Magnetic Properties
of
Organic Materials

Magnetic Properties
of
Organic Materials

edited by
Paul M. Lahti
University of Massachusetts
Amherst, Massachusetts

MARCEL DEKKER, INC.   NEW YORK · BASEL

Library of Congress Cataloging-in-Publication Data

Magnetic properties of organic materials / edited by Paul M. Lahti.
    p. cm.
   Includes bibliographical references and index.
   ISBN 0-8247-1976-X (alk. paper)
   1. Organic compounds—Magnetic properties.   2. Physical organic chemistry. I. Lahti, Paul M.
QD476.M24   1999
620.1'1791—dc21   99-26161
                                                      CIP

This book is printed on acid-free paper.

Headquarters
Marcel Dekker, Inc.
270 Madison Avenue, New York, NY 10016
tel: 212-696-9000; fax: 212-685-4540

Eastern Hemisphere Distribution
Marcel Dekker AG
Hutgasse 4, Postfach 812, CH-4001 Basel, Switzerland
tel: 41-61-261-8482; fax: 41-61-261-8896

World Wide Web
http://www.dekker.com

The publisher offers discounts on this book when ordered in bulk quantities. For more information, write to Special Sales/Professional Marketing at the headquarters address above.

Current printing (last digit):
10 9 8 7 6 5 4 3 2 1

PRINTED IN THE UNITED STATES OF AMERICA

To Eino and Helen Lahti—thanks for giving me the chance.

To Maureen, Jim, and Melanie—always my inspiration and my strength.

Preface

The aim of this book is to give a detailed overview of research into the design of magnetic materials that incorporate organic molecules. For centuries, magnetism has fascinated humans as a novelty, a guide to navigation, and eventually a technological phenomenon whose utilization has led to its incorporation into billion-dollar businesses such as information storage and computer logic production. Despite the massive use of magnetism in our society, it would be inapt to say that we have mastered this phenomenon. To the scientist, mastery suggests complete understanding and control. The inorganic ores and mixtures that yielded magnetic behavior to the first interested humans long ago are still the standard sources of today's technological materials. The genius of modern engineers has seemed—to the layperson, at least—to yield a never-ending succession of improvements to the use of standard materials. For example, in 1987, one could buy a 250-megabyte computer hard-drive device that was about the size of a typical business briefcase at a price of $20,000. In 1998, one can buy 17-gigabyte hard drive that will rest comfortably in one's hand for less than $1000. This trend of higher performance and lower cost will continue past the turn of the millenium. Given this, what is to be gained by intensive research into an area that seems already mature by the standards of its incorporation into the lifeblood of our present society?

The goal of such research is mastery of the field of magnetism. Ideally, one dreams not only of identifying the types of materials that would allow a particular technological innovation, but also of designing them from first principles and then synthesizing them from available substances. Although this goal still remains a dream, scientists have made tremendous advances

v

in this direction over the last decade. Just as the 19th century was a time of discovering the basic physical properties of atoms and molecules, and the 20th century one of mastering the basic chemistry of individual molecules and their reactions, so the 21st century will be one of aiming to master the assemblage of molecules into supramolecular arrays with known structure and properties. Therefore, it is no accident that as the 20th century comes to an end, the field of molecular magnetism has grown from a dream in the minds of a few scientists into a research thrust that involves groups from all over the world. It represents a great challenge to those who seek to implement theoretical models in the form of actual materials. In the Foreword to his wonderful book on this subject, Professor Olivier Kahn has noted that "quite spectacular advances have been achieved in this way." But much remains to be done.

For more than a century, chemists who work primarily with organic molecules have been fascinated by many of the physical properties of molecules and materials. Physical organic chemistry is even a well-recognized —although not always well-defined—subdiscipline of organic chemistry. However, as previously stated, the technologically useful magnetic materials are all inorganic metals and alloys. In addition, nature has blessed the numerous inorganic elements with the strong, natural magnetic moments desirable for construction of supramolecular arrays with large overall magnetism. Organic chemists are typically viewed as working to synthesize stable, often large molecules with elemental composition constrained to the first row or two of the periodic table. These elements are not generally perceived as conducive to the design of magnetic arrays. In fact, in the 1950s, the notion of a book about the magnetic properties of organic-based materials would have been quite fantastic.

Fortunately, much of the field of organic chemistry has been aimed at mastery of the synthesis and properties of carbon-containing molecules. Thanks to a century of work, organic chemists have a well-established array of synthetic tools and a firm understanding of structural relationships by which to design specific reactivity and properties for molecules. Since the time of Gomberg, organic chemists have been able to make molecules with unpaired electrons, hence with paramagnetic moments. By the 1960s, chemists could dream of utilizing the well-defined structure-property and synthetic tools of organic chemistry within the framework of paramagnetic open-shell organic molecules to become part of the nascent field of de novo designer molecular magnetic materials.

This book describes some of the basic background precepts of physical chemistry that define the paramagnetic natures of open-shell organic molecules, through a description of the synthetic and physical tools used to build and analyze such molecular building blocks; the design of oligomeric and polymeric organic molecules and crystals that exibit bulk magnetic ordering; and the melding of organic with inorganic spin-bearing units to create hybrid materials with fascinating magnetic properties and beautifully complex overall structures. The book recounts the field of organic-based molecular magnetism research from the 1960s to the present, through a series of monographs by scientists in this field of research. The chapters are divided into specific subsections so that one may more readily find topics within specific areas of theory, building block evaluation, and multispin assembly. The field of magnetic research is greatly diverse and could not be covered comprehensively by any single compendium. The boundary conditions used to limit the scope of this book must therefore be set forth, in order to understand the selection of authors from among so large a number of scientists performing critical work in this area.

A completely arbitrary decision was made to classify "organic-based magnetic materials" as those composed, in major part, of organic polymers or organic paramagnetic molecules. The field of molecular magnetism as it now stands has shown that boundaries of definition must be somewhat arbitrary, given its strong interaction of physics, inorganic chemistry, and organic chemistry. Therefore, this decision was made primarily to limit this book to a manageable scope. The decision is also scientifically justifiable, given the primary role that inorganic paramagnetic elements and clusters have played in numerous studies described elsewhere, wherein any organic fragments or components of such materials act as molecular scaffolding rather than as paramagnetic units in their own rights. This book is aimed at filling a gap in the compendia of this field, by presenting molecular magnetism research that utilizes organic polymers and spin-carrier units. These topical boundary conditions impose some limitations on what the reader will find in this book. The serious aficionado of molecular magnetism is advised to use other sources in addition to this one, and a brief suggested reading list is provided at the end of the frontmatter to whet one's curiosity further about this endlessly fascinating area.

Even within these boundary conditions, it was not easy to limit this book to a manageable size. Because the field continues to grow and attract new scientists, choosing authors was difficult, even during the time course of assembling and producing the book. I have done my best to select a mixture of well-experienced and young research groups, in order to show both established techniques as well as some of the multifold possibilities for future work. It has been a thing of joy to interact with these authors during production of the book. It has also been an honor and a pleasure for me to know many of the contributing authors both personally and professionally. The field of molecular magnetism is distinguished not only by its alluring subtlety, but by the unusual spirit of camaraderie among those who work in the field. As the book demonstrates, molecular magnetism is achievable only through strong cooperativity of numerous molecular interactions—fortunately, those working in this area are a wonderful demonstration on the human level of what they desire to achieve on the microscopic level. I specifically thank the contributors for their time and work, and also all others working in this area for making possible a level of scientific sharing that has made me quite proud to contribute to the level of interaction by editing this book.

Enough. If you have read this far, no further justification should be necessary. The beauty of molecular magnetism lies not solely in its history and its synergism, but in the rigor of its science. Let these experts in the area be your guides to its past, present, and future. Enjoy the trip. I have.

Paul M. Lahti

RECOMMENDED ADDITIONAL READING

Kahn, O. Molecular Magnetism (380 pp.) This book by a premier scientist in the field of molecular magnetism gives much detail about the physics of magnetism. Inorganic systems are covered to a considerable extent, make this an excellent complement to the present book.

Gatteschi D, Kahn O, Miller JS, Palacio F. Magnetic Molecular Materials, vol. 198E, Kluwer Academic Publishers: Dordrecht, The Netherlands, 1991 (411 pp.) This special collection of monographs details the presentations at a NATO Advanced Studies Institute (ASI) meeting at Il Ciocco, Italy. Many of the goals of present studies in molecular magnetism were suggested at this meeting.

Turnbull MM, Sugimoto T, Thompson LK, eds. Molecule-Based Magnetic Materials. Theory, Techniques, and Applications, vol. 644, American Chemical Society: Washington, DC, 1996 (340 pp.) This book on molecular magnetism summarizes presentations at the Pacifichem95 international meeting at Honolulu, Hawaii.

Proceedings of the International Conferences on Molecular Magnetism. This series in the journal Molecular Crystals and Liquid Crystals details the presentations of the biennual International Conferences on Molecular Magnetism. The series includes the following special journal issues from Gordon & Breach.

Miller JS, Dougherty DA (eds.) Mol. Cryst. Liq. Cryst., vol. 176. Proceedings of the Symposium on Ferromagnetic and High-spin Molecular Materials, 197th National Meeting of the American Chemical Society, Dallas, Texas, Fall 1989; 1989.

Iwamura H, Miller JS (eds.) Mol. Cryst. Liq. Cryst., vol. 232–233. Proceedings of the Symposium on the Chemistry and Physics of Molecular Based Magnetic Materials, Tokyo, Japan, October 25–30, 1992; 1993.

Miller JS, Epstein AJ (eds.) Mol. Cryst. Liq. Cryst., vol. 271–274, 1ff. Proceedings of the Fourth International Conference on Molecule-based Magnets, Salt Lake City, Utah, October 16–21, 1994; 1995.

Itoh K, Miller JS, Takui T (eds.) Mol. Cryst. Liq. Cryst., vol. 305–306, 1ff. Proceedings of the Fifth International Conference on Molecule-based Magnets, Salt Lake City, Utah, October 16–21, 1994; 1997.

Contents

EXCHANGE IN MODEL OPEN-SHELL SYSTEMS

EXCHANGE IN POLYMERIC ORGANIC SYSTEMS

Contents

SUMMATION

Contributors

Kunio Awaga, Ph.D. Associate Professor, Department of Basic Science, The University of Tokyo, Tokyo, Japan

Martin Baumgarten, Ph.D. Max-Planck-Institute for Polymer Research, Mainz, Germany

Jerome A. Berson, Ph.D. Department of Chemistry, Yale University, New Haven, Connecticut

Silas C. Blackstock, Ph.D. Assistant Professor, Department of Chemistry, The University of Alabama, Tuscaloosa, Alabama

Weston Thatcher Borden, Ph.D. Professor, Department of Chemistry, University of Washington, Seattle, Washington

Ronald Breslow, Ph.D. Professor, Department of Chemistry, Columbia University, New York, New York

Richard J. Bushby, D.Phil. School of Chemistry, University of Leeds, Leeds, England

Joan Cirujeda, Ph.D. Department of Molecular and Supramolecular Materials, Barcelona Institute of Material Science (CSIC), Bellaterra, Spain

Mercè Deumal, Ph.D. University of Barcelona, Barcelona, Spain

Fritz Dietz, Ph.D.   Professor, Wilhelm-Ostwald Institute of Physical and Theoretical Chemistry, University of Leipzig, Leipzig, Germany

Dante Gatteschi   Department of Chemistry, University of Florence, Florence, Italy

Béatrice Gillon, Ph.D.   Léon Brillouin Laboratory, Atomic Energy Commission—C.N.R.S., Gif-sur-Yvette, France

Klaus Griesar, Ph.D.* Institute of Physical Chemistry, Darmstadt University of Technology, Darmstadt, Germany

Wolfgang Haase, Ph.D.   Professor, Institute of Physical Chemistry, Darmstadt University of Technology, Darmstadt, Germany

Takao Hashiguchi, M.Sc.   Microcalorimetry Research Center, School of Science, Osaka University, Osaka, Japan

Hidenari Inoue, Ph.D.   Professor, Department of Applied Chemistry, Keio University, Yokohama, Japan

Koichi Itoh, D.Sci.   Professor, Department of Material Science, Graduate School of Science, Osaka City University, Osaka, Japan

Hiizu Iwamura, Ph.D.   Professor, National Institution for Academic Degrees, Nagatsuta, Japan

Takashi Kaneko, Ph.D.   Research Associate, Department of Chemistry and Chemical Engineering, Niigata University, Niigata, Japan

Piotr Kaszynski   Department of Chemistry, Vanderbilt University, Nashville, Tennessee

Takashi Kawakami, Ph.D.   Department of Chemistry, Graduate School of Science, Osaka University, Osaka, Japan

Douglas J. Klein, Ph.D.   Professor, Texas A&M University at Galveston, Galveston, Texas

---

Current affiliation: Department of R&D Strategy, SKW Trostberg AG, Trostberg, Germany.

Noboru Koga, Ph.D. Professor, Department of Pharmaceutical Sciences, Kyushu University, Fukuoka, Japan

Paul M. Lahti Professor, Department of Chemistry, University of Massachusetts, Amherst, Massachusetts

Michio M. Matsushita, Ph.D. Research Associate, Department of Applied Chemistry, Graduate School of Engineering, Tokyo Metropolitan University, Tokyo, Japan

Harden M. McConnell, Ph.D Professor, Department of Chemistry, Stanford University, Stanford, California

Yozo Miura, Ph.D. Professor, Department of Applied Chemistry, Faculty of Engineering, Osaka City University, Osaka, Japan

Yuji Miyazaki, D.Sc. Research Associate, Microcalorimetry Research Center, School of Science, Osaka University, Osaka, Japan

Kazuo Mukai, Ph.D. Professor, Department of Chemistry, Faculty of Science, Ehime University, Matsuyama, Japan

Jotaro Nakazaki Department of Basic Science, Graduate School of Arts and Sciences, University of Tokyo, Tokyo, Japan

Shigeaki Nimura, Ph.D. Research and Development Center, Ricoh Company, Ltd., Yokohama, Japan

Hiroyuki Nishide, Ph.D. Professor, Department of Polymer Chemistry, Waseda University, Tokyo, Japan

Juan J. Novoa, Ph.D. Professor, University of Barcelona, Barcelona, Spain

Akifumi Oda Department of Chemistry, Graduate School of Science, Osaka University, Osaka, Japan

Fernando Palacio Institute of Material Science, C.S.I.C.—University of Zaragoza, Zaragoza, Spain

Andrzej Rajca Department of Chemistry and Center for Materials Research and Analysis, University of Nebraska, Lincoln, Nebraska

Jeremy M. Rawson Department of Chemistry, University of Cambridge, Cambridge, England

Paul Rey, Ph.D. Director of Research, Department of Condensed Matter Sciences, CEA-DRMFC, Grenoble, France

Kazunobu Sato, D.Sci. Lecturer, Department of Chemistry, Graduate School of Science, Osaka City University, Osaka, Japan

Jacques Schweizer Atomic Energy Commission, Grenoble, France

Trent D. Selby Department of Chemistry, The University of Alabama, Tuscaloosa, Alabama

Daisuke Shiomi, D.Sci. Associate Professor, Department of Material Science, Graduate School of Science, Osaka City University, Osaka, Japan

David A. Shultz, Ph.D. Associate Professor, Department of Chemistry, North Carolina State University, Raleigh, North Carolina

Michio Sorai, D.Sc. Professor and Director, Microcalorimetry Research Center, School of Science, Osaka University, Osaka, Japan

Tadashi Sugawara, Ph.D. Professor, Department of Basic Science, Graduate School of Arts and Sciences, The University of Tokyo, Tokyo, Japan

Takeji Takui, D.Eng. Professor, Department of Chemistry, Graduate School of Science, Osaka City University, Osaka, Japan

Yoshio Teki, D.Sci. Associate Professor, Department of Material Science, Graduate School of Science, Osaka City University, Osaka, Japan

Philippe Turek, Ph.D. Lecturer, Department of Physics, Charles Sadron Institute, Louis Pasteur University, Strasbourg, France

Nikolai Tyutyulkov, Ph.D. Professor, Department of Physical Chemistry, University of Sofia, Sofia, Bulgaria

Jaume Veciana, Ph.D. Professor, Department of Molecular and Supramolecular Materials, Barcelona Institute of Material Science (CSIC), Bellaterra, Spain

Akira Yabe, Ph.D.   Chief Senior Researcher, National Institute of Materials and Chemical Research, Tsukuba, Japan

Kizashi Yamaguchi, D.Eng.   Professor, Department of Chemistry, Graduate School of Science, Osaka University, Osaka, Japan

Naoki Yoshioka, Ph.D.   Associate Professor, Department of Applied Chemistry, Keio University, Yokohama, Japan

Yasunori Yoshioka, Ph.D.   Associate Professor, Department of Chemistry, Graduate School of Science, Osaka University, Osaka, Japan

Kazunari Yoshizawa, Ph.D.   Associate Professor, Department of Molecular Engineering, Kyoto University, Kyoto, Japan

1
Intermolecular Ferromagnetic Spin Exchange

Harden M. McConnell
Stanford University, Stanford, California

I. BACKGROUND

The purpose of this chapter is to give a brief resumé and historical account of early proposals for achieving ferromagnetic exchange interaction between aromatic or olefinic free radicals. Under most circumstances one expects that two spin = 1/2 ($S = 1/2$) free radicals should interact, or react, so as to produce a diamagnetic ($S = 0$) product. However, in a 1963 proposal it was suggested that if two odd-alternant free radicals were geometrically constrained in their relative configurations, a high-spin ($S = 1$) ground state should result [1]. The geometrical constraint is one in which carbon atoms with positive spin density on one molecule face carbon atoms with negative spin density on the second molecule. The term *ferromagnetic* was used in connection with suggesting that a crystalline array of such molecules might show ferromagnetic properties. The prediction of this proposal with respect to the formation of high spin states by appropriately oriented pairs of alternant radicals was verified in elegant experiments some twenty years later by Iwamura and collaborators [2,3].

An unsaturated hydrocarbon is said to be alternant when the unsaturated carbon atoms can be labeled so that atoms of one type (the *starred atoms*) are surrounded by atoms of the other type (the *unstarred atoms*). In case there is an odd number of unsaturated atoms, the atoms are labeled so that there are more starred atoms than unstarred atoms. In the case of an odd-alternant molecule, Longuet-Higgins has shown that there is at least one zero-energy molecular orbital, in which the unpaired electron is confined to

1

atomic orbitals localized only on starred atoms [4]. Schematic representations of the unsaturated atoms in allyl and benzyl are given in Figure 1. In this figure valence bond structures are drawn so that light lines represent covalent bonds involving σ-orbitals, and dark lines represent bonds involving π-orbitals.

When the electronic structures of odd-alternant radicals are described by (nonionic) valence bond structures, one generally finds positive spin densities on the starred atoms and negative spin densities on the unstarred atoms [5,6]. That is, if the net z-component of the molecule's electron spin angular momentum is $+1/2$, this angular momentum is distributed over the molecule so that the z-components of spin are positive on the starred atoms and negative on the unstarred atoms. The 1963 proposal considers the consequences of placing one unsaturated free radical on top of the other so that atoms with opposite spin density face one another. Since spins on neighboring atoms in molecules tend to be antiparallel, the net spin on the first molecule should therefore tend to be parallel to the net spin on the second molecule. This corresponds to ferromagnetic coupling.

In 1989, Iwamura and collaborators prepared diphenylcarbene molecular groups incorporated into a [2.2] paracyclophane skeleton [2,3]. Their experiments are most easily discussed in terms of two benzyl radicals placed on top of one another so that the two benzene rings are coaxial (see Figure 2). In accordance with the proposal, when the positive-spin-density atoms (starred atoms) faced the negative-spin-density atoms (unstarred atoms), the ground state was found to be high spin (Figures 2a and 2c). When the spin densities on opposite atoms were the same (Figure 2b), the ground states were found to be low spin [2,3].

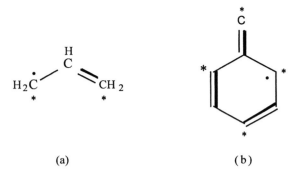

(a)   (b)

Figure 1 Valence bond resonance structures of the alternant free radicals allyl (a) and benzyl (b). The light lines represent σ-bonds, and the heavy lines represent bonds between π-orbitals.

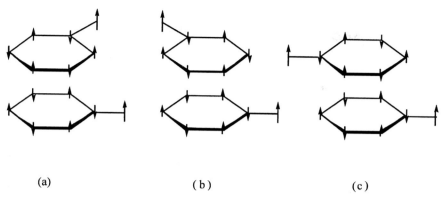

Figure 2 Complexes of benzyl radicals. In orientations (a) and (c) there are positive and negative spin densities on intermolecular neighboring carbon atoms, leading to ferromagnetic coupling (high-spin ground state of the complex). In (b), spin densities of the same sign are opposite one another, and the lowest energy state is low spin.

The proposal for ferromagnetic interaction between odd-alternant radicals in terms of interacting positive and negative spin densities presumes that the strongest intermolecular–interatomic interactions are between the intermolecular atom pairs that are closest to one another. This of course requires that configuration interaction within each in-plane π-electron system give rise to negative spin densities on unstarred atoms. These negative spin densities are well known both experimentally and theoretically. Calculations by Yamaguchi et al. [7] and by Buchachenko [8] have provided quantitative support for the proposal. Recently, however, Yoshizawa and Hoffmann [9] have pointed out that even if these negative spin densities were zero, the ground spin state of a pair of alternant radicals should be high spin for the described radical orientations. Instead of repeating their arguments, we describe a simple general proof.

II. THREE-DIMENSIONAL ALTERNANT MOLECULES

When alternant radicals are stacked on top of one another as in Figure 2, so that the starred atoms of one radical face the unstarred atoms of the second radical, one has a *three-dimensional alternant molecule*. Starred atoms are surrounded by unstarred nearest neighbors, in both two and three dimensions. We can apply the alternant orbital theorems of Longuet-Higgins [4] to this complex since there is nothing in the derivation of these theorems

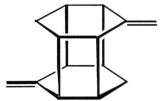

Figure 3 The complex of two benzyl radicals is treated as a three-dimensional alternant molecule, and in the principal resonance structure shown there are four intermolecular bonds between π-orbitals.

that requires the system to be two-dimensional, nor is there any requirement for all the molecular orbital resonance integrals to be equal to one another. One theorem states that there are at least $N - 2T$ zero-energy orbitals in an alternant molecule, where N is the number of carbon atoms and T can be defined as the number of bonds between π-electrons in any principal resonance structure. One of the principal resonance structures for a high-spin complex with two benzyl radicals is shown in Figure 3; this particular structure involves intermolecular bonds between π-orbitals. Thus, by Hund's rule the ground states of the benzyl radical pairs sketched in Figures 2a and 2c are high spin, $S = 1$. One sees that this scheme is easily generalized to other pairs of alternant free radicals. The radicals, of course, need not be identical.

The two reasons for achieving high-spin ground states involve distinct assumptions about the sizes of various matrix elements, but both depend in an essential way on the three-dimensional alternant molecular topology.

III. EARLY HISTORY

I have been asked to write a few words on the history of this 1963 proposal and related work. As noted elsewhere [10], after finishing graduate studies at Cal Tech in 1950 and postdoctoral work at Chicago in 1952, I felt quite frustrated with respect to the then-current theories of molecular electronic structure—valence bond and molecular orbital theories. My view at that time was that there was no satisfactory experimental method for testing the adequacy of these approximate theories. This situation changed dramatically with the discoveries of magnetic resonance—especially the paramagnetic resonance of free radicals. Proton nuclear hyperfine splittings in these free radicals—as well as zero-field splittings in states with spin $S > 1/2$—permitted quantitative tests of calculated spin distributions and correlations in molecules, with remarkably satisfying results. I was then anxious to test these theories even further by considering weak intermolecular forces. In this connection the first problem tackled was that of propagating triplet states in molecular crystals, triplet excitons [11]. In fact, triplet exciton motion involves just the same sort of integrals as those considered later in the 1963 proposal. Once again experimental confirmation of the basic ideas came only a number of years later [12]!

In conclusion, it is perhaps worth noting that a "second proposed mechanism" for intermolecular ferromagnetic coupling had its origin in a completely extemporaneous remark following a lecture by Robert S. Mulliken at a Welch Foundation Symposium [13]. Mulliken gave a long and rather difficult lecture on intermolecular charge transfer. As I recall, a rather long period of silence followed his lecture. In order to encourage some discussion I made a suggestion involving charge transfer and ferromagnetic coupling. I am pleased that both of my proposals have been critically evaluated experimentally and theoretically, and may have contributed to the field of molecular magnetism.

ACKNOWLEDGMENTS

I am indebted to Hiizu Iwamura, Kazunari Yoshizawa, and Roald Hoffmann for helpful correspondence. This work was supported by the National Science Foundation.

REFERENCES

1. McConnell HM. J Chem Phys. 1963; 39:1910.
2. Izuoka A, Murata S, Sugawara T, Iwamura H. J Am Chem Soc. 1985; 107: 1786–1787.
3. Izuoka A, Murata S, Sugawara T, Iwamura H. J Am Chem Soc. 1987; 109: 2631–2639.
4. Longuet-Higgins, HC. J Chem Phys. 1949; 18:265.
5. Brovetto P, Ferroni S. Nuovo cimento 1957; 5:142.
6. Lefebvre R, Dearman HH, McConnell HM. J Chem Phys. 1960; 32:176–181.
7. Yamaguchi K, Toyoda Y, Fueno T. Chem Phys Lett. 1989; 159:459.
8. Buchachenko AL. Russ Chem Rev. 1990; 59:307.
9. Yoshizawa K, Hoffmann R. J Am Chem Soc. 1995; 117:6921.
10. McConnell HM. In: Eaton GR, Eaton SS, and Salikhov KV, eds. Foundations of Modern ERP. London: World Scientific, 1997, pp. 236–241.
11. Sternlicht H, McConnell HM. J Chem Phys. 1961; 35:1793–1800.
12. Haarer D, Wolf HC. Mol Liq Cryst. 1970; 10:359.
13. McConnell HM. Proc Robert A. Welch Found Conf Chem Res. 1967; 11:144.

2

Structural Determinants of the Chemical and Magnetic Properties of Non-Kekulé Molecules

Jerome A. Berson
Yale University, New Haven, Connecticut

I. SCOPE

Space limitations do not permit comprehensive reviews of the several topics mentioned in this chapter. Therefore, at the direction of the editor, and with the risk (say, rather, the certainty!) of offending the community of workers in the field, this chapter omits many important contributions from other laboratories.

II. INTRODUCTION

π-Conjugated non-Kekulé molecules lack one or more bonds and cannot be represented by conventional resonance structures [1–5]. These species eventually might find practical use as the monomeric components of polymers with useful bulk ferromagnetic or electrically conductive properties [6]. On the other hand, the very existence of the monomeric non-Kekulé molecules as examples of an unconventional form of matter raises new fundamental questions about the interrelationships among molecular structure, spin, and reactivity. Although mindful of the potential for application, we cannot deny that curiosity is the principal motivation for our efforts to answer these questions.

III. TWO MAJOR STRUCTURAL DETERMINANTS OF THE SPIN OF NON-KEKULÉ MOLECULES

Non-Kekulé molecules with N missing bonds are biradicals (or oligoradicals) with N pairs of degenerate or nearly degenerate frontier molecular orbitals (FMOs). Only $2N$ electrons, half the number needed for full occupancy, are available to these orbitals. This deficiency can give rise to more than one low-lying spin multiplet: when $N = 1$, for example, a triplet and one or two singlets [7]. Much effort has been expended on the assignments of the energy ordering and spacing of the multiplets, since these relationships control the chemical and magnetic properties. The prediction of the spin of the ground state has been a difficult and even controversial problem from the earliest appearances of such molecules. Today, some sixty years after the first theoretical studies, it is recognized that *frontier orbital degeneracy or near-degeneracy* [1,7] plays an important but not exclusive role. A second major factor is the *connectivity of the π-orbital framework* [8–10]. The following section reviews some of the contributions of our laboratory to both of these issues.

In a seminal 1950 paper [1], Longuet-Higgins invoked an approximate form of Hund's rule [11,12] in considering the ground states of non-Kekulé molecules (Scheme 1), including trimethylenemethane (TMM) **1**, *meta-*benzoquinodimethane **2**, and tetramethyleneethane **3**. The prediction of a triplet ground state [1] for each was based upon the presence of degenerate half-filled molecular orbitals (MOs) at the highly approximate Hückel one-electron MO theoretical level. This degeneracy persists at higher levels of theory for **1** because its symmetry point group (D_{3h}) contains E representations. However, the point groups of compounds **2** (C_{2v}) and (planar) **3** (D_{2h}) do not have such representations, the nonbonding molecular orbital (NBMO) degeneracies are not symmetry-enforced, and their energies will split apart at higher levels of theory.

Trimethylenemethane **1**, synthesized years later by Dowd in 1966, displayed an electron spin resonance (ESR) spectrum consistent with a triplet

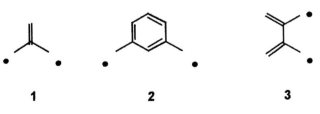

1   **2**   **3**

Scheme 1

molecule of D_{3h} symmetry [13,14]. It later was assigned a triplet ground state in recognition of the linear response of its ESR signal intensity to inverse temperature (Curie law plot) [15]. These results were nicely in accord with Longuet-Higgins' prediction.

We gradually began to recognize several new questions raised by this work: Would Hund's rule still apply to TMM derivatives in which the frontier orbital degeneracy was broken by distortion and/or substitution? More generally, does the molecular Hund's rule require a strict degeneracy of FMOs? Still more generally, why should there be a molecular Hund's rule at all? One could argue that such a rule should be a hard thing for a chemist to accept. It says that regardless of the structure of the non-Kekulé molecule, high spin will always be preferred. This seems to conflict with one of the most fundamental chemical postulates: molecular structure determines molecular properties. Why should spin be an exception? Even if we tell ourselves that there *must* be violations of a generalization as sweeping as Hund's rule, we are still faced with the problem of recognizing what kinds of structures will cause a violation.

IV. IS STRICT FMO DEGENERACY REQUIRED FOR A HIGH-SPIN NON-KEKULÉ GROUND STATE?

Practice has not been uniform on whether the molecular Hund's rule applies only when the frontier orbitals are exactly degenerate by symmetry, as in the classical cases O_2 and NH. Some theoreticians, including Hund himself, have kept the rule as a guide, even when the degeneracy is only approximate.

Our opportunity to address this question grew out of a different study [16], whose results prompted us to attempt the synthesis of the substituted TMM biradical **5a** from the diazene **4a** (Scheme 2). The photodeazetation of diazene **4** [17] in cold matrices produced biradical **5a** as a triplet species, which was characterized by its ESR spectrum, $|D/hc| = 0.027$ cm^{-1}, $|E/hc| = 0.0023$ cm^{-1}. The D-value is very similar to that of TMM **1** itself [13,15]. Note that **5a** not only lacks threefold symmetry because of the substituents, but it also lacks even *local* threefold symmetry because of the distortion from trigonality at C_2, which has an intracyclic C–C–C angle of less than 120°. This analysis is directly confirmed by the finite value of the ESR zero-field splitting parameter $|E/hc|$ which should be and is [13–15] zero for a truly threefold symmetrical case like the parent TMM **1**. Nevertheless, like **1**, biradical **5a** gave a linear Curie plot [17]. Barring an accidental degeneracy of the singlet and triplet, this result shows that **5a** has a triplet ground state *despite its nondegenerate frontier orbitals*. The triplet ground state also is confirmed by extensive chemical studies [18]. Further examples of ground

4   **5**

a: $X_1, X_2 = CH_3$ **d:** $X_1 = H, X_2 = Ph$

b: $X_1, X_2 = H$ **e:** $X_1, X_2 = OCH_3$

c: $X_1, X_2 = Ph$ **f:** $X_1 = H, X_2 = Cl$

Scheme 2

state triplet TMM derivatives without FMO degeneracies include the hydro-carbons **5b–d** and the heteroatom substituted derivatives **5e–f** [19].

Subsequently, a number of other non-Kekulé molecules with near rather than exact FMO degeneracies have been shown to have persistent ESR spectra, which in many cases follow the Curie law. These species probably have triplet ground states. The list includes hydrocarbons such as m-quinodimethane **2** [20–23], tetramethyleneethane **3** [24], and derivatives with heteroatoms in the π-system such as m-benzoquinomethane (Scheme 3) **6** [25], m-naphthoquinomethane **7** [26], and 1-methylene-8-imino-naphthalene **8** [27]. Clearly, the preference for the higher multiplet state in such molecules survives the FMO splitting caused by low symmetry and even by heteroatom substitution in or on the π-system.

V. CONNECTIVITY

The first non-Kekulé molecules were the hydrocarbons **9** and **10** prepared in 1915 by Schlenk and Brauns by reductive dehalogenation of the corre-sponding dichlorides **11** and **12** (Scheme 4) [28]. In 1936, Müller and Müller-Rodloff [29] showed by magnetic susceptibility measurements that preparations of compound **10** contained a paramagnetic component. Follow-ing a suggestion of Hund [30], who here made the first applications to organic molecules of his multiplicity rule [11,12], Müller and Bunge [31] ascribed the paramagnetism to the triplet state of **10**. Attempts to demon-strate paramagnetism in the other Schlenk-Brauns hydrocarbon **9** by mag-

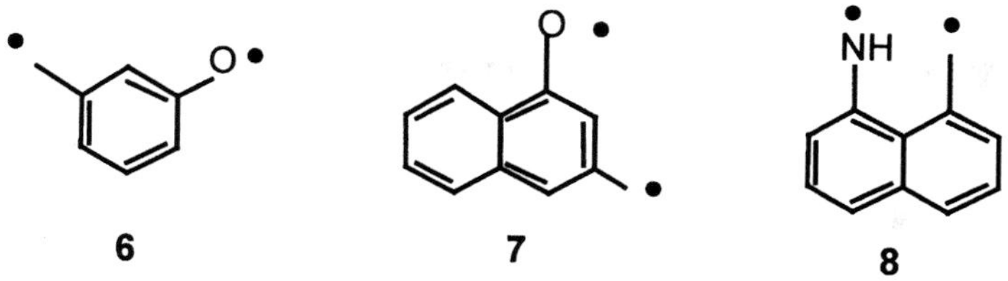

6   **7**   **8**

Scheme 3

netic susceptibility were unsuccessful at this early stage [31], but electron spin resonance spectroscopy eventually disclosed this triplet state also [32].

However, Erich Hückel soon gave a prescient warning [8a], which was effectively ignored for decades [8b,9e], that the rule might fail for **10** and related molecules. There is now strong theoretical and experimental evidence [9,10] that shows Hückel's doubts to have been well founded. The previous invariable high-spin rule must be modified to take into account more subtle structural factors: the *connectivity* of the atoms of the π-framework of a conjugated non-Kekulé species is a strong determinant of the energy separation and ordering of the lowest spin states.

Generally, the Hund's rule preference for the state of higher multiplicity is expected to operate when the non-Kekulé molecule can be imagined to be formed by the union of two radicals so that one or both of the points of union is "active," that is, bears a finite NBMO coefficient. However,

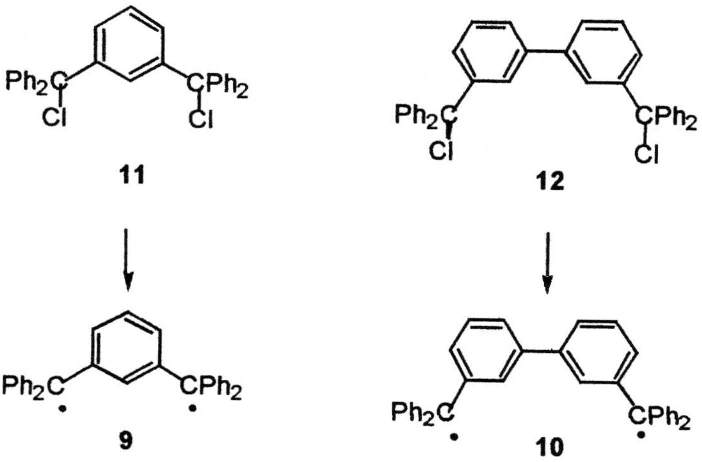

11   **12**

9   **10**

Scheme 4

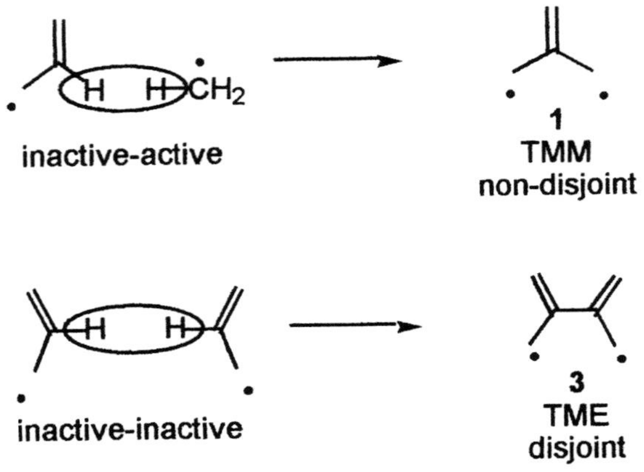

Scheme 5

when the non-Kekulé species can be imagined to be formed by union at two "inactive" sites, with NBMO coefficients both zero, the exchange energy approaches zero [8,9]; and since the singlet state is selectively stabilized by dynamic spin polarization, the Hund's rule high-spin preference is greatly reduced or even reversed [9a,d]. Terminology derived from Borden and Davidson [9a], gradually has been adopted to distinguish these categories: "nondisjoint" (example trimethylenemethane, TMM, 1, Scheme 5) and "disjoint" (example tetramethyleneethane, TME, 3), respectively.

As we already have seen, TMM 1, with a triplet ground state, behaves as expected for a nondisjoint species. Tetramethyleneethane TME 3 [24] and two of its derivatives, 13 [33] and 14 [34], are disjoint and therefore are expected to have little if any spin preference (Scheme 6). Nevertheless, they too have been assigned triplet ground states based upon linear Curie law plots [33,34]. It was not clear from these results alone whether the failure to observe singlets in some such disjoint cases means that the disjoint hypothesis is valid but the singlet-triplet gap, although small, still favors the

Scheme 6

triplet, or whether the hypothesis is simply wrong and the gap favoring the triplet is large. The answer to this conundrum was not resolved until recently for **13** [35]. Soon after the modern theoretical classification of the spin of non-Kekulé molecules in 1977 [9], we began a program to design and implement experimental tests of the disjoint hypothesis.

VI. TESTS OF CONNECTIVITY IN BIRADICALS RELATED TO *m*-QUINONOID COMPOUNDS

The original synthesis of the biradicals **9** and **10** by Schlenk and Brauns [28] generated them in fluid solutions from the corresponding dichlorides **11** and **12** (Scheme 4). Variations of this general method of forming 1,3-dimethylenebenzene or 3,3'-dimethylenebiphenyl derivatives from intact 1,3-disubstituted benzene or 3,3'-disubstituted biphenyl precursors have been used to prepare the parent hydrocarbon *m*-benzoquinodimethane **2** and related species [20,36].

A different approach is through ring-opening of *m*-quinonoid [25,26] precursors [21,23,25,26,38] (Scheme 7, Eqs. 1–4). Although the ground states of **7** and (probably) **6** are triplets, the chemistry seems to be largely that of the singlet [25,26].

Note that the Schlenk-Brauns compound **9** (Scheme 4) belongs to the nondisjoint category and its companion **10** to the disjoint category. Preparations of **10** are paramagnetic, as judged by direct magnetic moment measurements (Gouy balance [29,31], and ESR spectra have been recorded for both [32], but the instability of the species has so far not permitted an assignment of the ground state [39]. Highly hindered derivatives of **9** have been shown to have triplet ground states [39], despite the nonplanarity of the π-system. Hindered derivatives of **10** have been shown to have triplet ground states also, but the singlet is very close in energy [39]. A dicarbene related to **10** has been shown to have a singlet ground state [37a,b], and several closely related dinitrenes and dinitroxyls also have singlet ground states [9d,37c].

An early direct test of the disjoint vs. nondisjoint categorization was based on a comparison of the two isomeric bis-*m*-benzoquinomethane tetraradicals **24** and **27** (Scheme 8), species that are nondisjoint and disjoint, respectively [9e,40–42]. These tetraradicals can be generated by two-step photolysis of the pentacyclic precursors **22** and **25**, via the biradicals **23** and **26**, respectively. In the case of tetraradical **24**, its nondisjoint connectivity, combined with the proposal [9c] that the spin rules should apply even when some of the π-electron centers are heteroatoms, amounted to the prediction that this compound should have a quintet ground state, the highest spin

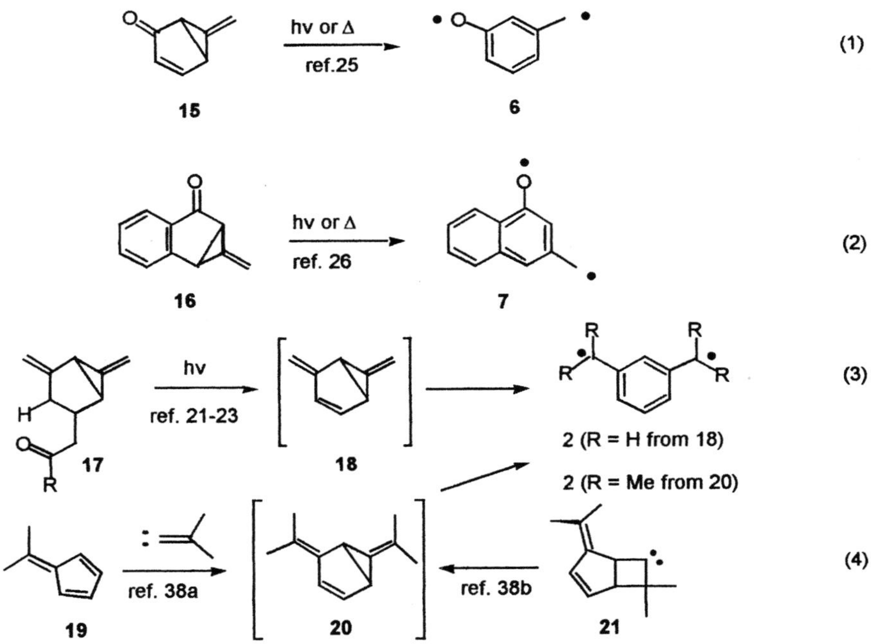

Scheme 7

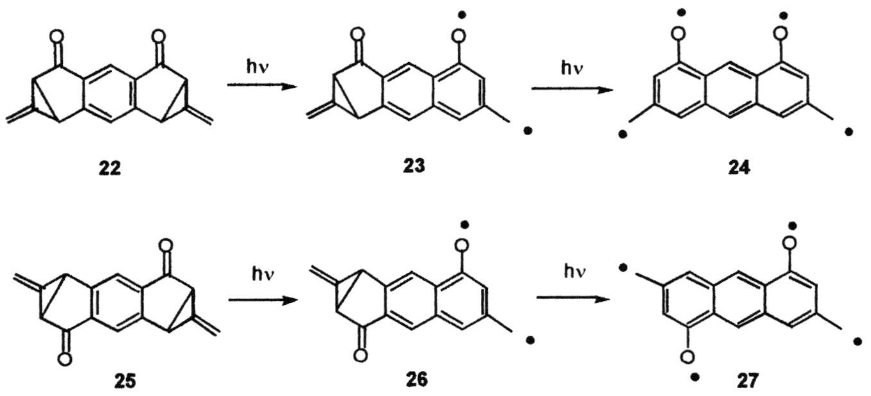

Scheme 8

available to a four-electron system. Experimental observations [9e,40,42] confirmed the prediction. The isomeric system **27**, on the other hand, is disjoint and hence might be a singlet, in violation of Hund's rule. Experimentally, however, **27** was found to be a triplet instead [9e,41,42], so that although it does not follow Hund's rule, neither does it manifest the potential consequence of disjoint character.

The hydrocarbon biradical tetramethylenebenzene (TMB) **28** (Scheme 9) promised a decisive test of the disjoint hypothesis. Formally, this species is the result of a union of two pentadienyl radicals at inactive sites and therefore belongs to the disjoint category. Both semiempirical [43] and ab initio [44,45] calculations predicted that the preference for the singlet state in TMB **28** should be even larger than that in the simplest disjoint hydrocarbon, TME **3**. The first published experimental studies [46] of TMB reported the photolysis of the bicyclic ketone **29** in low-temperature media to give a colored preparation that showed several bands in the UV-Vis spectrum and a narrow ESR spectrum. The authors [46] assigned both spectra to the triplet state of TMB **28**. Citing the presence in the ESR spectrum of a weak half-field transition and the adherence of the signal intensity to the Curie law, they concluded [46] that the triplet is the ground state of TMB, thus contradicting the theoretical predictions [43–45].

However, experimental re-examination [10,47] of the chemistry and spectroscopy of TMB ultimately showed that although the intensely purple carrier of the UV-Vis spectrum is the authentic TMB singlet, the carrier of the ESR spectrum is formed in a side reaction and is not associated with the TMB structure. Positive support for the *singlet* ground state of TMB is provided by the observation [10,47a] that the irradiation of the precursor ketone **29** labeled with ^{13}C in the exocyclic methylene groups gives a purple preparation whose cross-polarization magic angle spinning (CP MAS) ^{13}C nuclear magnetic resonance (NMR) spectrum in the solid state at 77 K shows an unbroadened, well-defined resonance at 113 ppm, a normal shift for an sp^2 trivalent carbon. A triplet would have shown a drastically broadened and shifted resonance that would not have been visible in the normal ^{13}C NMR region. The ground state of TMB thus is singlet, as predicted theoretically

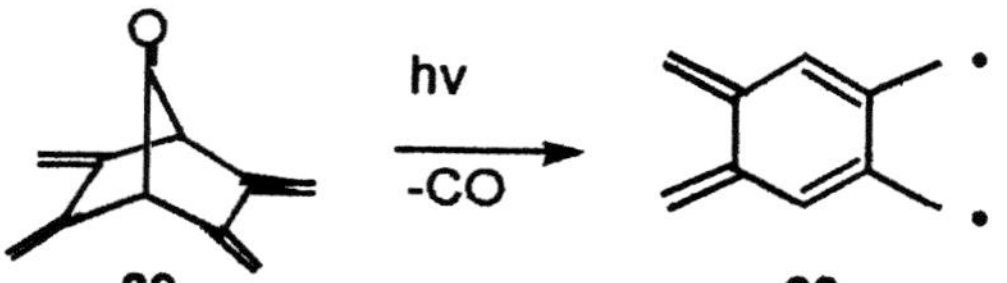

29   **28**

Scheme 9

[43–45]. This returns us from the idealized, icy elegance and generality of Hund's rule to the real-world warmth, the myriad detailed contingencies, the refractory nonuniformity characteristic of chemistry. It fortifies the chemist's belief that molecular structure determines molecular properties.

VII. A NON-KEKULÉ SERIES WITH TUNABLE MULTIPLET ENERGY SPACINGS

The objective of this study was to gain the power of choice over the energy separation between the singlet and triplet by designed structural variation, in other words, to learn how to *tune* the spacing. In the series of non-Kekulé heterocycles **31** (Scheme 10), the tetramethyleneethane hydrocarbon structure has been modified by the insertion of a heteroatom bridge between a pair of methylene groups. The interaction of the heteroatom lone pair of electrons with the NBMOs of the TMM unit perturbs the energy separation between the highest occupied molecular orbital (HOMO) and lowest unoccupied molecular orbital (LUMO) and in turn affects the singlet–triplet gap. This opens the way to adjust the gap gradually and at will by judicious choice of the group X. Since theory [9,48–52] strongly points to a small singlet–triplet gap in the parent hydrocarbon tetramethyleneethane (TME) **3**, one would expect that the electronic perturbations might suffice to switch the multiplicity within the series and thereby test a computationally based prediction of the relationship between molecular structure and spin state. Semi-empirical [43,48] and ab initio [49] calculations of the lowest-lying states of furan and thiophene biradicals **31a** and **31b** predict that the singlet should be preferred. In the pyrrole series, the parent compound **31c** (R = H)

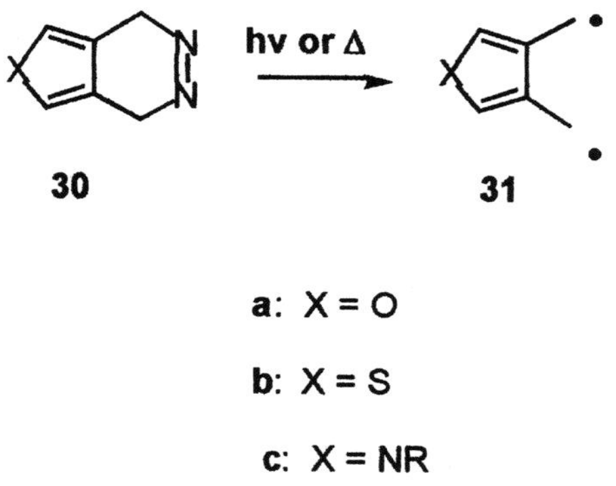

30   **31**

        a: X = O

        b: X = S

        c: X = NR

Scheme 10

again should be a singlet, but by an even larger preference than those in **31a** and **31b**. Thermal or photochemical deazetations of the diazenes **30a** and **30b** generate the corresponding biradicals **31a** and **31b** [53−57]. The photochemical reaction in low-temperature matrices gives rise to ESR-silent red-purple preparations that show strong absorption in the UV-Vis spectrum near 570 nm ($\lambda_{max} \sim 5000$ M^{-1}cm^{-1}).

These species also can be liberated by laser flash photolysis and their concentration monitored by nanosecond time-resolved spectroscopy. The relative rates at which the alkenes quench the visible absorption are in the same order and ratio as those determined by preparative competition experiments, in which the cycloadducts **32** and **33** are obtained stereospecifically (Scheme 11). This strongly argues that the carrier of the visible color and the species trapped in the cycloaddition experiments are the same and have the 3,4-dimethyleneheterocycle structure **31** [57−60].

The chemistry and the absence of ESR absorption point to the singlet nature of the biradicals **31a** and **31b**, but the assignment is strengthened by the CP MAS ^{13}C NMR spectra of solid matrices of isotopically enriched samples of **31a** or **31b**. These show the unbroadened lines at normal chemical shifts expected for singlet species [61−63]. Theory thus has correctly predicted the singlet ground states of **31a** and **31b**.

In the pyrrole series **31c** [6,64−68], variation of the substituent R offers a special opportunity to adjust the electronic nature of the heteroatom and hence to narrow the singlet−triplet gap by replacement of H with electron-withdrawing groups of increasing strength. These qualitative perturbational arguments are supported by AM1-CI semiempirical calculations, which suggest that relative to the singlet−triplet gap in TME **3** and the hydrocarbon TME derivatives **13** and **14**, $E_T − E_S$ should decline from a value near 8 kcal/mol in the parent pyrrole derivative **31c** (R = H) or its methyl analog **31c** (R = Me), to a small but definite preference for the singlet in an N-acyl derivative, to a value close to zero in derivatives in which the R group is as strongly electron-withdrawing as *p*-nitrobenzoyl. Since the hydrocarbon TMEs **3**, **13**, and **14** each have experimentally accessible triplet states

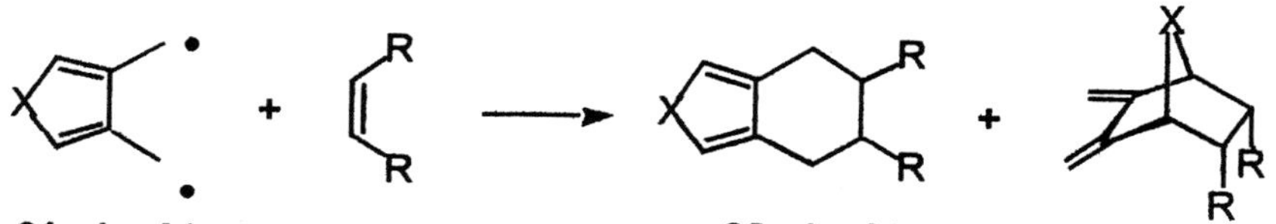

31a (or b)   **32a (or b)**   **33a (or b)**

Scheme 11

[13,33,34], one would expect that a triplet would be visible in the case **31c** (R = *p*-toluenesulfonyl). If the anticipated switch in spin state in response to changes in N-substituent actually occurred, it would be a direct confirmation of the theoretical ideas. In fact, this is essentially what we observed [64–66]: the derivatives **31c** (R = Me) and **31c** (R = *t*-BuCO) both are singlets as judged by their chemistry and by the absence of triplet ESR signals, but the derivative **31c** (R = Ts) shows a beautiful four-line triplet ESR signal in the $\Delta m_s = 1$ region, characterized by the zero-field splitting parameters $|D|/hc = 0.0226$ cm^{-1} and $|E|/hc \sim 0$ cm^{-1}, and a weak $\Delta m_s = 2$ transition, diagnostic of a triplet species. Similar triplets were observed in the related cases of the *N*-(*p*-bromobenzenesulfonyl, Bs)-3,4-dimethylene-pyrrole **31c** (R = Bs) ($|D|/hc = 0.0231$ cm^{-1}, $|E|/hc = 0.0005$ cm^{-1}) and *N*-tosyl-2-phenyl-3,4-dimethylenepyrrole ($|D|/hc = 0.0187$ cm^{-1}, $|E|/hc = 0.0005$ cm^{-1}) biradicals. However, nature provided a startling complication in the phenomenon of *long-lived spin isomerism*, one of the delightful surprises we encountered in our work on non-Kekulé compounds (see later).

VIII. CATENATION OF NON-KEKULÉ BIRADICALS TO TETRARADICAL PROTOTYPES OF CONDUCTIVE OR MAGNETIC POLYMERS [67,68]

In π-electron theory, each monomeric unit of the non-Kekulé series **31** embodies a pair of half-filled, nearly degenerate, nominally NBMOs. Catenation of singlet monomers such as **31a**, **31b**, and **31c** (R = electron-releasing or weakly electron-withdrawing group) through a spacer should lead to antiferromagnetic low-spin "nonclassical" [69,70] polymers **34a**, **34b**, or **34c** (R = ERG or weak EWG), in which the half-filled electron bands might confer the capability for metallic conduction [6] without doping. On the other hand, the triplet monomers, such as **31c** (R = Ts or Bs), might lead to conductively insulating high-spin polymers **34c** (R = strong EWG). Recently [67,68], we have synthesized the first member of this potentially large group of catenated multiradicals, the singlet tetraradical **35**, a green ESR-inactive transient that is stable in solid matrices (Scheme 12).

IX. SOME SURPRISES DISCOVERED IN THE STUDY OF NON-KEKULÉ MOLECULES

Much of the theory of organic chemistry was built to explicate the properties of molecules in terms of conventional structures. The structures of non-Kekulé molecules, by definition, lie outside this domain. Thus, the emern is large

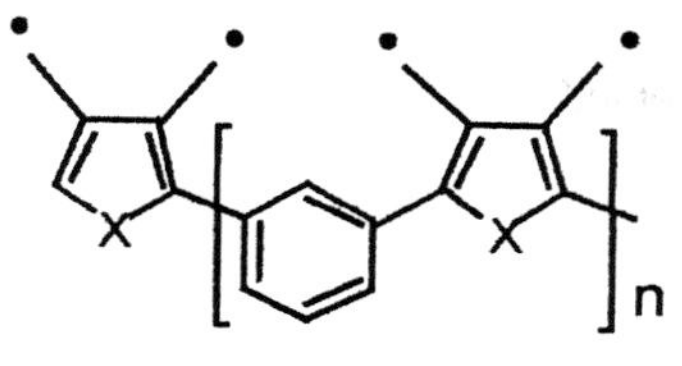

34a: X = O   34c: X = NR

34b: X = S

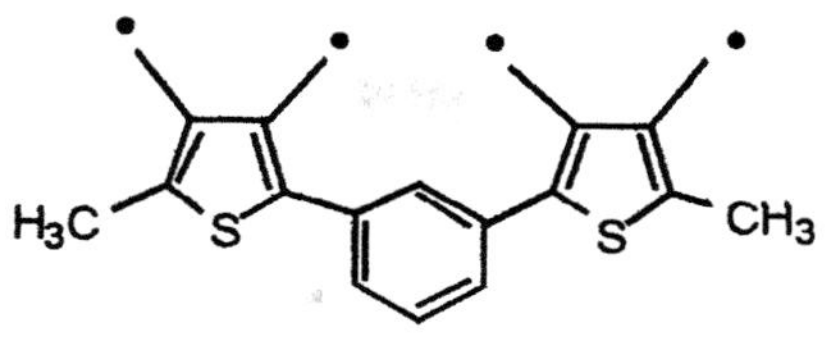

35

Scheme 12

gence of surprises should be no surprise. The following brief accounts illustrate the knowledge gained in the examination of a few of these.

X. LONG-LIVED SPIN ISOMERISM [64–68]

As we have already seen, it is possible to generate the triplet state of the *N*-tosyl-3,4-dimethylenepyrrole biradical **31c** (R = Ts) by 265-nm photolysis of the precursor diazene **30c** (R = Ts). When immobilized in a solid matrix, the triplet biradical persists for at least a month. A conventional interpretation of this result would be that the triplet is the ground state. However, photolysis of **30c** (R = Ts) *at a different wavelength, 370 nm*, produces an entirely different result, forming the blue (λ_{max} 590 nm), ESR-inactive *singlet* state of **31c** (R = Ts). Even more startling is the finding that the triplet and singlet forms of the biradical **31c** (R = Ts) do not interconvert during weeks of storage (Scheme 13). Similar results are obtained with the analog **31c** (R = Bs). Independent evidence [66] that both the singlet and the triplet in the tosyl series have the structure **31c** (R = Ts) forces us to the inference that the carrier of the ESR spectrum and the carrier of the blue color and UV-vis spectrum have the same connectivity, so that they are long-lived *spin isomers* whose rate of interconversion is many orders of magnitude slower than that observed in conventional molecules. The current explanation of this phenomenon is that the singlet and triplet have different rotational conformations about the N–Ts bond and that the barrier to internal rotation is high enough to make the rate very slow in the low-temperature matrices.

A partial analogy to this spin isomerism occurs in certain transition metal complexes [71], such as the Fe(dithiocarbamate)$_3$ species discovered

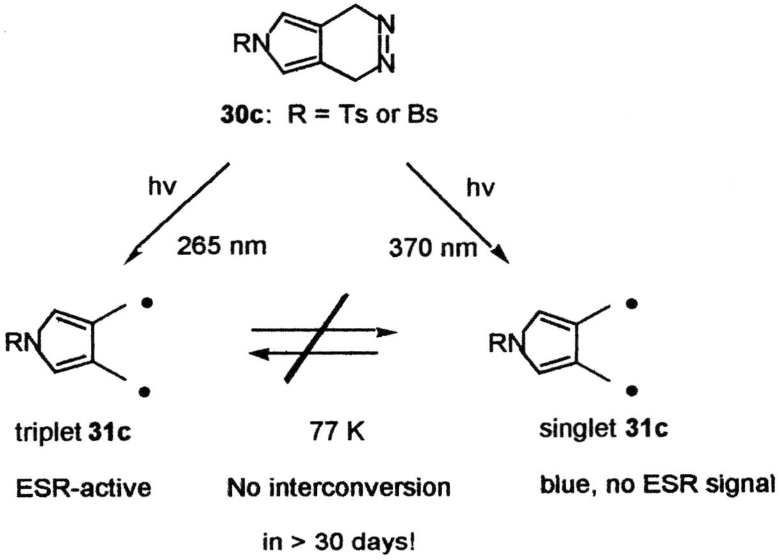

30c: R = Ts or Bs

triplet 31c   77 K   singlet 31c

ESR-active   No interconversion   blue, no ESR signal

in > 30 days!

Scheme 13

by Cambi [72] and subsequently studied extensively [73,74]. In those cases also, the atomic connectivity is the same in the low-spin and high-spin members of a pair. However, the structural difference between the isomers derives not from differences in rotational conformation but rather from differences in the geometrical disposition of the ligands around the metal atom. In several cases with hexavalent metal centers, for example, these structures correspond to local minima in the itinerary of distortions from octahedral symmetry.

XI. NEGATIVE BOND DISSOCIATION ENERGIES

Normally, the homolytic cleavage of a ring bond in a full-valency Kekulé molecule to form its biradical non-Kekulé isomer would be endothermic by the bond dissociation energy. Is it possible to imagine a system in which the thermodynamic relationship is reversed, so that the biradical is actually *more stable* than the full-valence isomer? This situation might be approached, for example, when drastic ring strain destabilizes the full-valence structure and when, simultaneously, conjugation and the singlet–triplet splitting stabilize the biradical. An example of such a set of relationships is seen

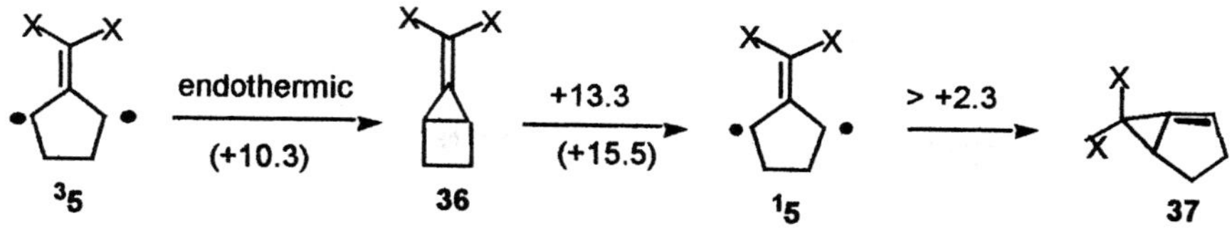

Scheme 14

in the four hydrocarbons ^{3}**5**, **36**, ^{1}**5**, and **37** (Scheme 14) [75,76]. The available experimental values [75,76] (in kcal/mol) of the heats of reaction where X = CH_3 are shown above the line, and the values calculated [77] at an MCSCF level of electronic structure theory for the parent analogs (X = H) are shown in parentheses below the line. In this series, the triplet biradical ^{3}**5** is the global minimum of the energy surface on which the four species reside. One bond in each of the full-valence Kekulé structures **36** and **37** has a *negative* dissociation energy. Thus, these compounds owe their existence as persistent species only to a kinetic barrier to ring cleavage, not to a thermodynamic advantage over the biradical.

XII. WHY DO FRONTIER ORBITAL ENERGY SEPARATIONS CORRELATE ENTROPY-CONTROLLED CYCLOADDITION REACTIONS OF NON-KEKULÉ MOLECULES?

There was early evidence [53,54], subsequently confirmed repeatedly [55,58–60,65,67,78], that the relative reactivities of 3,4-dimethyleneheterocycles **31a** or **31b** with alkenes (Scheme 11) ran parallel to those of the same alkenes in the conventional Diels-Alder reactions with conjugated dienes such as cyclopentadiene **38** (Scheme 15).

The relative reactivities in diene additions of the alkenes maleic anhydride, fumaronitrile, dimethyl fumarate, and acrylonitrile to give products **39** cover a range of ~5000-fold [79,80]. Surprisingly, this discriminatory

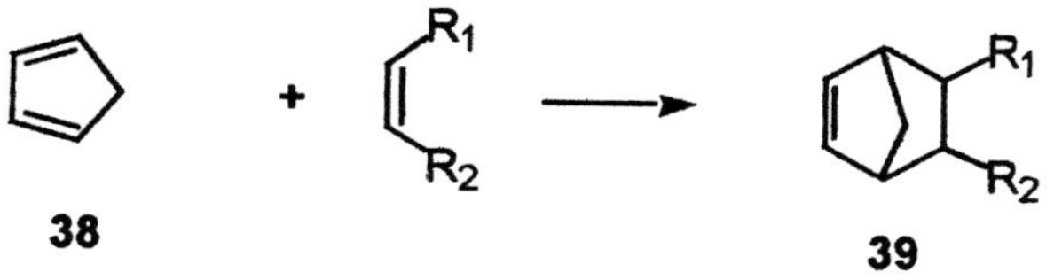

38   **39**

Scheme 15

ability of the group of alkenes persists even though the diyl reactions of Scheme 11 with the same alkenes are some 10^{10} times as fast as the diene reactions of Scheme 15. The orders and relative magnitudes of the rates of the additions of both series are essentially the same [58–60]. Superficially, one might have expected the so-called "reactivity-selectivity relationship" [81] to produce a much narrower range of reactivities in the more reactive diyl series.

There is another intriguing aspect of the data: in the diene reactions (Scheme 15) [79,80], the rates are strongly temperature dependent, $T\Delta\Delta S^{\ddagger}$ is essentially zero throughout the series, and the reactivity range thus is controlled entirely by $\Delta\Delta H^{\ddagger}$; in the diyl reaction series (Scheme 11) [59], there is little or no temperature dependence $\Delta H^{\ddagger}$ and $\Delta\Delta H^{\ddagger}$ both are about zero, and the reactivity range is controlled entirely by variations in $T\Delta S^{\ddagger}$. Nevertheless, not only do the rates of the diyl and diene reactions correlate with each other, but also, *both* correlate with the frontier MO (FMO) *energy separation* [59,60,65,78,80]. This suggests that the familiar Eyring activation energy and activation entropy terms have a common physical basis in the crossing of ground state and excited state reaction coordinate curves [59].

XIII. CONCLUSIONS

A major objective in our work on non-Kekulé molecules has been to elucidate the relationship between spin and molecular structure. In the pursuit of it, we found many strange and unfamiliar phenomena. A wide range of new non-Kekulé molecules now is imaginable and accessible to synthesis, and we do not doubt that the study of these species will bring new insights on reaction mechanisms and physical properties, new synthetic tools, and new surprises.

ACKNOWLEDGMENTS

I thank the many collaborators in our laboratory and in other institutions who have contributed to this work. Financial support was provided by the National Science Foundation, the National Institutes of Health, the Humphrey Chemical Company, the American Chemical Society's Arthur C. Cope Scholar program, and the Dox Foundation.

REFERENCES

1. Longuet-Higgins HC. J Chem Phys. 1950; 18;265.
2. Clar E, Stewart DG. J Am Chem Soc. 1953; 75:2667.

3. Clar E, Kemp W, Stewart DG. Tetrahedron. 1958; 3:325.

4. Clar E. Aromatische Kohlenwasserstoffe. 2nd ed. Berlin: Springer Verlag, 1952, pp. 93, 461.

5. Dewar MJS. The Molecular Orbital Theory of Organic Chemistry. New York: McGraw-Hill, 1969, p 232.

6. A list of references is given in refs. 3–5 in the following paper: Lu HSM, Berson JA. J Am Chem Soc. 1997; 119:1428.

7. (a) Salem L, Rowland C. Angew Chem Intl Ed Engl. 1972; 11:92. (b) Borden WT. In Borden WT, ed. Diradicals. New York: Wiley, 1982, Ch 1 and references cited therein.

8. (a) Hückel E. Z Phys Chem Abt B. 1936; 34:339. (b) Review: Berson JA. Angew Chem Intl Ed Engl. 1996; 35:2751.

9. (a) Borden WT, Davidson ER. J Am Chem Soc. 1977; 99:4587. (b) Misurkin IA, Ovchinnikov AA. Russ Chem Rev (Engl Transl). 1977; 46:967. (c) Ovchinnikov AA. Theor Chim Acta. 1978; 47:497. (d) Review: Borden WT, Iwamura H, Berson JA. Acc Chem Res. 1994; 27:109. (e) Review: Berson JA. In Patai S, Rappoport Z, eds. The Chemistry of the Quinonoid Compounds. New York: Wiley, 1988, Vol. II Ch 10. (f) Review: Klein, DJ. In Lahti PM, ed. Magnetic Properties of Organic Materials. New York: Marcel Dekker, 1998. Review: Borden WT. In Lahti PM, ed. Magnetic Properties of Organic Materials. New York: Marcel Dekker, 1998.

10. Reynolds JH, Berson JA, Kumashiro KK, Duchamp JC, Zilm KW, Scaiano JC, Berinstain AB, Rubello A, Vogel P. J Am Chem Soc. 1993; 115:8073.

11. (a) Hund F. Linienspektren Periodisches System der Elemente; Springer-Verlag, Berlin, 1927, p. 124ff. (b) Hund F. Z Phys. 1928; 51: 759. (c) Lennard-Jones, JE. Trans Faraday Soc. 1929; 25:668.

12. (a) See Lowe JP. Quantum Chemistry. Boston, MA: Academic Press, 1978, pp 112–127. (b) Review: Kutzelnigg W. Angew Chem Intl Ed Engl. 1996; 35: 573. (c) Kutzelnigg W, Morgan JD III, Z Für Physik D. 1996; 36:197.

13. Dowd P. J Am Chem Soc. 1966; 88:2587.

14. Review: Dowd P. Accts Chem Res. 1972; 5:242.

15. Baseman RJ, Pratt DW, Chow M, Dowd P. J Am Chem Soc. 1976; 98:5726.

16. Berson JA, Bauer W, Campbell MM. J Am Chem Soc. 1970; 92:7515.

17. Bushby RJ, McBride JM, Tremelling M, Berson JA. J Am Chem Soc. 1971; 93:1544.

18. (a) Review: Berson JA. Accts Chem Res. 1978; 11:446. (b) Review: Berson JA. In de Mayo P, ed. Rearrangements in Ground and Excited States. New York: Academic Press, 1980, p 311. (c) Review: Berson JA. In Borden WT, ed. Diradicals. New York: Wiley-Interscience, 1982, Ch 4.

19. (a) Platz MS, McBride JM, Little RD, Harrison JJ, Shaw A, Potter SE, Berson JA. J Am Chem Soc. 1976; 98:5725. (b) For theoretical discussion, see: Carpenter BK, Little RD, Berson JA. J Am Chem Soc. 1976; 98:5723.

20. (a) Wright BB, Platz MS. J Am Chem Soc. 1983; 105:628. (b) Review: Platz MS. In Borden WT, ed. Diradicals. New York: Wiley-Interscience, 1982, Ch 5.

21. Goodman JL, Berson JA. J Am Chem Soc. 1984; 106:1867.

22. Goodman JL, Berson JA. J Am Chem Soc. 1985; 107:5409.
23. Goodman JL, Berson JA. J Am Chem Soc. 1985; 107:5424.
24. (a) Dowd P. J Am Chem Soc. 1970; 92:1066. (b) Dowd P, Chang W, Paik YH. J Am Chem Soc. 1986; 108:7416.
25. (a) Rule M, Matlin AR, Dougherty DA, Hilinski EF, Berson JA. J Am Chem Soc. 1979; 101:5098. (b) Matlin AR, Inglin TA, Berson JA. J Am Chem Soc. 1982; 104:4954. (c) Inglin TA, Berson JA. J Am Chem Soc. 1986; 108:3394.
26. (a) Seeger DE, Hilinski EF, Berson JA. J Am Chem Soc. 1981; 103:720. (b) Goodman JL, Peters KS, Lahti PM, Berson JA. J Am Chem Soc. 1985; 107:276.
27. Platz MS, Burns JR. J Am Chem Soc. 1979; 101:4425.
28. Schlenk W, Brauns M. Chem Ber. 1915; 48:661, 716.
29. Müller E, Müller-Rodloff I. Justus Liebig's Ann der Chem. 1935; 517:134.
30. Hund F. Personal communication cited in ref 31.
31. Müller E, Bunge W. Chem Ber. 1936; 69:2168.
32. (a) Kothe G, Denkel K-H, Sümmermann W. Angew Chem Intl Ed Engl. 1970; 9:906. (b) Luckhurst GR, Pedulli GF, Tiecco M. J Chem Soc (B). 1971:329.
33. Dowd P, Chang W, Paik YH. J Am Chem Soc. 1987; 109:5284.
34. Roth WR, Kowalczik U, Maier G, Reisenauer HP, Sustmann R, Müller W. Angew Chem Intl Ed Engl. 1987; 26:1285.
35. Matsuda K, Iwamura H. J Am Chem Soc. 1997; 119:7412, have shown by magnetic susceptibility that the singlet and triplet of **13** are degenerate. This, and not the triplet ground state previously proposed [33] causes the linear Curie plot from ESR measurements.
36. (a) Migirdicyan E, Baudet J. J Am Chem Soc.1975; 97:7400 (b) Haider K, Platz MS, Despres A, Lejeune V, Migirdicyan E, Bally T, Haselbach E. J Am Chem Soc. 1988; 110:2318.
37. (a) Itoh K. Pure Appl Chem. 1978; 50:1251. (b) Teki Y, Takui T, Kitano M, Itoh K. Chem Phys Lett. 1987; 142:181. (c) Murata S, Iwamura H. J Am Chem Soc. 1991; 113:346.
38. (a) Gajewski JJ, Chang MJ, Stang PJ, Fisk TE. J Am Chem Soc. 1988; 110:2318. (b) Stadler H, Rey M, Dreiding AS. Helv Chim Acta. 1984; 67:1379.
39. (a) Review: Rajca A. Chem Rev. 1994; 94:871. (b) Rajca A, Utamapanya S, Smithhisler DJ. J Org Chem 1993; 58:5650.
40. Seeger DE, Berson JA. J Am Chem Soc. 1983; 105:5144.
41. Seeger DE, Berson JA. J Am Chem Soc. 1983; 105:5146.
42. Seeger DE, Lahti PM, Rossi AR, Berson JA. J Am Chem Soc. 1986; 108:1251.
43. Lahti PM, Rossi AR, Berson JA. J Am Chem Soc. 1985; 107:2273.
44. Lahti PM, Rossi AR, Berson JA. J Am Chem Soc. 1985; 107:4362.
45. Du P, Hrovat DA, Borden WT, Lahti PM, Rossi AR, Berson JA. J Am Chem Soc. 1986; 108:5072.
46. (a) Roth WR, Langer R, Bartmann M, Stevermann B, Maier G, Reisenauer HP, Sustmann R, Müller W. Angew Chem Intl Ed Engl. 1987; 26:256. (b) See also: Roth WR, Langer R, Ebbrecht T, Beitet A, Lennartz H-W. Chem Ber. 1991; 124:2751.

47. (a) Reynolds JH, Berson JA, Kumashiro KK, Duchamp JC, Zilm KW, Rubello A, Vogel P. J Am Chem Soc. 1992; 114:763. (b) Reynolds JH, Berson JA, Scaiano JC, Berinstain AB. J Am Chem Soc. 1992; 114:5866.
48. Lahti PM, Ichimura AS, Berson JA. J Org Chem. 1989; 54:958.
49. Du P, Hrovat DA, Borden WT. J Am Chem Soc. 1986; 108:8086.
50. Du P, Borden WT. J Am Chem Soc. 1987; 109:930.
51. Nachtigall P, Jordan KD. J Am Chem Soc. 1992; 114:4743.
52. Nachtigall P, Jordan KD. J Am Chem Soc. 1993; 115:270.
53. Review: Berson JA. Mol Cryst Liq Cryst. 1989; 176:1, and references cited therein.
54. Greenberg MM, Blackstock SC, Berson JA. Tetrahedron Lett. 1987; 28:4263.
55. Stone KJ, Greenberg MM, Blackstock SC, Berson JA. J Am Chem Soc. 1989; 111:3659.
56. Greenberg MM, Blackstock SC, Stone KJ, Berson JA. J Am Chem Soc. 1989; 111:3671.
57. Review: Berson JA. Acc Chem Res. 1997; 30:238.
58. Scaiano JC, Wintgens V, Bedell A, Berson JA, J Am Chem Soc. 1988; 110:4050.
59. Scaiano JC, Wintgens V, Haider K, Berson JA. J Am Chem Soc. 1989; 111:8732.
60. Heath RB, Bush LC, Feng XW, Berson JA, Scaiano JC, Berinstain AB. J Phys Chem. 1994; 97:13355.
61. Zilm KW, Merrill RA, Greenberg MM, Berson JA. J Am Chem Soc. 1987; 109:1567.
62. Zilm KW, Merrill RA, Webb GG, Greenberg MM, Berson JA. J Am Chem Soc. 1989; 111:1523.
63. Greenberg MM, Blackstock SC, Berson JA, Merrill RA, Duchamp JC, Zilm KW. J Am Chem Soc. 1991; 113:2318.
64. Bush LC, Heath RB, Berson JA. J Am Chem Soc. 1993; 115:9830.
65. Bush LC, Heath RB, Feng XW, Wang PA, Maksimovic L, Song A IH, Chung W-S, Berinstain AB, Scaiano JC, Berson JA. J Am Chem Soc. 1997; 119:1406.
66. Bush LC, Maksimovic L, Feng XW, Lu HSM, Berson JA. J Am Chem Soc. 1997; 119:1416.
67. Lu HSM, Berson JA. J Am Chem Soc. 1996; 118:265.
68. Lu HSM, Berson JA. J Am Chem Soc. 1997; 119:1428.
69. Tyutyulkov N, Karabunarliev S, Ivanov C. Mol Cryst Liq Cryst. 1989; 176:139, and references cited therein.
70. Tyutyulkov N, Kanev I, Polansky O, Fabian J. Theoret Chim Acta. 1977; 46:191.
71. I thank Professor R. Crabtree for pointing out this analogy.
72. Cambi L, Szego L. Ber Dtsch Chem Ges. 1931; 64:2591. (See also subsequent articles.)
73. Cotton FA, Wilkinson G. Advanced Organic Chemistry. 3rd ed. New York: Wiley-Interscience, 1972. p. 567 ff.
74. Gütlich P, Hauser A, Spiering H. Angew Chem Intl Ed Engl. 1994; 33:2024.

75. (a) Mazur MR, Berson JA. J Am Chem Soc. 1981; 103:684. (b) idem. ibid. 1982; 104:2217.
76. Berson JA. In Borden WT, ed. Diradicals. New York: Wiley, 1982, Ch 4 and references cited therein.
77. Dixon DA, Dunning TH Jr, Eades RA, Kleier DA. J Am Chem Soc. 1981; 103:2878.
78. Haider KW, Clites JA, Berson JA. Tetrahedron Lett. 1991; 32:5305.
79. Sauer J, Wiest H, Mielert A. Chem Ber. 1964; 97:3183.
80. Sauer J, Sustmann R. Angew Chem Int Ed Engl. 1980; 19:770.
81. For an instructive review of this topic and examples of the numerous exceptions to the relationship, see: Leffler JE, Grunwald E. Rates and Equilibria of Organic Reactions. New York: Wiley, 1963, p 162 ff.

3

Antiaromatic Triplet Ground State Molecules: Building Blocks for Organic Magnets

Ronald Breslow
Columbia University, New York, New York

I. TRIPLET GROUND STATES OF ANTIAROMATIC SYSTEMS

In contrast to the cyclic conjugated systems with $4n + 2\pi$ electrons that are classed as "aromatic," reflecting their conjugative stabilization, the properties of cyclic conjugated systems with $4n$ π-electrons are quite different. Many of them exhibit conjugative destabilization, a property we have called "antiaromaticity," particularly when referred to similar analogs lacking full cyclic conjugation [1]. For example, solvolytic evidence [2] makes it clear that the 4-π-electron cyclopentadienyl cation **1** is destabilized relative to cyclopentenyl cation **2**. Also, cyclopropenyl anion **3** is destabilized [3] relative to the allyl anion **4** (Scheme 1). Modern theoretical treatments account for such properties.

At least as interesting, it is found that some antiaromatic molecules exist as triplet ground state species, with two unpaired electrons. This is predicted by simple HMO theory, in which the two highest-energy electrons of cyclopentadienyl cation **1** or of cyclopropenyl anion **3** occupy a degenerate pair of molecular orbitals; higher levels of theory also make such a prediction.

The triplet is predicted to be the ground state for **1** and **3** only if they have full polygonal symmetry, at least C_3. With lower symmetry, the HMO treatment, and others, predict a singlet ground state with no unpaired elec-

1   2   3   4   5

Scheme 1

trons. Thus the real-life question about such potential triplet species is: Will they stay symmetric and be triplets, or instead distort and be singlets? Such distortion will stabilize a singlet state, so a Jahn–Teller distorted singlet could be lower in energy than an undistorted triplet, in spite of the general energy advantage of triplet states in which electron repulsion is lower.

As the rest of this chapter will demonstrate, some such species are indeed singlets, others are triplets. The general situation is that smaller derivatives are triplets, larger ones singlets. In the larger species the electrons are spread over more space so that the electron repulsion advantage of the triplet is diminished. Also, larger species find more ways to distort. As another consideration, Borden has shown that odd ring systems such as **1** and **3** are more likely to be triplets than are even ring systems such as cyclobutadiene **5**, because of the specific electronic distributions [4].

Some of our work in this area has been reviewed previously, with a discussion of its relationship to magnetic materials [5–7].

II. TRIPLETS AND MAGNETIC MATERIALS

Harden McConnell proposed a mechanism by which charge transfer could lead to ferromagnetism, provided one of the components of the charge transfer system were a ground state triplet species [8]. The idea was that electron transfer from a normal singlet molecule into a triplet would lead to a pair of radicals with parallel spins (Figure 1). An extension into a solid material, made up of equal numbers of singlet and triplet molecules, could lead to ferromagnetic ordering in which all the unpaired spins were parallel.

It was not clear whether this model for ferromagnetism would work in a real case. The impetus to test this theory furnished us with an incentive to prepare stable triplet state antiaromatic species, in addition to the desire to determine whether such species would indeed exist preferentially in their triplet states. As will be described, one such test of this approach was unsuccessful. Others have argued from theory that systems of this type will not lead to ferromagnetism [9], but it is fair to say that there is not yet enough experimental evidence on the point.

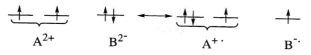

Figure 1 A charge transfer complex between a dication with a triplet ground state and a dianion with a normal singlet ground state. This is one of several models by which a triplet ground state species can align spins in solids, leading to ferromagnetism.

III. POTENTIAL TRIPLETS THAT PREFER THE SINGLET STATE

If cyclooctatetraene **6** (Scheme 2) were a flat regular octagon, it could be a triplet $4n$ π electron molecule, but it is a nonplanar singlet. The nonplanarity may be thought of as an extreme case of Jahn–Teller distortion, but of course it also relieves angle strain. The tetrapotassium salt of tetrahydroxybenzoquinone **7** could be a triplet, but is not [10]. It is likely that potassium coordination distorts the potential hexagonal symmetry of the species. Later we will describe a related stabilized benzene dictation derivative that is in fact a triplet.

We prepared the heptaphenylcycloheptatrienyl anion **8** from the corresponding cation [11–12]. Anion **8** was stable in solution, but electron spin resonance (ESR) and nuclear magnetic resonance (NMR) studies and magnetic susceptibility measurements showed that **8** was a singlet with no detectable equilibrating triplet state. Perhaps **8** is distorted from heptagonal symmetry.

IV. CYCLOPENTADIENYL CATIONS WITH DETECTABLE TRIPLET STATES

IV.A. Pentaphenylcyclopentadienyl Cation

The first example of a $4n$ π electron system with a detectable triplet state was the pentaphenylcyclopentadienyl cation **9**. When this species was prepared at room temperature, it rapidly rearranged [13]; but at lower temperature it gave stable solutions [14–15], which were frozen into glasses for triplet ESR studies. These showed clear triplet spectra with the expected values for the spectral parameters. In particular, the zero-field splitting parameters showed a species having polygonal symmetry, with the two unpaired electrons at a sensible average distance separation for **9** (Scheme 3).

For a triplet ground state, the ESR spectrum should follow the Curie law, increasing in intensity as the temperature is lowered and the lowest-

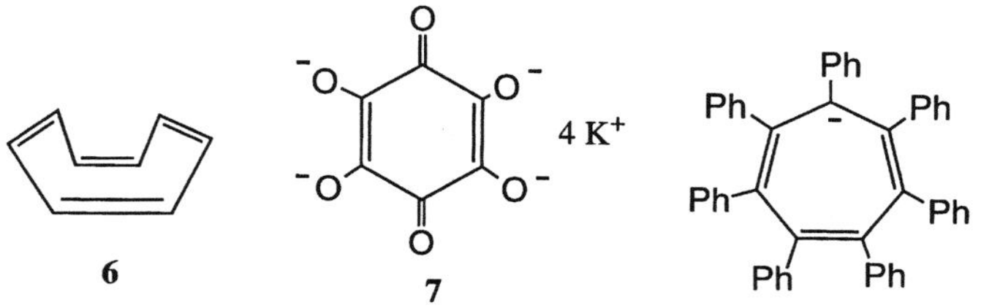

Scheme 2

energy triplet sublevel increases in population. In our studies of **9**, the spectral intensity **decreased** as the temperature was lowered, indicating an increased population of a singlet state. There are two possible interpretations: (1) the singlet is the lowest state for **9**, but the triplet is ca. 0.5 kcal/mole higher in energy; or (2) the triplet is the lowest state, but in frozen solution a significant population of **9** is in an anisotropic environment, for instance, as an unsymmetrical ion pair, in which distortion favors the singlet state. Both interpretations would account for the findings. For other pentaarylcyclopentadienyl cations without the full pentaphenyl symmetry, we also detected triplet states [15], but again as low-lying excited states with higher excitation energies than for the symmetric **9**.

IV.B.   Pentachlorocyclopentadienyl Cation

Treatment of hexachlorocyclopentadiene with SbF_5 afforded a solution containing the cation **10** [15–16]. This showed the triplet ESR spectrum expected for such a species, and temperature studies showed that the spectrum obeyed the Curie law down to 4.2 K (liquid helium). Thus **10** is almost certainly a ground state triplet species. The contrast with **9** reflects the smaller size of **10**. This leads to larger electronic interaction in the ring, favoring the triplet; and the small size of **10** makes distortion more difficult than for **9**. The yield of the cation was not large, and it was not stable at room temperature. Thus it had no serious potential for a test of the use of triplets to make ferromagnets.

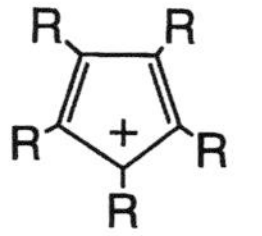

9: R = Ph
10: R = Cl
11: R = H

Scheme 3

IV.C. Unsubstituted Cyclopentadienyl Cation

We prepared the first examples of 5-halocyclopentadienes [2], and from the 5-bromo derivative it was possible to prepare a salt of the cyclopentadienyl cation **11** by using a simple molecular beam procedure with SbF_5 [17]. Again a triplet ESR spectrum was observed, with parameters consistent with structure **11**, and again Curie law behavior was observed down to liquid helium temperature. Thus as expected from the behavior of **10**, the even smaller derivative **11** is a ground state triplet. Again, however, it is not stable enough to be considered for constructing ferromagnetic systems.

Current work is aimed at preparing a stabilized fully symmetric derivative of **11**.

V. TRIPLET DICATIONS

V.A. A Triphenylene Dication Derivative

Hoijtinck [18] showed that some dianions derived from aromatic systems by adding two electrons were triplets. We explored stabilized dications produced by the removal of two electrons from aromatic rings. The first example [19] was the dication **12** of a hexaazatriphenylene derivative. The neutral compound was prepared from triphenylene by hexabromination, then reaction with ethylenediamine, then hexaacetylation, then borane reduction.

The neutral triphenylene was converted to dication **12** with one equivalent of Br_2, and its ESR spectrum in frozen solution showed the typical triplet pattern with reasonable spectroscopic parameters. The intensity followed the Curie law from 12 K to 150 K, so it is almost certainly a ground state triplet. The dication was stable at room temperature. Dication **12** (Scheme 4) is related to a less stable dication **13** derived from hexamethoxy-triphenylene by Bechgard and Parker [20]. They reported a triplet spectrum for that species, which we confirmed, but their dication is not stable above $-80°C$. Wasserman had also reported [21] that the chemically unstable dication **14** of hexachlorobenzene was a triplet.

The reversible electrochemical oxidation of our triphenylene derivative to the dication **12** required a remarkably low potential, 0.27 V vs. SCE. We found [19] that the system could be oxidized reversibly all the way to the tetracation, but beyond that the electrochemistry was irreversible.

We have prepared analogs of dication **12** with various electron-withdrawing groups to replace the ethyl groups of **12** [22]. These were prepared to modify the oxidation potential of the cation, as described later. We have also prepared some relatives of dication **12** with bulky groups replacing the ethyl groups of **12** [7]. They have not yet been fully evaluated.

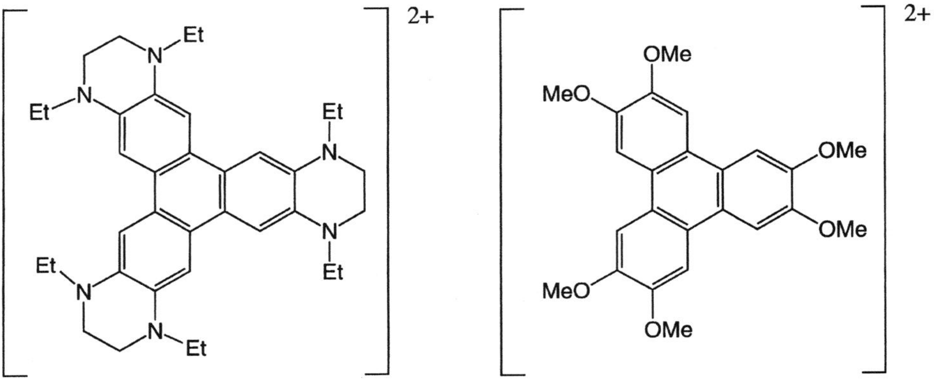

12   13

Scheme 4

V.B.   Stabilized Benzene Dications

We prepared hexakis-dimethylaminobenzene **15** (Scheme 5), by reductive dimerization of tris-dimethylaminocyclopropenyl cation, but found that it was not easily oxidized to the dication [5]. Apparently the dimethylamino groups are twisted out of enough conjugation to stabilize the dication. For this reason we were attracted to compound **16**, with fully conjugated amino groups [23–24]. We prepared **16** by a remarkable "zipper" reaction that is further described in the original paper.

As expected, **16** was easily oxidized, chemically and electrochemically. The reversible potential for conversion of **16** to its monocation radical was -0.46 V, while it was reversibly converted to the dication **17** at 0.03 V (vs. SCE in acetonitrile). Analytically pure samples of **17** as its bis I_3^- salt were prepared, and samples for ESR studies were prepared from **16** with NO^+ BF_4^- in various solvents.

The question of whether **17** is a ground state triplet is complex. In frozen solutions with CH_2Cl_2/CH_3CN there was a typical triplet ESR spectrum with the appropriate parameters. It followed the Curie law over the temperature range 109–156 K, but was not examined lower. However, in CH_3CN frozen solution the ESR spectrum had slightly different line positions for the triplet, and the temperature dependence indicated that the triplet was 0.9 kcal/mole above a singlet state. A DMSO frozen solution of the I_3^- salt showed no triplet ESR spectrum at all.

It seems likely that in some environments **17** is a ground state triplet;

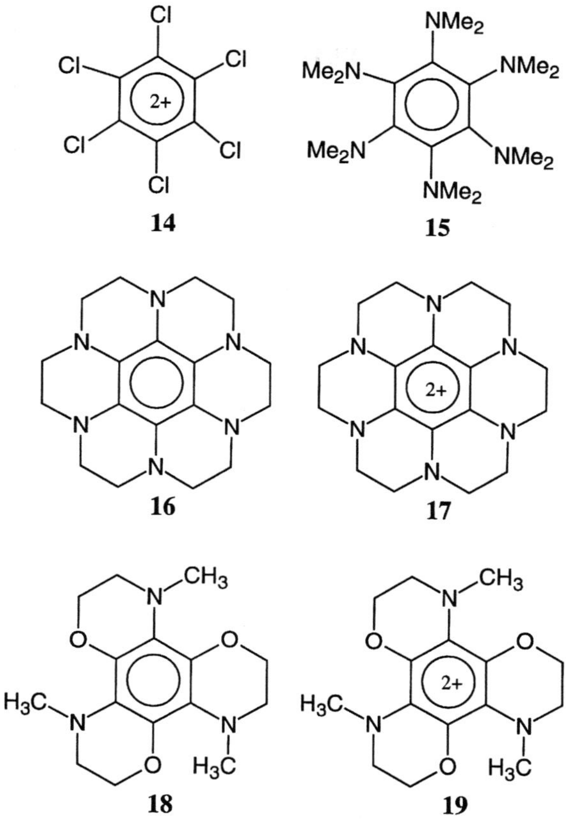

Scheme 5

but the singlet state is very close in energy, and small environmental per-turbations are enough to break the symmetry of the system and lower the singlet state below the triplet.

We also prepared the threefold symmetric compound **18** [24]. The idea here was that various distortions of **17** would break the sixfold symmetry but still leave the positive charges conjugated with nitrogen atoms. However, the dication **19** derived from **18** probably must retain threefold symmetry to keep the positive charges stabilized by nitrogen atoms. We were able to prepare the dication **19**, and saw that it had a typical triplet ESR spectrum, almost the same as for **17**. The intensity followed the Curie law from 104 K to 165 K, again suggesting a ground state triplet.

We prepared some other derivatives related to **18**, but do not yet know about the triplet question for their dications [7,24].

VI. CYCLOPROPENYL ANIONS

The simplest antiaromatic system is the cyclopropenyl anion, with 4π electrons. It is likely to have a triplet ground state, but the combination of antiaromaticity and ring strain makes it very reactive. We have studied the chemistry of the parent compound and various derivatives in an attempt to prepare stable substituted cyclopropenyl anions with threefold symmetry, to determine whether they are indeed triplets.

VI.A.   Unstabilized Cyclopropenyl Anions

Electrochemical reduction of simple cyclopropenyl cation affords the unsubstituted cyclopropenyl anion **3** as a reactive intermediate [25–27]. From the potential required and an thermodynamic cycle, we estimated [27] that the pKa of cyclopropene to form **3** is of the order of 60, so high that there is no prospect of preparing a stable solution of **3**. The estimated pKa's of 1,2,3-trimethylcyclopropene and 1,2,3-tri-t-butylcyclopropene are even higher [27]. Even though electrochemical reduction of the cations to the anions should lead to the lowest-energy electronic states, there is as yet no chemical or physical evidence whether that state is a triplet or not.

VI.B.   Triphenylcyclopropenyl Anion

The same electrochemical thermodynamic cycle [27] indicated that the pKa of 1,2,3-triphenylcyclopropene to form the anion **20** (Scheme 6) is of the order of 50. This is lower than that for the unsubstituted or alkyl derivatives of cyclopropene, but still too high for preparation of a stable solution. In a very early study [28] we treated 1,2,3-triphenylcyclopropene with strong base, but were not able to generate **20** even as a transient intermediate. Borden [29] has generated **20** transiently in solution by desilylation of triphenylcyclopropene carrying a trimethylsilyl group on C-3. There is no evidence about the electronic state of **20** at present.

VI.C.   Approaches to Stabilized Symmetric Cyclopropenyl Anions

Our earliest attempt in this field [30] was an approach to the tribenzoylcyclopropenyl anion **21**. We prepared halotribenzoylcyclopropanes **22**, with X

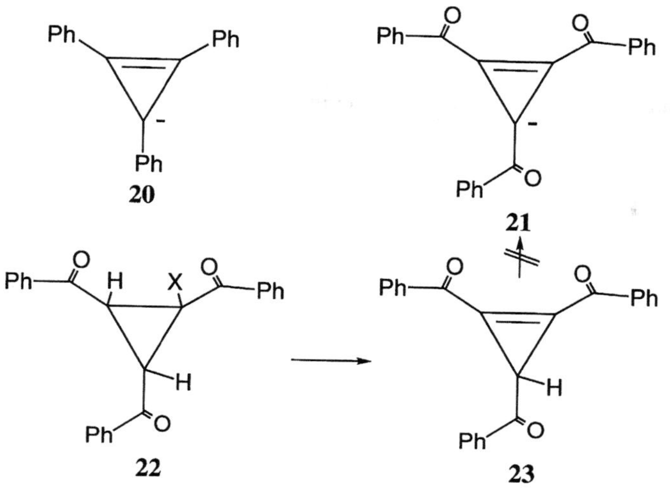

Scheme 6

as chlorine or bromine, and treated them with various bases. They eliminated HX to generate the cyclopropene derivative **23**, which could be trapped, but no base that we used converted this to a solution of anion **21**. Instead, various other reactions of reactive **23** occurred. While **21** remains an attractive target, other approaches to it are being pursued.

Another interesting target is the tricyanocyclopropenyl anion **24** (Scheme 7). Two approaches to this were tried [31]. In one, we chlorinated tricyanocyclopropane to the trichloro derivative **25**, and examined its reduction with metals or by electrochemistry. No evidence for the anion **24** was found, but instead a rearranged reductive dimer **26** was produced.

In a second approach, **25** was partially dechlorinated with tributyltin hydride to the monochloro derivative related to **23**. As with **23**, treatment with base led to HX elimination and the formation of a cyclopropene; but even with quite strong base, no anion was formed. In this case there was evidence that the cyclopropene could sit in solution with the strong base and eventually be trapped without having formed the anion **24**. It is apparent that the rate of deprotonation of tricyanocyclopropene is slow, even though the anion **24** should be quite stable, judging from the acidity of tricyanomethane.

A pyridinium cation should strongly stabilize a conjugated anion. Thus we made some approaches [32] to species **27**, which is a cyclopropenyl anion stabilized by three pyridinium cation groups. We were able to make and isolate cyclopropenes related to **27** carrying halogens or other leaving

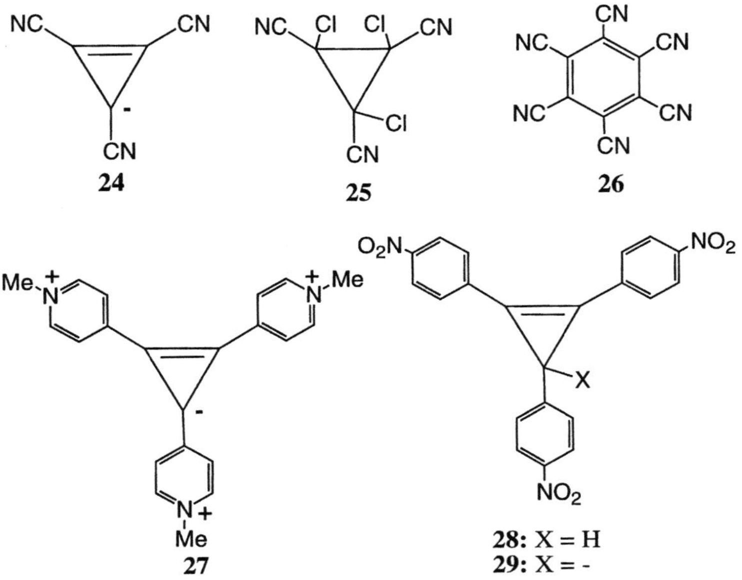

Scheme 7

groups on C-3, but all attempts to generate the anion **27** by reduction led to dimerization to a benzene ring carrying six pyridinium rings.

We have also made tris-*p*-nitrophenylcyclopropene **28**, and tried to deprotonate it to anion **29** [33]. From the effect of nitro groups on the pKa of triphenylmethane, and the pKa of triphenylcyclopropene, the pKa of **28** should be 32 or lower. However, no attempts to deprotonate **28**, even with very strong bases, have yet been successful. We are currently preparing the corresponding cyclopropene related to **28** but with a trimethylsilyl group in place of the C-3 proton. If the anion **29** is indeed as stable as our extrapolation predicts, desilylation should be able to generate it as a persistent species.

VII.   AN APPROACH TO A FERROMAGNET

With the finding that the triphenylene dication **12** is a stable triplet species, probably a ground state triplet, it was possible to try to test the idea that a charge transfer complex involving one triplet and one singlet species should lead to ferromagnetic ordering [22]. That could be the basis of bulk ferromagnetism in a solid.

There are several requirements for the singlet dianion partner of the dication. One is that its oxidation potential be such that the solid complex with **12**, or a relative of **12**, have partial charge transfer between the ions. That is, the solid must consist of a hybrid of dication/dianion and monocation/monoanion structures. It is also necessary that the solid consist of alternating cation and anion species, and in a complex with this much paired charge such alternation is extremely likely. Finally, it is necessary that the dianion not distort the environment of the dication so as to destabilize its triplet state relative to the singlet. This means in particular that the dianion should have threefold (or sixfold) symmetry.

The best candidate seemed to be the dianion **30** (Scheme 8) [34]. However, it is such a weak reducing agent that its salt with **12** showed no evidence of charge transfer to mix in a monocation/monoanion state. Thus we changed the dication [22]. Replacing the ethyl groups of **12** by β-trifluoroethyl groups led to a dication **31** that also showed a triplet ESR spectrum with Curie law behavior over the temperature range −155°C to −100°C. However, it was such a strong oxidizing agent that its salt with **30** showed no evidence of a mixed valence state, instead appearing to be a simple monocation/monoanion salt.

We then prepared dication **32**, with only two fluorine atoms on each ethyl group, and saw that it had the right oxidation potential to form a salt with **30** that indeed existed as a mixed-valence-charge transfer complex. The clearest evidence for this was in the infrared spectrum of the salt, which

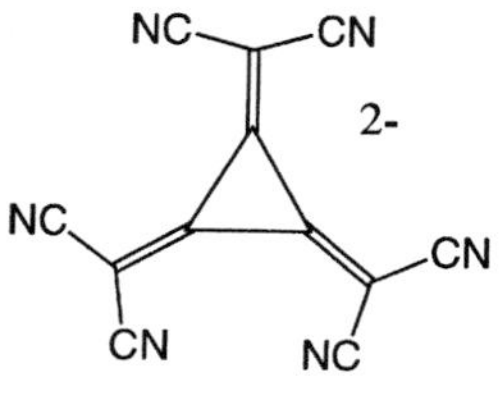

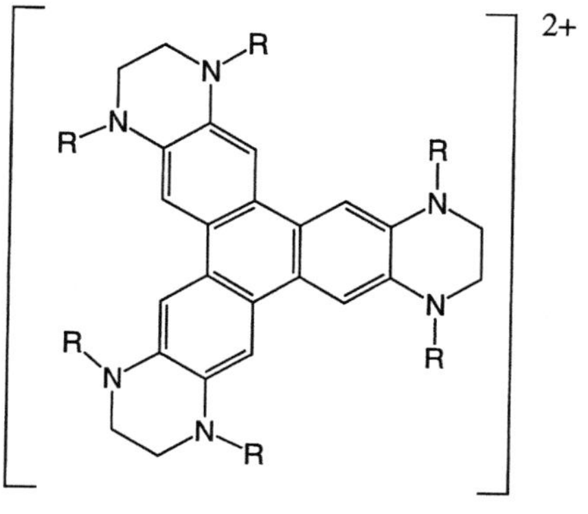

31: R = CH$_2$CF$_3$

32: R = CH$_2$CHF$_2$

Scheme 8

showed the C-N stretch of **30** intermediate in position between that for the dianion **30** and for the corresponding monoanion, and with a shape characteristic of that seen in related charge transfer complexes.

The **32/30** complex was extensively investigated. The details are published [22] and will not be repeated here. However, the major conclusion was that this complex is not ferromagnetic but has antiferromagnetic ordering.

There can be several explanations for this. One is that the cations **12**, **31**, and **32** may actually have singlet ground states, but with triplet states so low that the Curie law study does not detect this situation over the temperature range examined. Charge transfer involving two singlets will lead to antiferromagnetic ordering. Another is that environmental or electronic interaction between the cation and anion may lead to spin pairing even if the free cation were a ground state triplet. Furthermore, there are requirements for interactions in three dimensions that must be met for ferromagnetism, and there may be problems with this in the **32/30** complex.

It is fair to say that this single test of the charge transfer mechanism for producing a ferromagnet does not prove that the mechanism cannot work. However, it did not work in this particular case, which was perhaps not ideal.

REFERENCES

1. Breslow R. Antiaromaticity. Accts Chem Res. 1973; 6:393–398.
2. Breslow R, Hoffman JM Jr. Antiaromaticity in the parent cyclopentadienyl cation. Reaction of 5-iodocyclopentadiene with silver ion. J Am Chem Soc. 1972; 94:2110–2111.
3. Breslow R, Brown J, Gajewski JJ. Antiaromaticity of cyclopropenyl anions. J Am Chem Soc. 1967; 89:4383–4390.
4. Border WT. Diradicals. New York: Wiley, 1982.
5. Breslow R. Stable $4n$ pi electron triplet molecules. Pure Appl Chem. 1982; 54:927–938.
6. Breslow R. Approaches to organic ferromagnets. Mol Cryst Liq Cryst. 1985; 125:261–267.
7. Breslow R. $4n$ Pi electron triplet species and their potential use in organic ferromagnets. Mol Cryst Liq Cryst. 1989; 176:199–210.
8. McConnell H. Informal remarks. Proc Robert A Welch Found Conf Chem Res. 1967; 11:144.
9. Kollmar C, Kahn O. Ferromagnetic spin alignment in molecular systems: an orbital approach. Acc Chem Res.1993; 26:259–265.
10. West R, Niu HY. Symmetrical resonance stabilized anions, $C_nO_n^{m}$.II. $K_4C_6O_6$ and evidence for $C_6O_6^{-3}$. J Am Chem Soc. 1962; 84:1324–1325.

11. Breslow R, Chang HW. Heptaphenylcycloheptatrienyl anion, a singlet with potential spin degeneracy. J Am Chem Soc. 1962; 84:1484.
12. Breslow R, Chang HW. Heptaphenylcycloheptatrienyl anion. J Am Chem Soc. 1965; 87:2200–2203.
13. Breslow R, Chang HW. The rearrangement of the pentaphenylcyclopentadienyl cation. J Am Chem Soc. 1961; 83:3727–3728.
14. Breslow R, Chang HW, Yager WmH. A stable triplet state of pentaphenylcyclopentadienyl cation. J Am Chem Soc. 1963; 85:2033–2034.
15. Breslow R, Chang HW, Hill R, Wasserman E. Stable triplet states of some cyclopentadienyl cations. J Am Chem Soc. 1967; 89:1112–1119.
16. Breslow R, Hill R, Wasserman E. Pentachlorocyclopentadienyl cation, a ground state triplet. J Am Chem Soc. 1964; 86:5349–5350.
17. Saunders M, Berger R, Jaffe A, McBride JM, O'Neill J, Breslow R, Hoffman JM, Perchonock C, Wasserman E, Hutton RS, Kuck VJ. Unsubstituted cyclopentadienyl cation, a ground state triplet. J Am Chem Soc. 1973; 95:3017–3018.
18. Jesse RE, Biloen P, Prins R, van Voorst JDW, Hoijtinck GT. Hydrocarbon ions with triplet ground states. Mol Phys 1963; 6:633–635.
19. Breslow R, Jaun B, Kluttz RQ, Xia C-Z. Ground state pi-electron triplet molecules of potential use in the synthesis of organic ferromagnets. Tetrahedron. 1982; 38:863–867.
20. Bechgard K, Parker V. Mono-, di-, and trications of hexamoethoxytriphenylene. A novel anodic trimerization. J Am Chem Soc. 1972; 94:4749–4750.
21. Wasserman E, Hutton RS, Kuck VJ, Chandross EA. Dipositive ion of hexachlorobenzene. A ground state triplet. J Am Chem Soc. 1974; 96:1965–1966.
22. Lepage TJ, Breslow R. Charge transfer complexes as potential organic ferromagnets. J Am Chem Soc. 1987; 109:6412–6421.
23. Breslow R, Maslak P, Thomaides J. Synthesis of the hexaaminobenzene derivative hexaazaoctadecahydrocoronene (HOC) and related cations. J Am Chem Soc. 1984; 106:6453.
24. Thomaides J, Maslak P, Breslow R. Electron rich hexasubstituted benzene derivatives and their oxidized cation radicals, dications with potential triplet ground states, and polycations. J Am Chem Soc. 1988; 110:3970–3979.
25. Breslow R, Chu W. Thermodynamic determination of pK_a's of weak hydrocarbon acids using electrochemical reduction data. Triarylmethyl anions, cycloheptatrienyl anions, and triphenyl- and trialkylcyclopropenyl anions. J Am Chem Soc. 1973; 95:411–418.
26. Breslow R, Drury RF. Trapping of electrochemically generated cyclopropenyl and cycloheptatrienyl anions by charged reagents. J Am Chem Soc. 1974; 96:4702–4703.
27. Wasielewski MR, Breslow R. Thermodynamic measurements on unsubstituted cyclopropenyl radical and anion, and drivatives, by second harmonic alternating current voltammetry of cyclopropenyl cations. J Am Chem Soc. 1976; 98:4222–4229.
28. Breslow R, Dowd P. The dimerization of triphenylcyclopropene. J Am Chem Soc. 1963; 85:2729–2735.

29. Koser HG, Renzoni GE, Borden WT. Evidence for rapid pseudorotation in triphenylcyclopropenyl anion. J Am Chem Soc. 1983; 105:6359–6360.

30. Breslow R, Ehrlich K, Higgs T, Pecoraro J, Zanker F. Some reactions of 1,2,3-tribenzoylcyclopropene. Tetrahedron Lett. 1974; 13:1123–1126.

31. Breslow R, Cortes D, Jaun B, Mitchell R. Studies on the tricyanocyclopropenyl system. Tetrahedron Lett. 1982; 23:795–798.

32. Breslow R, Crispino GA. Tripyridylcyclopropene derivatives and their conversion to hexapyridylbenzene. Tetrahedron Lett. 1991; 32:601–604.

33. Klisic J, Rubin Y, Breslow R. Approaches to stable cyclopropenyl anions: tris-1,2,3-p-nitrophenylcyclopropene. Tetrahedron. 1997; 53:4129–4136.

34. Fukunaga T. Negatively substituted trimethylenecyclopropane dianions. J Am Chem Soc. 1976; 98:610–611.

4

Resonating Valence Bond Theory and Magnetic Properties

Douglas J. Klein
Texas A&M University at Galveston, Galveston, Texas

I. OVERVIEW

Magnetic properties of para-, ferro-, antiferro-, and ferrimagnetic materials typically are described in terms of the Heisenberg spin Hamiltonian, which derives [1] from an underlying quantum mechanical exchange interaction [2]. Though Heisenberg's description was "spin-free," the spin description seems to have become preferred, following influential expositions by Dirac [1] and by van Vleck [3]. General solutions of this Heisenberg model for magnetic properties are typically a challenging, many-body problem, so that rather often approximate solutions are presented in terms of simple, localized spins and the spin on such a site is viewed to respond to a mean "molecular" field due to all the other sites. Of course proper quantitative quantum mechanical detail may involve somewhat more elaborated computations, but the simplified uncorrelated mean-field picture is often the preferred picture. Especially in the ferromagnetic case, this makes immediate contact with the prequantum mechanical molecular field theory of Weiss [4]. For the antiferromagnetic case one has two different sets of sites, such that sites of one set have only neighbors that are in the other set—and for this circumstance one allows the possibility of different molecular fields on the two sublattices, as developed later in the 1930s by Néel [5].

Separately, Pauling and Wheland [6] proposed for neutral conjugated hydrocarbons a valence bond (VB) model that turns out to be representable as this same Heisenberg spin Hamiltonian for the case that the spin on each site is of spin 1/2 and the coupling parameter is of the antiferromagnet sign.

Originally the correspondence was somewhat disguised, since Pauling and Wheland represented this model instead in terms of matrix elements on a Rumer spin-pairing VB basis. The strongly correlated character is manifest in this description, with the exact solutions to be obtained via diagonalization of Hamiltonian matrices, which become larger rather quickly (i.e., exponentially fast with the number of sites). But also, Pauling and Wheland developed approximate solutions for the description of these conjugated hydrocarbons in terms of the more important VB basis functions, with singlet spin-pairings between near-neighbor sites, thereby leading to a qualitative "resonance-theoretic" picture, which matched neatly onto classical organic chemical descriptions of these compounds, e.g., even incorporating some aspects of "resonance" in prequantum mechanical work of Armit and Robinson [7], as later refined by Clar [8]. Thence now the resonance-theoretic description is often used with much success throughout organic chemistry (and some too in inorganic chemistry), although this use is usually quite qualitative and magnetic properties are usually not quite the main focus. See, e.g., Pauling's masterwork on *The Nature of the Chemical Bond* [9], or Wheland's book [10] *Resonance in Organic Chemistry*, with a little more quantum mechanical detail.

Work on the Heisenberg model in these two fields long had little overlap, and indeed at casual glance might even appear contradictory. In the chemical area the idea persisted widely that a full brute-force diagonalization over all covalent structures was needed. In physics, many-body "spin-wave" methods were developed [11,12], though again it is to be emphasized that [13] these are of a "mean-field" nature (albeit more sophisticated than the earlier molecular field ideas). But perhaps a more fundamental distinction is that the local (e.g., nearest-neighbor) spin-pairing ideas of Pauling and Wheland never lead to the long-range correlations between pairs of spins as found in the "molecular" or "spin-wave" mean-field solutions mentioned. Thus there naturally arise such questions as:

> What might be the relations between the resonance-theoretic and mean-field theories?
> Granted the fundamental distinctness of the two theories, which theory is more nearly correct?
> How might resonance theory be more neatly utilized to characterize magnetic properties?

Early on, such questions seem not to have been asked, evidently since quite distinct sets of people worked with each approach. But especially in the last two decades, interest in "molecular magnetism," and in particular in "organic magnetism," has intensified, e.g., even now with many reviews [14]. Thence such questions have become of more relevance, and here are addressed. Section II considers the first two questions, which turn out to be quite tightly related (with the basic argument being that the mean-field and resonance-theoretic views apply to different Heisenberg models for different structural circumstances). Section III, in considering the third question, describes a natural qualitative resonance-theoretic picture for magnetic properties, for the appropriate systems. Section IV recalls some more rigorous theorematic results for a general class of "bipartitic" Heisenberg models, such rigorous results being of relevance in that they should constrain whatever approximate picture is developed—further, they are noted to agree neatly under the appropriate circumstances with the resonance-theoretic picture of Section III. Finally, Section V discusses in the context of the preceding sections some general aspects of molecular and organic magnetism.

The (isotropic) Heisenberg model under consideration may be expressed as

$$H = J \sum_{i \sim j} 2\mathbf{s_i} \cdot \mathbf{s_j} \tag{1}$$

where the sum is over near-neighbor pairs of sites, $\mathbf{s_k}$ denotes a spin operator for site k, and J is an exchange-coupling parameter. Such parameters incorporating the effects of exchange of electrons between a pair of sites fall very rapidly with distance, so that the near-neighbor interaction is most important. More generally, farther distant (rapidly diminishing) exchange parameters (along with their associated interactions) might be included, and, further, the near-neighbor parameters may be different (so that J_{ij} instead appears under the $i \sim j$ summation) if not all near-neighbor pairs are equivalent. The "molecular" structure (or exchange-interaction pattern) is conveniently represented by a graph G defined in terms of the constituent sites and the set of near-neighbor pairs of sites in the structure. That is, G corresponds rather closely to what classically in chemistry is a "structural formula," and, moreover, the solution of H in Eq. (1) can be seen as a purely graph theoretic problem. A negative sign for the parameter $J < 0$ leads to ferromagnetic ordering in suitable three-dimensional systems G, whereas for the sign $J > 0$, antiferromagnet ordering can be obtained in three-dimensional lattices, most especially if G is bipartite in the sense that the set of sites of G may be divided into two sets, with all near-neighbor pairs comprised from one site of each set. In the solid-state literature, the two sets of sites in such a partitioning are termed the $\mathcal{A}$ and $\mathcal{B}$ sublattices, while in organic chemistry they are termed the starred and unstarred sets of sites. If the two sets $\mathcal{A}$ and $\mathcal{B}$ are to be of different sizes, then the starred set $\mathcal{A}$ may be chosen as the larger. Thence, e.g., for 1,3,5-trimethylenebenzene one has the circumstance of Figure 1.

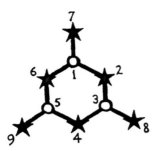

Figure 1   The case of 1,3,5-trimethylenebenzene, with starred set $\mathscr{A} = \{2, 4, 6, 7, 8, 9\}$ and unstarred set $\mathscr{B} = \{1, 3, 5\}$.

II. MOLECULAR FIELD THEORY AND RESONATING VALENCE BOND THEORY

When every site spin is a doublet (spin = 1/2) site and $J > 0$, there occurs a correspondence between solid-state and organic chemical models. But actually, molecular field theory in solid-state physics and resonating VB theory in organic chemistry have traditionally been applied to somewhat different structures. In solid-state magnetism traditionally the structures have been somewhat tightly packed, three-dimensional crystals with higher coordination number, whereas in organic chemistry the structures have been molecules with strong bonds but low in coordination number, with interactions between molecules considered no more than secondarily. Further, the solid-state structures typically involve (transition or rare-earth) metal ions frequently of spin $s > 1/2$, whereas the organic structures involve the π-electrons of conjugated carbon atoms with site spins $s = 1/2$. Perhaps one might then imagine that entirely different types of models might apply, say, Heisenberg models for the solid-state case and Hückel models for the organic case. But perhaps, might too the Heisenberg model simply have rather different types of solutions for the two characteristic types of structures? And indeed, something like this was advocated by Anderson and Fazekas [15] in the early 1970s when they proposed that for the particular case of the triangular lattice the Heisenberg model with an antiferromagnetically signed exchange interaction, one might better describe the ground state by a resonating VB wave function than by one associated with the molecular field theory. But the general type of argument may be made [16] much more generally for models with antiferromagnetically signed J and spin $s = 1/2$ sites. Indeed, one simply evaluates the ground state energy for the simplest wave functions appearing in these two types of theories.

The mean-field-theoretic case is considered briefly first. The ground state and excited state wave functions are all approximated to be of the simple product form

$$|\sigma(N)\rangle \equiv |\sigma_1, \sigma_2, \ldots, \sigma_N\rangle \equiv \sigma_1(1) \times \sigma_2(2) \times \cdots \times \sigma_N(N) \qquad (2)$$

with N being the total number of sites, σ_i labeling the spin assigned to site i being either up or down (α or β). For a bipartite graph G with antiferro-magnetically signed J, the ground state is the so-called Néel state of the form

$$\Psi_{\text{Néel}} \equiv \prod_{i \in \mathcal{A}} \alpha(i) \times \prod_{j \in \mathcal{B}} \beta(j) \qquad (3)$$

with $\sigma_i = \alpha$ if i is a starred site and $\sigma_i = \beta$ if i is an unstarred site. The approximation may be viewed to presume an effective mean-field Hamiltonian

$$H_{MF} = J \sum_{i \sim j} (\mathbf{s_i} \cdot \langle \mathbf{s_j} \rangle + \langle \mathbf{s_i} \rangle \cdot \mathbf{s_j}) \qquad (4)$$

with the expectations appearing here being statistical mechanical averages

$$\langle O \rangle \equiv \sum_{\sigma(N)} \langle \sigma(N)|O|\sigma(N)\rangle \cdot \exp\{-\langle \sigma(N)|H|\sigma(N)\rangle/kT\} \cdot Z^{-1} \qquad (5)$$

where T is the temperature and Z is the partition function $Z \equiv \Sigma_{\sigma(N)}$ $\exp\{-\langle \sigma(N)|H|\sigma(N)\rangle/kT\}$. That is, the overall mean-field Hamiltonian H_{MF} may be viewed as a sum over a set of single-site Hamiltonians $h_i \equiv \gamma\mathbf{B_j} \cdot \mathbf{s_i}$, each with a site-spin $\mathbf{s_i}$ in an effective magnetic or molecular field $\gamma\mathbf{B_i} \equiv \Sigma_{j \sim i} 2\langle \mathbf{s_j} \rangle$ whose strength is self-consistently mediated by the average relative orientations of the spins on adjacent sites. When $J < 0$, there is a preference for parallel orientation of neighbor spins; whereas for $J > 0$, (as in the VB-theoretic case), there is a preference for antiparallel orientation of neighbor spins.

Next, the resonating VB case may be considered. The wave functions are built up as combinations of VB structures, each of which involves prod-ucts of singlet spin-pairing functions of the form

$$[i, j] \equiv \alpha(i)\beta(j) - \beta(i)\alpha(j) \qquad (6)$$

for the one pairing spins associated to sites $i \in \mathcal{A}$ and $j \in \mathcal{B}$. Then each Kekulé structure K is to pair maximal numbers of such pairs with i and j neighbors. For a bipartite G with at least as many starred sites as unstarred, every spin from the unstarred set should then be paired with a neighbor site from the unstarred set while leaving any remaining starred sites unpaired; thusly:

$$|K\rangle \equiv \prod_{j \in \mathcal{B}} [K(j), j] \cdot \prod_{i \in U(\mathcal{A})} \alpha(i) \tag{7}$$

where the first product is over unstarred sites, $K(j)$ is the starred site to which j is paired, and the second product is over any (excess) unpaired starred sites. Because a pair-interaction term $2\mathbf{s}_i \cdot \mathbf{s}_j$ has a local ground state $[i, j]$ in the absence of interactions with other pairs, one anticipates that these Kekulé structures are more favored over other VB structures, in that they are composed of maximal numbers of near-neighbors such local singlet spin-paired ground states. But in general, there may be more than one such maximally neighbor-paired Kekulé structure, so that the ground state is plausibly approximated as a near uniform combination of all of them:

$$\Psi_{RVB} \equiv \sum_K |K\rangle \tag{8}$$

Indeed, this is the simplest resonating VB ground state Ansatz. Higher-order approximations would weight the different Kekulé structures differently and include VB structures with more distant spin-pairing.

Now the energies for the Néel state and a single Kekulé structure are readily evaluated rather generally to give

$$E(\text{Néel}) = \frac{-JzN}{4}$$

$$E(\text{Kekulé}) = \frac{-3JN_o}{2} \tag{9}$$

for a bipartite graph G with N sites, N_o unstarred sites, and $zN/2$ edges. Note particularly the different dependences on the average coordination number z. Thence [16] for a bipartite G with a matching number of starred and unstarred sites (i.e., $N_o = N/2$), one arrives at the plot of Figure 2. There the additional curve marked resonance theory is that obtained for a particular sequence of graphitic strips. Of course the Néel-state-theoretic picture can be improved over that obtained with just a single pure Néel state [17–20], and also the resonance-theoretic picture can improved [19–22]. Anyway, one sees that the two theories seem to have different structural regions of better application:

1. Resonance-theoretic descriptions are most appropriate for bipartite systems with low mean coordination number ($z \leq 3$) and with many Kekulé structures (per site).
2. Néel-state-based descriptions are most appropriate for (three-dimensional) structures with high mean coordination numbers and low extents of frustration (i.e., few small odd cycles of antiferromagnetically signed interactions).

energy per site

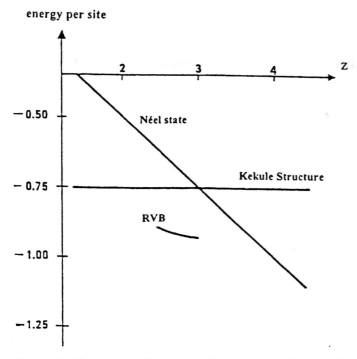

Figure 2   Plots of energies per site for several simple ground state approximants.

Notably, these rather different, largely disjoint structural circumstances are, first, that pertaining to organic chemistry and, second, that pertaining to the conventional solid-state magnetic materials. Possibly, the resonating VB theory is relevant [23] for high-T superconductivity.

III.   GENERAL PREDICTIONS FROM QUALITATIVE RESONANCE THEORY

Resonance theory in a qualitative manner has been applied [9,10] most widely for stable nonradical species. As already noted, Pauling's simple resonance theory is based upon the consideration of sets of different classical chemical-bonding patterns consistent with the considered bipartite graph. And the graph should preferably be of lower coordination number. For the (neutral) benzenoids there is one π-electron per site, and to each bonding pattern there is imagined a basis wave function with singlet spin pairings (corresponding to π-bonds), and such that the most important individual bonding patterns are those with a greater number of neighbor spin pairings (giving lower energies over the VB Hamiltonian). Further, the greater the number of low-energy such VB patterns, the greater the stabilization (because of mutual "configuration" interaction amongst them), this phenomenon being termed resonance. But even patterns with non-neighbor pairing can contribute to this resonance, especially if they are not too different from a maximally neighbor-spin-paired pattern and there are a great number of such structures. Thus, to achieve overall energetic stability the tendencies first to maximize the number of neighbor-paired sites and second to maximize the number of resonance (or VB) structures often are in competition. Then generally there is some contribution from non-neighbor bonding, and in some cases this may even be dictated by the structure (being without a fully neighbor-paired VB structure). But as an extension to the preference to neighbor pairing there is a secondary preference to a slightly more distant local "vicinity" pairing. If pairing between very distant sites makes a contribution, then the spin pairing is weak, in the sense that at a very slight energy above this there should be a state with the electron pair truly unpaired—i.e., an excited state with additional unpaired spin density and an overall spin greater by 1 than the ground state.

It is an important point that the non-neighbor pairing may be stipulated (in the ground state) to occur solely between sites in the starred and unstarred subsets (these subsets being such that any site from one subset has neighbors solely in the other subset). A double spin-pairing, one with $\mathcal{A}$ and one within $\mathcal{B}$, may be reexpressed as a linear combination of two interset double spin-pairings. Thus there is a basis of VB functions with any intraset spin-pairings confined solely to one of the sets $\mathcal{A}$ or $\mathcal{B}$. But VB functions with such an intraset spin-pairing do not gain energy stabilization due to such a spin-pairing, since there are no intraset (antiferromagnetic) exchange couplings in H for bipartite G, and such spin-pairing introduces constraints on the function that otherwise would be absent. Thence it seems that the ground state for bipartite G should have no such intraset spin-pairings.

The question of the competition between maximization of neighbor pairing and resonance may be rephrased in terms of "bond orders" and "free valences." The resonance bond order for a neighbor pair e of sites is putatively identified as the fraction p_e of ground state contributing structures that have the π-electrons on these two sites spin-paired; and the free valence for a π-center i is the fraction v_i of ground state contributing structures that have no spin-pairing to another π-electron in the vicinity of site i. What is meant by "vicinity" here may be given different interpretations. If the vicinity is taken as the limit of nearest neighbors with all spin-pairing between such neighbors (and each of the maximally spin-paied bonding patterns is taken to contribute equally to the ground state), then the p_e are called [24]

Pauling bond orders, and the associated v_i might be called Pauling free valences. For instance, in Figure 3 Pauling bond orders and free valences for three molecular species are indicated. It has been demonstrated [24,25] that for a variety of nonradicaloid benzenoids, the Pauling bond orders correlate closely with experimental bond lengths. Too, any remnant free valence may be indicative of unpaired spin density, as in the third case of Figure 3, where trimethylenemethane is known [26] to be a triplet ground state with spin density primarily on the end atoms.

The maximization of the number of neighbor-paired sites clearly minimizes the free valences. The maximization of resonance is also somewhat intuitively clear: resonance is greatest when the bonding patterns are as delocalized as possible. That is, for maximal resonance one might anticipate that the probability (or bond order) of a double bond along any one of the directions away from a site to its nearest neighbors is equally likely. For benzenoids the maximum number of neighbors is 3, so resonance would seem greatest the nearer the π-bond orders are to 1/3. Then even without explicit consideration of the Kekulé structures of a species, the maximization of resonance might lead one (tentatively) to assign 0-order bond orders of 1/3 and corresponding 0-order free valences as deficits of the sums of these 0-order bond orders incident at a site. That is, this 0-order free valence for a site i turns out to be just the deficit from 1 of one-third of the degree of site i. In many cases additional pairing between neighbor sites may further

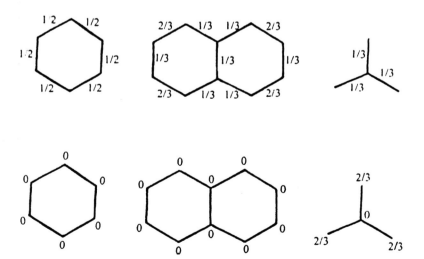

Figure 3 Pauling bond orders and free valences for benzene, naphthalene, and trimethylenemethane.

reduce the free valences, yielding first-order bond orders and first-order free valences. Thence such first-order free valences for the species of Figure 3 turn out to be the same as already illustrated. For the 1,3,5-trimethylene-benzene species of Figure 1, the first-order free valences turn out as in Figure 4. This result is, however, slightly different than that obtained with explicit listing and examination of the 18 Kekulé structures, whereby one finds free valences of 5/9 on each terminal carbon and 4/9 on each of the unstarred ring carbons with nonzero free valences in Figure 4.

But one may now introduce additional pairing between starred and unstarred sites, even if not a neighbor (though preferably as close as possible). That is, the free valences are (at least partly) satiated by other free valences located on the other type of sites (i.e., starred vs. unstarred sites). As a consequence it is just the difference between the net free valence on the two sublattices that results in unpaired spins (in the ground state). And this difference is the same for the full free valences as well as the 0-order ones, for which their sums on the starred and unstarred sites can be expressed in terms of the numbers a_z and b_z of degree-z sites on the $\star$- and $\circ$-sets of sites. That is, the overall ground state spin is predicted to be

$$S_{RT} = \frac{\left|\left\{\overset{\star}{\underset{i}{\sum}} v_i - \overset{\circ}{\underset{j}{\sum}} v_j\right\}\right|}{2} = \frac{|\{2a_1 + a_2 - 2b_1 - b_2\}|}{6} \tag{10}$$

Notice too that the (unpaired electron) spin density should appear primarily on the sites with an excess free-valence sum, especially for those such sites more well separated from opposite-type sites with nonzero free valence.

There are other rather qualitative resonating-VB-theoretic predictions [27] concerning the occurrence of "bond localization" and of the nature of excitations, as to whether they might be solitonic or bound soliton–antisoliton pairs. Indeed structural characteristics dictate the resonating-VB func-

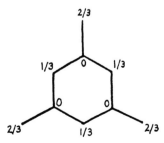

Figure 4   First-order free valences for 1,3,5-trimethylenebenzene.

tions that can occur; and if there is a tendency for the spin-pairings to be localized in these (Kekulé-like) functions, then there is corresponding π=bond localization, manifesting a shorter neighbor distance. Further, there can be different classes of VB structures to be associated to noninteracting sectors in configuration space, and thereby to give separate ground state phases, which if degenerate can only occur together in different regions, with a "soliton" separating the phases. If the phases are nondegenerate, then there is "confinement," and excitations consist of bound soliton–antisoliton pairs. Some of the manner in which molecular structures correlate with these different behaviors are understood. One case of some interest concerns the "3-leg" and ordinary "2-leg" inorganic (Cu-peroskovite) ladder structures, for which the spin-1/2 antiferromagnetically signed Heisenberg models should [28] respectively have soliton–antisoliton and single-soliton excitations, with the ground state in the former case being less well described by a resonance-theoretic wave function. Indeed, the picture was (later) supported by experimental observations [29] and by further quantitative computations [30]. In the organic realm, very much has been done (in a band-theoretic MO framework) with polyacetylene [31] and a few related species. But there are many further possibilities for novel conjugated organic polymers, as have theoretically been studied to some extent, primarily in MO frameworks [32] but also some in resonating VB pictures [18,25].

IV. THEOREMS FOR MORE GENERAL CIRCUMSTANCES

For suitable Heisenberg models, allowing general site spin as well as both antiferromagnetically and ferromagnetically signed exchange coupling (at least in certain patterns), there are a number of theorematic results. Many of these results arise from Marshall's identification [33] of a basis for which the ground state often turns out to be "nodeless." The result applies with a rather general set of site spins interacting with many different exchange parameters J_{ij} associated to interactions $2s_is_j$, and a generalized definition of bipartiteness, involving the division of the sites into two (disjoint) subsets such that any J_{ij} has a sign: $J_{ij} > 0$ or $J_{ij} < 0$ as i and j both in the same set or each in different sets of the bipartition. The two subsets might still be termed "starred" and "unstarred" or $\mathcal{A}$ and $\mathcal{B}$. Then the definitions

$$S_A \equiv \sum_{i \in \mathcal{A}} s_i \quad \text{and} \quad S_B \equiv \sum_{j \in \mathcal{B}} s_j \qquad (11)$$

prove of use. In particular, the ground state spin is

$$S_{\text{ground}} = |S_A - S_B| \qquad (12)$$

as noted elsewhere [34,35] and as conjectured by Ovchinnikov [36].

Ovchinnikov emphasized the ease with which this result enables one theoretically to design "high-spin" hydrocarbons. Further, there are some results [35] concerning point group (or translational) symmetries for the ground state. In addition, signs of ground state expectations are [35]

$$\langle \mathbf{s_i \cdot j_j} \rangle \begin{cases} < 0 & i\varepsilon\mathcal{A} \text{ and } j\varepsilon\mathcal{B}, \text{ or } i\varepsilon\mathcal{B} \text{ and } j\varepsilon\mathcal{A} \\ > 0 & i, j\varepsilon\mathcal{A}, \text{ or } i, j\varepsilon\mathcal{B} \end{cases} \tag{13}$$

and ground state spin densities are [37]

$$\langle s_i^z \rangle \begin{cases} \geq 0 & i\varepsilon\mathcal{A} \\ \leq 0 & i\varepsilon\mathcal{B} \end{cases} \tag{14}$$

where $\mathcal{A}$ and $\mathcal{B}$ have been chosen such that $S_A \geq S_B$, and the expectation is (as standard) with respect to the $M_S = S = S_{\text{ground}}$ z-component (of spin) of the ground state manifold.

These various theorems can be seen to have a degree of consonance with the Néel-state-based ideas. Equations (13) and (14) predict the same sign relations as for the Néel state $\Psi_{\text{Néel}}$ of Eq. (3). And indeed, in application to conjugated organics, Malrieu and coworkers [38] and also Yamaguchi [39] have emphasized the utility of this Néel-state-based idea.

For the case of spin-1/2 sites, these theorematic results are in consonance with the resonance-theoretic results of the preceding section. In this case

$$S_A = \frac{|\mathcal{A}|}{2} \quad \text{and} \quad S_B = \frac{|\mathcal{B}|}{2} \tag{15}$$

while also, in the notation of Eq. (10),

$$|\mathcal{A}| = a_1 + a_2 + a_3 \quad \text{and} \quad |\mathcal{B}| = b_1 + b_2 + b_3 \tag{16}$$

so that

$$S_{\text{ground}} = \frac{\{(a_1 + a_2 + a_3) - (b_1 + b_2 + b_3)\}}{2} \tag{17}$$

Yet further, since all the antiferromagnetically signed couplings $J_{ij} 2\mathbf{s_i \cdot s_j}$ occur between sites in different sets $\mathcal{A}$ and $\mathcal{B}$, one sees that the number $\#_c$ of such couplings can be written in two ways:

$$a_1 + 2a_2 + 3a_3 = \#_c = b_1 + 2b_2 + 3b_3 \tag{18}$$

Thence this can be used to eliminate a_3 and b_3 from the ground state spin expression to establish agreement:

$$S_{ground} = \frac{\{[a_1 + a_2 + (\#_c - a_1 - 2a_2)/3] - [b_1 + b_2 + (\#_c - b_1 - 2b_2)/3]\}}{2}$$

$$= S_{RVB} \tag{19}$$

Notably, too, Lieb has established [40] that the ground state spin of the Hubbard model (with all on-site repulsions equal and the only nonzero hopping parameters being between nearest neighbors such that all are equal) is given for bipartite G as $(|\mathcal{A}| - |\mathcal{B}|)/2$ when the number of electrons matches the number of sites. Indeed, in a numerical test with a slightly more elaborate Parisier–Parr–Pople model, this same result is also obtained [41] for all (~90) bipartite "chemical" G with up to eight sites. Further, the qualitative resonance-theoretic results concerning location of ground state unpaired spin density are in concert with Eq. (14) and also with quantitative numerical spin-density computations [42] for Heisenberg models. Thus the resonance-theoretic view seems quite reliable.

Finally it may be noted, even for purely organic molecules, there are cases where some of the neighbor exchange couplings are of ferromagnet sign. One such case involves conjugated π-network molecules, including carbene groups, whereat there is a dangling σ-bond. Then the exchange coupling between σ- and π-orbitals on the same atom is of ferromagnet sign, while the coupling between neighbor π-orbitals (still) is of antiferromagnet sign, and the generalized definition of bipartiteness at the beginning of this section is relevant. Indeed, a graph might be drawn out for the consequent Heisenberg model for the considered set of exchange-coupled sites, (or orbitals) with solid and dotted edges, respectively, indicating antiferromagnetically and ferromagnetically signed couplings, as in Figure 5. For this species, $|\mathcal{A}| = 13$ and $|\mathcal{B}| = 9$, so $S_{ground} = 2$, in agreement with experiment [43]. In addition, in Figure 5 first-order free valences are shown, indicating a spin-density distribution not unlike that obtained [44] via a brute-force matrix diagonalization. There is a list [45] of about a dozen such polyradical species exhibiting uniform agreement for ground state spin with experiment.

V. MOLECULAR MAGNETS

By "molecular magnets" we here understand a molecular crystal that exhibits cooperative behavior and, in particular, some sort of magnetic hysteresis. Then a molecular magnet may be a single cluster molecule, say, perhaps with as few as a dozen spin sites, as for the novel $[Mn_{12}O_{12}(CH_3COO)_{16}(H_2O)_4] \cdot 2CH_3COOH \cdot 4H_2O$ cluster that has been recently discovered [46]. But more often a molecular magnet may be an extended array of molecules (e.g., a molecular crystal), each a high-spin mol-

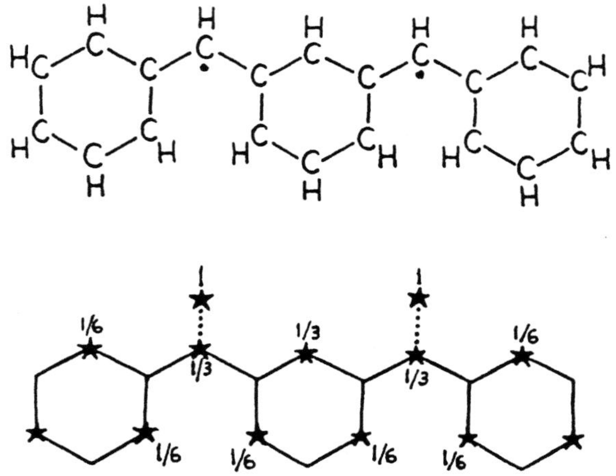

Figure 5 Formula for *m*-bis(phenylcarbenyl)benzene and the related exchange-coupling graph *G*, with first-order free valences.

ecule (or atom or ion) that then interacts with others in some "cooperative" manner. The ideas from the preceding sections can then be used in two manners:

1. In the standard context of organic chemical resonance theory to predict characteristics of (either small or macroscopically large) covalent networks
2. In the standard solid-state-physics Heisenberg modeling context to predict results of the interactions between (smaller) molecules

For the case of molecular magnetism with smaller molecules, resonating-VB theory is but one conceivable method for obtaining a prediction of the ground state spin of a molecule.

Sometimes even when a molecular orbital Hartree–Fock wave function is considered, Heisenberg models arise. For example, in the case of a conjugated molecule, a simple MO wave function might yield an open-shell circumstance: a set of (near) nonbonding orbitals, each to be occupied but once, whence one is interested in the exchange coupling between these singly occupied orbitals. Indeed, the more qualitative ideas of Borden and Davidson [47] may be neatly formulated in this manner, with ferromagnet or antiferromagnetic exchange coupling, respectively, occurring, since two such MOs have negligible or non-negligible differential overlap. In the cases that Borden and Davidson considered there were but two nonbonding orbitals and the Heisenberg model was not explicitly noted. But for greater numbers

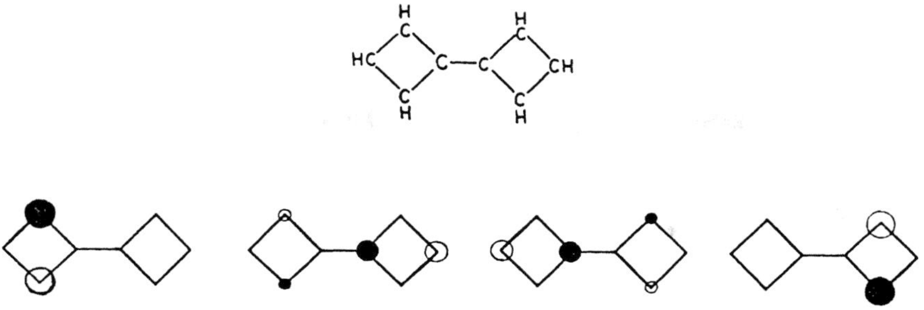

Figure 6 Structure and nonbonding MOs for bis(cyclobutadienyl).

of nonbonding orbitals, a more explicit consideration of a Heisenberg model seems worthwhile. For example, for a π-bonded pair of cyclobutadiene rings, one finds four nonbonding Hückel MOs, as indicated in Figure 6, where circles of different sizes on the different sites indicate the magnitude of MO amplitudes in proportion to the circle radii, and the shading vs. emptiness of a circle indicates the sign of the MO amplitude. Evidently there is differential overlap between the first and second nonbonding MOs as well as the third and fourth, so that antiferromagnetically signed coupling occurs between these two pairs of orbitals. On the other hand there is no differential overlap between the second and third, so one anticipates ferromagnetically signed coupling. The pairs (1,3), (2,4) and (1,4) also do not overlap, but have even less density on adjacent sites than for the (2,3) case, so one might imagine neglecting these couplings. Then the exchange-coupling graph for these four nonbonding orbitals (with solid and dashed edges, respectively, indicating antiferromagnetically and ferromagnetically signed couplings) is as in Figure 7. Starring and unstarring the sites in accord with the generalized definition of bipartiteness, one then obtains the prediction that $S_{ground} = 0$, in concert with resonance theory and with the exact theorems for the Heisenberg [34,35] and Hubbard [40] models for the 8-site π-network. Indeed, this result is in agreement with the high quality ab initio quantum

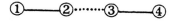

Figure 7 A plausible exchange-coupling graph for the nonbonding MOs of Figure 6.

chemical computations [48]. But a word of caution is in order: the same type of high-quality ab initio computations yield [48] a triplet ground state for a string of three cyclobutadiene rings, although the usual Heisenberg and Hubbard models for this system evidently exhibit singlet ground states. The deficiency of the nearest-neighbor model for systems with 4-membered rings is perhaps not surprising (in that it has been realized [49] for some time that there are additional cyclic 4-site exchange couplings of relevance), but the discrepancy with the Hubbard model is more surprising. (For the sequence or 1-, 2-, 3-, . . . ring strings of cyclobutadiaenes Raos et al. [48] indicate a modified sort of resonance picture that accords with their computational results.)

Regardless of the single-molecule methodology, one often is faced with a Heisenberg model whose "sites" are the different molecules (at least if the individual molecules are of high spin). The size, and particularly the signs, of the intersite exchange couplings are of special interest, and depend on the relative orientations of the molecules and of the singly occupied single-molecule "magnetic" orbitals. Indeed, this has been much considered, e.g., as reviewed by Kahn [50].

Granted the model, the long-range ordering behavior for the consequent Heisenberg model is important, for it dictates whether the material is ferromagnetic, ferrimagnetic, antiferromagnetic, paramagnetic, or whatever. And as concerns such behavior it may be mentioned that there are some qualitative theorematic results, first due to Mermin and Wagner [51]. The long-range order that they consider concerns statistical-mechanical expectations:

$$\langle(s_i^z - \langle s_i^z \rangle)(s_j^z - \langle s_i^z \rangle)\rangle, \qquad \text{for } i, j \text{ distant} \qquad (20)$$

For ferromagnetic ordering this expectation approaches a constant that is nonzero even as the distance between i and j approaches infinity. For ferrimagnetic or antiferromagnetic ordering this expectation approaches fluctuating nonzero positive and negative values, respectively, associated to pairings between the two ($\mathcal{A}$ and $\mathcal{B}$) sublattices, with the magnitudes of these values being different for the ferrimagnetic case and the same for the antiferromagnetic case. Then for a Heisenberg model solely with ferromagnetically signed couplings, it is proved [51] that there is no ferromagnetic long-range ordering at any nonzero temperature so long as G is a translationally symmetric regular lattice of one or two dimensions. For a Heisenberg model solely with antiferromagnetically signed couplings all between two sublattices ($\mathcal{A}$ and $\mathcal{B}$), it is proved [51] that there is no long-range ferrimagnetic or antiferromagnetic ordering at any nonzero temperature so long as G is a translationally symmetric regular lattice of one or two dimensions. Indeed, both these results may be extended [52] to allow G to be a general (connected) subgraph of the earlier-considered regular lattices. Thus the pattern of interactions and its dimension are crucial to the possibility of cooperative long-range ordering. But perhaps it should be mentioned that in these proofs the dimension 2 appears just at the "borderline" (so they do not apply to fractal systems with a dimension just ever so slightly greater than 2). Thence one might expect the possibility of magnetic ordering in two dimensions if there is relaxation of some of the assumptions involving the exchange couplings (and not only their signs to different sublattices, but also whether anisotropic spin–spin couplings may occur). Indeed, there evidently are two-dimensional systems experimentally observed [53] to exhibit long-range magnetic ordering—whence, the question arises as to which of the assumptions might be violated. The "no long-range order" predictions for the case of one-dimensional systems should be much more robust, the idea being that at any nonzero temperature the spoiling of long-range ordering through even very few local excitations would give a notable lowering in free energy due to the entropy increase. A rigorous result incorporating this idea is found in Ruelle's fundamental book [54], though the result there applies for a class of general lattice-gas-like models (that do not encompass the Heisenberg model).

Regardless of whether there is long-range ordering, quantitative finite-temperature solutions to Heisenberg models and the associated properties are relevant. But the solution of the Heisenberg model is a nontrivial many-body problem. Classical mean-field and spin-wave approximations are rather readily available, and seem to do best in higher dimensions with higher coordination numbers and for higher-spin sites, most especially for the ferromagnetic case. With the advent of high-temperature superconductivity, some powerful methods have been developed—e.g., perturbative (high-temperature) expansions can be made [55] to quite high order (and consequent accuracy), and the "density-matrix renormalization group" method [56] seems to be quite powerful, at least for quasi-one-dimensional systems.

On occasion, models other than the Heisenberg model are relevant. Most particular in this regard are the cases of partly filled bands or mixed-valence compounds. Again, sizes and signs of intersite couplings are of special interest, and depend on the relative orientations of the molecules and "magnetic" orbitals. And again Kahn's work can be consulted [50].

VI. CONCLUSIONS

Overall, one sees that the general theorematic results for the Heisenberg model and resonating-VB theory both seem to be quite useful. Further, if formulated appropriately there is a remarkable degree of agreement between the resonating-VB and other many-body methods common to physics. Perhaps the qualitative resonance theory may find more utility and be more widely melded with quantitative resonating-VB methods.

ACKNOWLEDGMENT

Acknowledgment for research support is made to the Welch Foundation of Houston, Texas.

REFERENCES

1. Dirac PAM. Quantum Mechanics. 3rd ed. Oxford: Clarendon Press, 1949, Sec. 58.
2. Heisenberg W. Zeit Physik. 1928; 49:619.
3. van Vleck JH. Theory of Electric and Magnetic Susceptibilities. Oxford: Clarendon Press, 1932.
4. Weiss P. J Phys Radium. 1907; 4:661.
5. Néel L. Ann Phys (Paris). 1936; 5:232.
6. Pauling L, Wheland GW. J Chem Phys. 1933; 1:362.
7. Armit JW, Robinson R. J Chem Soc (London). 1922; 38:827.
8. Clar E. The Aromatic Sextet. New York: Wiley, 1972.
9. Pauling L. The Nature of the Chemical Bond. Ithaca, NY: Cornell University Press, 1939.
10. Wheland GW. Resonance in Organic Chemistry. New York: Wiley, 1955.
11. Holstein T, Primakoff H. Phys Rev. 1940; 58:1098.
12. (a) Anderson PW. Phys Rev. 1952; 84:694. (b) Kubo R. Phys Rev. 1952; 87: 568.
13. Lieb E, Schultz T, Mattis D. Ann Phys. 1961; 16:407.
14. (a) Breslow R. Pure Appl Chem. 1982; 54:927. (b) Miller JS, Epstein AJ, Reiff WM. Science 1988; 240:240. (c) Dougherty DA. Acc Chem Res. 1991; 24:88. (d) Rajca A. Chem Rev. 1994; 94:871. (e) Kinoshita M. Japan J Appl Phys. 1994; 33:5718. (f) Gatteschi D. Adv Mat. 1994; 6:635. (f) Borden WT, Iwamura H, Berson JA. Acc Chem Res. 1994; 27:109. (g) Miller JS, Epstein AJ. Angew Chem Intl Ed Engl. 1994; 33:5718. (h) Berson JA. Acc Chem Res. 1997; 30:238.
15. (a) Anderson PW. Mat Res Bull. 1973; 8:153. (b) Anderson PW, Fazekas P. Magnetism. 1974; 30:423.
16. Klein DJ, Alexander SA, Seitz WA, Schmalz TG, Hite GE. Theor Chim Acta. 1986; 69:393.
17. (a) Vroelant C, Daudel R. Bull Soc Chim. 1949; 16:36. (b) Nebanzahl I. Phys Rev. 1969; 177:1001. (c) Bartowski RR. Phys Rev B. 1972; 55:4536. (d) Klein DJ. J Chem Phys. 1976; 64:4868. (d) Suzuki MA. J Stat Phys. 1986; 43:883.

18. (a) Garcia-Bach MA, Klein DJ. Intl J Quantum Chem. 1977; 12:273. (b) Bishop RF, Parkinson JB, Xian Y. Phys Rev B. 1991; 44:9425. (c) Harris FE. Phys Rev B. 1993; 47:7903. (d) Bishop RF, Hale RG, Xian Y. Phys Rev Lett. 1994; 73:3157.
19. Liang S, Doucet N, Anderson PW. Phys Rev Lett. 1988; 61:365.
20. (a) Garcia-Bach MA, Penaranda A, Klein DJ. Phys Rev B. 1992; 45:10891. (b) Garcia-Bach MA, Valenti R, Klein DJ. Phys Rev B. 1997; 56:1751.
21. (a) Klein DJ, Garcia-Bach MA. Phys Rev B. 1979; 19:877. (b) Baskaran G, Zou Z, Anderson PW. Solid State Commun. 1987; 69:973.
22. (a) Affleck I, Marston JB. Phys Rev B. 1988; 37:316. (b) Zeng C, Parkinson JB. Phys Rev B. 1995; 51:11609.
23. Anderson PW. Science. 1987; 235:1196.
24. Ham NS, Ruedenberg K. J Chem Phys. 1958; 29:1215.
25. (a) Pauling L. sec. 7.5 and 7.6 of 3rd ed of ref 9. (b) Herndon WC, Ellzey ML. J Am Chem Soc 1974; 16:6631. (c) Pauling L. Acta Cryst B. 1980; 36:1898.
26. (a) Dowd P. J Am Chem Soc. 1966; 88:2587. (b) Yarkony DR, Schaeffer HF III. J Am Chem Soc. 1974; 96:3754. (c) Auster SB, Pitzer RM, Platz MS. J Am Chem Soc. 1982; 100:3812. (d) Berson JA. In: WT Borden, ed. Diradicals. New York: Wiley, 1982. (e) Weinhold PG, Hu J, Squires RR, Lineberger WC. J Am Chem Soc. 1996; 118:475.
27. Klein DJ, Schamalz TG, Seitz WA, Hite GE. Int J Quantum Chem. 1986; 19:707.
28. Zivković TP, Sandleback BL, Schmalz TG, Klein DJ. Phys Rev B. 1990; 41:2249.
29. (a) Tranquada TM, Sternlieb BJ, Axe JD, Nakamura Y, Uchida S. Nature. 1995; 375:561. (b) Hirol Z, Takano M. Nature. 1995; 377:41.
30. (a) Dagotto E, Rice TM. Science. 1996; 271:618. (b) White SR, Noack RM, Scalapino DJ. Phys Rev Lett. 1994; 73:886. (c) Syljuåsen OF, Chakravarty S, Greven M. Phys Rev Lett. 1997; 78:4115.
31. Chien JWC. Polyacetylene. New York: Academic Press, 1984.
32. (a) Tyutyulkov NN, Schuster P, Polansky OE. Theor Chim Acta. 1983; 63:291. (b) Chance RR, Boudreaux DS, Eckhardt H, Elsenbaumer RL, Frommer JE, Brédas JL, Silbey R. In: Ladik J, ed. Quantum Chemistry of Polymers—Solid State Aspects. Reidel, 1984, pp 221–248. (c) Nasu K. Phys Rev B. 1986; 33:330. (d) Lahti PM, Ichimura AS. J Org Chem. 1991; 56:3030. (e) Gao Y-D, Kumazaki H, Terai J, Chida K, Hosoya H. J Math Chem. 1993; 12:279. (f) Klein DJ. Chem Phys Lett. 1994; 217:261. (g) Fujita M, Wakabayashi K, Nakada K, Kusakabe K. J Phys Soc Japan. 1996; 65:1920. (h) Minato M, Lahti PM. J Am Chem Soc. 1997; 119:2187. (i) Tyutyulkov NN, Madjarova G, Dietz F, Baumgarten M. Int J Quantum Chem. 1998; 66:425.
33. Marshall W. Proc Roy Soc (London) A. 1955; 232:48.
34. Lieb EH, Mattis DC. J Math Phys. 1962; 3:749.
35. Klein DJ. J Chem Phys. 1982; 77:3098.
36. Ovchinnikov AA. Theor Chim Acta. 1978; 47:297.

37. (a) Cheranovskii VO. Teor Eksp Kh. 1980; 16:147. (b) Klein DJ, Alexander SA. In: King RB, Rouvray DH, eds. Graph Theory and Topology in Chemistry. Amsterdam: Elsevier, 1987, pp 404–419.
38. Maynau D, Said M, Malrieu JP. J Am Chem Soc. 1983; 105:5244.
39. (a) Yamaguchi K, Yoshioka Y, Fueno T. Chem Phys Lett. 1977; 46:360. (b) Yamaguchi K, Fueno T. Chem Phys Lett. 1989; 159:465.
40. Lieb EH. Phys Rev Lett. 1989; 62:1201.
41. Alexander SA, Klein DJ, Randic M. Mol Cryst Liq Cryst. 1989; 176:109.
42. (a) Jiang Y, Zhu H, Wang G. J Molec Struct. 1993; 297:327. (b) Li S, Ma J, Jiang Y. Chem Phys Lett. 1995; 146:221. (c) Li S, Ma J, Jiang Y. J Phys Chem. 1997; 101:5567.
43. (a) Itoh K. Chem Phys Lett. 1967; 1:235. (b) Takui T, Itoh K. Chem Phys Lett. 1973; 19:120.
44. (a) Iwamura H, Izuoka A. J Chem Soc Japan. 1987:595. (b) Teki Y, Takui T, Kitano M, Itoh K. Chem Phys Lett. 1987; 142:181.
45. Alexander SA, Klein DJ. J Am Chem Soc. 1988; 110:3401.
46. (a) Friedman JR, Sarachik MP, Tejada J, Ziolo R. Phys Rev Lett. 1996; 76: 3830. (b) Thomas L, Lionti F, Ballou R, Gatteschi D, Sessoli R, Barbara B. Nature. 1996; 383:145.
47. (a) Borden WT. J Am Chem Soc. 1975; 97:5968. (b) Borden WT, Davidson E. J Am Chem Soc. 1977; 99:4587; Acc Chem Res. 1981; 14:16.
48. Raos G, McNicholas SJ, Gerratt J, Cooper DL, Karadakov PB. J Phys Chem B. 1997; 101:6688.
49. (a) Mulder JJC, Oosterhoff LJ. Chem Commun. 1970; 305:307. (b) van der Hart WJ, Mulder JJC, Oosterhoff LJ. J Am Chem Soc. 1972; 94:5724. (c) Kuwajima S. J Am Chem Soc. 1984; 106:6496. (d) Poshusta RD, Schmalz TG, Klein DJ. Mol Phys. 1989; 66:317.
50. Kahn O. Molecular Magnetism. New York: VCH Pub. 1993.
51. Mermin ND, Wagner H. Phys Rev Lett. 1966; 17:1133.
52. Klein DJ, Nelin CJ, Alexander SA, Matsen FA. J Chem Phys. 1982; 77:3101.
53. (a) van Amstel WD, deJongh LJ. Sol State Comm. 1972; 11:1423. (b) Heger G, Henrich E, Kanellakopulos B. Sol State Comm. 1973; 12:1157. (c) Gerstein BG, Chang K, Willet RD. J Chem Phys. 1974; 60:3454. (d) Losee DB, McGregor KT, Estes WE, Hatfield WE. Phys Rev B. 1976; 14:4100.
54. Ruelle D. Statistical Mechanics. New York: W.A. Benjamin, 1969, sec 5.6
55. (a) Weihong Z, Oitmaa V, Hamer CJ. Phys Rev B. 1991; 43:8321. (b) Oitmaa J, Hamer CJ, Weihong CJZ. Phys Rev B. 1992; 45:9834.
56. (a) White SR. Phys Rev Lett. 1992; 69:2863. (b) White SR. Phys Rev B. 1993; 48:10345.

5

Qualitative and Quantitative Predictions and Measurements of Singlet-Triplet Splittings in Non-Kekulé Hydrocarbon Diradicals and Heteroatom Derivatives

Weston Thatcher Borden
University of Washington, Seattle, Washington

I. INTRODUCTION

Non-Kekulé molecules are conjugated molecules for which no classical Kekulé structures can be written [1]. Since the late Paul Dowd's synthesis of trimethylenemethane (TMM) in 1966 [2], non-Kekulé hydrocarbons have been of great interest to both theoreticians and experimentalists. A major reason for this interest is that, like carbenes, but unlike almost all other organic molecules with even numbers of electrons, non-Kekulé hydrocarbons can have high-spin as well as low-spin ground states.

The first question about such a hydrocarbon that must be addressed is, therefore, which electronic state is lowest in energy. Next, one would like to know qualitatively whether the energy separation between the ground state and the first excited state is large (e.g., on the order of 10 kcal/mol) or small (e.g., on the order of 1 kcal/mol). In the latter case, if equilibration between the high- and low-spin states is fast, both states will contribute to the observed physical and chemical properties of the hydrocarbon. Most challenging of all is the quantitative prediction and/or measurement of the exact size of the energy difference between the lowest two spin states.

Although non-Kekulé hydrocarbons with large numbers of unpaired electrons have been generated and studied [3], this chapter focuses on those molecules in which just two electrons occupy two orbitals of nearly equal energy, either with parallel spins, to give a triplet state, or with spins paired, to give a singlet state. Such molecules are known as diradicals [3,4]. The non-Kekulé hydrocarbon diradicals that have been prepared and most thoroughly studied over the past 30 years are shown in Figure 1.

The tetraphenyl derivative of *m*-benzoquinodimethane (MBQDM) was the first non-Kekulé hydrocarbon diradical to be prepared. Its synthesis was reported by Schlenck and Brauns in 1915 [5], more than 50 years before Dowd published the synthesis of TMM [2], the simplest non-Kekulé hydrocarbon diradical. However, it was not until nearly 70 years after the synthesis of the Schlenck–Brauns hydrocarbon that Platz characterized the parent MBQDM diradical by electron paramagnetic resonance (EPR) [6].

Four years after Dowd reported the preparation of TMM [2], he published the synthesis of tetramethyleneethane (TME) [7], the second molecule in Figure 1 to be prepared, unadorned by phenyl substituents. Dowd also published a synthesis of 2,4-dimethylenecyclobutane-1,3-diyl (DMCBD) [8a], albeit a year after Dougherty reported the first preparation of this diradical [8b]. The last of the diradicals in Figure 1 to be prepared was 1,2,4,5-tetramethylenebenzene (TMB). In 1987 Roth and coworkers reported the first synthesis of this diradical [9a,b], which was also prepared and studied by the Berson group at Yale [9c–e].

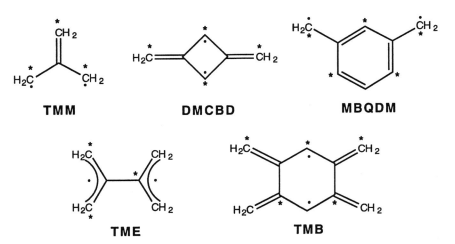

Figure 1   Some non-Kekulé alternant hydrocarbon (AH) diradicals. Starred and unstarred atoms are shown.

Sixteen years before Dowd published the synthesis of TMM, Longuet-Higgins proved an important theorem about non-Kekulé hydrocarbons, which, like those in Figure 1, contain no odd- or $4n$-membered rings [1]. He showed that the number of electrons that do not form bonds in any Kekulé structure is equal to both the number of nonbonding molecular orbitals (NBMOs) and the number of electrons that must be accommodated in them. Using this theorem, it is easy to see, simply by inspection of the structures in Figure 1, that each of these hydrocarbons is a diradical.

According to Hund's rule [10], in a diradical the lowest energy state is the triplet, in which one electron occupies each NBMO and the electrons have parallel spins. Hund's rule was actually derived for atoms; but, if it is valid for the molecules in Figure 1, all of them should have triplet ground states. This was what Longuet-Higgins expected [1].

In 1977–78, two theoretical papers appeared that predicted that violations of Hund's rule might be found in non-Kekulé hydrocarbons. The paper by Borden and Davidson [11] arrived at this prediction by use of molecular orbital (MO) theory. The paper by Ovchinnikov [12] employed a Heisenberg Hamiltonian in what chemists term a valence bond (VB) approach.

Although, as we will see, MO and VB theory eventually arrive at the same set of qualitative predictions regarding the ground states of non-Kekulé hydrocarbons, they do so from diametrically different starting points. In the former approach, the starting point is the set of π-MOs that maximize bonding. Based on whether or not the NBMOs for a diradical have atoms in common, the effect of spin state on the Coulomb repulsion energy between the pair of electrons in the NBMOs is then predicted.

In contrast, the VB approach that was used by Ovchinnikov begins by assuming that electron repulsion is minimized by placing one electron in the p-π atomic orbital (AO) on each carbon. It is then easy to predict which distribution of electron spins provides maximum π-bonding between the carbons; and Ovchinnikov showed how to determine to which electronic state this distribution of spins belongs.

These two different methods for predicting ground states are discussed in some detail in the next section. It will be shown that the MO approach has the advantage of being able to predict qualitatively whether the size of the singlet-triplet splitting (ΔE_{ST}) in a diradical is likely to be large or small. In contrast, the VB method makes no prediction about the size of ΔE_{ST}. However, the VB method is always able to make a prediction as to whether the singlet or triplet is the ground state, even in cases where the MO approach cannot predict the sign of ΔE_{ST} but can predict only that its magnitude is small. Thus, the two approaches are complementary.

The third section of this chapter reviews the current state of computational and experimental research on the relative energies of the lowest singlet and triplet states of the five diradicals shown in Figure 1. In order to show how well the qualitative methods, discussed in the second section, do when applied to non-Kekulé hydrocarbons that have more than two non-bonding electrons, recent computational results on a triradical, which is currently being studied experimentally, are also presented in the third section.

Although calculated values of ΔE_{ST} for all five of the non-Kekulé hydrocarbons in Figure 1 have been available for more than ten years, it is only very recently that photoelectron (PE) spectra of the radical anions of three of them have been obtained. The PE spectra allow very precise measurements of ΔE_{ST} in the corresponding neutral diradicals; and, as will be discussed, the calculated and experimental values of ΔE_{ST} are in very good agreement for these three diradicals. For all five of the non-Kekulé hydrocarbons in Figure 1, qualitative theory successfully predicts whether the ground state is a singlet or a triplet.

If theory does well in predicting singlet-triplet splittings in these five diradicals, what does it predict will happen when heteroatoms are present? For instance, when an oxygen atom is substituted for one or more of the methylene groups in each of these diradicals that has a triplet ground state, is it possible that a singlet will be the ground state of some or all of the oxa analogs? The fourth section of this chapter is concerned with the effects of heteroatom substitution on changing the values of ΔE_{ST} from those in the hydrocarbon diradicals.

The fifth and final section briefly summarizes the current state of experimental research on the singlet-triplet splittings in non-Kekulé diradicals. This section looks backward, to see what progress has been made during the past 30 years. It also attempts to look forward, to see where research in this area appears to be headed.

II. QUALITATIVE METHODS FOR PREDICTING GROUND STATES AND THE SIZE OF ΔE_{ST}

II.A. Ovchinnikov's Equation

As already noted, in the VB model, used by Ovchinnikov, the π-electrons are assumed to be perfectly correlated so that one electron occupies the p-π AO on each carbon. A π-bond can form between adjacent carbons only if their p-π electrons have opposite spin. Thus, in the VB model the ground state of any molecule is determined by which set of possible distributions of α- and β-electron spins allows the formation of the maximum number of bonds.

Ovchinnikov derived a simple mathematical expression for predicting the ground state of any alternate hydrocarbon (AH), whether or not it is a non-Kekulé molecule. An AH is a conjugated hydrocarbon in which the carbons can be divided into two sets, traditionally called starred and un-starred, such that no atoms belonging to the same set are nearest neighbors. As shown in Figure 1, this division is possible for all the molecules in this figure, so they are all AHs [1].

The problem of determining the spin distribution that gives the most bonding in an AH has a simple solution: put a p-π electron of one spin, for instance α, at each starred atom and one of opposite (β) spin at each un-starred atom, thus ensuring that each carbon will be able to form π-bonds to each of its nearest neighbors. If the number of starred atoms, N^*, is equal to the number of unstarred atoms, N, as is the case in most AHs (e.g., ethylene, cyclobutadiene, benzene), the number of α and β spin electrons will also be the same, and the ground state will be a singlet. However, if there are more starred than unstarred atoms in an AH, in the ground state the number of α spin electrons will exceed the number of β spin electrons by $N^* - N$. Ovchinnikov presented a proof that the excess of α spin is equal to the spin quantum number, S. It then follows that for the ground state of any AH [12],

$$S = \frac{N^* - N}{2} \tag{1}$$

Application of this formula to predicting the spin of the ground states of the non-Kekulé hydrocarbons in Figure 1 simply involves counting starred and unstarred atoms. Noting that in TMM, DMCBD, and MBQDM, $N^* - N = 2$, these molecules are all predicted to have triplet ($S = 1$) ground states. However, when Eq. (1) is applied to predicting the ground state of TME and TMB, since $N^* - N = 0$, both diradicals are predicted to have a singlet ground state and thus to violate Hund's rule.

Klein and coworkers have provided a mathematically more rigorous derivation of Eq. (1) [13] and shown how it can be extended to predicting the ground states of conjugated polycarbenes [14]. Since vinyl and aryl monocarbenes have triplet ground states, the preferred spin of an electron in the σ-orbital of such a carbene is the same as that of an electron in the p-π AO on the carbenic carbon. Thus, Eq. (1) remains valid for predicting the ground state of a conjugated polycarbene if each carbenic center is given a value of 2 in summing up N^* and N.

For example, in each repeating unit of the oligomeric carbene **A**, shown in Figure 2, there is one more starred than unstarred atom. Therefore, the spin quantum number of the ground state of the oligomer with m re-

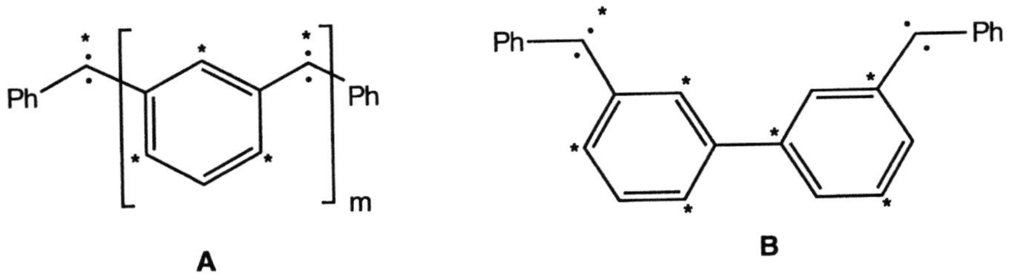

Figure 2   Non-Kekulé carbenes with $N^* \neq N$ (**A**) and $N^* = N$ (**B**).

peating subunits should be given by the formula $S = m + 1$. Experimental confirmation of the predictions implicit in this expression have been provided by the research of Itoh and Wasserman for $m = 1$ [15] and by Itoh, Iwamura, and coworkers for $m = 2$, 3, and 4 [16]. Branched variants of **A** with 1,3,5-trisubstituted benzene rings and six [17a,b] or nine [17c,d] carbenic centers have been prepared by Iwamura and coworkers and found, as expected, to have high-spin ground states with, respectively, $S = 6$ and 9.

In contrast to **A**, bis-carbene **B** in Figure 2 has $N^* = N$. Therefore, it should have $S = 0$. In agreement with the expectation of $S = 0$ in B, in 1978 Itoh found that B does, indeed, have a singlet ground state [18a]. Subsequently, other examples of biscarbenes [19a], non-Kekulé hydrocarbons [19b], bisnitrenes [20b,c], and bisnitroxides [20c,21b] with $N^* = N$ and singlet ground states have been reported. Thus, as predicted by Eq. (1), all of these molecules may be viewed as violating Hund's rule [22].

Itoh explained the difference between the ground states of **A** and **B**, not on the basis of Eq. (1) and the VB model that underlies it, but rather in terms of the form of the π-NBMOs of these two carbenes [18a]. He pointed out that in **A** the two π-NBMOs both are confined to only the starred carbons, whereas in B the two NBMOs are confined to disjoint sets of atoms, one to the starred set and the other to the unstarred set.

In the previous year, Borden and Davidson had anticipated Itoh's experimental result by predicting that non-Kekulé hydrocarbon diradicals with disjoint NBMOs might have singlet ground states and thus violate Hund's rule [11]. When Itoh published his 1978 paper [18a], he apparently did not know of this paper by Borden and Davidson or, for that matter, of the paper by Ovchinnikov [12]. It is almost certain that none of these authors was aware of the 1936 paper in which Hückel had pointed out that, because the π-NBMOs of **B** do not have any atoms in common, Hund's rule might fail for the m,m'-Schlenck−Brauns hydrocarbon, the diradical that has this π-system [18b].

II.B. Disjoint and Nondisjoint NBMOs—The MO Approach

Borden and Davidson chose Hückel theory as their starting point [11]. As noted earlier, VB theory assumes that electrons are correlated perfectly, thus minimizing their Coulomb repulsion energy; and bonding is conceived of as a pertubation on this distribution of electrons. In contrast, Hückel theory totally ignores Coulomb's law and focuses only on maximizing the buildup of electron density in the regions where the AOs on the atoms overlap most strongly.

Borden and Davidson considered how the form of the two Hückel NBMOs of an AH diradical affects the mutual Coulomb repulsion energy of the two electrons that occupy these MOs. The Hückel NBMOs of any AH can easily be found, without actually doing a Hückel calculation, by using the zero-sum rule [1]. In the NBMOs of an AH the coefficients of the starred atoms must sum to zero about each unstarred atom. The coefficients of the unstarred atoms must also sum to zero about each starred atom, but in some diradicals the coefficients of the unstarred atoms may all be identically zero.

If the coefficients of the unstarred atoms in one of the NBMOs of a diradical are nonzero, the NBMOs can be chosen so that one spans just the starred set of atoms and the other spans just the unstarred set. Since NBMOs of this type have no atoms in common, they were described by Borden and Davidson as being *disjoint*.

On the other hand, if the coefficients of the unstarred atoms are identically zero in all the NBMOs, the NBMOs are obviously confined to only the starred atoms. Although the NBMOs may still be disjoint (vide infra), much more frequently they have atoms in common. If they do have atoms in common, the NBMOs are termed *nondisjoint*.

It is possible to tell whether or not the NBMOs of an AH are all confined to the starred set of atoms, simply by counting the number of atoms that are starred and unstarred. By convention, if there are unequal numbers of starred and unstarred atoms, the starred set is always chosen to be the more numerous (i.e., $N^* \geq N$). It can be proven [1,11] that the difference between the number of starred and unstarred atoms, $N^* - N$, is equal to the difference between the number of NBMOs that span just the starred set of atoms and the number that span just the unstarred set. Thus, $N^* - N = 0$ means that equal numbers of NBMOs span just the starred and just the unstarred sets of atoms; so $N^* - N = 0$ is a sufficient condition for showing that one or more pairs of NBMOs in a non-Kekulé hydrocarbon are disjoint.

II.C.   How the Form of the NBMOs Affects ΔE_{ST}

Whether the NBMOs of a non-Kekulé molecule are disjoint or nondisjoint has a profound effect on the energy difference between high- and low-spin states. In the state of highest spin, where all the electrons in the NBMOs have parallel spins, the Pauli principle (which is embodied mathematically in the antisymmetrization of the electronic wave function) gives zero probability of finding any of these electrons in the same AO simultaneously. Consequently, even if the NBMOs are nondisjoint, the spatial wave function for the highest spin state does not contain ionic terms, corresponding to the simultaneous occupancy of the same AO by any of these electrons.

In contrast, since the Pauli principle does not apply to electrons of opposite spin, wave functions for lower spin states *usually* contain such ionic terms. These terms have high Coulomb energies, since, when two electrons occupy the same AO, the distance between them is obviously smaller than when the electrons occupy different AOs. This is the physical reason why high spin states usually have substantially lower energies than states of lower spin and why Hund's rule is generally successful.

However, Borden and Davidson pointed out that in an AH diradical with $N^* - N = 0$, since the NBMOs are disjoint, the nonbonding electrons will, *regardless of spin*, be confined to different sets of atoms (starred for one electron, unstarred for the other). Therefore, in diradicals with disjoint NBMOs, there is no probability that two electrons in different NBMOs will simultaneously occupy the same AO in either the singlet or the triplet state. Thus, at least to a first approximation, the singlet and triplet will have the same energy.

This physical picture is embodied in the mathematical expression $\Delta E_{ST} = 2K_{ij}$ for the energy difference between the singlet and triplet coupling of the spins of two electrons, one of which occupies MO ψ_i and the other which occupies ψ_j [4a]. K_{ij} is the symbol for the exchange integral, which corrects for the fact that two electrons of the same spin in ψ_i and ψ_j cannot simultaneously appear in the same region of space.* The mathematical expression for K_{ij} is

$$K_{ij} = \int \psi_i(1)\psi_j(1) \left[\frac{e^2}{r_{12}}\right] \psi_i(2)\psi_j(2) \, d\tau \tag{2}$$

The size of K_{ij} depends on the extent to which ψ_i and ψ_j both have large amplitudes in the same region of space. If ψ_i and ψ_j are confined to different sets of atoms, K_{ij} may not be exactly zero, but it will certainly be

---

*In VB theory "exchange" has a different meaning. It is the name given to the bonding interaction between two electrons of opposite spin.

quite small, certainly much smaller than when ψ_i and ψ_j have atoms in common. Therefore, when ψ_i and ψ_j are disjoint, $\Delta E_{ST} = 2K_{ij} \approx 0$, because the exclusion principle provides no energetic advantage for the triplet over the singlet.

In contrast, when ψ_i and ψ_j are nondisjoint, $\Delta E_{ST} = 2K_{ij} \gg 0$. In this case a large singlet-triplet splitting is expected, because the exclusion principle keeps the two nonbonding electrons from appearing simultaneously in the region of space where the NBMOs both have large amplitudes.

II.D. Non-Kekulé Diradicals with $N^* = N$, Which Therefore Have Disjoint NBMOs

Tetramethyleneethane (TME) and TMB each have $N^* - N = 0$; so Eq. (1) predicts a singlet ($S = 0$) ground state for both. Since $N^* - N = 0$, both diradicals must also have disjoint NBMOs; so $\Delta E_{ST} \approx 0$. Valence bond theory unequivocally predicts a singlet ground state, but makes no prediction about the size of the splitting. In contrast, solely from the disjointness of the NBMOs, it can be predicted that the singlet-triplet splitting is small; but the prediction of a singlet ground state cannot be made. However, using both methods simultaneously leads to the prediction that TME and TMB each have small singlet-triplet splittings, with the singlet below the triplet in energy.

The reason why TME and TMB are both expected to have singlet ground states may be seen by following the VB prescription for maximum bonding in these two diradicals. This involves placing an α spin electron at each starred atom and a β spin electron at each unstarred atom, as shown for TME in structure **A** of Figure 3. It can then be easily seen that this spin distribution allows π-bonding to occur across the C–C single bond that connects the two three-carbon fragments, and it also allows long-range π-

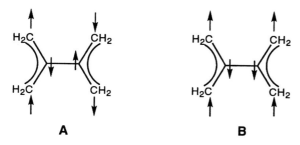

Figure 3 Valence bond depictions of the lowest singlet (**A**) and triplet (**B**) state of TME.

bonding interactions between the terminal methylene groups of the two fragments. In contrast, in the lowest triplet state of TME, which is depicted in structure **B** of Figure 3, the parallel spins at these pairs of carbons prevent π-bonding between them.

The prediction of singlet ground states for both TME and TMB can be made using MO theory, albeit not nearly as simply as through the use of VB theory. The disjoint NBMOs in TME are the localized NBMOs for each of the two allyl moieties; one NBMO spans the two starred methylene groups and the other NBMO the two methylene groups that are unstarred. In TMB the disjoint NBMOs are the localized NBMOs for each of the two pentadienyl moieties.

Because TME and TMB are disjoint diradicals, in both molecules the lowest singlet and triplet states are expected to have comparable energies. However, two types of bonding, which are not included in simple Hückel theory, cause the singlet to fall below the triplet in energy [4a,11]. The first of these is bonding between nonnearest-neighbor p-π AOs. These long-range interactions cause ψ_S—the in-phase combinations of the localized, allylic NBMOs in TME and pentadienylic NBMOs in TMB—to have a lower energy than ψ_A—the out-of-phase combination in each diradical.

The wave function for the lowest triplet state is

$$\Psi(T) = |\cdots \psi_S^\alpha y_A^\alpha\rangle \tag{3}$$

Since it places one electron in ψ_S and one electron in ψ_A, it cannot profit from the lower energy of the former NBMO.

The wave function for the lowest singlet state is

$$\Psi(S) = c_S |\cdots \psi_S^\alpha \psi_S^\beta\rangle - c_A |\cdots \psi_A^\alpha \psi_A^\beta\rangle \tag{4}$$

It is easy to show that this wave function places one electron in $\sqrt{c_S}\psi_S + \sqrt{c_A}\psi_A$ and the other in $\sqrt{c_S}\psi_S - \sqrt{c_A}\psi_A$ [4]. These are called the *generalized valence bond* (GVB) orbitals. When $c_S = c_A$, the GVB orbitals are the disjoint pair of localized allylic NBMOs in TME and pentadienylic NBMOs in TMB.

However, since nonnearest neighbor interactions between AOs results in ψ_S being lower in energy than ψ_A in both TME and TMB, $c_S > c_A$ in Eq. (4). Because more electron density is placed in ψ_S than in ψ_A, the singlet can take maximal advantage of the long-range bonding between the AOs of the localized NBMOs, whereas, as already noted, the triplet cannot. Thus, nonnearest-neighbor bonding interactions selectively stabilize the singlet.

The second effect that selectively stabilizes the lowest singlet state of TME and TMB involves π-bonding between the carbons where the two localized radical fragments are joined by single bonds [11]. The NBMOs have nodes at these carbons; so one might expect that the spins of the electrons in the NBMOs would not have any effect on the π-bonding between these pairs of nodal carbons. However, when electron correlation is included, the spins in the NBMOs of allyl and pentadienyl radicals cause an uneven distribution of the α and β spin electrons in the bonding π-orbitals; and this spin polarization results in some unpaired spin density appearing at the nodal carbons [4a].

In triplet TME and TMB, the spins that are induced at the nodal carbons in each of the radical fragments are parallel, since the spins that induce them are parallel. This is depicted in Figure 3B for TME. In contrast, in the lowest singlet state of each on these diradicals, the spins that are induced at the nodal carbons in each of the radical fragments are paired, since the spins at these carbons are induced by electrons of opposite spins in the NBMOs. This is depicted in Figure 3A.

In a singlet diradical electrons of α and β spin have equal probability of appearing in each NBMO. Thus, Figure 3A depicts the distribution of electron spins in just half the wave function for the lowest singlet state of TME. In the other half, all the spins are reversed. Consequently, unlike the case in monoradicals and in triplet diradicals, spin polarization in singlet diradicals is not static but dynamic.

By inducing opposite spins at the nodal carbons in the lowest singlet states of TME and TMB, *dynamic spin polarization* [4a] allows π-bonding between these carbons, which is absent from the triplet, where the spins induced at the nodal carbons are parallel. Consequently, dynamic spin polarization selectively stabilizes the lowest singlet state over the triplet in TME and TMB [11] and thus contributes to the violation of Hund's rule that is predicted computationally for both of these non-Kekulé, AH diradicals [22].

The specific examples of TME and TMB serve to illustrate the strengths and weaknesses of the VB and MO approaches to predicting the ground states of diradicals. Equation (1) predicts that both molecules should have singlet ground states; but, by itself, Eq. (1) gives no indication of whether the energy separation between the singlet and triplet in these diradicals is expected to be large or small. Of course, the size of ΔE_{ST} can be assessed by drawing VB structures like those in Figure 3 for TME and noting that the nearest-neighbor interactions that result in the prediction of a singlet ground state for TME and TMB by Eq. (1) involve the small, negative spin densities at the nodal carbons.

On the other hand, using the MO approach, the prediction of small energy separations between the lowest singlet and triplet state in TME and TMB is easily made, because the NBMOs are disjoint. However, the prediction of which state is actually lowest in energy requires additional considerations, such as the effects of long-range bonding and dynamic spin polarization. The former effect always stabilizes the singlet; and, except in a few cases where the latter effect is very weak anyway, it too selectively stabilizes the singlet. Therefore, if the NBMOs of a diradical are disjoint, it can be predicted that not only is the singlet-triplet splitting likely to be small, but also that the singlet is likely to be the ground state.

II.E. Non-Kekulé Diradicals That Have $N^* \neq N$ but Whose NBMOs Are Disjoint

If $N^* - N \geq 2$, at least two of the NBMOs will span just the starred atoms. However, it is still possible that the NMBOs may, nevertheless, be disjoint [11]. As an example, Borden and Davidson cited pentamethylenepropane (PMP), whose structure is depicted in Figure 4. In PMP, $N^* - N = 2$; and, hence, both NBMOs in this diradical are confined to just the starred atoms. However, it is clear from inspection of the structure of PMP that the NBMOs of PMP are the NBMOs of the two allyl moieties that PMP contains. These NBMOs are disjoint; and, moreover, unlike the disjoint NBMOs in TME, those in PMP are separated by a vinylidene group. Iwamura has suggested the term "doubly disjoint" to describe this type of NBMO topology [20c,21b].

Equation (1) predicts a triplet ground state for PMP, but not the size of the singlet-triplet energy gap. However, as a result of the interposition of a vinylidene group between the allyl moieties in PMP, one might anticipate that the amount of interaction between the small spin density at the central carbons of the allyl moieties would be much less in PMP than in TME. This is indicated by comparison of the VB structure for triplet PMP, which is shown in Figure 4, with that for the singlet ground state of TME, which is shown in Figure 3.

Figure 3 suggests that in TME the energy difference between the singlet and triplet states, resulting from the interaction of electrons on the two

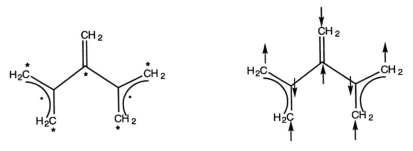

Figure 4   Bonding in PMP and VB depiction of the triplet state.

central carbons, might be approximated by the single-triplet energy difference in ethylene, scaled by some factor involving the amount of unpaired spin density at the central carbons of the two allyl moieties. Figure 4 indicates that, using the same type of model, the energy difference between the singlet and triplet states of PMP should be no greater than the singlet-triplet energy difference in TMM times the same scaling factor. Since ΔE_{ST} is a factor of about 5 smaller in TMM than in ethylene, this model suggests that in PMP there should be a substantial reduction in the already small singlet-triplet energy difference that is expected in TME.

Studies by the groups of Iwamura [20c] and Lahti [20d] of isomeric, conjugated bisnitrenes and by Iwamura of the corresponding bisnitroxides [20c,21b] provide support for this qualitative prediction. The effect of spin polarization in the "doubly disjoint" isomer was found to be so small that, in contrast to the prediction of Eq. (1), the ground state of the doubly disjoint diradical was actually found to be a singlet.

This finding is not really surprising, because Eq. (1) considers only nearest-neighbor interactions, which in PMP should confer only a tiny energy advantage on the triplet. Long-range bonding interactions between the two allyl moieties in PMP, which selectively stabilize the singlet, are presumably responsible for the latter being found to be the ground state of the doubly disjoint diradical.

II.F. Non-Kekulé Diradicals That Have $N^* - N = 2$ and Nondisjoint NBMOs

Non-Kekulé AHs with no four-membered rings have $N^* - N$ Hückel NBMOs that are confined to only the starred set of atoms [1,11]. As in PMP, some of these NBMOs may be disjoint, but whether or not this is the case in a particular molecule can be quickly ascertained by using the zero-sum rule [1] to find the NBMOs.

In a molecule that does have $N^* - N \geq 2$ NBMOs that span common atoms, Hund's rule is expected to apply to the $N^* - N$ electrons in them [4a,11]. Consequently, the ground state of a non-Kekulé AH with $N^* - N$ nondisjoint NBMOs is expected to have this number of electrons with parallel spins. Thus, its spin quantum number is predicted to be $S = (N^* - N)/2$, in agreement with Eq. (1).

As already discussed, Eq. (1) predicts only the multiplicity of the ground state, not the size of the energy difference between it and the lowest excited state. However, if a molecule is found to have nondisjoint Hückel NBMOs, it can also be predicted that there is a substantial energy difference between the high-spin ground state and excited states with lower spin [4a,11]. Only in the high-spin ground state does the Pauli principle prevent the electrons in the Hückel NBMOs from appearing simultaneously on the atoms that the NBMOs have in common and thus giving rise to high-energy ionic terms in the wave function. Consequently, ΔE_{ST} is expected to be considerably larger in non-Kekulé molecules with nondisjoint NBMOs than in those in which the NBMOs are disjoint or "doubly disjoint." As discussed in Section III of this chapter, this qualitative prediction is supported by the results of both calculations and experiments on the diradicals in Figure 1, as well as by the studies of isomeric bisnitrenes [20] and bisnitroxides [20c,21] with nondisjoint, disjoint, and doubly disjoint NBMOs.

A third prediction, made possible by the finding that a non-Kekulé AH has nondisjoint Hückel NBMOs, is that low-spin excited states will have very different equilibrium geometries from the geometry of the high-spin ground state [4a,11]. When the Hückel NBMOs are nondisjoint, the very large Coulombic repulsion between the nonbonding electrons in low-spin excited states can be reduced if the NBMOs for these states are confined to different sets of atoms. However, localization of the NBMOs to disjoint sets of atoms comes at a price, because the MOs used by the low-spin excited states provide less π-bonding than the more delocalized Hückel orbitals that are utilized by the high-spin ground state. The more localized MOs of the low-spin states result in their having C–C bond lengths that are more unequal than those of the high-spin ground state.

II.G. Dependence of the Equilibrium Geometry on the Spin State—The Example of TMM

A very well-studied example of different spin states of a nondisjoint diradical having very different equilibrium geometries is provided by TMM [4a,11,23]. The bonding in each of the three lowest electronic states of TMM is depicted in Figure 5. The bonding in the triplet ground state is fully delocalized, with π-bonds of equal strength between the central carbon and each of the three methylene groups and equal spin densities on all three.

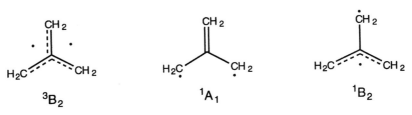

Figure 5   Bonding in the three lowest electronic states of TMM.

However, the bonding in each of the two lowest singlet excited states is much more localized.

The reason for this difference is that, as shown in Figure 6, the Hückel NBMOs, which are utilized by the $^3A_2'$ state, are nondisjoint. Consequently, if singlet wave functions, either the two-configuration wave function for $^1A_1$,

$$^1A_1 = \frac{|\cdots 2b_1^\alpha 2b_1^\beta\rangle - |\cdots a_2^\alpha a_2^\beta\rangle}{\sqrt{2}} \tag{5}$$

or the open-shell singlet wave function for $^1B_2$

$$^1B_2 = |\cdots 2b_1 a_2(\alpha\beta - \beta\alpha)/\sqrt{2}\rangle \tag{6}$$

utilized the Hückel NBMOs, these wave functions would contain high-energy, ionic terms, corresponding to the simultaneous occupancy of the same AOs by both nonbonding electrons.

A lower energy is obtained for each singlet state with a set of MOs that are different from the Hückel orbitals and that confine the two electrons of opposite spin to different atoms [23b]. For example, in the $^1B_2$ state, confining the $2b_1$ NBMO to the carbon where the a_2 NBMO has a node prevents the electrons that occupy these orbitals from ever appearing in the same AO. However, the resulting reduction in Coulomb repulsion energy in $^1B_2$ is achieved only at the expense of a reduction in π-bonding from that in $^3A_2'$. As shown in Figure 5, the π-bonding in $^1B_2$ resembles that in an allyl radical, with the second nonbonding electron localized at the remaining methylene group.

Consequently, this singlet state prefers a different equilibrium geometry than the D_{3h} geometry of the triplet, a geometry with two medium-length and one long C–C bond. In contrast, the $^1A_1$ state has a strong π-bond to one methylene carbon and little π-bonding to the other two; so this

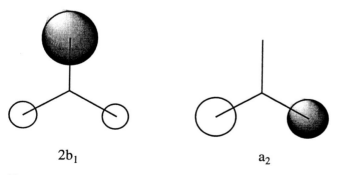

$$2b_1 \qquad\qquad a_2$$

Figure 6 Hückel NBMOs for TMM. Only one lobe of each p-π AO is shown.

state is calculated to prefer a geometry with one short and two long C–C bonds. These two states are predicted to have nearly the same energy, with $^1A_1$ slightly lower in energy than $^1B_2$ [23].

Since TMM has three equivalent methylene groups, there must be three equivalent $^1A_1$ states and three equivalent $^1B_2$ states. Singlet TMM can pseudorotate between three equivalent $^1A_1$ minima via $^1B_2$ transition states on the potential surface for the planar diradical [4a,23b]. However, lower in energy than both of these planar TMM singlet states are three equivalent nonplanar $^1B_1$ states, in which one methylene group is twisted out of conjugation. Twisting the unique methylene group out of conjugation relieves the small antibonding interaction between the $p\text{-}\pi$ AO on this carbon and the $p\text{-}\pi$ AO on the central carbon in the wave function for the planar $^1B_2$ state [23].

The prediction of differences between the bonding in the triplet and in the two lowest singlet states of TMM can also be made from VB theory. Valence bond structures for TMM are shown in Figure 7. The first structure, which has α spin electrons at the three methylene groups and a β spin electron at the central carbon, corresponds to the $S_z = 1$ component of the triplet ground state. In this state, π-bonding occurs between the central carbon and all three methylene groups.

The next three structures, in which a second β spin electron is located on one methylene group, obviously all have $S_z = 0$. The sum of all three structures represents the second of the three components of the triplet state. The out-of-phase combination of just the first two represents $^1A_1$, in which there is a strong π-bond to one methylene carbon and nonbonding electrons of opposite spin localized, one each, on the other two methylene groups. The third $S_z = 0$ structure, together with a smaller contribution from the sum of the first two, represents $^1B_2$, in which the central carbon is π-bonded to two of the methylene groups but not to the third.

Like TMM, DMCBD and MBQDM have nondisjoint NBMOs. Consequently, like the former non-Kekulé hydrocarbon, the latter two can be

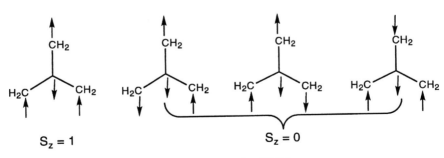

Figure 7 $S_z = 1$ and $S_z = 0$ VB structures for TMM.

predicted to have (a) triplet ground states, (b) large singlet-triplet splittings, and (c) singlet excited states in which the bonding is much more localized than that in the triplet ground state. The next section discusses the results of calculations and experiments on these three nondisjoint diradicals, as well as on TMM and TMB, the two disjoint diradicals in Figure 1.

III. CALCULATIONS AND MEASUREMENTS OF SINGLET-TRIPLET ENERGY DIFFERENCES

III.A. Trimethylenemethane (TMM)

As noted in the introduction, TMM was the first of the non-Kekulé diradicals in Figure 1 to be synthesized [2]. It was also the first for which (a) the triplet was shown to be the ground state by the finding that the intensity of the EPR signal gave a linear Curie plot [24]; (b) the lowest singlet was found experimentally to prefer a geometry different from the triplet [25]; and (c) the singlet-triplet energy difference was measured accurately by PE spectroscopy on the radical anion [26]. The results of these experiments are all in agreement with the qualitative expectations that are based on the NBMOs of TMM being nondisjoint [11]. In fact, it was really the experimental data on TMM and the results of the calculations that were subsequently performed on this non-Kekulé hydrocarbon [23] that provided the basis for the qualitative expectations for diradicals with nondisjoint MOs.

Of the many calculations that have been performed in order to compute the singlet-triplet splitting in TMM, only the two most recent will be discussed here. Cramer and Smith carried out (10/10)CASSCF calculations to correlate all the electrons in the C–C bonds and then used second-order perturbation theory (CASPT2) to recover the rest of the correlation energy [23i]. For the energy difference between $^1A_1$ and $^3A_2'$, they obtained $\Delta E_{ST} = 17.5$ kcal/mol. Hrovat and Borden used (4/4)CASSCF to correlate the π-electrons and performed CI with single and double excitations [23j]. After addition of corrections for quadruple excitations and calculated differences in zero-point energies and heat capacities, they also obtained $\Delta E_{ST} = 17.5$ kcal/mol for the energy difference between $^1A_1$ and $^3A_2'$. Both groups calculated that $^1B_1$ lies ca. 2 kcal/mol below $^1A_1$.

The experimental measurement of the singlet-triplet splitting in TMM was made possible by the development by Squires and coworkers of a very clever method for the generation of the radical anion in the gas phase [27]. Reaction of $H_2C{=}C[CH_2Si(CH_3)_3]_2$ with a mixture of F^- and F_2 formed the radical anion of TMM, and this synthesis allowed the measurement of ΔE_{ST} in the neutral from the PE spectrum of the radical anion [26].

Since the TMM radical anion is expected to be planar, Franck–Condon factors should favor formation of planar TMM on photodetachment of an electron. Therefore, the PE spectrum was interpreted as measuring the adiabatic energy difference between $^1A_1$ and $^3A_2'$. The experimental value of $\Delta E_{ST} = 16.1$ kcal/mol and the calculated value of 17.5 kcal/mol for the adiabatic energy difference between $^1A_1$ and $^3A_2'$ are in good agreement.

Although ΔE_{ST} in TMM has now been measured, an experiment performed by Dowd, long before this measurement was made, suggested a value less than half as large. Dowd found that the EPR signal for TMM disappeared with $E_a = 7.8$ kcal/mol [28]. He attributed the disappearance as being due to closure of triplet TMM to methylenecyclopropane (MCP), and he interpreted the E_a as measuring the energy necessary to reach the lowest-energy intersection of the singlet and triplet potential energy surfaces. If this intersection occurs at a singlet energy minimum, for instance, at the geometry of $^1B_1$, E_a would be equal to the adiabatic energy difference between $^1B_1$ and $^3A_2'$.

However, the lowest-energy minima on the singlet potential surface are those that occur at the three equivalent geometries for MCP. Since MCP is lower in energy than triplet TMM, it is conceivable that, as ring closure occurs, the steeply descending singlet potential surface intersects the more slowly rising triplet surface at a geometry that is lower in energy than that of $^1B_1$. Although this hypothesis provides an attractive rationalization of Dowd's experimental results, the existence of such an intersection has received no support from calculations that have attempted to find it [23g,h]. The process that is responsible for the disappearance of the EPR signal with $E_a = 7.8$ kcal/mol currently remains an unsolved mystery.

III.B.   Tetramethyleneethane (TME)

There is general agreement that the size of the singlet-triplet energy separation in TME is small, as expected for a disjoint diradical [11]. However, whether a singlet or triplet is the ground state of this non-Kekulé hydrocarbon has proven to be controversial and is still an active area of research.

Following Dowd's synthesis of TME in 1970 [7], a Curie plot of the temperature dependence of the EPR signal intensity showed that the population of the triplet is invariant with temperature [29]. This finding has three possible interpretations [30]: (a) the triplet lies well below the singlet in energy at the equilibrium geometry of the triplet; (b) the singlet and triplet have the same energy to within several cal/mol; or (c) the singlet and triplet do not interconvert on the time scale (hours) of the experiment.

Two cyclic analogs of TME—the 2,2-dimethyl derivative of 4,5-dimethylenecyclopentane-1,3-diyl (DMCPD) [31] and 2,3-dimethylenecyclohexane-1,3-diyl (DMCHD) [32] (Figure 8)—have also been prepared and found to have triplet EPR spectra whose intensities give linear Curie plots. However, recent magnetization measurements by Iwamura have shown that the linear Curie plot for DMCHD [32b] is apparently due to nearly degenerate energies for the lowest singlet and triplet states [32c,d], as had previously been predicted for DMCHD by ab initio calculations [33].

The finding that $\Delta E_{ST} \approx 0$ in DMCHD, a cyclic derivative of TME, is also in accord with the qualitative prediction, based on the disjointness of the NBMOs. However, as discussed in the previous section, two effects— (a) non-nearest-neighbor interactions between the AOs of the two localized, allylic NBMOs and (b) weak bonding between the p-π AOs of the two nodal carbons, due to dynamic spin polarization—should selectively stabilize the lowest singlet of planar TME and make it the ground state. All the ab initio calculations that have been performed on *planar* TME confirm that the sign of ΔE_{ST} is negative (i.e., the singlet is computed to lie below the triplet) at this geometry of the diradical [34].

However, the calculations also find that the planar (D_{2h}) geometry of TME is not an energy minimum for either the lowest singlet or triplet state. The lowest-energy singlet prefers a D_{2d} geometry, which, as shown in Figure 9, allows the p-π AO on each terminal carbon to undergo long-range bonding interactions with the p-π AOs on both terminal carbons of the other allylic radical. At a D_{2d} geometry, it is only these long-range bonding interactions that stabilize the singlet, relative to the triplet. Dynamic spin polarization has no effect on ΔE_{ST} at this geometry, because the p-π AOs on the central carbons are orthogonal. Nevertheless, the D_{2d} singlet is also computed to lie below the D_{2d} triplet by all the calculations [34], although by less energy than at the D_{2h} geometry.

As shown in Figure 9, the combination of localized allylic NBMOs that is stabilized by long-range interallylic bonding changes on going from D_{2h} to D_{2d}. Near a dihedral angle of 45°, the two combinations have the same energy, because there are no net long-range bonding interactions be-

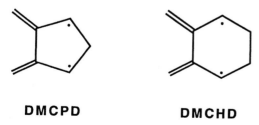

DMCPD   **DMCHD**

Figure 8 Two cyclic derivatives of TME.

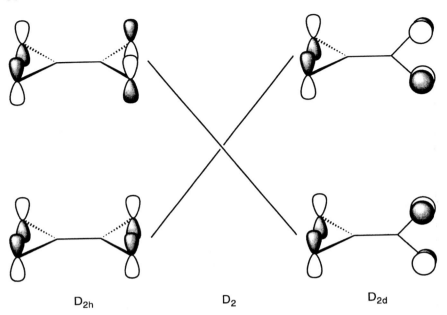

D_{2h} D_2 D_{2d}

Figure 9 Schematic depiction of how the energies of the two combinations of localized allylic NBMOs change on rotation about the central C–C bond in TME.

tween the localized allylic NBMOs. Since interallylic bonding is actually slightly destabilizing for the triplet, it prefers an equilibrium geometry near the dihedral angle where the energies of the two NBMOs cross.

The question that remains to be answered unequivocally by calculations is: What are the predicted sign and size of ΔE_{ST} at the equilibrium geometry of the triplet? Calculations by Nachtigall and Jordan at the (6/6)CASSCF level found $\Delta E_{ST} \approx -1$ kcal/mol; but at the CISD level, after a correction for the effect of quadruple excitations, $\Delta E_{ST} \approx +1$ kcal/mol was obtained [34c]. Both calculations have shortcomings that would be remedied by a (6/6)CASCISD calculation (i.e., a CI calculation that includes single and double excitations from *all* of the configurations that result from distributing the six π-electrons in TME among the six π-MOs). Unfortunately, although (4/4)CASCISD calculations have been carried out for TMM [23j], even with modern computational hardware, (6/6)CASCISD is still too big to have been performed for TME. When such a calculation is carried out, it should provide a reliable theoretical prediction of at least the sign of the very small singlet-triplet splitting in TME at the equilibrium geometry of the triplet.

Unfortunately, there is no clear-cut experimental answer to the question of the sign, much less the precise value, of ΔE_{ST} at the equilibrium geometry of the triplet. Reaction of 2,3-dimethylbutadiene with O^- in the gas phase has allowed the generation of the radical anion of TME [35a], and its PE spectrum has been obtained [35b]. Since the radical anion of TME is calculated to prefer a D_{2d} geometry, the Franck–Condon factors for formation of the D_2 equilibrium geometry of the triplet upon electron loss are expected to be very small. Therefore, the value of $\Delta E_{ST} = -3.0 \pm 0.3$ kcal/mol that was obtained from the spectrum was interpreted as corresponding to the energy difference between the lowest singlet and triplet states at the D_{2d} equilibrium geometry of the singlet, not at the D_2 equilibrium geometry of the triplet.

However, an estimate of ΔE_{ST} at the D_2 equilibrium geometry of the triplet can be obtained by combining the computational results of Nachtigall and Jordan [34d] with the experimental results from PE spectroscopy [35b]. The CISD/TZP value calculated for ΔE_{ST} increases by 2.0 kcal/mol on rotation of TME from D_{2d}, where $\Delta E_{ST} = -1.9$ kcal/mol is calculated, to D_2 where $\Delta E_{ST} = 0.1$ kcal/mol is computed. Since the experimental value of $\Delta E_{ST} = -3.0$ kcal/mol is about 1 kcal/mol more negative than the CISD/TZP value at D_{2d}, an estimated value of $\Delta E_{ST} = -1$ kcal/mol at the D_2 equilibrium geometry of the triplet is obtained. Based on the results of these calculations, or of any of the others that have been performed to date [34], there is every reason to believe that, if $\Delta E_{ST} = -3.0$ kcal/mol at the D_{2d} equilibrium geometry of the singlet, the singlet also lies below the triplet at the D_2 equilibrium geometry of the latter state.

If the singlet really does lie below the triplet at the D_2 equilibrium geometry of the latter, it supports the claim [34a] that there is no obvious physical mechanism by which the triplet can fall below the singlet. Even if there is no long-range bonding between the NBMOs to stabilize the singlet selectively at this geometry, the overlap between the p-π AOs on the central carbons of TME is reduced by only a factor of cos $\phi \approx 0.7$ at a dihedral angle of $\phi \approx 45°$ between the allyl moieties. Therefore, at this geometry, the bonding between these AOs in the singlet, due to dynamic spin polarization, should act to stabilize the singlet, relative to the triplet.

A singlet ground state for TME at the D_2 equilibrium geometry of the triplet would constitute a violation of the strictest form of Hund's rule [22]. It should be recalled that Hund's rule was formulated for atoms, where orbitals with the same principle quantum number and angular momentum necessarily have exactly the same energy. However, in most non-Kekulé diradicals, the two NBMOs are not degenerate by symmetry; hence, they do not have exactly the same energies. For example, as already discussed, non-nearest-neighbor interactions result in an energy difference between the

NBMOs of TME at D_{2h} and D_{2d} geometries. This energy difference contributes to the singlet's being predicted to lie below the triplet at the planar geometry and is wholly responsible for this energy ordering at the $\phi = 90°$ geometry, where no bonding is possible between the nodal carbons.

However, at the D_2 equilibrium geometry of the triplet state of TME, the two NBMOs have essentially the same energy (see Figure 9). Consequently, if the singlet really is lower in energy than the triplet at this geometry, this finding would prove that, as predicted [11,34a], dynamic spin polarization can result in violations of Hund's rule in diradicals, even when the two NBMOs have exactly the same energy [22]. The predicted violation of the strictest form of Hund's rule in D_{8h} cyclooctatetraene [36a] has been recently confirmed experimentally by the PE spectrum of the radical anion [36b].

If the singlet really is the ground state of TME at the D_2 equilibrium geometry of the triplet, an obvious question is: Why did Dowd obtain a linear Curie plot? One possible answer is that Dowd did his experiments in frozen solutions [29], whereas the PE spectroscopy was done in the gas phase [35b]. If solvent were to stabilize the triplet selectively, this state could become degenerate with or fall below the singlet; and the frozen medium might also prevent rotation of the D_2 singlet to its D_{2d} equilibrium geometry. As attractive as this rationalization may be, there is computational evidence against interactions with solvent providing sufficient stabilization of the triplet [23h].

Another possible explanation of Dowd's linear Curie plot is that the singlet and triplet did not interconvert on the time scale of his study of the temperature dependence of the EPR signal intensity. This explanation would seem highly improbable, had not Berson and coworkers reported immeasurably slow intersystem crossing (ISC) in arylsulfonyl derivatives of 2-azaDMCPD [37]. Berson suggested that rotamers about the bond to the sulfonyl group might have different ground states and that different, non-interconverting rotamers might be populated by different modes of diradical generation in frozen solutions. However, there is no compelling evidence that this really is the correct explanation for the immeasurably slow rate of ISC in the diradicals studied by Berson and coworkers [37]; and, if another factor causes ISC to be slow in these cyclic TME derivatives, the same factor may make ISC very slow in TME itself.

Magnetization studies on TME, of the type done on DMCHD [32c,d], could, at least in principle, reveal the relative amounts of singlet and triplet diradicals present under the conditions used by Dowd for his Curie study. Until the results of such a study become available or until the PE spectrum is shown to be compatible with a positive value of ΔE_{ST} at the D_2 equilibrium geometry of triplet TME, the reason why Dowd obtained a linear Curie plot for TME will remain an unsolved mystery.

III.C. 2,4-Dimethylenecyclobutane-1,3-diyl (DMCBD)

There are fewer mysteries associated with ΔE_{ST} in DMCBD than in TME or even TMM, because the value of ΔE_{ST} in DMCBD has not been measured experimentally. Nevertheless, Curie studies [8a,c] do indicate a triplet ground state. Because the NBMOs of DMCBD are nondisjoint [11], a triplet ground state and a large energy difference ($\Delta E_{ST} = 18-20$ kcal/mol) between $^3B_{2u}$ and the lowest excited singlet state, ^1A_g, have been computed by CI calculations [38].

The bonding in $^3B_{2u}$, ^1A_g, and the next lowest singlet state, $^1B_{2u}$, are depicted schematically in Figure 10. Since the NBMOs of DMCBD are nondisjoint, both singlets use different sets of MOs than the triplet, and, consequently, the bonding in the singlets is more localized than in the triplet. The reason why ^1A_g is computed to lie well below $^1B_{2u}$ becomes rather obvious after inspection of Figure 10, because this figure shows that the π-bonding in the four-membered ring of DMCBD resembles more closely that of cyclobutadiene in the $^1B_{2u}$ state than in the ^1A_g state.

An experimental value for ΔE_{ST} in DMCBD from PE spectroscopy should be available in the near future, because Hill and Squires have succeeded in generating the radical anion in the gas phase by reaction of 1,3-dimethylenecyclobutane with O^- [39]. Interestingly, they find that, in forming triplet DMCBD from 1,3-dimethylenecyclobutane, the second C–H BDE is ca. 10 kcal/mol larger than the first, even after correcting for the fact that breaking the first C–H bond creates a conjugated pentadienyl radical from a molecule with unconjugated double bonds. Quite reasonably, Hill and Squires attribute the difference between the BDEs to the fact that breaking the second C–H bond creates a triplet DMCBD diradical, which has some of the antiaromatic character of cyclobutadiene.

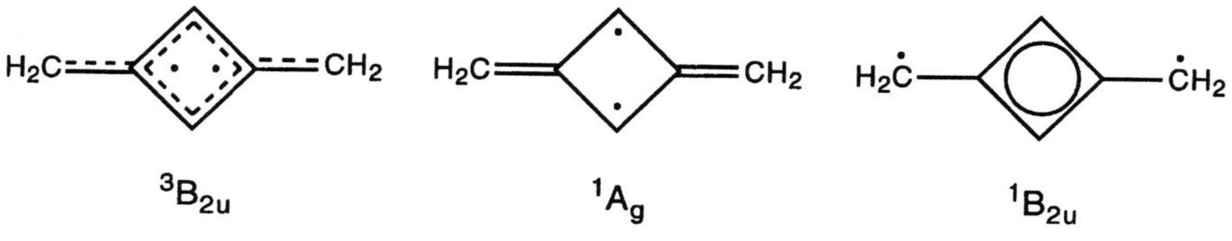

$^3B_{2u}$ $\qquad$ ^1A_g $\qquad$ $^1B_{2u}$

Figure 10 Bonding in the lowest electronic states of DMCBD. There is somewhat more π-bonding to the exocyclic methylene groups in $^1B_{2u}$ than is suggested by the structure drawn here, but there is less π-bonding in $^1B_{2u}$ than in $^3B_{2u}$.

III.D. *meta*-Benzoquinodimethane (MBQDM)

meta-Benzoquinodimethane (MBQDM) has been shown to have a triplet ground state by the Curie study performed by Platz [6a]. Using the terminology of Dougherty [4d], the *m*-phenylene group in MBQDM is a strong ferromagnetic coupler of the unpaired π-electron on each methylene group, and *m*-phenylene has been found to result in high-spin coupling of the unpaired π-electrons at a variety of radical [3,5,6], diradical [4d,40], triplet carbene [4c,e,15–17,19], nitrene [20a], and nitroxide [21a] centers. A few bisnitroxides have been found to exhibit antiferromagnetic coupling [41], but in these diradicals steric effects cause the *tert*-butylnitroxide moieties to be strongly twisted out of conjugation with the π-orbitals of the benzene ring. In such cases, through-bond interactions have been shown to favor antiferromagnetic coupling [42].

As in TMM and DMCBD, the two other nondisjoint diradicals in Figure 1, minimization of the Coulombic repulsion energy in the two lowest singlet states of MBQDM causes their NBMOs to be more localized than those of the triplet state. Consequently, as shown in Figure 11, the bonding in the singlets is less delocalized than in the triplet. The $^1A_1$ state preserves the aromaticity of the benzene ring by confining the two nonbonding electrons largely to the exocyclic methylene groups. Hence, $^1A_1$ is expected to be lower in energy than $^1B_2$, whose π wave function is essentially that for an allyl radical plus a pentadienyl radical [43].

Since the NBMOs of MBQDM are nondisjoint, the energy separations between the triplet and both singlets are expected to be substantial. (8/8)CASSCF/6-31G* calculations place $^1A_1$ and $^1B_2$, respectively, 12.2 and 22.4 kcal/mol above $^3B_2$ [43c]. Inclusion of dynamic correlation through the CASPT2 method stabilizes $^1A_1$ by 1.2 kcal/mol and destabilizes $^1B_2$ by 0.5 kcal/mol, relative to $^3B_2$.

Experimental values for these two energy differences are now available. Wenthold and Lineberger generated the radical anion of MBQDM by reaction of O$^-$ with *m*-xylene and obtained the PE spectrum of the ion [44].

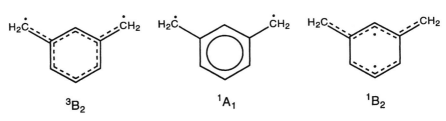

$^3B_2$ $^1A_1$ $^1B_2$

Figure 11   Bonding in the lowest electronic states of MBQDM.

Values of ΔE_{ST} = 9.6 ± 0.2 and ≤21.5 kcal/mol were measured. These experimental values for ΔE_{ST} are each 1.4 kcal/mol less than the corresponding CASPT2N/6-31G* values. Comparison of the best calculated values of ΔE_{ST} in TMM [23i,j], TME [34c], and MBQDM [43c] with those obtained experimentally by PE spectroscopy [26,35b,44] shows that the energies of the singlets, relative to the triplets, are all overestimated, but only by ca. 1−2 kcal/mol.

III.E.   1,2,4,5-Tetramethylenebenzene (TMB)

Like TME, TMB has disjoint NBMOs; and, as in TME, the lowest singlet state of TMB is predicted to be stabilized, relative to the triplet, by a combination of long-range π-bonding between nonadjacent carbons and dynamic spin polarization. However, unlike the case with TME, where rotation about the central C−C bond can modulate the size of both of these interactions between the two allyl fragments, the two single bonds that join the pentadienyl moieties in TMB keep them coplanar. This difference makes both theoretical predictions and the interpretation of experiments much less equivocal for TMB than for TME.

Calculations unequivocally predict a singlet ground state for TMB [45]. At the TCSCF-ROHF level of theory, the effect of dynamic spin polarization on ΔE_{ST} is not included, so the calculated value of ΔE_{ST} = −2 kcal/mol is due only to stabilization of the singlet by long-range interactions between the two pentadienyl fragments. When all the π-electrons are correlated, ΔE_{ST} = −5 kcal/mol [45d,e]. Thus, the π-bonding between the two sets of nodal carbons that results from dynamic spin polarization appears to stabilize the singlet selectively by an additional 3 kcal/mol. Evidence for the presence in the singlet of π-bonding between these carbons, which is absent in the triplet, comes from the finding that, when all the π-electrons are correlated, the C−C bonds that join the two pentadienyl fragments are found to be 0.01 Å shorter in the former state than in the latter [45d,e].

As noted in the introduction, Roth and coworkers were the first to prepare TMB [9a]. In disagreement with the prediction of a singlet ground state [45], they found that TMB has a triplet EPR spectrum, whose intensity gives a linear Curie plot. Consequently, they assigned a triplet ground state to the diradical. Further, they noted,

> Support for this assignment [of a triplet ground state for TMB] is provided by the UV/VIS spectrum . . . with its [weak] long-wave[length] bands between 550 and 600 nm. The position of these bands is in good agreement with the long-wave[length], forbidden transition at 570 nm calculated for the triplet [45d], these calculations do not lead one to expect any bands for the singlet in this region.

Consequently, Roth and coworkers concluded, "Contrary to the theoretical expectations the singlet . . . can lie 0.02 kcal/mol at most below the lowest triplet state, i.e., the states are [sic] degenerated, or—what could be more likely—the diradical has a triplet ground state'' [9a].

At the time that Roth et al. published their paper, Berson and his coworkers at Yale had also generated TMB and were busy studying its spectroscopy and chemistry. These studies led Berson and coworkers to the following conclusions [9c,d,22a]: (a) The EPR spectrum observed when TMB is generated does not belong to TMB. (b) The UV-VIS spectrum does belong to TMB. (c) Since a sharp resonance is seen in the CPMAS ^{13}C NMR spectrum of TMB, enriched with ^{13}C, TMB has a singlet ground state. If the ground state were a triplet, dipolar broadening, due to the unpaired electrons, would have been observed.

Although these studies appeared to confirm the predictions of a singlet ground state for TMB [45], Berson's findings also raised an important question. If the singlet is the ground state of TMB, why does the longest-wavelength absorption in the UV-VIS spectrum of this state correspond to a forbidden transition, a spectral feature that π CI calculations had predicted for the triplet but not for the singlet [45d]? The answer to this question was subsequently given by the results of CASPT2N calculations, which showed the inclusion of correlation between σ- and π-electrons is essential for computing the UV spectrum of the singlet [45e].

The long-wavelength band in the UV-VIS spectrum of TMB was identified as a forbidden transition from the $^1A_{1g}$ ground state to the 2^1A_{1g} excited state, and the calculations were able to rationalize the width and extensive vibrational structure in this band. The energies of the absorptions in the UV-VIS spectrum were duplicated to within 0.2 eV by the CASPT2N calculations for excitations from the lowest singlet state, but the observed spectrum did not fit at all that computed for excitations from the lowest triplet state. It appears that comparing the observed UV-VIS spectra of non-Kekulé diradicals with those calculated by CASPT2N for the lowest singlet and triplet may provide a generally applicable method for identifying which is the ground state [22b].

Although there is now little doubt that the singlet is the ground state of TMB, the size of the singlet-triplet splitting has not yet been measured. By analogy with the generation of the radical anion of TME [35a], reaction of 1,2,4,5-tetramethylenecyclohexane [46] with O^- appears to be a promising method for forming the radical anion of TMB in the gas phase; or the F_2/F^- method of Squires [27] could be applied to tetrakis(trimethylsilyl) durene. If successful, formation of the radical anion should allow measurement of ΔE_{ST} in TMB by photoelectron spectroscopy.

III.F. Trisdehydromesitylene (TDM)—A Triradical

The initials "TMB" are already in common use for 1,2,4,5-tetramethyl-enebenzene. Therefore, it seems preferable to use the alternative name, 1,3,5-trisdehydromesitylene, rather than 1,3,5-trimethylenebenzene, so that TDM can be used for the non-Kekulé hydrocarbon in Figure 12. As shown in the figure, any resonance structure for TDM leaves three carbons without π-bonds. Hence, TDM is a triradical [1]. Also as shown in Figure 12, in TDM $N^* - N = 3$, so Eq. (1) predicts a quartet ground state.

$N^* - N = 3$ also means that all three NBMOs are confined to the starred set of carbons [1,11]; and the zero-sum rule can be used to find the NBMOs and to confirm that they are all disjoint. Consequently, it can be predicted that the quartet ground state lies well below any excited doublet state. It can also be predicted that in the lowest double states, minimization of Coulombic repulsions between electrons of opposite spin in the NBMOs will result in the NBMOs for these states being localized to different sets of atoms to a greater extent than the NBMOs for the quartet. Consequently, the π-bonding in these doublet states should be more localized than in the quartet ground state.

The results of recent calculations on TDM confirm these qualitative expectations [47]. The bonding in the quartet and in the two lowest doublet states is depicted in Figure 12. At D_{3h} geometries, the doublet states comprise the two components of a $^2E''$ state, and the localized bonding in the doublets causes them to undergo the first-order e' distortions from D_{3h} symmetry that are predicted by the Jahn–Teller theorem [48].

The adiabatic doublet-quartet splitting in TDM is calculated to be 14.9 kcal/mol by (9/9)CASSCF/6-31G* calculations, 12.7 kcal/mol at the CASPT2N level, and 15.5 kcal/mol by multireference CI calculations [47].

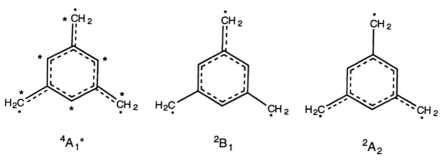

$^4A_1''$ $\qquad$ $^2B_1$ $\qquad$ $^2A_2$

Figure 12 Bonding in the lowest electronic states of TDM. Starred and unstarred atoms are shown.

Squires has successfully generated the radical anion of TDM by reaction of a tris(trimethylsilyl) precursor with F^-/F_2 [49]; and PE spectroscopy on the radical anion should provide an experimental value of ΔE_{DQ} to compare with these calculated values.

IV. NON-KEKULÉ DIRADICALS WITH O SUBSTITUTED FOR CH₂

IV.A. Qualitative Predictions

Ovchinnikov conjectured that heteroatom substitution would not invalidate the arguments that produced Eq. (1) [12]. However, substitution of oxygen for a methylene group in a non-Kekulé hydrocarbon diradical has two effects, both of which can stabilize a singlet, relative to a triplet state [50]. Consequently, even if a diradical is nondisjoint and, hence, has a triplet ground state, it is possible, at least in principle, that substitution of O for CH_2 will result in a singlet ground state for the heteroatom derivative.

The first effect that can selectively stabilize a singlet state on substitution of O for one or more CH_2 groups has its origin in the fact that C–O π-bonds are stronger than C–C π-bonds. Consequently, states that have large π-bond orders to the methylene group(s) where substitution occurs should be stabilized, relative to those that do not. In non-Kekulé diradicals whose NBMOs are nondisjoint, the π-bonding in the two lowest singlet states is more localized than in the triplet ground state. Consequently, the singlet state with the larger amount of π-bonding to a CH_2 group for which oxygen is substituted should be selectively stabilized, relative to both the triplet and the other singlet.

The second effect that can decrease ΔE_{ST} comes from oxygen's being more electronegative than carbon. The energy of the NBMO with the greater density at the CH_2 group that is replaced by oxygen will be lowered the most. In the wave functions, $\Psi(T)$ for the triplet and $\Psi(S_O)$ for the "open-shell" singlet,* one electron occupies the lower-energy (ψ_L) and one the higher-energy (ψ_H) NBMO:

$$\Psi(T, S_O) = |\cdots \psi_L \psi_H (\alpha\beta \pm \beta\alpha)/\sqrt{2}\rangle \qquad (7)$$

Consequently, neither of these two states can take maximal advantage of the difference in energy between the NBMOs.

In contrast, in the two-configuration wave function, $\Psi(S_C)$, for the "closed-shell" singlet,

---

*In Eq. (7) the positive sign gives the wave function for the $m_s = 0$ component of the triplet. The spin wave functions for the two other components are, of course, $\alpha\alpha$ and $\beta\beta$.

$$\Psi(S_C) = c_L \left| \cdots \psi_L^\alpha \psi_L^\beta \right\rangle - c_H \left| \cdots \psi_H^\alpha \psi_H^\beta \right\rangle \tag{8}$$

the coefficient, c_L, for the lower-energy configuration can be larger than that, c_H, for the higher-energy configuration. As a result, the electron occupation number, $2c_L^2$, of ψ_L can be larger than that, $2c_H^2$, of ψ_H. Consequently, unlike the triplet and "open-shell" singlet, the "closed-shell" singlet can take maximal advantage of the energy difference between the NBMOs. Thus, for a sufficiently large energy difference between ψ_L and ψ_H, the flexibility of the two-configuration wave function in Eq. (8) allows the "closed-shell" singlet to fall below the triplet in energy.

A very nice example of the operation of this latter effect has been provided by Dougherty and coworkers [51]. They showed computationally that, unlike MBQDM, which is both calculated [43] and found [6,44] to have a triplet ground state, the pyridinium analogs can have either a singlet or a triplet ground state, depending on where the protonated nitrogen is located in the six-membered ring.

The degree to which the singlet state is stabilized, relative to the triplet, was found to depend on how effective the high electronegativity of the protonated nitrogen is at lifting the degeneracy of the NBMOs of MBQDM. The same predictions could also be made by assessing the relative stabilities of VB structures in which a pair of π-electrons is localized on the pyridinium nitrogen. Dougherty and coworkers also reported experimental confirmation of their own predictions [51].

The effects of the less highly perturbing substitution of O for CH_2, rather than N^+ for C, is the subject of the following subsections of this chapter. The next two subsections show that a combination of both effects, just discussed, allows the "closed-shell" singlet to become the ground state when one CH_2 group is replaced by oxygen in some derivatives of TMM and when both CH_2 groups are replaced by oxygens in DMCBD. The third subsection shows that, in contrast, calculations predict little effect on ΔE_{ST} when both methylene groups in MBQDM are replaced by oxygens, and the reason why this is the case is discussed. The final subsection is concerned with the effect on ΔE_{ST} of replacing just one CH_2 group in MBQDM by oxygen.

IV.B. Oxyallyl (OXA)

If one methylene group in TMM is replaced by an oxygen atom, the bonding in the lowest three states of TMM (Figure 5) can be used to obtain the first-order description of the bonding in the lowest three states of OXA, which is shown in Figure 13. Since, as noted in the introduction to this section, C–O π-bonds are considerably stronger than C–C π-bonds, it is trivial to

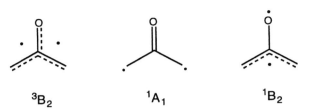

$^3B_2$ $^1A_1$ $^1B_2$

Figure 13   Bonding in the lowest electronic states of OXA.

predict from Figure 13 the changes that occur in the relative energies of the lowest three states of TMM on replacing a CH_2 group by oxygen to form OXA.

The $^1A_1$ state should be stabilized the most by this substitution, since it has the largest amount of C–O π-bonding. Only one of the three resonance structures for the triplet ($^3B_2$) contains a π-bond to oxygen, so $^3B_2$ should be stabilized less; and the $^1B_2$ state should not be stabilized at all, since it has no π-bonding to oxygen. Therefore, on going from TMM to OXA, the near degeneracy of $^1A_1$ and $^1B_2$ in TMM [23] should be strongly lifted, and $^1A_1$ should also be stabilized with respect to $^3B_2$.

Ab initio calculations do, in fact, predict that substitution of oxygen for one methylene group in TMM should result in the lowest singlet ($^1A_1$) and triplet ($^3B_2$) states of oxyallyl (OXA) being nearly isoenergetic [52a,b]. Alkyl substituents at the remaining two methylene groups of OXA are unequivocally predicted to make $^1A_1$ the ground state [52b–d]. Experimental results indicate that alkyl derivatives of OXA do, indeed, have singlet ground states [53].

However, the greater strength of a C–O than a C–C π-bond is not the only effect that is responsible for causing $^1A_1$ to be close to or lower in energy than $^3B_2$ in OXA and derivatives. The second effect, discussed in the introduction to this section, also plays an important role in selectively stabilizing $^1A_1$ in OXA. The lowering of the energy of the $2b_1$ NBMO of TMM on substitution of O for CH_2 also contributes to the selective stabilization of the $^1A_1$ state of OXA.

In singlet TMM, the $2b_1$ and a_2 NBMOs in Figure 6 have almost exactly the same energy. Consequently, in the wave function in Eq. (5) for the $^1A_1$ state of this non-Kekulé hydrocarbon, the two configurations have almost exactly the same coefficients [i.e., $c_L \approx c_H \approx 1/\sqrt{2}$ in Eq. (8)]. Thus, $2b_1$ and a_2 are each occupied by essentially one electron, not only in the $^3B_2$ and $^1B_2$ states of TMM, but also in the $^1A_1$ state.

As shown in Figure 6, only the $2b_1$ NBMO of TMM has density on the CH_2 group where oxygen is substituted in OXA, because the a_2 NBMO has a node at this atom. Consequently, the $2b_1$ NBMO is considerably lower in energy than the a_2 NBMO in OXA. As a result, $c_L > c_H$ in the wave function for the "closed-shell" $^1A_1$ state in Eq. (8); so in this state the electron occupancies of the NBMOs are $2b_1 > 1 > a_2$. Since the NBMOs are each occupied by one electron in both the $^3B_2$ and $^1B_2$ states, the $^1A_1$ state of OXA is stabilized, relative to both B_2 states, by the greater occupancy of the lower energy of the two NBMOs.

If $c_L \gg c_H$ in Eq. (8) for the wave function for the lowest singlet state of OXA, the zwitterionic resonance structure for OXA, shown in Figure 14, would provide a good depiction of this state. However, $c_L < 1.5c_H$ [52a], and, as a result, the dipole moment of singlet OXA is computed to be only 25% greater than that of cyclopropanone [52e]. It is for this reason that the solvent effects on ring opening of cyclopropanone to OXA are both calculated [52e] and found [54] to be modest. Consequently, singlet OXA is best regarded as a diradical to which a zwitterionic resonance structure makes only a modest contribution [52].

Another indication of the dominance of the diradical rather than the zwitterionic resonance structure for OXA is the frequency of 1730 cm^{-1}, calculated for the C=O stretch [52f]. This frequency indicates a strong C=O π-bond in OXA. If the zwitterionic resonance structure were important, one would expect a lower C=O stretching frequency. Interestingly, in tetramethyl-OXA the C=O stretching frequency is calculated to drop to 1670 cm^{-1} [52f], suggesting that the methyl groups act to stabilize the zwitterionic resonance structure shown in Figure 14 and thus lower the C=O stretching frequency.

Although substitution of oxygen for CH$_2$ in TMM lowers substantially the energy of the $^1A_1$ state of OXA, relative to that of $^3B_2$, the energy difference between the $^1B_2$ and $^3B_2$ states increases on going from TMM to OXA [52a,b]. This change in relative energies is not due to lifting the degeneracy of the NBMOs in OXA. Since the $^1B_2$ and $^3B_2$ states of TMM each have one electron in the $2b_1$ and one electron in the a_2 π-MO, lifting of the degeneracy of these MOs by replacement of a CH$_2$ in TMM by O in OXA does not have a significant effect on the relative energies of the $^1B_2$ and $^3B_2$ states of these two diradicals. Instead, the increase in the energy difference

Figure 14 Resonance structures for the lowest singlet state of OXA.

between these two states is due to the larger amount of π-bonding to oxygen in the triplet (see Figure 13).

IV.C. Cyclobutane-2,4-dione-1,3-diyl (CBDOD)

Inspection of Figure 10, which depicts the bonding in the low-lying states of DMCBD, allows one to predict that substitution of oxygens for both methylene groups* should provide selective stabilization for the ^1A_g state of CBDOD, which is depicted in Figure 15. Moreover, one would expect greater stabilization of this state than substitution of oxygen for one methylene group in TMM provides for the $^1A_1$ state of OXA. This expectation has been confirmed computationally. Although ΔE_{ST} is computed to be slightly larger in DMCBD ($\Delta E_{ST} = 18-20$ kcal/mol) [38c] than in planar TMM [23], the lowest singlet state is calculated to be far below the triplet in energy ($\Delta E_{ST} < -20$ kcal/mol) in CBDOD [38c], rather than at nearly the same energy, as in OXA [52a,b]. Unfortunately, CBDOD has not yet been prepared, so this prediction has not received even a qualitative experimental test.

The reason that ΔE_{ST} is computed to be nearly zero in OXA but very negative in CBDOD is that the energy difference between the $2b_1$ and a_2 orbitals in OXA is not nearly as large as the energy difference between the corresponding two MOs, respectively, $2b_{1u}$ and b_{3g} in CBDOD. One contributor to the large energy difference between the latter pair of MOs is the fact that the in-phase combination of AOs at C-1 and C-3 is stabilized by interacting with the low-lying π* orbital of one carbonyl group in the $2b_1$ MO of OXA but with the π* orbitals of two carbonyls in the $2b_{1u}$ MO of CBDOD [38c,50]. In addition, the small distance between C-1 and C-3 in CBDOD results in the cross-ring interaction between the p-π AOs at these two carbons being rather large. Since this interaction is bonding in $2b_{1u}$ but antibonding in b_{3g}, it also stabilizes $2b_{1u}$, relative to b_{3g}.

As a consequence of the very large energy difference between the $2b_{1u}$ and b_{3g} MOs of CBDOD, the occupation number of the former MO is much

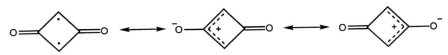

Figure 15   Resonance structures for the lowest singlet state of CBDOD.

---

*The effect of substituting oxygen for just one methylene group in DMCBD is to leave the triplet still substantially below the singlet in energy, and the reason for this has been discussed [38c,43c].

larger than that of the latter in the wave function for the ^1A_g state [38c]. In fact, the difference between the occupation numbers of these two MOs is so large that the two ionic resonance structures in Figure 15 really do contribute substantially to the lowest singlet state. Comparing the resonance structures in Figures 14 and 15, one also can see why the two carbonyl groups and the four-membered ring in CBDOD both make ionic resonance a much greater contributor to the lowest singlet state of this molecule than of OXA.

IV.D. *meta*-Benzoquinone (MBQ)

Unlike the case with DMCBD, where substitution of oxygen for both methylene groups is unequivocally predicted to result in a singlet ground state for CBDOD [38c], the same substitution of O for both CH_2 groups in MBQDM is predicted, just as unequivocally, to have little effect on the size of the singlet-triplet splitting in *m*-benzoquinone (MBQ) [43b,50]. The reason for this difference between the four- and six-membered rings is that in the lowest singlet state ($^1A_{1g}$) of DMCBD, the two methylene groups that are replaced by oxygen atoms in CBDOD have large π-bond orders to them. In contrast, as shown in Figure 16, in the lowest singlet state ($^1A_1$) of MBQDM, the CH_2 groups that are replaced by O in MBQ have little π-bonding to them.

Unlike the $^1A_1$ state, the $^3B_2$ state of MBQDM does have partial π-bonds to the exocyclic CH_2 groups; and the strengths of these π-bonds are increased upon substitution of O for CH_2 to form MBQ. Consequently, as is the case in forming the $^1B_2$ state of OXA from $^1B_2$ TMM [52a,b], substitution of O for CH_2 in the $^1A_1$ state of MBQDM actually increases the energy difference between $^1A_1$ and $^3B_2$ in MBQ [43b,50].

Unlike the $^1A_1$ state of MBQDM, the $^1B_2$ state does have substantial π-bonding to the exocyclic methylene groups. Upon substitution of O for CH_2, the strength of the C–O π-bonds in the $^1B_2$ state of MBQ reduces the CASPT2N energy difference between $^1B_2$ and $^3B_2$ from 22.9 kcal/mol in

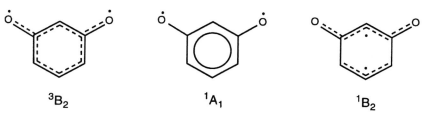

$^3B_2$ $^1A_1$ $^1B_2$

Figure 16 Bonding in the lowest electronic states of MBQ.

MBQDM to 11.9 kcal/mol in MBQ [43c]. In addition, the strong C–O π-bonds in the $^1B_2$ state of MBQ also result in this state being calculated at the CASPT2N level to go from being 11.9 kcal/mol above $^1A_1$ in MBQDM to being below $^1A_1$ by 10.0 kcal/mol in MBQ [43c].

In both the $^1B_2$ and $^3B_2$ states of MBQ, one electron occupies each NBMO. Thus, any difference between the energies of the NBMOs cannot have a substantial effect on the relative energies of these two states of MBQ. Since $^1B_2$ is the lowest singlet state of MBQ and since the singly occupied MOs of $^3B_2$ are nondisjoint, it is possible to make the qualitative prediction that the triplet is *unequivocally* the ground state. It may be hoped that the predictions about both the sign and the magnitude of ΔE_{ST} in MBQ will be tested experimentally in the near future.

IV.E.   *meta*-Benzoquinomethane (MBQM) and 1,3-Napthoquinomethane (NQM)

Although resonance structures can easily explain why the lowest singlet state is $^1A_1$ in MBQDM but $^1B_2$ in MBQ, it is harder to use resonance structures to predict the electronic structure of the lowest singlet state in the less symmetrical *m*-benzoquinomethane (MBQM), which is shown in Figure 17. Since no resonance structure can be drawn for MBQM that simultaneously possesses both an aromatic ring and a carbonyl π-bond, the question arises as to the electronic structure of the lowest singlet state. The ground state of MBQM is known to be a triplet [6b–e, 55]; and one might wonder whether the inability of MBQM to adopt the electronic structure of the lowest singlet state of either MBQDM or MBQ causes ΔE_{ST} in MBQM to be larger than that in either of these other two diradicals.

Experiments on 1,3-naphthoquinomethane (1,3-NQM), the benzo analog of MBQM, suggest that this might in fact be the case. Like MBQM, 1,3-NQM has been found to have a triplet ground state [6b–e,56]; and sev-

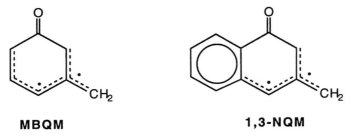

MBQM   **1,3-NQM**

Figure 17   Structures of MBQM and 1,3-NQM.

eral years before photoelectron spectroscopy was used to measure the singlet-triplet splittings in TMM [26], TME [35], and MBQDM [44], Goodman and Kahn used photoacoustic calorimetry to obtain the value $\Delta E_{ST} = 18.5$ kcal/mol in 1,3-NQM [57]. This energy separation between the lowest singlet and triplet states of 1,3-NQM is nearly twice as large as the value measured for MBQDM [44] and over 50% larger than those calculated for both MBQDM and MBQ at the CASPT2N level of theory [43c].

As shown in Figure 17, MBQM and 1,3-NQM may be viewed as carbonyl-bridged derivatives of, respectively, vinyl-TMM and phenyl-TMM. Similarly, OXA can be analyzed as two isolated radical centers connected to a carbonyl group [50]. Since two isolated radical centers should have $\Delta E_{ST} \approx 0$, one can ask what the effect of the carbonyl group is that connects the two radical centers in OXA and then use this information to estimate the size of ΔE_{ST} in MBQM [43c].

In fact, as discussed in Section IV.B, calculations on OXA find $\Delta E_{ST} \approx 0$ [52a,b]. The unchanged value of ΔE_{ST} on joining two radical centers by a carbonyl group is due to the near cancellation of two effects, one that stabilizes the triplet selectively and one that favors the singlet. Because the NBMOs of OXA, like those of TMM, are nondisjoint, correlation between electrons of the same spin that occupy these MOs stabilizes the triplet over the singlet. On the other hand, the singlet is selectively stabilized by the fact that the in-phase (b_1), but not the out-of-phase (a_2), combination of the two radical centers is stabilized by mixing with the π^* orbital of the carbonyl group in OXA. The $^1A_1$ state, in which the electron occupation number in the resulting $2b_1$ MO is greater than that in the a_2 NBMO, has an advantage over the $^3B_2$ state, in which one electron must occupy each MO.

If it were the case in MBQM and in 1,3-NQM, as in OXA, that the carbonyl group has little *net* effect on altering ΔE_{ST} from that in the diradical that lacks the carbonyl bridge, ΔE_{ST} in MBQM should be similar to that computed for vinylTMM [58]. In addition, ΔE_{ST} in 1,3-NQM should be similar to that in phenylTMM, which, in turn, should have a value of ΔE_{ST} similar to that in vinylTMM, since phenyl is known to be only slightly less radical stabilizing than vinyl [59].

Very recent ab initio calculations have provided quantitative support for these qualitative predictions [43c]. (8/8)CASSCF calculations on MBQM gave essentially the same value of ΔE_{ST} as that computed at the (6/6)CASSCF level for vinylTMM; and both the CASSCF and CASPT2N values for ΔE_{ST} in 1,3-NQM were only 2 kcal/mol larger than the values, calculated at the same levels of theory for MBQM [43c].

The CASPT2N/6-31G* value of $\Delta E_{ST} = 11.6$ kcal/mol for 1,3-NQM is 7 kcal/mol lower than that measured by Goodman and Kahn, using photoacoustic calimetry [57]. Thus, there is a sizable discrepancy between theory and experiment in the size of ΔE_{ST} for this diradical. This conflict will probably be resolved only when a second experimental value becomes available through PE spectroscopy on the radical anion of 1,3-NQM, although PE spectroscopy on the radical anion of MBQM should at least indicate how accurate the computed values of ΔE_{ST} are for such non-Kekulé quinomethanes.

V. SUMMARY AND PROGNOSTICATIONS

Experimental studies have made it clear that qualitative theory—either the use of Eq. (1) or the classification of NBMOs as being disjoint or nondisjoint—provides an adequate method for predicting the ground states of non-Kekulé hydrocarbon diradicals. In addition, experiments support the predictions of the MO-based method regarding the relative size of the singlet-triplet splittings in diradicals with nondisjoint, disjoint, and doubly disjoint NBMOs.

The recent use of photoelectron spectroscopy on the radical anions of three non-Kekulé hydrocarbon diradicals—TMM, TME, and MBQDM—has made quantitatively accurate experimental values of ΔE_{ST} available for them. These values are close to, but all 1–2 kcal/mol lower than, those predicted, in advance of the experiments, by the best ab initio calculations. It seems likely that in the near future, photoelectron spectroscopy of radical anions will also make available experimental values of ΔE_{ST} for the DMCBD and TMB diradicals and ΔE_{DQ} for the TDM triradical, against which the predictive ability of ab initio calculations will be further tested.

Values of ΔE_{ST} for heteroatom derivatives of several non-Kekulé hydrocarbon diradicals have now been computed at levels of ab initio theory that, for the hydrocarbons themselves, afford rather accurate predictions. Of course, this does not guarantee that the same levels of theory will do as well in calculating ΔE_{ST} for diradicals, such as OXA, CBDOD, MBQ, MBQM, and 1,3-NQM, in which one or two methylene groups in the corresponding non-Kekulé hydrocarbons have been replaced by oxygen. Therefore, the accurate measurement of ΔE_{ST} in these heteroatom derivatives is of particular interest.

More than thirty years after Dowd reported the synthesis of TMM, the study of non-Kekulé hydrocarbon diradicals has entered a new era. During the first thirty years, EPR spectroscopy provided a means for determining (albeit, not always reliably) the ground state of a diradical. However, except in rare cases, when ΔE_{ST}, was so small that curvature in Curie plots could be seen, there was no way to measure the size of singlet-triplet energy differences.

During this thirty-year period, the singlet-triplet splittings in many non-

Kekulé diradicals were computed, but without any guarantee that it would ever be possible to measure a single one of the calculated values of ΔE_{ST}. Whether this state of affairs frustrated or emboldened the research groups that performed these calculations is an interesting question.

For most of the past thirty years, the Lineberger laboratory has been providing accurate values of ΔE_{ST} by PE spectroscopy of radical ions, but not of the radical ions of non-Kekulé molecules. Organic chemists are good at making molecules, and it was the ability of organic chemists, such as Squires, Wenthold, and Grabowski, to synthesize the radical ions of, respectively, TMM, MBQDM, and TME that has brought about the beginning of a new era in the study of non-Kekulé hydrocarbons.

During this next era, it seems likely that accurate experimental values of the energy splittings between the low-lying states will become available for many di-, tri-, and polyradicals. The ability of experimentalists to measure these energy gaps by PE spectroscopy and the apparent ability of ab initio theory to calculate them with reasonable accuracy will almost certainly encourage computational chemists to perform calculations on additional non-Kekulé hydrocarbons and derivatives of them. The existence of these new predictions will surely stimulate experimentalists to test them.

Thus, at this writing, it seems, at least to this author, that we are entering what may prove to be the "golden age" of the study of singlet-triplet splittings in non-Kekulé hydrocarbons and in molecules derived from them. Whether this era will also last for thirty years will depend on factors such as the ability of organic chemists to synthesize the necessary radical ions, the availability of spectrometers and physical chemists to measure the PE spectra of these ions, the success of computational chemists in predicting the energy separations in the non-Kekulé molecules produced by photodetachment of these ions, and the willingness of funding agencies to support this research. The end of the "golden age" will come when the state separations in so many non-Kekulé molecules are known that, as is currently the case in the study of closed-shell organic molecules by PE spectroscopy, it is no longer of great interest to make measurements on yet another molecule.

This chapter has undoubtedly betrayed its author's delight that qualitative theory has, at least to date, proven very successful in predicting the ground states of non-Kekulé hydrocarbon diradicals and that quantitative calculations have had equally good success in computing the size of the energy differences between the low-lying states in these molecules. Perhaps unfortunately, the continued success of theory could hasten the end of the "golden age" of the study of singlet-triplet splittings in non-Kekulé molecules.

If the energy gaps between the electronic states of these molecules can be computed with such reliability that future experiments are unlikely to provide data that cannot be obtained more quickly and cheaply by calculations, experimentalists will turn their attention to more exciting areas of chemistry. The author of this chapter believes that the end of the current synergism between theory and experiment in the study of non-Kekulé molecules would be a great loss. Therefore, it is his hope that future measurements of the energy separations between the low-lying states of these molecules will reveal that theory has less predictive ability than currently appears to be the case.

ACKNOWLEDGMENTS

This chapter is dedicated to the memories of Professors Paul Dowd and Wolfgang Roth, both of whom made very important contributions to the study of non-Kekulé molecules and both of whom the author was fortunate to know and be inspired by at the start of his career. I wish to thank my coworkers, whose names appear on the papers on non-Kekulé molecules that we published together, many of which can be found in the references for this chapter. Two of my collaborators deserve special thanks. Professor Ernest Davidson convinced me nearly a quarter of a century ago that ab initio calculations are a more powerful method than back-of-the-envelope, Pariser–Parr–Pople calculations for making predictions about diradicals. Dr. David Hrovat has, for many years, taught the students and postdocs in my group how to do those calculations on non-Kekulé molecules that he did not do himself. During this period he has also found all the mistakes I missed in almost every manuscript that I have written, including this one. This manuscript also benefited from the comments of Professors Jerome A. Berson and Hiizu Iwamura, two experimentalists whose names appear frequently in the references for this chapter. I am grateful to Professor Rolf and Frau Gertrude Gleiter, Professor Peter Hofmann, Dr. Peter Bischof, and Frau Christiane Eckert for their help and hospitality while I was in Heidelberg, where this chapter was written. A Senior Scientist Award from the Alexander von Humboldt Foundation made my stay in Heidelberg possible, and the National Science Foundation has generously supported my work during the more than thirty years that I have been doing research on non-Kekulé molecules.

REFERENCES

1. Longuet-Higgins HC. J Chem Phys. 1950; 18:265.
2. (a) Dowd PJ. Am Chem Soc. 1966; 88:2587; (b) Experimental work on TMM is reviewed by Berson JA. In: Diradicals, Borden WT, ed. New York: Wiley-Interscience, 1982, p. 152.

3. Review: Rajca A. Chem Rev. 1994; 94:871.
4. Reviews: (a) Borden WT. In: Diradicals, Borden WT, ed. New York: Wiley-Interscience, 1982, p. 1; (b) Kinetics and Spectroscopy of Carbenes and Biradicals, Platz MS, ed. New York: Plenum Press, 1990; (c) Iwamura H. Adv Phys Org Chem. 1990; 26:179; (d) Dougherty DA. Acc Chem Res. 1991; 24: 88; (e) Iwamura H, Koga N. Acc Chem Res. 1993; 26:346; (f) Borden WT. In: Encyclopedia of Computational Chemistry, von R Schleyer P, ed. New York: Wiley, 1998.
5. (a) Schlenck W, Brauns M. Ber Deutsch Chem Ges. 1915; 48:661. More than half a century later, EPR studies showed the ground state to be a triplet: (b) Kothe G, Denkel KH, Sümmerman W. Angew Chem Int Ed Engl. 1970; 9: 906; (c) Luckhurst GR, Pedulli GF. J Chem Soc. 1971; 329.
6. (a) Wright BB, Platz MS. J Am Chem Soc, 1983; 105:628. Experimental work on MBQDM and other quinodimethanes has been reviewed by (b) Platz MS. In: Diradicals, Borden WT, ed. New York: Wiley-Interscience, 1982, p. 195; (c) Rule M, Matlin AR, Seeger DE, Hilinski EF, Dougherty DA, Berson JA. Tetrahedron. 1982; 38:787; (d) Berson JA. In: The Chemistry of the Quinonoid Compounds, Patai S, Rappoport Z. eds. New York: Wiley, 1988, vol. 2, p. 455; (e) Berson JA. Mol Cryst Liq Cryst. 1989; 17:1.
7. Dowd PJ. Am Chem Soc. 1970; 91:1066.
8. (a) Dowd P, Paik YH. J Am Chem Soc. 1986; 108:2788; (b) Snyder GJ, Dougherty DA. J Am Chem Soc. 1985; 107:1774; (c) Snyder GJ, Dougherty DA. J Am Chem Soc. 1986; 108:299; (c) Snyder GJ, Dougherty DA. J Am Chem Soc. 1989; 111:3927; (d) Snyder GJ, Dougherty DA. J Am Chem Soc. 1989; 111:3942.
9. (a) Roth WR, Langer R, Bartmann M, Stevermann B, Maier G, Reisenauer HP, Sustmann R, Müller W. Angew Chem Int Ed Engl. 1987; 26:256; (b) Roth WR, Langer R, Ebbrecht T, Beitat A, Lennartz H-W. Chem Ber. 1991; 124: 2752; (c) Reynolds JH, Berson JA, Kumashiro KK, Duchamp JC, Zilm KW, Rubello A, Vogel PJ. Am Chem Soc. 1992; 114:763; (d) Reynolds JH, Berson JA, Scaiano JC, Berinstain, AB. J Am Chem Soc. 1992; 114:5866; (e) Reynolds JH, Berson JA, Kumashiro KK, Duchamp JC, Zilm KW, Scaiano JC, Berinstain AB, Rubello A, Vogel P. J Am Chem Soc. 1993; 115:8073.
10. Hund F. Linienspektren und periodisches System der Elemente. Berlin: Springer-Verlag, 1927, p. 124.
11. Borden WT, Davidson ER. J Am Chem Soc. 1977; 99:4587.
12. Ovchinnikov AA. Theor Chim Acta. 1978; 47:297.
13. (a) Klein DJ, Nelin CJ, Alexander S, Matsen FA. J Chem Phys. 1982; 77:3101. (b) See also Tyutyulkov N, Karabunarliev S, Ivanov C. Mol Cryst Liq Cryst. 1989; 176:139 and references cited therein.
14. Alexander SA, Klein DJ. J Am Chem Soc. 1988; 110:3401.
15. (a) Itoh K. Chem Phys Lett. 1967; 1:235; (b) Wasserman E, Murray RW, Yager WA, Trozzolo AM, Smolinsky G. J Am Chem Soc. 1967; 89:5076.
16. (a) For $m = 2$: Teki Y, Takui T, Yagi H, Itoh K, Iwamura H. J Chem Phys. 1985; 83:539. (b) For $m = 3$; Teki Y, Takui T, Yagi H, Itoh K, Iwamura H, Kobayashi K. J Am Chem Soc. 1986; 108:2147 and references cited therein.

(c) For $m = 4$: Fujita I, Teki Y, Takui T, Kinoshita T, Itoh K, Miko F, Sawaki Y, Iwamura H, Izuoka A, Suguwara T. J Am Chem Soc. 1990; 112:4074.

17. (a) Nakamura N, Inoue K, Iwamura H. J Am Chem Soc. 1992; 114:1484; (b) Matsuda M, Nakamura N, Takahashi K, Inoue K, Koga M, Iwamura H. J Am Chem Soc. 1995; 117:5550; (c) Nakamura N, Inoue K, Iwamura H, Angew. Chem Int Ed Engl. 1993; 32:872; (d) Matsuda M, Nakamura N, Takahashi K, Inoue K, Iwamura H. Bull Chem Soc Jpn. 1996; 69:1483.

18. (a) Itoh K. Pure Appl Chem. 1978; 50:1252. (b) Hückel E. Z Phys Chem. 1936; 34:339 and discussed in Berson JA. Angew Chem Int Ed Engl. 1996; 35:2751.

19. (a) Murata S, Sugawara T, Iwamura H. J Am Chem Soc. 1987; 109:1266; (b) Tukada H. J Am Chem Soc. 1991; 113:8991.

20. (a) Wasserman E, Murray RW, Yager WA, Trozzolo AM, Smolinsky GJ. J Am Chem Soc. 1967; 89:5076; (b) Murata S, Iwamura H. J Am Chem Soc. 1991; 113:5547; (c) Matsumoto T, Ishida T, Koga N, Iwamura H. J Am Chem Soc. 1992; 114:9952; (d) Ling C, Minato M, Lahti PM, van Willigen H. J Am Chem Soc. 1992; 114:9959.

21. (a) Calder A, Forester AR, James PG, Luckhurst GR. J Am Chem Soc. 1969; 91:3724; (b) Matsumoto T, Koga N, Iwamura H. J Am Chem Soc. 1992; 114: 5448.

22. For reviews of violations of Hund's rule in molecules see: (a) Borden WT, Iwamura H, Berson JA. Acc Chem Res. 1994; 27:109; (b) Hrovat DA, Borden WT. J Mol Struct (Theochem). 1997; 398–399:211; (c) Hrovat DA, Borden WT. In: Modern Electronic Structure Theory and Applications in Organic Chemistry, Davidson ER, ed. Singapore: World Scientific, 1997, p. 171.

23. (a) Yarkony DR, Schaefer HF III. J Am Chem Soc. 1974; 96:3754; (b) Davidson ER, Borden WT. J Am Chem Soc. 1977; 99:2053; (c) Davis JH, Goddard WA III. J Am Chem Soc. 1977; 99:4242; (d) Hood DM, Schaefer HF III, Pitzer RM. J Am Chem Soc. 1978; 100:4587; (e) Dixon DA, Dunning TH Jr, Eades RA, Kleier DA. J Am Chem Soc. 1981; 103:2878; (f) Auster SB, Pitzer RM, Platz MS. J Am Chem Soc. 1982; 104:475; (g) Feller D, Tanaka K, Davidson ER, Borden WT. J Am Chem Soc. 1982; 104:967; (h) Ma B, Schaefer HF III. Chem Phys. 1997; 207:31; (i) Cramer CJ, Smith BA. J Phys Chem. 1996; 100: 9664; (j) Hrovat DA, Borden WT. unpublished results, cited by Bally T, Borden WT. In: Reviews in Computational Chemistry, Lipkowitz KB, Boyd DB, eds. New York: VCH Publishers, 1999, in press.

24. (a) Baseman RJ, Pratt DW, Chow M, Dowd P. J Am Chem Soc, 1976; 98: 5726. (b) The first Curie law studies were performed on TMM derivatives, rather than on the parent hydrocarbon, by Platz MS, McBride JM, Little RD, Harrison JJ, Shaw A, Berson JA. J Am Chem Soc. 1976; 98:5725.

25. von E, Doering W, Roth HD. Tetrahedron. 1970; 26:2825.

26. Wenthold PG, Hu J, Squires RR, Lineberger WC. J Am Chem Soc. 1996; 118: 475.

27. Wenthold PG, Hu J, Squires RR. J Am Chem Soc. 1994; 116:6961.

28. Dowd P, Chow M. J Am Chem Soc. 1977; 99:8507.

29. Dowd P, Chang W, Paik YH. J Am Chem Soc. 1985; 108:7416.

30. For a discussion of Curie plots, see ref. 6d, pp. 462–469.
31. Roth WR, Kowalczik U, Maier G, Reisenauer HP, Sustmann R, Müller W. Angew Chem Int Ed Engl. 1987; 26:1285.
32. (a) Roth WR, Biermann M, Erker G, Jelich K, Gerhartz W, Görner H. Chem Ber. 1980; 113:586; (b) Dowd P, Chang W, Paik YH. J Am Chem Soc. 1987; 109:5284; (c) Matsuda K, Iwamura H. J Am Chem Soc. 1997; 119:7413; (d) Matsuda K, Iwamura H. J Chem Soc Perkin. 1998; 2:1023.
33. Nash JJ, Dowd P, Jordan KD. J Am Chem Soc. 1992; 114:10071.
34. (a) Du P, Borden WT. J Am Chem Soc. 1987; 109:930; (b) Nachtigall P, Jordan KD. J Am Chem Soc. 1992; 114:4743; (c) Nachtigall P, Jordan KD. J Am Chem Soc. 1993; 115:270.
35. (a) Lee J, Chou PK, Dowd P, Grabowski JJ. J Am Chem Soc. 1993; 115:7902; (b) Clifford EP, Wenthold PG, Lineberger WC, Ellison GB, Wang CX, Grabowski JJ, Vila F, Jordan KD. J Chem Soc Perkin 2. 1998:1015.
36. (a) Hrovat DA, Borden WT. J Am Chem Soc. 1992; 114:5879; (b) Wenthold PG, Hrovat DA, Borden WT, Lineberger WC. Science. 1996; 272:1456.
37. (a) Bush LC, Maksimovic L, Feng XW, Lu HSM, Berson JA. J Am Chem Soc. 1997; 119:1416, and references cited therein; (b) Berson JA. Acc Chem Res. 1997; 30:238.
38. (a) Davidson ER, Borden WT, Smith J. J Am Chem Soc. 1978; 100:3299; (b) Feller D, Davidson ER, Borden WT. J Am Chem Soc. 1982; 104:1216; (c) Du P, Hrovat DA, Borden WT. J Am Chem Soc. 1989; 111:3773.
39. Hill BT, Squires RR. J Chem Soc Perkin 2. 1998:1027.
40. Jacobs SJ, Schultz DA, Jain R, Novak J, Dougherty DA. J Am Chem Soc. 1993; 115:1744.
41. (a) Dvolaitzky M, Chiarelli R, Rassat A. Angew Chem Int Ed Engl. 1992; 31: 180; (b) Kanno F, Inoue K, Koga N, Iwamura H. J Am Chem Soc. 1993; 115: 844.
42. Fang S, Lee M-S, Hrovat DA, Borden WT. J Am Chem Soc. 1995; 117:6727.
43. (a) Kato S, Morokuma K, Feller D, Davidson ER, Borden WT. J Am Chem Soc. 1983; 105:1791; (b) Fort RC Jr, Getty SJ, Hrovat DA, Lahti PM, Borden WT. J Am Chem Soc. 1992; 114:7549; (c) Hrovat DA, Murcko MA, Lahti PM, Borden WT. J Chem Soc Perkin 2. 1998:1037.
44. Wenthold PG, Kim JB, Lineberger WC. J Am Chem Soc. 1997; 119:1354.
45. (a) Lahti PM, Rossi A, Berson JA. J Am Chem Soc. 1991; 113:2318; (b) Lahti PM, Ichimura AS, Berson JA. J Org Chem. 1989; 54:958; (c) Lahti PM, Rossi A, Berson JA. J Am Chem Soc. 1985; 107:4362; (d) Du P, Hrovat DA, Borden WT, Lahti PM, Rossi A, Berson JA. J Am Chem Soc. 1986; 108:5072; (e) Hrovat DA, Borden WT. J Am Chem Soc. 1994; 116:6327.
46. (a) Bailey WJ, Fetter EJ, Economy J. J Org Chem. 1962; 27:3479; (b) Longone DT, Boettcher, F-P. J Am Chem Soc. 1963; 85:3436.
47. Kemnitz CR, Squires RR, Borden WT. J Am Chem Soc. 1997; 119:6564.
48. Jahn HA, Teller E. Proc R Soc London, Ser A. 1937; 161:220. The distortion of the $^1E'$ state of TMM from D_{3h} symmetry is also a first-order Jahn–Teller effect.
49. Hu J, Squires RR. J Am Chem Soc. 1996; 118:5816.

50. (a) Borden WT. Mol Cryst Liq Cryst. 1993; 232:195. (b) See also the discussion in ref. 43c.
51. West AP Jr, Silverman SK, Dougherty DA. J Am Chem Soc. 1996; 118:1451.
52. (a) Osamura Y, Borden WT, Morokuma K. J Am Chem Soc. 1984; 106:5212; (b) Coolidge MB, Yamashita K, Morokuma K, Borden WT. J Am Chem Soc. 1990; 112:1752; (c) Ichimura AS, Lahti PM, Matlin AR. J Am Chem Soc. 1990; 112:2868; (d) Powell HK, Borden WT. J Org Chem. 1995; 60:2654; (e) Lim D, Hrovat DA, Borden WT, Jorgensen WL. J Am Chem Soc. 1994; 116: 3494; (f) Hrovat DA, Sander W, Borden WT, unpublished results.
53. Hirano T, Kumagai T, Miyashi K. Akiyama, Ikegami Y. J Org Chem. 1991; 56:1907.
54. (a) Sclove DB, Pazos JF, Camp RL, Greene FD. J Am Chem Soc. 1970; 92: 7488. (b) Solvent effects on the ring opening of sterically unhindered cyclopropanones are complicated by interactions between solvents with lone pairs of electrons and the very electrophilic carbonyl carbon of the cyclopropanone; Cordes MHJ, Berson JA. J Am Chem Soc. 1996; 118:6241.
55. Rule M, Matlin AR, Hilinski EF, Dougherty DA, Berson JA. J Am Chem Soc. 1979; 101:5098.
56. Seeger DE, Hilinski EF, Berson JA. J Am Chem Soc. 1981; 103:720.
57. Khan MI, Goodman JL. J Am Chem Soc. 1994; 116:10342.
58. Davidson ER, Gajewski JJ, Shook CA, Cohen T. J Am Chem Soc. 1995; 117: 8495.
59. Recent references are: (a) Hrovat DA, Borden WT. J Phys Chem. 1994; 98: 10460; (b) Davico GE, Bierbaum VM, DePuy CH, Ellison GB. Int J Mass Spectrom, Ion Proc. 1996; 156:109.

6

Conformational Exchange Modulation in Trimethylenemethane-Type Biradicals

David A. Shultz
North Carolina State University, Raleigh, North Carolina

I. *META*-PHENYLENE AS A FERROMAGNETIC EXCHANGE COUPLING FRAGMENT

There are two reasons that *meta*-phenylene is commonly used as a ferromagnetic coupling unit. Arguably, the first is because there are a wide variety of synthetic approaches to creating bonds to benzene rings. Second, and more important, *meta*-phenylene is a good choice for coupling unpaired electrons because the parent biradical, *meta*-xylylene, is a robust triplet ground state biradical. This is not to say that *meta*-xylylene is unreactive, rather that the triplet state lies well below the singlet state, in fact by ca. 9 kcal/mol [1]. Since *meta*-phenylene is such an important, popular coupler we discuss it first before discussing trimethylenemethane-type high-spin molecules.

High-spin coupling of the unpaired electrons in *meta*-xylylene to give a triplet ground state is due to the topology of the π-system. To exemplify the effects of connectivity, let us consider the union of a benzyl radical and a methylene radical fragment to yield a xylylene biradical. There are three possible positions for attaching the methylene to the phenyl ring—*ortho*, *meta*, and *para*. Two of these (*ortho* and *para* are active positions (those with positive spin density), while *meta* is a nodal position. Union at either the *ortho*- or *para*-positions leads to closed-shell (low-spin) singlets. Only

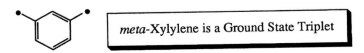

para- and *ortho-*Xylylene are Ground State Singlets

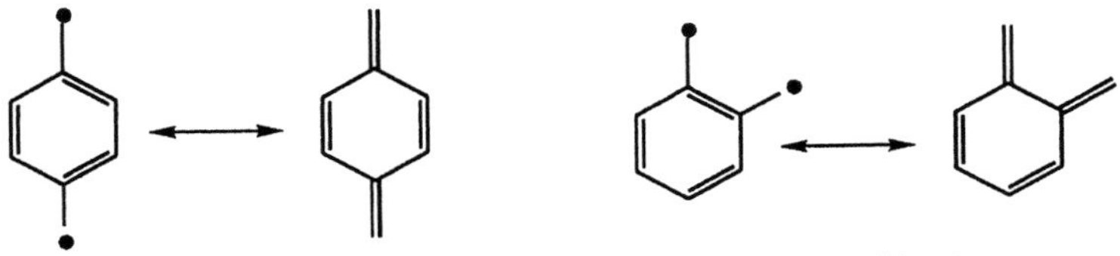

Scheme 1

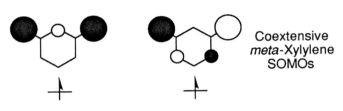

*meta-*Xylylene is a Ground State Triplet

Scheme 2

union between the nodal *meta-*position and methyl radical leads to a ground state triplet biradical (Schemes 1 and 2) [2].

Molecular orbital (MO) theory supports these claims. The key to predicting the multiplicity of an organic biradical from an MO viewpoint is to determine whether the singly occupied MOs (SOMOs) are coextensive (SOMOs have coefficients on a common set of atoms) or disjoint (SOMOs do not have coefficients on a common set of atoms) [3]. If the SOMOs are coextensive (or nondisjoint), then exchange interactions that stabilize high-spin states are significant. On the other hand, if the SOMOs are disjoint, then exchange interactions are negligible and so is stabilization of the high-spin state (Scheme 3).

The origin of the relationship between SOMO coextensivity and ground spin state can be ascribed to the fact that electrons of like spin avoid each other in a region of shared space (the Pauli exclusion principle). By comparison, spin-paired electrons behave as if they are "attracted" to each other in this shared volume element and electron–electron repulsion is increased. The overall effect is that nondisjoint biradicals and related species

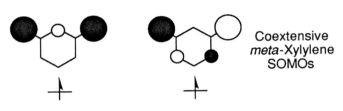

Coextensive
*meta-*Xylylene
SOMOs

Scheme 3

are predicted to be ground state triplet biradicals [4]. As a rule of thumb, any species having at least two degenerate or nearly degenerate (by symmetry or accidental degeneracy) SOMOs, for which the $2p$ AO coefficients *cannot be isolated on separate atoms*, is a candidate for having the maximum spin multiplicity as the ground electronic state (i.e., Hund's rule may be applied). Consequently, any steric interaction that destroys conjugation of the π-system separates the SOMO coefficients and narrows the singlet-triplet gap, and in extreme cases can result in a singlet ground state. Documenting such cases of "conformational J-modulation" is the goal of the following section [5,6].

II. EFFECT OF BOND TORSION ON EXCHANGE COUPLING IN *META*-PHENYLENE-TYPE BIRADICALS AND RELATED SPECIES

Elucidating the effects of conformation on spin–spin coupling in high-spin ($S = 1$) organic compounds is critical to understanding the magnetic properties of materials incorporating such species [7]. There have been numerous studies of the effect of torsion on exchange coupling in *meta*-phenylene-based bi- and triradicals [8].

For some time it was known that nondisjoint dinitroxide **1"** possessed a triplet ground state but decomposed rapidly [9]. In 1992, Rassat and coworkers prepared the isolable biradical **2**, a methylated derivative of **1"** [10]. The added methyl groups stabilize **2** with respect to undesired bimolecular decomposition. Interestingly, **2"** was isolated in two isomeric forms, both exhibiting rather strong antiferromagnetic spin–spin coupling, resulting in a ground state singlets with exchange parameters, J indicated in Scheme 4. $J < 0$ is characteristic of antiferromagnetic coupling; $J > 0$ is characteristic of ferromagnetic coupling. This result suggests that steric interactions that result in the inhibition of resonance delocalization of the spin-containing nitroxide unit onto the π-system of the ferromagnetic coupling unit can confound the topological rules for spin–spin coupling in organic biradicals and related species.

Isomer **2A"**: $J \approx -70$ cal/mol
Isomer **2B"**: $J \approx -83$ cal/mol

1"   **2"**

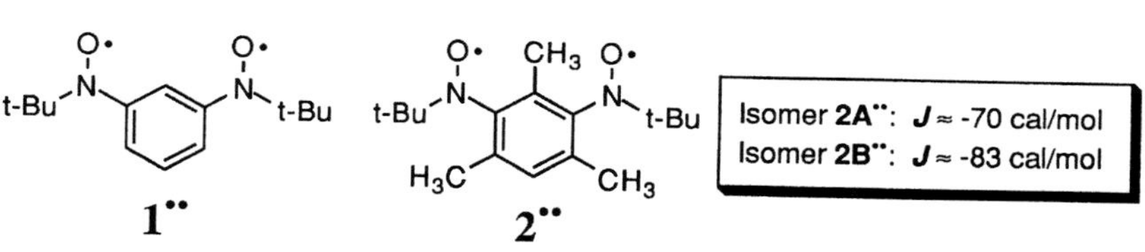

Scheme 4

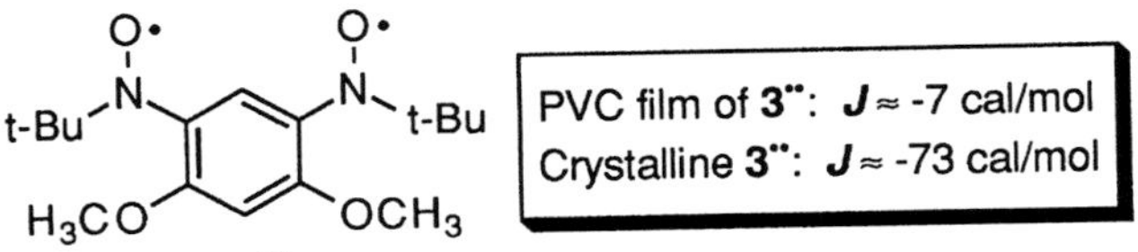

Scheme 5

Not long after Rassat's paper, Iwamura published a paper on a similar dinitroxide, **3** [11]. The exchange parameter calculated from a sample dispersed in PVC was -7 cal/mol; that calculated from a crystalline sample was -73 cal/mol (Scheme 5). Interestingly, the room-temperature EPR spectrum was consistent with $|J| \sim |a|$.

The x-ray crystal structure of **3"** showed that the *t*-butylnitroxide groups were twisted 65.1° and 75.3° from the plane of the benzene ring, experimentally suggesting nitroxide-benzene ring torsion as the culprit for reversing the predicted exchange coupling.

A few years later, Iwamura and coworkers followed up this work with a description of the exchange coupling in the fascinating, related trinitroxide, **4"'** [12]. In this triradical, the nitroxides are again twisted relative to the benzene ring with torsions of 90° and 79°. The variable-temperature susceptibility was fit with two different models, and representative *J*-values are given in Scheme 6. The exchange parameter J_B is more strongly antiferromagnetic than J_A. This is explained by the fact that the antiferromagnetic nature of J_A is attenuated by polarization due to the larger antiferromagnetic J_B. Thus, J_A has an antiferromagnetic term originating from torsion of the nitroxides with respect to the *meta*-phenylene coupling fragment, and a polarization-induced ferromagnetic component due to the large antiferro-

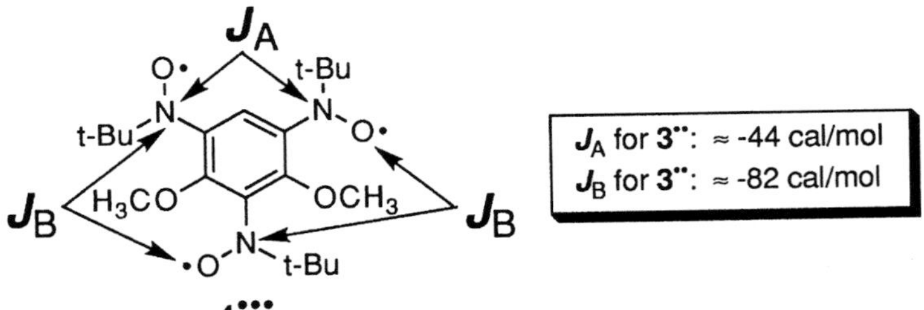

Scheme 6

magnetic J_B. This is the first example of competing exchange interactions in an organic triradical.

In 1995, Borden and coworkers published a computational paper that provided a mechanism for antiferromagnetic coupling in highly twisted *meta*-phenylene biradicals and related species [13]. The calculations clearly showed that for *meta*-xylylene, **5**, and the dinitroxide, **6**, the origin of the singlet ground states for species with radical centers twisted 90° relative to the benzene ring is the energetic splitting of the two SOMOs through mixing with benzene ring σ-orbitals of appropriate symmetry as shown in Scheme 7. The antisymmetric (**A**) combination is destabilized to a greater extent than the symmetric (**S**) combination.

Other structures have been helpful in illustrating the change in ground spin state when conjugation of spin-containing units with ferromagnetic coupling fragment is disrupted. For example, 10,10'-(*meta*-phenylene)di-

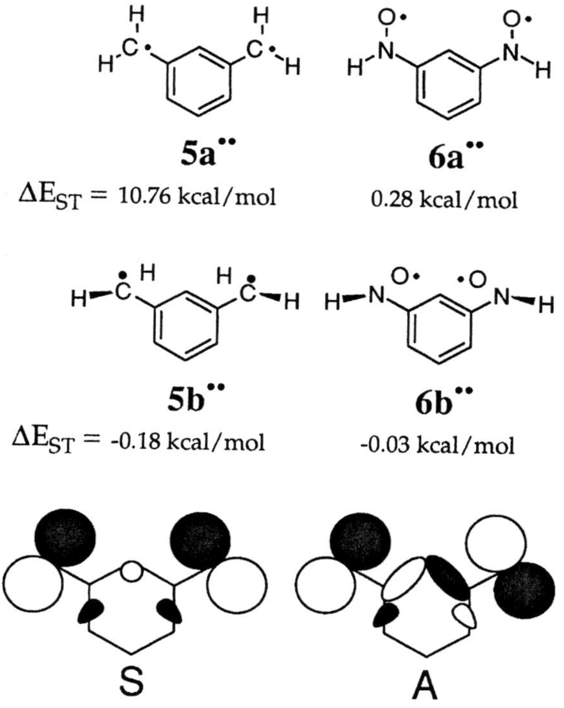

5a
ΔE_{ST} = 10.76 kcal/mol

6a
0.28 kcal/mol

5b
ΔE_{ST} = -0.18 kcal/mol

6b
-0.03 kcal/mol

SOMOs for **5b** when methylenes are twisted 90°

Scheme 7

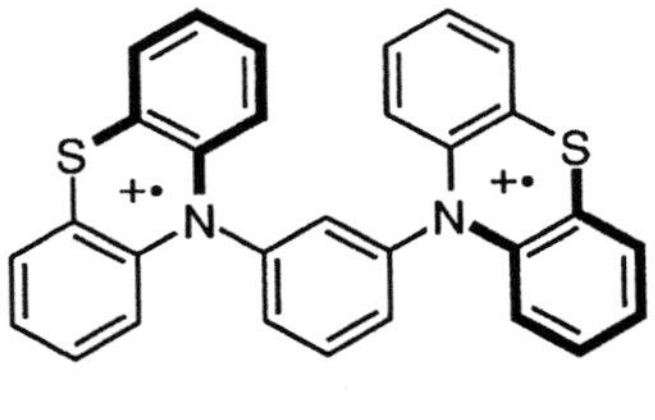

$7^{2+\cdot\cdot}$ 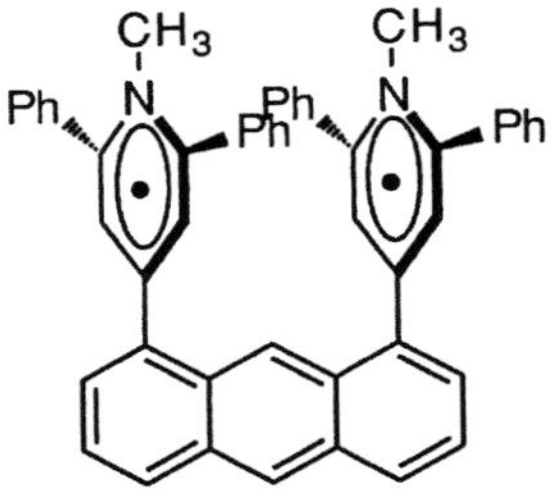 $8^{\cdot\cdot}$

Scheme 8

phenothiazine dication ($7^{2+\cdot\cdot}$), was shown to be a ground state singlet with $J = -14$ cal/mol [14]. An anthracene-coupled dipyridinyl biradical, $8^{\cdot\cdot}$, is also a singlet ground state with $J = -6.3$ cal/mol (Scheme 8) [15].

There are several noteworthy exceptions to these findings, some of which might be solvent matrix–induced geometries that do not display large bond torsions. For example, ferromagnetic coupling was observed for sterically congested tetraradical $9^{\cdot\cdot\cdot\cdot}$ (Scheme 9) [16].

A unique case for exchange coupling through a *meta*-phenylene coupler was described by Dougherty and Silverman [17], who studied bis-TMM tetraradicals. They found that in $11^{\cdot\cdot\cdot\cdot}$, the structure where severe bond torsions were expected, two noninteracting triplets are produced from bond torsions rather than the expected antiferromagnetic coupling, which would have produced a singlet ground state. To our knowledge, noninteracting spin centers connected to a π-system has been reported in only one other case—*disjoint* tetraradical $12^{\cdot\cdot\cdot\cdot}$ [18]. Dougherty and Silverman went on to conclude that moderate torsions, as in $10^{\cdot\cdot\cdot\cdot}$, still gave ferromagnetic coupling. Spin polarization was invoked as the factor enforcing ferromagnetic coupling in $10^{\cdot\cdot\cdot\cdot}$ (Scheme 10).

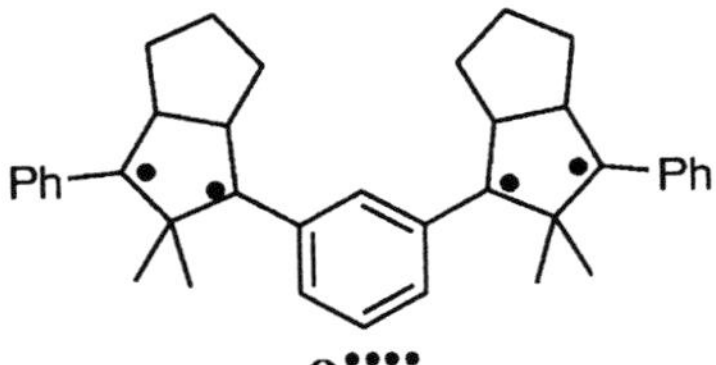

$9^{\cdot\cdot\cdot\cdot}$

Scheme 9

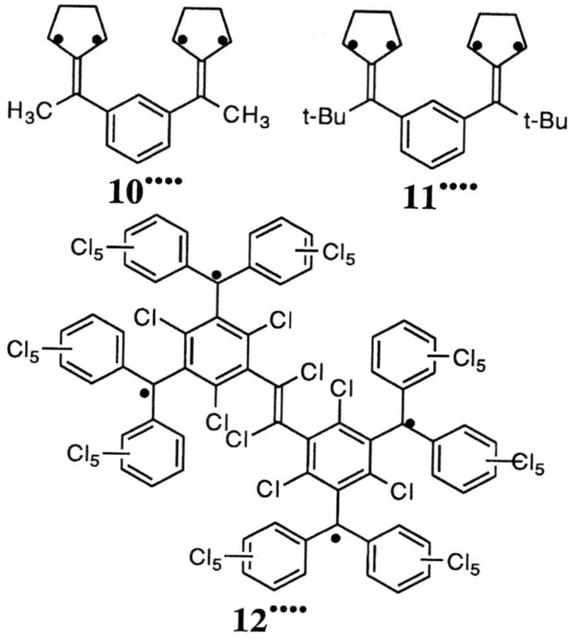

Scheme 10

That moderate torsion existed in **10""** was based on the fact that the crystal structure of **13** showed a 37° torsion between the fulvene and aryl groups, while nearly orthogonal fulvene and aryl groups were observed in **14**, the model for **11""** (Scheme 11) [17]. Evidence for similar twisting in the biradical versions is seen in the zero-field splitting parameters, **D**. Since **D** is inversely proportional to the cube of the interelectronic distance [19], the larger **D** for **16"**, as compared to **15"**, is consistent with less delocalization into the phenyl substituent due to phenyl-TMM torsion for **16"** than

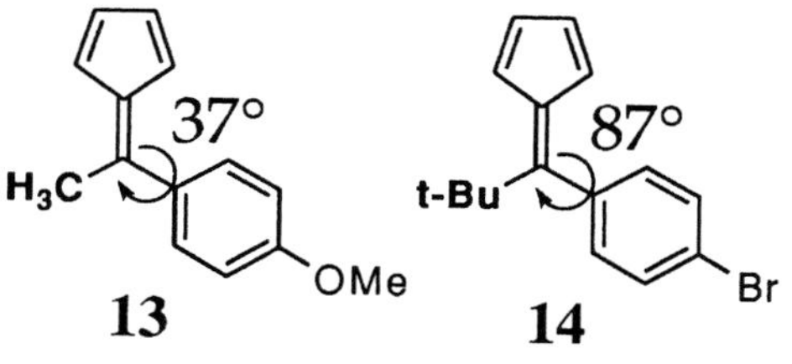

Scheme 11

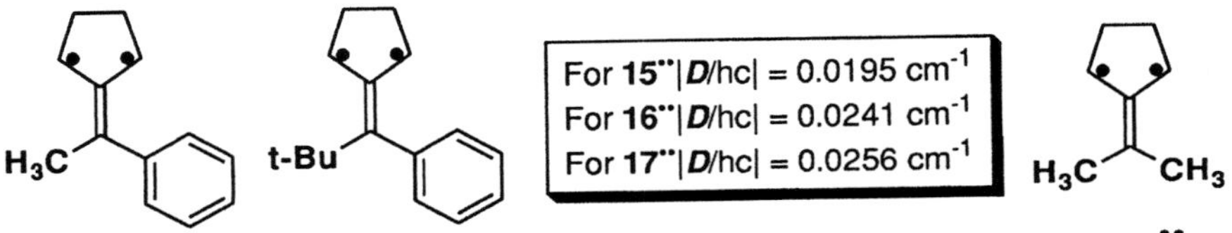

For **15"** $|D/hc| = 0.0195$ cm^{-1}
For **16"** $|D/hc| = 0.0241$ cm^{-1}
For **17"** $|D/hc| = 0.0256$ cm^{-1}

15"   **16"**   **17"**

Scheme 12

for **15"**. In fact, the *D*-value for **16"** is nearly equal to that of **17"**, which lacks a phenyl substituent (Scheme 12) [17].

Ferromagnetic coupling has been observed in structures having severe torsion of spin-containing units with respect to the coupling unit. For example, Rajca has shown that *meta*-phenylene appears to be a ferromagnetic coupler in the 2,4-dimethyl-, 2,4-diisopropyl-, and 2,4,6-trimethyl-modifications, as in compounds **18"**—**20"** [20,21], while Veciana and co-workers have shown that tetrachloro-*meta*-phenylene (**21"**) is a ferromagnetic coupler (Scheme 13) [22,23].

Obviously, the ability of *meta*-phenylene to couple attached spin-containing units is sensitive to bond torsions, but structures where severe torsions are expected do not always give rise to the expected antiferromagnetic coupling. Some of these observations might be the result of matrix

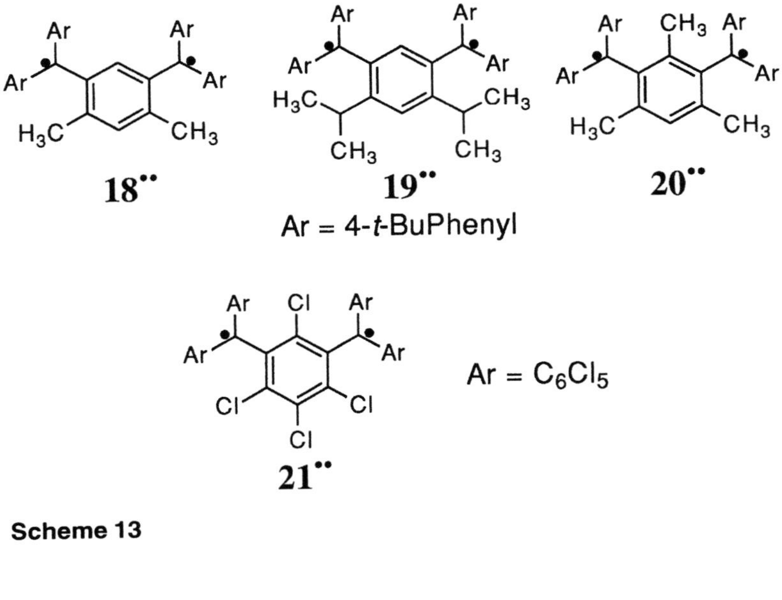

18"   **19"**   **20"**

Ar = 4-*t*-BuPhenyl

Ar = C$_6$Cl$_5$

21"

Scheme 13

effects, and additional study—both theoretical and experimental—is warranted.

III. DESIGN CRITERIA FOR TRIMETHYLENEMETHANE (TMM)-BASED HIGH-SPIN ORGANIC MOLECULES

Dougherty has used trimethylenemethane (TMM)-containing spin-containing units to explore exchange coupling through saturated coupling fragments [24,25]. However, high-spin molecules having a TMM-type ferromagnetic coupling unit are rare compared to those based on *meta*-xylylene. The major advantage to using the TMM design is that the parent biradical has a larger singlet-triplet gap (14.5 kcal/mol) [26] than *meta*-xylylene (9.2 ± 0.2 kcal/mol) [1].

The design elements of a TMM-based biradical are shown in Scheme 14. First, geminal attachment of spin-containing units to an ethylene moiety completes the TMM topological requirement. Second, stable delocalized radicals are very helpful for fundamental studies of exchange coupling. As discussed later, phenyl-*t*-butylnitroxides are somewhat less attractive in such studies, since spin density delocalization onto the phenyl ring is small compared to that in phenoxy radicals. The resulting spin–spin coupling is weak in the nitroxides, and consequently is less likely to be sensitive to structural variations [27]. Third, a capping group that sterically protects unpaired electrons is necessary for biradicals having spin delocalization into the TMM functionality. However, this necessity creates a potential problem: steric interactions between atoms in the capping group and those in the spin-containing unit might disrupt conjugation between the spin-containing units and the TMM functionality. As already indicated in *meta*-xylylene analogs, this inhibition of resonance delocalization attenuates the effectiveness of the coupling fragment to high-spin couple the unpaired electrons.

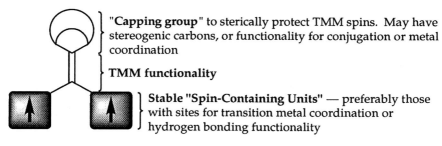

"**Capping group**" to sterically protect TMM spins. May have stereogenic carbons, or functionality for conjugation or metal coordination

TMM functionality

Stable "**Spin-Containing Units**" — preferably those with sites for transition metal coordination or hydrogen bonding functionality

Scheme 14

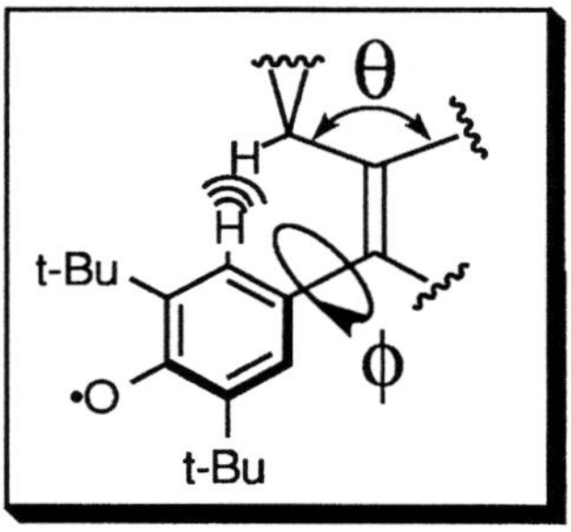

Scheme 15

Currently, we are studying phenoxyl and nitroxyl radicals as spin-containing units. We are interested in the phenoxy radicals primarily as models for semiquinone biradicals [28,29], but also because their ring spin density is large and their spin delocalization is expected to be greater than in nitroxides. Nitroxide radicals are attractive, however, due to their stability and their utility in forming coordination polymers with transition metals [30–32].

Possible structures we consider for the capping group include: heteroatom groups, aryl groups, and bicyclic or tricyclic ring systems. Simple primary or secondary alkyl groups were not chosen, since they provide allylic hydrogens that are excellent candidates for abstraction and loss of spin [33]. Difficulties in forming carbon–carbon double bonds having attached *tert*-butyl groups are well known. Consequently, we chose to examine bicyclic or tricyclic ring systems as capping groups first. As shown in Scheme 15, the bridgehead hydrogens are orthogonal to the TMM π-system, a geometric consequence of a bicyclic capping group. However, these same hydrogens experience van der Waals repulsions with phenoxy rings when the phenoxy rings approach coplanarity with the TMM functionality. This steric interaction could result in conformational attenuation of exchange coupling, and should decrease as the bond angle θ decreases. To test this hypothesis, we are preparing biradicals **22··–24··**. Biradicals **23··** and **24··** have been prepared, and most of the results discussed next will focus on these two biradicals.

IV. MOLECULAR MECHANICS CALCULATIONS ON DIPHENYLMETHYLENE COMPOUNDS

Biradicals **22··–24··** differ in the ring size of the capping ring system attached to the TMM functionality. Compound **22··** has a bicyclic ring structure in-

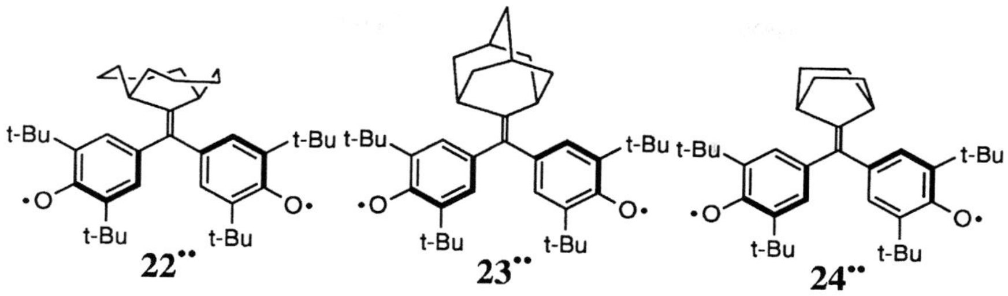

22"   **23"**   **24"**

Scheme 16

corporating two seven-membered rings, biradical **23"** has six-membered rings attached to the C=C, while **24"** has two five-membered rings attached in a bicyclic form (Scheme 16). These different ring sizes control the bond angles, θ, and are expected therefore to affect torsion angles, ϕ. To check the correlation between θ and ϕ computationally, we studied the [$n.m.1$] alkenes shown in Scheme 17. The structures were constrained to geometries having C_2 symmetry, and the phenyl rings were rotated in a conrotatory fashion until a minimum energy was reached.

The results of the MM2 [34] energies for various diphenylmethylene compounds, [$n.m.1$], in Figure 1 show that phenyl torsion, ϕ, is indeed directly proportional to the bond angle, θ, with the greatest conjugation of phenyl and C=C occurring at small values of θ. As can be seen in Figure 1, torsions increase as θ increases. The obvious conclusion is that larger ring systems introduce steric elements that cause torsion of the phenyl rings.

These results suggest a natural propensity for TMM-based biradicals to exhibit conformational J-modulation, but the attenuation of the expected ferromagnetic J can be adjusted by using capping groups with small ring sizes.

V. X-RAY CRYSTALLOGRAPHY OF DI(4-BROMOPHENYL)METHYLENE COMPOUNDS

As opposed to other related molecules, dibromides **25–27** were found to crystallize quite easily from ethanol solutions, and were used to test the results of the molecular mechanics calculations. The predictions of our MM2 calculations are substantiated by the results of the x-ray crystallography. Chem-3D©®™ depictions of the crystal structures for model compounds **25–27** are shown in Figure 2. Average aryl torsions and bond angles are

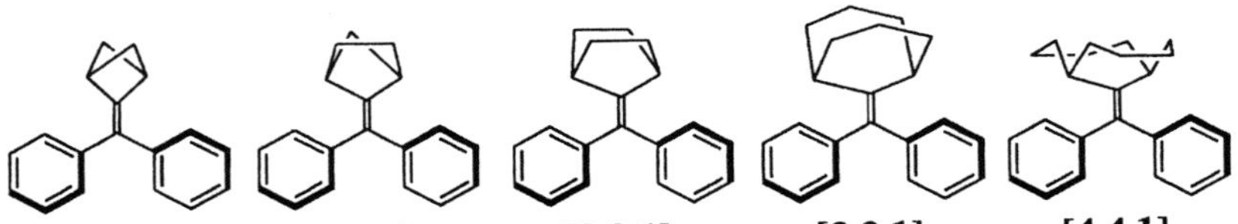

[1.1.1]   **[2.1.1]**   **[2.2.1]**   **[3.3.1]**   **[4.4.1]**

Scheme 17

given in Table 1. The measured decrease in θ in going from an adamantyl cap to a norbornyl cap is 14°, identical to the calculated decrease of 14°. Both **25** and **26** displayed solid-state structures having two different aryl torsion angles. The average angle for **25** is 44.9°; the average aryl torsion for **26** is 52.3°. These average angles differ by 7.4°, very close to the MM2-predicted difference of 8.6°.

The observed average torsion angles are lower than the calculated values, but the experimental trend follows the computations. Interestingly, the structure of **27** is quite close to that of **26**, indicating that potentially destructive steric interactions are not limited to those species having bicyclic ring systems as the capping group.

VI. SYNTHESIS OF COMPOUNDS 23–27

Schemes 18 and 19 are indicative of our general approach to the synthesis of TMM-based molecules and were adapted from the procedure of Iwamura [27]. Biradicals **23$^{\cdot\cdot}$** [35] and **24$^{\cdot\cdot}$** were prepared as shown in Schemes 18 and 19 (Eqs. 1 and 2). The biradicals gave blue-green solutions, with λ_{max} ~ 600 nm. Both **23$^{\cdot\cdot}$** and **24$^{\cdot\cdot}$** are quite photosensitive and oxygen sensitive. The infrared spectrum of the isolated solids are devoid of –OH stretches due to the phenolic precursons (ca. 3650 cm^{-1}, sharp) as well as the "normal" phenolic C–O stretches (ca. 1230 cm^{-1}). Several attempts were made at crystallizing **23$^{\cdot\cdot}$** and **24$^{\cdot\cdot}$**, but they were unsuccessful.

Ester **30** was prepared by metalation of 7-bromonorbornane [36] using lithium di-4,4′-t-butylbiphenylide [37], followed by reaction with CO_2 and subsequent Fisher esterification.

Dibromides **25**, **26**, and **27** [27] were prepared in a similar fashion, starting from condensation of esters **30, 28**, and methyl isobutyrate, respectively, with 4-bromophenyllithium (Scheme 20). Dehydration of the resultant tertiary alcohols provided the alkenes in excellent yields.

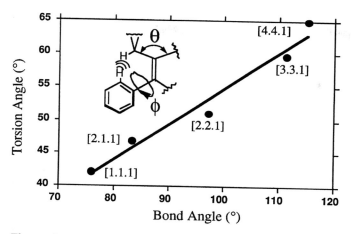

Figure 1 MM2-calculated torsion angles, φ, plotted against MM2-calculated bond angles, θ.

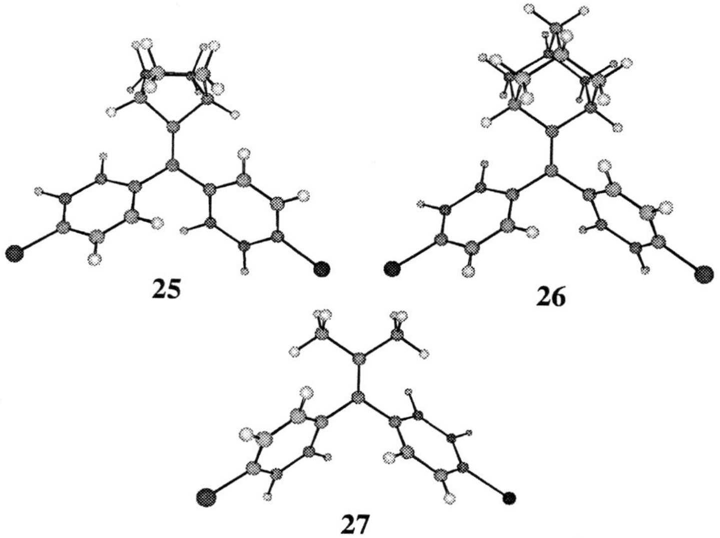

25

26

27

Figure 2 Chem-3D®©™ depictions of x-ray crystal structures of dibromides **25–27**.

Table 1 Comparison of MM2 and X-Ray Structural Parameters

Compound	MM2-calculated bond angle, θ	Observed bond angle, θ^a	MM2-calculated torsion angle, ϕ	Observed torsion angle, ϕ
25/[2.2.1]	97.3	97.0	51.0	44.9[a]
26/[3.3.1]	111.3	111.0	59.6	52.3[a]
27[b]	113.0	111.3	54.7	51.6[a]

[a]Average value.
[b]MM2 structure was 1,1-diphenyl-2-methylpropene.

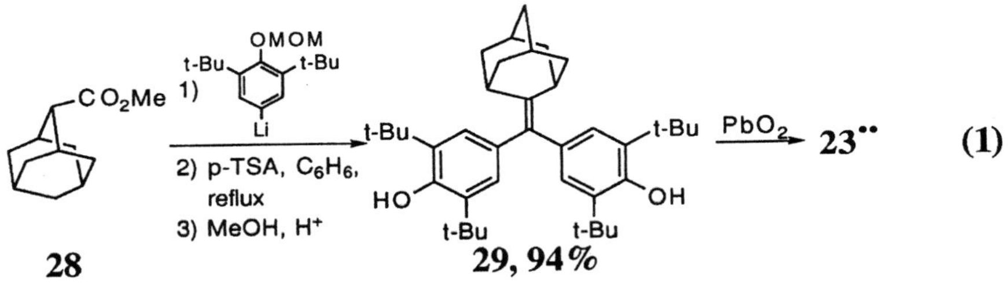

Scheme 18

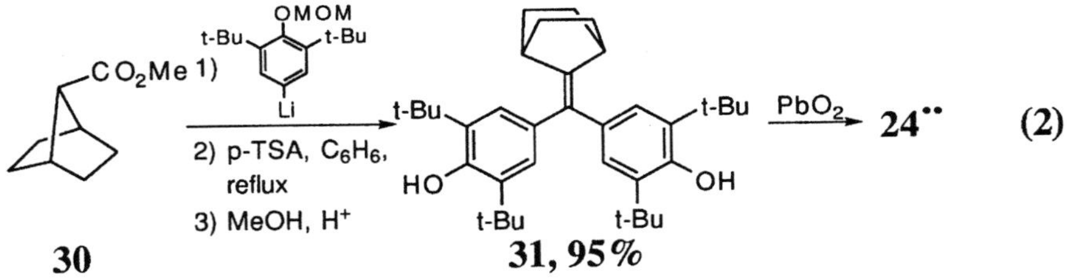

$$(1)$$

Scheme 19

$$(2)$$

VII. ELECTRON PARAMAGNETIC RESONANCE SPECTROSCOPY OF BIRADICALS 23·· [35] AND 24··

The X-band EPR spectra of biradical **23··** in frozen toluene are shown in Figure 3A (77 K) and B (4 K). Generally, the spectrum is characteristic of a randomly oriented triplet biradical with monoradical impurity [38]. As can be seen, the spectral features at 77 K include additional peaks, marked with

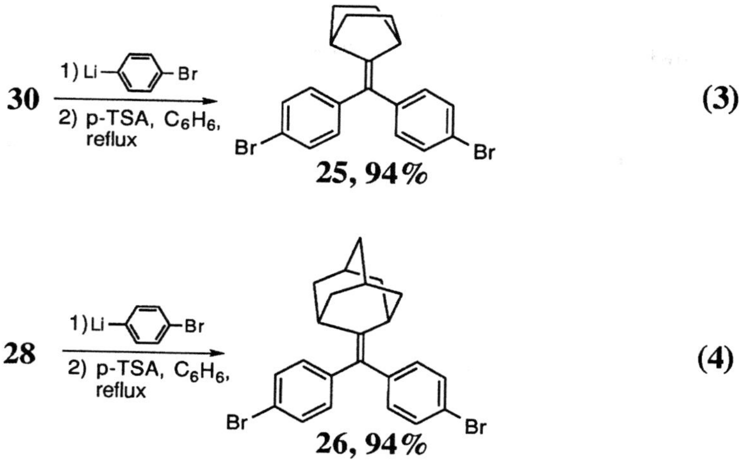

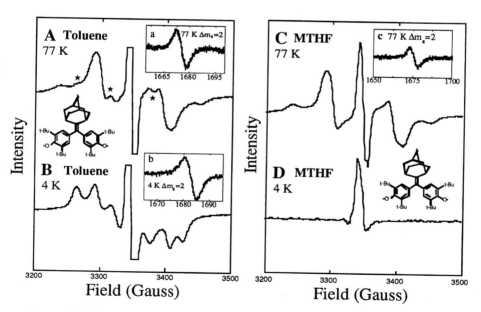

Scheme 20

Figure 3 77 K EPR spectra of biradical **23** in toluene (left) and MTHF (right). Insets **a–c**: $\Delta m_s = 2$ transitions.

a ★. As the temperature is lowered, these features increase in intensity; the remaining features decrease in intensity until at 4 K (Figure 3B) a new spectrum is apparent. Zero-field splitting parameters determined by spectral simulation are given in Table 2. At both temperatures, a $\Delta m_s = 2$ transition is apparent, linking the spectral features with a triplet biradical. We attribute these two different $S = 1$ spectra to two rotamers, **23A** and **23B**.

We believe that the size of the molecules and the rigidity of the solvent matrix precludes a dynamic equilibrium between rotamers **A** and **B**. Instead, hindered rotations in **23**, steric bulk, and, most importantly, matrix site effects combine to cause a freezing-out of the two rotamers (Scheme 21).

Below ca. 35 K, only spectral features of rotamer **B** are visible. The intensity of the $\Delta m_s = 2$ signal was measured as a function of temperature, and the Curie plot for the doubly integrated signal of **23B** is shown in Figure 4. Biradical **23B** exhibits a linear Curie plot, consistent with either a triplet ground state or a singlet/triplet degeneracy [40].

Spectra of **23** recorded in frozen 2-methyltetrahydrofuran (MTHF), are shown in Figure 3C and D. In this solvent, only rotamer **23A** is present (Scheme 22). As in the case for **23A**, when toluene is the solvent the intensity of the spectrum in Figure 3C is diminished as the temperature is lowered until, at 4K, no fine structure is observed.

The Curie plot for the spectra obtained in MTHF are shown in Figure 5 and were fit according to Eq. 5 [40]:

$$I_{EPR} = \frac{C}{T} \left[\frac{3 \exp \left(\dfrac{-2J}{RT} \right)}{1 + 3 \exp \left(\dfrac{-2J}{RT} \right)} \right] \tag{5}$$

where C is a constant and J is the exchange parameter. Best fit results give $J = -165 \pm 5$ cal/mol.

Table 2 Zero-Field Splitting Parameters for Biradicals **23** and **24**

Biradical		$|D/hc|/cm^{-1}$	$|E/hc|/cm^{-1}$
23	Rotamer A:	0.0102	0.0002
	Rotamer B:	0.0077	0.0010
24		0.0079	0.0007

Zfs parameters determined by simulation [39].

$23A^{\cdot\cdot} \rlap{\,/\!/}{=\!=} 23B^{\cdot\cdot}$ $23^{\cdot\cdot} \xrightarrow[\text{Toluene}]{\text{Freeze}} 23A^{\cdot\cdot} + 23B^{\cdot\cdot}$

Scheme 21

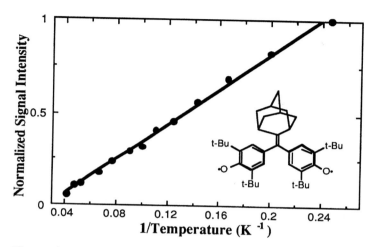

Figure 4   Curie plot for biradical **23**·· in a frozen toluene matrix (rotamer **B**). The data points were collected between 4 and 25 K and represent normalized signal intensities obtained by double integration of the $\Delta m_s = 2$ signals from three separate experiments.

As indicated by the loss of triplet signal at low temperature, rotamer **A** is a singlet ground state; the increase in signal intensity at low temperatures for rotamer **B** is consistent with either a triplet ground state or a singlet/triplet degeneracy for **B** [40]. In support of a dominant matrix effect, we conducted experiments in which one solvent was removed in vacuo and replaced with the other solvent. In every case, two rotamers appeared in toluene and only one rotamer was present in MTHF. This result also excludes matrix-dependent chemical reactions. The findings for biradical **23**·· are summarized in Scheme 23.

Possible structures for rotamers A and B are shown in Scheme 24. The rotamer having a singlet ground state could have two large phenoxy torsions (left) or exocyclic C=C torsion (middle).

$23^{\cdot\cdot} \xrightarrow[\text{MTHF}]{\text{Freeze}} 23A^{\cdot\cdot}$

Scheme 22

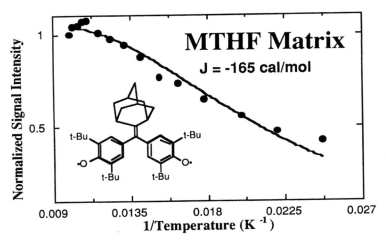

Figure 5 Curie plot for biradical **23··** in MTHF (rotamer **A**). The data points were collected between 40 and 100 K and represent normalized signal intensities obtained by double integration of the $\Delta m_s = 2$ signals from three separate experiments.

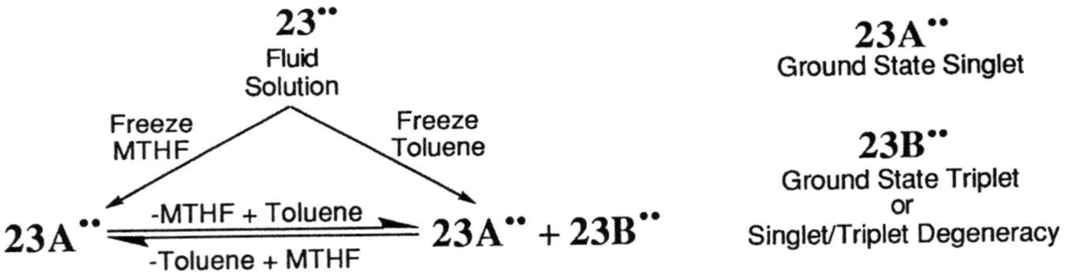

Scheme 23

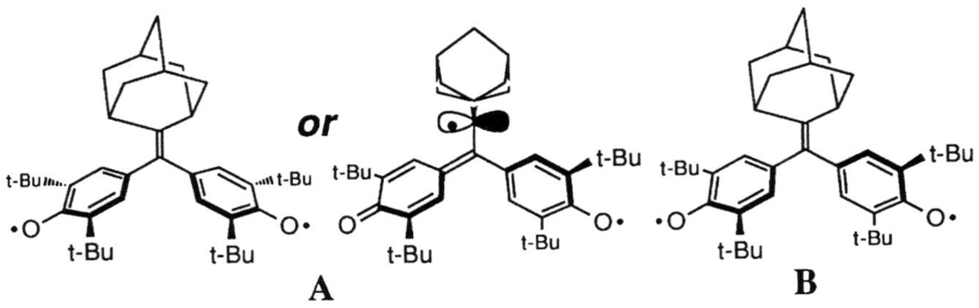

Scheme 24

    The EPR spectral properties of **23**" are the result of its structure and its interaction with solvent molecules within the frozen matrix. It might prove difficult to determine the origin of the matrix dependence; however, the relationship between the structural elements of **23**" and its spectral properties can be probed. We believe that the properties of **23**" can, in part, be attributed to large phenoxy torsions (ϕ) induced by steric interactions between bridgehead hydrogens and nearby atoms in the phenoxy rings, as discussed previously. A contraction of the angle θ should lessen these steric interactions by moving the bridgehead hydrogens away from the phenoxyl rings. This decrease in steric congestion should consequently allow a decrease in ϕ to enhance conjugation, and this less sterically congested geometry manifests itself in different EPR spectral parameters and in a change in exchange coupling.

    The 77 K EPR spectra of **24**" in toluene and MTHF are shown to the right in Figure 6. As can be seen, the spectral appearance (as determined by the zero-field splitting parameters) is independent of solvent, in contrast to the corresponding spectra for **13**" (Figure 6, left). In fact, the spectral features attributed to **23B**" (marked ♦ in Figure 6, left) are quite similar to the spectral features of **24**". Zero-field splitting parameters are given in Table 2.

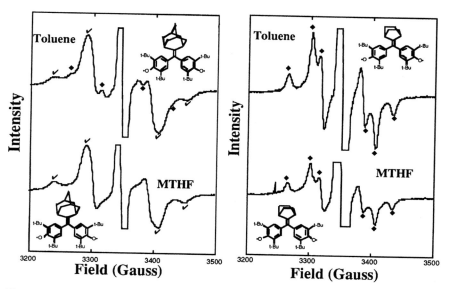

Figure 6 A comparison of X-band EPR spectra of biradical **23**" in toluene and MTHF at 77 K (left); and biradical **24**" recorded at 77 K in toluene and MTHF (right).

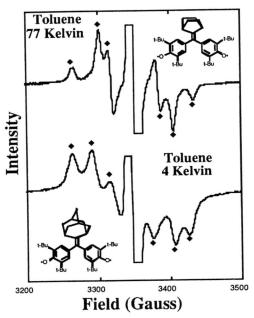

Figure 7 A comparison of X-band EPR spectra of biradical **24··** in toluene at 77 K (top) and biradical **23B··** in toluene at 4 K (bottom).

The spectral similarities between **23B··** and **24··** are more evident in the spectra presented in Figure 7. Not only is the spectrum of **24··** similar to that of **23B··**, but the temperature dependence of the spectral intensity is identical—consistent with either a ground state triplet or a singlet/triplet degeneracy for **24··**. The Curie plots for **24··** in both toluene and MTHF are given in Figure 8.

These spectra provide convincing evidence that **24··** is not as susceptible to the matrix effects and to conformational exchange modulation as is **23··**. This is consistent with the conformational differences in the model compounds imposed by the different capping groups.

VIII. CONCLUSIONS

Methyleneadamantane is a good TMM-type coupling unit to study the effects of conformation on spin–spin coupling. Biradical **23··** exhibits a type of spin isomerism, since geometry changes result in a change in ground state multiplicity. The origin of the spin isomers is a combination of large phen-

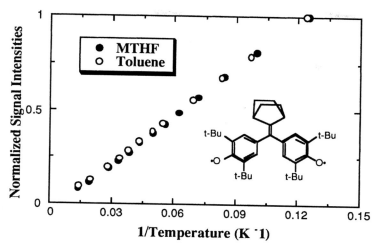

Figure 8 Curie plot for biradical **24ᐧᐧ** in MTHF and toluene. The data points were collected between 8 and 85 K and represent normalized signal intensities obtained by double integration of the $\Delta m_s = 2$ signals from three separate experiments.

oxyl ring torsions, bulky phenoxyl ring substituents, and matrix site effects. Similar conclusions have been drawn for another biradical [41]. For achieving the predicted sign of spin–spin coupling, methylenenorbornane is a better TMM-type coupling unit. Further studies of the properties of the dinitroxide biradical analogs of **23ᐧᐧ** and **24ᐧᐧ** and related species are under way.

ACKNOWLEDGMENTS

We thank the National Science Foundation (CHE-9634878) for support of this work. We also thank the National Science Foundation for funding a Research Experience for Undergraduates program.

REFERENCES AND NOTES

1. Wenthold PG, Kim JB, Lineberger WC. J Am Chem Soc. 1997; 119:1354–1359. The singlet triplet gap, $\Delta E_{ST} = 2J$, where J is the exchange parameter. The exchange parameter has a ferromagnetic term, proportional to the exchange integral, and an antiferromagnetic term, proportional to the overlap integral for the monoradical halves of a biradical.
2. Wright BB, Platz MS. J Am Chem Soc. 1983; 105:628.
3. Borden WT, Davidson ER. J Am Chem Soc. 1977; 99:4587.

4. Michl J, Bonacic-Koutecky V. Electronic Aspects of Organic Photochemistry. New York: Wiley-Interscience, 1990.
5. Atherton NM. Principles of Electron Spin Resonance. New York: Prentice Hall, 1993.
6. Wertz JE, Bolton JR. Electron Spin Resonance. New York: Chapman and Hall, 1986.
7. For general texts on molecular magnetism, see the following: (a) Gatteschi D. Molecular Magnetic Materials. Amsterdam: Kluwer Academic, 1991; (b) Kahn O. Molecular Magnetism. New York: VCH, 1993; (c) Molecular Magnetism: From Molecular Assemblies to the Devices. Coronado E, Delhaes P, Gatteschi D, Miller JS, eds. Dordrecht, The Netherlands: Kluwer Academic, 1996, vol. 321; (d) Turnbull MM, Sugimoto T, Thompson LK, eds. Molecule-Based Magnetic Materials. Theory, Technique, and Applications. Washington, DC: ACS Symposium Series 644, 1996.
8. Rajca A. Chem Rev. 1994; 94:871.
9. Calder A, Forrester AR, James PG, Luckhurst GR. J Am Chem Soc. 1969; 91: 3724.
10. Dvolaitzky M, Chiarelli R, Rassat A. Angew Chem Int Ed Engl. 1992; 31:180.
11. Kanno F, Inoue K, Koga N, Iwamura H. J Am Chem Soc. 1993; 115:847.
12. Fujita J, Tanaka M, Suemune H, Koga N, Matsuda K, Iwamura H. J Am Chem Soc. 1996; 118:9347.
13. Fang S, Lee M-S, Hrovat DA, Borden WT. J Am Chem Soc. 1995; 117:6727–6731.
14. Okada K, Imakura T, Oda M, Murai H, Baumgarten M. J Am Chem Soc. 1996; 118:3047–3048.
15. Okada K, Matsumoto K, Oda M, Murai H, Akiyama K, Ikegami Y. Tetrahedron Lett. 1995; 36:6693–6694.
16. Adam W, van Barneveld C, Bottle SE, Engert H, Hanson GR, Harrer HM, Heim C, Nau WM, Wang D. J Am Chem Soc. 1996; 118:3974–3975.
17. Silverman SK, Dougherty DA. J Phys Chem. 1993; 97:13273.
18. Carilla J, Juliá L, Riera J, Brillas E, Garrido JA, Labarta A, Alcalá R. J Am Chem Soc. 1991; 113:8281.
19. Gleason WB, Barnett RE. J Am Chem Soc. 1976; 98:2701.
20. Rajca A, Utamapanya S, Xu J. J Am Chem Soc. 1991; 113:9235–9241.
21. Rajca A. J Am Chem Soc. 1990; 112:5890.
22. Veciana J, Rovira C, Crespo MI, Armet O, Domingo VM, Palacio F. J Am Chem Soc. 1991; 113:2552.
23. Veciana J, Rovira C, Armet O, Domingo VM, Crespo MI, Palacio F. Mol Cryst Liq Cryst. 1989; 176:77.
24. Jacobs SJ, Shultz DA, Jain R, Novak J, Dougherty DA. J Am Chem Soc. 1993; 115:1744.
25. Dougherty DA. Acc Chem Res. 1990; 24:88.
26. Wenthold PG, Squires RR, Lineberger WC. J Am Chem Soc. 1996; 118:475.
27. Matsumoto T, Ishida T, Koga N, Iwamura H. J Am Chem Soc. 1992; 114: 9952; Shultz DA, Boal AK, Lee H, Farmer GT. J Org Chem. 1998, submitted.

28. Shultz DA, Boal AK, Driscoll DJ, Farmer GT, Hollomon MG, Kitchin JR, Miller DB, Tew GN. Mol Cryst Liq Cryst. 1997; 305:303–310.
29. Shultz DA, Boal AK, Driscoll DJ, Kitchin JR, Tew GN. J Org Chem. 1995; 60:3578.
30. Inoue K, Hayamizu T, Iwamura H, Hashizume D, Ohashi Y. J Am Chem Soc. 1996; 118:1803–1804.
31. Iwamura H, Inoue K, Hayamizu T. Pure & Appl Chem. 1996; 68:243.
32. Nakatsuji S, Anzai H. J Mat Chem. 1997; 7:2161–2196.
33. Anderson KK, Shultz DA, Dougherty DA. J Org Chem. 1997; 62:7575–7584.
34. Allinger NL, Yuh YH. QCPE. 1981; 13:395.
35. Shultz DA, Boal AK, Farmer GT. J Am Chem Soc. 1997; 119:3846.
36. Kwart H, Kaplan L. J Am Chem Soc. 1954; 76:4072–4077.
37. Cohen T, Jeong I, Mudryk B, Bhupathy M, Mohamed MA. J Org Chem. 1989; 55:1528–1536.
38. Platz MS, ed. Kinetics and Spectroscopy of Carbenes and Biradicals. New York: Plenum Press, 1990.
39. For powder simulations, the Shareware version of Winepr Simfonia (Version 1.25 © 1994–96 Brüker analytische Messtechnik GmbH) was used.
40. Berson JA, ed. The Chemistry of Quinonoid Compounds, vol. II. New York: Wiley, 1988, p 482.
41. Berson JA. Acc Chem Res. 1997; 30:238.

7

Use of Dinitrenes as Models for Intramolecular Exchange

Shigeaki Nimura
Ricoh Company, Ltd., Yokohama, Japan

Akira Yabe
National Institute of Materials and Chemical Research, Tsukuba, Japan

I. INTRODUCTION

Nitrenes are primary reactive intermediates generated as a result of the elimination of molecular nitrogen by the photolysis or thermolysis of azides. The chemistry of the nitrenes has attracted considerable attention from the viewpoints of reactive intermediates in organic syntheses [1]. In particular, most past studies on dinitrenes were focused mainly on photosensitive materials [2]. With increasing interest in organic magnetic materials of high-spin molecules, extensive studies of dinitrenes were begun in order to clarify their spin–spin interaction by using electron spin resonance (ESR) spectroscopy in the 1990s.

Nitrenes are heteroanalogs to carbenes. Among many recent studies aiming to design and synthesize molecular-based magnetic materials by using high-spin molecules such as polycarbenes and polynitrenes [3,4], most studies of dinitrenes have paid attention to understanding the basic interactions between two nitrenes through linkers, as shown in Scheme 1. To achieve organic magnets consisting of super-high-spin molecules, we should optimize their topology, conformation, and linkers. In the study of dinitrenes, the relationship between these parameters and exchange interactions has also been investigated.

Ferromagnetic
Coupling

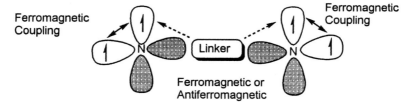

Ferromagnetic
Coupling

Ferromagnetic or
Antiferromagnetic

Scheme 1

In this chapter, we discuss the interelectronic spin–spin interactions of dinitrenes in terms of ESR spectroscopy. Dinitrenes can be divided into two classes, bis(arylnitrenes) (Table 1) and quinonoidal dinitrenes (Table 2).

In the bis(arylnitrenes), the spin–spin interaction of the π-spins of the two nitrene units is relatively weak. Most spin–spin interactions of the bis(arylnitrenes) are basically explained by a spin-polarization mechanism. Herein, however, we also describe bis(arylnitrenes) that cannot be explained simply by the spin-polarization mechanism: spin–spin interactions through an oxygen and a sulfur atom and a carbonyl group.

In the quinonoidal dinitrenes, the spin–spin interactions between two π-spins on terminal nitrogens are strong. They covalently couple, and leave two localized unpaired spins on the terminal nitrogens. The exchange interactions between the two nitrenes are thus antiferromagnetic and are not suitable for achieving the super-high-spin molecules. However, studies of the interactions are very important to understand the basic mechanisms of spin polarization in π-systems.

II. BIS(ARYLNITRENES)

In 1967 the first quintet ground state dinitrene, 1,3-dinitrenobenzene (**1**), was reported by Wasserman et al. [5]. The quintet ground state was also predicted from a UHF-INDO calculation by Olsen in 1969 [6]. After that, however, there was no experimental study of quintet dinitrenes until Iwamura et al. reported spin–spin interactions in ethynylenebis (phenylnitrenes) (**2**) and 1,3-butadiyne-1,4-diylbis(phenylnitrenes) (**3**) in 1989 [7]. In 1991 Lahti and Ichimura studied several types of spin–spin interactions between two nitrenes by using semiempirical AM1-CI calculations [8]. Since then many studies of spin–spin interactions in dinitrenes have been reported as interest in organic magnetism increased.

Table 1   Ground State Multiplicities of Bis(arylnitrenes)[a]

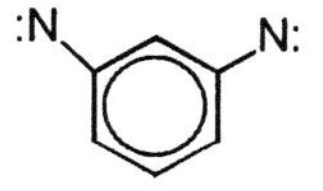

1[5]: Q

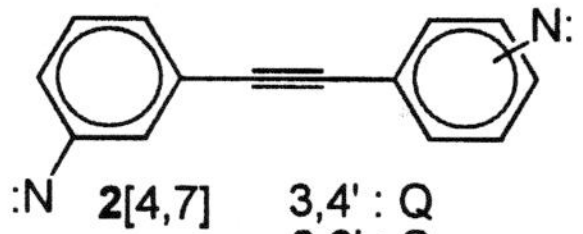

2[4,7]   3,4' : Q
3,3' : S
2,3' : Q

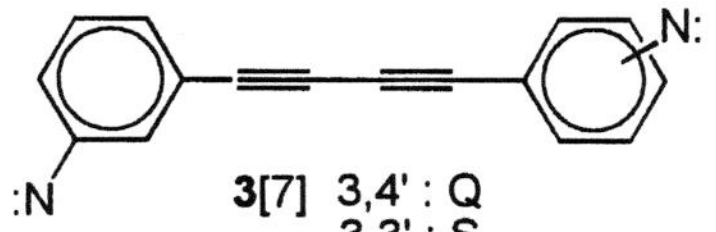

3[7]   3,4' : Q
3,3' : S

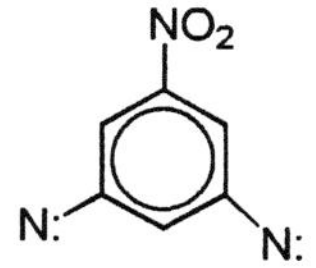

4[10]: Q

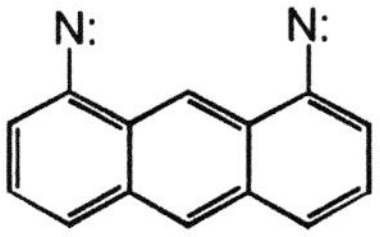

5[11]: Q

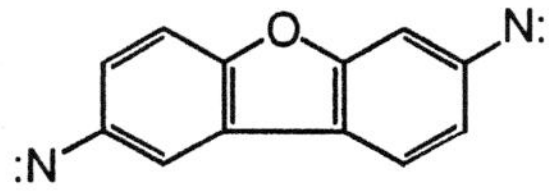

6[11]: Q

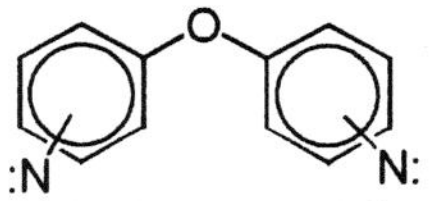

7[12,13]   4,4' : NS
3,4' : Q
3,3' : S

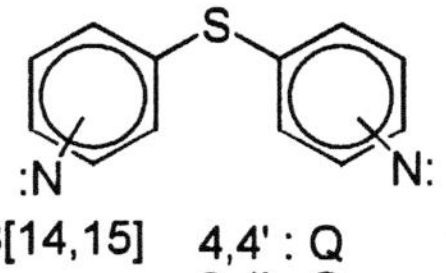

8[14,15]   4,4' : Q
3,4' : Q
3,3' : S, Q

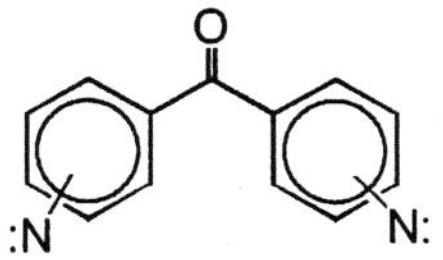

9[13,16,17,20]   4,4' : Q, S[b]
3,3' : S

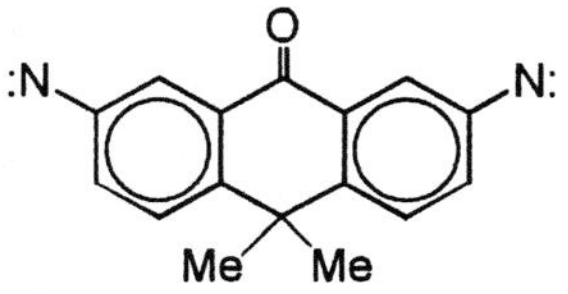

10[18]: S

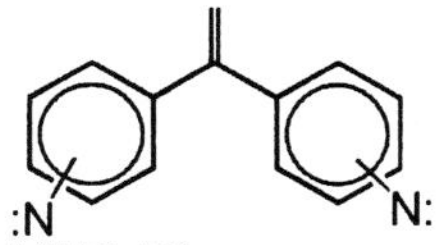

11[16,19]   4,4' : Q
3,4' : S
3,3' : S

[a]References are given in parentheses. Q, S, and NS denote quintet, singlet, and no signal, respectively.
[b]This result was obtained by a single-crystal ESR study in ref. 20.

Table 1 Continued[a]

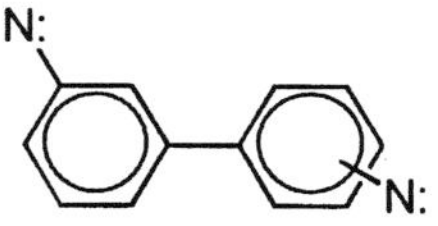

N:

12[22] 3,4' : Q
          3,3' : NS

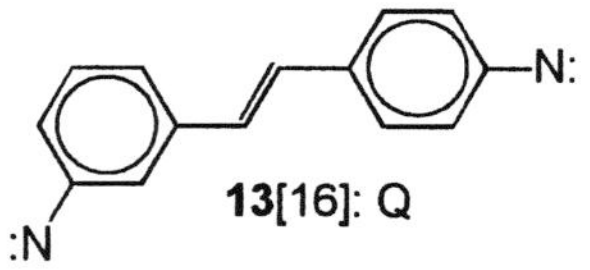

13[16]: Q

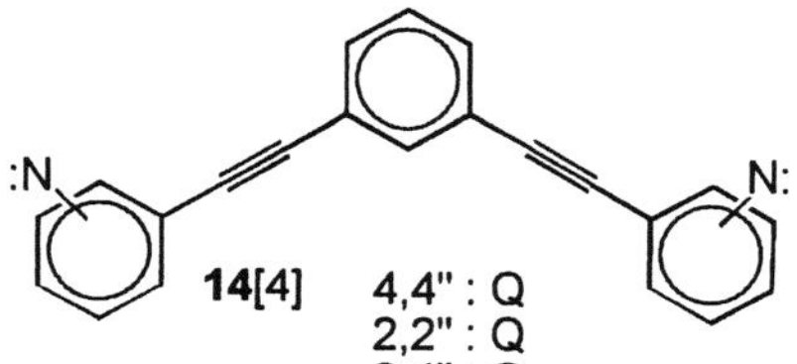

14[4] 4,4" : Q
          2,2" : Q
          2,4" : Q
          3,4" : S
          3,3" : NS

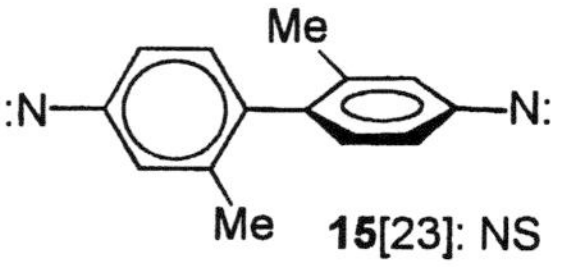

Me

15[23]: NS

Me

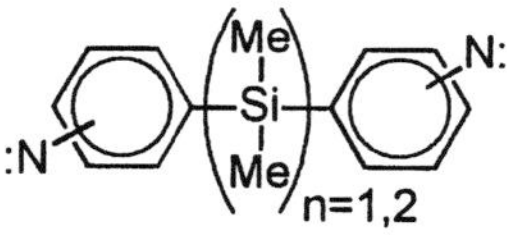

16[24]   n=1 n=2
          4,4' : NS   S
          3,3' : NS   S
          3,4' : VW   Q

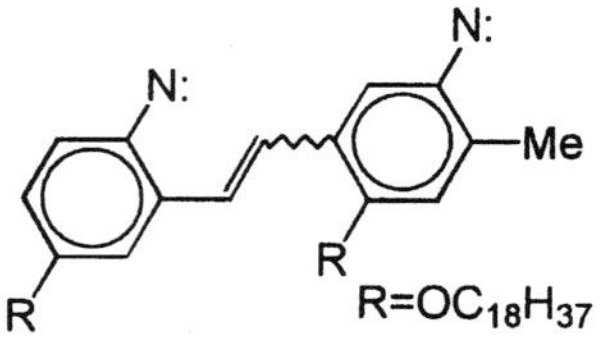

17[25] cis, trans : Q   R=OC$_{18}$H$_{37}$

R=H,Me

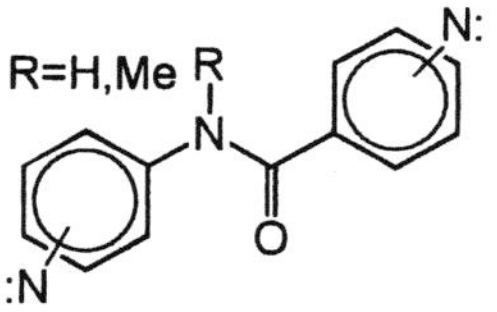

18[26] 4,4' (R=H) : S
          3,4' (R=H) : Q
          4,3' (R=H) : Q
          3,3'(R=H,Me) : S

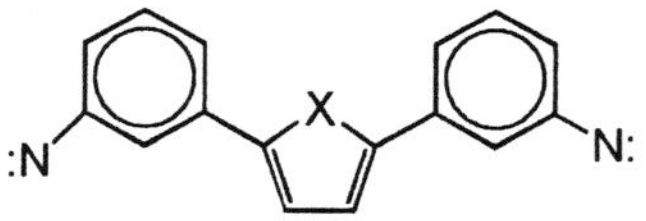

19[27] X=O : S
          S : S
          NH : S

[a]References are given in parentheses. Q, S, VW, and NS denote quintet, singlet, very weak signal, and no signal, respectively.

Table 2 Singlet-Triplet Energy Gaps and zfs D Values of Quinonoidal Dinitrenes[a]

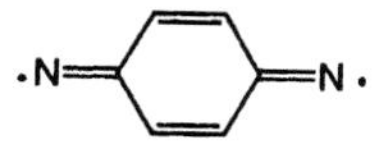

24:
-	0.171	[33]
574	0.169	[35,38]
722 ± 7	0.169	[39]
820	0.1539	[20][b]

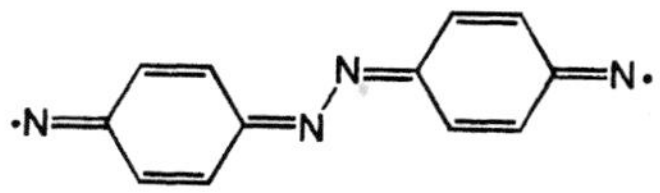

25:
-	0.059	[33]
251 ± 2	0.058	[39]

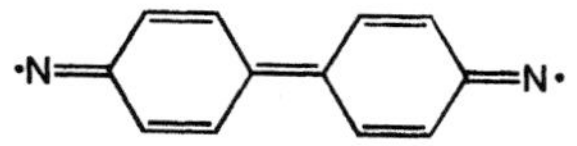

26:
580	0.188	[22,35,38]
750 ± 17	0.191	[39]

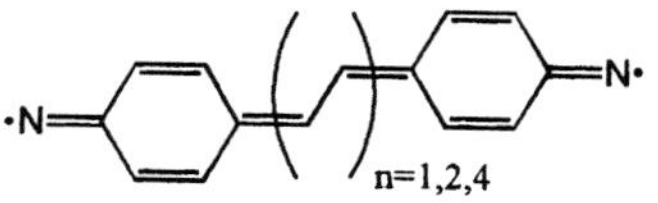

27: n=1,2,4
n=1	480	0.122	
2	677	0.0865	[35,38]
4	145	0.0442	
n=1	472 ± 8	0.123	[39]

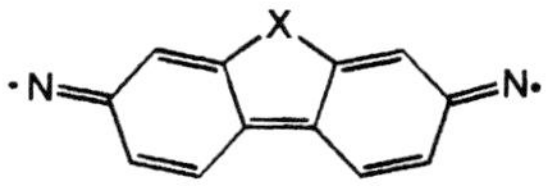

28[40]:
X=		
C_2H_2	0.93 ± 0.05	0.286
S	0.88 ± 0.02	0.236
O	0.85 ± 0.01	0.205
CH_2	0.78 ± 0.02	0.188
CO	0.79 ± 0.01	0.174

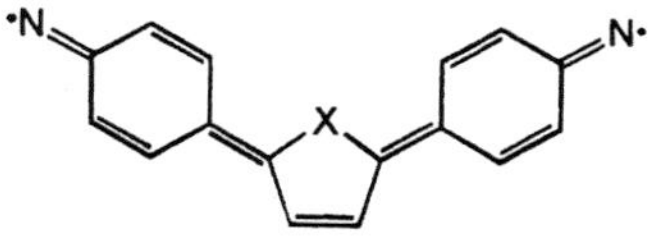

29[27]:
X=		
S	-	0.190
O	-	0.137
NH	-	0.183

[a] The first value for each entry is the energy gap, in kcal/mol, and the second is the D value, in cm^{-1}. References are given in parentheses.
[b] This value was determined in a p-dinitrobenzene single crystal at 288 K.

In the following, we describe a weakly interacting model and ESR analyses of quintet dinitrenes, and then discuss interactions through linkages of oxygen and sulfur atoms and a carbonyl group.

II.A. Low-Lying Spin Multiplicities in Two Weakly Interacting Nitrenes

Now let us consider low-lying spin multiplicities that are generated from an interaction between two nitrenes. In the case of bis(arylnitrenes), we can apply a weakly interacting model of two triplet units [9]. This model gives rise to a quintet, a triplet, and a singlet state for the exchange spin–spin coupling of two triplet nitrenes. Energy-level diagrams of these states (Scheme 2) depend on whether the sign of exchange coupling (J) is positive or negative. If J is positive, ferromagnetic coupling occurs and a quintet state is the ground state; if J is negative, antiferromagnetic coupling occurs and a singlet is the ground state.

The weakly interacting model also gives Eq. (1), which suggests that the quintet zero-field splitting parameters depend on the relative orientation of the two nitrenes:

$$D_q = \frac{1}{6}(D_a + D_b) \tag{1}$$

where D_q is the quintet **D** tensor of the dinitrene, and D_a and D_b are the **D** tensors of each mononitrene. Therefore, if the relative orientation and mononitrene **D** tensors are similar, then the ESR spectra of the related quintet states are also similar. This has been an important point in analyzing the ESR spectra of dinitrenes. Full details are given elsewhere by Itoh [9].

Simulations of quintet ESR spectra of dinitrenes have not been available until recent years because of their large zero-field splitting (zfs) values. As just described, if the relative orientation and **D** tensors of the interacting mononitrenes are similar, the ESR spectra of the quintet dinitrenes are expected to also be similar. In the case of bis(arylnitrenes), **D** tensors of mononitrenes are usually large and similar. Therefore, in analyses of the quintet ESR spectra of dinitrenes, the spectrum of the first quintet dinitrene **1** has been used as a benchmark. For example, a weak peak near 800 mT has previously been assigned as the highest-field z-transition, a key signal of quintet dinitrenes [4,5,7,12–27]. The relative orientation angles between the C–N bonds in most dinitrenes in Table 1 are almost 120° which are comparable to the relative orientation angle of dinitrene **1**. Hence, it is expected that the ESR spectra of these quintet dinitrenes are similar. Using the assumption that the weak signal near 800 mT was the highest-field z-transition, the zfs parameters for these dinitrenes were determined by perturbation approaches, and most estimated $|D/hc|$ to be approximately 0.16 cm^{-1}.

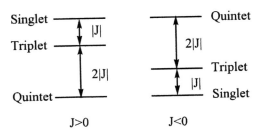

Scheme 2

However, this procedure has proven to be ambiguous, because one cannot reproduce these ESR spectra in a full magnetic field range using the estimated zfs parameters. Recently, Fukuzawa et al. reported the first ESR lineshipe simulation of a quintet dinitrene, 1,3-dinitreno-5-nitrobenzene (**4**), in a magnetic field range from 80 to 800 mT, using an eigenfield method [10]. The simulated spectrum fits well to an experimental spectrum using zfs parameters of $|D/hc| = 0.224$, $|E/hc| = 0.038$ cm^{-1}, which are larger than the zfs parameters estimated by the earlier perturbation approaches. This study suggested also that the weak peak near 800 mT was not the highest z-transition but an x-transition, and, thus, all reported zfs values estimated by the earlier perturbation procedures do not fit experimental spectra. The same ESR lineshape simulation technique was also applied to ESR spectra of two other quintet dinitrenes, 1,8-dinitrenoanthracene (**5**) and 2,6-dinitrenodibenzofuran (**6**), by Kalgutkar and Lahti [11]. The fitted zfs parameters of **5** and **6** are $|D/hc| = 0.243$, $|E/hc| = 0.003$ cm^{-1}, and $|D/hc| = 0.292$, $|E/hc| = 0.009$ cm^{-1}, respectively. The zfs parameters obtained by the simulations are consistent with those estimated from Eq. (1) using the corresponding mononitrene **D** tensors. These studies represent significant progress in dinitrene research, since only these three zfs values seem well established to date. Reinvestigations of zfs values of the known dinitrenes may reveal other interesting electronic features.

II.B. Superexchange Interactions Through Oxygen and Sulfur Atoms

To expand the variety of known linker groups, it is useful to examine the oxygen and sulfur heteroatom linkers. These interactions are not explained simply by a spin-polarization mechanism.

For linkage with an oxygen atom, three topological isomers, 4,4'-, 3,4', and 3,3'-dinitrenodiphenyl ethers (**7**), were examined by ESR spectroscopy [12,13]. The results showed that the ground states were singlet for the 3,3'-isomer and quintet for the 3,4'-isomer. On the other hand, the 4,4'-isomer did not show ESR signals besides mononitrene and radical peaks. The lack of dinitrene ESR signals suggests that either the aryl mononitrene units are isolated by the oxygen linkage or the dinitrene has a singlet ground state. A further study by higher-temperature measurements or by using a planar-constrained dinitrene may be required to discuss the details of the interaction in the 4,4'-isomer.

The exchange interactions of two triplet diphenylcarbene units linked by an oxygen atom in bis(phenylmethylene)diphenyl ether (**20**) (Scheme 3) have been also studied in a single crystal of benzophenone [28]. Antiferromagnetic coupling and a singlet ground state were demonstrated in the 4,4'-

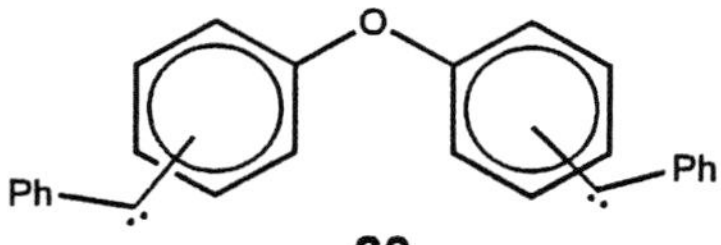

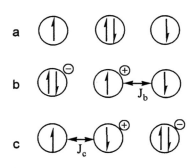

20

Scheme 3

and 3,3′-isomers of **20**, while a ferromagnetic interaction and a quintet ground state were observed in the 3,4′-isomer. These interactions are comparable to those of the corresponding dinitrenes, excepting the 4,4′-isomer.

The results for the dinitrenes **7** and dicarbenes **20** have been attributed to superexchange interactions through a lone pair of the oxygen atom [28,29]. The simplest model for superexchange involves the coupling of two odd electrons through an intervening electron pair (Scheme 4). The constants J_b and J_c in the excited configurations b and c control the exchange interaction between the two odd electrons. The excited configurations b and c make the two spins on the bridged carbons in **7** and **20** interact antiferromagnetically [28]. Accordingly, the overall coupling results in antiferromagnetic interaction in the 4,4′- and 3,3′-isomers and ferromagnetic interaction in the 3,4′-isomer by applying the spin-polarization mechanism on the phenyl rings. This is pretty much in accord with experimental results.

Next let us turn to linkage by the sulfur atom, which belongs to the same group in the periodic table as the oxygen atom. In the case of the 3,4′-isomer of dinitrenodiphenyl sulfide (**8**), a quintet ground state was demonstrated in agreement with the case of oxygen linkage [14,15]. On the other hand, the 4,4′-isomer of **8**, which was expected to be a singlet ground state by the superexchange model, was also proved to have a quintet ground state

Scheme 4

[14,15]. Furthermore, in the case of the 3,3'-isomer of **8**, both excited and ground quintet states were found [15].

Semiempirical molecular orbital calculations (PM3-CI) were performed to understand the effects of conformations on the singlet-quintet energy gap of the dinitrene **8** [14,15]. The energy gaps between the quintet and singlet states were calculated while changing the torsional twist angles between the phenyl rings and the C–S–C plane, assuming C_2 symmetry. For the 4,4'-isomer, although the ground state is a singlet in a planar conformation, the ground state becomes quintet, in agreement with the experimental results, when the phenyl–phenyl torsional angle is larger than 50°, as shown in Figure 1. For the 3,3'-isomer, the singlet and quintet are nearly degenerate, which suggests that ground spin-state inversion is easy, even if the conformational change is small. For this reason, the 3,3'-isomer has both excited and ground quintet states, presumably conformational isomers. For the 3,4'-isomer, inversion of the ground state does not occur. These results led to the conclusion that the observed violation of the superexchange mechanism in **8** is due to the torsion of the phenyl rings.

A similar study that used carbenes as spin centers, 4,4'-bis(phenylmethylene)diphenyl sulfide (**21**) (Scheme 5), found this system to have a quintet ground state in accord with the corresponding dinitrene study [30].

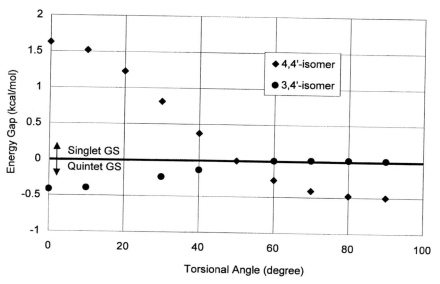

Figure 1 Torsional angle dependence of the singlet-quintet energy gaps in 4,4'- and 3,4'-isomer of **8**. (From Ref. 15.)

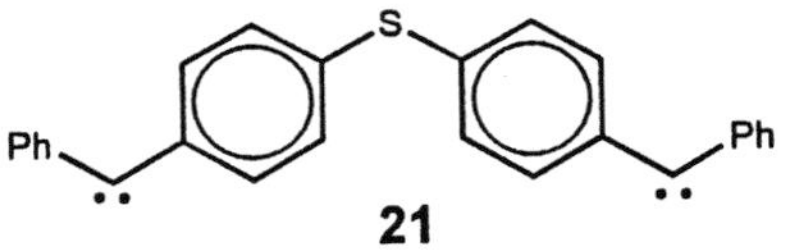

21

Scheme 5

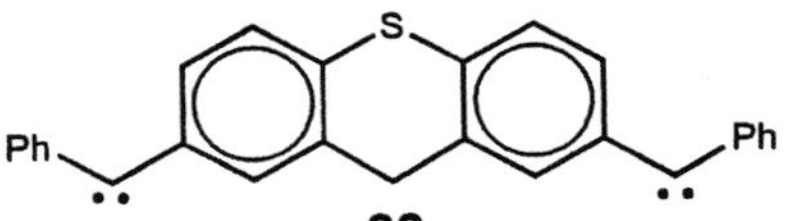

22

Scheme 6

The dicarbene **21** was also predicted to have a twist between the phenyl rings and the C–S–C plane by ab initio calculation. Moreover, this study demonstrated that a planar-constrained **21**, 2,7-bis(phenylmethylene)thiaxanthene (**22**) (Scheme 6) had a *singlet* ground state. These results also suggest that the torsional angle between the phenyl planes and the C–S–C plane can alter the exchange interaction between the two carbenes, in qualitative agreement with the case of dinitrene **8**.

It is suggested that conformational effects on electronic nature, as well as by topology, can be crucial to achieve a super-high-spin ground state by through-bond coupling.

II.C. Exchange Coupling Through Cross-Conjugated Linkage

Next let us look at another interesting coupling linkage, the carbonyl group. Usually, predictions of ground states on the basis of the spin-polarization mechanism and the Borden–Davidson criterion [31] are comparable and agree well with experimental results. In the case of 1,1,2,3,3-pentamethylenepropane (**23**) (Scheme 7), however, predictions of these mechanisms

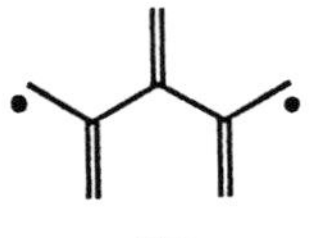

23

Scheme 7

are in formal contradiction: the former predicts a triplet ground state, while the latter predicts a singlet ground state [31]. There is still uncertainty about its experimental ground state multiplicity.

The 3,3′-isomer of dinitrenobenzophenone (**9**) belongs to the same connectivity type as **23** [16]. Based on the spin-polarization mechanism, the 3,3′-isomer of **9** is predicted to be a quintet ground state. On the other hand, on the basis of the Borden–Davidson criterion a singlet ground state is predicted.

As a result of ESR measurements of the 3,3′-isomer of **9**, two conformationally different species were observed, both of which were found to be excited quintet states, in accord with the prediction of the Borden–Davidson criterion [13,16,17]. Furthermore, in order to exclude the effects of conformation on the exchange interaction, a planar-constrained dinitrene, 10,10-dimethylanthrone-2,7-dinitrene (**10**), was examined [18]. The ESR study of dinitrene **10** suggested that this ground state was also a singlet.

An analogous coupling linkage with a 1,1-ethenediyl group was also examined using 1,1-bis(3-nitrenophenyl)ethene (**11**) [16,19]. These results were consistent with those of the carbonyl linkage. Accordingly, the Borden–Davidson criterion proved to be effective to explain the ground state multiplicity of these dinitrenes.

For the 4,4′-isomer of **9**, predictions of the spin polarization and the Borden–Davidson criterion are the same, viz, the quintet ground state, and also are in agreement with experimental results of ESR studies in a 2-methyltetrahydrofuran (MTHF) glassy solution [13,16]. On the other hand, Ichimura et al. reported an ESR spectrum of a quintet state of the 4,4′-isomer of **9** in a 4,4′-dinitrobenzophenone single crystal [20]. The 4,4′-isomer of **9** was stable at room temperature for months. Interestingly, the ESR study of the single crystal suggests that a singlet state is the ground state with a singlet-quintet energy gap of 90 cal/mol, while the ESR spectrum of this dinitrene in glassy MTHF solution shows no evidence that the quintet is a thermally excited state. These results suggest that the magnetic interactions in the MTHF glassy solution and the single crystal are different from each other. In addition, a big difference between the two randomly oriented ESR spectra is a strong signal at 48.6 mT in the single-crystal sample, which does not appear in the glassy solution ESR in the temperature range 8–80 K [13,16,21]. Therefore, the conformations and electronic structures of the 4,4′-isomer of **9** in the single crystal and the solution are apparently different from each other. For this reason, apparently, the reversal of the exchange interaction occurs, as mentioned earlier for the case of **8**.

Inversions of exchange interaction that are due to conformational changes imply that conformational rigidity is very important in the molecular modeling of super-high-spin molecules, together with the overall topologies and natures of linkage groups. This fact also suggests the possibility of magnetic switching by controlling conformations or environmental matrix: e.g., by photoisomerization.

III. QUINONOIDAL DINITRENES

Unlike bis(arylnitrenes), quinonoidal dinitrenes have singlet ground states and are not directly useful to achieving super-high-spin organic materials. However, since they have two localized σ-spins that interact through π-systems, their excited triplet states are observable by means of ESR spectroscopy. In this section, interesting features of their exchange interactions are discussed.

1,4-Dinitrenobenzene (**24**) is the simplest quinonoidal dinitrene. In 1963, the first study of this triplet dinitrene by ESR spectroscopy reported the zfs parameter $|D/hc| = 0.0675$ cm^{-1} and suggested that the triplet was the ground state of this species [32]. However, a later reinvestigation found that this first reported $|D/hc|$ value was due to 4,4'-dinitrenoazobenzene (**25**) derived by dimerization of **24** [33]. The latter report gave the $|D/hc|$ value of **24** as 0.171 cm^{-1}.

Until 1993 there was no other experimental evidence of the ground state multiplicity of **24**. In that year, for quinonoidal dinitrenes **24**, **26**, and **27**, temperature dependences of triplet ESR signal intensities were examined and singlet ground states demonstrated for all [34,35]. At present, all quinonoidal dinitrenes in Table 2 are known to have singlet ground states and thermally populated triplet states.

A simple application of Hund's rule suggests that the quinonoidal dinitrenes would have triplet ground states. However, all quinonoidal dinitrenes observed by means of ESR spectroscopy have singlet ground states. The experimental results are explained by the spin-polarization mechanism. The localized spins on the terminal nitrogen atoms polarize the π-spins on the nitrogen by one-center interactions. For example, a π-spin polarization in the triplet state of **24** is shown in Scheme 8. A mismatching of the spin polarization occurs in the π-system and destabilizes the triplet state. Consequently, the singlet ground states are expected. The analogous spin-polarization mechanism for p-phenylenebis(phenylmethylene) (**30**) was also supported by an ENDOR study [36,37], which found the observed spin distribution in the π-system of **30** (Scheme 9) to be similar to that shown in Scheme 8.

One interesting feature of the quinonoidal dinitrenes is their large zfs $|D/hc|$ values. For a simple point-dipole approximation to $|D/hc|$ caused by interaction of two localized spins, the $|D/hc|$ values are proportional to the

mismatching

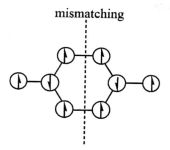

Scheme 8

inverse cubic root of the distance (r) between two spins, and is expressed by

$$|D/hc| = \frac{1.299g}{r^3} \tag{2}$$

where g is the free electron g value. The $|D/hc|$ value should decrease sharply as the distance increases. For example, the experimental $|D/hc|$ values of **27** decrease as the distance between the two terminal nitrogen increases [35,38]. However, if one assumes that the two unpaired electrons are at the nitrogen atoms, Eq. (2) yields $|D/hc| = 0.0015$, 0.0009, and 0.0004 cm^{-1} for **27** ($n = 1$, 2, and 4), respectively [35,38]. These values are quite small compared to the experimental ones.

Recently, a multipoint-dipole approximation, which includes one-center interactions using a 6-31G* SCF-MO-SDTQ-CI spin-density distribution, estimates the zfs $|D/hc|$ value of **24** to be 0.2 cm^{-1}, in reasonable agreement with the observed values [38]. This result implies that the large $|D/hc|$ value of **24** is due to the one-center interactions on the terminal nitrogens [37]. For the other, longer quinonoidal dinitrenes, unfortunately, quantitatively reliable values of spin-density distributions are still difficult to estimate at present. However, the large $|D/hc|$ values of other quinonoidal dinitrenes are also quite likely to be attributable to large one-center interactions.

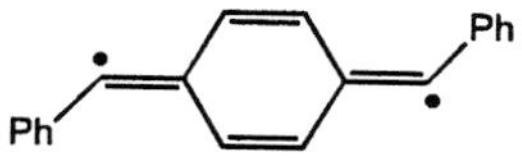

30

Scheme 9

Now let us consider the energy gaps between the singlet and triplet states in the quinonoidal dinitrenes. To determine the energy differences (ΔE) between the singlet and triplet states, one can apply a Curie law analysis. In the quinonoidal dinitrenes, the near-degenerate states are singlet and triplet. Therefore the temperature dependence of the ESR signal intensities (I) of the triplet states is fitted by

$$I = \left(\frac{C}{T}\right) \frac{3 \exp(\Delta E/RT)}{1 + 3 \exp(\Delta E/RT)} \tag{3}$$

where T, C, and R are absolute temperature, an arbitrary constant, and the gas constant, respectively.

The singlet-triplet energy gaps of the dinitrenes **24** and **26** reported by two groups are somewhat different from each other [22,38,39]. There are probably two reasons: (1) the groups could not observe the maxima of the ESR signal intensities, because they could not increase temperature above ca. 90 K without matrix thawing; and (2) there were differences of determination procedures for the sample temperatures. More sophisticated procedures are required in order to obtain more exact energy gaps. However, the relative energy differences of dinitrenes **24** and **26** in different groups were the same.

For dinitrene **24**, an ESR spectrum at room temperature was obtained in a p-dinitrobenzene single crystal [20]. In this case, one may obtain a reliable singlet-triplet energy gap, because the maximum of the ESR signal intensity is observable. The singlet-triplet energy gap ΔE was determined to be 820 cal/mol by a Curie analysis of the ESR signal in a wide temperature range (ca. 100–300 K), which was larger than the values determined in glassy organic solvents at low temperature. However, zfs parameters of the ESR at room temperature were $|D/hc| = 0.1539$ cm^{-1} and $|E/hc| = 0.00683$ cm^{-1}, which are slightly different from the values in low-temperature matrices. Moreover, an increase of $|D/hc|$ and decrease of $|E/hc|$ were observed with decreasing temperature. Although the origin of this phenomenon in **24** is not clear, one can suppose they could be caused by changes of the host crystal or vibronic effects in the dinitrene with temperature variation.

The relation between electronic and geometric structures is also very interesting in these biradicals. Recently, the correlation that zfs **D** values increased with increasing singlet-triplet energy gaps was found in planar-constrained, doubly linked quinonoidal dinitrenes **28**, as shown in Table 2 [40]. It should be noted that the distances between the two nitrene spin centers are almost the same in each dinitrene, but the zfs $|D/hc|$ values are large and widely different from each other. These large magnitudes of the $|D/hc|$ values appear to be due also to one-center interactions on the terminal

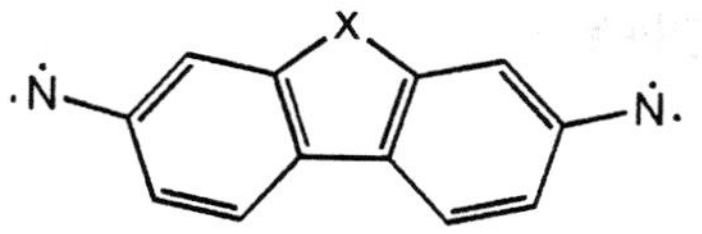

benzenoid structure

Scheme 10

nitrogens. Therefore, the wide differences in $|D/hc|$ values are ascribed to the difference in contribution of the benzenoid resonance forms, which have two unpaired π-spins, in the total wave functions. Moreover, it was found that the benzenoid contribution was correlated with the stability of each benzenoid structure, as estimated from the resonance energies of the intervening π-systems linking the two nitrogens. On the other hand, spin polarization in the benzenoid structure (Scheme 10) destabilizes the triplet states, because of spin-polarization mismatching. For this reason, the benzenoid nature relates also to the singlet-triplet energy gaps; hence a correlation between energy gaps and zfs $|D/hc|$ values appears. PM3-CI molecular orbital calculations found a similar relation between the benzenoid structure combinations and the energy gaps. This study demonstrated a "fine-tuning" control of the exchange interactions while maintaining a constant distance between spin centers.

IV. SUMMARY AND OUTLOOK

In the last decade, many valuable results have been obtained in dinitrene research (Tables 1 and 2). Most research, however, has been characterized only by low-temperature ESR spectroscopy in MTHF matrices. Recently, some papers have reported the single-crystal ESR work at ambient temperature. Such studies that use different methods or matrices are meaningful to investigate environmental effects on open-shell state properties and multiplicities.

A remarkable advance of experimental technique in dinitrene studies has been the ESR lineshape simulation of quintet spectra. Such simulations may throw a clearer light on details of the exchange interactions of dinitrenes and polynitrenes. Of particular interest is the excited dinitrene triplet state that is apparently observed in some experiments. Simulation distinguishes quintet and triplet states; thus one should be able to distinguish the triplet state peaks clearly.

Finally, it might be very interesting to study exchange interaction switching by means of conformational control, given the definite effect on exchange interaction that we saw in this chapter.

ACKNOWLEDGMENTS

We thank Dr. Tsuguyori Ohana (NIMC), Dr. Masahiro Kaise (NIMC), Professor Osamu Kikuchi (University of Tsukuba), and Dr. Koji Ujiie (RICOH Co., Ltd.) for their helpful advice and assistance.

REFERENCES

1. Scriven EFV, ed. Azides and Nitrenes. Reactivity and Utility. New York: Academic Press, 1984.
2. For instance: Reiser A, Wagner HM, Marley R, Bowes G. Photolysis of aromatic azides. Part 2. Formation and spectra of dinitrenes. Trans Faraday Soc. 1967; 63:2403–2410. Tsunoda T, Yamaoka T, Nagamatsu G. Spectral sensitization of bisazide compounds. Photogra Sci Eng. 1973; 17:390–393.
3. For instance: Matsuda K, Nakamura N, Inoue K, Koga N, Iwamura H. Toward dendritic two-dimensional polycarbenes: Syntheses of "starburst"-type nona- and dodecadiazo compounds and magnetic study of their photoproducts. Bull Chem Soc Jpn. 1996; 69:1483–1494. Matsuda K, Nakamura N, Inoue K, Koga N, Iwamura H. Design and synthesis of a "starburst"-type nonadiazo compound and magnetic characterization of its photoproduct. Chem Eur J. 1996; 2:259–264.
4. Sasaki S, Iwamura H. Approaches toward high-spin poly[m-(nitrenophenylene)ethynylenes]. Magnetic interaction in the oligomers containing two, three, and five nitrene units. Chem Lett. 1992; 1759–1762.
5. Wasserman E, Murray RW, Yager WA, Trozzolo AM, Smolinsky G. Quintet ground states of m-dicarbene and m-dinitrene compounds. J Am Chem Soc. 1967; 89:5076–5078.
6. Olsen JF. m-Phenylenedicarbene and m-phenylenedinitrene: Compounds having quintet ground states. Spectrosc Lett. 1969; 2:217–223.
7. Iwamura H, Murata S. Magnetic coupling of two triplet phenylnitrene units joined through an acetylenic or a diacetylenic linkage. Mol Cryst Liq Cryst. 1989; 176:33–47. Murata S, Iwamura H. Magnetic interaction between the triplet centers in ethynylenebis(phenylnitrenes) and 1,3-butadiyne-1,4-diylbis(phenylnitrenes). J Am Chem Soc. 1991; 113:5547–5556.
8. Lahti PM, Ichimura AS. Semiempirical study of electron exchange interaction in organic high-spin π-systems. Classifying structural effects in organic magnetic molecules. J Org Chem. 1991; 56:3030–3042.
9. Itoh K. Electronic structures of aromatic hydrocarbons with high spin multiplicities in the electronic ground state. Pure Appl Chem. 1978; 50:1251–1259.

10. Fukuzawa TA, Sato K, Ichimura AS, Kinoshita T, Takui T, Itoh K, Lahti PM. Electronic and molecular structures of quintet bisnitrenes as studied by fine-structure ESR spectra from random orientation: All the documented zfs constants correct? Mol Cryst Liq Cryst. 1996; 278:253–260.

11. Kalgutkar RS, Lahti PM. Rigid geometry bis(arylnitrenes) as definitive tests for angular dependence of zero-field splitting in high-spin molecules. J Am Chem Soc. 1997; 119:4771–4772.

12. Minato M, Lahti PM. Intramolecular exchange coupling of arylnitrenes by oxygen. J Phys Org Chem. 1994; 7:495–502.

13. Kaise M, Ohana T, Yabe A, Nimura S, Kikuchi O. 64th National Meeting of the Chemical Society of Japan, Abstract 3-A2-13 (Japanese), 1992.

14. Nimura S, Kikuchi O, Ohana T, Yabe A, Kaise M. ESR study of intramolecular magnetic interactions in bis(nitrenophenyl) sulfides. Chem Lett. 1994; 1679–1682.

15. Nimura S. ESR study of spin–spin couplings in dinitrenes. PhD dissertation, University of Tsukuba, Japan, 1996.

16. Ling C, Minato M, Lahti PM, van Willigen H. Models for intramolecular exchange in organic π-conjugated open-shell systems. A comparison of 1,1-ethenediyl and carbonyl linked bis(arylnitrenes). J Am Chem Soc. 1992; 114: 9959–9969. Lahti PM, Ling C, Yoshioka N, Rossitto FC, van Willigen H. Theory and experiment of investigation of exchange interactions in organic molecules and materials. Mol Cryst Liq Cryst. 1993; 233:17–32.

17. Nimura S, Kikuchi O, Ohana T, Yabe A, Kaise M. ESR of excited quintet states in 3,3′-dinitrenobenzophenone. Chem Lett. 1993; 837–838.

18. Ling C, Lahti PM. 10,10-Dimethylanthrone-2,7-dinitrene. A planar-constrained, doubly disjoint high-spin molecule with a singlet ground state. Chem Lett. 1994; 1357–1360.

19. Matsumoto T, Ishida T, Koga N, Iwamura H. Intramolecular magnetic coupling between two nitrene or two nitroxide units through 1,1-diphenylethylene chromophores. Isomeric dinitrenes and dinitroxides related in connectivity to trimethylenemethane, tetramethyleneethane, and pentamethylenepropane. J Am Chem Soc. 1992; 114:9952–9959.

20. Ichimura AS, Sato K, Kinoshita T, Takui T, Itoh K, Lahti PM. An electron spin resonance study of persistent dinitrenes. Mol Cryst Liq Cryst. 1995; 272:57–66.

21. Nimura S, Kikuchi O, Ohana T, Yabe A, Kaise M. Unpublished results.

22. Minato M, Lahti PM. Biphenyl-3,4′-dinitrene. J Phys Org Chem. 1991; 4: 459–462. Minato M, Lahti PM, van Willigen H. Models for intramolecular exchange in organic π-conjugated open-shell systems: A comparison of three non-Kekulé biphenyldinitrenes. J Am Chem Soc. 1993; 115:4532–4539.

23. Ling C, Lahti PM. 2,2′-Dimethylbiphenyl-4,4′-dinitrene. Chem Lett. 1993; 769–772.

24. Doi T, Ichimura AS, Koga N, Iwamura H. Magnetic interaction between two triplet nitrene units through diphenylsilane and 1,2-diphenyldisilane couplers. J Am Chem Soc. 1993; 115:8928–8932.

25. Mitsumori T, Koga N, Iwamura H. Magnetic interaction between the photochemically generated triplet centers through the π-conjugated skeleton of PPV. Mol Cryst Liq Cryst. 1994; 253:51–57.

26. Ichimura AS, Ochiai K, Koga N, Iwamura H. Isomeric N-(nitrenophenyl)nitrenobenzamides. Comparison of the energy differences between their singlet and quintet states obtained computationally and experimentally. J Org Chem. 1994; 59:1970–1972.

27. Ling C, Lahti PM. Models for intramolecular exchange in organic π-conjugated open-shell systems: 3-Nitrenophenyl and 4-nitrenophenyl units connected by 2,5-furandiyl, 2,5-thiophenediyl, and 2,5-pyrrolediyl nonalternant exchange linkers. J Am Chem Soc. 1994; 116:8784–8792. Lahti PM, Minato M, Ling C. Experimental investigation of exchange in organic open-shell molecular building blocks for magnetic materials. Mol Cryst Liq Cryst. 1995; 271:147–154.

28. Itoh T, Takui T, Teki Y, Kinoshita T. Spin alignment in a model compound of organic ferrimagnets. Mol Cryst Liq Cryst. 1989; 176:49–66.

29. Kramers HA. The interaction of magnetogenic atoms in paramagnetic crystals. Physica. 1934; 1:182–192. Anderson PW. Antiferromagnetism. Theory of superexchange interaction. Phys Rev. 1950; 79:350–356.

30. Matsuda K, Yamagata T, Iwamura H. Effect of the oxidation state of the sulfur atom on the exchange interaction between two triplet carbene units through diphenyl sulfide p,p'-diyl couplers. Chem Lett. 1995; 1085–1086. Matsuda K, Yamagata T, Seta T, Iwamura H, Hori K. Control of the "superexchange" interaction through diphenyl sulfide 4,4'-diyl magnetic coupler by changing the oxidation state and conformation of the sulfur atom. J Am Chem Soc. 1997; 119:8058–8064.

31. Borden WT, Davidson ER. Effects of electron repulsion in conjugated hydrocarbon diradicals. J Am Chem Soc. 1977; 99:4587–4594.

32. Trozzolo AM, Murray RW, Smolinsky G, Yager WA, Wasserman E. The e.p.r. of dicarbene and dinitrene derivatives. J Am Chem Soc. 1963; 85:2526–2527.

33. Singh B, Brinen JS. Low-temperature photochemistry of p-diazidobenzene and 4,4'-diazidoazobenzene. J Am Chem Soc. 1971; 93:540–542.

34. Ohana T, Kaise M, Nimura S, Kikuchi O, Yabe A. The photolysis of 4,4'-diazidobiphenyl in rigid matrices at low temperature. Chem Lett. 1993; 765–768.

35. Minato M, Lahti PM. Zero-field splitting versus interelectronic distance in triplet electron spin resonance spectra of localized dinitrenes. J Phys Org Chem. 1993; 6:483–487.

36. Teki Y, Sato K, Okamoto M, Yamashita A, Yamaguchi Y, Takui T, Kinoshita T, Itoh K. Topology and spin alignment in organic high-spin molecules. Bull Magn Reson. 1992; 14:24–29.

37. Yamaguchi Y, Sato K, Teki Y, Kinoshita T, Takui T, Itoh K. Electronic and molecular structures of p-phenylenebis(phenylmethylene) in its thermally excited triplet state as studied by single-crystal ^{1}H-ENDOR. Mol Cryst Liq Cryst. 1995; 271:67–78; in this paper, unlike the dinitrene **24**, it was concluded that π–π interaction is a dominant origin for a large D value of the dicarbene **30**.

In this case, however, the induced π-spins by the one-center interaction is also a prerequisite factor for the large D value.

38. Minato M, Lahti PM. Characterizing triplet states of quinonoidal dinitrenes as a function of conjugation length. J Am Chem Soc. 1997; 119:2187–2195.

39. Nimura S, Kikuchi O, Ohana T, Yabe A, Kaise M. Singlet-triplet energy gaps of quinonoidal dinitrenes. Chem Lett. 1996; 125–126.

40. Nimura S, Kikuchi O, Ohana T, Yabe A, Kondo S, Kaise M. Effects of additional linkers in biphenyl-4,4'-dinitrene on the low-lying singlet-triplet energy gap and zero-field splitting. J Phys Chem A. 1997; 101:2083–2088.

8

Radical-Ion Building Blocks for Organic Magnetic Materials: Computational and Experimental Approaches

Martin Baumgarten
Max-Planck-Institute for Polymer Research, Mainz, Germany

I. INTRODUCTION

In recent years we have combined synthetic, spectroscopic, and theoretical approaches in order to obtain deeper insight into mechanisms of electron spin–spin interactions, focusing mainly on radical ions of organic π-systems and their possible use for new high-spin molecules [1–3]. We tried to test different approaches, such as direct linkage with close-to-orthogonal alignment, linkage through nodal planes of frontier orbitals, phenylene bridging of π-radical ions, and intermolecular aggregation of radical ions.

II. DIRECTLY LINKED BI- AND OLIGORADICAL IONS

The conjugative interaction in bi- and oligoarylenes strongly depends on the topology and geometry of linkage. The linkage can lead either to fully extended conjugated polymers such as polyparaphenylene or to systems of almost-independent redox active subunits such as poly(9,10-anthrylene)s. While for the former class charged excitations extend over several repeat units, the latter class allows the formation of radical ions on each separate subunit (Scheme 1).

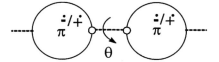

Scheme 1

An early example of a conjugation-separated biradical was given by Veciana et al. [4] with the perchlorobiphenyl dication **1**, where the large chloro-substituents lead to a strong twisting of the biphenyl unit and also decrease the oxidation potential. For exploring whether pure hydrocarbons can also be used for oligoradical formation upon charging, we started to study other congested bi- and oligoarylenes based on the anthracene unit **2–7** (Scheme 2) [5–8].

Bianthryl can be oxidized with $SbCl_5$ to the biradical dication (zero-field splitting $|D/hc| = 21.0$ mT) or reduced with alkali metal to the biradical dianion ($|D/hc| = 16.5$ mT), and therefore higher oligomers were synthesized. Temperature-dependent studies of the zero-field splittings (zfs) in the frozen-solution electron spin resonance (ESR) or electron paramagnetic resonance (EPR) spectra, revealed that the multiple charged oligo(9,10'-anthrylene)s **2, 5, 6** exhibit thermally exited high-spin multiplicities with the maxima of signal intensities at 20, 40, and 60 K for the bi-, tri- and tetraradicals, respectively, corresponding to 60, 120, and 180 cal/mol [7]. The low-temperature measurements of the bianthryl dianion additionally revealed that the average orthogonal alignment in liquid solution is lost, with a clear ortho-

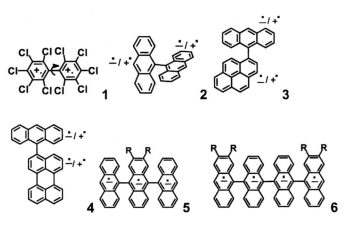

Scheme 2

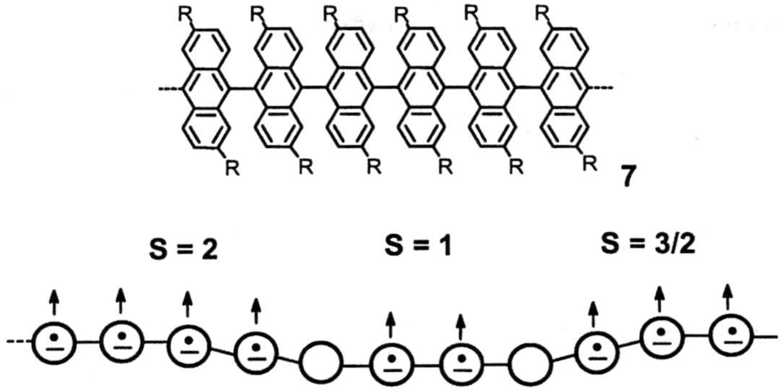

Scheme 3

rhombicity of the zero-field splitting parameters $|D/hc|$ (17 mT) and $|E/hc|$ (0.9 mT) at 20 K. These correlate to an angle θ of roughly 80° when judging from the two extremes $|E/hc| = 0$ ($\theta = 90°$) and $|E/hc|_{max} = |D/hc|/3$ ($\theta = 0°$), assuming a cosine dependence of $|E/hc|$. Also, the dianions and dications of **3** and **4** form thermally activated biradicals [8], leading to the conclusion that weakly coupled biaryls **2–4** with strong intramolecular twisting (orthogonal approach) enable biradical formation in either redox process (oxidation, reduction). In the polymer **7**, charging led to discrete higher-spin states of $S = 1$, $S = 3/2$, $S = 2$ (see Scheme 3), comparable to those in the small oligomers, but then precipitation occurred and the spin states are either separated through a neutral anthrylene unit or a doubly charged one.

   Linking of spin sites through a nodal plane of the frontier orbitals as in the bipyrenyl **8a** and terpyrenyl **8b** enables bi- and triradical formation upon multiple charge formation (Scheme 4) [9]. But again these spin states were found to be thermally populated with maximum EPR signal intensity for the triplet and quartet state at 40 and ca. 100 K corresponding to ca. 120 and to several hundred calories per mole of thermal activation of the higher-

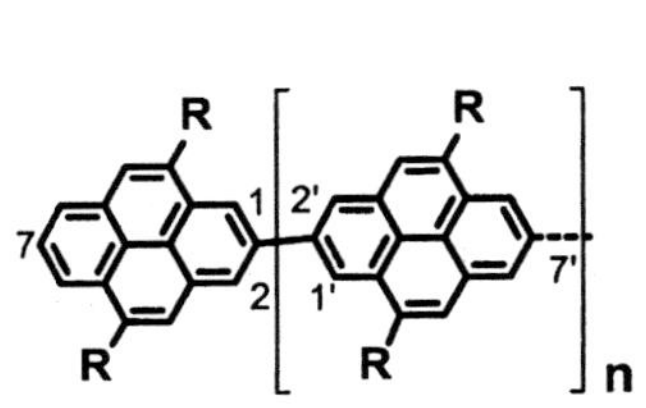

8 a,b ; n = 1,2

Scheme 4

spin states, respectively. For the stabilization of the low-spin state in these systems, spin polarization is responsible. Due to spin polarization, small but nonneglible spin densities propagate to the nodal π-sites, including the bridging 2 position. The direct overlap between the π-spin densities of the same sign in the 2 and 2′ positions destabilizes the triplet [9,10]. The antiferromagnetic spin exchange between the bridging sites may be avoided only when the conjugative interaction is completely broken in an exact orthogonal alignment. Thus, stronger steric hindrance would certainly be needed for the promotion of ferromagnetic coupling in the bi- and oligoradicals **2–8**, a task that seems still difficult to achieve synthetically.

Recently we proposed a novel approach toward stabilization of high-spin states in directly linked biaryl radicals. Conjugated monoradicals combine large positive and small negative para-site densities in an alternating fashion. The nominally antiferromagnetic interaction between adjacent π-sites could lead to parallel spin alignment in a biradical when the two monoradicals are linked through π-sites of opposite spin densities. A typical example of such a ferrimagnetic π-spin system is 1,2′-biperinaphthenyl **9**. Semiempirical calculations with configuration interaction to account for spin polarization suggest stabilization of the triplet state relative to the singlet, whereas for the 2,2′ bridged perinaphthenyls the predicted ground state is singlet. The conceptual idea proposed here can be extended to a large class of biaryls, where the triplet state should be a stable ground state entity.

The general prerequisite for a molecule to serve as a building block in a high-spin bi- or polyradical structure is the presence of alternating spin density in the neutral or charged monoradical form. Such candidates are azulene, s-indacene, pyrene, other molecules whose frontier π-orbitals possess a nodal plane, and some nonalternant aromatic hydrocarbons. Along this line, Gompper et al. [12] synthesized, 1,2′-bipyrenyl **10** and 2(9-anthryl)-pyrene **11** (Scheme 5). The latter molecules are predicted triplet biradicals in their dication and dianion (2+ and 2−) forms. They are already characterized as persistent biradicals, and currently we are studying the temperature dependence of their zero-field splittings to establish the singlet-triplet energy difference.

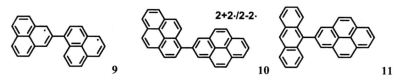

9   **2+2·/2-2·**   **10**   **11**

Scheme 5

Since this direct combination should allow ferromagnetic electron spin–spin interaction, it follows also that 1,2-ethylene or 1,4-phenylene (*para*) bridging should result in triplet state molecules [11] and *meta*-phenylene bridging should destabilize the triplet state, compared to the symmetrical substitution. And this important prediction will be tested further and leads us to the next section, where the phenylene bridging of radical ions is considered.

III. PHENYLENE BRIDGING OF RADICAL-ION SITES: KEKULÉ VS. NON-KEKULÉ STRUCTURES

After it was shown that the degree of rotation between the conjugation planes of two bridged monoradicals plays an important role for the effective spin–spin coupling, we studied this problem in more depth, starting with calculations of ground state multiplicities of typical neutral or charged biradicals **12–18a** [13] (Scheme 6). Thereby additional fixation to ladder type structures of the Schlenk hydrocarbon derivatives as in **12–18 b–f** was included. The results clearly indicate that the triplet state of the Schlenk hydrocarbon should be stabilized upon planarization of the backbone through a saturated methylene (−CH$_2$−) bridge, with the extra phenyl substituents being twisted out of conjugation. This effect certainly enhances the spin density on the radical centers and in the bridging *meta*-phenylene unit.

If, on the other hand, the neutral carbon radicals are exchanged by nitrogen, the situation changes completely. Cationic nitrogen centers lead to a large increase in the twisting angle θ to the *meta*-phenylene coupling unit and results in a strong/dramatic destabilization of the triplet state (Figure 1, Table 1).

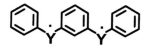

12a–18a	12–18 b–e

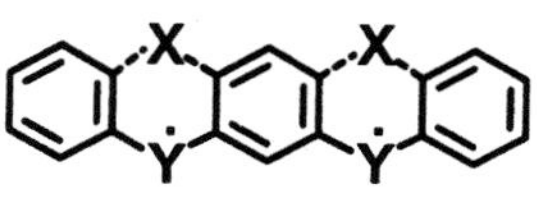

Y = CH (**12a**), CPh(**13a**), CCH$_3$(**14a**)   X= CH2 (**b**), CO (**c**), O (**d**), S(**e**), NH (**f**)

$^+$NPh(**15a**), $^+$NH(**16a**), N(**17a**) NO(**18a**)   Y = **12 - 18**

Scheme 6

Figure 1   The POLO density distributions of singly linked biradicals **12a** and **17a**.

Table 1   Semiempirical Calculated (AM1) Singlet Triplet (ΔE_{ST}/eV) Splittings of **12–18a–e**

·Y\X	-\- (a)	CH$_2$ (b)	C=O (c)	O (d)	S (e)	NH (f)	S=O (g)	SO$_2$ (h)
·CH (**12**)	0.45	0.40	0.28	0.14	0.14	−0.20	0.003	0.18
·CPh (**13**)	0.20$_{ex1}$	0.36$_{ex2}$	0.24	0.14	0.14	−0.18	0.003	0.11
·CCH$_3$ (**14**)	0.48	0.32	0.23	0.10	0.12	−0.18	0.003	0.16
·+NPh (**15**)	0.02	0.05	0.07	−0.12	−0.21	−0.37	0.002	0.04
·+NH (**16**)	0.03	0.03	0.08	−0.15	−0.20$_{ex3}$	−0.44	0.001	0.04
·N (**17**)	0.86	0.53	0.31	0.16	0.06$_{ex3}$	−0.30	0.003	0.13
NO· (**18**)	0.22	0.19	0.15	0.16	0.22	0.10	0.01	0.09

ex = experimentally verified: ex1 (Ref. 14), ex2 derivative of **13b** (Ref. 15), ex3 derivatives of **16e** and **17e** (Ref. 16).

III.A.   Radical Anions

In 1992/1993 we started to study different phenylene-bridged redox centers **19–24** (Scheme 7), since not many charged π-radical sites had been used before. The 1,3,5-(tris-1-naphthyl)benzene **20**, however, could be reduced with K in THF only to the biradical dianion, with very similar properties as the dinaphthylbenzene **19** reported by Tukada [17] (Na/THF), indicating that

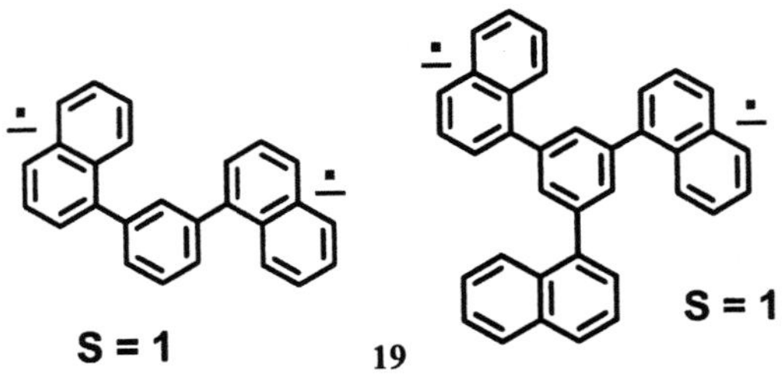

S = 1   **19**

S = 1   **20**

Scheme 7

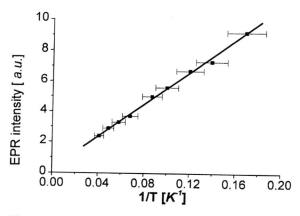

Figure 2 EPR signal intensity of the $\Delta M_s = 2$ transition vs. reciprocal temperature for $\mathbf{21}^{2.2-}$.

the third naphthyl unit does not contribute much to the spin-density delocalization in the doubly charged state.

The fulvenes **21a,b** and the diboride **22** were studied in collaboration with Profs. M. Oda and K. Okada (Osaka, Japan) [18] and for $\mathbf{21}^{2.2-}$, for our first time; we found a linear behavior of EPR signal intensity vs. reciprocal temperature (Figure 2), indicating a triplet ground state of the dianion (Scheme 8).

On the other hand, the diboride **22** (Scheme 9) showed only a very small zero-field splitting of approximately 3.7 mT (0.0034 cm^{-1}) when reduced with K in THF, and we hesitated to publish this result. A year later, Rajca et al. [19] showed a relatively strong zero-field splitting of $D_{max} = 0.022$ cm^{-1} for this same compound, which differs strongly with the counter ion conditions in mixtures of MTHF/THF. Whether intermolecularly coupled aggregates also are formed and are responsible for these dramatic differences, as in the case of ketyl radical anions discussed in the next section, has not been established so far.

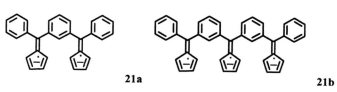

21a   **21b**

Scheme 8

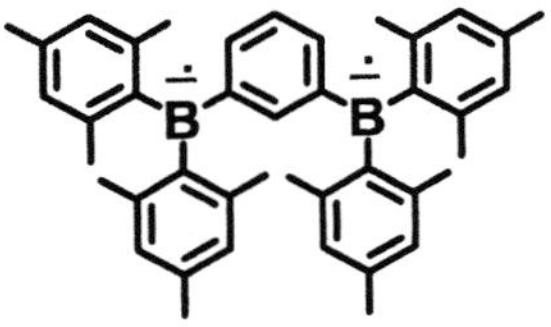

22

Scheme 9

The *meta*-distyrylbenzene **23** showed a striking behavior upon charging. In the monocharged form, a strong CT absorption in the near infrared region ($\lambda_{max} = 1700$ nm) occurs, suggesting that the monoradical species is localized on a single stilbene unit, with charge transfer to a neutral styryl part. The dianion also absorbs at long wavelength ($\lambda_{max} = 1400$ nm) but turns out to be diamagnetic instead of yielding a biradical with the *meta*-phenylene as a ferromagnetic coupler [20]. This result puts a question mark to the generalization of the Fukutome model [21], also intensively studied by Dougherty and his group [22,23]. The bis-stilbenylbenzene **24**, on the other hand, can be doubly charged to the biradical state, and semiempirical calculations also suggest a triplet ground state for this molecule (Scheme 10) [24].

III.B. Radical Cations

There are not many dications of aromatic hydrocarbons in the literature. One example was given by Tukada [17] for the 1,3-bis-(9-anthryl)benzene **25**, which according to their results should possess a ground state triplet multiplicity (Scheme 11).

For polymeric amino cation radicals **26a–c** (Scheme 12) and different derivatives, we predicted ferromagnetic coupling between the spin centers [25,26], and further bandstructure considerations are outlined in Chapter 18. But care should be taken to compare small subunits, as presented earlier, such as **15** and **16** [13], since very different spin-density distribution can

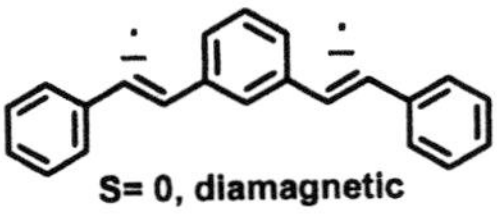

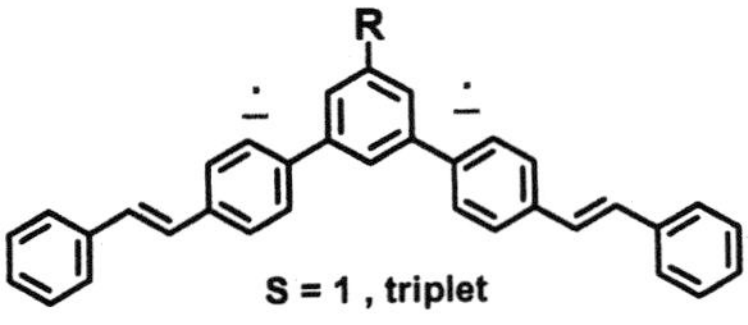

Scheme 10

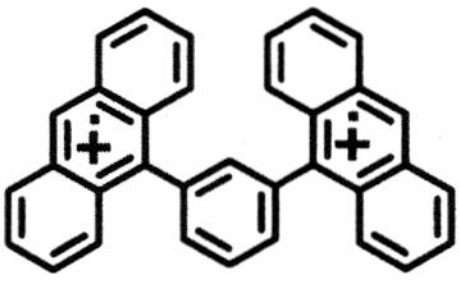

25

Scheme 11

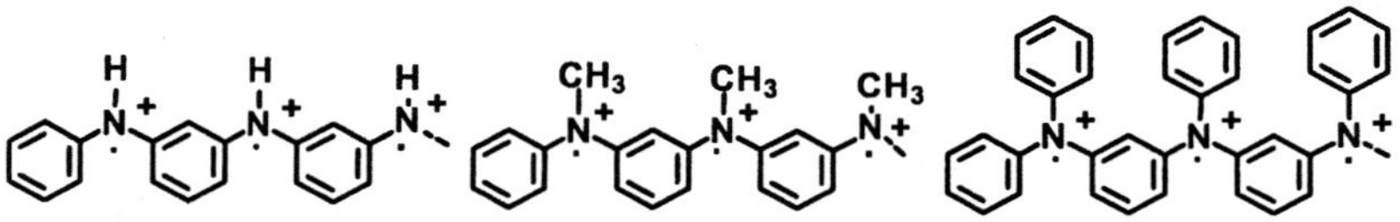

26a-c

Scheme 12

occur and depending on the substituents the ground state multiplicity can be changed. Similar findings have been reported by Stickley and Blackstock [28] and Yano et al. [29]. Instead of phenylene, 1,3-substituted triazines [29] can also be used, where the triplet state for diamino dications is stabilized compared to the former.

Recent reports showed that *meta*-phenylene does not always act as ferromagnetic coupling unit, especially when due to strong steric hindrance the conjugation planes of the radical centers are twisted close to perpendicular to the phenylene unit. Such examples are reported for bisnitroxides by Dvolaitsky et al. [30] and Kanno et al. [31] and have been confirmed and explained in ab initio calculations by Borden [32]. In cooperation with K. Okada (Osaka, Japan) we studied and compared some extremely hindered cases, namely the *meta-, para-, ortho*-phenylene, and the 9,10-anthrylene-bridged bisphenothiazines **27a–d** (Scheme 13) and bisphenoxazines (S = O) as well as 1,3,5-trisubstituted benzenes [33–35].

The phenothiazine and phenoxazine units can easily be oxidized with sulfuric acid to their corresponding cation radicals. All the compounds

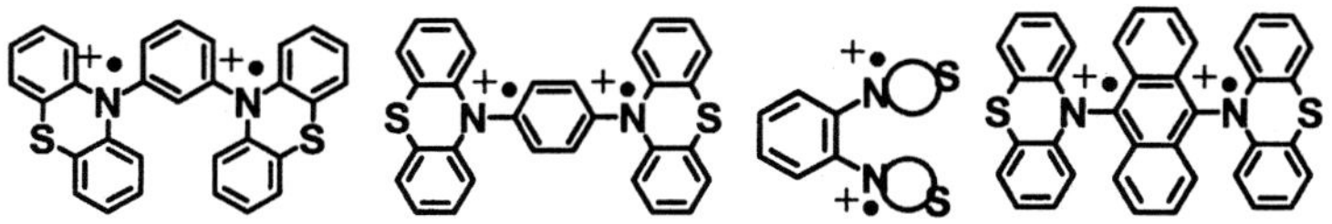

27a   27b   27c   27d

Scheme 13

yielded biradicals with typical zero-field splittings in the $\Delta M_s = 1$ region, closely reflecting the size of spin density on nitrogen and the averaged distance between the radical centers. In addition, strong $\Delta M_s = 2$ signals have been measured that are further split due to ^{14}N $(I = 1)$ hyperfine coupling, corresponding to the one also found in liquid solution of the monoradical cations. To elucidate the ground state spin multiplicity, temperature-dependent studies of the signal intensities were performed. To our surprise, the *para*-phenylene-bridged biradicals **27b** behaved Curie-like with linear dependence, while *meta*-phenylene-connected **27a** yielded a maximum at around 10–15 K, clearly demonstrating thermal activation (28 cal/mol) of the triplet state. To our knowledge, these are the first clearcut experiments demonstrating that *para*-phenylene-connected biradicals might be more stable triplets than *meta*-phenylene-connected ones [36]. These examples also demonstrate that one should not combine the orthogonal approach together with *meta*-phenylene bridging. Even the *ortho*-phenylene-bridged bisphenothiazinyl, which is somewhat more difficult to synthesize due to the strong steric hindrance, has now been shown to form biradicals in sulfuric acid, but not upon electrochemical oxidation [37].

For a better understanding of the role of bridging units for twisted biradicals we used AM1-CI calculations on the triplet stabilities for model benzoquinodimethane (xylylene)-type biradicals **28** (Scheme 14) [38]. For θ between 60° and 90° it is shown that *para*- and *ortho*-phenylene-coupled biradicals may be more stable triplets than the *meta*-isomers, which may even yield singlets. Preliminary ab initio calculations [39] support these findings.

IV. INTERMOLECULAR AGGREGATION

High-spin molecules have been considered valuable building blocks for organic magnetic materials as far as they can be coupled ferromagnetically in three dimensions [40,41]. Three principles for intermolecular spin exchange have been considered so far: (1) metal bridging [42], (2) hydrogen bonding [43,44], and (3) stack-type alignment [45,46]. In this vein we reconsidered some basic principles, starting with the bridging of ketyl radicals by their countercations. The carbonyl function thereby seemed very promising, since

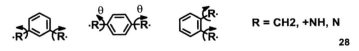

R = CH2, +NH, N

28

Scheme 14

even by the 1960s Hirota and Weissman [47,48] had demonstrated that benzophenone forms strongly intermolecular-coupled biradicals upon alkali metal reduction. This principle, on the other hand, has not been used further for obtaining higher-spin states. We thus used the easily accessible *meta*-dibenzoylbenzenes **29a** (R = 4-*tert*-butyl) and **29b** (R = 2,4,6-trimethyl) [2,49], for testing our prediction (Scheme 15). As with benzophenone, the monoanions of **29a**$^-$ exhibit strongly coupled radical pairs, where the large zfs components (11.3 and 20.0 mT) by far exceed the anisotropic hyperfine couplings.

In the mixture of mono- and dianionic states, indicated also by a raise in optical absorptions at 340 and 682 nm, the EPR spectrum changes dramatically due to formation of the next higher spin state, with $S = 3/2$, which can only arise upon formation of dimeric diketones to give trianion triradicals $(29a)_2$. The identification of this spin state is evidenced by detection of a $\Delta m_S = 3$ transition at one-third of the typical resonance field, measurable only at very low temperatures, and can be reproduced by computer simulations with zfs values $|D/hc| = 7.125$ mT and $|E/hc| = 0.25$ mT. Since intermolecular-coupled radical centers are often aligned antiferromagnetically, the temperature dependence of the EPR signal intensity was investigated. We found typical Curie behavior down to 10 K, supporting strongly a high-spin ground state multiplicity [2,49]. Continued reduction even allowed the detection of the quintet state $S = 2$, with maximum zfs splittings of 40.1 and 35.0 mT.

In the charged mesitylene derivative **29b**, the quintet state from dimeric dianions with zfs components of 45.0 and 38.0 mT seems to be favored over the quartet state. The smaller zfs components in the case of $(29a)_2^{4.-}$ compared to $(29b)_2^{4.-}$ can easily be understood in view of enhanced steric hindrance and larger donating effects of the methyl groups in **29b**, increasing the charge and spin density in the central part of the molecule. Up to now, however, we have not succeeded in obtaining even higher-spin states in extended phenylketone oligomers.

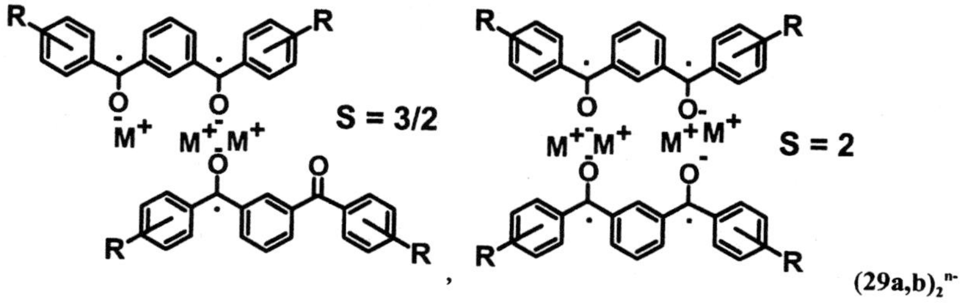

Scheme 15

Recently some stable radicals have been bridged through hydrogen bonds [43,44], but it was not clear if hydrogen bonds in these cases just fix the spin centers in a stack-type arrangement—which according to McConnell's model should allow ferromagnetic interaction—or whether exchange interactions are possible through hydrogen bonds. Starting first with semiempirical calculations on the spin–spin interaction of many different hydrogen-bridged radicals, we then depicted the diaminotriazine dication **30** as a basic unit, after we found an even stronger triplet stabilization in **30** than in the corresponding diaminophenylene dication [27,50]. Diaminotriazine can be hydrogen-bridged to itself $(30)_2$ or to neutral cyanuric acid **31a**, where the charging should yield the tetraradicals (Scheme 16) [27].

It was found that when combining two biradicals hydrogen-bonded through a neutral cyanuric acid molecule **31a**, the quintet is strongly stabilized ($\Delta E_{QT} = 0.04$ eV) in contrast with the self-aggregate $(30^{2.2+})_2$ and the spaced dimer without the cyanuric acid ($31b^{4.4+}$), where no stabilization is found ($\Delta E_{QT} < 0.001$ eV). The former motif clearly resembles the one of covalently linked biradicals through a *meta*-phenylene unit, since the biradicals span an angle of 120° and there is an odd number of centers between the spin-carrying units. Further synthetic efforts are now directed toward this issue.

In a recent paper, the stacked alignment of aminotriazine cations **32** and diaminotriazine dications **33** was calculated, and the high-spin stabilization compared to results for the hydrogen-bonded structures mentioned earlier (Scheme 17) [51]. It was found that the ferromagnetic coupling through space strongly depends on the distance (d, calculated for 0.1-Å steps from 2.5 to 4.0 Å along z) and alignment (rotational angles θ for every 10° from 0° to 180° and movement along x and y calculated in 0.5-Å steps) of

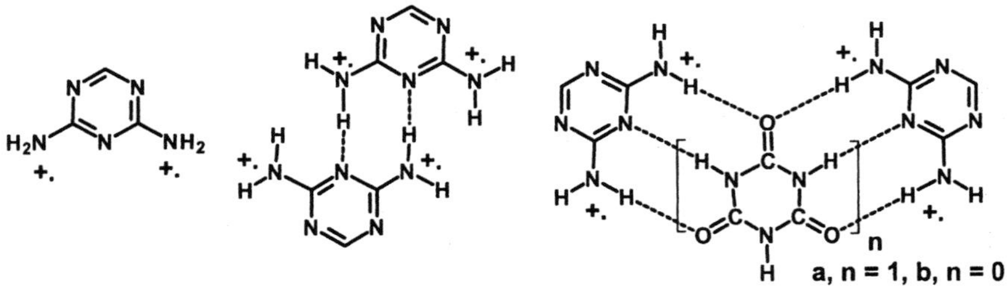

| 30 | (30)₂ | 31a,b |

Scheme 16

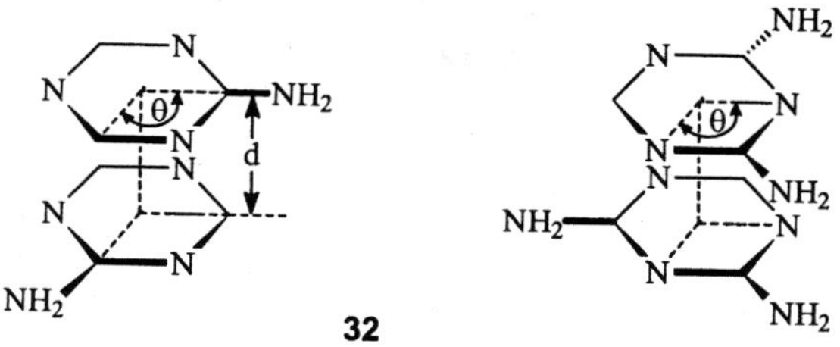

Scheme 17

32

33

the radicals, as predicted earlier (Scheme 18) [45]. Usually the stack-type interaction is stronger than that through hydrogen bonding, and only for distances $d > 4.0$ Å do these interactions become of comparable size.

V. CONCLUSION

In this review it is shown that radical ions of organic π-systems can be used for the synthesis of novel high-spin molecules, either by direct linkage or by connection through organic spacers such as phenylene. Some rules are given, along with some exceptions where the simple predictions are not followed. The high-spin molecules, which should be created to the maximum extent, may then be considered as building blocks for new intermolecular aggregates, and some possibilities for such spin alignment in larger entities are discussed in the final section.

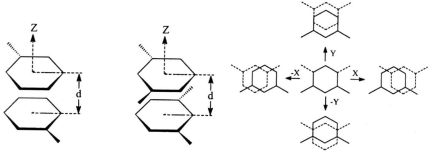

Scheme 18

ACKNOWLEDGMENTS

I wish to thank all the contributors who helped to achieve these results and are cited in the text and references, especially Prof. Dr. Stoyan Karabunarliev (Burgas University, BG), Dipl.-Ing. Lileta Gherghel and Prof. Keiji Okada. Financial support by the Volkswagenstiftung, the Deutsche Forschungsgemeinschaft, and the Fond der chemischen Industrie is gratefully acknowledged.

REFERENCES

1. Baumgarten M. Control of Conjugation and High Spin Formation in Radical Ions of Extended π-Systems. Acta Chem Scand. 1997; 51:193–198.
2. Baumgarten M. Noel Oligoradicals and High Spin Formation. Mol Cryst Liq Cryst. 1995; 272:109–116.
3. Baumgarten M, Müllen K. Radical Ions: Where Organic Chemistry Meets Materials Science. Top Curr Chem. 1994; 169:1–103.
4. Veciana J, Vidal J, Jullian N. Is the Orthogonality of Partially Filled Orbitals in a Regular Chain a Proper Strategy Towards High-Spin Molecules? Mol Cryst Liq Cryst. 1989; 176:443–450.
5. Baumgarten M, Müller U, Bohnen A, Müllen K. Oligo(9,10)anthrylenes as Substrate for Organic High Spin States. Angew Chem Int Ed Engl. 1992; 31: 448–451.
6. Baumgarten M, Müller U. Magnetic and Optical Properties of Anthrylenes. Synth Met. 1993; 55–57:4807–4812.
7. Baumgarten M, Müller U. Novel Oligo-(9,10-anthrylene)s: Models for Electron Transfer and High Spin Formation. J Am Chem Soc. 1995; 117:5840–5850.
8. Friedrich J, Gherghel L, Baumgarten M, Retting W, Jurczok M. Electronic Decoupling in Ground and Excited States of Asymmetric Biaryls. Chem Phys., submitted.
9. Kreyenschmidt M, Baumgarten M, Tyutyulkov N, Müllen K. 2,2′-Bipyrenyl and para-Terpyrenyl: A new type of Electronically Decoupled Electrophores. Angew Chem Int Ed Engl. 1994; 33:1957–1959.
10. Tyutyulkov N, Dietz F, Baumgarten M. Organic Polymers with Indirect Magnetic Interaction Caused by the Symmetry of the Elementary unit. Int J Quantum Chemistry 1997. In press.
11. Baumgarten M, Karabunarliev S. New Principle for High Spin formation. Chem Phys, submitted.
12. Gompper, Harfmann, Gherghel L, Karabunarliev S, Baumgarten M. 1,2′bipyrenyl and 2-(9-anthryl)pyrene: Triplet State Biradical Biaryls. To be published.
13. Zhang J, Karabunarliev S, Baumgarten M. The Ground State Spin Multiplicity of Schlenk-type Biradicals and the Influence of Linkage to Ladder Type Structures. Chem Phys. 1996; 206:339–351.

14. Schlenk W, Brauns M. Title Ber Dt Chem Ges. 1915; 48:661–669.
15. Rajca A, Utamapanya S. π-Conjugated Systems with Unique Electronic Structure: A Case of "Planarized" 1,3-Connected Polyarylmethyl Carbodianion. J Org Chem. 1992; 57:1760–1767.
16. Kistenmacher A, Baumgarten M, Enkelmann V, Pawlik J, Müllen K. The Novel Donor 1,4-Benzothiazino[2,3-b]phenothiazine. J Org Chem. 1994; 59:2743–2747.
17. Tukada H. Diradical Diions of *meta*-Bis(naphthyl) and (anthryl)Phenylenes as New High Spin Molecules. J Chem Soc Chem Commun. 1994:2293–2294.
18. Oda M, Okada K, Gherghel L, Baumgarten M. Fulvenes as High Spin Building blocks. Unpublished.
19. Rajca A, Rajca S, Desai SR. Boron-centered Diradical Dianion: A New Triplet State Molecule. J Chem Soc Chem Comm. 1995:1957–1958.
20. Gregorius H, Baumgarten M, Reuter R, Tyutyulkov N, Müllen K. *Meta*-Phenylene units as Conjugation Barrier in Phenylenevinylene Chains. Angew Chem Int Ed Engl. 1992; 104:1621–1623.
21. Fukutome H, Takahashi A. Ozaki M. Design of Conjugated Polymers with Polaronic Ferromagnetism. Chem Phys Lett. 1987; 133:34–38.
22. Dougherty DA, Kaisaki DA. New Designs For Organic Molecules and Materials with Novel Magnetic Properties. Mol Cryst Liq Cryst. 1990; 183:71–79.
23. Kaisaki DA, Chang W, Dougherty DA. Novel Magnetic Properties of a Doped Organic Polymer. A Possible Prototype for a Polaronic Ferromagnet. J Am Chem Soc. 1991; 113:2764–2766.
24. Yüksel T, Gherghel L, Baumgarten M. Novel Blue Fluorescent Conjugated Polymers. Makromol Chem Phys. In prep.
25. Baumgarten M, Müllen K, Tyutyulkov N, Madjarova G. On the Nature of the Spin Exchange Interaction in Poly(*meta*-aniline). Chem Phys. 1993; 169:81–84.
26. Madjarova G, Baumgarten M, Müllen K, Tyutyulkov N. Structure, Energy Spectra and Magnetic Properties of Poly(*meta*-anilines). Makromol Theory Simul. 1994; 3:803–815.
27. Zhang J, Baumgarten M. Using Triazine as Coupling Unit for Intra- and Intermolecular Ferromagnetic Coupling. Chem Phys. 1997; 214:291–299.
28. Stickley KR, Blackstock SC. Triplet Dication and Quartet Trication of a Triaminobenzene. J Am Chem Soc. 1994; 116:11576–11577.
29. Yano M, Furuichi M, Sato K, Shiomi D, Ichimura A, Abe K, Takui T, Itoh, K. Models for Positive Charge Fluctuations vs. Spin Polarization in Organic Systems: Synthesis and Cyclic Voltammetry of 2D and 1D Hyperbranched π-Aryl-Based Amines. Synth Met. 1997; 85:1665–1666.
30. Dvolaitzky M, Chiarelli R, Rassat A. Stable N,N′-Di-tert-butyl-*meta*-phenylenbisnitroxides-unexpected Singlet Ground States. Angew Chem Int Ed Engl. 1992; 31:180–183.
31. Kanno F, Inoue K, Kago N, Iwamura H. 4,6-Dimethoxy-1,3-phenylenebis(N-tert-butyl nitroxide) with a Singlet Ground State. J Am Chem Soc. 1993; 115: 847–850.

32. Fang BoS, Lee MS, Hrovat DA, Borden WT. Ab-Inito Calculations Show Why *meta*-Phenylene Is Not Always a Ferromagnetic Coupler. J Am Chem Soc. 1995; 117:6727–6731.

33. Okada K, Imakura T, Oda M, Murai M, Baumgarten M. 10,10′-(*para*-Phenylene) bis(phenothiazinyl)bis(radical ions): A Triplet Species in the Ground State in a Kekulé System. J Am Chem Soc. 1996; 118:3047–3048.

34. Okada K, Imakura T, Oda M, Kajiwara A, Kamachi M, Sato K, Shiomi D, Takui T, Itoh K. Gherghel L, Baumgarten M. 10,10′,10″-(1,3,5-Benzenetriyl)triphenothiazine Trication. J Chem Soc Perkin Trans II. 1997:1059–1060.

35. Okada K, Imakura T, Oda M, Kajiwara A, Kamachi M, Sato K, Shiomi D, Takui T, Itoh K, Gherghel L, Baumgarten M. Invalidity of Simple Application of Spin-Prediction Rule To Heteroaromatic π-Conjugated Systems. Mol Cryst Liq Cryst. 1997; 306:301–306.

36. Rassat A. Stable Nitroxide-Based Magnetic Materials. Mol Cryst Liq Cryst. 1997; 305:455–478.

37. Friedrich. PhD dissertation, Mainz, 1997; Okada K, Imakura T, Oda M, Kajiwara A, Takui T, Itoh K, Yüksel T, Friedrich J, Gherghel L, Baumgarten M. Manuscript in preparation.

38. Baumgarten M, Zhang J, Okada K, Tyutyulkov N. Intramolecular Exchange Interaction in Xylylene-Type Biradicals. Mol Cryst Liq Cryst. 1997; 305: 509–514.

39. Dietz F, Schleitzer A, Vogel H, Tyutyulkov N, Baumgarten M. Can *para*-Phenylene Act as a Ferromagnetic Coupling Unit? Manuscript in preparation.

40. Iwamura H. High-Spin Organic Molecules and Spin Alignment in Organic Molecular Assemblies. Adv Phys Org Chem. 1990; 26:179–253.

41. Miller JS, Epstein AJ. Organic and Metalorganic Molecular Magnetic Materials: Designer Magnets. Angew Chem Int Ed Engl. 1994; 33:385–418.

42. Inoue K, Iwamura H. Ferro- and Ferrimagnetic Ordering in a Two-Dimensional Network Formed by Manganese (II) and 1,3,5-Tris(*para*-(N-tert-butyl-N-oxyamino)phenyl)benzene. J Am Chem Soc. 1994; 116:3173–3174.

43. Cirujeda J, Hernandez-Gasio E, Rovira C, Stanger J-L, Turek P, Veciana J. Role of Hydrogen Bonding in the Propagation of Ferromagnetic Interactions in Organic Molecular Solids. J Mater Chem. 1995; 5:243–252.

44. Izuoka A, Kumai R, Sugawara T. Crystal Designing of Organic Ferrimagnets. Adv Mater 1995; 7:672–674.

45. McConnell HM. Ferromagnetism in Solid Free Radicals. J Chem Phys. 1963; 33:1910.

46. Breslow R, Jaun B, Klutiz RQ, Xia C-Z. Ground state Pi-Electron Triplet Molecules of Potential Use in The Synthesis of Organic Ferromagnets. Tetrahedron 1981; 38:863–867.

47. Hirota N, Weissman SI. Rate of Oxidation-Reduction Reactions between Ketones and Ketyls. J Am Chem Soc. 1964; 86:2538–2545.

48. Hirota N. Electron Paramagnetic Resonance Studies of Ion Pairs Metal Ketyls. J Am Chem Soc. 1967; 89:32–42.

49. Baumgarten M, Gherghel L, Wehrmeister T. Ketyl Radicals as Building Blocks for Intra- and Intermolecular Ferromagnetic Coupling. Chem Phys Lett. 1997; 267:175–178.
50. Zhang J, Wang R, Baumgarten M. Role of Hydrogen Bonds in the Propagation of Ferromagnetic Intermolecular Coupling between High Spin Molecules. Mol Cryst Liq Cryst. 1997; 306:119–123.
51. Zhang J, Baumgarten M. Novel Supramolecular Architecture of High Spin Aggregates? The Comparison of Ferromagnetic Coupling through Space and through Hydrogen Bond. Chem Phys. 1997; 222:1–7.

9
Polyradical Cations of High Spin

Silas C. Blackstock and Trent D. Selby
The University of Alabama, Tuscaloosa, Alabama

I. INTRODUCTION

The fields of organic high-spin molecules and organic magnetic materials are historically and phenomenologically connected, with many lessons from the former aiding the development of the latter. Understanding the intramolecular interactions of multiple electron spins is a beginning point for the design and preparation of ferromagnetic or ferrimagnetic organic materials. Stable high-spin organic molecules are envisioned as potential components of molecular-based magnetic materials. Also, the structure/property relationships afforded from studies of organic polyradicals are the cornerstones of attempts to build molecule-based bulk organic magnets.

   This report focuses on a particular set of organic polyradicals, the polyradical cations. Traditionally, the field of organic polyradical chemistry [1] has not included many studies of polyradical ion structures. But as interest in developing organic magnetic materials has grown in recent years, polyradical ions have gained more attention. Several features of such ions are attractive from a materials science viewpoint. First, redox activation of spin sites provides a very fast and generally reversible means of introducing spins into a molecule and potentially into a material. Second, the ionic character of the paramagnetic species provides a means for manipulation of intermolecular interactions between spin-active units within the macroscopic solid. Most of the work on high-spin polyradical cations in recent years has focused on characterization of molecular redox properties and the resultant spin multiplicities of various oxidation states. A brief review of some highlights in this area is the focus of this chapter, with other details concerning some topics being given in Chapters 3 and 10. No attempt has been made to discuss anions or the growing area of *polymeric* high-spin materials based on radical cation spin units, although studies of molecular and macromolecular polyradical cations are clearly relevant to the latter area; the reader is referred to Chapter 10 for this material.

II. OXIDIZED AROMATIC POLYRADICAL CATIONS

The simplest high-spin polyradical cations are two-spin triplets. In principle, two-electron oxidation of aromatic π-systems will give biradicals (i.e., molecules with doubly degenerate or nearly doubly degenerate orbitals containing two electrons) [2] that should have low-lying triplet states. Several examples of this class of triplet cations are well-established as some of the first cases of high-spin organic polyradical ions.

II.A. Cyclopentadiene-Based Cations

Some time ago, a series of cyclopentadienyl cations (formally the two-electron oxidation products of aromatic cyclopentadienide anions) were investigated as potential high-spin organic ions (Scheme 1). Pentaphenylcyclopentadienyl cation [3], 1^+, was found to be a ground state singlet with a low-lying, thermally populated triplet excited state that was detectable by electron spin resonance spectroscopy (ESR). By comparison, pentachlorocyclopentadienyl cation [4], 2^+, has a triplet ground state, as does the parent cyclopentadienyl (cp) cation structure, 3^+ [5]. These cations were generated from cyclopentadiene derivatives in strong Lewis acid media at low temperatures and observed by ESR in frozen solutions, typically at 77 K and often at 4 K. The cp triplet D-values suggest that phenyl and chloro substituents delocalize the spin/charge in these cations. The dication of hexachlorobenzene, 4^{2+}, has also been studied [6] and found to be a triplet molecule. This dication was prepared by matrix photolysis at 77 K of the

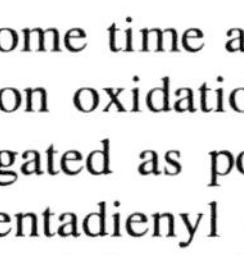

		ground state	D (triplet, cm^{-1})
1^+	R=Ph	singlet	0.1050
2^+	R=Cl	triplet	0.1495
3^+	R=H	triplet	0.1868

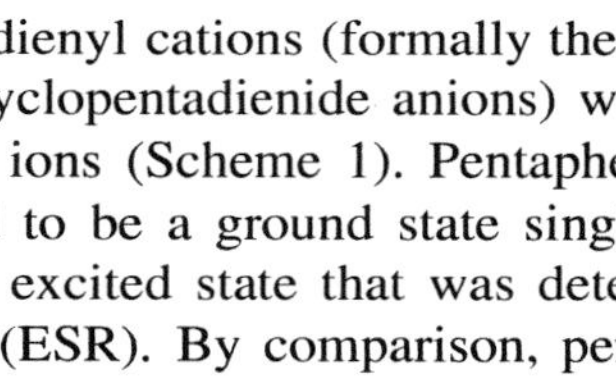

4^{2+} triplet ground state
D = 0.1012 cm^{-1}

Scheme 1

corresponding radical cation in SbF_5Cl_2. The lower D-value for 4^{2+} by comparison to that of 2^{2+} suggests a greater average spin separation in the larger π-system.

II.B. Triphenylene Dications

The oxidation of substituted triphenylenes to triplet dications has also been investigated. Hexamethoxytriphenylene dication [7] 5^{2+} (Scheme 2) was generated electrochemically and chemically (by $SbCl_5$ oxidation) at low temperature. The oxidation potentials in $9:1$ $CH_2Cl_2:TFA$ (Bu_4NBF_4 electrolyte) were $E^{\circ\prime}(1) = 0.55$ and $E^{\circ\prime}(2) = 0.85$ V vs. standard calomel electrode (SCE). In TFA-HFSO$_3$ at $-70°C$, a third chemically reversible oxidation of **5** (the $5^{2+}/5^{3+}$ couple) was observed by cyclic voltammetry (CV) at ~450 mV anodic of $E^{\circ\prime}(2)$, although 5^{3+} has not been further characterized. Even the dication 5^{2+} is unstable above $-50°C$; but its triplet ESR spectrum has been measured in frozen media, and a Curie analysis down to 10 K [8] suggests that it is a ground state triplet or within a few calories of such.

Amino-substituted triphenylenes have also been prepared [8] and are more easily oxidized than **5**. The ethyl-substituted hexaazatritetralin **6** shows four chemically reversible oxidations by CV in CH_3CN (25°C, 0.1 M Bu_4NClO_4) at 0.022, 0.271, 0.475, and 0.790 V vs. SCE. In order to tune the oxidation potential of the hexaazatritetralin for making charge-transfer salts appropriate for testing the McConnell/Breslow electron-exchange spin-coupling mechanism for ferromagnetism, derivatives of **6** in which CH_2CH_3 is replaced by CH_2CF_3 and CH_2CHF_2 were prepared [9]. These derivatives have higher oxidation potentials than **6** by about 0.4 and 0.3 V, respectively. For all the hexaazatritetralin dications studied, triplet ESR signals have been observed, and Curie plots suggest low-lying triplet states, either as ground states or as very weakly (significantly populated) excited states. None of the substituted triphenylene dications so far prepared has substantial chemical stability at room temperature.

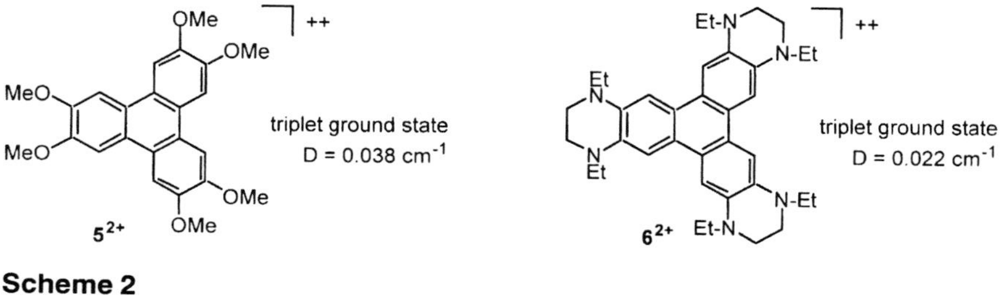

Scheme 2

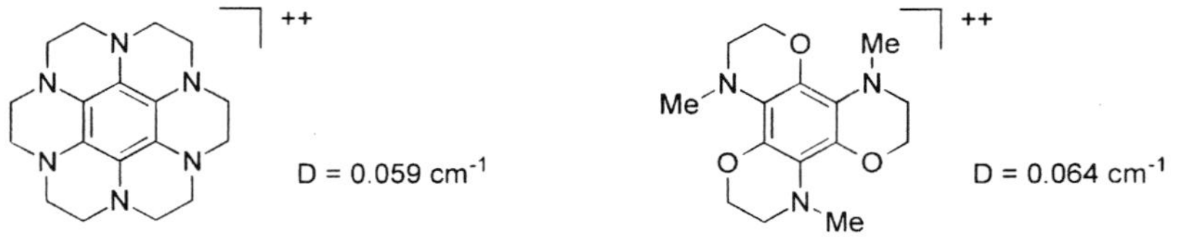

D = 0.059 cm^{-1} D = 0.064 cm^{-1}

7^{2+}   **8^{2+}**

Scheme 3

II.C. Hexaaminobenzene Polycations

A more easily oxidized substituted aromatic system with a much more stable dication is hexaazaoctadecahydrocoronene **7** [10], which has been studied in some detail [11–13] since first reported in 1984. The hexaaminobenzene derivative **7** shows four chemically reversible oxidations by CV, giving $E^{\circ\prime}$ values (CH$_3$CH, 0.1 M Bu$_4$NClO$_4$) of -0.44, 0.06, 0.52, and 0.92 V vs. SCE for its $n/+$, $+/2+$, $2+/3+$, and $3+/4+$ oxidations, respectively. The cation, dication, trication, and tetracation states of **7** have all been isolated and analyzed by means of x-ray diffraction [13]. Of these, only the dication **7^{2+}** (Scheme 3) has a propensity for electron spin alignment. A triplet ESR spectrum of some **7^{2+}** salts was observed in some solvents, such as CH$_2$Cl$_2$/CH$_3$CN mixtures, but not in others, such as DMSO [11]. In frozen CH$_2$Cl$_2$/CH$_3$CN, the half-field ESR signal intensity followed the Curie law from 109 to 156 K, consistent with a triplet ground state; but in CH$_3$CN, an inverse relation was found, suggesting a singlet ground state for **7^{2+}** [11]. Electron spin resonance data for neat powdered samples of **7^{2+}** salts showed this triplet to be a populated excited state, lying 3–5 kcal mol^{-1} above the singlet state [13]. So it appears that the **7^{2+}** triplet and singlet states are close in energy, with their precise ordering being medium dependent.

The dication of the triaminotrialkoxybenzene derivative **8** was prepared at low temperature by NOSbF$_6$ oxidation and its triplet ESR signal observed. The ESR signal intensity for **8^{2+}** obeyed the Curie law from 104 to 165 K, but at room temperature this dication rapidly decomposed [11].

III. POLYAMINE RADICAL CATIONS

Another approach to obtaining high-spin polyradical cations is to prepare isoelectronic radical ion analogs of the corresponding neutral non-Kekulé molecules (see Chapter 2). This approach relies heavily on integrating knowledge of organic radical ion chemistry with that of high-spin bi- and polyradicals, especially the structure/spin-content relationships that have

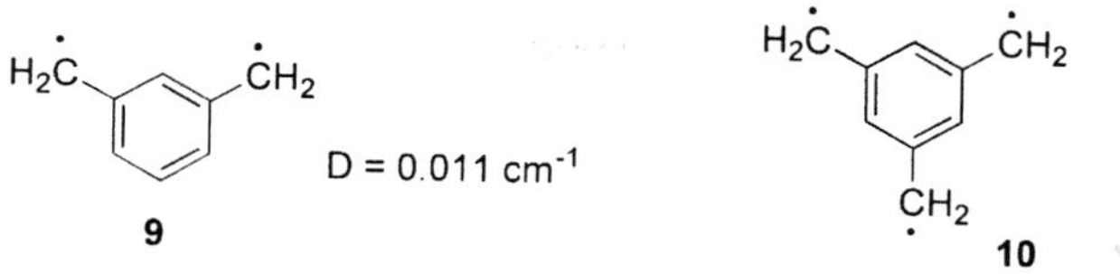

D = 0.011 cm^{-1}

9

10

Scheme 4

emerged over the past several decades. Structural motifs with a strong preference for high-spin ground states are desirable so that spin-dilution effects or other electronic perturbations that may be introduced in the radical ion analogs will not reverse the high-spin predisposition of the polyradical cation structure.

III.A. Tris(diarylamino)benzene Di- and Trications

Meta-substituted polyamines are potential substrates for making high-spin polyradical cations that are analogs of triplet *meta*-xylylene, **9**, or 1,3,5-trimethylenebenzene, **10** (Scheme 4). Biradical **9** has been prepared in a low-temperature matrix and analyzed by ESR [14]. It is a ground state triplet molecule for which ab initio calculations (6-31G*/π-SDCI) suggest a triplet preference of ∼10 kcal mol^{-1} [15]. Triradical **10** is unknown, but likewise is expected to have a strong preference for a high-spin (quartet) ground state.

The first report of a 1,3,5-triaminobenzene trication was for the putative hexaphenyl-substituted system **11** (Scheme 5) [16]. We were not able to obtain a clearly assignable quartet spectrum under the conditions described by these workers, and we believe that the spectrum originally attributed to a quartet state of **11**$^{3+}$ was caused by some other structure(s). However, anisyl-substituted analog 1,3,5-tris(di-*p*-anisylamino)benzene **12** has been shown to possess solution-stable dication and trication oxidation

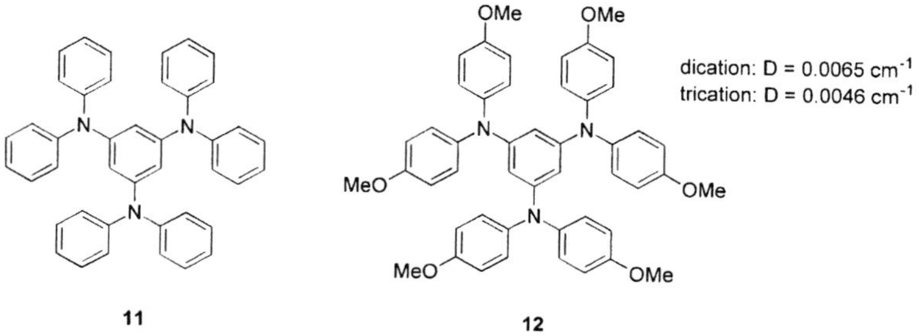

dication: D = 0.0065 cm^{-1}
trication: D = 0.0046 cm^{-1}

11

12

Scheme 5

states long-lived enough for analysis [17]. The CV oxidation of **12** gave three chemically reversible waves at low temperature, with $E^{\circ\prime}$ values of 0.67, 0.87, and 0.98 V vs. SCE (PrCN, 0.1 M Bu$_4$NBF$_4$, $-78°$C) for $n/+$, $+/2+$, and $2+/3+$ redox couples, respectively. Monocation **12**$^+$ gave a 10-line ESR spectrum assigned to have hyperfine constants of a(3N) = 3.4 G and a(3H) = 2.6 G, typical of a number of tris(arylamino)benzene radical cations [18]. For **12**$^{2+}$ and **12**$^{3+}$, triplet and quartet ESR signals were observed with D-values of 0.0065 and 0.0046 cm^{-1}, respectively [17]. The **12**$^{3+}$ D-value of 0.0046 cm^{-1} is close to that (0.0049 cm^{-1}) observed for the related neutral hydrocarbon triradical quartet 1,3,5-tris(diphenylmethylene)benzene [19].

Ab initio calculations have predicted a strong high-spin ground state preference for 1,3,5-triaminobenzene trication, with a doublet-to-quartet energy gap $\Delta E_{D,Q}$ = 10.6 kcal mol^{-1} [20]. The ESR signal intensity of the half-field absorptions for **12**$^{2+}$ and for **12**$^{3+}$ over 90–140 K was found to obey the Curie law, consistent with either high-spin grounds states for these structures or near degeneracy of the singlet/triplet and doublet/quartet states [17]. More recently, analysis of **12**$^{3+}$ by ESR/electron spin transient nutation spectroscopy has demonstrated conclusively that **12**$^{3+}$ is a quartet ground state trication [21].

III.B. Bis(diarylamino)benzene and Bis(diarylamino)biphenyl Dications

The oxidation of a number of aryl-substituted diamines has been studied to investigate the bis(aminium radical cation) exchange couplings. A diamino analog of **12**, **13**, has been prepared and its dication (**13**$^{2+}$) was found to be a ground state triplet by an ESR/spin nutation study, with a D-value very close to that observed for **12**$^{2+}$ [21]. The biphenyl-linked isomeric diamines **14** and **15** have also been studied (Scheme 6) [22]. The ΔE values for the difference between first and second oxidation potentials for **14** and **15** in benzonitrile are 27 and 66 mV, respectively. The larger ΔE for **15** is consistent with its coextensive structure, which allows greater interaction between the redox groups. Triplet ESR signals were observed for both **14**$^{2+}$ and **15**$^{2+}$ [22]. The triplet signal for **14**$^{2+}$ diminished upon lowering of the temperature, suggesting that is is a thermally populated triplet; hence **14**$^{2+}$ is probably a singlet ground state species.

III.C. Tris(triaryl-*p*-phenylenediamino)benzene Di- and Trications

In an effort to prepare more stable poly(radical cations), some tris(phenylenediamine)benzene substrates, compounds **16** [23] and **17** [23–

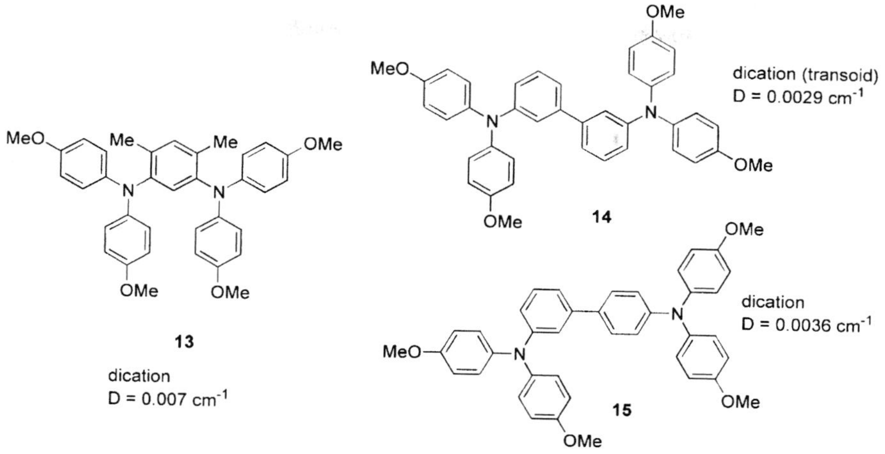

Scheme 6

25], have been prepared and their oxidations studied (Scheme 7). These substrates gave six chemically reversible oxidations by CV at room temperature. For **16**, the $E^{\circ\prime}$ values were $(n/+)$ 0.41, $(+/2+)$ 0.54, $(2+/3+)$ 0.61, $(3+/4+)$ 0.97, $(4+/5+)$ 1.01, and $(5+/6+)$ 1.08 V vs. SCE in CH_2Cl_2 (0.1 M Bu_4NBF_4). Chemical oxidation by stoichiometric amounts of $NO^+PF_6^-$ afforded **16**$^+$, **16**$^{2+}$, and **16**$^{3+}$ as isolable PF_6^- salts. ESR measurements showed a broad singlet for **16**$^+$ in liquid solution and characteristic triplet and quartet ESR spectra with half-field absorptions for **16**$^{2+}$ and **16**$^{3+}$ in frozen solutions, respectively. The D values for **16**$^{2+}$ and **16**$^{3+}$ were 0.0026 and 0.0018 cm^{-1}, respectively. Limited temperature range studies of the half-field signals (90–120 K) obeyed the Curie law, showing that the high-spin states are ground states or fairly close to being ground states. The extreme stability of these cations allowed the measurement of their effective magnetic susceptibility values (μ_{eff}) in solution by nuclear magnetic resonance (NMR) analysis using the Evans method [26]. From the derived μ_{eff} values and correction for the statistical advantage afforded to the high-spin states, we concluded that **16**$^{2+}$ is a ground state triplet by ~1.4 kcal mol^{-1} and that **16**$^{3+}$ is actually a ground state doublet, with its quartet state being a low-lying, highly populated excited state at 25°C.

A linearly meta-linked tris(phenylenediamine) **18** and its solution-stable cations have also been reported (Scheme 8) [25]. Compound **18** undergoes chemically reversible two-electron oxidation at $E^{\circ\prime}(1,2) = 0.56$ V and a third oxidation at $E^{\circ\prime}(3) = 0.75$ V vs. SCE in CH_2Cl_2 (0.1 M Bu_4NPF_6). The dication **18**$^{2+}$ gives a 9-line solution ESR spectrum with a(4N) = 2.9

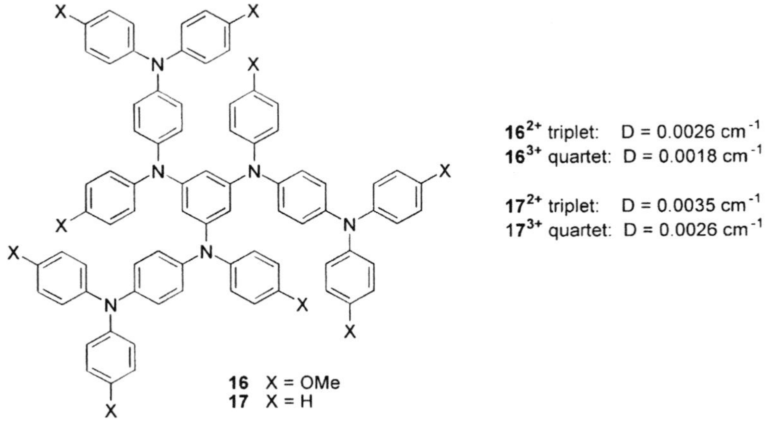

16^{2+} triplet: D = 0.0026 cm^{-1}
16^{3+} quartet: D = 0.0018 cm^{-1}

17^{2+} triplet: D = 0.0035 cm^{-1}
17^{3+} quartet: D = 0.0026 cm^{-1}

16   X = OMe
17   X = H

Scheme 7

G, indicating weak dipolar spin–spin interactions between radical cations located on the terminal phenylenediamine groups. Trication **18**$^{3+}$ shows a quartet ESR signal ($D = 0.0026$ cm^{-1}), which appears to be superimposed on a doublet signal at 100 K in frozen CH$_2$Cl$_2$/TFA (99:1). The intensity of the **18**$^{3+}$ half-field signal was monitored over 4–100 K and found to obey the Curie law, indicating that the population of the **18**$^{3+}$ triplet state did not change significantly in this temperature range.

III.D.   Bis(*p*-phenylenediamino)benzene Dications

The first reported poly(phenylenediamine) radical cations were bis-derivatives **19** and **20** [27] and later **21** [28] and **22** (Scheme 9) [29]. Oxidation of **19** and **20** was performed with TCNQF$_4$ to make the **19**$^{2+}$(TCNQF$_4$)$_2$ and **20**$^{2+}$(TCNQF$_4$)$_2$ solids. These solids could not be crystallized but were otherwise characterized, and their magnetic susceptibilities were measured over

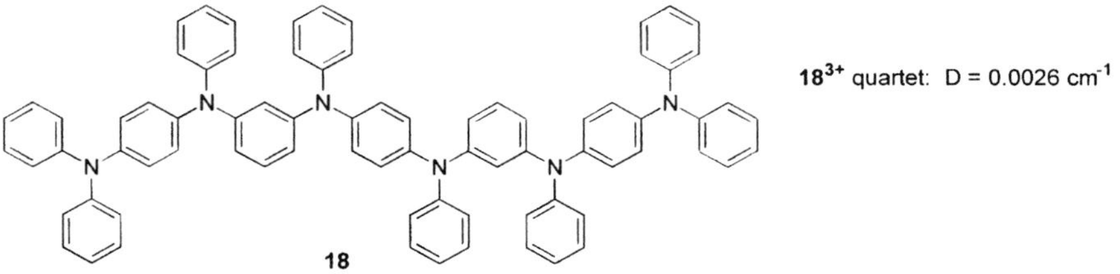

18^{3+} quartet: D = 0.0026 cm^{-1}

18

Scheme 8

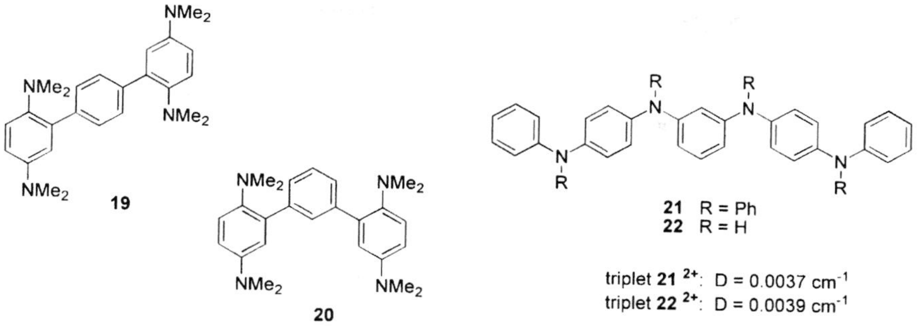

Scheme 9

19

20

21 R = Ph
22 R = H

triplet 21 $^{2+}$: D = 0.0037 cm^{-1}
triplet 22 $^{2+}$: D = 0.0039 cm^{-1}

5–300 K by SQUID magnetometry. The intermolecular interaction between the phenylenediamine radical cation spins was found to favor low-spin electron pairing in **19** (2J/k = −42 K or 83 cal mol^{-1}) and high-spin alignment in **20** with 2J/k = 240 K or 477 cal mol^{-1} [27].

The meta-linked, phenylated bis(p-phenylenediamine) **21** [28] gave chemically reversible oxidations with $E^{\circ\prime}$ values of (n/+) 0.54 and (+/2+) 0.68 V vs. SCE in CH$_2$Cl$_2$ (0.1 M Bu$_4$NPF$_6$). For **21**$^+$, a 5-line solution ESR spectrum is observed and is assigned as a(2N) = 5.7 G. The dication **21**$^{2+}$ prepared by thianthrenium perchlorate oxidation in CH$_2$Cl$_2$/TFA was stable for days at room temperature in solution. A triplet ESR signal for **21**$^{2+}$ was observed in frozen solutions (D = 0.0037 cm^{-1}), and its half-field signal intensity obeyed the Curie law from 4 to 100 K, indicating a triplet ground state for this dication or near singlet-triplet state degeneracy.

The N–H derivative of **21**, **22**, has been observed to give triplet dication **22**$^{2+}$ when treated with the N-dehydro analog plus acid. This process has been referred to as *acid doping* [29]. The reaction occurs formally between **22**$^{4+}$ and **22** as a conproportionation to give **22**$^{2+}$. A triplet ESR spectrum (D = 0.0039 cm^{-1}) is found for **22**$^{2+}$, and again the ESR signal intensity obeys the Curie relation from 4 to 100 K, indicating no change in the triplet population over this range. Thus, **22**$^{2+}$ is concluded to be either a triplet with no appreciable singlet population at 100 K or a degenerate singlet, triplet molecule.

IV. DIPHENOTHIAZINE AND DIHYDRAZINE RADICAL CATIONS

Although the bulk of spin-coupled polyradical cation work has focused on amine or phenylenediamine redox systems, similar studies employing other stable organic radical cation moieties have also been reported, and some are outlined next.

IV.A. Bis(phenothiazino)benzene Dications

The phenothiazine group is formally a 16 π-electron system that is easily oxidized to give radical cations having long solution lifetimes. By linking these groups via their N-atoms to a benzene core, both meta- and para-coupled bis(phenothiazino)benzenes **23** and **24** have been prepared for study of their electron exchange interactions in the dication state (Scheme 10) [30]. Oxidation to the dication state was accomplished by dissolving the neutral substrates in concentrated sulfuric acid. Triplet ESR spectra for **23**$^{2+}$ and **24**$^{2+}$ were observed; they had the D-values of 0.00597 and 0.00439 cm^{-1}, respectively. The temperature dependence of the spectra from 6 to 50 K indicated that the meta-linked isomer **23**$^{2+}$ is a ground state singlet, with the populated triplet $\sim$28 cal mol^{-1} higher in energy. The corresponding study for **24**$^{2+}$ showed Curie law behavior in this temperature region, indicating it to be a triplet ground state by more than $\sim$100 cal mol^{-1} or to have essentially degenerate singlet and triplet states to within $\pm$10 cal mol^{-1}. The low-spin ground state for the meta-linked isomer and the apparent lack thereof for the para-linked isomer are explained by assuming that the phenothiazine cation rings are twisted nearly orthogonal to the benzene core group plane. If so, sigma conjugation of the spin-bearing orbitals can lift the formal degeneracy of the SOMOs, giving preferential stabilization of the singlet state.

IV.B. Bis(hydrazino)benzene Dications

Hydrazine radical cations are solution-stable radicals that have been systematically studied in some depth [31]. When properly structured, such radical cations are isolable and stable. Recently, the set of bis(hydrazino)benzenes **25** and **26** with isolable diradical dication states has been prepared and their magnetic properties evaluated (Scheme 11) [32]. Chemically reversible ox-

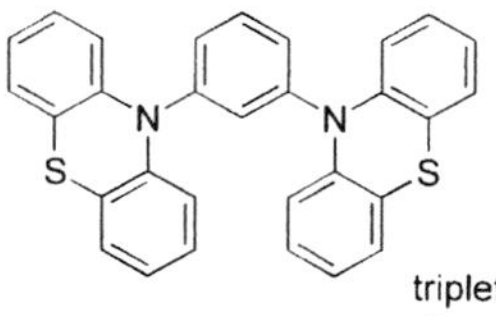

	triplet **23**$^{2+}$		triplet **24**$^{2+}$
23	D = 0.00597 cm^{-1}	**24**	D = 0.00439 cm^{-1}

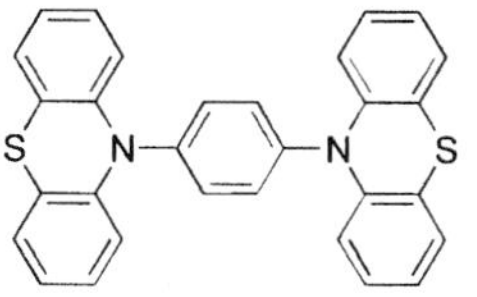

Scheme 10

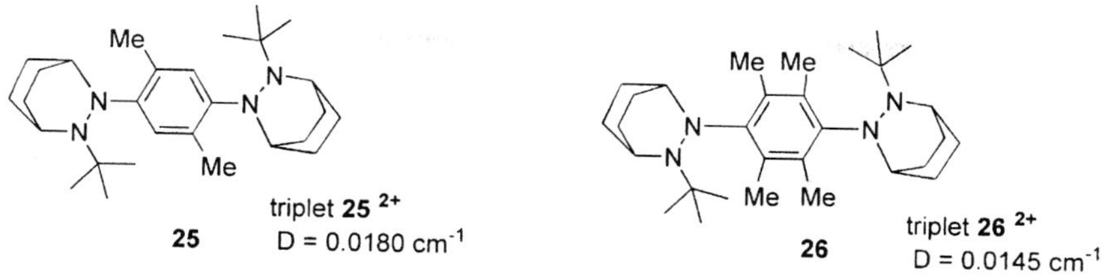

triplet 25 $^{2+}$

25   D = 0.0180 cm^{-1}

triplet 26 $^{2+}$

26   D = 0.0145 cm^{-1}

Scheme 11

idations by CV of **25** occurred, with $E^{\circ\prime}$ values of $(n/+)$ 0.03 and $(+/2+)$ 0.33 V vs. SCE in CH_3CN (0.1 M Bu_4NCl). The corresponding oxidation of **26** occurred, with $E^{\circ\prime}$ values of $(n/+)$ -0.17 and $(+/2+)$ 0.09 V. Both 25^{2+} and 26^{2+} had thermally populated triplets that were observable by means of ESR, with D-values of 0.0180 and 0.0145 cm^{-1}, respectively. Because the crystal structures of these dication salts are known, molecular dication geometries may be correlated with the ESR parameters, assuming a negligible effect of solution versus solid state. Analysis of 26^{2+} triplet ESR intensities over $4-113$ K gave a singlet energy preference 2J = 320 cal mol^{-1} [33]. Of additional interest in this study is the correlation drawn between the singlet-triplet energy gap 2J and the electron transfer parameters for 26^+ and 26^{2+} deduced from their optical spectra. The reader is referred to the original work for further details.

V.   OTHER POLYRADICAL CATION COUPLED SYSTEMS

V.A.   Bis- and Tris(methylenephosphorano)benzene Polycations

High-spin radical cations of methylenephosphoranes have been prepared [34]. Oxidation of the bis- and tris(methylenephosphorano)benzene **27** by CV gave $E^{\circ\prime}$ values of $(n/+)$ $-$ 0.14 and $(+/2+)$ 0.23 V for **27** vs. SCE in THF (0.1 M Bu_4NPF_6). The corresponding study for the analog **28** found $(n/+)$ -0.29, $(+/2+)$ -0.01, and $(2+/3+)$ 0.27 V (Scheme 12). Cold chemical oxidation by Ag^+ produced dication 27^{2+} that showed triplet ESR signals with $D = 0.0117$ cm^{-1} and a linear Curie plot from 3.8 to 110 K. Cold chemical oxidation of **28** produced a mixture of 28^{2+} and 28^{3+} (and possibly 28^+), which displayed doublet, triplet, and quartet ESR spectra in superposition. The presence of two $\Delta m_s = 2$ transitions and a discernible $\Delta m_s = 3$ transition at 4 K validated the simultaneous presence of triplet and quartet species. Curie law behavior of the quartet signal in the $\Delta m_s = 1$ spectral

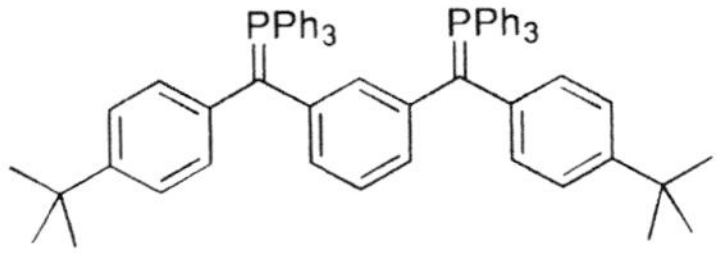

27

triplet **27** $^{2+}$:   D = 0.0117 cm^{-1}

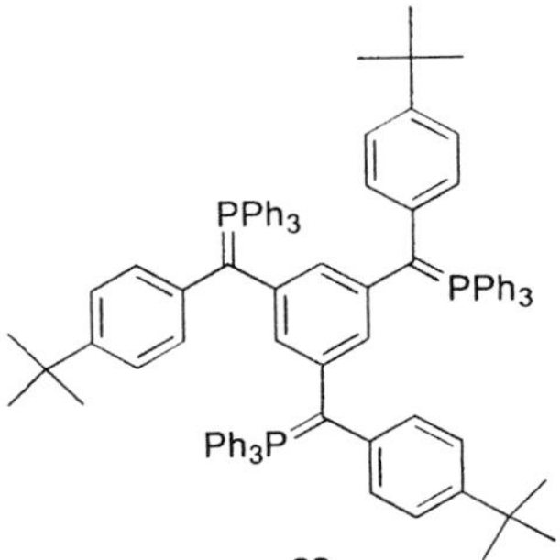

28

triplet **28** $^{2+}$:   D = 0.0117 cm^{-1}
quartet **28** $^{3+}$:   D = 0.0088 cm^{-1}

Scheme 12

region was found, indicating a quartet ground state or very nearly degenerate doublet-quartet states for **28**$^{3+}$.

V.B.   Radical–Radical Cation Structures

It is possible to prepare high-spin radical cations in which a radical cation spin couples to other neutral radical site(s), as in the oxidized forms of compounds **29** [35] and **30** (Scheme 13) (see also Chapter 26) [36]. Oxidation of **29** by I$_2$ in methyl-THF gave a green solution showing a triplet ESR signal at 100 K, with D = 0.0214 and E = 0.0022 cm^{-1}, respectively. The temperature dependence of this signal indicated a singlet ground state for **29**$^+$ with a thermally populated triplet state at ~200 cal mol^{-1} above the ground state [35]. Oxidation of **30** by I$_2$ gave **30**$^+$, with a triplet ESR signal in frozen tetrahydrofuran having D = 0.0276 and E = 0.0016 cm^{-1}, respectively. Variable temperature analysis of these triplet signals over 7.5–100 K showed linear Curie law behavior, indicating that the triplet population remained unchanged in this temperature range. The authors concluded that **30**$^+$ has a triplet ground state, with the nearest singlet being several hundred

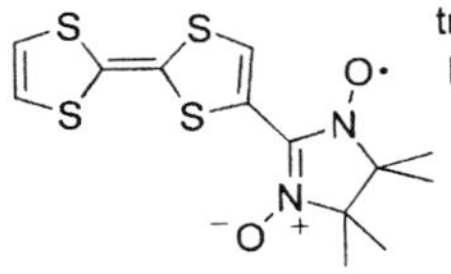

triplet **29** $^+$
D = 0.0214 cm^{-1}

29

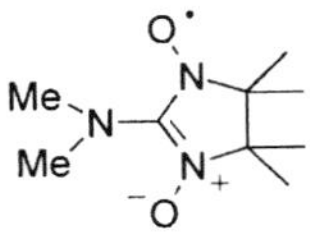

triplet **30** $^+$
D = 0.0276 cm^{-1}

30

Scheme 13

cal mol^{-1} higher in energy, or that the singlet and triplet states are within ~10 cal mol^{-1} of each other. Thus, 30^+ has been regarded as a hetero analog of a TMM-type spin-coupled diradical (comparable to other analogs given in Chapter 2) [36].

VI. CONCLUSIONS

As the foregoing discussions demonstrate, the field of polyradical cations is blossoming, especially in recent years, after a strong beginning in the 1960s and a period of slow but steady growth in the 1970s and '80s. Because of the ease of introducing spin centers by one-electron oxidation and the great stability possible for some of the resulting polyradical cation systems, one can expect the next decade of polyradical ion work to produce even more impressive results than already obtained in the late 1990s. In addition, the implications of this work for spin-coupled oligomeric radical ions will surely aid the continued progress being made in doped polymeric systems and other strategies for preparing polyradical ion–based organic magnetic materials.

REFERENCES

1. For recent reviews of organic polyradicals see: (a) Iwamura H. Adv. Phys. Org. Chem. 1990; 26:179–253. (b) Dougherty DA. Acc. Chem. Res. 1991; 24:88. (c) Miller JS, Epstein AJ. Angew. Chem. Int. Ed. Engl. 1994; 33:385. (d) Rajca, A. Chem. Rev. 1994; 94:871.
2. (a) Salem L, Rowland C. Angew. Chem. Int. Ed. Engl. 1972; 11:92. (b) Borden WT, ed. Diradicals. New York: Wiley, 1982, ch. 1.
3. Breslow R, Chang HW. J. Am. Chem. Soc. 1963; 93:2033.
4. Breslow R, Hill R, Wasserman E. J. Am. Chem. Soc. 1964; 86:5349.
5. Saunders M, Berger R, Jaffe A, McBride JM, O'Neill J, Breslow R, Hoffman JM Jr., Perchonock C, Wasserman E, Hutton RS, Kuck VJ. J. Am. Chem. Soc. 1973; 95:3017.
6. Wasserman E, Hutton RS, Kuck VJ, Chandross EA. J. Am. Chem. Soc. 1974; 96:1965.
7. Bechgaard K, Parker VD. J. Am. Chem. Soc. 1972; 94:4749.
8. Breslow R, Juan B, Kluttz RQ, Xia C-Z. Tetrahedron 1982; 38:863.
9. LePage TJ, Breslow R. J. Am. Chem. Soc. 1987; 109:6412.
10. Breslow R, Maslak P, Thomaides, JS. J. Am. Chem. Soc. 1984; 106:6453.
11. Thomaides J, Maslak P, Breslow R. J. Am. Chem. Soc. 1988; 110:3970.
12. Miller JS, Dixon DA, Calabrese JC. Science 1988; 240:1185.
13. Miller JS, Dixon DA, Calabrese JC, Vazquez C, Krusic PJ, Ward MD, Wasserman E, Harlow RL. J. Am. Chem. Soc. 1990; 112:381.
14. Wright BB, Platz MS. J. Am. Chem. Soc. 1983; 105:628.

15. Fort RC Jr, Getty SJ, Hrovat DA, Lahti PM, Borden, WT. J. Am. Chem. Soc. 1992; 114:7549.

16. Yoshizawa K, Chano A, Ito A, Tanaka K, Yamabe T, Fujita H, Yamauchi J, Shiro M. J. Am. Chem. Soc 1992; 114:5994.

17. Stickley KR, Blackstock SC. J. Am. Chem. Soc. 1994; 116:11576.

18. Stickley RK, Blackstock SC. Tetrahedron Lett. 1995; 36:1585.

19. Wilker W, Kothe G, Zimmerman H. Chem. Ber. 1975; 108:2124.

20. Yoshizawa K, Hatanaka M, Matsuzaki Y, Tanaka K, Yamabe T. J. Chem. Phys. 1994; 100:4453.

21. Sato K, Yano M, Furuichi M, Shiomi D, Takui T, Abe K, Itoh K, Higuchi A, Katsuma K, Shirota Y. J. Am Chem. Soc. 1997; 119:6607.

22. Bushby RJ, McGill DR, Ng KM, Taylor N. J. Chem. Soc. Perkin Trans. 2 1997:1405.

23. Stickley KR, Selby TD, Blackstock SC. J. Org. Chem. 1997; 62:448.

24. Ishikawa W, Noguchi K, Kuwabara Y, Shirota Y. Adv. Mater. 1993; 5:559.

25. Wienk MM, Janssen RAJ. J. Am. Chem. Soc. 1997; 119:5398.

26. (a) Evans DF. J. Chem. Soc. 1959:2003. (b) Live DH, Chan SI. Anal. Chem. 1970; 42:791.

27. Nakamura Y, Iwamura H. Bull. Chem. Soc. Jpn. 1993; 66:3724.

28. Wienk MM, Janssen RAJ. Chem. Commun. 1996:267.

29. Wienk MM, Janssen RAJ. J. Am. Chem. Soc. 1996; 118:10626.

30. Okada K, Imakura T, Oda M, Murai H, Baumgarten M. J. Am. Chem. Soc. 1996; 118:3047.

31. Nelsen SF. Acc. Chem. Res. 1981; 14:131.

32. Nelsen SF, Ismagilov RF, Powell DR. J. Am. Chem. Soc. 1997; 119:10213.

33. Nelsen SF, Ismagilov RF, Teki Y. J. Am. Chem. Soc. 1998; 120:2200.

34. Wienk MM, Janssen RAJ. J. Am. Chem. Soc. 1997; 119:5398.

35. Kumai R, Matsushita MM, Izuoka A, Sugawara T. J. Am. Chem. Soc. 1994; 116:4523.

36. Sakurai H, Kumai R, Izuoka, A, Sugawara T. Chem. Lett. 1996:879.

10

The Design of Stabilized High-Spin Spin-Bearing Blocks

Richard J. Bushby
University of Leeds, Leeds, England

I. INTRODUCTION

Almost thirty years ago, Mataga suggested that it would be possible to create ferromagnetic organic polymers by exploiting the strong ferromagnetic spin-coupling mechanism found in triplet ground state π-diradicals [1]. His idea was that a conjugated polymer in which the repeat unit was one of these triplets would possess a band structure in which a superdegenerate band occupied by ferromagnetically coupled unpaired electrons would lie between the valence and conduction bands [2]. Most triplet ground state π-diradicals—and hence most potential building blocks—belong to one of the three main families of non-Kekulé hydrocarbons [3]. These are shown in Figures 1–3. They are the non-Kekulé quinodimethanes (Figure 1) [4,5], of which the simplest is **meta**-benzoquinodimethane **1**, the non-Kekulé polyenes (Figure 2) [6–10], of which the simplest is trimethylenemethane **9**, and the non-Kekulé polynuclear aromatics (Figure 3) [3,11–13], of which the simplest is compound **19**, which is variously known as triangulene or Clar's hydrocarbon. Hückel Molecular Orbital (HMO) calculations predict that all of the hydrocarbon diradicals shown in Figures 1–3 have degenerate pairs of non-bonding π-molecular orbitals, which need to accommodate two electrons. In most cases these orbitals are orthogonal but coextensive [14] (non-disjoint [15]). There is a zero-overlap integral, but they overlap in their spatial distributions. In such cases there is a strong exchange interaction, and, just as with pairs of singly occupied cocentered atomic orbitals, Hund's Rule applies and a triplet ground state results [16]. In many cases the triplet

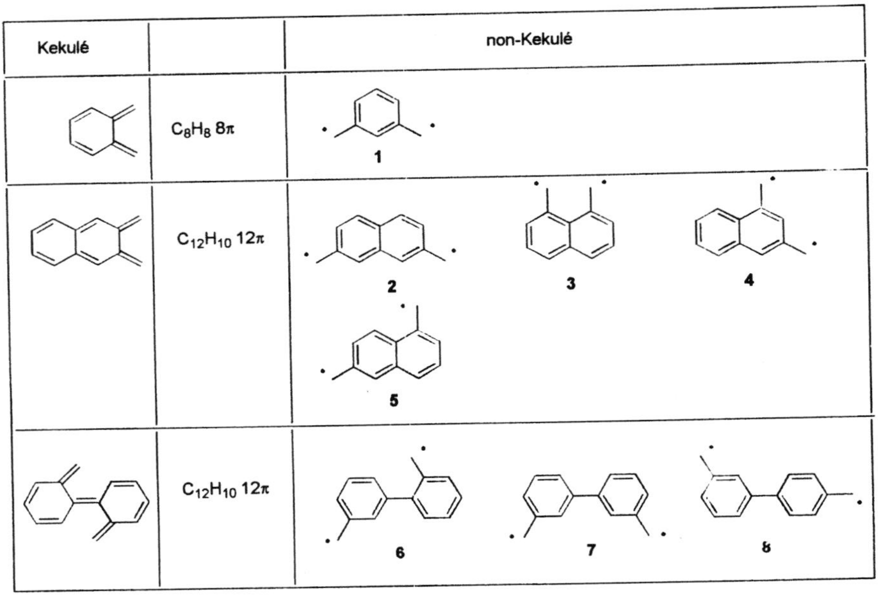

Kekulé		non-Kekulé
	C_8H_8 8π	**1**
	$C_{12}H_{10}$ 12π	**2** **3** **4** **5**
	$C_{12}H_{10}$ 12π	**6** **7** **8**

Figure 1 Representative Kekulé and non-Kekulé quinodimethanes based on the benzene, napthalene, and biphenyl nuclei. Note that, in a formal sense, each is derived from the parent hydrocarbon by substituting two hydrogens on the aromatic nucleus for two methylenes (CH_2). All of the diradicals (or simple derivatives) have been made. (From Refs. 4 and 5.)

ground state nature of these diradicals has been demonstrated experimentally [3–13].

Mataga's idea for exploiting these triplet diradicals, like many of the similar proposals of Ovchinnikov [17], looks attractive, but no experimentalist has produced a ferromagnetic polymer through this approach. The biggest problem is that all of the non-Kekulé hydrocarbons shown in Figures 1–3 are unstable. Indeed, this instability problem bedevils all attempts to produce organic materials with exploitable magnetic properties, whether based on this approach or on one of the two approaches suggested by McConnell [18]. Hence, the search for stable, organic, spin-carrying building blocks has become a major research industry.

Organic radicals can be stabilized in a variety of ways, including the use of bulky, conjugating donor and acceptor substituents. These approaches apply equally well to multiradicals. Alternatively, once can exploit the fact that radical-ion spin-bearing moieties are generally more stable than their

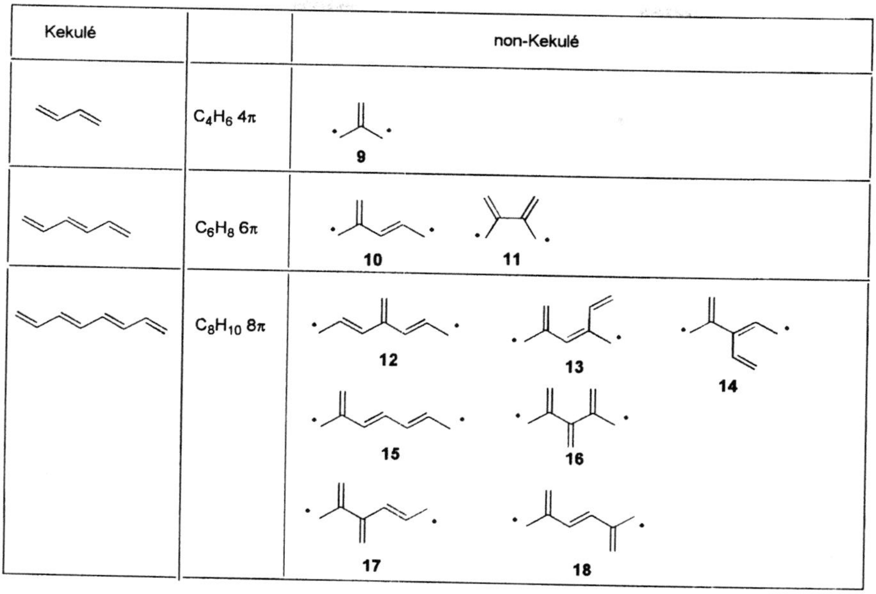

Kekulé		non-Kekulé
	C_4H_6 4π	 **9**
	C_6H_8 6π	 **10** **11**
	C_8H_{10} 8π	 **12** **13** **14** **15** **16** **17** **18**

Figure 2 The first ten members of the non-Kekulé polyene family **9–18**. Note that each non-Kekulé polyene is isomeric with a classical polyene. Hence TMM **9** is isomeric with the Kekulé polyene, butadiene, and also has four π-electrons. Only the first four non-Kekulé polyenes (or derivatives thereof) have been made so far. (From Refs. 6–10.)

neutral carbon-centered counterparts. Figure 4 illustrates some of the ways one might develop these ideas to create analogs of *meta*-benzoquinodimethane **1** that would be more stable.

Until recently, trimethylenemethane **9** was the only one of the π-diradicals shown in Figures 1–3 for which the energetic triplet to singlet gap was experimentally known with some degree of confidence [19]. The magnitude of this splitting proves to be important. As shown in Figure 4, the substitution of the two methylenes (CH_2^-) of *meta*-benzoquinodimethane **1** with stable entities like nitroxide **25**, amminium cation **26**, carbonyl radical anion **24, 27**, or aryl radical ion **28–31** can lift the degeneracy of the putative singly occupied molecular orbitals (SOMOs). If the exchange interaction in the parent system were small, this could result in a reversion to a singlet ground state. If, however, the exchange interaction is large, even when strongly perturbed the triplet will remain the ground state. Such systems have been termed "robust triplets" [14].

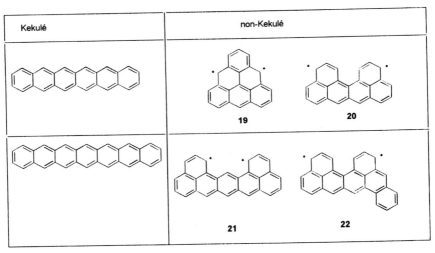

Figure 3 The four simplest non-Kekulé polynuclear aromatic compounds compared to Kekulé polynuclear aromatic structures. The majority of structures that can be generated by edgewise fusion of benzene nuclei are of the Kekulé type in which each π-electron can be formally paired with one on the neighboring carbon. For compounds **19–22**, however, this is not possible. Derivatives of compounds **19–21** have been reported. (From Refs. 3 and 11–13.)

The preference for the triplet state in diradical **1** is expected to be substantial, but until recently was not known. Wenthold and Lineberger have now used modern photoelectron ion spectroscopic techniques to estimate the preference to be 9.6 ± 0.2 kcal/mol [20]. Since this exchange interaction is moderately large, even strong substituent perturbation was not expected to reverse the triplet ground state preference. Yet, even in the most "robust" system, there must be limits. Hence not all *meta*-benzoquinodimethane analogs have triplet ground states; and although it was once argued that *meta*-phenylene was a "universal ferromagnetic spin-coupling moiety," this is now known not to be the case [21,22].

A considerable number of different stabilized systems have been explored, including (recently) ones based on boron [23] and phosphorus [24,25]. More thoroughly explored are systems in which carbon-based radicals are stabilized by multiple phenyl substitution [26], and the most notable successes have been achieved using systems exploiting nitroxide [27] and nitronylnitroxide spin-bearing units [28]. These are discussed elsewhere in this book. This chapter concentrates on those systems that we have explored

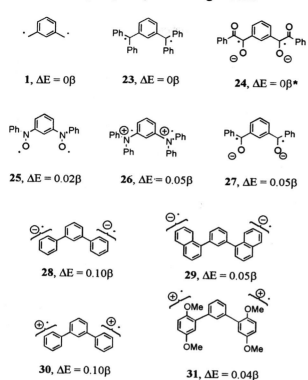

Figure 4 Stabilized building blocks based on metaquinodimethane and HMO values for the splitting between the putative singly occupied molecular orbitals (SO-MOs). The standard parameterization scheme was used. *In the case of compound **24**, a nonzero splitting is obtained with some parameterization schemes.

based on carbonyl radical-anion spin-bearing units, aryl radical-ion spin-bearing units, and amminium cation spin-bearing units.

II. CARBONYL RADICAL-ANION SPIN-BEARING UNITS

Although many different synthetic routes had been developed to the non-Kekulé quinodimethanes [4,5] (Figure 1) and the non-Kekulé polyenes (Figure 2) [6–10], the non-Kekulé polynuclear aromatic family (Figure 3) has been largely unknown and unexplored. A number of ways in which they could be used as the basis of molecular magnets had been proposed theoretically [13,17], but synthetic entry to this group of compounds looked particularly difficult. The problem was overcome by making the carbonyl radical anion analogs **33**, **35**, and **37**, each of which can be obtained by a two-electron reduction of the corresponding dione, **32**, **34**, or **36** [3,11–13]. Triangulene **19** (Figure 3) possesses a degenerate coextensive pair of singly occupied molecular orbitals. The introduction of the three O⁻ substituents in the derivative **33** (Figure 5) preserves the threefold symmetry of the system, the degeneracy, and the coextensive nature of the singly occupied molecular orbitals [3]. However, this is a special case. In the analogous bis-O⁻ derivative **35** of hydrocarbon **20** and derivative **37** of hydrocarbon **21**, the presence of the two substituents lifts the degeneracy of the putative singly occupied molecular orbitals by 0.15β and 0.09β, respectively, at the simple Hückel level of theory. In this series it was found that the diradical **33** has a triplet ground state [3,11] but that **35** and **37** have singlet ground states [13].

Similar problems accompany attempts to create carbonyl radical-anion analogs of *meta*-benzoquinodimethane **1**. In compound **27** the presence of the substituents lifts the degeneracy of the putative singly occupied molec-

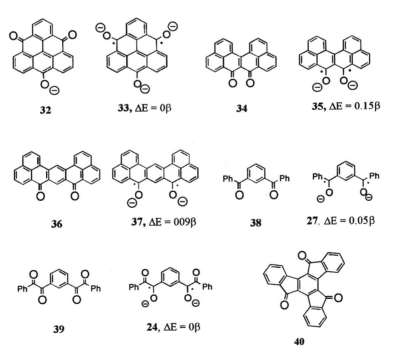

Figure 5   High-spin systems based on carbonyl radical-anion spin-bearing units.

ular orbitals by 0.05β. So far we have not observed triplet species by reduction of the diketone **38** [16,29], but Baumgartem et al. have obtained $S = 3/2$ and $S = 2$ ground state species from equivalent 4-*tert*-butyl and 2,4,6-trimethyl substituted systems [30–32]. In the case of the corresponding bis-α-dione product **24**, HMO calculations suggest a degenerate pair of singly occupied molecular orbitals that are orthogonal and coextensive [16]. Reduction of **39** gave a species showing a strong triplet EPR spectrum; but on cooling to liquid helium temperatures, the spectrum disappeared, suggesting that the triplet state was thermally populated and not the ground state. Despite the fact that the observed zero-field splitting was close to that expected for the desired dianion, it remains possible that this triplet spectrum was not due to this species but to some kind of ion pair aggregate: perhaps **39**$^{\cdot -}$, 2M$^+$, **39**$^{\cdot -}$ [16,29]. This kind of aggregation is particularly prevalent with carbonyl radical anions, and at least some of the "high-spin" electron spin resonance (ESR) spectra obtained on reduction of the trione **40** are due to this kind of aggregate [33].

Although exploitation of carbonyl radical-anion spin-bearing units provided the first entry to the non-Kekulé polynuclear aromatic family, in other ways this seems to be an unpromising approach. As all organic chemists who dry and degas THF using benzophenone radical anion know, these radicals are very thermally stable but also oxygen sensitive. Hence, although the triplet diradical **33** is indefinitely stable at room temperature under vacuum, it is instantly destroyed by atmospheric oxygen [3]. Furthermore, the strong tendency of carbonyl radical anions to form aggregates in which the spins couple weakly through space but in which antiferromagnetic coupling often (but not always [30–32]) dominates would make it very difficult to prepare a material with bulk ferromagnetic coupling using this approach.

III. ARYL RADICAL-ION SPIN-BEARING UNITS

Most studies of model systems of this type (Figure 6) have been based on aryl radical anions. Hence the dianion of the dinaphthyl derivative **29** [34] has a triplet ground state, as does the dianion of the triphenyl derivative **41** [16,35–38]. The dianion of the tris(biphenyl)-derivative **42** [16,35], like the dianion of the trinaphthyl derivative **43** [16,35], shows a strong triplet ESR spectrum, although the relative order of the singlet and triplet states is unknown. Unlike the dianion **41**, for which the zero-field splitting parameters are solvent- and counterion-dependent but for which the triplet is always the ground state [35–38], the ordering of the spin states for the equivalent triazine **44** is itself a function of the nature of the solvent and counterions

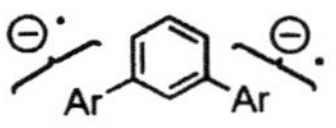

29, Ar = α-naphthyl, ΔE = 0.05β

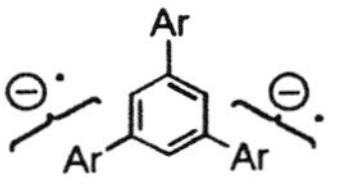

41, Ar = phenyl, ΔE = 0.05β
42, Ar = p-biphenyl, ΔE = 0.05β
43, Ar = α-naphthyl, ΔE = 0.05β

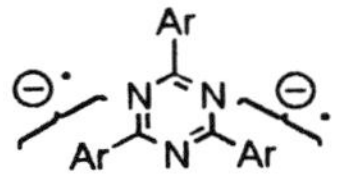

44, Ar = phenyl, ΔE = 0.05β

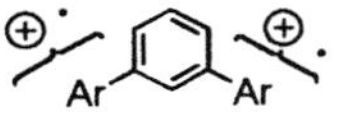

45, Ar = 9-anthryl, ΔE = 0.03β

46, Ar = 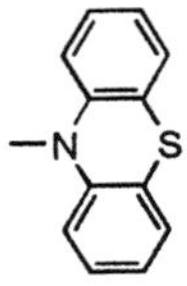   ΔE = 0.03β

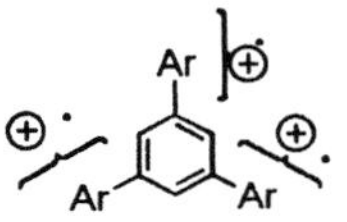

47, Ar =

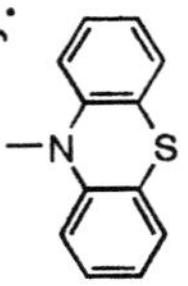

Figure 6 High-spin systems based on aryl radical-anion spin-bearing units.

employed [16,35,38]. The ground state of **44** in poorly solvating solvents is the singlet, and in strongly solvating solvents is the triplet.

In radical anions, electron–electron repulsion causes the electrons to be less strongly bound, and, since the counterions are also generally smaller, radical-anion-based species are more readily polarized than radical-cation-based species. Therefore, counterion sensitivity of the ordering of the spin states is likely to be more of a problem with radical ions.

There have been relatively few model studies of radical cations in this category. The dianthryl dication **45** has a triplet ground state. The equivalent heterocyclic systems **46** and **47** have singlet and doublet ground states, respectively. The loss of high-spin character in this case is attributed (at least in part) to their nonplanar nature [39].

Attempts to build high-spin polymer systems have exploited the radical cations. We have shown that the doped polymer **48** (Figure 7) gives magnetic behavior up to the equivalent of an $S = 1$ system [16,40], roughly equivalent to that shown by the pyrrole derivative **49** (Figure 8) reported by Nakazaki et al. [41]. Dougherty has made a variety of systems showing higher-spin

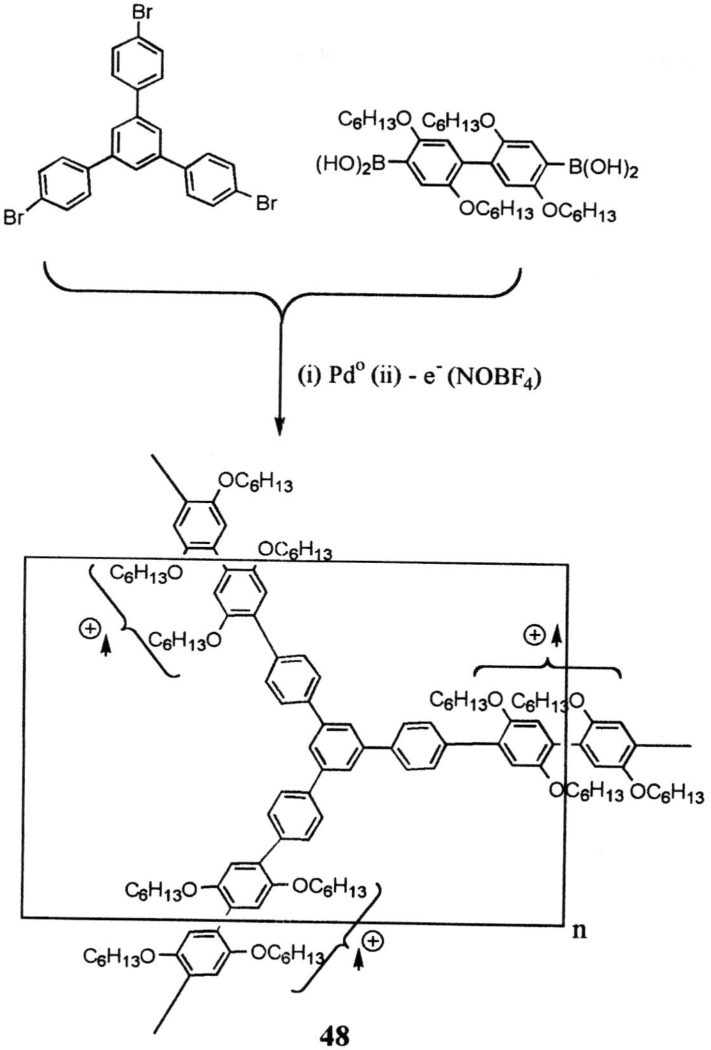

Figure 7 Route to the high-spin polymer **48** based on a tetramethoxybiphenyl radical-cation spin-bearing unit. The repeat structure is shown in the inner box. The extended structure is shown to indicate the essential ferromagnetic spin-coupling pathway. (From Refs. 16 and 40.)

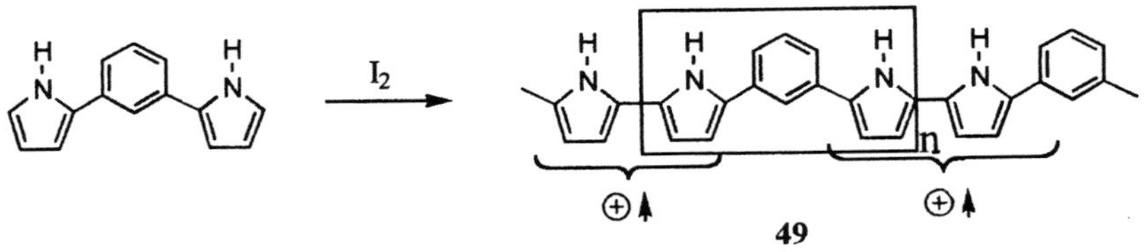

Figure 8 Route to the high-spin polymer **49** based on a dipyrrole radical-cation spin-bearing unit. The repeat structure is shown in the inner box. The extended structure is shown to indicate the essential ferromagnetic spin-coupling pathway. (From Ref. 41.)

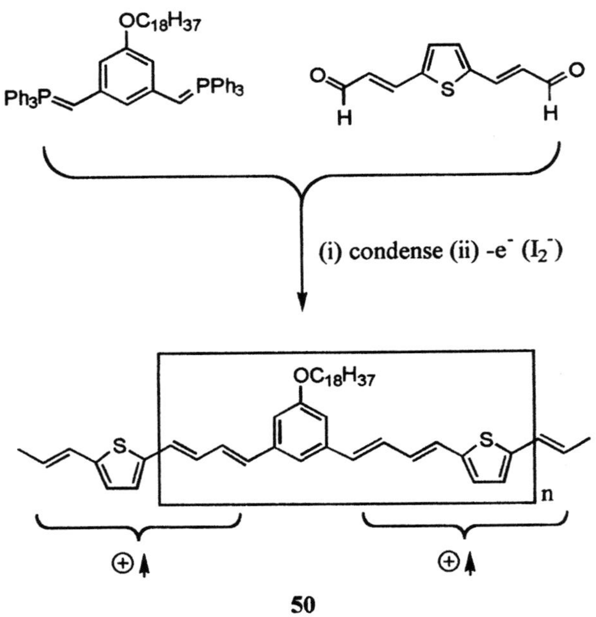

Figure 9 Route to the high-spin polymer **50** based on a dibutadienylthiophene radical-cation spin-bearing unit. The repeat structure is shown in the inner box. The extended structure is shown to indicate the essential ferromagnetic spin-coupling pathway. (From Ref. 42.)

behavior [42]. Polymer **50** (Figure 9), for example, is reported to show behavior up to the equivalent of $S = 4.5$ [42].

IV. AMMINIUM CATION SPIN-BEARING UNITS

Replacement of neutral methylene spin-bearing units (CH_2^-) with cationic amminium spin-bearing ($N^{\cdot+}$) centers gives isoelectronic structures but lifts the degeneracy of the putative SOMOs. Hence, in the hydrocarbon **23** (Figure 4) these are strictly degenerate, but in the analogous nitrogen-containing system **26** they are split by 0.05β. Nonetheless, the spin states in the amminium cation series seem to follow those of the corresponding hydrocarbons. Hence, the dications **51** [43] and **52** [44,45] (Figure 10) both have triplet ground states, and the trications **53** [46–48] and **54** [44,45] both have quartet ground states. These systems are based on a *meta*-benzoquinodimethane **1** motif. The radical cations **55** and **56** are based on the dimethylenebiphenyls **7** and **8** [49]. They illustrate the fact that not quite all of the diradicals shown in Figures 1–3 can be used as building blocks, but only those that are SOMO-coextensive. The radical cation **55** is based on the disjoint hydrocarbon **7**. Its SOMOs are spatially separated (Ψ_7 and Ψ_8 in Figure 10) and hence interact weakly. Its triplet ESR spectrum disappears on cooling to liquid helium temperatures, indicating that it is not a triplet ground state species. In such diradical-dication species, the spin–spin and hole–hole interactions parallel each other, and cyclic voltammetry studies of the dependence of the oxidation potentials on the dielectric constant of the solvent show that the hole–hole repulsion is also relatively weak. However, the isomeric radical cation **56** is based on the coextensive hydrocarbon **8**. Its SOMOs are orthogonal but overlap in their spatial distributions (Ψ_7 and Ψ_8 in Figure 10). It has a triplet ground state and a relatively large hole–hole (electrostatic) repulsion [49]. The stability of amminium radical cations is highly structure dependent. Hence tris(*o*-bromophenyl)amminium pentachloroantimonate is a commercially available stable solid that is used as a one-electron oxidant. However, the stabilities of some amminium cations investigated have proved to be disappointing [50]. This has led to the synthesis of even more stable radical cation species based on a *p*-phenylene diamine radical-cation spin-bearing unit. Hence, the dication **57** has a triplet ground state [51,52] and, most remarkably, the trication **58** is an isolable solid that has a quartet ground state [53].

Attempts to create ferromagnetic polymers based on amminium-ion spin-bearing units have a long history and go back to the pioneering work of Torrance et al. [54]. More recently, Yoshizawa et al. have developed a

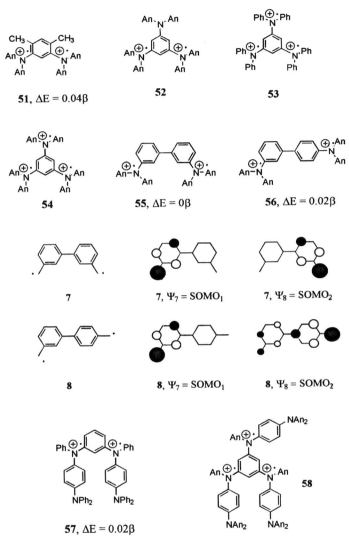

Figure 10 High-spin systems based on amminium radical-cation spin-bearing units. Ph = phenyl (C_6H_5). An = anisyl (p-$CH_3OC_6H_4$).

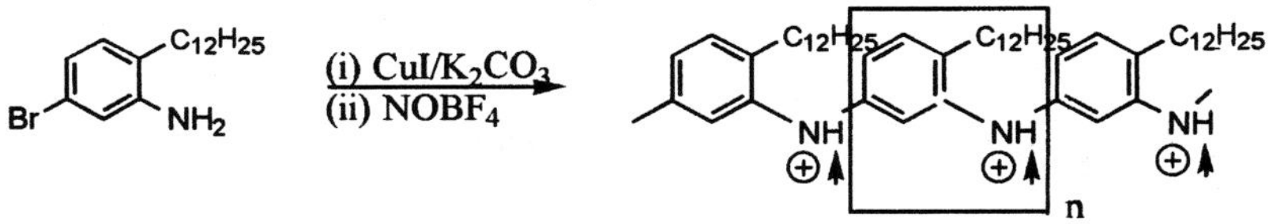

Figure 11 Route to the high-spin polymer **59** based on an alkylmetaphenylene-diamine radical-cation spin-bearing unit. The repeat structure is shown in the inner box. The extended structure is shown to indicate the essential ferromagnetic spin-coupling pathway. (From Refs. 53–58.)

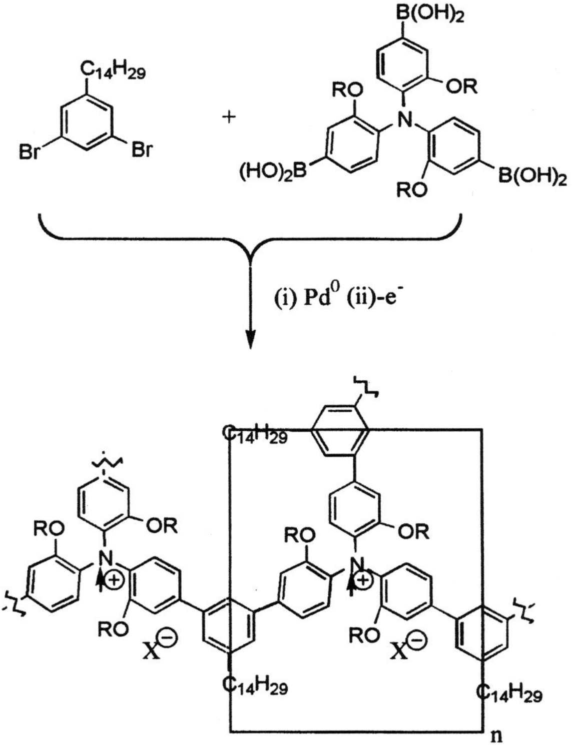

60a R = C_6H_{13}
60b R = C_4H_9

Figure 12 Route to the high-spin polymer **60** based on an extended metapheny-lenediamine-cation spin-bearing unit. The repeat structure is shown in the inner box. The extended structure is shown to indicate the essential ferromagnetic spin-coupling pathway. (From Refs. 16, 40, and 59–62.)

rational poly-*meta*-aniline synthesis (Figure 11). The doped polymer **59** shows high-spin behavior up to the equivalent of $S = 1$ [55–58].

We have also made a range of polymers in which the *meta*-phenylenediamine motif is exploited to create π-systems containing clusters of ferromagnetically coupled spins [16,40,59–61]. For example, when the polymer **60a** (Figure 12) is oxidatively doped in dichloromethane solution using NOBF$_4$, ca. 15% of the nitrogens are oxidized to the amminium-ion level. Nutation resonance spectroscopy shows the presence of spin clusters from $S = 1$ to $S = 3$ [59,60]. A Brillouin-function fit of the field dependence of the magnetization of this doped polymer at 2 K corresponds to behavior equivalent to an average $S = 2.5$ [16,40]. Clearly, if a higher proportion of the sites could be doped, higher-spin behavior would result. In theory, the size of the largest spin cluster should show a percolation limit with explosive growth beyond ca. 75% doping. This has not yet been achieved, but some improvements have been made. Hence, when a thin film (ca. 1 μ thick) of the polymer **60b** was exposed to gaseous antimony pentachloride ca. 40–60% of the potentially dopable sites were oxidized to the amminium ion ($N^{\bullet+}$) level. A Brillouin-function fit of the field dependence of the magnetization of this doped polymer at 2 K corresponds to average behavior equivalent to a nonet ($S = 4$) π-multiradical [61]. This remains the current limit, but it will certainly prove possible to design systems that can be doped to a higher level [62].

REFERENCES

1. Mataga N. Possible ferromagnetic states of some hypothetical hydrocarbons. Theoret Chim Acta. 1968; 10:372–376.
2. Hughbanks T, Yee KA. Superdegeneracies and orbital delocalisation in extended organic systems. In: Gatteschi D, Kahn O, Miller JS, Palacio F, eds. Magnetic Molecular Materials. Dordrecht: Kluwer, 1990, pp 133–144.
3. Allinson G, Bushby RJ, Paillaud J-L. Organic molecular magnets—The search for stable building blocks. J Mater Sci Mater Electronics. 1994; 5:83–88.
4. Platz MS. Quinodimethanes and related diradicals. In: Borden WT, ed. Diradicals. New York: Wiley, 1982, pp 195–258.
5. Berson JA. *m*-Qinonoid compounds. In: Pati S, Rappoport Z, eds. The Chemistry of the Quinonoid Compounds. Vol. 2. Part 1. New York: Wiley, 1988, pp 455–536.
6. Dowd P. Trimethylenemethane. J Am Chem Soc. 1966; 88:2587–2589.
7. Dowd P. Tetramethyleneethane. J Am Chem Soc. 1970; 92:1066–1068.
8. Berson JA, Bushby RJ, McBride JM, Tremelling M. 2-Isopropylidenecyclopentane-1,3-diyl. Preparation, properties and reactions of a distorted trimethylenemetane. Direct evidence of a triplet reaction. J Am Chem Soc. 1971; 93: 1544–1546.

9. Bushby RJ, Jarecki C. Making a vinyl-trimethylenemethane precursor through the addition of diethyl azodicarboxylate to tropone. J Chem Soc Perkin Trans 1. 1990:2335–2343.

10. Bushby RJ, Jarecki C. Generation of eight-π non-Kekulé polyenes through the $2\pi + 4\pi$ diethyl azodicarboxylate + tropone cycloadduct. J Chem Soc Perkin Trans 1. 1992:2215–2222.

11. Allinson G, Bushby RJ, Paillaud J-L, Oduwole D, Sales K. ESR spectrum of a stable triplet π biradical: Trioxytriangulene. J Am Chem Soc. 1993; 115: 2062–2064.

12. Allinson G, Bushby RJ, Paillaud J-L, Thornton-Pett M. Synthesis of a derivative of triangulene: The first non-Kekulé polynuclear aromatic. J Chem Soc Perkin Trans 1. 1995:385–390.

13. Allinson G, Bushby RJ, Jesudason MV, Paillaud J-L, Taylor N. The synthesis of singlet ground state derivatives of non-Kekulé polynuclear aromatics. J Chem Soc Perkin Trans 2. 1997:147–156.

14. Dougherty DA. Spin control in organic molecules. Acc Chem Res. 1991; 24: 88–94.

15. Borden WT, Davidson ER. Effects of electron repulsion in conjugated hydrocarbon diradicals. J Am Chem Soc. 1977; 99:4587–4594.

16. Bushby RJ, McGill DR, Ng KM. Organic magnetic polymers. In: Kahn O, ed. Magnetism a Supramolecular Function. Dordrecht: Kluwer, 1996, pp 181–204.

17. Ovchinnikov AA. Multiplicity of the ground states of large alternant organic molecules with conjugated bonds. Theor Chim Acta. 1978; 47:297–304.

18. McConnell HM. Ferromagnetism in solid free radicals. J Chem Phys. 1963; 39:1910. McConnell HM. In printed discussions at the end (pp 144–145) of Mulliken RS. Electron-donor acceptor interactions and charge transfer spectra. Proc R A Welch Found Chem Res. 1967; 11:109–143.

19. Mazur MR, Berson JA. Formal kinetic proof of reversible unimolecular transformation to a biradical as the obligatory first step in the mechanism of cycloaddition of 5-isopropylidenebicyclo[2.1.0]pentane to olefins. J Am Chem Soc. 1982; 104:2217–2222.

20. Wenthold PG, Kim JB, Lineberger WC. J Am Chem Soc. 1997; 119:1354.

21. Kanno F, Inoue K, Koga N, Iwamura H. 2,4-Dimethoxy-1,3-phenylenebis (*N*-*tert*-butyl nitroxide) with a singlet ground state. Formal violation of the rule that *m*-phenylene serves as a robust ferromagnetic coupling unit. J Am Chem Soc. 1993; 115:847–850.

22. West AP, Silverman SK, Dougherty DA. Do high-spin topology rules apply to charged polyradicals? Theoretical and experimental evaluation of pyridiniums as magnetic coupling units. J Am Chem Soc. 1996; 118:1452–1463.

23. Rajca A, Rajca S, Desai SR. Boron-centred diradical dianion: a new triplet state molecule. Chem Commun. 1995:1959–1960.

24. Wienk MM, Janssen RAJ, Meijer EW. Triplet-state phosphoryl biradicals. Syn Met. 1995; 71:1833–1834.

25. Weink MM, Janssen RAJ. High-spin cation radicals of methyenephosphoranes. J Am Chem Soc. 1997; 119:5398–5403.

26. Rajca A. Organic diradicals and polyradicals. From spin coupling to magnetism? Chem Rev. 1994; 94:871–893.

27. Chiarelli R, Novak MA, Rassat A, Tholence JL. A ferromagnetic transition at 1.48 K in an organic nitroxide. Nature. 1993; 363:147–149.

28. Tamura M, Nakazawa Y, Shiomi D, Nozawa K, Hosokoshi Y, Ishikawa M, Takahashi M, Kinoshita M. Bulk ferromagnetism in the β-crystal of the *p*-nitrophenyl nitronyl nitroxide radical. Chem Phys Lett. 1991; 186:401–404.

29. Halliwell C. Polaronic Magnets MS thesis, University of Leeds, 1994.

30. Baumgarten M. Novel oligoradicals and high spin formation. Mol Cryst Liq Cryst. 1995; 272:109–116.

31. Baumgarten M, Gherghel L, Wehrmeister T. Novel high-spin states in reduced dibenzoylbenzenes. Chem Phys Lett. 1997; 267:175–178.

32. Baumgarten M. Control of conjugation and high-spin formation in radical ions of extended π-systems. Acta Chem Scand. 1997; 51:193–198.

33. Celina M, Lazana RLR, Luisa M, Franco TMB, Shohoji MCBL. Electron paramagnetic resonance characterization of high spin states of tribenzoylenebenzene radical ions. J Chem Res S. 1996:48–49.

34. Tukada H. Diradical dianions of *m*-bis(naphthyl) and (anthryl)phenylenes as new high-spin molecules. Chem Commun. 1994:2293–2294.

35. McGill DR. The exploration of novel routes to molecular magnets. PhD dissertations, University of Leeds, 1997.

36. Jesse RE, Biloen P, Prins R, van Voorst JDW, Hoijtink GJ. Hydrocarbon ions with triplet ground states. Mol Phys. 1963; 6:633–635.

37. van Brockhoven JAM, van Willigen H, de Boer E. Counterion effects on the spin density distribution in triplet aromatic dianions. Mol Phys. 1968; 15:101–103.

38. van Broekhoven JAM, Sommerdijk JL, de Boer E. An ESR study of the triplet dianions of 1,3,5-triphenylbenzene and 2,4,6-sym-triazine. Mol Phys. 1971; 20:993–1003.

39. Okad K, Imakura Y, Oda M, Kajiwara A, Kamachi M, Sato K, Shiomi D, Takui T, Itoh K, Gherghel L, Baumgarten M. Invalidity of simple application of spin-prediction topology rule to heteroatomic π-congugated systems. Mol Cryst Liq Cryst. 1997; 306:301–308.

40. Bushby RJ, McGill DR, Ng KM, Taylor N. *p*-Doped high-spin polymers. J Mater Chem. 1997; 7:2343–2354.

41. Nakazaki J, Matsushita MM, Izuoka A, Sugawara T. Ground state spin multiplicity of cation diradicals derived from pyrroles carrying nitronyl nitroxide. Mol Cryst Liq Cryst. 1997; 306:81–88.

42. Murray MA, Kaszynski P, Kasisaki DA, Change W-H, Dougherty DA. Prototypes for the polaronic ferromagnet. Synthesis and characterization of high-spin organic polymers. J Am Chem Soc. 1994; 116:8152–8161.

43. Sato K, Yano M, Furuichi M, Shiomi D, Takui T, Abe K, Itoh K, Higuchi A, Katsuhiko K, Shirota Y. Polycation high-spin states of one- and two-dimensional (diarylamino)benzenes, prototypical model units for purely organic ferromagnets as studied by pulsed ESR/electron spin transient nutation spectroscopy. J Am Chem Soc. 1997; 119:6607–6613.

44. Stickley KR, Blackstock SC. Triplet dication and quartet trication of a tri-aminobenzene. J Am Chem Soc. 1994; 116:11576–11577.
45. Yoshizawa K, Hatanaka M, Ago H, Tanaka K, Yamabe T. Magnetic properties of 1,3,5-tris[bis(p-methoxyphenyl)amino]benzene cation radicals. Bull Chem Soc Japan. 1996; 69:1417–1422.
46. Yoshizawa K, Chano A, Ito A, Tanaka K, Yamabe T, Fujita H, Yamauchi J. Electron spin resonance of the quartet state of 1,3,5-tris(diphenylamino)benzene. Chem Lett. 1992:369–372.
47. Yoshizawa K, Chano A, Ito A, Tanaka K, Yamabe T, Fujita H, Yamauchi J, Shiro M. ESR of the cationic triradical of 1,3,5-tris(diphenylamino)benzene. J Am Chem Soc. 1992; 114:5994–5998.
48. Yoshizawa K. Molecular orbital study of quartet molecules with trigonal axis of symmetry. Mol Cryst Liq Cryst. 1993; 233:323–332.
49. Bushby RJ, McGill DR, Ng KM, Taylor N. Disjoint and coextensive diradical diions. J Chem Soc Perkin Trans 2. 1997:1405–1414.
50. Stickley KR, Blackstock SC. Cation radicals of 1,3,5-tris(diarylamino)benzene. Tet Lett. 1995; 36:1585–1588.
51. Wienk MM, Janssen RAJ. Stable triplet state of di(cation radicals) of an N-phenylaniline oligomer. Chem Commun. 1996:267–268.
52. Wienk MM, Janssen RAJ. Stable triplet state di(cation radicals) of a meta-para aniline oligomer by acid doping. J Amer Chem Soc. 1996; 118:10626–10628.
53. Stickley KR, Selby TD, Blackstock SC. Isolable polyradical cations of poly-phenylenediamines with populated high-spin states. J Am Chem Soc. 1997; 62:448–449.
54. Torrance JB, Oostra S, Nazzal A. A new simple model for organic ferromag-netism and the first organic ferromagnet. Syn Met. 1987; 19:709–714.
55. Yoshizawa K, Tanaka K, Yamabe T. Synthesis of poly(m-aniline) by dehydro-halogenation of m-chloroaniline. Chem Lett. 1990:1311–1314.
56. Ito A, Saito T, Tanaka K, Yamabe T. Syntheis of oligo(m-aniline). Tet Lett. 1995; 36:8809–8812.
57. Yoshizawa K, Tanaka K, Yamabe T, Yumauchi J. Ferromagnetic interactions in poly(m-aniline): Electron spin resonance and magnetic susceptibility. J Chem Phys. 1992; 96:5516–5524.
58. Ito A, Ota K, Tanaka K, Yamabe T, Yoshizawa K. n-Alkyl group-substituted poly(m-aniline)s: Synthesis and magnetic properties. Macromolecules. 1995; 28:5618–5623.
59. Shiomi D, Sato K, Takui T, Itoh K, McGill DR, Ng KM, Bushby RJ. FT pulsed-ESR/electron spin transient nutation of hyperbranched polycationic or-ganic high-spin polymers. Syn Met. 1997; 85:1721–1722.
60. Shiomi D, Sato K, Takui T, Itoh K, McGill DR, Ng KM, Bushby RJ. High-spin states of hyperbranched polycationic organic polymers as studied by FT pulsed-ESR/electron spin transient nutation. Mol Cryst Liq Cryst. 1997; 306: 513–520.
61. Bushby RJ, Gooding D. Higher-spin pi multiradical sites in doped polyaryl-amine polymers. J Chem Soc Perkin Trans 2. 1998; 1067–1075.
62. Bushby RJ, McGill DR, Ng KM, Taylor N. Coulombic effects in radical cation based high-spin polymers. Chem Commun. 1996:2641–2642.

11

Continuous-Wave ESR and Fourier-Transform-Pulsed Electron Spin Transient Nutation Spectroscopy in Molecule-Based Magnetics and Molecular Spinics: Theory and Applications

Takeji Takui, Kazunobu Sato, Daisuke Shiomi, and Koichi Itoh
Osaka City University, Osaka, Japan

I. INTRODUCTION

The last decade has seen organic-molecule-based magnetics [1–10] become a rapidly growing interdisciplinary field in the pure and applied natural sciences [11–23]. This is not only due to the rich variety of novel physical phenomena and properties that synthetic-molecule-based materials are anticipated to exhibit (both macro- and mesoscopically), but also due to their underlying potential applications as future molecular-device technology in molecular quantum materials science. These hopes are based on both their anticipated multiple supramolecular functionality and "system" properties [18–23]. Spin-mediated electronics, which we termed "spinics," has become an important issue in the interdisciplinary field between physics and chemistry [18–21].

Among the diverse topics of organic-molecule-based magnetic materials, molecular high-spin chemistry has continued to be an underlying theme. Impressive progress has been made in molecular design and syntheses—both organic and inorganic—and has led to the realization of

197

various types of organic-molecule-based magnets [11–17]. Efforts to synthesize low-dimensional (zero- to two-dimensional) molecular magnetic systems include extremely large spins (superparamagnets and super high-spin polymeric systems). These are motivated not only by the device application possibilities of novel "systems" or "soft" magnetism, but also by pure scientific interest in the spin manipulation chemistry and modulated spin structures of molecular orbital (crystal orbital) bands in neutral or ionic polymeric open-shell systems. Interest around the world continues to grow and to encourage us in the research of such low-dimensional molecular magnetic systems. More controlled attempts to generate extremely high-spin ground states of neutral or charged (polycationic or polyanionic) organic molecular systems continue, while extremely high-spin large clusters composed of transition-metal ions have also emerged [11–17,24]. Molecular spinics has thus become a focus of attention in molecular spin science, with relation to spin-mediated quantum tunneling as well as magnetic quantum well effects.

Continuous-wave electron spin resonance/electron nuclear multiple resonance (cw-electron magnetic resonance: cw-EMR) has been used to explore microscopic details of high-spin molecules and molecular spin assemblies, focusing on electronic spin/molecular structural analyses and structure-magnetism relationship in solids. This chapter presents an overview of recent developments in the rapidly growing interdisciplinary area of organic/molecular magnetics in terms of cw and Fourier-transform (FT)-pulsed EMR spectroscopy. The main emphasis of this work is the spectral analysis of random-orientation fine-structure ESR spectroscopy, which is frequently invoked to identify molecular high-spin states and with high sensitivity and to determine fine-structure parameters with high precision. Some important recipes and guidelines for spectral simulation procedures—rather than merely tedious mathematical expressions—are given and exemplified graphically. This chapter discusses the potential capability of the high-field/high-frequency (W-band) EMR spectroscopy in comparison to X-band spectroscopy. This chapter also deals with late-breaking results from FT-pulsed electron spin resonance/electron spin transient nutation (ESR/ESTN) spectroscopy applied to high-spin systems.

Electron nuclear multiple resonance spectroscopy gives us valuable information about spin-alignment mechanisms in terms of spin-density distributions of molecular high-spin systems. Nevertheless, multiple resonance spectroscopy is not included in this chapter because of space restrictions. In ordered regimes, electron magnetic resonance spectroscopy can probe many important aspects concerning low-dimensional magnetics inherent in organic-molecule-based materials. Related issues are only briefly discussed here, including super-high-spin magnetics and the possible occurrence of radiative cooperative phenomena in terms of dynamic magnetizations.

II. PRACTICAL IMPORTANCE OF NONORIENTED FINE-STRUCTURE ESR SPECTROSCOPY IN HIGH-SPIN MOLECULAR SCIENCE

In order to characterize the spin structures of open-shell components or building units of molecular magnetic systems and assemblages in microscopic details, fine-structure (FS) ESR spectroscopy from random orientation or in nonoriented media is a powerful spectroscopic technique and has been frequently invoked. Successful analyses of FS spectra can intensify investigations a great deal. Nonetheless, some "fine-structure" ESR spectra from high-spin origins appear to be structureless single-absorption peaks, which are subject to inhomogeneous line-broadening or exchange-narrowing processes, while others feature complex and apparently bizarre lineshapes. Thus, a facile spectroscopic method and a practical, easy-to-access interpretive approach (in addition to well-established ones) for high-spin identification are required to stimulate work in this particular research area.

The aim of this chapter is to carry readers who do not specialize in EMR spectroscopy from a sort of qualitative understanding of high-spin FS-ESR spectra, to the stage where they are able to extract spin Hamiltonian parameters from observed FS spectra with the help of spectral simulation or directly without simulation procedures. The spin Hamiltonian parameters include spin quantum numbers (S's), g-values, zero-field splitting (zfs) parameters (D and E values), hyperfine splitting (hfs) parameters, and nuclear electric quadrupole splitting parameters, as follows:

$$H = \sum_i \left\{ \beta \tilde{\mathbf{B}} \cdot \mathbf{g}_i \cdot \mathbf{S}_i + \tilde{\mathbf{S}}_i \cdot \mathbf{D}_i \cdot \mathbf{S}_i + J_{il} \mathbf{S}_i \cdot \mathbf{S}_l + \sum_j \left[\tilde{\mathbf{I}}_j \cdot \mathbf{A}_j \cdot \mathbf{S}_i \right. \right.$$

$$\left. \left. - g_I \beta_N \tilde{\mathbf{I}}_j \cdot \mathbf{B} + \tilde{\mathbf{I}}_j \cdot \mathbf{P}_j \cdot \mathbf{I}_j + \sum_k \tilde{\mathbf{I}}_j \cdot \mathbf{J}_{jk} \cdot \mathbf{I}_k \right] \right\} \qquad (k \neq j, \, l \neq i) \qquad (1)$$

where group-theoretically allowed quartic or higher-order terms of even numbers such as BS_m^3, $S_m^2 S_n^2$, and $S_m^3 S_n^3$ are neglected. It should be noticed that these terms take part in the interpretation of FS spectra if S becomes enormous and the spin–orbit interaction becomes significantly large. A simulation procedure is not necessarily required in order to interpret observed FS spectra and to extract, with high accuracy, the first four parameters mentioned, as will be described later. For ESR-allowed transitions ($\Delta M_S = \pm 1$

and $\Delta M_I = 0$), the other terms contribute to second- and higher-order corrections in terms of perturbation theory.

Due to space limitations, this chapter does not deal with single-crystal EMR spectroscopy, but presents only random-orientation EMR spectroscopy. In general, random-orientation spectroscopy is easy to access from the experimental side, but the spectra are not readily interpretable from the theoretical side. This is particularly the case for random-orientation spectra in noncrystalline media such as organic rigid glasses. We review mostly cw FS-ESR spectroscopy with random orientations, exemplifying conventional cw X-band (~9 GHz) FS spectra obtained both theoretically and experimentally. At the same time, we attempt to present cw W-band (~95 GHz) FS spectra for comparison's sake, illustrating what makes the X-band FS spectra complex or difficult of interpretation.

Also, we present recent breakthrough results from FT-pulsed ESR/ESTN spectroscopy applied to high-spin systems, showing the possibilities this new method augurs in high-spin molecular science. This methodological advance could further intensify the study of high-spin systems, organic or inorganic, and their quantum-constrained effects in low-dimensional molecular magnetic systems, semimacroscopically reduced dimensional magnetic systems, and charged high-spin polymeric systems created through laser excitation.

III. APPROACHES TO RANDOM-ORIENTATION X-BAND FS SPECTRAL ANALYSES

III.A. Exchange-Coupled Systems and Quantum Spin Mixing

If the definition is broadly based, there are two kinds of molecular FS spectra. One arises from noninteracting high-spins S and their states that do not have energetically nearby electronic states. In this case S is a good quantum number; thus, no quantum spin mixing takes place. The other case involves the exchange coupling of spins, such that the resulting ESR transition fields and probabilities are interrelated between different spin states. Readers can refer to a comprehensive treatise on this issue by Bencini and Gatteschi [25].

Figure 1 exemplifies the quantum spin mixing in FS-ESR spectra, showing typical FS-ESR spectra at K-band (~25 GHz) calculated for an exchange-coupled triplet pair in a static magnetic field $\mathbf{B}_0$ along a given orientation, as described by the FS spin Hamiltonian (equation 1 where i = 1, 2 and $\mathbf{D}_1 = \mathbf{D}_2$). As the spin quantum mixing grows, new transitions arise with intensity borrowing, and the shifting of transition fields becomes appreciable. In the case of complete mixing, the salient spectroscopic features

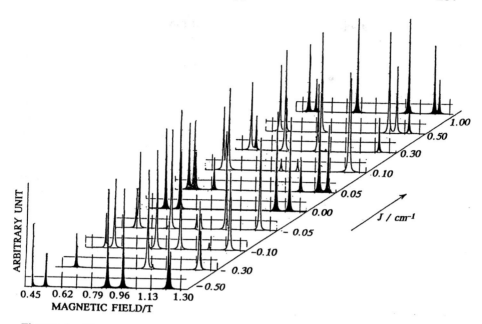

Figure 1 Fine-structure spectra arising from quantum spin mixing in a weakly interacting triplet pair.

arising from a quintet state and a triplet state disappear, giving the appearance of *only* two triplet FS spectra, even though there are not two independent triplets. In this example, singlet-quintet complete mixing takes place due to the group-theoretical symmetry requirement of a pair of the equivalent $S_i = 1$ spins ($i = 1, 2$), while the triplet state is isolated because of symmetry requirements. The permutation symmetry of Bose ($S_i = 1$) particles is symmetric, preventing mixing between the triplet and the other spin multiplets.

   Thus, a drastic spectral change from the "typical case" is anticipated when complete spin quantum mixing occurs with a nearly vanishing exchange term J_{il}, as illustrated in Figure 1. In X-band FS-ESR spectroscopy, the transition probabilities that gain intensity due to the good high-field approximation in high-frequency ESR spectroscopy are reduced a great deal, and, as a result, the intensity distribution characteristic of the high-field approximation with high-spin states is destroyed. In this sense, the practical advantages of X-band ESR spectroscopy are limited for cases of high quantum spin mixing. High-field/high-frequency ESR spectroscopy is desirable for the detection of spin quantum mixing in terms of the transition probability of intermediate spin states. For these reasons, molecular design and fine-tuning of the quantum spin mixing with potential device applications are the current topics of molecular magnetic science.

III.B. Salient Features of FS Spectra from Noninteracting High Spins and Spectral Simulation

Fine-structure ESR spectral analysis methods for isolated molecular triplet states are well established. Assuming that the zfs parameters not larger than the microwave transition energy ($h\nu$) employed at X-band and that small g-anisotropy is present, there appear six group-theoretically allowed resonance peaks (singularities of absorption intensity) arising from canonical orientations where resonance fields correspond to B_0 oriented along the principal axes (X, Y, and Z) of the $\mathbf{D}$ tensor. Thus, the number of the allowed transitions is given by $2S \times 3(X, Y, Z)$ for an arbitrary spin, S. For $|\mathbf{D}| \ll h\nu$, these absorption peaks occur in the $g \sim 2$ region. In addition, peaks due to group-theoretically forbidden transitions can appear at low magnetic fields, such that the maximum total number of possible peaks for an arbitrary S is given by $S(2S - 1) \times 3(X, Y, Z)$.

III.B.1. Inevitable Appearance of Off-Principal-Axis Extra Lines

In general, lineshapes of the forbidden transition peaks are anomalous, compared with those of the allowed ones. This is because the angular dependence (anisotropy) of forbidden transitions is constrained and thus subject to angular anomaly, which corresponds to stationary points with B_0 oriented off the principal axis. A typical example is $B_{\min}$ appearing in the FS spectra from triplet states. The resonance field of $B_{\min}$ is given by

$$B \geq B_{\min} = \frac{1}{2g\beta} \sqrt{(h\nu)^2 + 4(XY + YZ + ZX)}$$

$$= \frac{1}{2g\beta} \sqrt{(h\nu)^2 - \frac{4}{3}(D^2 + 3E^2)} \tag{2}$$

$$X \sin^2\theta \cos^2\phi + Y \sin^2\theta \sin^2\phi + Z \cos^2\theta = XYZ/(g\beta B_{\min})^2 \tag{3}$$

where

$$-X = \frac{D}{3} + E$$

$$-Y = \frac{D}{3} - E$$

$$-Z = \frac{-2D}{3}$$

and $(\theta, \phi)_{min}$ is calculated by substituting B_{min} obtained from Eq. (2) into Eq. (3).

The B_{min} anomaly underlies the stationary behavior of the off-principal-axis orientation in FS spectra from high spins $(S_i > 1)$ [26,27]. In order to understand straightforwardly the anomalous behavior of the forbidden transition $|M_S\rangle \leftrightarrow |M_S + 2\rangle$, the general expression for the resonance field (B_0) and transition probability $(I_{M_S M_{S+2}})$ in an arbitrary coordinate-axis system are given by the following:

$$hv = 2g\beta B_0 - 6D_0(M_S + 1) - \frac{1}{2g\beta B_0}\left\{2|D_1|[4S(S + 1)\right.$$

$$\left. - 8M_S(M_S + 1) - 33] - \frac{1}{2}|D_2|^2[2S(S + 1) - 2M_S(M_S + 1) - 9]\right\} \quad (4)$$

with

$$D_0 = -(\tilde{\mathbf{u}} \cdot \mathbf{D} \cdot \mathbf{u})$$
$$|D_1|^2 = (\tilde{\mathbf{u}} \cdot \mathbf{D} \cdot \mathbf{D} \cdot \mathbf{u}) - (\tilde{\mathbf{u}} \cdot \mathbf{D} \cdot \mathbf{u})^2$$
$$|D_2|^2 = 2Tr(\mathbf{D} \cdot \mathbf{D}) + (\tilde{\mathbf{u}} \cdot \mathbf{D} \cdot \mathbf{u})^2 - 4(\tilde{\mathbf{u}} \cdot \mathbf{D} \cdot \mathbf{D} \cdot \mathbf{u})$$

and

$$I_{M_S M_{S+2}} \propto (2g\beta B_1)^2 |G|^2 |\langle M_S|S_i|M_S + 2\rangle|^2 \cdot |\langle M_I(M_S)|M_I'(M_S + 2)\rangle|^2 \quad (5)$$

with

$$|\langle M_S|S_i|M_S + 2\rangle|^2 = \left[\frac{|D_1|^2}{(2g\beta B_0)^2}\right]$$
$$\times [(S + 1)^2 - (M_S + 1)^2] \times [S^2 - (M_S + 1)^2]$$

$$|G|^2 = (\tilde{\mathbf{h}}_1 \cdot \tilde{\mathbf{g}} \cdot \mathbf{g} \cdot \mathbf{h}_1) - \frac{(\tilde{\mathbf{h}} \cdot \tilde{\mathbf{g}} \cdot \mathbf{g} \cdot \mathbf{h}_1)^2}{g^2}$$

$$g^2 = \tilde{\mathbf{h}} \cdot \tilde{\mathbf{g}} \cdot \mathbf{g} \cdot \mathbf{h}$$

$$\mathbf{u} = \frac{\mathbf{g} \cdot \mathbf{h}}{g}$$

$$\mathbf{h} = \frac{\mathbf{B}_0}{B_0}$$

$$\mathbf{h}_1 = \frac{\mathbf{B}_1}{B_1}$$

where $\mathbf{u}$ stands for a unit vector, S_i is the component of S along the unit vector $\mathbf{i}$. For these equations, $\mathbf{i} \perp \mathbf{u}$, $\mathbf{B}_1$ is the magnetic field of microwave irradiation, hv stands for the microwave transition energy, and $M_S = S - 2$,

$S - 3, \ldots, -S$. Additional peaks due to angular anomaly corresponding to off-principal-axis orientations ($\mathbf{B}_0//X$, Y, or Z) are called, in general, off-axis extra lines (or simply extra lines). The first identified hfs extra line was observed in the spectrum of copper(II)phthalocyanine in an H_2SO_4 glass [28]. Physical origins of hfs extra lines have been elucidated by several authors [29,30].

Fine-structure extra lines have been studied extensively in terms of higher-order perturbation treatments [27]. Angular anomaly is not observed for the allowed transitions of the triplet state within the framework of the second-order perturbation treatment [27]. The appearance of extra line features in FS-ESR spectra from high spins larger than $S_i = 1$ inevitably are due to half-integer spins. In terms of perturbation theory, the angular anomaly originates in second- and higher-order corrections for resonance fields. For half-integer spins, first-order terms of the resonance field corresponding to an $M_S = -1/2 \leftrightarrow M_S = 1/2$ transition are vanishing, leading to a condition where the second- and higher-order terms dominate, as described in the following;

$$h\nu = E(M_S + 1, M_I) - E(M_S, M_I)$$

$$= g\beta B_0 + \frac{3}{2}(\tilde{\mathbf{u}} \cdot \mathbf{D} \cdot \mathbf{u})(2M_S + 1) + [K(M_S + 1)M_I(M_S + 1)$$

$$- K(M_S)M_I(M_S)] + [\text{second- and higher-order terms}] \qquad (6)$$

with

$$K^2(M_S) = \tilde{\mathbf{h}} \cdot \tilde{\mathbf{K}}(M_S) \cdot \mathbf{K}(M_S) \cdot \mathbf{h}$$

$$\mathbf{K}(M_S) = \left(\frac{\tilde{\mathbf{A}} \cdot \mathbf{g}}{g}\right) M_S - g_I \beta_N B_0 \mathbf{E}$$

where B_0 stands for the resonance field of the ESR allowed $|M_S, M_I\rangle \leftrightarrow |M_S + 1, M_I\rangle$ transition. Thus, in order to generate simulated FS spectra, first-order treatment is *not* a priori sufficient to interpret observed FS spectra when $S_i > 1$. Recipes for the analyses of FS extra lines and their appearance conditions are given in detail in Ref. 27.

For smaller zfs parameters, first-order perturbation treatment has been invoked for the sake of simplicity in order to extract zfs parameters and g-values. It has turned out that most of the documented analyses based on the first-order treatment of molecular high spins—where $\pi-\pi$ spin–spin interactions dominate in the contribution to the zfs parameters—do not fulfill the appearance condition for FS extra lines. During the simulation procedure, readers are strongly recommended to check the appearance condition by substituting any trial zfs parameters and the microwave frequency employed into the appropriate equation given in the literature [27]. Whether or not extra lines appear also depends on the linewidth of the single transition used for the simulation. If the difference in resonance fields between extra lines and principal-axis lines is comparable to or smaller than the linewidth, the extra line will *not* be distinctively seen, although asymmetry in the lineshape or intensity anomaly will be appreciable.

III.B.2. Higher-Order Perturbation Approaches

In order to reproduce overall FS spectra—including low-field absorption peaks from forbidden transitions—the second- or higher-order perturbation treatments are recommended for depicting the angular dependence of all the resonance fields and corresponding transition probabilities. Taking the Boltzmann distribution into account for the peak intensities of the FS spectra observed at low temperature is in principle trivial, but it is sometimes necessary even for X-band spectroscopy. In addition, when the perturbation approach is applied in order to simulate the low-field X-band FS spectra including forbidden transitions, close attention must be paid. The appearance of a somewhat peculiar-looking lineshape (e.g., out-of-phase lineshape) in the simulated spectrum strongly suggests a possible breakdown of the perturbation approach.

All the mathematical expressions required for spectral simulation are available in analytical forms with respect to an arbitrary coordinate-axis system [27,31,32], and the complete FORTRAN program software packages based on the second-order perturbation approach, including hfs terms, were intended to be distributed by various authors previously [33,34]. The perturbation approach has been developed to the third-order stage only in terms of FS terms [27]. A program package based on the second-order perturbation approach to spectral simulation for an arbitrary electron spin, S, and arbitrary nuclear spins, I's, is commercially available, but its efficiency is hampered by a failure to simulate forbidden transition peaks [35]. The perturbation-approach programs are efficient in terms of computation time, but a weakness is that they still fail to reproduce low-field peaks in X-band FS spectra. Still, the programs that enable one to simulate overall FS spectra should be useful for cases when $|D| \ll h\nu$ under the experimental condition of X-band or high-frequency/high-field spectroscopy (K-, Q-, and W-band).

III.B.3. Program Software Packages

The program software packages that include ESR powder-pattern (random-orientation) spectral simulation based on higher-order perturbation treatment are available on request [36]. The program package for random-orientation

ESR spectral simulation with second-order FS terms and anisotropic line-width variations is commercially available [35].

III.C. Exact Numerical Diagonalization Approaches

The breakdown of the perturbation approach based on analytical solutions can be avoided by invoking the exact numerical diagonalization of the spin Hamiltonian matrix, either the $n \times n$ eigen-energy matrix or the $n^2 \times n^2$ eigenfield matrix [37,38], where $n = 2S + 1$. The former method is subject to notorious nonconvergence problems while matching $h\nu$ to the difference between the energies involved in the transitions, if the calculation involves avoided energy crossings. For example, this difficulty takes place in the simulation procedure of the transitions appearing in the low-field region of X-band FS spectra. Therefore, we concentrate on the latter method.

III.C.1. Eigenfield Approach

The difficulty of nonconvergence can be overcome by direct numerical diagonalization of eigenfield matrices; note that the *exact analytical* solution of the eigenfield for triplet states can be derived in an arbitrary orientation of the static magnetic field, $\mathbf{B}_0$ [39]. The only weakness of the eigenfield method is the sizable dimension of the $n^2 \times n^2$ eigenfield matrix. The eigenfield method must solve generalized eigenvalue problems, which give rise to imaginary eigenfield values. During the numerical convergence procedure, elaborate mathematical techniques are necessary. In addition, the method needs a large amount of CPU computational time, which usually increases in proportional to the third power of the dimension of the matrix. Thus, the eigenfield method becomes impractical when the dimension of the original eigenenergy matrix is large, but to the best of our knowledge of this restriction, required CPU time is tolerable even for a $2S + 1$ multiplet larger than quintet.

Once exact eigenfields are obtained, the corresponding transition probabilities can be numerically calculated by diagonalizing either the eigenfield transition probability matrix [38,40] or the eigenenergy transition probability matrix, after substituting the already-obtained exact transition (resonance) field and calculating the appropriate eigenfunctions [41,42]. The latter procedure for the transition probability is a sort of a "hybrid approach" enabling one to save a great deal of CPU time without numerically technical difficulty.

Figure 2 illustrates the X-band FS spectrum and corresponding angular dependence of the transition fields and probabilities for a quintet state ($S_i = 2$) with relatively large zfs parameters at which the higher-order perturbation

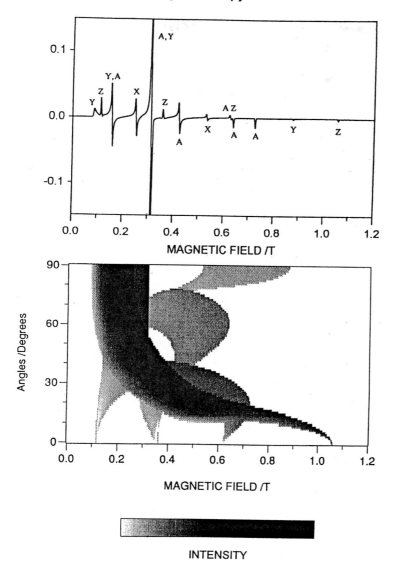

Figure 2 X-band fine-structure spectrum (top) from a molecular quintet state, and the angular dependence (bottom) of the transition fields and intensities.

approach breaks down; $g = 2.003$, $D = 0.224$ cm^{-1}, and $E = 0.038$ cm^{-1}. The X-, Y-, and Z-labels denote the three canonical orientations of the molecule, i.e., the orientations of $\mathbf{B}_0$ parallel to the three principal axes of the zfs tensor $\mathbf{D}_i$ with respect to each transition involved. Variable A designates the off-axis extra absorption peaks that arise from the stationary point (angular anomaly) that occurs when neither $\theta = 0$, $\pi/2$ nor $\phi = 0$, $\pi/2$, where θ and ϕ stand for the polar angles of $\mathbf{B}_0$ from the Z- and X-axis, respectively. Quite a few angular anomalous points and low transition probabilities occur throughout the angular dependence of the transition fields and probabilities, showing that the high-field approximation does not hold any more in this spin-quintet system. All the transition fields and probabilities have been computed in terms of the numerical direct diagonalization based on the eigenfield approach.

III.C.2. W-Band Quintet FS Spectrum

An intermediate case with $|D| < h\nu$ gives rise to a most complex FS spectrum, as seen in Figure 2. The perturbation approach will not enable us to extract accurate zfs or other spin Hamiltonian parameters from the observed FS spectra of the intermediate case. This is the reason that all documented zfs data on spin-quintet nitrenes are not correct, leading to a failure to interpret their electronic and molecular structures in a reasonable way.

Figure 3 exemplifies the W-band FS spectrum (top) at 3 K, corresponding to the same spectrum in Figure 2. Also, the angular dependence of the transition fields and intensities with Boltzmann distribution at 3 K taken into account are depicted at the bottom of the same figure. The W-band FS spectrum illustrates in a straightforward manner all the salient features anticipated for randomly oriented quintet states in the high-field approximation. The allowed transition probability shows up strongly, as expected, all the extra lines disappear, and the forbidden transitions appear in the low-field region with only weak intensities. The spectral assignment is straightforward, facilitating the extraction of the spin Hamiltonian parameters with high accuracy, even by means of the perturbation approach.

It should be noted in addition that the absolute sign of the D-value can be determined experimentally here. The positive sign for the present case is concluded directly from the much stronger intensity of the outermost Z-peak in the high field than that of the outlying Z-peak in the low field.

III.C.3. Program Software Package

The program software package for random-orientation FS spectral simulation based on the eigenfield approach is available on request via diskette media [44]. The package allows 2D representation for the angular depen-

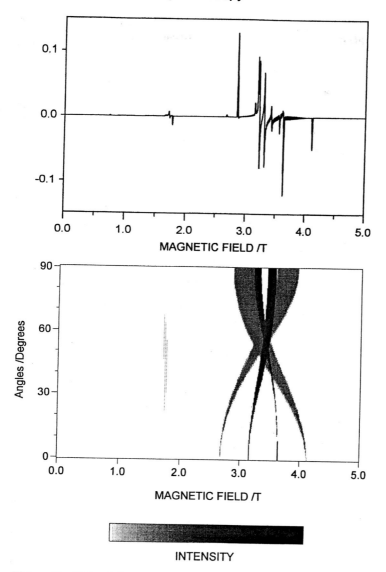

Figure 3 W-band fine-structure spectrum (top) from the quintet state, and the angular dependence (bottom) of the transition fields and intensities.

dence of both transition fields and probabilities, and that of transition intensities with Boltzmann statistics taken into account. Anisotropic linewidth variations as a function of the orientation of B_0 and the g and hfs tensors are incorporated in the program. The package optionally incorporates the eigenenergy approach based on the numerical direct diagonalization of the energy matrix of the spin Hamiltonian, i.e., Eq. (1) ($i = 1$).

III.D. Use of the Exact Analytical Expressions for the Resonance Fields and Eigenenergies of High Spins up to $S = 3$ and $S = 4$

It is well known that equations up to quartic can be analytically solved, while general analytical expressions for the exact solutions of equations higher than quartic are not available. Nonetheless, the eigenfields of B_0 along the principal axes of the D_i tensor *can* be solved exactly. The exact analytical expressions for the transition fields have been derived for the first time from the eigenfield FS equation up to $S_i = 3$, and the exact analytical expressions for the eigenenergies in the principal-axis orientation have been derived up to $S_i = 4$ [40]. The lineshapes characteristic of canonical orientations enable us to discriminate between the X-, Y-, and Z-lines. Thus, the exact analytical expressions for the transition fields serve to determine the spin Hamiltonian parameters without numerical spectral simulation in many cases.

Figure 4 illustrates the transition assignment of the spin quartet FS spectrum (top) of Cr^{3+} in a cage complex of the ligand, 1,8-diamino-3,6,10,13,16,19-hexaazobicyclo-[6,6,6]eicosane [43,45]. The angular dependence of the transition fields and probabilities are also shown (bottom). The complete spectrum has been simulated by the eigenfield approach. Spectral assignments were carried out using the exact analytical solution, which enables us to extract the following set of spin Hamiltonian parameters: $S = 3/2$, $g = 1.99$, $D = 7134$ MHz, and $E = 422$ MHz. The anomalous behavior of the transition fields and intermediate transition probabilities are shown in Figure 4. The label A denotes off-axis extra absorption peaks; the label F designates an apparent forbidden transition. No perturbation approach would be applicable in order to extract these spin Hamiltonian parameters with reasonably good accuracy.

Figure 5 shows the simulated W-band FS spectrum of Cr^{3+} in the cage complex at 3 K when assuming a positive D-value, exemplifying the inevitable appearance of the off-axis extra line from the $|M_S = +1/2\rangle \leftrightarrow |M_S = -1/2\rangle$ transition for a half-integer high-spin system, even in the high-field approximation. The strongest central peak in the high-field side is attributable to the angular anomaly. The transition intensities from the forbidden transitions were diminished a great deal, and the allowed transitions gained

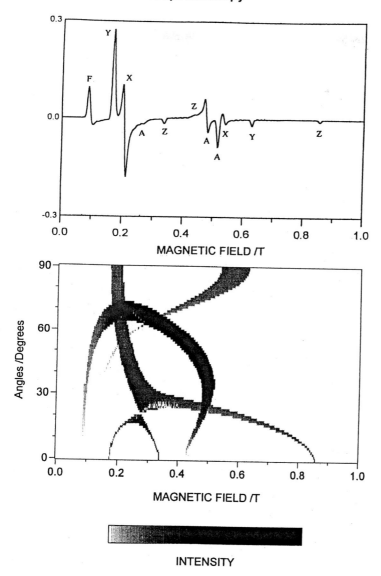

Figure 4 Transition assignments (top) of the quartet fine-structure spectrum of Cr^{3+} in a cage complex, and the angular dependence (bottom) of the transition fields and intensities.

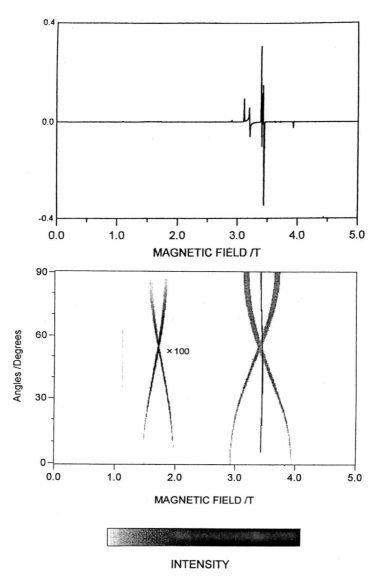

Figure 5   Simulated W-band fine-structure spectrum (top) of Cr^{3+} in the cage complex, and the angular dependence (bottom) of the transition fields and intensities.

intensity and dominated, as expected (Figure 5, bottom). In the W-band FS spectrum, the assignments for the canonical orientations are straightforward, facilitating accurate determination of the spin Hamiltonian parameters.

III.D.1. Program Software Package

The exact analytical expressions described herein are tedious to reproduce explicitly here. They are available on diskette media [46]. The diskette contains the program package that generates the transition fields and probabilities, the eigenenergies in the principal-axis orientations of B_0 for a given set of spin Hamiltonian parameters, i.e., a spin quantum number $S(\leq 4)$, g-values, and zfs parameters. The transition probabilities in the principal-axis orientations are calculated by means of the "hybrid" approach mentioned earlier [42].

IV. STATISTICAL MOLECULAR STRUCTURAL FLUCTUATION AND LINEWIDTH VARIATION AS A FUNCTION OF THE ORIENTATION OF B_0

Statistical molecular structural fluctuation shows up prominently in organic rigid glasses. Spectral simulation assuming random orientation will not reproduce observed FS spectra when structural fluctuation takes place. The most typical example is the existence of geometrical isomers, conformers of high-spin molecules or various degrees of molecular clustering. Specific absorption peaks arising from canonical orientations are subject to line broadening in some cases. Thermally stimulated relaxation processes that occur due to annealing will change the ratio of the conformers, or favor particular conformations if they are persistent during the annealing of organic rigid glass states.

In order to interpret the occurrence of statistical molecular fluctuation due to a continuous distribution of conformers, structurally reinforced molecular design has been invoked, such as bridging to suppress the angular fluctuation of dihedral angles. Rigid planar or round molecules, large or small, give less statistical fluctuation in FS-ESR spectra that do not exhibit intermolecular clustering. On the other hand, linear high-spin systems have ample opportunity to be subject to the fluctuation of various kinds, as do hyperbranched π-aryl systems.

IV.A. Simulation Programs

In the case of linewidth variation originating in hyperfine interactions, spectral simulation procedures should incorporate the anisotropic contribution to overall FS lineshapes. Simulation software that includes anisotropic line-width variation are available, based on both the numerical direct diagonal-ization (eigenenergy approach) and higher-order perturbation treatments. Phenomenological treatments on the linewidth variation are also available. A program package based on the second-order approximation for fine-struc-ture terms and the first-order−to−hyperfine interaction terms with some re-strictions is commercially available [35].

Finally, from the viewpoint of general theory and electron magnetic resonance spectroscopy, Pilbrow et al. have developed theoretical treatments on linewidth variations [47].

V. A NEW EMR SPECTROSCOPIC APPROACH TO MOLECULAR HIGH-SPIN SYSTEMS: FT-PULSED ESR/ELECTRON SPIN TRANSIENT NUTATION SPECTROSCOPY

With increasing spin quantum S of molecular high-spin systems, the spectral density of the central regions increases rapidly ($\propto \sim S^2$), even if the spin−spin or spin−orbit interactions are kept similar, as exemplified in Figure 6 for a nonet state (top: $S = 4$; $g = 2.003$, $D = 0.06$ cm^{-1}, and $E = 0.004$ cm^{-1}) and a decet state (bottom: $S = 9/2$; $g = 2.003$, $D = 0.05$ cm^{-1}, and $E = 0.003$ cm^{-1}). The absorption peaks spread out in both outlying directions of the magnetic field. The peaks appearing in the outermost field lose their intensity most rapidly. This arises from the $|M_S\rangle \leftrightarrow |M_S + 1\rangle$ allowed tran-sition probability, as given the zero-order transition probability for the $|M_S\rangle \leftrightarrow |M_S + 1\rangle$ transition in the following:

$$I_{M_S M_{S+1}} \propto (2g\beta B_1)^2 |G|^2 \left(\frac{1}{4}\right) [S(S + 1) - M_S(M_S + 1)]$$

$$|\langle M_I(M_S)|M_I'(M_S + 1)\rangle|^2 \tag{7}$$

When extremely large molecular spins or polymeric ones are designed and synthesized, their observed FS spectra are often composed of a mixture of different high spins or conformers. In addition, the spectra are inhomo-geneously broadened with poor resolution. Spin state discrimination and unequivocal identification are thus very difficult to make for a high-spin mixture in nonoriented media such as organic glasses because of the spectral high density mentioned earlier.

On the other hand, high-spin polymeric systems are frequently subject to inter- or intramolecular exchange interactions, giving rise to structureless FS spectra. Cw FS-ESR spectroscopy alone hardly offers valid interpretation for such cases. These experimental difficulties encountered in studying the

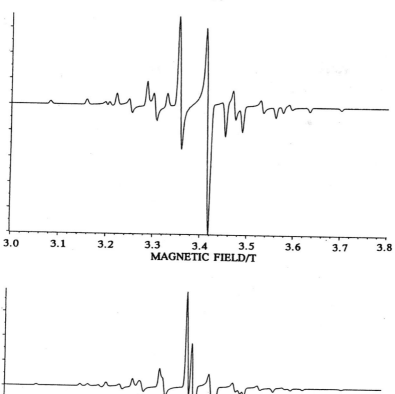

Figure 6 Simulated W-band fine-structure ESR spectra from random orientation of a nonet spin state (top), and a decet spin state (bottom).

most complex molecular spin system can be eliminated by FT-pulsed ESR spectroscopy operating in the time domain instead of the cw frequency domain. For this reason, in the last part of this chapter we concentrate on FT-pulsed ESR/ESTN (electron spin transient nutation) spectroscopy, which has emerged recently [48,49] and has been developed for application to non-oriented molecular and polymeric high-spin systems [50]. Fourier-transform-pulsed ESR/ESTN spectroscopy enables us to discriminate and identify unequivocally the spin quantum number of a system under consideration. The spin identification procedure does not include any spectral simulation. The method is straightforward and does not require well-resolved cw FS spectra of high quality.

In the strict sense, ESR spectroscopy has been considered too specialized when compared with NMR spectroscopy. This is in part due to the many different terms to be handled in the ESR spin Hamiltonian and in part due to the exploitation by ESR of the "notorious" static magnetic field-swept scheme instead of the energy(frequency)-swept scheme, because of technical restrictions. The latter reason forces spectral analyses and eigenvalue-eigenfunction problems to be mathematically complex. In this context, the spin discrimination analyses involved in ESTN spectroscopy are straightforward, and are easy to understand when using modern NMR-like ESTN experiments. It is natural that ESTN spectroscopy based on FT-pulsed electron magnetic resonance (EMR) methods give rise to new aspects of EMR spectroscopy and to simplifications of the anisotropy problems intrinsic to FS hfs spectra and to time-dependent phenomena of high spins or molecular high-spin assemblies, whether they are exchange coupled or not.

VI. VECTORIAL PICTURE OF ESTN SPECTROSCOPY AND NUTATION FREQUENCY OBTAINED BY QUANTUM MECHANICAL TREATMENT

VI.A. What ESTN Spectroscopy Tells Us

Electron spin transient nutation spectroscopy is based on electron spin resonance to measure the spin Hamiltonian in terms of the rotating frame. The time evolution of the electron spin system in the presence of a microwave field B_1 and a static magnetic field B_0 is observed in the rotating frame. As a result, ESTN spectroscopy yields pseudo-"zero-field" or pseudo-"nearly zero-field" FS spectra with high sensitivity due to the use of high-field ESR spectroscopy (with an offset parameter $x \sim 0$; see later for the physical meaning of x). Nutation spectroscopy provides us with extremely high resolution in terms of spin multiplicity discrimination, the magnitude of zfs parameters, and thus, also, the asymmetric nature of the zfs tensor and its surroundings.

One of the advantages of FT-pulsed ESR/ESTN spectroscopy is the use of two-dimensional (2D) representations inherent in time-domain spectroscopy. The 2D representation of FT-pulsed ESR/ESTN spectroscopy includes field-swept ($\mathbf{B_0}$) vs. ESTN spectra and $\mathbf{B_1}$ vs. ESTN phenomena. The resolution enhancement intrinsic to ESTN spectroscopy does *not* require high-quality cw FS spectra. On the contrary, ESTN spectroscopy shows its real ability in the spectroscopic study of complex mixtures of high spins, displaying clear-cut experimental evidence of magnetic properties for spin systems in microscopic detail.

All aspects inherent in and characteristic of ESTN spectroscopy have not been fully apparent until recently. Thus, ESTN spectroscopy is still developing. In particular, application to nonoriented spin systems such as organic glasses has for the first time been carried out and developed from the methodological viewpoint in this work. For example, we emphasize that ESTN spectroscopy has proven the existence of exchange coupling in neutral high-spin organic polymers [50] and enabled us to determine experimentally and evaluate extremely small zfs parameters attributable to departure from cubic or octahedral symmetry environments of the high-spin state of a transition-metal ion. The high resolution of the latter experiment is due to the intrinsic nature of pseudo-zero-field spectroscopy in time-domain ESTN. Such a small magnitude ($\sim 10^{-4}$ cm^{-1}) of the zfs parameter has never before been experimentally detected and evaluated. In addition, the nutation behavior of electron spin multiplets has been evaluated from the theoretical side, suggesting in particular the possible occurrence of multiple quantum nutation phenomena in high-spin systems and their usefulness in evaluating zfs parameters (see Table 1).

VI.B. Vectorial Picture of ESTN Spectroscopy

In Figure 7 we depict the nutation of a single quatum transition in terms of a classical vectorial picture for the motion of the spin magnetization. The magnetization $\mathbf{M_0}$ in the presence of a static magnetic field $\mathbf{B_0}$ precesses at a nutation angle ϕ from the initial direction around an effective field $\mathbf{B_e} = \mathbf{B_0} + \mathbf{B_1}$ when applying a microwave field ($\mathbf{B_1}$) pulse with width t_1. Then, $\mathbf{B_0}$ in thermal equilibrium undergoes free induction decay (FID) when the excitation pulse is turned off: $B_e = \sqrt{(B_0 - \omega/\gamma)^2 + B_1^2}$, $\tan \theta = B_1/(B_0 - \omega/\gamma)$, where we can define x as an offset parameter such that $x = (B_0 - \omega/\gamma)/B_1$. ϕ is defined as the nutation angle of $\mathbf{M_0}$ around $\mathbf{B_e}$ in time t, and ϕ_0 is defined as the nutation angle of $\mathbf{M_0}$ in the same duration time t around $\mathbf{B_1}$. Thus $\phi = \gamma B_e t$, $\phi_0 = \gamma B_1 t$, and $\phi = \phi_0 \sqrt{1 + x^2}$ holds. ϕ_0 is the rotation

Table 1 On-Resonance Nutation Frequencies for Various Cases

Condition	$\omega_n,\ \omega_n^{dq},\ \omega_n^{tq}$
$H_D = 0$	$\omega_n = \omega_1$
$H_D \ll H_1$	$\omega_n \sim \omega_1$
$H_D \sim H_1$	not a single ω_n
$H_D \gg H_1$	$\omega_n = \omega_1 \sqrt{S(S + 1) - M_S M_S'}$ with $M_S' = M_S - 1$

$$\omega_n = \omega_1 \left(S + \frac{1}{2} \right) \quad \text{for the } M_S = 1/2 \leftrightarrow M_S' = -1/2 \text{ transition}$$
$$(S = 3/2,\ 5/2,\ 7/2, \ldots)$$

$$\omega_n = \omega_1 \sqrt{S(S + 1)} \quad \text{for the } M_S = 0 \leftrightarrow M_S' = -1 \text{ or}$$
$$M_S = -1 \leftrightarrow M_S' = 0 \text{ transition } (S = 1, 2, 3, \ldots)$$

$$\omega_n^{dq} = \omega_1 \left(\frac{\omega_1}{\omega_D} \right) \quad \text{for } S = 1$$

$$\omega_n^{dq} = \omega_1 \left(\frac{7\omega_1}{4\omega_D} \right) \quad \text{for } S = 3/2$$

$$\omega_n^{tq} = \omega_1 \left(\frac{3\omega_1}{8\omega_D} \right)^2 \quad \text{for } S = 3/2$$

ω_n^{dq} and ω_n^{tq} denote the nutation frequency for double and triple quantum transitions ($S \geqq 1$), respectively.

angle on exact resonance ($x = 0$). In this classical vectorial picture, the nutation frequency ω_n is given as $\omega_n = \gamma B_e$ off exact resonance ($x \neq 0$) and $\omega_n = \omega_1 = \gamma B_1$ on exact resonance.

VI.C. Quantum Mechanical Description of ESTN

In order to describe ESTN phenomena quantum mechanically, the equation of motion of the density matrix governed by the Liouville–von Neumann equation is invoked. For the vanishing FS term (H_D), the second term of Eq. (1) ($i = 1$), or $H_D \ll H_1$, the ensemble of high spins nutates at a frequency of $\omega_1 = -\gamma B_1$ under the on-resonance condition, where ω_1 is independent of S. H_1 stands for the interaction term between microwave field $\mathbf{B_1}$ and an electron high spin.

For nonvanishing H_D, the nutation is modified due to the presence of H_D in the rotating frame, and is not described in terms of a single frequency.

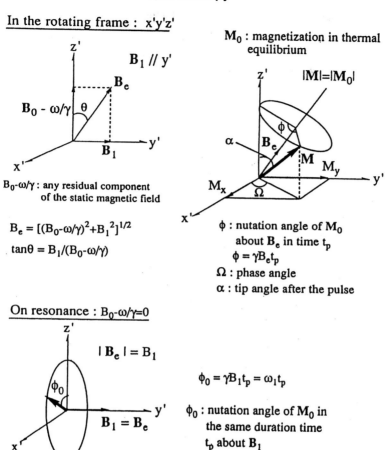

In the rotating frame : x'y'z'

$B_1 // y'$

$B_0 - \omega/\gamma$ θ

B_e

B_1

$B_0-\omega/\gamma$: any residual component of the static magnetic field

$B_e = [(B_0-\omega/\gamma)^2 + B_1^2]^{1/2}$

$\tan\theta = B_1/(B_0-\omega/\gamma)$

M_0 : magnetization in thermal equilibrium

$|M|=|M_0|$

B_e

M M_y

α

M_x Ω

ϕ : nutation angle of M_0 about B_e in time t_p
   $\phi = \gamma B_e t_p$

Ω : phase angle

α : tip angle after the pulse

On resonance : $B_0-\omega/\gamma=0$

$|B_e| = B_1$

ϕ_0

$B_1 = B_e$

$\phi_0 = \gamma B_1 t_p = \omega_1 t_p$

ϕ_0 : nutation angle of M_0 in the same duration time t_p about B_1

Figure 7   Vectorial picture of ESTN and definitions of various quantities.

In the weak extreme limit of $H_D \gg H_1$, however, the nutation frequency ω_n is expressed simply as

$$\omega_n = \omega_1 \sqrt{S(S + 1) - M_S(M_S - 1)} \qquad (8)$$

where M_S denotes the electron spin sublevel involved in the ESR-allowed transition [48–50]. The rotating-frame matrix element corresponding to the transition is given in the first order as

$$\langle S, M_S | H_{1,R} | S, M_S' \rangle = -\omega_1 \sqrt{(S + M_S)(S - M_S')}$$

$$= -\omega_n \tag{9}$$

where $M_S' = M_S - 1$ for the allowed transition, showing that ω_n is equivalent to the transition frequency of coherent microwave excitation. Thus, for the weak extreme limit, $H_D \gg H_1$, the nutation spectrum ω_n depends on S and M_S.

For integer spins $S = 1, 2, 3, \ldots$, $\omega_n = \omega_1 \sqrt{S(S + 1)}$ for the $|S, M_S = 1\rangle \leftrightarrow |S, M_S' = 0\rangle$ or $|S, M_S = 0\rangle \leftrightarrow |S, M_S' = -1\rangle$ transition. Therefore, even if the ESR transitions involving the $|S, M_S = 0\rangle$ level overlap due to small zfs parameters, the spin quantum number S can be discriminated in the nutation spectrum by time-domain ESTN spectroscopy. Practically, the offset-frequency effect on nutation must be carefully considered in some cases in order to carry out $\mathbf{B_0}$-swept 2D nutation spectroscopy.

For half-integer spins, $S = 3/2, 5/2, \ldots$, the fine-structure term $\omega_D(2M_S - 1)$ to the first order with allowed $M_S \leftrightarrow M_S - 1$ transitions is vanishing for the $|S, M_S = 1/2\rangle \leftrightarrow |S, M_S' = -1/2\rangle$ transition, and higher-order corrections due to the fine-structure term contribute only as off-axis extra absorption peaks in the random-orientation FS spectrum if ω_D is large [27]. The corresponding nutation frequency ω_n is given simply as $\omega_n = \omega_1(S + 1/2)$. Thus, the nutation spectrum is distinguishable between $S = 1/2$ and other S's even if the fine-structure splitting is not apparent in the cw-ESR spectrum because of inhomogeneous line broadening, large ω_D values, and so on. For intermediate cases, i.e., $H_D \sim H_1$, the nutation spectrum appears more or less complicated, but the spectrum is interpretable in terms of the rotating-frame spin Hamiltonian.

In addition, multiple quantum transitions are observable in nutation spectroscopy in the weak extreme limit. It has been found in theoretical treatments that the nutation frequency arising from a multiple quantum transition is considerably reduced due to the scaling effect of the effective field that spin ensembles experience in the rotating frame [50]. Table 1 summarizes nutation ω_n on resonance for various cases of S and H_D vs H_1 at experimentally typical discretions.

VI.D. Experimental Methods

A nutation experiment can be carried out by observing either the FID or the electron spin or rotary echo (ESE) signal, $s(t_1, t_2)$, as time-domain spectroscopy (time axis t_2) by incrementing the time interval t_1 of the microwave pulse excitation with the field strength $\propto \omega_1 (\omega_1 = -\gamma B_1)$ parametrically, as depicted in Figure 8. The time variables t_1 and t_2 are independent. In the echo-detected scheme, a typical two-pulse sequence can be invoked for the

FID-Detected Nutation

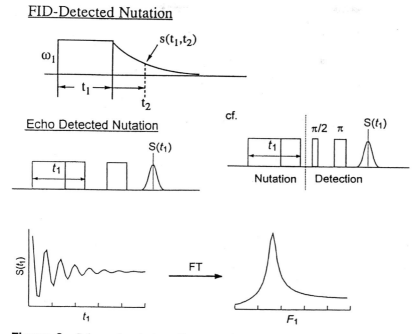

Figure 8 Schematic timing diagrams for the FID-detected and ESE-detected ESTN.

detection period by exploiting particular phase cyclings in order to detect the desired echo signal. The signal $s(t_1, t_2)$ measured as a function of t_1 and t_2 is converted into a 1D or 2D frequency domain spectrum—i.e., $S(f_1, t_2)$ [or $S(t_1, f_2)$] or $S(f_1, f_2)$—by Fourier transformation. The nutation spectrum $S(f_1, t_2)$ (measured also as a function of the static magnetic field $\mathbf{B_0}$) gives a 2D $\mathbf{B_0}$-swept nutation spectrum, which enables us to discriminate directly between ESR transition assignments in FS spectra and to differentiate spin states spectroscopically.

VII. ESTN SPECTROSCOPY APPLIED TO NONORIENTED MEDIA

VII.A. Two-Dimensional ESTN Spectroscopy: An Example of Field-Swept 2D ESTN

Figure 9 exemplifies 2D ESTN spectroscopy illustrating direct spin discrimination and ESR transition assignments for a spin-quartet state observed in

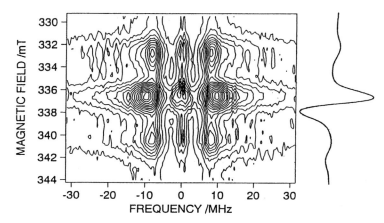

Figure 9 Contour plot of the field-swept 2D ESTN spectra of the tricationic state of N,N,N',N',N'',N''-hexa-4-anisyl-1,3,5-triaminobenzene and the conventional field-swept ESR spectrum (on the right) observed at 6 K. ω_1 corresponds to about 5 MHz.

an organic glass, where lower-spin states ($S = 1/2$, 1) from intermediate oxidation states or by-products are expected to contaminate the chemical reaction. The central peak (ω_b) and the two "outermost" peaks (ω_a and ω_c) along the magnetic field $\mathbf{B_0}$ are unequivocally assigned to the $|S = 3/2, M_S = +1/2\rangle \leftrightarrow |S = 3/2, M_S = -1/2\rangle$ transition and $|S = 3/2, M_S = \pm 3/2\rangle \leftrightarrow |S = 3/2, M_S = \pm 1/2\rangle$ transition, respectively. This is because $\omega_b{:}\omega_a$ (or ω_c) $= 10.0$ MHz:8.1 MHz $\cong 2\omega_1{:}\sqrt{3}\omega_1$, the ratio expected theoretically from Eq. (8) for $S = 3/2$ in the weak extreme limit.

Also, Figure 9 apparently shows that no trace peak attributable to $S = 1/2$ or $S = 1$ species was observed at 7 K, demonstrating that the cw FS spectrum at 7 K originates only from the highest-spin quartet species in the electronic ground state, with the lower-spin states being located higher in energy. It should be emphasized that the central peak, whose transition (nutation) frequency is ω_b, does not contain any appreciable amount of spin-doublet species, and thus the FS spectrum is purely from the spin quartet state.

Figure 10 illustrates the observed and simulated FS hfs spectra for the spin quartet species just described. The simulation was carried out by incorporating hyperfine interaction terms only for the pure spin quartet state. The agreement is essentially perfect, enabling us to extract the spin Hamiltonian parameters with high accuracy.

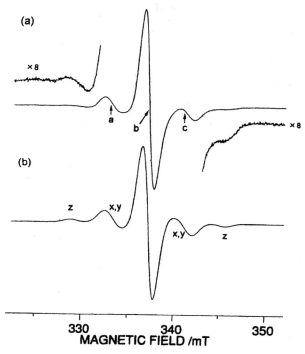

Figure 10 (a) Powder-pattern ESR quartet spectrum of the tricationic state observed at 7 K. The microwave frequency was 9.44751 GHz. The letters a, b, and c correspond to the peaks ω_a, ω_b, and ω_c, respectively, appearing in the 2D ESTN spectra in Figure 9. (b) Simulated ESR spectrum with $S = 3/2$, $g = 2.0023$, $|D| = 0.004$ cm^{-1}, $|E| = 0.0002$ cm^{-1}. The letters x, y, and z indicate the canonical orientation.

VII.B. An Example of 1D ESTN

Organic high-spin polymeric systems are frequently subject to inter- or intramolecular exchange interactions, giving rise to structureless FS spectra. For such cases cw-ESR FS spectroscopy is useless for discriminating between spin states. Comparatively, magnetic susceptibility measurements and Brillouin-function fittings of observed magnetization curves give only an average value of the spin multiplicity for the polymeric high-spin mixture under study. In the procedure for obtaining the average value, technical assumptions such as technical saturation magnetization must be invoked. The latter drawbacks are intrinsic to the nonspectroscopic, bulk measurements.

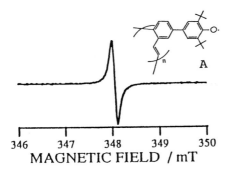

Figure 11   cw-ESR spectrum of a quasi-1D organic high-spin polymer A in the solid state observed at 6 K.

Figures 11 and 12 show the first example of 1D ESTN spectroscopy applied to organic high-spin polymers in order to identify the high-spin components in the polymer mixture [51]. Figure 11 shows a typical cw-ESR spectrum for a π-conjugated high-spin polymer in the solid state, observed at 6.7 K. The ESR lineshape was Lorentzian, indicating that exchange narrowing was taking place in the system. Figure 12 shows the dependence of many nutation-frequency components on microwave amplitude. In the weak extreme limit of $H_D \gg H_1$, the distinguishable nutation peaks $\omega_b(S_i)$ occurring in addition to the strong one near zero frequency were evaluated by Eq. (8). Table 2 summarizes the peak assignments for effective molecular

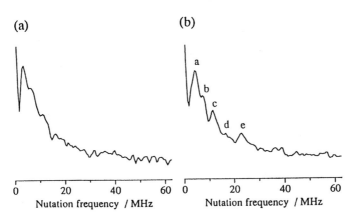

Figure 12   Nutation spectra of a solid-state polymer A observed at 6 K. The microwave amplitude dependence of the on-resonance nutation is shown: (a) 30 dB (b) 25 dB.

Table 2 Nutation Peak Assignments to Effective Spin Quantum Numbers Appearing in Nutation Spectra Observed from the Quasi-1D Polymer A at 6.7 K

$\omega_n{}^a$	ω_n/ω_1	S	$[S(S + 1)]^{1/2}$
a. 4.84 $(=\omega_1)$	1	1/2	
b. 7.88	1.63	1	$[1(1 + 1)]^{1/2} = 1.41$
c. 11.47	2.37	2	$[2(2 + 1)]^{1/2} = 2.45$
d. 16.73	3.45	3	$[3(3 + 1)]^{1/2} = 3.46$
e. 22.95	4.74	4	$[4(4 + 1)]^{1/2} = 4.47$

^{a}a−c correspond to the nutation peaks a−c appearing in Figure 12b.

spin quantum numbers, assuming that $M_S = 0 \leftrightarrow M_S' = \pm 1$ transitions are dominant in the nutation spectra. It was concluded that the high-spin polymer contains high spins up to at least nonet species. The magnetization curve fitting described in the previous paragraph is insensitive to an ensemble of various spin quantum numbers S_i's, due to the increasing number of S_i's expected for extended spin networks of magnetic polymers and high-spin molecular clusters. The present nutation experiment demonstrates that the ESTN method is more suitable for the spectroscopic discrimination of different spins for high-spin ensembles, accepting that one of the drawbacks of the nutation technique is the limited time resolution in an X-band experiment. High-frequency/high-field pulsed ESR spectroscopy has potential capability of improving this disadvantage because of nearly zero deadtime that may be employed.

A comprehensive treatment will be published elsewhere on both experimental and theoretical techniques for random-orientation ESTN spectroscopy for high-spin 3d block transition and lanthanide ions, molecular high-spin clusters, ionic high-spin polymers, and low-dimensional molecular magnetic assemblies.

VIII. ELECTRON MAGNETIC RESONANCE IN ORDERED REGIME

VIII.A. General Survey

Magnetic materials in ordered regimes continue to provide diverse interest for solid-state physics and chemistry. Organic-molecule-based magnetics particularly show salient features as subjects of this rapidly growing interdisciplinary field. They exhibit a variety of types of interesting types of behavior due to different design strategies: (a) low-symmetry nature in their spin-carrying sites due to multicentered "spin-in-molecule" natures, (b) manifestation of quantum nature of spin units (molecular spins), (c) weak spin–orbit interactions due to light atoms, (d) low-dimensionality nature of magnetic interactions in molecular assemblages, (e) molecularly super-high-spin nature, (f) through-bond topological control of exchange behavior, and (g) molecularly designing natures with multiple functionality [18–21].

Molecule-based magnetics is an important testing ground for a variety of theoretical models, whether established or not. For example, organic ferrimagnetics demonstrates a breakdown of classical pictures for ferrimagnetics [52]. Organic systems also give exceptional diversity as subjects of novel quantum magnetic phenomena or functionalities, as well as potential applications in future technology. Organic super-high-spin magnetics is closely related to conceptual advances in magnetics that underlie novel molecular devices such as genuine liquid-phase magnets, molecular quantal magnets, and magnetic quantum well effects.

In magnetically ordered substances, the macroscopic magnetization as cooperative property is described in terms of dynamics of microscopic details (spins). There are a variety of magnetic excitations, where the excitation does not remain localized at a given spin site but propagates in the form of coherent waves. The collectivized coherency (collective excitation) typically originates in quantum mechanical exchange interactions, forming the simplest type of magnetic excitation as spin waves or magnons. In terms of spin carriers, all documented spin waves are atomic spin waves. Additionally, there exist standing spin waves originating in gigantic magnetic dipolar interactions. This novel magnetic excitation occurring in artificial superlattices or molecularly superstructured lattices is at a semimicroscopic scale and has interesting group-theoretical symmetry as well. The creation of highly excited and concentrated spin assemblages in terms of parametric microwave excitation of the standing spin wave receives serious attention in spinics. Such magnetic phenomena or functionalities are essentially nonlinear properties. Dynamic magnetization is basic for nonlinear magnetic properties. The exploitation of dynamic magnetization as well as solitary waves is an important issue for spinics, as well as for further development of magnetics in materials science.

VIII.B.   Magnetic Relaxation and Equation of Motion of Magnetization

In order to highlight the cooperative nature of ordered spin assemblages in terms of microscopic details, the magnetic dynamics of the spins is briefly treated here, with the spins taken to be coupled to give macroscopic magnetization. In an ordered regime, each electron spin is in local magnetic surroundings with thermal fluctuation, while macroscopic magnetization $\mathbf{M}$ is treated in terms of the resulting spin dynamics of spin assemblages. The magnetization dynamics can be treated in two different manners. One describes the dynamics when the magnitude of $\mathbf{M}$ remains invariable during the relaxation process, and the other when $\mathbf{M}$, such as polarization in the presence of an external magnetic field $\mathbf{B}$, is not conserved. In the latter, Maxwell relaxation terms are introduced.

The former includes the equation of magnetization motion by Landau–Lifshitz [54] as given by

$$\mu_0 \frac{d\mathbf{M}}{dt} = \gamma[\mathbf{M} \times \mathbf{B}] - \lambda[\mathbf{M} \times [\mathbf{M} \times \mathbf{B}]] \tag{10}$$

and the Gilbert type of the equation [55] as given by

$$\frac{d\mathbf{M}}{dt} = \gamma \left\{ \mathbf{M} \times \left[\frac{\mathbf{B}}{\mu_0} - \left(\frac{\lambda}{\gamma^2} \right) \frac{d\mathbf{M}}{dt} \right] \right\} \tag{11}$$

where $\gamma\hbar = -g\beta$. The Kittel equation [56] is covered in Eq. (11). Bloch equations [57] represent the latter approach, where a longitudinal relaxation with a characteristic time T_1 and a transverse relaxation with T_2 are assumed to occur independently. The Bloch approach is defined by two equations;

$$\frac{dM_z}{dt} = \left(\frac{\gamma}{\mu_0} \right) [\mathbf{M} \times \mathbf{B}]_z - (M_z - M_0)/T_1 \tag{12-1}$$

$$\frac{dM_{x,y}}{dt} = \left(\frac{\gamma}{\mu_0} \right) [\mathbf{M} \times \mathbf{B}]_{x,y} - M_{x,y}/T_1 \tag{12-2}$$

where T_1 is defined as a spin-lattice relaxation time of dissipating Zeeman magnetic energy to lattices as a thermal bath and T_2 is a spin–spin relaxation (or phase memory) time in which rotational phases of magnetic moments $\mu_i = -g\beta S_i$ are disordered due to interactions between the moments and $M_{x,y} = \Sigma_i(\mu_i)_{x,y}$ vanishes. The relaxation of the transverse magnetization is driven by any interactions that do not commute with $\mathbf{M} = \Sigma_i\mu_i$, quantum mechanically, such as spin dipolar or anisotropic exchange interactions and local magnetic interactions due to thermal fluctuations.

VIII.C. Microscopic Spin Dynamics Underlying Magnetization Dynamics

The magnetization dynamics of ordered spin assemblages is underlain by microscopic spin dynamics [58]. Both Kubo–Tomita [59] and Mori (and Tokuyama–Mori) [60] theories treat magnetization dynamics in terms of spin assemblages in microscopic detail. The former theory describes the transverse relaxation by invoking a relaxation function $\phi(t)$ of a macroscopic magnetization as

$$\phi(t) = \frac{\langle M_x(t)M_x(0)\rangle}{k_B T} \tag{13}$$

where $\langle M_x(t)M_x(0)\rangle$ stands for a time correlation function of M_x under thermal fluctuation and therefore symbol $\phi(t)$ describes the magnetization, classically. A response $\chi(\omega)$ for an oscillating magnetic field $\mathbf{B}_1(\omega)(\mathbf{B}_{1x}(t) = B_1 e^{i\omega t})$, such as microwave irradiation, is given, according to fluctuation-dissipation theorem, as

$$\chi(\omega - \omega_0) = \left(\frac{1}{2\pi}\right) \int_{-\infty}^{\infty} \phi(t)\exp(-i\omega t)\, dt \tag{14}$$

where the imaginary part of Eq. (14) gives a magnetic resonance spectrum. Noncommutable local magnetic fluctuations are expressed in terms of a time correlation function $\varphi(\tau)$ for the local magnetic field as

$$\varphi(\tau) = \frac{\langle \omega_1(\tau)\omega_1(0)\rangle}{\langle \omega_1^2(0)\rangle} \tag{15}$$

with $\varphi_1(t) = \gamma B_{loc}(t)$. In terms of the time correlation function for the local fluctuation field, $\phi(t)$ is expressed by

$$\phi(t) = \exp\left[-\langle\omega^2(0)\rangle \int_0^t (t - \tau)\phi(\tau)\, d\pi\right] \tag{16}$$

Noting that the phase memories of individual spins $\mathbf{S}_i$'s are vanishing in $h/2\pi J$ due to commuting interactions with surrounding spins, the motion of $\mathbf{M}$ is conserved for a long time because $\mathbf{M}$ commutes with isotropic exchange interactions. Dynamics of $\mathbf{S}_i$ may be expressed in terms of a spin self-time-correlation function $\Phi(t)$ as

$$\Phi(t) = \frac{3\langle S_{xi}(t)S_{xi}(0)\rangle\}}{S(S + 1)} \tag{17}$$

In the rotating frame, the first three terms of $\Phi(t)$ are approximated to be

$$\Phi(t) \cong \exp\left[\frac{-(t/\tau_e)^2}{2}\right] \tag{18}$$

with $t < \tau_e$ and $\tau_e = \{h/|2\pi J|\}\{3/2qS(S + 1)\}^{1/2}$, where J stands for an exchange integral of a pair of nearest-neighbor spins and q for the number of the nearest-neighbor spins. For long times, $t > \tau_e$,

$$\Phi(t) \propto \left[\frac{S(S + 1)}{3} \right] t^{-d/2} \qquad (19)$$

holds, where d stands for the dimension of the spin systems. Thus, $\Phi(t)$ decays slowly for low-dimensional spin systems ($d = 1, 2$), leading to the occurrence of a long transverse relaxation called a long time tail (LTT). An LTT arises from increases in the contributing weights of the zero-mode Fourier component of the wave vector in $\Phi(t)$ for smaller d. Compared with three-dimensional spin systems, whose spin correlations decay in times much shorter than periods of microwave frequency $\omega_0/2\pi \sim 10 - 10^2$ GHz, free induction decays with transverse relaxations characteristic of LTT or satellite peaks due to spin dipolar interactions appear for low-dimensional organic spin systems in pulsed ESR experiments. In this context, high-field/high-frequency electron spin resonance spectroscopy such as W-band spectroscopy is potentially capable of extending vital information on magnetization dynamics in microscopic detail. In particular, it should be noted that microscopic spin dynamics characteristic of low dimensionality in organic magnetics reflect the contribution of nearly-zero-mode Fourier components in the process of $T \rightarrow T_C$ (or T_N). The difference in the contributions arises from the difference in temperature evolution of short-range interactions between ferromagnetics and antiferromagnetics. Pulsed-FT-based time domain spectroscopy is direct and sensitive to the difference.

VIII.D.   Electron Spin Resonance of Low-Dimensional Magnetics

In ESR measurements of low-dimensional magnetic materials, subtle changes or abnormalities appear in absorption lineshapes and/or resonance fields near the critical temperature. When $T \gg T_C, T_N$, there is no difference in the characteristics of absorption lines between ferromagnetics and antiferromagnetics. As already mentioned, however, with T approaching T_C or T_N the contribution to the lineshapes from the nearly-zero-mode components is different. Assuming that spin assemblages of low dimensionality are described by a spin Hamiltonian as

$$H = H_{eZ} + H_{ex} + H' \qquad (20)$$

where H_{eZ} stands for electron Zeeman interactions, H_{ex} for exchange interactions between spins, and H' for two-center spin interactions such as spin dipolar interactions. The latter two terms are subject to the restrictions of the dimensionality under study. For isotropic Heisenberg-type antiferromagnetic systems, Eq. (20) is expressed as

$$H = g\beta\mathbf{B_0} \sum_i S_{iz} + \sum_i (-2J)\mathbf{S}_i \cdot \mathbf{S}_{i+1}$$

$$+ \sum_i \sum_j (g\beta)^2/r_{ij}^3 \left[S_i \cdot S_j - \left(\frac{3}{r_{ij}^2}\right)(\mathbf{r}_{ij} \cdot \mathbf{S}_i)(\mathbf{r}_{ij} \cdot \mathbf{S}_j) \right] \qquad (21)$$

where no interchain interactions are considered and only nearest-neighbor interactions between spins and spin dipolar interactions $H' = H_{dip}$ are assumed. H_{dip} depends on the angle θ between the direction of a one-dimensional magnetic chain and the $z(\|\mathbf{B_0})$-axis. With decreasing temperature, resonance fields observed with $\mathbf{B_0}$ along the magnetic chain shift to low field, while ones with $\mathbf{B_0}$ perpendicular to the chain shift to high field. The temperature dependence is similar to phenomena observed in antiferromagnetic magnetics. Quantitative interpretations in terms of semiclassical treatments of spins are available [61]. Useful equations to predict the behavior of g-values in ESR experiments are given as

$$B_r^i = \frac{[(\chi_j\chi_k)^{1/2}/\chi_i]\hbar\omega}{g_i\beta} \qquad (22)$$

where B_r^i stands for the resonance observed with $\mathbf{B_0}$ along the i-axis, χ_l and g_l stand for the anisotropic susceptibility, and the g-value is with $\mathbf{B_0}$ along the ℓ-direction ($\ell = i, j, k$). Thus,

$$B_r^i B_r^j B_r^k = \left(\frac{\hbar\omega}{\beta}\right)^3 (g_i g_j g_k)^{-1} \qquad (23)$$

holds independent of temperature. When the i-axis is along the direction of the one-dimensional magnetic chain, $\chi_i > \chi_j$, χ_k holds for both ferromagnetics and antiferromagnetics, leading to positive g-shifts, $\Delta g_i > 0$. On the other hand, for ferrimagnetics $\chi_i < \chi_j$, χ_k holds, leading to $\Delta g_i < 0$ and $\Delta g_{j,k} > 0$, and $\Delta g_\ell = 0$ for the magic angle $\theta = 54.7°$ [62,63]. For two-dimensional magnetic systems with square-planar symmetry and a uniform spin structure, Eq. (23) holds. But the equation does not hold for random magnetic materials, in which spin interactions are truncated and cover only finite lengths or sizes. Random magnetic materials have the characteristic of nonuniformity in spin structures.

Magnetic relaxation phenomena reflect spin dynamics and interactions in microscopic detail. Anomalies appearing in magnetic relaxations are vital for organic magnetics because of the low dimensionality inherent in organic systems. The anomalies originate in dimension d and the low symmetry of spin interactions. Related ESR phenomena are described by a relaxation function for magnetization, $\phi(t) = \langle M_+(t)M_-(0)\rangle/\langle M_+M_-\rangle\}$ (a spin self-time-correlation function for transverse magnetization) and its Fourier transform. Time dependence of $M_{+/-}$ is expressed by the slow modulation due to H'.

Table 3 Relations Between the Dimension of Magnetics and Relaxation Function $\phi(t)$ vs. Lineshapes and Linewidths Appearing in ESR Spectra

Dimension d	Relaxation function $\phi(t)$	Lineshape	Linewidth $\Delta B_{1/2}(\theta) \propto$
1	$\exp(-\Gamma t^{3/2})$	FT of $\phi(t)$	$(3\cos^2\theta - 1)^{4/3}$
2	$\exp(-\Gamma t \ln t)$	FT of $\phi(t)$	$(3\cos^2\theta - 1)^2$
3	$\exp(-\Gamma t)$	FT of $\phi(t)$	$(\cos^2\theta + 1)$

For $d = 1$ and 2, the lineshapes are non-Lorenzian except for $\theta = 54.7°$, where the linewidth is minimized. For $d = 1$, $\theta =$ the angle between the direction of a one-dimensional magnetic chain and $\mathbf{B_0}$. For $d = 2$, $\theta =$ the angle between the direction of a normal axis and $\mathbf{B_0}$.

When H' stands for the third term in Eq. (21), absorption linewidths as a function of θ depend on the dimension of the systems under study, where θ stands for the angle between the normal axis of the plane and $\mathbf{B_0}$ in two-dimensional square-planar systems. For reduced symmetric spin systems, the appearance of additional shifts in resonance fields is predicted; the shifts are due to the topology of J-connectivity when the contribution to magnetic relaxation from LTT occurs. Table 3 briefly summarizes the relations between ESR lineshapes, linewidths vs. dimensions, and spin functions, $\phi(t)$ [53].

VIII.E. Electron Spin Resonance of Random Magnetic Materials

Random magnetic materials having nonuniformity in spin arrangements show positive shifts in resonance fields at any orientation of $\mathbf{B_0}$ with decreasing temperature in ESR measurements. This is due to the appearance of long-range ordered phases at low temperature. Contributions from LTT are expected, but propagation and diffusion of spin correlation in low-dimensional organic magnetics are influenced enormously by magnetic impurities. If there are nonmagnetic impurities in the systems, spin assemblages are comprised of magnetic chains or spin clusters with finite lengths or sizes. Below thresholds of the length or size, Eq. (15) becomes constant. Thus, the lineshapes abruptly approach Gaussian nature.

VIII.F. Ferromagnetic and Spin-Wave, Antiferromagnetic, and Ferrimagnetic Resonance

In contrast to those of conventional electron spin resonance spectroscopy, dynamics of magnetic moments in ferromagnetically ordered states are described by invoking the inclusion of the molecular field $\mathbf{B_m}$ and demagnetization field $\mathbf{B_d}$ due to free magnetic moments appearing on the surface of magnetic materials, where $\mathbf{B_d}$ depends on the shape of the material and $\mathbf{B_d}$ is assumed to be uniform inside the material. Contributions from anisotropic energy can be incorporated in $\mathbf{B_d}$. $\mathbf{B_m}$ does not contribute vectorial products in equations, including the magnetic moment. It should also be noted that the usual assumption of the saturation of magnetization in the presence of relatively large $\mathbf{B_0}$ is not valid for the dynamics of magnetization in low field.

If magnetizations of sublattices $\mathbf{M_a}$ and $\mathbf{M_b}$ are cancelled ($\mathbf{M_a} = -\mathbf{M_b}$) by the formation of antiferromagnetically ordered states, no precession of spins is induced in the presence of a microwave oscillating field $\mathbf{B_1}$. When the two magnetizations make an angle against a magnetic easy axis, spin resonance is induced. While $\mathbf{B_m}$ does not contribute spin resonance in ferromagnetics, exchange interactions and anisotroic energy play a vital role in antiferromagnetic resonance. An antiferromagnetic resonance frequency ω is given as

$$\frac{\omega}{\gamma} = \left\{ 2\lambda K + \left(\frac{K}{M_0}\right)^2 + \left(\frac{1}{4}\right) \lambda \chi_\parallel B_0^2 \right\}^{1/2} \pm B_0 \left(1 + \frac{K\chi_\parallel}{2M_0} - \frac{\lambda\chi_\parallel}{2} \right) \quad (24)$$

where λ stands for a molecular-field parameter, K the constant for the anisotropic energy in a uniaxial crystal, and M_0 for saturation magnetization. Setting $\alpha = \dfrac{(\chi_\perp - \chi_\parallel)}{\chi_\perp}$ and further approximation gives

$$\frac{\omega}{\gamma} = \left\{ 2\lambda K + \left(\frac{1}{4}\right) (1 - \alpha)^2 B_0^2 \right\}^{1/2} \pm \frac{B_0(1 + \alpha)}{2} \quad (25)$$

With T approaching zero, α becomes zero, leading to

$$\frac{\omega}{\gamma} = (2\lambda K)^{1/2} \pm B_0 \quad (26)$$

Equation (26) indicates that spin resonance occurs even in the absence of $\mathbf{B_0}$ because $(2\lambda K)^{1/2} > B_0$. A critical applied magnetic field B_0^c changes the precession direction of spins along the easy axis to that of $\mathbf{B_0}$, and is given from Eq. (25) as a solution of the following equation:

$$\frac{B_0^c(1 + \alpha)}{2} = \left\{ 2\lambda K + \left(\frac{1}{4}\right) (1 - \alpha)^2 (B_0^c)^2 \right\}^{1/2} \quad (27)$$

Ferrimagnetic resonance is similar to ferromagnetic resonance as far as $\mathbf{M_a}$ and $\mathbf{M_b}$ ($M_a \neq M_b$) are in antiparallel arrangements. If an angle between the two magnetizations is considered, an additional resonance appears in the high-frequency region. This resonance, called an exchange-mode resonance, is expected, for organic magnetics, to appear in the microwave frequency region.

The variety of magnetically ordered spin assemblages gives rise to characteristic types of magnetic excitations. The corresponding spin resonances have their own related excitations. Nevertheless, spin waves (or magnons) are important in searching for novel quantal phenomena on a mesoscopic scale, such as in low dimensionally patterned superlattices. On account of magnetic interactions other than atom-based exchange interactions, novel standing spin waves are expected to appear in microstructured ferromagnetic thin films. The possible appearance of self-oscillation and geometrical degeneracy on a semimacroscopic scale is of great interest in standing spin-wave resonance spectroscopy. They will be vital in studies of novel molecular devices and spinics.

IX. FINAL REMARKS

There have been many review articles and books dealing with electron magnetic resonance in ordered regimes. Magnetic excitations in low-dimensional organic magnetics are quite different from those of atomic-based magnetics in many respects. A comprehensive treatise, nevertheless, has not yet appeared on such topics of organic-molecule-based magnetics in ordered regime [53]. Similarly, this chapter cannot be complete, even in terms of selected topics of methodology. Interested readers should refer to basic treatises on electron magnetic resonance of magnetically ordered substances in quantum terms [53].

ACKNOWLEDGMENTS

This work has been supported by Grants-in-Aid for Scientific Research on Priority Area "Molecular Magnetism" (Area Nos. 228/04 242 103 and 04 242 105) from the Ministry of Education, Culture, and Science, Japan, and also by the Ministry of International Trade and Industries, Japan (NEDO Project "Organic Magnets"). The authors (K. S. and D. S.) acknowledge the Ministry of Education, Culture, and Science, Japan, for Grants-in-Aid for Encouragement of Young Scientists (Grant Nos. 07740468 and 07740553, respectively).

REFERENCES

1. Morimoto S, Tanaka F, Itoh K, Mataga N. Preprints of Symposium on Molecular Structure. Chem Soc Japan. 1968:67.
2. Mataga N. Theor Chim Acta. 1968; 10:372.
3. Itoh K. Bussei. 1971; 12:635.
4. Itoh K. Pure Appl Chem. 1978; 50:1251.
5. Ovchinnikov AA. Theor Chim Acta. 1978; 47:297.
6. Takui T. PhD. dissertation, Osaka University, 1973.
7. McConnell, HM. J Chem Phys. 1963; 39:1910.
8. McConnell HM. Proc R.A. Welch Found Chem Res. 1967; 11:144.
9. Itoh K. Chem Phys Lett. 1967; 1:235.
10. Wasserman E, Murray RW, Yager WA, Trozzolo AM, Smolinsky G. J Am Chem Soc. 1967; 89:5067.
11. For a recent overview, see Refs. 12–17 as well as Miller JS, Dougherty DA, eds. Mol Cryst Liq Cryst. 1989; 176:1–562.
12. Chiang LY, Chaikin PM, Cowan DO. eds. Advanced Organic Solid State Materials. Mater Res Soc. 1990:1–92.
13. Gatteschi D, Kahn O, Miller JS, Palacio F, eds. Molecular Magnetic Materials. Dortrechdt: Kluwer, 1991.
14. (a) Iwamura H, Miller JS, eds. Mol Cryst Liq Cryst. 1993; 232/233:1-360/1-366. (b) Miller JS, Epstein AJ, eds., Mol Cryst Liq Cryst. 1995; 272/273:1-222/1-216. (c) Ibid., 274/275:1-218/1-226.
15. Turnbull MM, Sugimoto T, Thompson LK, eds. Molecule-Based Magnetic Materials: Theory, Technique, and Applications. New York: American Chemical Society, 1996.
16. Itoh K, Miller JS, Takui T, eds. Mol Cryst Liq Cryst. 1997; 305/306:1–586/1–520.
17. (a) Miller JS, Epstein AJ, Reiff WM. Acc Chem Res. 1988; 22:144 and references therein. (b) Ibid., Chem Rev. 1988; 88:201 and references therein.
18. Takui T, Itoh K. Polyfile. 1990; 27:49.
19. Takui T, Itoh K. J Mat Sci Japan. 1991; 28:315.
20. Takui T. Polyfile. 1992; 29:48.
21. Takui T. Chemistry. 1992; 47:167.
22. Miller JS, Epstein AJ. Chemtech. 1991; 21:168.
23. Landee CP, Melville D, Miller JS. In: Gatteschi D, Kahn O, Miller JS, Palacio F, eds. Molecular Magnetic Materials. Dortrechdt: Kluwer, 1990.
24. Kahn O, ed. Magnetism: A Supramolecular function. Dortrechdt: Kluwer, 1996.
25. Bencini A, Gatteschi D. EPR of Exchanged Coupled Systems. Berlin: Springer-Verlag, 1990, pp. 1–287.
26. de Groot MS, van der Waals JH. Mol Phys. 1960; 3:190.
27. Teki Y, Takui T, Itoh K. J Chem Phys. 1988; 88:6134.
28. Neiman R, Kivelson D. J Chem Phys. 1961; 35:156.
29. Ovchinnikov V, Konstantinov VN. J Magn Reson. 1978; 32:179.
30. Weltner Jr W. Magnetic Atoms and Molecules. New York: Scientific and Academic Editions, 1983, pp. 1–422.

31. Iwasaki M. J Magn Reson. 1974; 16:417.
32. Rockenbauer A, Simon P. J Magn Reson. 1973; 11:217.
33. Toriyama K, Iwasaki M. Personal communication, 1974.
34. Shimokoshi K. Personal communication, 1975.
35. Bruker Report: Bruker Analytishe Messtechnik GMBH, D-7512 Rheinstetten, FRG.
36. Sato K, Takui T, Itoh K. Program Software Package (ESR Spectral Simulation Based on Higher-Order Perturbation Approach). Osaka City University.
37. Banwell CN, Primas H. Mol Phys. 1963; 6:225.
38. Belford GG, Belford RL, Burkhalter JF. J Magn Reson. 1973; 11:251.
39. Wasserman E, Snyder LC, Yager WA. J Chem Phys. 1964; 41:1763.
40. Sato K. PhD dissertation, Osaka City University, 1994.
41. Teki Y, Fujita I, Takui T, Kinoshita T, Itoh K. J Am Chem Soc. 1994; 116: 11499.
42. Sato K, Takui T, Itoh K. Program Software Package (ESR Spectral Simulation Based on Eigenfield Approach). Osaka City University.
43. Strach SJ, Bramley R. J Chem Phys. 1988; 88:7380.
44. Sato K, Takui T, Itoh K. Program Software Package (Exact Analytical Expressions for Eigenfields up to $S = 3$ and Eigenenergies up to $S = 4$). Osaka City University.
45. Sato K, Takui T, Itoh K. Unpublished work.
46. Sato K, Takui T, Itoh K. Program Software Package (Exact Analytical Expressions for Eigenfields up to $S = 3$ and Eigenenergies up to $S = 4$), Ver. 2.0: Graduate School of Science, Osaka City University.
47. Pilbrow JR. Abstracts of Twelfth Conference of ISMAR. Sydney: ISMAR, 1995. L 18.1.
48. Isoya J, Kanda H, Norris JR, Tang J, Bowman MK. Phys Rev. 1990; B41: 3905.
49. Astashkin AV, Schweiger A. Chem Phys Lett. 1990; 174:595.
50. Sato K, Shiomi D, Takui T, Itoh K, Kaneko T, Tsuchida E, Nishide H. J Spectrosc Soc Japan. 1994; 43:280.
51. Nishide H, Kaneko T, Nii T, Katoh K, Tsuchida E, Yamaguchi K. J Am Chem Soc. 1995; 117:548.
52. Shiomi D, Nishizawa M, Sato K, Takui T, Itoh K, Sakurai H, Izuoka A, Sugawara T. J Phys Chem. 1997; 101:3342.
53. Takui T. Organic Molecule-Based Magnetic Materials. In: Horie K., Taniguchi A, eds. Handbook of Opto-Electronic Organic Functionality Materials. Tokyo: Asakura, 1994, Ch. 9.
54. Landau L, Lifshitz E. Phys Zei'ts Soviet Union. 1935; 8:153.
55. Gilbert TL. Phys Rev. 1995; 100:1243.
56. Kittel C. Phys Rev. 1950; 80:918.
57. Bloch F. Phys Rev. 1946; 70:460.
58. (a) Nagata K. Bussei. 1972; 13:149. (b) Nagata K. Physics for Random Systems. Tokyo: Barifuhkar, 1981, Ch. 13.
59. Kubo R, Tomita K. J Phys Soc Japan. 1954; 9:888.

60. (a) Mori H. Prog Theor Phys. 1965; 33:423. (b) ibid., 34:399. (c) Tokuyama M, Mori H. Prog Theor Phys. 1976; 55:411.
61. (a) Nagata K, Tazuke Y. J Phys Soc Japan. 1972; 32:337. (b) Tazuke Y, Ngata K. J Phys Soc Japan. 1971; 30:285.
62. Caneschi A, Gatteschi D, Rey D, Sessoli R. Inorg Chem. 1988; 27:1756.
63. Boucher J-P. J Mag Mat. 1980; 15:687.

12

Design and Experimental Investigation of High-Spin Organic Systems

Yoshio Teki and Koichi Itoh
Osaka City University, Osaka, Japan

I. INTRODUCTION

This chapter deals with the design and experimental investigation of high-spin organic systems, with focus on high-spin polycarbenes and their topological isomers [1–14]. We define a π-topological isomer as a molecule that differs from others only in the topology of its π-electron network, i.e. in the linking positions of its π-bonds. Polycarbenes are ideal model compounds for the study of intramolecular spin correlation and the resulting spin alignment in the open-shell organic spin systems.

High-spin polycarbenes as organic high-spin systems are, in spite of their high chemical reactivity, very important from the viewpoint of organic magnetism as well as of spin ordering/controlling in chemistry, for the following reasons:

1. One of the most prominent features of high-spin polycarbenes is the presence of multielectron open-shell systems in the ground and/or low-lying excited states. These high-spin states arise from the spins occupying degenerate delocalized π-orbitals and dangling σ-orbitals localized at divalent carbon atoms, the latter orbitals being nearly degenerate with the highest half-filled π-orbitals.

2. The degeneracy of their π-orbitals is governed by a particular connectivity of the π-electron network, i.e. by the topology of the π-electron network.
3. The number of their degenerate π-orbitals is in principle unlimited within the framework of the π-electron molecular-orbital approximation.
4. The spin correlation of their unpaired electrons via the delocalized π-orbital network is most important for determining the spin states of these high-spin polycarbenes and their topological isomers.

In this chapter, we describe spin alignment and the role of the spin correlation on the intramolecular spin alignment in the high-spin ground states or the low-spin ground states with low-lying high-spin excited states of the topological isomer. The organic high-spin molecules and their topological isomers **1–9** appearing in this review are shown in Figure 1. Their spin density distributions provide the most direct information about the spin correlation among the unpaired electrons and the mechanism of the spin alignment in these organic molecules. Their spin density distributions were, therefore, determined by ^{1}H- and ^{13}C-ENDOR (electron nuclear double resonance) spectroscopies as well as by the theoretical calculations based on two model Hamiltonian approaches (the Hubbard model and the valence bond Heisenberg model).

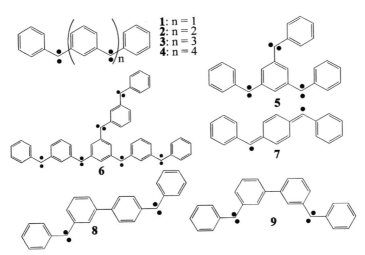

Figure 1 Chemical structures of high-spin polycarbenes and their topological isomers described in the text.

II. SIMPLE SPIN-PREDICTION THEORIES AND HUBBARD HAMILTONIAN APPROACH FOR ORGANIC OPEN-SHELL SYSTEMS

For π-electron networks of alternant hydrocarbons (AHs), there are two simple rules that have been proposed for predicting the ground state spin: one by Longuet-Higgins using molecular orbital (MO) theory [15], and one by Ovchinnikov and Klein et al. using valence bond (VB) theory [16–18]. The Hubbard model Hamiltonian approach presented here is more sophisticated than these, in that it gives the physical picture of the spin alignment as well as the spin state prediction.

II.A. Simple Spin-Prediction Theories

Molecular orbital theory gives a simple prediction for the number of non-bonding MOs, which in turn gives the number of degenerate singly occupied MOs (SOMOs) for neutral AHs. This rule states that an AH has at least $(N - 2T)$-fold degenerate SOMOs, where N is the number of carbon atoms in the π-network and T is the possible maximum number of double bonds. The $(N - 2T)$-fold degeneracy is due to topological symmetry of the π-electron network. By simply combining Hund's rule with this topological nature, one can predict the maximum spin S^π in the ground state as follows:

$$S^\pi = \frac{|N - 2T|}{2} \tag{1}$$

As described in Section III, this guiding principle is useful for designing high-spin hydrocarbons such as **1–6**. This simple spin prediction has, however, been violated sometimes in the cases of "disjoint" AHs [19], in which possible open-shell spin states are nearly degenerate with each other, while for "non-disjoint" AHs the highest spin state predicted from Eq. (1) lies well below the other, lower-spin states and Eq. (1) gives reliable spin prediction in the ground state. The first example of the breakdown of this simple ground state prediction based on Eq. (1) was biphenyl-3,3'-bis-(phenylmethylene) **9** [5], which is discussed in more detail in Section V.

Simple VB theory has proved to give better predictions, since the theory takes spin correlation into account. It gives

$$S^\pi = \frac{|n^* - n|}{2} \tag{2}$$

where n^* and n denote the number of starred and unstarred carbon atoms, respectively, in the π-network. This approach gives the correct ground state spin for both disjoint and nondisjoint AHs. These simple spin predictions, however, give no detailed information about the mechanism and the physical picture of the intramolecular spin alignment in the ground spin state as well as in the low-lying excited spin states.

II.B. Generalized Hubbard Model

The Hubbard model provides the simplest model Hamiltonian that takes into account both degenerate SOMOs and spin correlation. This approach gives a clear physical picture of the intramolecular spin alignment of AHs discussed hereafter, as well as the spin prediction of low-lying excited states generated as the result of correlation among the unpaired electron spins in degenerate SOMOs. The following generalized Hubbard model takes into account the additional nonbonding ("n") electrons in the dangling σ-bonds at divalent carbon sites. This model Hamiltonian approach has provided satisfactory and complementary descriptions for the spin-alignment and spin-density distributions of organic high-spin polycarbenes and their topological isomers [14,20].

The generalized Hubbard model Hamiltonian is given by [20–23].

$$
H = -T \sum a_{m'\sigma}^+ + a_{m\sigma} + \left(\frac{1}{2}\right) U \sum n_{m-\sigma} n_{m\sigma}
$$
$$
- J \sum \left[S_k^z S_m^z + \frac{S_k^+ S_m^- + S_k^- S_m^+}{2} \right] \tag{3}
$$

where $n_{m\sigma} = a_{m'\sigma}^+ a_{m\sigma}$ and T is a π-electron transfer integral (i.e., minus Hückel β) between the adjacent carbon sites. U and J are the effective electron repulsions between two π-electrons and the intramolecular exchange integral between the n- and π-electrons at the divalent carbon atoms, respectively. The sign of J should be positive (ferromagnetic), since the σ- and π-orbitals are mutually orthogonal on each divalent carbon atom. This model Hamiltonian has been solved by employing the Hartree–Fock approximation [22],

$$
n_{m-\sigma} n_{m\sigma} = \langle n_{m-\sigma} \rangle n_{m\sigma} + n_{m-\sigma} \langle n_{m\sigma} \rangle - \langle n_{m-\sigma} \rangle \langle n_{m\sigma} \rangle \tag{4}
$$

Here, $\langle \ \rangle$ denotes the average with respect to a desired spin state. In this calculation, each n-electron is assumed to be localized in each dangling σ-orbital of the divalent carbon site. Taking the parameters to be $U/T = 2.0$, $J/T = 0.25$, and $T = 2.4$ eV [14], the spin multiplicities of a series of high-spin polycarbenes in their ground states (as well as their spin-density distributions) have been well reproduced. The physical picture of intramolecular spin alignment and the design of very high-spin polycarbenes are described in the next section on the basis of the generalized Hubbard model.

The spin alignment in the topological isomers is also discussed in this review.

III. EXPERIMENTAL INVESTIGATION OF VERY HIGH-SPIN ORGANIC SYSTEMS

A series of *meta*-conjugated polyphenylcarbenes such as **1–6** are expected to have *m*-fold degenerate π-SOMOs according to Eq. (1), where *m* is the number of the divalent carbon atoms. The unpaired π-electrons in the *m*-fold degenerate π-SOMOs are conjugated with the π-systems of the benzene rings. The other unpaired "*n*"-electrons remain in the dangling σ-bonds at each divalent carbon atom. All these unpaired electrons are expected to become spin parallel with each other, leading to a high-spin ground state with $S = m$.

We have investigated the electronic ground states of these polyphenylcarbenes **1–6** [1,3,5–10,12,14] by means of electron-spin resonance (ESR) experiments. The highest spin multiplicity expected from the total number of the unpaired electrons ($2m$ from m sets of unpaired π- and *n*-electrons) was actually found for all the cases examined. Thus, both the spin prediction for the ground state from Eqs. (1) and (2) and the results of the generalized Hubbard model calculations agree well with experiments. Figure 2a gives a typical example of the high-spin ESR spectrum observed for the nonet ($S = 4$) polycarbene **3** [6,8]. This spectrum was observed at 4.2 K after photolysis of the corresponding diazo precursors (Scheme 1), which were diluted as a guest molecule in the benezophenone host crystals.

The spectrum consists of four well-resolved pairs A±, B±, C± and D±; the relative separations of each pair are nearly (A$_-$–A$_+$):(B$_-$–B$_+$): (C$_-$–C$_+$) (D$_-$–D$_+$) = 7:5:3:1 and the relative integrated intensities are nearly A±:B±:C±:D± = 4:7:9:10. These ratios are just what one would expect for the $\Delta M_S = \pm 1$ allowed transitions between the fine-structure sublevels of the nonet ($S = 4$) spin manifold in the high-field limit. Figure 2b shows the observed angular dependence of these four pairs, A±, B±, C±, and D±. The angular dependence of the resonance fields and the relative

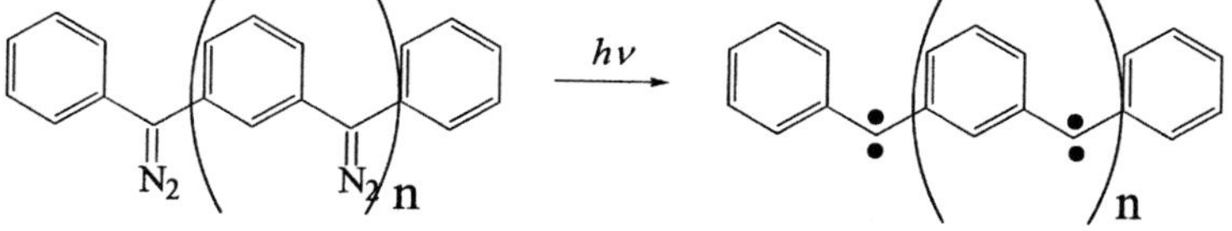

Scheme 1

(a)

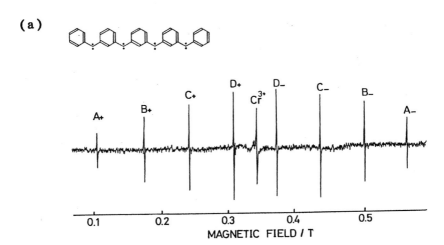

(b)

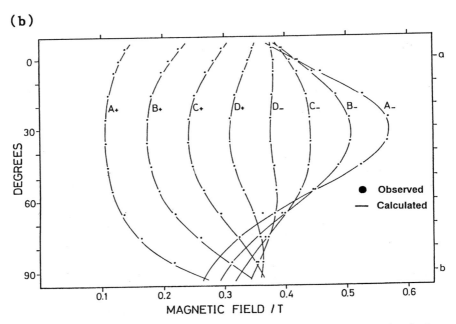

Figure 2 ESR spectrum of the nonet polycarbene **3** observed after photolysis at 4.2 K. (a) The ESR spectrum of the eight allowed transitions with the external magnetic field along the direction 26° from the a-axis in the ab-plane of the host crystal. The central line is due to Cr^{3+}, used as a reference. (b) The angular dependence of the resonance fields. The microwave frequency $\nu = 9550.6$ MHz, and the ESR parameters are $g = 2.002$, $D = +0.03161$ cm^{-1}, and $E = -0.00394$ cm^{-1}.

intensities were analyzed using the following spin Hamiltonian with a high-spin state of $S = 4$:

$$H'_{spin} = \beta_e H.g.S + S.D.S$$

$$= \beta_e H.g.S + D\left[S_z^2 - \frac{S(S+1)}{3}\right] + E(S_X^2 - S_Y^2) \quad \text{for } S = 4 \quad (5)$$

From the angular dependence of the resonance fields, the spin-Hamiltonian parameters were evaluated as given in Table 1. The solid curves in Figure 2b were obtained by the exact diagonalization of spin-Hamiltonian Eq. (5) using the experimentally determined g-value and fine-structure parameters. The spin-Hamiltonian parameters for the other high-spin polycarbenes, i.e. undecet ($S = 5$) **4** [10] and tridecet **6** ($S = 6$) molecules [12] as well as the quintet **1** [1] and septet molecules **2**, **5** [3,7] are also listed in Table 1. The D-value apparently decreases with increasing S. However, if we compare $(2S - 1)D$ instead of D for the high-spin polycarbenes, this quantity is nearly a constant value between 0.2 and 0.25. For the high-spin polycarbenes, D is determined predominantly from the large, one-center $n-\pi$ interactions on the divalent carbon sites. In such a case, the quantity $(2S - 1)D$ measures the averaged $n-\pi$ interaction in the molecule, where the factor $2S - 1$ comes from the projection theorem of angular momentum. Thus, a series of the polycarbenes with similar electronic structure are expected to have a similar magnitude of $(2S - 1)D$.

For a spin system with eight unpaired electrons, such as **3**, five kinds of energy levels with $S = 0, 1, 2, 3$, and 4 are permissible. Figure 2a, however, shows that only the ESR spectrum of the nonet state was observable at 4.2 K. Moreover, no triplet, quintet, and septet signals could be detected up to the temperature where **3** decomposed, i.e., 160 K. Given the sensitivity of the spectrometer, it can therefore be concluded that the spin

Table 1 D, E, and $(2S - 1)D$ Compared among the High-Spin Polycarbenes with Similar Electronic Structures

Polycarbenes	S	D	$\lvert E \rvert$	$(2S - 1)D$	Ref.
1	2	0.0713	0.0190	0.2139	1
2	3	0.0487	0.0089	0.2437	7
3	4	0.0316	0.0039	0.2213	6,8
4	5	−0.0168	0.0036	0.1513	10
5	3	0.0416	0.0103	0.2079	3
6	6	0.0191	0.0019	0.2101	12
8	2	0.1250	0.0065	0.3750	14

states other than the singlet (which is undetectable by ESR) are located at least 300 cm^{-1} above the nonet state. In order to estimate the location of the singlet state, the temperature dependence of the total signal intensity of the nonet signals $I_4(T)$ was measured down to 2 K. The total signal intensity is plotted as a function of inverse temperature in Figure 3. The solid curves are plots of the total intensity calculated for the cases of $\Delta E = -5, -3, -1$, and $+\infty$ cm^{-1}, where ΔE is given by the energy separation between the singlet and the nonet spin states. If the singlet state were located below the nonet state ($\Delta E < 0$), a considerable decrease in the total intensity would occur below 4.2 K. However, since Figure 3 shows no appreciable negative curvature, it is concluded that the nonet high-spin state is the electronic ground state, and that other spin states are located at least 300 cm^{-1} above the nonet ground state. For other *meta*-conjugated polycarbenes, **1, 2, 4–6**, a similar temperature dependence was also observed. Therefore, one can conclude that their high-spin ground states are electronic ground states with a large energy separation ($\Delta E > 300$ cm^{-1}) between the ground states and other spin states. These findings clearly show that an organic spin system with a very large number of parallel spins can be obtained if its molecular structure is properly designed on the basis of its topological symmetry. Extension of the *meta*-conjugated high-spin hydrocarbons reported here may

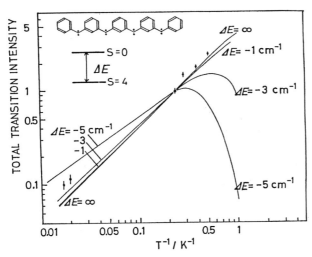

Figure 3 Plots of the observed total signal intensity of **3** as a function of inverse temperature. The solid curves represent the calculated total intensities. ΔE is the energy gap between the singlet and the nonet states.

be one of the most promising approaches to organic polymer ferromagnets or superparamagnets.

IV. PHYSICAL PICTURE OF THE SPIN ALIGNMENT IN HIGH-SPIN ORGANIC SPIN SYSTEMS

In order to clarify the physical picture of the intramolecular spin alignment and the role of intramolecular spin correlation, we have investigated the spin-density distribution of two typical organic high-spin dicarbenes, biphenyl-3,4'-bis(phenylmethylene) **8** [14,24] and biphenyl-3,3'-bis(phenylmethylene) **9** [24,25] by means of ^{1}H-ENDOR experiment [26] as well as theoretical calculations using the generalized Hubbard model [20–23]. Molecule **8** is the topological isomer of an important dicarbene **9** [26], described in detail in Section V. Figures 4a and 4b show typical ESR and ^{1}H-ENDOR

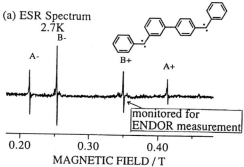

(a) ESR Spectrum
2.7K
B-
A-
B+
A+
monitored for
ENDOR measurement

0.20   0.30   0.40
MAGNETIC FIELD / T

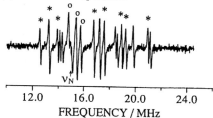

(b) ^{1}H-ENDOR Spectrum

ν_N

12.0   16.0   20.0   24.0
FREQUENCY / MHz

Figure 4 Typical ESR spectrum (a) and ^{1}H-ENDOR spectrum (b) of the quintet ($S = 2$) ground state of **8**. The external magnetic field is along the b-axis of the host benzophenone crystal. The ^{1}H-ENDOR signals indicated by asterisk were assigned to the signals arising from the adjacent protons on the biphenyl group by using the partially deuterium-labeled compound.

spectra of the quintet ($S = 2$) ground state of **8** [24]. All the ^{1}H-hyperfine tensors were determined from analysis of the angular dependence of their ^{1}H-ENDOR frequencies. The π-spin densities on the carbon atoms adjacent to the hydrogen atoms were obtained from the isotropic terms A_F's of their hyperfine tensors with the help of McConnell's equation $A_{Fi} = Q\rho_i^\pi/2S$ ($Q = -66.9$ MHz [27,28]). The spin densities on the carbon atoms without adjacent protons could not be obtained from the ^{1}H-ENDOR experiment. The observed π-spin densities and the calculated ones based on the generalized Hubbard model are given in Figures 5a and 5b, respectively. The calculated spin-density distribution in the case of the typical high-spin dicarbene **8** consistently explains the observed spin-density distribution, the number of the positive and negative π-spin densities, and the magnitude of each spin density.

These spin-density distributions give an excellent physical picture of the intramolecular spin alignment of high-spin polycarbenes. We present here the case of **8** using unrestricted Hartee-Fock (UHF) calculations based on the generalized Hubbard model. The Hubbard calculation of **8** revealed that the quintet high-spin state is the most stable electronic spin state. A similar situation also holds for molecule **1** [26]. Although the total spin is not generally conserved in a UHF calculation, the spin contamination from other spin states is not so significant for the high-spin ground state. Figure 5b shows the calculated spin-density distribution in the quintet ground state. On the basis of this spin distribution, we can understand the spin alignment of **8**. The unpaired π-spins are distributed over the carbon skeleton with the sign of the spin density alternating from carbon to carbon, thus leading to the formation of a pseudo π-spin-density wave (pseudo π-SDW) in the π-electron network, as schematically shown in Figure 6.

The localized unpaired n-spins in the dangling σ-bonds are exchange-coupled ferromagnetically [$J > 0$ in Eq. (3)] to the unpaired π-spins at each divalent carbon atom, causing the n- and π-spins to be parallel to each other. Since the signs of the π-spin densities at both divalent carbon atoms are the same, due to the pseudo π-SDW, this molecule has four parallel spins, leading to a quintet ground state. Thus, the highest-spin state comes from the π-SDW that is energetically most favorable in view of the total spin-exchange correlation energy. The formation of a pseudo π-SDW state depends strongly on the topology of the π-electron network, since the first term of the Hubbard Hamiltonian [Eq. (3)] expresses the transfer of π-spins between the adjacent carbon atoms, indicating the important role of π-topology. This example shows that the π-topology plays a key role in intramolecular spin-exchange correlation as well as for the number of degenerate SOMOs [($N - 2T$)-fold degeneracy of the unpaired π-orbitals]. For m-polyphenylcarbene systems such as **1–4**, Hubbard model calculations of the one-dimensional infinite system have been carried out, and the stability of these high-spin ground states has been shown by Nasu [23].

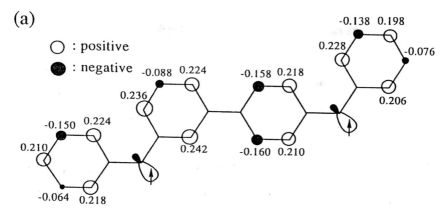

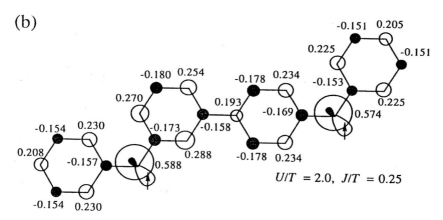

Figure 5 π-Spin density distribution of the quintet ground state of **8**. (a) Observed values obtained from the ^{1}H-ENDOR experiment. (b) Calculated values obtained from the generalized Hubbard model calculation ($U/T = 2.0$ and $J/T = 0.25$). The alternation of the sign of the π-spin density occurs from carbon to carbon.

V. BREAKDOWN OF THE SIMPLE MOLECULAR ORBITAL SPIN PREDICTION AND THE MECHANISM OF INTRAMOLECULAR SPIN ALIGNMENT IN LOW-LYING EXCITED SPIN STATES

V.A. Spin States of Biphenyl-3,3′- and -3,4′-bis(phenylmethylene)

Figure 7 shows the energy diagrams of the ground states and the low-lying excited states of the typical high-spin dicarbenes, biphenyl-3,4′- and -3,3′-

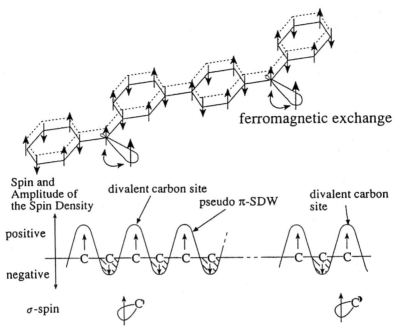

Figure 6 Mechanism of the intramolecular spin alignment of the high-spin poly-carbene **8** in terms of the pseudo π-SDW. The up-and-down network of the π-spins is established. Note that the π-spin densities at the two divalent carbon atoms have the same positive sign. The π-spin densities were calculated using the generalized Hubbard model.

bis(phenylmethylene), **8** and **9**. As already mentioned in Section II, 3,3'-isomer **9** was the first example of a disjoint alternant hydrocarbon that violated the ground state spin prediction based on the Longuett-Higgins theory [5,25]. Thus, the lowest energy levels of biphenyl-3,3'-bis(phenylmethylene), **9**, were found to be nearly degenerate singlet, triplet, and quintet states, in the order of increasing energy, i.e., a singlet ground state [5] as shown in Figure 7. On the other hand, its topological isomer, 3,4'-isomer **8**, has a single high-spin ($S = 2$) ground state [14]. The low-lying triplet and quintet states of **9** were determined from the temperature dependence of the ESR signal intensity to be located above the ground state by 20 and 60 cm^{-1}, respectively [5]. The fine-structure parameters of the low-lying triplet and quintet states are $D_T = -0.2986$ cm^{-1}, $E_T = 0.0603$ cm^{-1}, and $D_Q = 0.1035$ cm^{-1}, $E_Q = 0.0146$ cm^{-1}, respectively [5]. It should be noted that a particular relationship $\mathbf{D}^T = 3\mathbf{D}^Q$ holds between the observed fine-structure tensor of

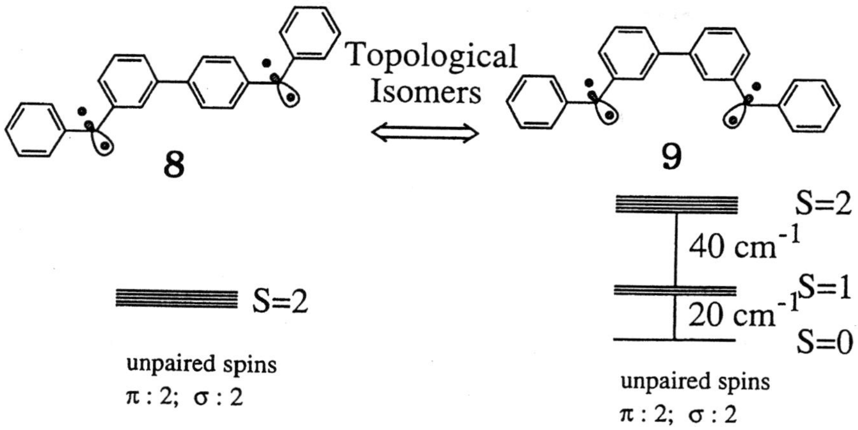

Figure 7   Ground states and the low-lying excited states of the topological isomers **8** and **9**, whose ground states are high-spin and low-spin, respectively.

the low-lying triplet state, $\mathbf{D}^T$, and that of the nearly degenerate quintet state, $\mathbf{D}_Q$.

V.B.   Breakdown of Spin State Prediction

Both molecules **8** and **9** have four unpaired electron spins in their quadruply nearly degenerate nonbonding π- and σ-orbitals, as easily shown by simple MO calculation. It can also be shown that more sophisticated molecular orbital calculations lead qualitatively to the same findings. The simple spin prediction of Eq. (1) gives a high-spin quintet ground state for both of molecules **8** and **9**, as follows: Equation (1) predicts two parallel spins in the two π-SOMOs, as determined from $N - 2T = 2$. In addition, the σ- and π-spins on the same divalent carbon atom should be parallel, since the one-center exchange integral between the σ- and π-spins on the same atom is ferromagnetic. As a result, the spin prediction based on simple MO theory plus Hund's rule leads to quintet ($S = 2$) high-spin ground states for both of the molecules. Molecule **8** actually has a high-spin ($S = 2$) ground state, in agreement with the foregoing spin prediction. However, **9** has a low-spin singlet ($S = 0$) ground state with low-lying excited higher spin states, against the prediction.

The difference in their ground state spin multiplicities demonstrates clearly that the intramolecular spin alignment in organic molecules is highly dependent on the topological nature of the π-electron network. Thus, it is expected that the different ground state spins arise from the different pathways of the intramolecular spin correlation among unpaired spins. As a typical example showing the importance of spin correlation, we hereafter deal with the intramolecular spin alignment of biphenyl-3,3′-bis-(phenylmethylene), **9**. This molecule is peculiar in that it violates the MO spin prediction of Eq. (1) [5], whereas the VB spin prediction of Eq. (2) gives the correct spin state for the ground states.

V.C. Simple Valence Bond Explanation of the Spin Alignment in Biphenyl-3,3′-bis(phenylmethylene)

As already mentioned in Section IV, the pseudo π-spin-density wave (the up-and-down spin network of the π-spins) can be recognized by identifying the sign of the spin densities at the divalent carbon sites in the case of **8**, in agreement with the simple valence bond prediction of Eq. (2). The physical picture of the spin alignment in **8** via the pseudo π-SDW is shown in Figure 6a. On the other hand, in the case of **9**, the pseudo π-SDW has opposite signs at the two divalent carbon sites in the pathway of the spin correlation, according to the topology of its π-electron network, as shown in Figure 8. Thus, this pseudo π-SDW leads to a singlet ground state whose spin densities, however, cannot be detected because of its ESR silence. This simple picture gives a simple qualitative reason why **9** has a low-spin ground

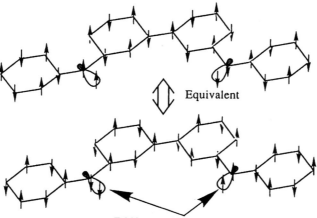

Different Signs of Spin Densities

Figure 8   Two equivalent pseudo π-SDW states with equal energies but opposite phases of the spin densities. Superposition of these two gives the singlet ground state without net spin densities. Note that the two divalent carbon atoms have opposite signs of spin density.

state, in contrast to **8**. It should be mentioned that these two pathways shown in Figure 8 correspond to the two most important ones among many other spin configurations contributing to the singlet ground state [34].

The foregoing simple approach gives no detailed understanding of the spin alignment for the low-lying higher-spin states. The topological isomers of the high-spin polycarbenes always have a low-spin ground state and low-lying excited higher-spin states. Clarifying the spin-density distribution of the low-lying excited higher-spin state is especially important in order to understand the intramolecular spin correlation among the unpaired spins in organic multi-open-shell systems. Since one can obtain experimentally no detailed information of the spin correlation in the singlet ground state without any net spin density, we discuss the intramolecular spin correlation in the low-spin polycarbenes on the basis of the spin-density distribution in the low-lying excited high-spin state.

V.D. Weakly Interacting Multiplet–Multiplet Model

In addition to the experimental work, we have calculated the spin-density distribution and the electronic state of **9** by a model Hamiltonian approach that is referred to as a weakly interacting multiplet–multiplet model [5,24,25]. Using the model described next, we have successfully dealt with the ground and low-lying excited states that are formed by a magnetically weak interaction between two high-spin units, as shown in Figure 9. In what follows, the Hubbard and Heisenberg model Hamiltonian approaches and a weakly interacting multiplet–multiplet model are briefly described. The useful relationships between the spin densities (the fine-structure parameters) of the whole molecule and those of the isolated moieties are also derived. The energy relationship among the ground and the low-lying excited spin

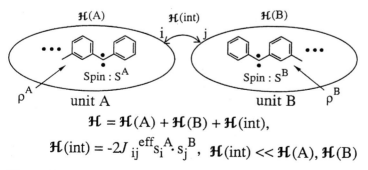

Figure 9 Weakly interacting multiplet–multiplet model. The molecule is regarded as composed of units A and B, which are connected at the sites i and j, respectively.

states is discussed within the framework of the weakly interacting multiplet–multiplet model described next.

In this model, we regard the whole molecule as being composed of two unit moieties weakly interacting with each other, as shown in Figure 9. This model is a good approximation for molecule **9**, in which the two high-spin diphenylmethylene molecules are connected to each other at the carbon sites where the unpaired π-orbitals (π-SOMOs) have a node in the restricted Hartee-Fock (RHF) one-electron MO or the simple Hückel MO [5]. Thus, molecule **9** corresponds to the disjoint case in the disjoint–nondisjoint theory by Borden and Davidson [19].

The whole Hamiltonian for the system consisting of the two weakly interacting units A and B with the spins S^A and S^B, respectively, may be written as

$$H = H(A) + H(B) + H(\text{int}), \qquad H(\text{int}) \ll H(\text{A}), H(\text{B}) \tag{6}$$

where the first (second) term is the electronic Hamiltonian consisting of both orbital and spin parts for separate units A (B) and the third term is that for the interaction between them. We assume the interaction term as being a Heisenberg exchange type:

$$H(\text{int}) = -2J_{ij}^{\text{eff}} s_i^A . s_j^B \tag{7}$$

This term arises from the effective exchange through the chemical bond in the case of **9**. The term s_i^A (s_j^B) is the spin operator of the π-electron spin on the carbon site $i(j)$ belonging to unit A (B). As discussed later, the relation $H(\text{int}) \ll H(\text{A}), H(\text{B})$ is valid for disjoint-type molecules. Under this condition (weak-interacting limit), we derived the following relationship for the spin densities of the whole molecule with the total spin S. In this weak interaction limit, the wave function of the whole molecule $|\Psi(S, M)\rangle$, is given by the direct product of the wave functions of the two isolated unit moieties, $|\chi(S^A, m_A)\rangle$ and $|\chi(S^B, m_B)\rangle$, namely:

$$|\Psi(S, M)\rangle = \sum C(S^A S^B S; m_A m_B M)|\chi(S^A, m_A)\rangle|\chi(B^B, m_B)\rangle \tag{8}$$
$$+ \text{ (higher-order terms)}$$

where $C(S^A S^B S; m_A m_B M)$ is the Clebsch–Gordan coefficient. The spin density ρ_i^A on the carbon site i in the interacting A unit of the whole molecule can be expressed in terms of the corresponding spin density ρ_{i0}^A in the isolated A-unit molecule with the spin S^A and of the Clebsch–Gordan coefficient $C(S^A S^B S; m_A M - m_A)$ [24,25]:

$$\rho_i^A(S, M) = \left(\frac{1}{S^A}\right) \rho_{i0}^A \sum m_A C(S^A S^B S; m_A M - m_A)^2 \tag{9}$$

where $S = S^A + S^B, S^A + S^B - 1, \ldots, |S^A - S^B|$. A similar expression also holds for unit B. Table 2 shows the relationships derived from Eq. (9) for the several cases of $S^A = S^B$ (homo-spin system) and of $S^A \neq S^B$ (hetero-spin system). Molecule **9** corresponds to the homo-spin system, with $S^A = S^B = 1$. We can, therefore, calculate the spin densities in the whole molecule from those of the isolated unit moieties. In this weak-interacting limit, we can also derive the energy relationship among the ground and the low-lying excited spin states formed from the two interacting high-spin units, as given by

$$H = E_A + E_B - J_{ex}[S(S + 1) - S^A(S^A + 1) - S^B(S^B + 1)] \qquad (10)$$

The effective exchange interaction J_{ex} is expressed in terms of the spin densities ρ_{i0}^A and ρ_{j0}^B, on the connecting sites i and j in Figure 9, and of the effective exchange integral J_{ij}^{eff} in Eq. (7), yielding

$$J_{ex} = J_{ij}^{eff} \frac{\rho_{i0}^A \cdot \rho_{j0}^B}{4S^A S^B} \qquad (11)$$

This equation corresponds to a generalization of the relation given by McConnell [29] for the arbitrary spins, S^A and S^B, in the intermolecular case.

The special relationships between the fine-structure tensor, D^S, of the whole molecule with total spin S and those of the isolated moieties, D^A and D^B, in the spin Hamiltonian can be derived within the weakly interacting model, using the operator technique of angular momentum [30]. Then, the

Table 2 Relationship of Spin Densities between Resultant Spin (S) and Two Constituent Spins (S^A, S^B)

(i) $S^A = S^B = 1$ (homo-spin system)					
$S = 2$		$S = 1$		$S = 0$	
Unit A	Unit B	Unit A	Unit B	Unit A	Unit B
ρ_i^A	ρ_i^B	$\frac{1}{2}\rho_i^A$	$\frac{1}{2}\rho_i^B$	0	0

(ii) $S^A = 2$, $S^B = 1$ (hetero-spin system)					
$S = 3$		$S = 2$		$S = 1$	
Unit A	Unit B	Unit A	Unit B	Unit A	Unit B
ρ_i^A	ρ_i^B	$\frac{5}{6}\rho_i^A$	$\frac{1}{3}\rho_i^B$	$\frac{3}{4}\rho_i^A$	$-\frac{1}{2}\rho_i^B$

spin Hamiltonian for the whole system can be written in terms of the total spin S in the case of the isotropic g-factor of $g^A = g^B = g$ as

$$H_{\text{spin}} = -J_{\text{ex}}[S(S + 1) - S^A(S^A + 1) - S^B(S^B + 1)] + g\beta_e H.S + S.DS$$

(12)

Each spin Hamiltonian for the isolated unit moieties, A and B, can be explicitly given by

$$H_{\text{spin}}^A = g^A\beta_e H.S^A + S^A.D^A.S^A$$

(13)

and

$$H_{\text{spin}}^B = g^B\beta_e H.S^B + S^B.D^B.S^B$$

(14)

The relationship among the fine-structure tensors, D^S, D^A, and D^B, is given by

$$D^S = [P(S, S^A, S^B)D^A + P(S, S^B, S^A)D^B] + Q(S, S^A, S^B)D^{AB}$$

(15)

In the present case of $S^A = S^B = 1$ for molecule **9**, $P(S, S^A, S^B)$ and $Q(S, S^A, S^B)$ are rewritten as

$$P(S, 1, 1) = \frac{3S^2 + 3S - 11}{2(2S - 1)(2S + 3)}$$

(16)

and

$$Q(S, 1, 1) = \frac{S^2 + S + 8}{2(2S - 1)(2S + 3)}$$

(17)

Thus, the fine-structures tensors are $D^T = -(1/2)(D^A + D^B) + D^{AB}$ for the triplet spin state ($S = 1$), and $D^Q = (1/6)(D^A + D^B) + (1/3)D^{AB}$ for the quintet spin state ($S = 2$), respectively. Since $D^{AB} \ll D^A, D^B$, these equations lead to the particular relationship $D^T = -3D^Q$. This particular relationship of the fine-structure tensors holds for the experimentally determined fine-structure tensors of the present system **9** [5], as already described in Section V.A.

V.E. Spin-Density Distribution and Spin Alignment of the Low-Lying Triplet Excited State

Figure 10a shows a typical ESR spectrum of **9** observed at 34 K with the external magnetic field applied parallel to the crystallographic a-axis of the benzophenone-d_{10} host single crystal [24,25]. For the ENDOR experiments, the corresponding diazo precursor was diluted in the host crystal and **9** was generated at 2 K by photolysis. ^{1}H- and ^{13}C-ENDOR were measured at 10–15 K for the low-lying excited triplet state of **9** due to thermal excitation

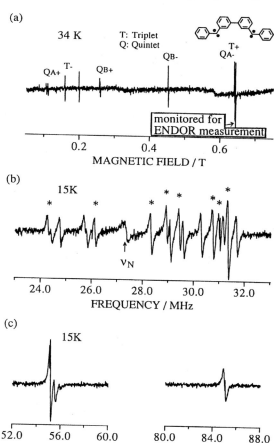

Figure 10 Single-crystal ESR and ENDOR spectra of **9** with the external magnetic field along the *a*-axis of the host benzophenone crystal. (a) ESR spectrum. The signals denoted by *T* and *Q* belong to the thermally populated excited triplet and quintet states, respectively. (b) ^{1}H-ENDOR spectrum and (c) ^{13}C-ENDOR spectrum of the excited triplet state. The ^{1}H-ENDOR signals indicated by asterisks were assigned to the signals arising from the adjacent protons on the biphenyl group by using the partially deuterium-labeled compound.

of this state. Figures 10b and 10c show typical ^{1}H- and ^{13}C-ENDOR spectra observed at 15 K, by monitoring the EPR transition T_+ in Figure 10a [24,25].

The angular dependence of the ENDOR transitions was measured for the rotation of the external magnetic field in the three crystallographic planes, *ab*, *bc*, and *ca*, of the benzophenone host single crystals (space group: $P2_12_12_1$, $z = 4$). The spin Hamiltonian used for the analysis of the ENDOR spectra is given by

$$H'_{\text{spin}} = \beta_e H.g.S + S.D.S + \sum (S.A.I_i - g_N\beta_N H.I_i) \tag{18}$$

where g, D, and A are the electron g tensor, the fine-structure tensor, and the hyperfine tensor, respectively. g_N, β_e, and β_N are the nuclear g-factor, the electron Bohr magneton, and the nuclear magneton, respectively. The analysis was carried out using the numerical fitting of the observed ENDOR frequencies to those calculated by the exact numerical diagonalization of the preceding spin Hamiltonian matrix [24,25]. From the angular dependence of the ENDOR transitions, the hyperfine tensors for all the 18 protons and the ^{13}C hyperfine tensors of both the divalent carbons were determined. The π-spin densities on the carbon sites adjacent to the hydrogen atoms and the π- and σ-spin densities on both divalent carbon atoms thus obtained from the hyperfine data are shown in Figure 11a. The spin densities on the six carbon atoms without circles could not be determined, since they have no adjacent protons.

The following characteristic features should be noted from this experimentally determined spin-density distribution.

1. As shown in Figure 11a, the observed spin-density distribution of the low-lying excited triplet state of **9** is roughly symmetrical with respect to the broken line inserted in this figure.
2. The unpaired π-spins are distributed over the whole molecule, in which the alternation of the signs of the spin densities occurs within each diphenylmethylene unit moiety.
3. Both of the divalent carbon atoms have positive spin densities similar in magnitude.
4. The magnitudes of both of the π- and σ-spin densities at both divalent carbon atoms are about one-half of those for the isolated triplet diphenylmethylene molecule reported by Hutchison and Kohler [28].

The calculation of the spin densities shown in Figure 11b was carried out with the help of the weakly interacting model described previously and with the unrestricted Hartree–Fock approximation [Eq. (4)] based on the generalized Hubbard model. Thus, we regard molecule **9** as composed of the two diphenylmethylene moieties A and B weakly interacting with each

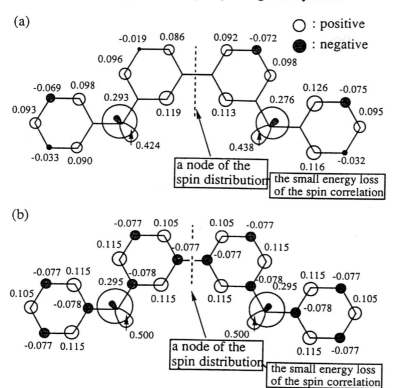

Figure 11 π-Spin-density distribution of the low-lying excited triplet state of **9**. (a) Observed values obtained from ^{1}H- and ^{13}C-ENDOR spectra. (b) Theoretical values obtained from the Hubbard model and the weakly interacting multiplet– multiplet model.

other. As discussed in the previous section, the spin densities $\rho_i^A(S, M)$ of unit A in **9** can be derived from $\rho_{i0}^A(S^A = 1, m_A = 1)$ of the isolated triplet diphenylmethylene moiety using Eq. (9). Thus,

$$\rho_i^A(S = 1, M = 1) = \left(\frac{1}{2}\right) \rho_{i0}^A(S^A = 1, m_A = 1) \qquad (19)$$

for the triplet state, and

$$\rho_i^A(S = 2, M = 2) = \rho_{i0}^A(S^A = 1, m_A = 1) \qquad (20)$$

for the quintet state. Similar expressions hold also for unit B, since A and B are equivalent. The spin densities used for the isolated diphenylmethylene moieties were obtained from the UHF calculation based on the generalized Hubbard model.

This Hubbard calculation using the relationship of Eq. (19) interprets well the aforementioned salient points of our ENDOR results shown in Figure 11b. The observed and calculated spin-density distributions agree well in both sign and magnitude. Figure 11 shows that the sign of the π-spin density is alternately distributed on the carbon sites within the diphenylmethylene moiety, thus forming the up-and-down network of the π-spin. The pseudo π-SDW resulting from an up-and-down network is most favorable in view of the total spin energy (the sum of the spin-exchange correlation energy). However, in the low-lying triplet excited state of **9**, the pseudo π-SDW state is not formed in the whole molecule, but only within the unit moieties, as shown in Figure 11. The breakdown of the formation of pseudo π-SDW (appearance of the node of the pseudo π-SDW), i.e., the existence of the adjacent carbon sites with the same sign of spin density, is the key to the spin structure of the low-lying excited high-spin states.

The existence of the spin-density distribution nodes clarifies the reason that these low-lying higher-spin states lie above the ground state from the viewpoint of the spin-exchange correlation energy. Since the singlet spin state has zero spin density, such a node does not exist, resulting in no energy loss due to a node. Thus, the presence of the node in the spin-density distribution of the observed triplet state, which destabilizes the spin-correlation energy, makes the triplet state an excited state above the spinless, nodeless ground state. A similar spin distribution is also expected for the quintet state from the comparison of Eqs. (19) and (20). The exchange correlation energy through the π-bond between the bridged carbons can be estimated roughly from Eq. (10). Apart from a factor of S^2, the correlation energy is given by J_{ex} in Eq. (11), where J_{ex} is related to the magnitude of the product of the spin densities, ρ_{i0}^A and ρ_{j0}^B, at the bridged carbon sites i and j in the isolated moieties. Thus, the small spin densities on the bridge carbon sites lead to the weak interaction between the two moieties, and to the small energy loss arising from the existence of the node in the spin distribution. In the case of **9**, in which the two high-spin moieties are connected at disjoint positions, the effective exchange interaction through the π-bond between the bridged carbons is expected to be weak, since the loss of the spin correlation energy arising from the node at the center of the bridge is small owing to the small spin densities at the bridged carbon sites. These are the reasons that the low-lying excited higher-spin states for molecule **9** lie above its low-spin (spinless) ground state.

These findings give a clear physical picture for the spin alignment as well as the mechanism of the small energy splitting among the low-lying spin states of **9**. This may be summarized as follows:

1. All their low-lying spin states are formed from the unpaired spins within the four nonbonding molecular orbitals with nearly degenerate energy levels.
2. The energy difference, therefore, comes mainly from the spin-exchange correlation among their unpaired electrons via the π-electron network.
3. The small energy loss of the spin correlation due to the small spin densities at the bridged carbon atoms makes the observed triplet and the quintet high-spin states into low-lying excited states above the spinless ground state.

VI. VALENCE BOND PICTURE OF THE SPIN ALIGNMENT OF ORGANIC HIGH-SPIN MOLECULES

In the previous section, we focused on the low-lying excited high-spin state and clarified the mechanism of the spin alignment giving the low-lying excited spin state. In this section, we describe the exact numerical treatment of the valence bond Heisenberg–Dirac Hamiltonian as a more reliable theoretical method for dealing with the ground and low-lying excited electronic states of the organic high-spin molecules just mentioned [31]. This Hamiltonian has been applied to π-electron networks by Ovchinnikov [16] and Klein [17,18] for prediction of their ground state spin multiplicity on the basis of a theorem derived by Lieb and Mattis [32].

As a theoretical model that takes the spin correlation of many open-shell electrons in the polycarbenes into account, we have chosen the following valence bond Heisenberg–Dirac Hamiltonian:

$$H = -2 \sum J_{ij}^{\text{eff}} s_i . s_j \tag{21}$$

where J_{ij}^{eff} is the effective exchange integral given approximately by

$$J_{ij}^{\text{eff}} = -\frac{2|T_{ij}|^2}{U} + K_{ij} \tag{22}$$

This Hamiltonian can be regarded as a limiting case of $U \gg T$ of the generalized Hubbard Hamiltonian [20–23]. The effective exchange integrals were estimated to be $J_{\pi\pi}^{\text{eff}} = -1.5$ eV and $J_{\sigma\pi}^{\text{eff}}/J_{\pi\pi}^{\text{eff}} = -0.2$ eV [31,34]. This model Hamiltonian has been solved by an exact numerical diagonalization of the matrix derived from the full-basis set of the spin system $|s_1^z \ s_2^z \ s_3^z \ldots s_N^z\rangle$, in which the subscript denotes carbon sites. The technique of the numerical computation employed here is based on that by Oguchi et al. using the Lanczos method [33].

Figure 12 depicts the results of the present calculation using the Heisenberg model for o-, m-, and p-phenylenebis(methylene) (PBM), which are the simplest examples to demonstrate the role of topology in the spin alignment of organic high-spin polycarbenes. Their electronic structures are reasonably expected to be similar to those of the corresponding isomers of phenylenebis(phenylmethylene) (PBPM). The features of the electronic states of PBM are consistent with the experimental results: the *meta* compound has a high-spin quintet ($S = 2$) ground state, while the *para* isomer has a singlet ground state with a low-lying triplet ($S = 1$) excited state, as shown in Figure 12. In addition, the energy separation between the quintet ground state and the first excited triplet state of m-BPM is about three times larger than those between the singlet ground state and the first excited triplet state of o- and p-PBM, in quantitative agreement with experiment. Furthermore, Figure 12 reveals that the spin multiplicity of the ground state remains unchanged if the ratio of $J_{\sigma\pi}^{\text{eff}}/J_{\pi\pi}^{\text{eff}}$ is varied. This shows that the spin alignment in the ground state is uniquely determined by the topology of the π-electron network and is independent of the magnitude of the effective exchange integrals. This finding predicts that the polycarbenes **1–6**, which have m-phenylene as their spin coupler, give robust high-spin ground states, in agreement with experimental results [1–14].

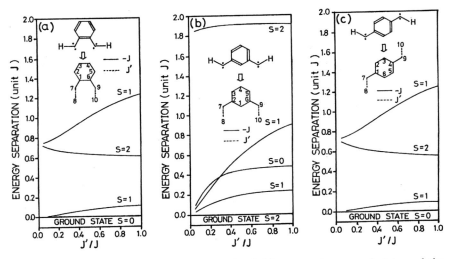

Figure 12   Energy separations in the unit of J between the ground state and the low-lying excited states of the isomers of PBM as a function of J'/J, where J and J' are $J_{\pi\pi}^{\text{eff}}$ and $J_{\sigma\pi}^{\text{eff}}$ in the text, respectively: (a) o-PBM, (b) m-PBM, and (c) p-PBM.

Figures 13a and 13b show the spin-density distributions on each carbon site in the quintet ground state of m-PBM and in the first excited triplet state of p-PBM, respectively, as a function of $J^{\text{eff}}_{\sigma\pi}/J^{\text{eff}}_{\pi\pi}$. The unpaired π-electrons in the quintet state are distributed over the carbon skeleton with alternate sign of the spin density from carbon to carbon, in agreement with the simple VB picture. This behavior is consistent with our ENDOR experiment for m-PBPM [26]. In addition, the spin-density distribution of m-PBM is almost independent of the $J^{\text{eff}}_{\sigma\pi}/J^{\text{eff}}_{\pi\pi}$ ratio.

In contrast to m-PBM, the spin distribution of p-PBM in the first excited triplet state changes depending on the $J^{\text{eff}}_{\sigma\pi}/J^{\text{eff}}_{\pi\pi}$ ratio, i.e., with increasing ratio, the n-spin densities decrease while the π-spin densities increase. This dependence on the ratio may be understood as follows: Contrary to m-PBM, the net unpaired spin in the π-network of p-PBM is zero, from top-

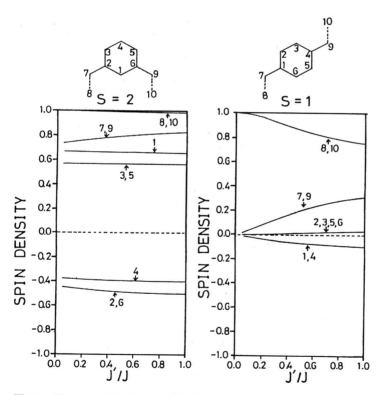

Figure 13 Spin-density distributions on each site in the quintet ground state of m-PBM and in the low-lying excited triplet state of p-PBM as a function of J'/J, where J and J' are $J^{\text{eff}}_{\pi\pi}$ and $J^{\text{eff}}_{\sigma\pi}$ in the text, respectively.

ological symmetry. Thus, the π-spin densities of *p*-PBM arise from the spin polarization induced by the exchange interaction between the *n*- and π-electrons at each divalent carbon site. Hence the spin distribution depends directly on the magnitude of the n–π exchange integral $J_{\sigma\pi}^{eff}$. The magnitude of the zero-field splitting fine-structure parameter D of *p*-PBPM can also be explained by taking the induced π-spin into account [31].

Figures 14a and 14b show the energy location of the ground and low-lying excited spin states and the spin density distribution in the first excited triplet state of biphenyl-3,3′-bis(methylene) (BP-3,3′-BM). They were calculated via the Heisenberg model Hamiltonian approach, where we replaced the end phenyl groups of biphenyl-3,3′-bis(phenylmethylene) **9** with hydrogen atoms to reduce the dimensionality of the Hamiltonian matrix. The results were obtained without the help of the weakly interacting model mentioned already. Figure 14a clearly shows that **9** has a low-spin singlet ground state with low-lying triplet and quintet excited states, in quantitative agreement with experiment. Thus, the pseudo π-SDW is formed only within the

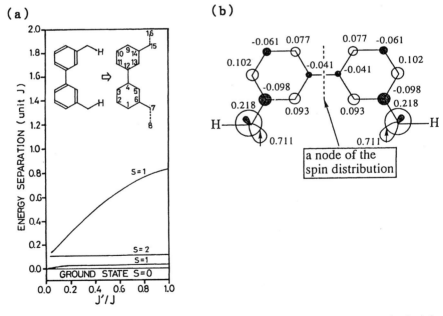

Figure 14 Energy levels and the spin-density distribution in the first excited triplet state of BP-3,3′-BM. The spin densities were calculated by means of the Heisenberg model Hamiltonian approach. (a) The energy location of the ground and low-lying excited spin states. (b) The spin density distribution.

phenylmethylene moieties in this case, and the two bridged carbon sites have the same signs of spin densities in the low-lying triplet excited state. Although this model is a limiting case of $U \gg T$ of the generalized Hubbard model Hamiltonian, the spin correlation among all the π- and n-spins is taken exactly into account within the model Hamiltonian, leading to the correct ordering of the ground and low-lying excited states [24,31,34]. This finding confirms the physical picture for spin alignment, as well as the formation of low-lying excited spin states, as was derived in Section V with the help of both the weakly interacting multiplet–multiplet model and the generalized Hubbard model.

VII. SUMMARY

The spin alignment in the ground state and the low-lying excited states of the high-spin polycarbenes and their topological isomer are reviewed. The generalized Hubbard Hamiltonian and the valence bond Heisenberg–Dirac Hamiltonian approaches are presented as useful methods to treat the high-spin polycarbenes and their topological isomers, i.e., to give a correct interpretation of their spin-density distribution as well as their ground state multiplicity. A series of m-phenylene-bridged polycarbenes that have been designed according to their topological symmetry are shown to have robust high-spin ground states located more than 300 cm^{-1} below the other lower-spin states. Their electronic structures and intramolecular spin alignment have been clarified by ENDOR experiments and generalized Hubbard calculations. It is shown that the formation of pseudo π-SDW plays an important role on their spin alignment. As a typical example showing the breakdown of the simple spin-prediction rule according to Eq. (1), the physical picture of the spin alignment of biphenyl-3,3′-bis(phenylmethylene) is presented. Electron spin resonance, ^{1}H-, and ^{13}C-ENDOR experiments as well as generalized Hubbard and Heisenberg model calculations have revealed that the formation of pseudo SDWs accounts well for the singlet ground state and the low-lying excited high-spin states of this molecule. A weakly interacting multiplet–multiplet model is also presented, in combination with the generalized Hubbard model, as an excellent procedure to interpret the spin alignment and the spin-density distribution of this molecule.

REFERENCES

1. Itoh K. Electron spin resonance of an aromatic hydrocarbon in its quintet ground state. Chem Phys Lett. 1967; 1:235–238.

2. Wasserman E, Murray RW, Yager WA, Trozzolo AM, Smolinsky G. Quintet ground states of *m*-dicarbene and *m*-dinitorene compounds. J Am Chem Soc. 1967; 89:5076–5078.
3. Takui T, Itoh K. Detection of an aromatic hydrocarbon in its septet electronic ground state by electron spin resonance. Chem Phys Lett 1973; 19:120–124.
4. Trozzolo AM, Murray RW, Smolinsky G, Yager WA, Wasserman E. The E.p.r. of dicabene and dinitrene derivatives. J Am Chem Soc. 1963; 85:2526–2527.
5. Itoh K. Electronic structures of aromatic hydrocarbons with high-spin multiplicities in the electronic ground state. Pure Appl Chem. 1978; 50:1251–1259.
6. Teki Y, Takui T, Itoh K., Iwamura H, Kobayashi K. Design, preparation, and ESR detection of a ground-state nonet hydrocarbon as a model for one-dimensional organic ferromagnets. J Am Chem Soc. 1983; 105:3722–3723.
7. Teki Y, Takui T, Yagi H, Itoh K, Iwamura H. Electron spin resonance line shapes of randomly oriented molecules in septet and nonet states by a perturbation approach. J Phys Chem. 1985; 82:539–547.
8. Teki Y, Takui T, Itoh K, Iwamura H, Kobayashi K. Preparation, and ESR detection of a ground-state nonet hydrocarbon as a model for one-dimensional organic ferromagnets. J Am Chem Soc. 1986; 108:2147–2156.
9. Sugawara T, Bandow S, Kimura K, Iwamura H, Itoh K. Magnetic behavior of nonet tetracarbene as a model for one-dimensional ferromagnets. J Am Chem Soc. 1986; 108:368–371.
10. Fujita I, Teki Y, Takui T, Kinoshita T, Itoh K, Miko F, Sawaki Y, Iwamura H, Izuoka A, Sugawara T. Design, preparation, and electron spin resonance detection of a ground-state undecet ($S = 5$) hydrocarbon. J Am Chem Soc. 1990; 112:4074–4075.
11. Nakamura N, Inoue K, Iwamura H, Fujioka T, Sawaki Y. Synthesis and characterization of a branched-chain hexacarbene in a tridecet ground state. An approach to supermagnetic polycarbenes. J Am Chem Soc 1992; 114:1484–1485.
12. Furukawa K, Matsumura T, Teki Y, Kinoshita T, Takui T, Itoh K. Molecular design, synthesis, and electron spin resonance detection of a ground-state tridecet ($S = 6$) hydrocarbon as a model for 2D organic ferro- and superparamagnets. Mol Cryst Liq Cryst. 1993; 232:251–260.
13. Nakamura N, Inoue K, Iwamura H. A branched-chain nonacarbene with a nonadecet ground state: A step nearer to superparamagnetic polycarbenes. Angew Chem Int Ed Engl. 1993; 32:872–874.
14. Teki Y, Fujita I, Takui T, Kinoshita T, Itoh K. Topology and spin alignment in a novel organic high-spin molecule, 3,4′-bis(phenylmethylene)biphenyl, as studied by ESR and a generalized UHF Hubbard calculation. J Am Chem Soc. 1994; 116:11499–11505.
15. Longuett-Higgins HC. Some studies in molecular orbital theory I. Resonance structures and molecular orbitals in unsaturated hydrocarbons. J Chem Phys. 1950; 18:265–274.
16. Ovchinnikov AA. Multiplicity of the ground state of large alternant organic molecules with conjugated bonds. Theoret Chim Acta. 1978; 47:297–304.

17. Klein DJ, Nelin CJ, Alexander S, Matsen A. High-spin hydrocarbons. J Chem Phys. 1982; 77:3101–3108.

18. Alexander SA, Klein DJ. High-spin carbenes. J Am Chem Soc. 1988; 110: 3401–3405.

19. Borden WT, Davidson ER. Effects of electron repulsion in conjugated hydrocarbon diradicals. J Am Chem Soc. 1976; 99:4587–4594.

20. Teki Y, Takui T, Kinoshita T, Ichikawa S, Yagi H, Itoh K. Spin alignment in organic high-spin molecules as studied by ESR and the generalized Hubbard Hamiltonian approach. Chem Phys Lett. 1987; 141:201–205.

21. Hubbard J. Electron correlation in narrow energy bands. Proc Roy Soc. 1963; 276A:238–257.

22. White RM. Quantum Theory of Magnetism. 2nd ed. Berlin: Springer-Verlag, 1983.

23. Nasu K. Periodic Kondo-Hubbard model for a quasi-one-dimensional organic ferromagnet m-polyphenylcarbene: Cooperation between electron correlation and topological structure. Phys Rev B. 1986; 33:330–338.

24. Teki Y, Takui T, Sato K, Yamashita A, Okamoto M, Kinoshita T, Itoh K. Role of the spin correlation and weakly interacting model for the spin alignment in organic high-spin molecules. Mol Cryst Liq Cryst. 1993; 232:261–270.

25. Okamoto M, Teki Y, Takui T, Kinoshita T, Itoh K. Spin distribution of an organic high-spin molecule, biphenyl-3,3'-bis(phenylmethylene), as studied by ^{1}H-ENDOR and a UHF Hubbard calculation. Chem Phys Lett. 1990; 173:265–270.

26. Takui T, Kita S, Ichikawa S, Teki T, Kinoshita T, Itoh K. Spin distribution of organic high-spin molecules as studied by ENDOR/TRIPLE. Mol Cryst Liq Cryst. 1989; 176:67–76.

27. Hirota H, Hutchison CA Jr, Palmer P. Hyperfine interactions and electron spin distribution in triplet-state naphthalene. J Chem Phys. 1964; 40:3717–3725.

28. Hutchison CA Jr, Kohler BE. Electron nuclear double resonance in an organic molecule in a triplet ground state. Spin densities and shape of diphenylmethylene molecules in diphenylethylene single crystals. J Chem Phys. 1969; 51: 3327–3335.

29. McConnell HM. Ferromagnetism in solid free radicals. J Chem Phys. 1963; 39:1910.

30. Rose ME. Elementary theory of angular momentum. New York: Wiley, 1957.

31. Teki Y, Takui T, Kitano M, Itoh K. Spin alignment in organic high-spin molecules. A Heisenberg Hamiltonian approach. Chem Phys Lett. 1987; 142: 181–186.

32. Lieb E, Mattis D. Ordering energy levels of interacting spin systems. J Math Phys. 1962; 3:749–751.

33. Oguchi T, Nishimori H, Taguchi Y. Ground state of antiferromagnetic quantum spin systems on the triangular lattice. J Phys Soc Japan. 1986; 55:323–330.

34. Teki Y, Takui T, Itoh K. Theoretical study of spin alignment of organic high-spin molecules in terms of VB description. Mol Cryst Liq Cryst. 1995; 274: 213–222.

13

Synthesis and Properties of Organic Conjugated Polyradicals

Yozo Miura
Osaka City University, Osaka, Japan

I. INTRODUCTION

Though synthetic polymer chemists have long investigated polyradicals as polymeric antioxidants or organic conductors [1], polyradical chemistry is now inspired by theoretical predictions that ferromagnetic materials might be made from exclusively organic components [2,3]. Possible models for polymeric ferromagnets based on the molecular orbital (MO) or valence bond (VB) theories of π-electron networks of the alternant hydrocarbons have been proposed, and theoretical considerations have spurred the design and preparation of a variety of π-conjugated polyradicals and examination of their magnetic properties. Since polyradical ferromagnets promise many interesting and useful properties, different from those of "inorganic" magnets, they are the focus of current research and development in many fields, and a variety of organic conjugated polyradicals have already been synthesized [4].

II. FUNDAMENTAL BASIS OF MOLECULAR DESIGN FOR ORGANIC CONJUGATED FERROMAGNETIC POLYRADICALS

Theoretical design studies are based on the topological symmetry of the π-electron networks in alternant hydrocarbons. To illustrate the essence of topologically controlled orbital degeneracy, three topological isomers of quinodimethane are considered in terms of the simple MO and VB methods (Figure 1). According to the π-nonbonding MO (NMBO) theory of alternant hydrocarbons by Longuet-Higgins [5], the number of degenerate π-NBMOs is given by $N - 2T$, and S is expressed by Eq. (1), where S is the total spin quantum number, N is the number of carbon sites, and T is the number of double bonds. For 1,3-benzoquinodimethane, $S = (N - 2T)/2 = 1$. Since each NBMO is singly occupied with parallel spins in the ground state, according to Hunt's rule, 1,3-benzoquinodimethane is predicted to be in the triplet ground state $(S = 1)$. On the other hand, $(N - 2T)$ is 0 for 1,2- and 1,4-benzoquinodimethane, the isomers being predicted to be in the singlet ground state $(S = 0)$:

$$S = \frac{N - 2T}{2} \tag{1}$$

According to the VB theory of Ovchinnikov [6], ground spin states

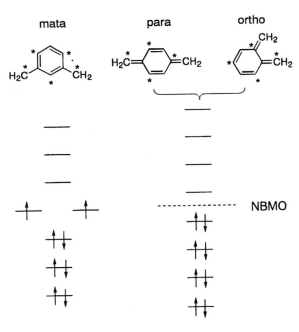

$$S = (N - 2T)/2 = (8 - 2 \times 3)/2 = 1 \quad (8 - 2 \times 4)/2 = 0$$

$$S = (n^* - n)/2 = (5 - 3)/2 = 1 \quad\quad (4 - 4)/2 = 0$$

Figure 1   Spin states of quinodimethanes predicted by MO and VB theories.

are predicted by Eq. (2), where n^* is the number of starred carbon atoms and n is the number of unstarred carbon atoms (where adjacent atoms are starred and unstarred, and identically denoted atoms are not adjacent to each other). For 1,3-benzoquinodimethane, n^* is 5 and n is 3, which predicts that the ground spin state of this isomer is triplet ($S = 1$), as shown by Figure 1. On the other hand, for 1,2- and 1,4-benzoquinodimethanes, n^* and n are both 4, which implies that the ground spin states are singlet ($S = 0$):

$$S = \frac{n^* - n}{2} \tag{2}$$

Although both methods are understandable to synthetic organic or polymer chemists, they do not address the strength of interactions between the unpaired electron spins. Furthermore, there is some disagreement between theory and experiment; for example, when π-conjugated systems are nonplanar or have so-called disjoint system [7], magnetic interactions between the unpaired electron spins can be negligibly small. In this case polyradicals are only paramagnetic.

When polyradicals are designed and synthesized according to theoretical prediction, one-dimensional spin parallel-ordering π-systems could be obtained by intramolecular (through-bond) ferromagnetic coupling between the unpaired electron spins (intramolecular exchange interactions $J_1 > 0$). If the magnetic interactions between the one-dimensional spin parallel-ordering π-systems are ferromagnetic (intermolecular spin-exchange interaction $J_2 > 0$), bulk ferromagnetic polyradicals will be achieved (Figure 2). We should note that intermolecular (through-space) magnetic interactions are, however, generally antiferromagnetic. It is therefore important that the one-dimensional spin parallel-ordering π-systems must be linked so that magnetic cou-

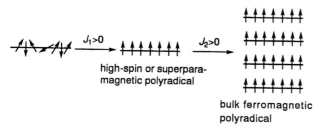

high-spin or superpara-
magnetic polyradical

bulk ferromagnetic
polyradical

J_1: intramolecular spin-exchange interaction
J_2: intermolecular spin-exchange interaction

Figure 2 Spin ordering of high-spin or superparamagnetic polyradicals and bulk ferromagnetic polyradicals.

plings between them are ferromagnetic. Construction of this three-dimensional ferromagnetic π-network is therefore essential for the realization of polymer ferromagnets.

As models for one-dimensional spin parallel-ordering π-systems based on theoretical considerations, some potentially synthesizable or synthesized polyradicals are shown in Scheme 1. The first group is represented by the structures having the unpaired electron spin sites pendent from their main chains, and the second group is represented by the structures having the unpaired electron spins incorporated in their conjugated main chain. Although the latter structures are expected to have stronger magnetic interactions than the former ones, the defects on spin centers will cause an interruption of the π-conjugating systems, leading to a significant reduction in S of the polyradicals. This disadvantage can be overcome by π-network structures of polyradicals [3].

Although poly(phenylacetylene)-based polyradicals are the most widely investigated of the conjugated polyradicals, none has shown a ferromagnetic interaction between the unpaired electron spins, because there is serious twisting in the conjugated π-systems that reduces significantly the magnetic interactions between the unpaired electron spins [4]. This twisting in the conjugated π-systems has often been a serious problem in the design and synthesis of ferromagnetic polyradicals.

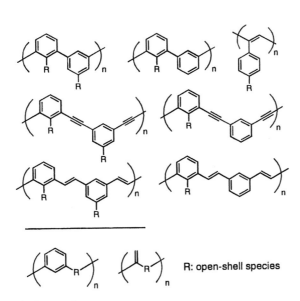

R: open-shell species

Scheme 1

In this chapter, recent advances in syntheses and magnetic properties of poly(1,3-phenyleneethynylene)-based, poly(2,3-thienyleneethynylene)-based, poly(1,3-phenylene)-based, and poly(thiophene-2,5-diyl)-based polyradicals, and poly(1,3-phenelene iminoxyl) polyradicals are described.

III. POLY(1,3-PHENYLENEETHYNYLENE)-BASED POLYRADICALS

III.A. Synthesis and Characterization

Poly(1,3-phenyleneethynylene)s can be readily obtained by the reaction of 1,3-phenylenediethynylenes with 1,3-diiodo- or 1,3-dibromobenzenes in basic solvents such as Et$_3$N or Et$_2$NH in the presence of (PPh$_3$)$_2$PdCl$_2$–CuI [8]. In particular, when 1,3-diiodobenzenes are employed as a dihalo compound, the polycondensation proceeds smoothly at room temperature. By using this condensation reaction, a variety of poly(1,3-phenyleneethynylene)-based polyradicals have been prepared.

Condensation reaction of 2-(3,5-diethynylphenyl)nitronyl nitroxide (**1**) with 3,5-diiodo-1-hexylbenzene or 3,5-diiodo-1-*tert*-butylbenzene in Et$_3$N-pyridine at 20°C in the presence of (PPh$_3$)$_2$PdCl$_2$–CuI gave polyradicals **2** as a light blue powder in ~100% yield (Scheme 2) [9]. 2-Arylnitronyl nitroxides are known as oxygen-insensitive, isolable, stable, free radicals and can be prepared by condensation of arylaldehydes with 2,3-dimethyl-2,3-bis(hydroxyamino)butane, followed by oxidation [10]. The number average molecular weights (M_n) determined by gel permeation chromatography (GPC) are 4300–4800, corresponding to 9–11 repeating units. The elemental analyses of the polyradicals showed the presence of 7.53–7.89% iodide, indicating that the polycondensation is terminated by 3,5-diiodobenzenes. This was supported by the fact that infrared (IR) spectrum showed no ab-

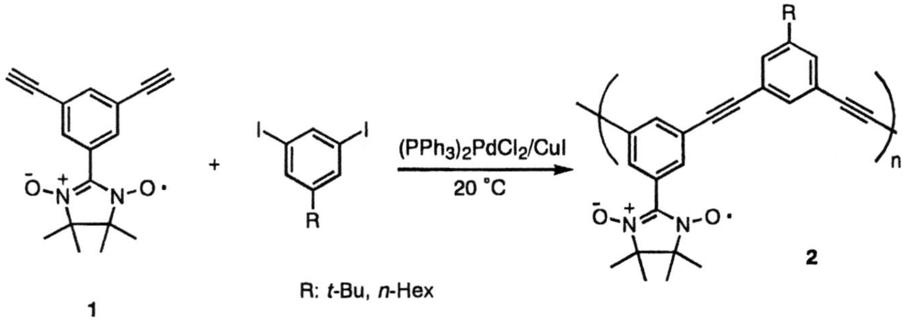

Scheme 2

sorption peak due to stretching vibration of the $\equiv$C—H. The spin concentrations of **2** determined by electron spin resonance (ESR) using 1,3,5-triphenylverdazyl as reference are 1.12×10^{21} to 1.33×10^{21} spins g^{-1} (0.82–0.91/repeating unit), indicating that no significant decomposition of nitronyl nitroxide radical moieties takes place during the condensation reaction.

Polycondensation of *N-tert*-butyl-1,3-diethynylphenyl nitroxide (**3**) with 3,5-diiodo-1-*tert*-butylbenzene in Et$_3$N-pyridine at 20°C in the presence of (PPh$_3$)$_2$PdCl$_2$–CuI gave polynitroxide **4** as a light red powder in 87% yield (Scheme 3) [11]. Although *N-tert*-butyl aryl nitroxides are familiar as isolable, stable, free radicals [12], they are not as stable as 2-arylnitronyl nitroxides. However, *N-tert*-butyl aryl nitroxides have a delocalized unpaired electron spin, which is advantageous for introduction of strong ferromagnetic coupling between the unpaired electron spins.

Polyradical **4** showed M_n of 3000 by GPC measurements. The spin concentrations changed in the range 0.85×10^{21} to 1.51×10^{21} spins g^{-1}, depending on the solvent used. When 1:1 Et$_3$N–pyridine was used as solvent, the spin concentration of **4** was 0.85×10^{21} spins/g (0.49/repeating unit), and when 1:4 Et$_3$N–pyridine was used, it was 1.51×10^{21} spins/g (0.86/repeating unit), indicating that the nitroxide radical is less stable in an Et$_3$N-rich solvent. However, in 1:4 Et$_3$N–pyridine, no significant decomposition of the nitroxide radical was found to occur during the polycondensation.

Therefore, polycondensation of **3** with 2-(3,5-diiodophenyl)nitronyl nitroxide (**5**) was carried out in 1:4 Et$_3$N–pyridine, giving **6** in 78–83% yield as a light blue powder (Scheme 4) [13]. The M_n values determined by GPC were 2670–3030 (5.5–6.9 repeating units), and the spin concentrations determined by ESR were 1.91×10^{21} to 2.02×10^{21} spins g^{-1} (1.40–1.48 spins/repeating unit). Since the theoretical spin number per repeating unit is

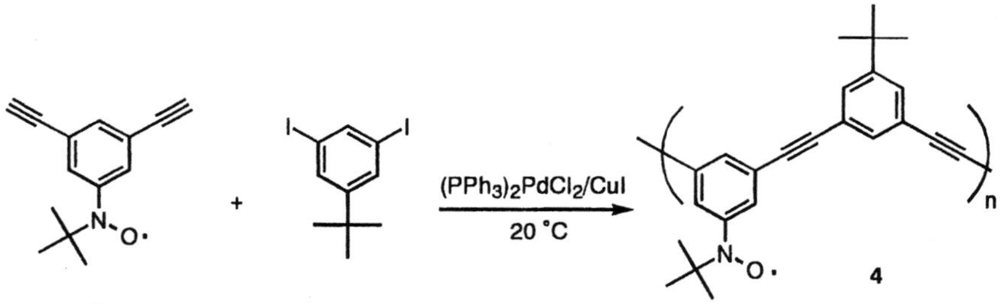

Scheme 3

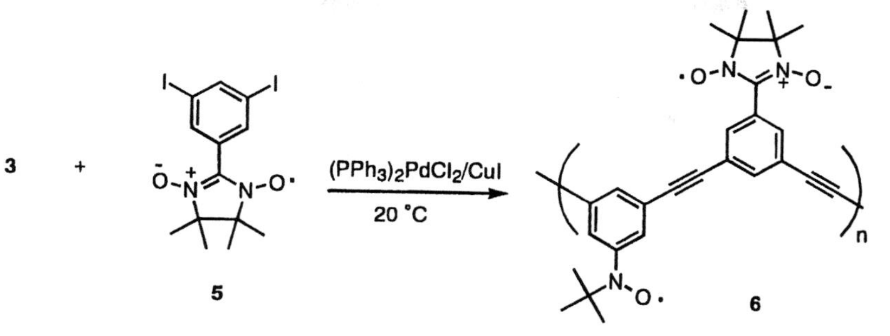

Scheme 4

2, the obtained spin concentrations indicate that some decomposition of the free radical moieties occurs during the polycondensation.

III.B. Electron Spin Resonance Spectra

Nitronyl nitroxides **1** and **5** give an ESR spectrum consisting of a 1:2:3: 2:1 quintet with $a_N \cong 0.75$ mT ($g = 2.0067$), characteristic of 2-arylnitronyl nitroxides, while nitroxide **3** gave a 1:1:1 triplet ESR spectrum, with $a_N = 1.210$ mT ($g = 2.0061$), characteristic of N-*tert*-butyl aryl nitroxides. In most cases the 1:1:1 triplet is further split by the interaction with aromatic protons. Indeed, each component of the 1:1:1 triplet of **3** was further split into a 1:3:3:1 triplet by the interaction with ortho and para protons. In contrast, those of polyradicals are different from those of the monoradicals when their spin concentrations are high. For example, the ESR spectrum of **6** gave a single line of $\Delta H_{pp} = 0.65$ mT with no resolved hyperfine splitting, even in a dilute solution, indicating that spin-exchange narrowing is taking place.

III.C. Magnetic Characterization

The magnetic properties of polyradicals have been studies with a superconducting quantum interference device (SQUID) magnetometer in the temperature range 1.8–300 K using the polyradical powder samples.

Figure 3 shows χT vs. T plots for **6**. Above 30 K, the χT values are constant, indicating that the polyradical is paramagnetic. From the χT value of 0.570 emu K/repeating unit in this temperature region, the spin concentration of **6** was determined to be 1.52 spins per repeating unit: the theoretical value is 0.752 emu K/repeating unit, provided that S is 1/2. This value is in good agreement with those determined by ESR (1.40–1.48). In the temperature region below 30 K, the χT vs. T plots showed a downward turn,

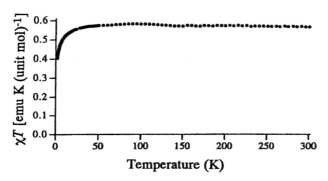

Figure 3 χT vs. T plots of **6**.

indicating that the magnetic coupling between the unpaired electron spins is antiferromagnetic with a Weiss temperature of $\theta = -1.5$ K. This antiferromagnetic interaction is probably due to typical through-space interactions between the unpaired electron spins. Consequently, it is assumed that the expected through-bond ferromagnetic interactions between the unpaired electron spins [14] are very weak and are masked by the antiferromagnetic through-space interaction. SQUID magnetic susceptibility measurements for **2** and **4** showed a similar antiferromagnetic coupling in the low-temperature region below 15 K.

IV. POLY(2,5-THIENYLENEETHYNYLENE)-BASED POLYRADICALS: SYNTHESIS AND CHARACTERIZATION

Magnetism of conductive polyradicals may be one of the important and challenging subjects in the area of polymeric magnetism. Coupling between conduction electrons in the main chain and the unpaired electrons in the

Scheme 5

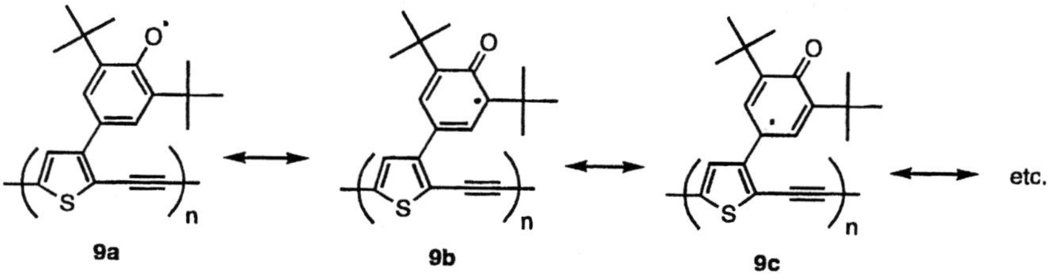

Scheme 6

side radical moieties may yield a strong ferromagnetic interaction between the unpaired electron spins. Selective oxidation of donor main chain, like poly(2,5-thienyleneethynylene), may produce conductive polyradicals.

Palladium-catalyzed polycondensation of 2-bromo-5-ethynylthiophene **7** with a hindered phenolic group gave a regioregular π-conjugated poly(2,5-thienyleneethynylene), **8** [15]. Polymer **8** had [η] of 0.15 dL g^{-1} (in CHCl$_3$ at 30°C), and the GPC measurements gave M_n of 7700. Oxidation of this polymer with K$_3$[Fe(CN)$_6$]/NaOH yielded polyradical **9** (Scheme 5). The IR spectrum of **9** showed the complete disappearance of the sharp OH stretching absorption peak at 3630 cm^{-1} and the appearance of a new, strong absorption at 1730 cm^{-1} in the carbonyl region. The appearance of the new absorption in the carbonyl region suggests contribution from the resonance forms represented by **9a–9c**, etc. (Scheme 6). The polyradical gave a strong and somewhat broad ESR signal with $\Delta H_{pp} \cong 0.50$ mT and a g-value of 2.0079. The broad ESR signal was explained in terms of the extensive delocalization of the unpaired electron due to its coupling with many protons.

Preparation of a poly(2,5-thienyleneethynylene) with nitronyl nitroxides was also attempted by Pd-catalyzed polycondensation of **10** (Scheme 7) [16]. Although monomer **10** could be prepared in eight steps, starting from 3-thiophenealdehyde, it was not as stable as expected. In the Pd-catalyzed polycondensation, **10** immediately decomposed to give brown nonradical compounds, and the expected polyradical **11** was not obtained.

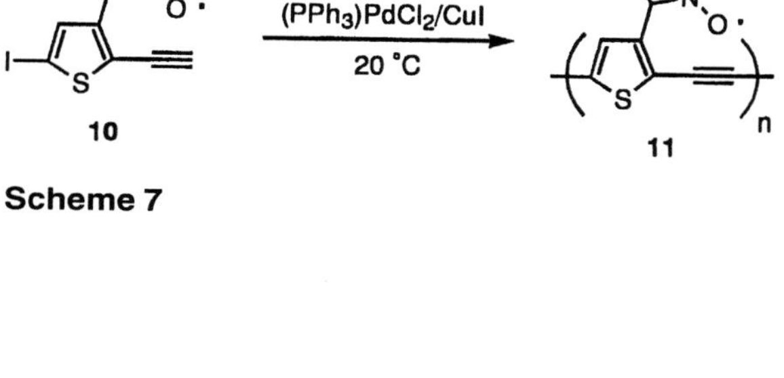

Scheme 7

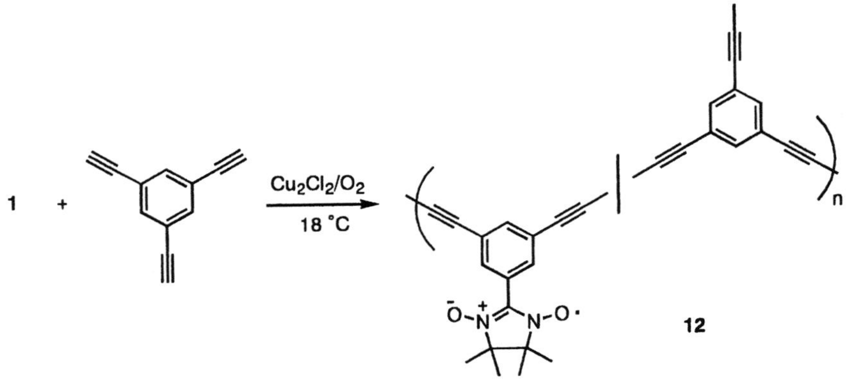

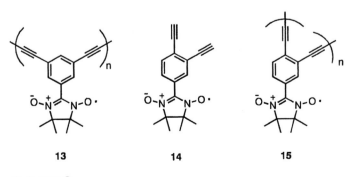

Scheme 8

V. POLY(1,3- OR 3,4-PHENYLENEDIETHYNYLENE)-
BASED POLYRADICALS

V.A. Synthesis and Characterization

Oxidative coupling of diethynylbenzenes at room temperature in the presence of Cu_2Cl_2 under bubbling oxygen gives poly(phenylenediethynylene)s in quantitative yields [17]. For example, oxidative coupling of **1** and 1,3,5-triethynylbenzene in the presence of Cu_2Cl_2 under bubbling oxygen gave polyradical **12** as a dark blue powder in quantitative yield (Scheme 8) [18]. Oxidative homocoupling reaction of **1** and **14** gave polyradicals **13** and **15**, respectively, in 83–91% yield (Scheme 9) [19–21].

Their IR spectra showed the complete disappearance of the strong absorption peak due to $\equiv C-H$ (3250 cm^{-1}) and the appearance of a new

Scheme 9

absorption peak at 2200 cm^{-1} due to —C≡C—. GPC measurements showed 6190 and 8930 for **13** and **15**, 22 and 32 units, respectively. Since the nitronyl nitroxide radicals are very stable and the reaction conditions are quire mild, the spin concentrations of the polyradicals obtained are expected to be very high. Indeed, the spin concentrations $(1.17 \times 10^{21} - 1.28 \times 10^{21}$ spins/g) of **12** determined by ESR indicated that 83–90% of the nitronyl nitroxide radicals survived under such mild polymerization conditions, and those of **13** and **15** estimated from the Curie values determined by a Faraday balance indicated no decomposition of the nitronyl nitroxide moieties.

V.B. Magnetic Characterization

The magnetic susceptibility measurements for polyradicals **13** and **15** were studied in the temperature range of 78–300 K with a Faraday balance. Although polyradical **15** was paramagnetic or antiferromagnetic, depending on the precipitation conditions of the polyradical, **13** was reported to show a weak ferromagnetic interaction between the unpaired electron spins. Since theoretical prediction shows that the magnetic interaction between the unpaired electron spins through π-conjugation is antiferromagnetic, the observed ferromagnetic interaction may be ascribable to the through-space interaction between the unpaired electron spins. It is proposed that the magnetic properties of polyradicals depend not only on the structures of polyradicals, but also on the precipitation conditions of polyradicals.

VI. POLY(1,3-PHENYLENE)-BASED POLYRADICALS

VI.A. Synthesis and Characterization

Poly(1,3-phenylenes) can be conveniently obtained by the Pd-catalyzed cross-coupling reaction of 1,3-dibromobenzenes with 1,3-phenylenediboronic acids or 1,3-phenylene-bis(trimethylene boronate) (**17**) [22]. Using this Pd-catalyzed cross-coupling reaction, poly(1,3-phenylene)-based nitroxide polyradical, **19**, was prepared according to Scheme 10 [23]. Thus, the cross-coupling reaction of *N-tert*-butyl-2,4-dibromoaniline (**16**) with **17** in benzene–H_2O containing *n*-Bu$_4$N$^+$Cl$^-$ in the presence of Pd(PPh$_3$)$_4$ gave a poly(1,3-phenylene) with *tert*-butylamino groups, **18**, in 70% yield as a light yellow powder. The M_n value of **18** determined by GPC was 4070 (18 repeating units) after reprecipitation (CH$_2$Cl$_2$/hexane). Oxidation of **18** with 3-chloroperbenzoic acid in CH$_2$Cl$_2$ gave nitroxide polyradical **19** in 50% yield as a light red powder. Interestingly, the spin concentration of the polyradical determined by ESR was 0.75 spin/repeating unit. Therefore, the ESR spectrum of the polyradical showed a broad single line due to the spin-exchange

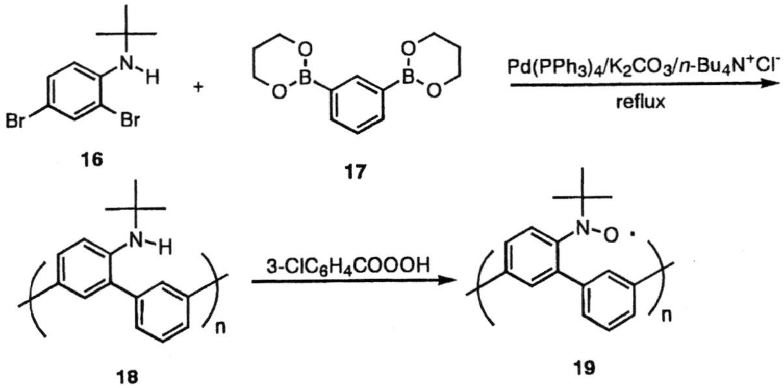

Scheme 10

narrowing process. The polyradical was very stable under the ambient conditions and showed no decomposition over a long period.

VI.B. Magnetic Characterization

The magnetic properties of **19** were measured with a SQUID magnetometer in the temperature range 1.8–300 K. The χ_gT vs. T plots showed a downward turn below 50 K. Therefore, in the high-temperature region, the polyradical is paramagnetic, and below 50 K, the magnetic interaction between the unpaired electron spins is antiferromagnetic. This result indicates that there is no strong ferromagnetic coupling between the unpaired electron spins and that the observed antiferromagnetic interaction is produced by the through-space mechanism. If the N–O· group is significantly twisted due to steric effects from the poly(1,3-phenylene) π-conjugated system, we cannot expect strong ferromagnetic interaction through the π-conjugated system. In this view, one ideal structure for poly(1,3-phenelene)-based polyradicals may be **20** (Scheme 11).

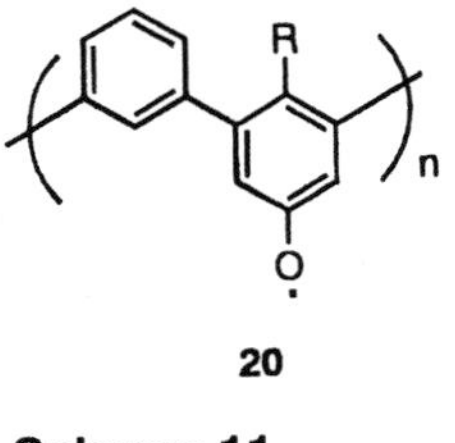

20

Scheme 11

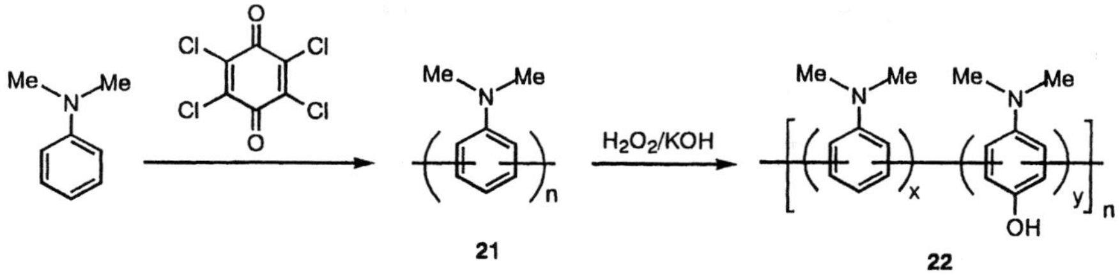

Scheme 12

Such a polyradical was recently prepared according to Scheme 12 [24]. Thus, the reaction of N,N-dimethylaniline with tetrachloro-p-benzoquinone at 120°C gave polymer **21**. ^{1}H NMR showed that its backbone was composed of 73% meta-linked units and 27% para-linked units, and its M_w determined by GPC was 5600. After the residual radical cations in **21** were quenched with KBH$_4$, treatment of **21** with H$_2$O$_2$/KOH gave **22**. A powder of **22** gave a broad ESR signal with g = 2.0042, and its spin concentration was determined to be 3.8×10^{19} g^{-1}. This value corresponds to <1% per repeating unit. Therefore, occurrence of ferromagnetic coupling between the unpaired electron spins cannot be expected for **22** because of its low-spin concentration. However, if high-spin concentration polyradicals are successfully prepared, we could expect a large positive exchange interaction for **22**.

VII.   POLY(THIOPHENE-2,5-DIYL)-BASED POLYRADICALS

Ni(0)-catalyzed polycondensation of **23** in DMF at 60°C or oxidative polycondensation of **23** with FeCl$_3$ in CHCl$_3$ at room temperature gave poly(thiophene-2,5-diyl) with hindered phenolic groups, **24**, in 73–89% yields (Scheme 13) [25]. The trimethylsilyl groups were completely removed from **24** during the polycondensation or by treatment of the resulting polymer with HCl-acidic MeOH. The η_{sp}/c value of the polymer obtained was 0.07 dL g^{-1} in CHCl$_3$ (c = 0.22 g dL^{-1}), and light-scattering analysis indicated an M_w of 13,000.

Oxidation of **25** was carried out with PbO$_2$ in benzene. The IR spectrum showed almost complete disappearance of the OH stretching absorption peak. Although the spin concentration of **26** was not measured, its ESR spectrum indicated that the unpaired electron spin is extensively delocalized onto the thiophene ring.

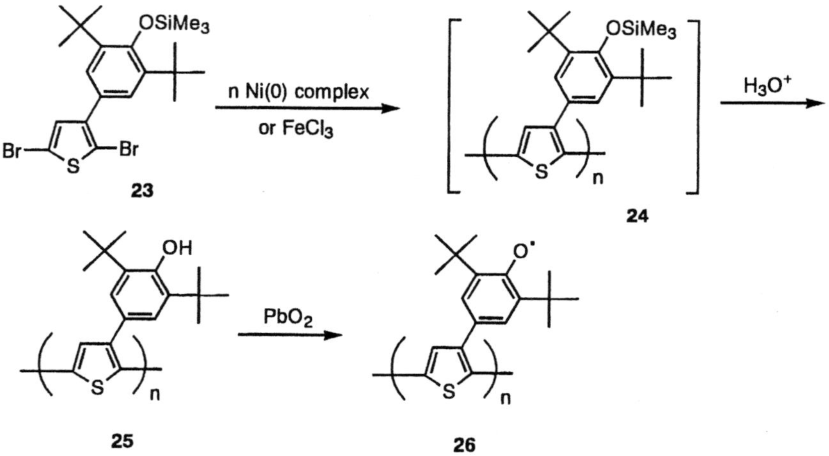

Scheme 13

VIII. POLY(1,3-PHENYLENEIMINOXYL)S

VIII.A. Synthesis and Characterization

This type of polyradical is structurally similar to 1,3-phenylene-connected high-spin polycarbenes [26] or 1,3-phenylene connected high-spin triarylmethyl polyradicals [3] and belongs to the second group of polyradicals (see Scheme 1). A model triradical of poly(1,3-phenyleneiminoxyl), **27** (Scheme 14), showed a large ferromagnetic coupling of $J/k_B = 240 \pm 20$ K between the neighboring unpaired electron spins [27].

Poly(1,3-phenyleneiminoxyl)s were prepared by oxidation of poly(1,3-phenyleneimine)s (**28**) (Scheme 15) [28]. Poly(1,3-phenyleneimine)s were prepared by the Ullmann condensation of 3-bromoanilines in nitrobenzene at 210°C in the presence of Cu_2I_2 and K_2CO_3. Although a clear IR absorption peak due to NH was not observed, GPC of the polymers showed M_w values of 2200–18,000 (50–100 repeating units). Oxidation of **28** (R = n-dodecyl)

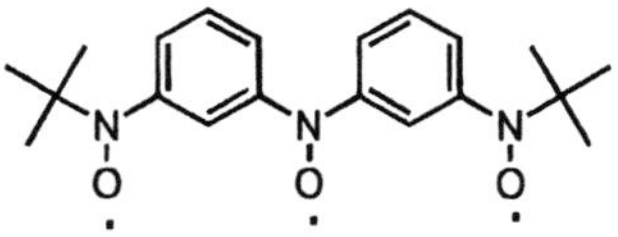

27

Scheme 14

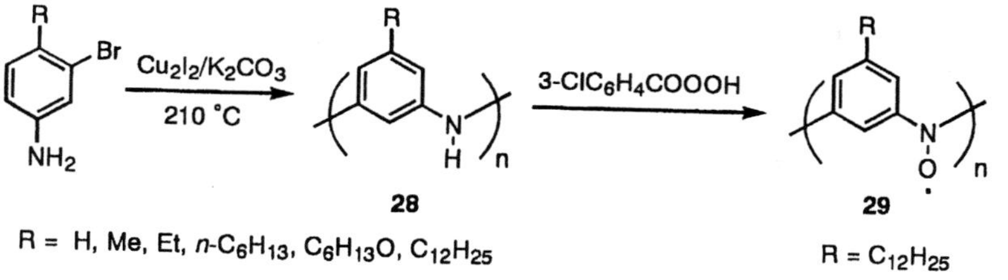
R
Br
NH_2
Cu_2I_2/K_2CO_3
210 °C
R
N
H
n
28
3-ClC_6H_4COOOH
R
N
O
n
29
R = H, Me, Et, n-C_6H_{13}, $C_6H_{13}O$, $C_{12}H_{25}$
R = $C_{12}H_{25}$

Scheme 15

with 3-chloroperbenzoic acid in CH_2Cl_2 gave a nitroxide polyradical, **29**, with $g = 2.0049$. The spin concentration of **29** determined by ESR was low (2.6×10^{19} spins g^{-1}), corresponding to ~1.0% spin/repeating unit.

VIII.B. Magnetic Characterization

Magnetic susceptibility measurements were carried out with a SQUID magnetometer in the temperature range 2–260 K. In the $\chi_g T$ vs. T plots the $\chi_g T$ were constant in the high-temperature region, and in the low-temperature region the $\chi_g T$ vs. T plots showed a downward turn, indicating antiferromagnetic interaction between the unpaired electron spins. The observed antiferromagnetic coupling was ascribed to through-space antiferromagnetic interaction. Therefore, the expected strong ferromagnetic coupling was not induced between the unpaired electron spins, and this was ascribed to the nonplanar conformation of the polyradical main chain. The low-spin concentration of the polyradical may also make the expected through-bond ferromagnetic interaction difficult.

An excellent method for preparation of poly(1,3-phenyleneimine)s was recently reported [29]. The Pd-catalyzed polycondensation of 1,3-phenylenediamine with 1,3-dibromobenzene or of 3-bromoaniline gave poly(1,3-phenyleneimine) with M_n of 10,500–20,400 in 30–86% yields. The ^{1}H and ^{13}C NMR spectra of the polymers obtained are clear, and the IR spectrum showed a strong NH absorption peak at 3384 cm^{-1}. Using this method, poly(5-*tert*-butyl-1,3-pheneleneimine) (**30**) with M_n of 2500–2800 (17–19 repeating units) was obtained from 5-*tert*-butyl-1,3-dibromobenzene and 5-*tert*-butyl-1,3-phenylenediamine (Scheme 16) [30]. Although oxidation of this polymer with 3-chloroperbenzoic acid was performed, no stable polyradical was obtained. To confirm this result, oxidation of **32** with 3-chloroperbenzoic acid was attempted, and no **33** was obtained. Nitroxides such as **31** and **33** may not be sufficiently persistent to be useful as magnetic materials.

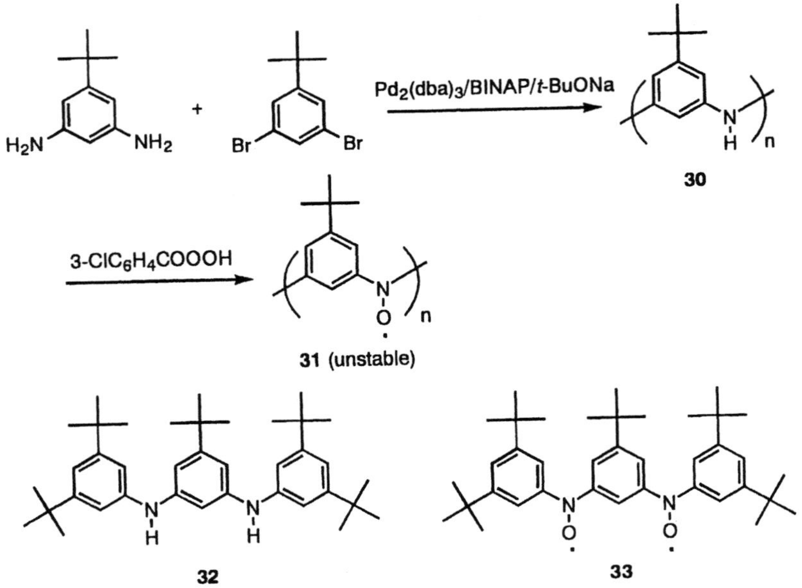

Scheme 16

REFERENCES

1. Rånby B, Rabek JF. ESR Spectroscopy in Polymer Research. Berlin: Springer-Verlag, 1977; Crayston JA, Iraqi A, Walton JC. Polyradicals: Synthesis, spectroscopy, and catalysis. Chem Soc Rev. 1994:147–153.
2. Mataga N. Possible "ferromagnetic states of some hypothetical hydrocarbons. Theor Chim Act. 1968;10:372–376; Miller, JS, Epstein AJ, Reiff WM. Ferromagnetic molecular charge-transfer complexes. Chem Rev. 1988;88:201–220; Dougherty DA. Spin control in organic molecules. Acc Chem Res. 1991; 24:88–94; Iwamura H, Koga N. Studies of organic di-, oligo-, and polyradicals by means of their bulk magnetic properties. Acc Chem Res. 1993;26:346–351; Turnbull MM, Sugimoto T, Thompson LK, eds. Molecule-based magnetic materials. Theory, techniques, and application. ACS symposium Series 644. Washington, DC: ACS, 1995.
3. Rajca A. Organic diradicals and polyradicals: From spin coupling to magnetism? Chem Rev. 1994;94:871–893.
4. Miura Y. Polyradicals (synthesis and magnetic properties). In: Salamone JC, ed. Polymeric Materials Encyclopedia. Vol. 9. Boca Raton, FL: CRC Press, 1996, pp 6686–6695; Lahti PM. Conjugated Polyradicals. In: Salomone JC, ed. Polymeric Materials Encyclopedia. Vol. 2. Boca Raton, FL: CRC Press, 1996, pp 1484–1494.

5. Longuet-Higgins HC. Some studies in molecular-orbital theory I. Resonance structures and molecular orbitals in unsaturated hydrocarbons. J Chem Phys. 1950; 18:265–274.

6. Ovchinnikov AA. Multiplicity of the ground state of large alternant organic molecules with conjugated bonds. Theor Chim Acta. 1978;47:297–304.

7. Borden WT, Davidson ER. Effects of electron repulsion in conjugated hydrocarbon diradicals. J Am Chem Soc. 1977;99:4587–4594.

8. Sonogashira K, Tohda Y, Hagihara N. A convenient synthesis of acetylenes: Catalytic substitutions of acetylenic hydrogen with bromoalkenes, iodoarenes, and bromopyridines. Tetrahedron Lett. 1975;4467–4470; Takahashi S, Kuroyama Y, Sonogashira K, Hagihara N. A convenient synthesis of ethynylarenes and diethynylarenes. Synthesis. 1980;627–630.

9. Miura Y, Ushitani Y, Inui K, Teki Y, Takui T, Itoh K. Syntheses and magnetic characterization of poly(1,3-phenyleneethynylene) with pendant nitronyl nitroxide radicals. Macromolecules. 1993;26:3698–3701.

10. Ullman EF, Osiecki JH, Boocock DGB, Darcy R. Studies of stable free radicals. X. Nitronyl nitroxide monoradicals and biradicals as possible small molecular spin labels. J Am Chem Soc. 1972;94:7049–7059.

11. Miura Y, Ushitani Y. Synthesis and characterization of poly(1,3-phenyleneethynylene) with pendant nitroxide radicals. Macromolecules. 1993;26:7079–7082.

12. Forrester AR, Hay JM, Thomson HR. Organic Chemistry of Stable Free Radicals. New York: Academic Press, 1968; Rozantsev EG. Free Nitroxide Radicals. New York: Plenum Press, 1970; Volodarsky LB, Reznikov VA, Ovcharenko VI. Synthetic Chemistry of Stable Nitroxides. Boca Raton, FL: CRC Press, 1994.

13. Miura Y, Issiki T, Ushitani Y, Teki Y, Itoh K. Synthesis and magnetic behavior of polyradical: Poly(1,3-phenyleneethynylene) with π-toporegulated pendant stable aminoxyl and imine N-oxide-aminoxyl radicals. J Mater Chem. 1996;6:1745–1750.

14. Sasaki S, Iwamura H. Approaches toward high-spin poly[m-(nitrenophenylene)ethynylenes]. Magnetic interaction in the oligomers containing two, three, and five nitrene units. Chem Lett. 1992;1759–1762.

15. Hayashi H, Yamamoto T. Synthesis of regioregular π-conjugated poly(thienyleneethynylene) with a hindered phenolic substituent. Macromolecules. 1997;30:330–332.

16. Miura Y, Hirato S. Unpublished results.

17. Hay AS. Oxidative coupling of acetylene. J Org Chem. 1960; 25:1275–1276; Hay AS. Oxidative coupling of acetylene. II. J Org Chem. 1962;27:3320–3321.

18. Miura Y, Inui K. Synthesis of polyradicals by the oxidative coupling copolymerization of 2-(3,5-diethynylphenyl)-4,4,5,5-tetramethyl-3-oxo-2-imidazolin-1-yloxyl and 1,3,5-triethynylbenzene catalyzed by CuO/O$_2$, and their magnetic properties. Makromol Chem. 1992;193:2137–2147.

19. Saf R, Swoboda P, Hummel K, Czaputa R. New derivatives of 4,4,5,5-tetra-

methyl-4,5-dihydro-1*H*-imidazol-1-oxyl-3-oxide as monomers for polyradicals with special magnetic properties. J Hetrocyclic Chem. 1993;30:425–428.

20. Saf R, Hengg D, Hummel K, Gatterer K, Fritzer HP. Influence of the precipitation method on the magnetic properties of a polyradical with conjugation in the backbone and nitronyl nitroxide side groups. Polymer. 1993;34:2680–2683.

21. Swoboda P, Saf R, Hummel K, Hofer F, Czaputa R. Syntehsis and characterization of a conjugated polymer with stable radicals in the side groups. Macromolecules. 1995;28:4255–4259.

22. Miyaura N, Suzuki A. Palladium-catalyzed cross-coupling reactions of organoboron compounds. Chem Rev. 1995;95:2457–2483.

23. Oka H, Tamura T, Miura Y. J Mater Chem. (in press).

24. Zhang R, Zheng H, Shen J. A new polyphenylene derivative bearing two stable radicals. Macromol Rapid Commun. 1997;18:961–965.

25. Yamamoto T, Hayashi H. π-Conjugated soluble and fluorescent poly(thiophene-2,5-diyl)s with phenolic, hindered phenolic and *p*-$C_6H_4OCH_3$ substituents. Preparation, optical properties, and redox reaction. J Polym Sci, Polym Chem. 1997;35:463–474.

26. Fujita I, Teki Y, Takui T, Kinoshita T, Itoh K, Miko F, Sawaki Y, Iwamura H, Izuoka A, Sugawara T. Design, preparation, and electron spin resonance detection of a ground-state undecet ($S = 5$) hydrocarbon. J Am Chem Soc. 1990; 112:4074–4075; Zuev PS, Sheridan RS. Organic polycarbenes: Generation, characterization, and chemistry. Tetrahedron. 1995; 51:11337–11376.

27. Ishida T, Iwamura H. Bis[3-*tert*-butyl-5-(*N*-oxy-*tert*-butylamino)phenyl] nitroxide in a quartet ground state: A prototype for persistent high-spin poly-[(oxyimino)-1,3-phenylenes]. J Am Chem Soc. 1991;113:4238–4241.

28. Ito A, Ota K, Tanaka K, Yamabe T. *n*-Alkyl group-substituted poly(*m*-aniline)s: Syntheses and magnetic properties. Macromolecules. 1995;28:5618–5625.

29. Kanbara T, Izumi K, Nakadani Y, Narise T, Hasegawa K. Preparation of poly(imino-1,3-phenylene) and its related polymer by palladium-catalyzed polycondensation. Chem Lett. 1997;1185–1186.

30. Nakatsuji M, Miura Y. Unpublished results.

14

Pendant and π-Conjugated Organic Polyradicals

Hiroyuki Nishide
Waseda University, Tokyo, Japan

Takashi Kaneko
Niigata University, Niigata, Japan

I. INTRODUCTION

One of the potential candidates to realize a very high-spin and purely organic-derived molecule is a π-conjugated linear polymer bearing multiple pendant radical groups that are attached to a π-conjugated backbone to yield a non-Kekulé structure (Scheme 1). For some of these pendant polyradicals, theoretical studies have predicted an intrapolymer and through-bond ferromagnetic interaction between the pendant spins, since such a π-conjugated backbone acts as a ferromagnetic coupler (FC) [1]. The pendant-type polyradicals possess the following advantages over the cross-conjugated-type polyradicals that are described in other chapters of this book, such as oligo(phenylenecarbene)s (Chapters 11 and 12) [2] and oligo(triarylmethyl)s (Chapter 17) [3], whose FCs were divided by the spin sites. First, a ferromagnetic spin interaction occurs through the π-conjugated backbone, and potentially even works over a long conjugation length (Scheme 1). That is, the spin coupling is not sensitive to the spin defects that are unavoidable for macromolecular polyradicals. This is in contrast to the cross-conjugated polyradicals, where the spin defect breaks down their π-conjugations. Second, a great number of spins is accumulated along one backbone or within one molecule. The spins are expected to interact among not only their neighboring spins but also with more remote spins. Third, various chemically stable radical species can be introduced to a polyradical as the pendant

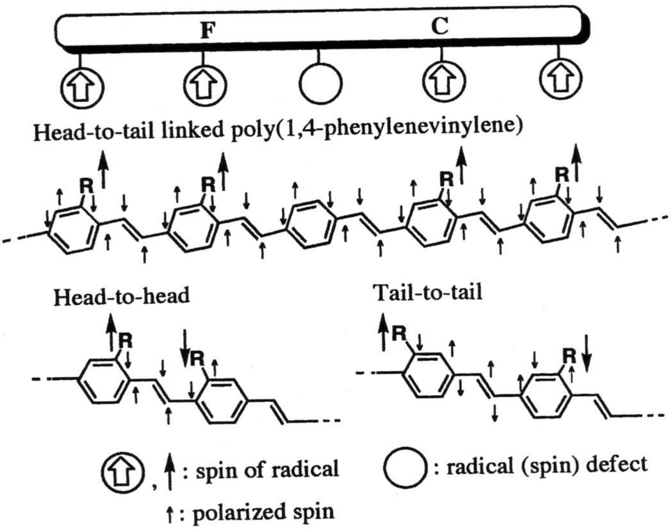

Head-to-tail linked poly(1,4-phenylenevinylene)

Head-to-head   Tail-to-tail

⊕, ↑ : spin of radical   ◯ : radical (spin) defect
↑ : polarized spin

Scheme 1

group. The resultant polyradical has substantial stability and is easily handled, e.g., at room temperature and under air. Such feasibility would be indispensable for the future application of magnetic organics as materials. Additionally, precursors of this type of polyradical may be synthesized via one-pot polymerization and are soluble in common solvents; these characteristics are very favorable for obtaining and processing well-characterized samples to study.

II. MOLECULAR DESIGN AND MAGNETIC PROPERTIES OF PENDANT POLYRADICALS

Polyacetylene, poly(diacetylene), poly(arylene)s, poly(arylenevinylene)s, and poly(aryleneethynylene)s are available as π-conjugated FC backbones of pendant polyradicals (Scheme 2) [4]. These polyradicals are formed via polymerization of the corresponding monomeric precursors, where monomer linkages involve three possible structures, head-to-tail, head-to-head, and tail-to-tail linkages in their dimer units. One can state from the theoretical and experimental studies of diradicals described in the previous chapters that the head-to-tail monomer linkage in the polyradicals is essential for ferromagnetic coupling. By applying spin-polarization rules to a poly-(phenylenevinylene)-based radical, Scheme 1 exemplifies mixtures of

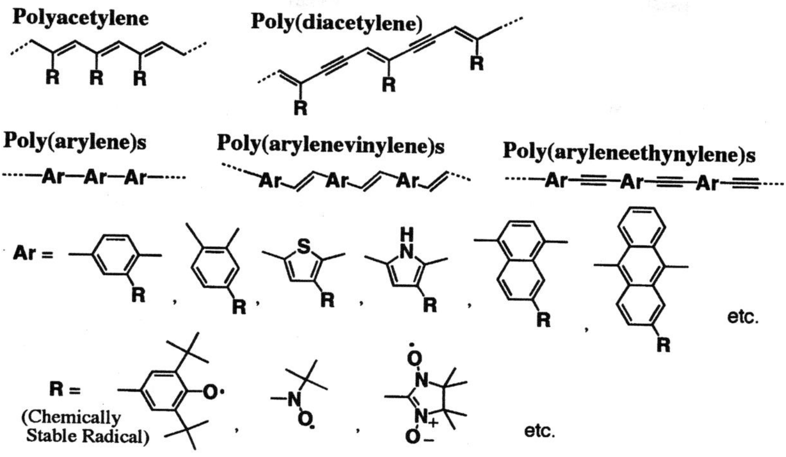

Scheme 2

head-to-tail (*o,m'*-isomers in a stilbene diradical), head-to-head (*o,o'*-isomer in a stilbene diradical), and tail-to-tail (*m,m'*-isomer in a stilbene diradical) linkages that cancel the expected ferromagnetic interaction along a conjugated backbone, because the spin quantum number (S) of the ground state (GS) is estimated to be $S = 2/2$ for head-to-tail and $S = 0$ for the head-to-head and tail-to-tail linked dimer units. Thus, much effort has been devoted to designing both appropriate monomer structures *and* precision polymerization with catalysts to yield completely head-to-tail linkage of the monomer. In principle, this strategy can bring about spin multiplicity values in proportion to the degree of polymerization.

Polyacetylene-based polyradicals have been extensively studied as the most accessible pendant polyradicals, because ferromagnetic through-bond interactions between the pendant spins have been predicted theoretically by using simple polyene models [1,4b]. The poly(phenylacetylene)-based polyradical **1** was synthesized in 1988 by the authors [5]. Following **1**, many pendant polyradicals **2–16** were synthesized via polymerization of the corresponding phenylacetylenic monomers by catalysis of group 5 or 6 transition-metal chlorides or Rh complexes (Table 1) [6–15].

Even the acetylenic monomers with bulky pendant radical precursors yielded the corresponding polymers, such as **3**, **13**, and **14**, with a 10^4–10^5 molecular weight, in which the hydroxy groups were converted to the oxy-radicals in a high-spin concentration due to their stable structures [7,14]. Some Rh complexes were not poisoned by the radical species and directly catalyzed the polymerization of the radical monomers to yield **5**, **11**, and **12**

Table 1 Magnetic Characterization of the Polyacetylene-Based Pendant Polyradicals[e]

R =	Mol wt[a] ($\times 10^4$)	Spin concen. [spin/g(spin/unit)]	Curie const. (emuK/g)	Magnetic property [Weiss temp (K)]	Ref.
1	1.8	2.6×10^{20} (0.1)		Antiferro (-0.2)	5
2	4[b]	4.4×10^{19} (0.022)[c]		Para	6
3	2.0	5.8×10^{20} (0.51–0.59)		Antiferro (-0.7)	7
4	3[b]	1.3×10^{21} (0.4)[c]		Para	6
	0.81	2.1×10^{21} (0.56–0.65)		Antiferro (-3.3)	8
5	0.92–1.1	2.02×10^{21} (0.82)	1.26×10^{-3}	Antiferro (-1.0 ± 0.3)	9
6	14[b]	1.9×10^{21} (0.75)		Para	6
7	20[b]			Para	10
8	0.16	2.1×10^{21d} (0.64)	1.3×10^{-3}	Para	11
9					11
a: R' =	m-1.9	1.5×10^{21d} (0.79)	0.91×10^{-3}	Para	
	p-3.6	1.3×10^{21d} (0.73)	0.83×10^{-3}	para	
b: R' =	0.18	1.3×10^{21d} (0.69)	0.82×10^{-3}	para	
c: R' =	8.1	1.6×10^{21d} (1.5)	1.0×10^{-3}	Para	

Table 1 Continued

R =	Mol wt[a] ($\times 10^4$)	Spin concen. [spin/g(spin/unit)]	Curie const. (emuK/g)	Magnetic property [Weiss temp (K)]	Ref.
10					
a: R' =		1.6×10^{21d} (0.73)	0.97×10^{-3}	Para	11
b: R' =	0.21	2.3×10^{21d} (1.86)	1.4×10^{-3}	Para	
c: R' =	m-0.10	1.5×10^{21d} (0.68)	0.91×10^{-3}	Para	
	p-0.17	1.7×10^{21d} (0.80)	1.1×10^{-3}	Para	
11		2.36×10^{21} (1.01)[c]	1.48×10^{-3}	Antiferro (−1.5)	12
	DP = ~100	1.1×10^{21} (0.5)			13
	18^b			Para	10
12	15^b	2.31×10^{21} (0.99)[c]	1.45×10^{-3}	Antiferro (−1.7)	12
13	2.0	4.5×10^{20} (0.33)		Para	14
14	1.7	1.3×10^{20} (0.09)		Para	14
15	2.0	$10^{16}-10^{17}$ (10^{-3})			15
16	7^b	2.3×10^{20} (0.12)		Para	6

[a] mol wt = $\overline{Mn}$.
[b] mol wt = $\overline{Mw}$.
[c] Calculated from spin/g.
[d] Calculated from Curie const.
[e] Pendant Radical Group **R** in Scheme 2.

with high-spin concentrations [9,12]. The nitroxide polyradicals **4–12** have been obtained with high-spin concentrations, and relatively large Weiss temperatures (θ) (e.g., $\theta = -3.3$ K for **4**) were reported.

However, the expected ferromagnetic interactions have not been observed to date for any polyacetylene-based pendant polyradicals. All the magnetic measurements revealed that the polyradicals were paramagnetic or weakly (probably through-space) antiferromagnetic. Mismatching between the theoretical ferromagnetic prediction and the experimental results is discussed using computational modeling and semiempirical molecular orbital (MO) calculations [16] in the next section.

Besides the polyacetylene-based polyradicals, poly(diacetylene)-based polyradicals have often been studied. These efforts have in part been inspired by the unusual magnetic properties reported for thermally polymerized 4,4′-(butadiyne-1,4-diyl)-bis(2,2,6,6-tetramethyl-4-hydroxypiperidine-N-oxyl) [17]. For example, 2-chloro-5-butadiynylphenyl *tert*-butyl nitroxide was thermally polymerized; however, most of the remaining spins were paramagnetic [18].

Poly(arylene)s are strong candidates for an effective backbone of the pendant polyradicals, as represented in Scheme 2. The poly(arylene)s, such as poly(pyrrole) and poly(thiophene), have been well characterized for conductive and electronic polymer materials. Their HOMO-LUMO energy gap can be tuned by electron-withdrawing or -donating substituents. For example, poly(isothianaphthene) has a very small HOMO-LUMO gap; such a backbone would be of interest for the pendant polyradicals. However, steric hindrances between the hydrogens and/or the substituents of the neighboring arylene ring reduce the π-conjugation length in the poly(arylene)s. Additionally, difficulty in achieving the head-to-tail linkage through polymerization disturbs the study of the poly(arylene)-based polyradicals, and no report has yet been published on their magnetic study.

Poly(arylenevinylene), on the other hand, possesses coplanarity and an extended conjugation length, even after substitution on the arylene ring. The pendant polyradicals based on poly(phenylenevinylene) (PPV) **17–22** have been synthesized by the authors; for the first time, they have succeeded in realizing a through-conjugated backbone bond and long-range ferromagnetic exchange interaction between the pendant unpaired electrons (Scheme 3) [19–21].

Experimental results of model stilbene-diradicals [22] indicated that the *o,m′*- and *m,p′*-substituted stilbenes are effective FCs of the radical sites. These molecular connectivities as dimer units are extended to polyradical structures: head-to-tail linked poly(1,4-phenylenevinylene) (1,4-PPV) and poly(1,2-phenylenevinylene) (1,2-PPV)–2– and –4–substituted with built-in and pendant radical groups, respectively. The 4- (or 2-) bromo-2- (or 4-)

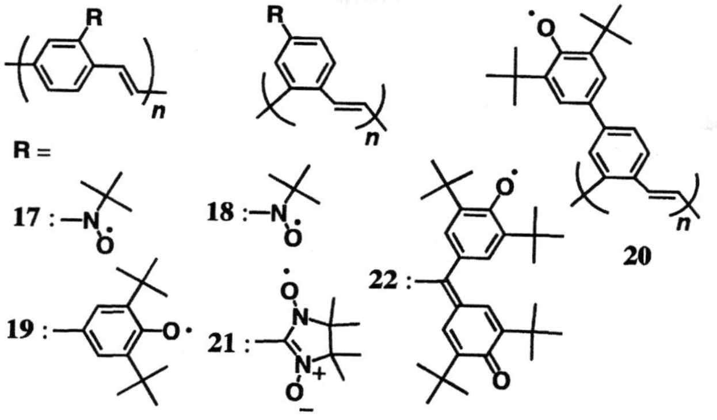

Scheme 3

radical precursor substituted styrene was designed as the starting monomer to yield the 1,4- and 1,2-PPV-based polyradicals, respectively. The Heck reaction of bromostyrene derivatives (β-phenylation of the styrenic vinyl group with phenyl bromide catalyzed with a Pd complex of tolylphosphine) was selected as the polymerization route to yield PPVs with a relatively high molecular weight, a specific head-to-tail linked phenylenevinylene connectivity, and an all *trans*-vinylene structure for increased planarity.

The polymers **17−22** were obtained as yellow powders with a 10^3−10^4 molecular weight, and were soluble in common solvents such as CHCl$_3$, benzene, and tetrahydrofuran. The UV-Vis absorption maximum (λ_{max}) of **17**, **19** (1,4-PPV) and **18**, **20** (1,2-PPV) at ca. 400 and 360 nm suggested a developed π-conjugation in the PPVs.

The precursor polymers were deprotected and chemically or electrochemically oxidized to yield the corresponding polyradicals **17−22**. The polyradicals **18−22** were stable, even in air at room temperature, and the spin concentrations could be increased by the oxidative conditions up to ca. 0.8 spin/unit for **18−20** and beyond 0.9 spin/unit for **21** and **22**. Side reactions such as the intramolecular oxygen migration probably prevented any increase in the spin concentration of **17** in which every *tert*-butyl-nitroxide group was neighboring the vinylene bridge. The polyradicals **18−22** could even be isolated as neat powders using a careful freeze-drying procedure. The spin concentration was decreased to half after 1, 7, and 10 days for **17**, **21** and **22**, respectively, in the powder state. Static magnetic susceptibility of the PPV-based polyradicals **18−22** (neat powder) essentially followed the Curie−Weiss law [$\chi = C/(T - \theta)$] in the temperature range of 10−280 K [20b], and their positive θ values (e.g., θ = 5.7 K for **19**) indicated ferromagnetic interactions in the PPV-based polyradicals.

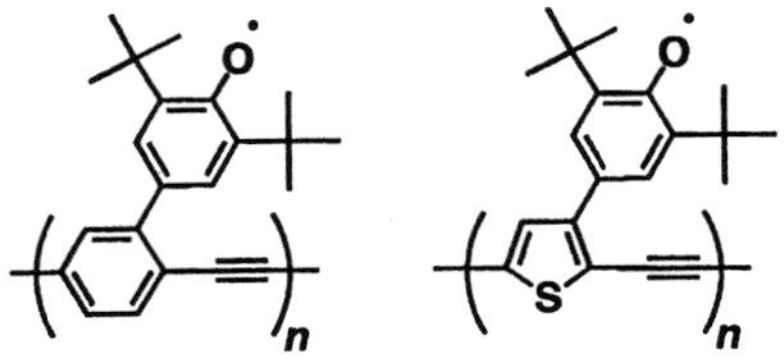

Scheme 4

The ethynylene bridge is characterized by a sterically compact and a hydrogen-free structure. It involves two orthogonal π-bonds. A poly-(aryleneethynylene) looks like a favorable choice as a π-conjugated FC backbone of the polyradicals, which are sterically crowded and often chemically unstable and whose conjugations suffer considerable twisting. The head-to-tail linked poly(aryleneethynylene)s have been synthesized from bromoarylacetylenes using a Pd catalyst: poly(1,4-phenyleneethynylene) **23** [23] and poly(2,5-thienyleneethynylene) **24** [24] are 2- and 3-substituted with phenoxyl, and have $10^3 - 10^4$ molecular weight (Scheme 4). The polyradical **23** showed an intrapolymer ferromagnetic interaction similar to that of the corresponding PPV derivative **19**, while **24** resulted in an unstable radical formation. The *m*-linked poly(phenyleneethynylene)-based polyradicals were also synthesized [25], though their connectivity was considered as a cross-conjugation system (see Chapter 13, by Miura, in this book).

To realize a ferromagnetic interaction through the π-conjugated backbone for pendant-type polyradicals, not only the FC structure or the molecular connectivity of the polyradicals but their conformations and spin-density distributions also need to be considered, as will be described in the following sections.

III. PLANARIZED π-CONJUGATION IN THE POLYRADICALS AFFECTING THE SPIN COUPLING

The triradical **25** [26] and diradical **26** [27] can be considered as trimer and dimer models of the poly(phenylacetylene)-based polyradical **4** and 1,2-PPV-based polyradical **18**, respectively, as shown in Figure 1. They had quartet and triplet GSs whose exchange coupling constants $2J$ were estimated to be 7.4 and 67 cm^{-1}, respectively (positive values show ferromagnetic exchange). However, the magnetic measurement of **4** resulted in a weak antiferromagnetic behavior [8], while the polyradical **18** had a ferromagnetic

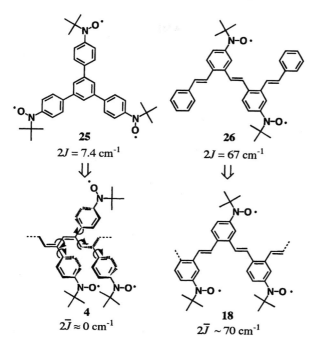

Figure 1 Magnetic interaction of the model oligoradicals and the corresponding polyradicals.

interaction and the exchange-coupling constant comparable with $2J$ of the diradical **26**, or an average value in the polyradical of $2J \sim 70$ cm^{-1} [19].

The incapability of poly(phenylacetylene) as an FC backbone has been discussed using computational modeling and semiempirical MO calculations [16]. Because many connectivity-based theoretical predictions are based on the assumption of planar structures in the π-conjugated hydrocarbons, the effects of the nonplanar that may arise for a polymer, and of heteroatoms, should be taken into account. Insight into this consideration was obtained using MMX force-field optimizations on poly(phenylacetylene). The phenyl-C=C dihedral twisting angle was calculated to be 35–55° for many of these systems, suggesting a strong reduction in the conjugated nature of the electronic structures for the poly(phenylacetylene)-based polyradicals.

Proton hyperfine coupling constants (a_H) at the polyene chain in **1** and **4** were estimated by INDO (intermediate neglect of differential overlap) calculations. They significantly decreased with the torsional angle of the pendant group vs. the polyene backbone. Torsional angles for the substituted poly(phenylacetylene)s, such as **1** and **4**, are presumably larger than those of the simple poly(phenylacetylene) calculated by MMX, because of the steric hindrance of the bulky *tert*-butyl substitutions at the pendant group. The ESR spectrum of **1** as oxidized in solution gave an $a_{\parallel}$ value of 0.02–0.05 mT ascribed to the polyene proton hyperfine coupling, which coincided with INDO calculations for a highly twisted pendant phenoxyl-C=C dihedral angle (ca. 70–90°) [16,28]. The electron spin resonance (ESR) spectrum of the nitroxide polyradical **4** indicated that the spin density at the polyene backbone was obviously not large enough to cause a ferromagnetic coupling or to overcome through-space antiferromagnetic interactions between the nitroxide groups stacked along the backbone. In conclusion, unavoidable problems in the poly(phenylacetylene)-based pendant polyradicals are the nonplanar polyene backbone itself, the highly twisted dihedral angle between the polyene backbone and the pendant radical groups, and possible intramolecular through-space antiferromagnetic interactions between the accumulated pendant radical groups.

On the other hand, the force-field optimization of the 1,2-PPV-based precursor polymer derivatives suggested a planarized form in which the pendant radical groups were located in an alternating fashion relative to the backbone [20b] which was consistent with the notion that good π-conjugation favored high-spin coupling in a pendant polyradical. Actual 1,2-PPV-based polyradicals such as **18** are found to have planarized conformations that display an intramacromolecular ferromagnetic coupling between their pendant radicals, in accord with the experimental results.

The ESR hyperfine spectra of monoradical model compounds corresponding to the unit structure of the PPV-based polyradicals provided more information about the spin-density distribution over the molecules or the related polyradical structures (Table 2) [29]. The proton hyperfine splitting of the nitroxide monoradical of the 1,4-PPV-based **17** was hardly detected, indicating localization of the spin density on the pendant N–O bond, probably because of significant twisting of the pendant group relative to the 1,4-PPV backbone. On the other hand, the hyperfine coupling constants attributed to the 5–7 protons of the vinylene and/or the two phenyl rings for the phenoxyl monoradical of the 1,4-PPV-based **19** imply a sterically less hindered structure for the pendant phenoxyl group. In contrast to the 1,4-PPV-based radicals, the monoradicals of the 1,2-PPV-based **18** and **20** behave as if their pendant radical moieties do not suffer such steric interactions with the *m*- and *p*-substituted PPV backbones, and yield the larger proton hyperfine coupling constants ascribed to the phenylene and the vinylene or vinyl. Similar changes were seen in the broad hyperfine ESR spectra of the polyradicals **18** and **20** themselves at low-spin concentrations. These results re-

Table 2 ESR Hyperfine Coupling Constants (mT) Estimated by the Spectral Simulation for Monoradical Model Compounds

Radical species	Corresponding polyradical	Phenylene ring substituted with the radical			Vinylene or vinyl	Radical moiety	
		o-a_H	m-a_H	p-a_H	a_H	a_N	a_H
Nitroxide	1,4-PPV-based **17**	None	None	None	None	1.42	
	1,2-PPV-based **18**	0.19	0.09		0.12, 0.02	1.15	
Phenoxyl	1,4-PPV-based **19**	0.11	0.06	0.11	0.03		0.17
	1,2-PPV-based **20**	0.17	0.10		0.09		0.18

Source: Ref. 29.

veal an effectively delocalized spin distribution into the 1,2-PPV backbone for **18** and **20**.

Ultraviolet photoelectron spectroscopy and the aforementioned UV/Vis absorption of the polyradicals also supported a significant π-conjugation length along the poly(phenylenevinylene) backbone [30]. It is concluded for the pendant and π-conjugated polyradicals that a total planarized π-conjugation over the macromolecule is a critical prerequisite to yield a significant magnetic interaction. If a pendant polyradical structure can be pictured on a flat paper without any overlapping bond lines, such as poly-(aryleneethynylene)-based polyradicals besides 1,2-PPV, it should be an effective backbone.

IV. EFFECTS OF DEGREE OF POLYMERIZATION AND SPIN DEFECT ON THE HIGH-SPIN MULTIPLICITY AND COUPLING CONSTANT

$\chi_{mol}T$ of the PPV-based polyradicals **18–22** (neat powder) exceeded 0.375 emu·K/mol (the value of $S = 1/2$) in the temperature range of 2–300 K though slightly decreased at low temperature ($< \sim 10$ K). The overall magnetic interactions of **18–22** were ascribed to both a ferromagnetic interaction through the intrapolymer π-bond and a weak antiferromagnetic through-space interpolyradical interaction. On the other hand, for the 1,3-PPV-based polyradical **27**, whose low-spin state was expected from the theoretical prediction based on its molecular connectivity, the $\chi_{mol}T$ curve, even for a diluted sample, significantly deviated downward from the value of $S = 1/2$, indicating a singlet GS for **27** (Scheme 5) [20b].

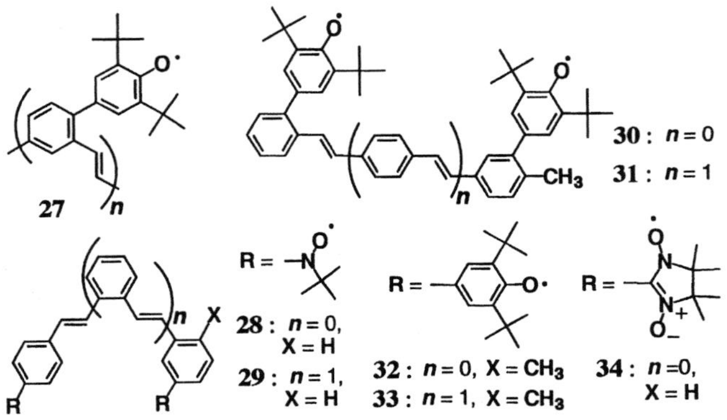

Scheme 5

The pendant type of polyradical involves the contribution of various arrangements of the radical sites along the backbone and has some spin-exchange-coupling constants. Therefore, the exchange-coupling constant in the polyradical was estimated approximately in this chapter as the average value ($\bar{J}$). $\bar{J}$ was estimated from the curve-fitting of the $\chi_{mol}T$ vs. T plots by a model that assumes a mixture of three-, two-, and one-spin systems, for polyradical samples with spin concentrations of 0.2–0.4 spin/unit and an average GS spin quantum number of $\bar{S} < 3/2$.

The $2\bar{J}$ values are listed in Table 3 with the $2J$ values of the model diradical compounds **26** and **28–34** [19–21,27]. The average GS spin quantum number ($\bar{S}$) was determined by comparing the magnetization M/M_S vs. the effective temperature ($T - \theta$) plots with the theoretical Brillouin curves, and are also given in Table 3. $2\bar{J}$ increases with the degree of polymerization (DP), e.g., for the 1,2-PPV-based nitroxide polyradicals in the first paragraph of Table 3, because the spin-exchange interaction works cooperatively from both sides along the conjugated backbone, and/or the high-spin state of the polyradical is stabilized by a decrease in the HOMO-LUMO gap or the availability of more degenerate nonbonding molecular orbitals (NBMOs) due to a better developed π-conjugation as the DP increases. This result is consistent with the polymeric effects expected from the semiempirical MO calculations.

A comparison of the 1,4-PPV-based **19** and the 1,2-PPV-based **20** reveals a stronger interaction in the 1,2-derivative, in which the aforementioned ESR result suggested a more effective spin distribution through the pendant phenoxyl to the backbone conjugation due to its more rigid coplanarity in the π-conjugation. The semiempirical MO calculations also suggest reduction in the $2\bar{J}$ value for the sterically hindered **19**.

The spin localization on the pendant nitronyl nitroxide and galvinoxyl residue for the polyradical **21** and **22**, respectively, resulted in smaller $2J$ values of **21** and **22**. Besides being a planarized structure, a pendant radical species needs to be carefully designed to provide sufficient spin delocalization into the backbone conjugation.

$\bar{S}$ is related to the DP and the spin concentration, i.e., to the number of spins in the polyradical. The 1,2-PPV-based phenoxyl polyradical **20** (DP of 17), even with a spin concentration of 0.67 spin/unit, revealed an $\bar{S}$ value equal to or greater than 4/2 [20], which is expected to be further enhanced by higher DPs and spin concentrations.

The $2J$ values of **29**, **31**, and **33** are reduced to 5- to 10-fold (from **28** to **29**, from **30** to **31**, and from **32** to **33**, respectively, as shown in Table 3) in response to the conjugated but more widely spaced phenylenevinylene unit. The semiempirical MO calculation data of the model compounds supported this reduction [20,27]. However, this 5- to 10-fold reduction of $2J$ with a spacing or a spin defect is not fatal. For a cross-conjugated type of polyradical, which involves a radical in its backbone, the $2J$ value was decreased by more than 100-fold by an increase of one unit of spacing (for oligo(1,3-phenylenemethylene)) [31].

A Monte Carlo simulation of the experimental magnetization plots of the 1,4-PPV-based phenoxyl radical **19** ($\bar{S} \sim 3/2$) suggested a contribution of the spin coupling among the pendant radicals separated by one nonradical phenylenevinylene spacing [32], which coincides with the quantitative $2J$ value of **31**. The electron spin transient nutation method based on pulsed ESR spectroscopy has been successfully applied to characterize the polyradical **20** as the mixture of molecular high-spin (mainly $S = 2/2, 4/2, 6/2,$ and 8/2 species) assemblages [33].

V. TWO-DIMENSIONAL EXTENSION OF THE PENDANT POLYRADICALS

In the previous chapters, descriptions were given about extending the cross-conjugated type of polyradicals to their pseudo-two-dimensional (2D) homologies, i.e., branched, dendritic, macrocyclic, annelated macrocyclic, and pseudo-ladder polyradicals [3,34]. The aim of such 2D extensions, especially the macrocyclic ones, is to diminish the effects of a spin defect, which can be fatal for cross-conjugated polyradicals, by guaranteeing a spin-coupling interaction with multi-FC conjugated pathways in the 2D frameworks.

For the pendant polyradicals, the 2D-extension strategy first means an increase in their molecular weight or DP, which enhances S in GS. For example, a star-shaped and PPV-based polyradical **35** with DP $\sim$ 70 [35]

Table 3 Average Spin Quantum Number ($\bar{S}$) and Average Exchange-Coupling Constants ($\bar{J}$) of the PPV-Based Polyradicals and Diradicals

Polyradical	DP[a]	n[b]	Spin concen. (spin unit^{-1})	$\bar{S}$ or S[c]	$2\bar{J}$ or $2J$[d] (cm^{-1})	MO calcd $2J$ (cm^{-1})
28	2	0		2/2	41 ± 1	115[e]
29	3	1		2/2	3 ± 2	25[e]
26	4	0		2/2	67 ± 11	
18	4					143[e]
18	7		0.50–0.59	1/2–2/2	$28 \leq 2J \leq 40$	153[e]
18	15		0.53–0.62	1/2–3/2	$33 \leq 2J \leq 70$	
30	2	0		2/2	18 ± 1	40[e,f], 200[g]
31	3	1		2/2	4 ± 6	8[e,f]
19	5					53[e,f]
19	6					55[e,f]
19	9		0.46	2/2–3/2	20 ± 2	
19	21		0.23	2/2–3/2	32 ± 2	
32	2	0		2/2	50 ± 6	68[e], ≥ 350[g]

Compound						
33	3	1	2/2		6 ± 1	14[e]
20	6					93[e]
20	8		1/2–2/2	0.25	48 ± 6	
20	17		2/2–3/2	0.26	47 ± 6	
20	17		4/2–5/2	0.68		
27	8		0–1/2	0.35	−90 ± 4	≤100[g]
34	2	0	2/2		5 ± 1	
21	7		1/2–2/2	0.80	29 ± 3	
22	4		2/2	0.97	12 ± 5	

[a]Degree of polymerization of the precursor polymers for **18–22** and **27**.

[b]Number of the spacing phenylenevinylene unit for the diradicals as shown in Scheme 5.

[c]GS spin quantum number (S) for **26, 28**, and **29–34**, and average spin quantum number ($\bar{S}$) for spin state of **18–22** and **27**.

[d]Exchange-coupling constant ($2J$) for **26, 28, 29–34**, and its average value ($2\bar{J}$) for **18–22** and **27**.

[e]Estimated by the INDO-CI method from refs. 20a, 27.

[f]The MO calculation after the 40° rotation for the phenyl-phenylenevinylene bond, corresponding to steric hindrance between the o-phenyl proton and the vinylene proton.

[g]Estimated by the AM1-CI method from ref. 20b.

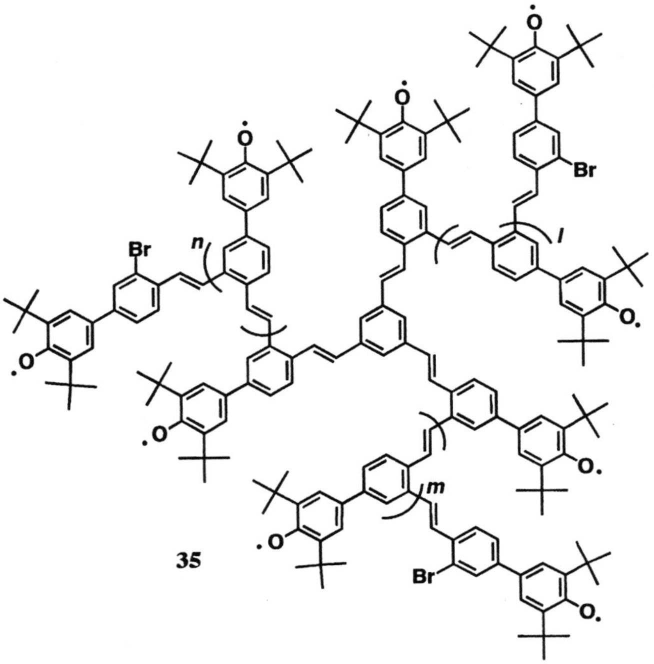

Scheme 6

revealed $\bar{S} = 7/2$, even with a spin concentration of 0.6 spin/unit (Scheme 6). Additional 2D effects on $\bar{S}$ of the pendant polyradicals are still not realized, despite the principal requirement that only a magnetically 3D interaction could realize a bulk ferromagnetism. Still, pendant polyradicals possess a powerful advantage of reduced sensitivity to the spin-defect damage for designing a DP-correlated, very high-spin polyradical.

REFERENCES

1. (a) Ovchinnikov AA. Multiplicity of the Ground State of Large Alternant Organic Molecules with Conjugated Bonds. Theoret Chim Acta. 1978; 47:297–304. (b) Klein DJ, Nelin CJ, Alexander S, Matsen FA. High-spin hydrocarbons. J Chem Phys. 1982; 77:3101–3108. (c) Tyutyulkov NN, Karabunarliev SH. Structure and Properties of Nonclassical Polymers. IV. Magnetic Properties of Polymers with Superexchange Interaction. Chem Phys. 1987; 112:293–299. (d) Nishide H. High-Spin Alignment in π-Conjugated Polyradicals: A Magnetic Polymer. Adv Mater. 1995; 7:937–941.

2. Iwamura H, Koga N. Studies of Organic Di-, Oligo-, and Polyradicals by Means of Their Magnetic Properties. Acc Chem Res. 1993; 26:346–351.
3. Rajca A. Organic Diradical and Polyradicals: From Spin Coupling to Magnetism? Chem Rev. 1994; 94:871–893.
4. (a) Yamaguchi K, Toyoda Y, Fueno T. A Generalized MO (GMO) Approach to Unstable Molecules with Quasi-degenerate Electronic States: Ab-initio GMO Calculations of Intramolecular Effective Exchange Integrals and Designing of Organic Magnetic Polymers. Synth Met. 1987; 19:81–86. (b) Lahti PM, Ichimura AS. Semiempirical Study of Electron Exchange Interaction in Organic High-Spin π Systems. Classifying Structural Effects in Organic Magnetic Molecules. J Org Chem 1991; 56:3030–3042. (c) Nishide H, Kaneko T, Kuzumaki Y, Yoshioka N, Tsuchida E. Poly(phenylvinylene) and Poly(phenylenevinylene) with Nitroxide Radicals. Mol Cryst Liq Cryst. 1993; 232:143–150.
5. (a) Nishide H, Yoshioka N, Inagaki K, Tsuchida E. Poly[3,5-di-*tert*-butyl-4-hydroxyphenyl)acetylene]: Formation of a Conjugated Stable Polyradical. Macromolecules. 1988; 21:3119–3120. (b) Nishide H, Yoshioka N, Inagaki K, Kaku T, Tsuchida E. Poly[(3,5-di-*tert*-butyl-4-hydroxyphenyl)acetylene] and Its Polyradical Derivative. Macromolecules. 1992; 25:569–575.
6. Iwamura H, Murata S. Magnetic Coupling of Two Triplet Phenylnitrene Units Joined Through an Acetylenic or a Diacetylenic Linkage. Mol Cryst Liq Cryst. 1989; 176:33–48.
7. (a) Nishide H, Yoshioka N, Kaneko T, Tsuchida E. Poly(*p*-ethynylphenyl)galvinoxyl: Formation of a New Conjugated Polyradical with an Extraordinarily High Spin Concentration. Macromolecules. 1990; 23:4487–4488. (b) Yoshioka N, Nishide H, Kaneko T, Yoshiki H, Tsuchida E. Poly[(*p*-ethynylphenyl)hydrogalvinoxyl] and Its Polyradical Derivative with High Spin Concentration. Macromolecules. 1992; 25:3838–3842.
8. Nishide H, Kaneko T, Yoshioka N, Akiyama H, Igarashi M, Tsuchida E. Poly[[4-(*N-tert*-butyl-*N*-hydroxyamino)phenyl]acetylene] and the Magnetic Property of Its Radical Derivative. Macromolecules. 1993; 26:4567–4571.
9. (a) Miura Y, Matsumoto M, Ushitani Y, Teki Y, Takui T, Itoh K. Magnetic and Optical Characterization of Poly(ethynylbenzene) with Pendant Nitroxide Radicals. Macromolecules. 1993; 26:6673–6675. (b) Miura Y, Matsumoto M, Ushitani Y. Synthesis of Poly(ethynylbenzene) with Pendant Nitroxide Radicals by Rhodium-Catalyzed Polymerization of Ethynylphenyl Nitroxide. Macromolecules. 1993; 26:2628–2630.
10. Saf R, Hummel K, Gatterer K, Fritzer HP. Paramagnetic conjugated polymers with stable radicals in side goups. Polym Bull. 1992; 28:395–402.
11. Dulog L, Lutz S. Polymerization of acetylene derivatives with nitroxyl radical pendant groups. Makromol Chem Rapid Commun. 1993; 14:147–153.
12. Fujii A, Ishida T, Koga N, Iwamura H. Syntheses and Magnetic Properties of Poly(phenylacetylenes) Carrying a (1-Oxido-3-oxy-4,4,5,5-tetramethyl-2-imidazolin-2-yl) Group at the Meta or Para Position of the Phenyl Ring. Macromolecules. 1991; 24:1077–1082.
13. Miura Y, Inui K, Yamaguchi F, et al. Molecular Design, Synthesis, and Magnetic Characterization of Poly(phenylacetylene) with Pi-Toporegulated Pendant

Nitronyl Nitroxide Radicals as Models for Organic Superpara- and Ferromagnets. J Polym Sci Polym Chem Ed. 1992; 30:959–966.

14. Nishide H, Yoshioka N, Saitoh Y, Gotoh R, Miyakawa T, Tsuchida E. Synthesis and Magnetic Property of Polyacetylene Bearing π-Conjugated Bis-(diphenylene)phenylallyl Radical. J Macromol Sci. 1992; A29:775–786.

15. Alexander C, Feast W. An approach to the preparation of conjugated polyradicals. Polym Bull. 1991; 26:245–252.

16. Nishide H, Kaneko T, Igarashi M, Tsuchida E, Yoshioka N, Lahti PM. Magnetic Characterization and Computational Modeling of Poly(phenylacetylene)s Bearing Stable Radical Groups. Macromolecules. 1994; 27:3082–3086.

17. Korshak UV, Medvedeva TV, Ovchinnikov AA, Spector VN. Organic polymer ferromagnet. Nature. 1987; 326:370–372.

18. Inoue K, Koga N, Iwamura H. An Approach to Organic Ferromagnets. Synthesis and Characterization of 1-Phenyl-1,3-butadiyne Polymers Having a Persistent Nitroxide Group on the Phenyl Ring. J Am Chem Soc. 1991; 113:9803–9810.

19. (a) Kaneko T, Toriu S, Kuzumaki Y, Nishide H, Tsuchida E. Poly-(phenylenevinylene) Bearing Built-in *tert*-Butylnitroxide: A Magnetically Coupled Polyradical in the Intrachain. Chem Lett. 1994; 2135–2138. (b) Nishide H, Kaneko T, Toriu S, Kuzumaki Y, Tsuchida E. Synthesis of and Ferromagnetic Coupling in Poly(phenylenevinylene)s Bearing Built-in *t*-Butyl Nitroxides. Bull Chem Soc Jpn. 1996; 69:499–508.

20. (a) Nishide H, Kaneko T, Nii T, Katoh K, Tsuchida E, Yamaguchi K. Through-Bond and Long-Range Ferromagnetic Spin Alignment in a π-Conjugated Polyradical with a Poly(phenylenevinylene) Skeleton. J Am Chem Soc. 1995; 117: 548–549. (b) Nishide H, Kaneko T, Nii T, Katoh K, Tsuchida E, Lahti PM. Poly(phenylenevinylene)-Attached Phenoxyl Radicals: Ferromagnetic Interaction through Planarized and π-Conjugated Skeletons. J Am Chem Soc. 1996; 118:9695–9704.

21. Nishide H, Hozumi Y, Nii T, Tsuchida E. Poly(1,2-phenylenevinylene)s Bearing Nitronyl Nitroxide and Galvinoxyl at the 4-Position: π-Conjugated and Non-Kekulé-Type Polyradicals with a Triplet Ground State. Macromolecules. 1997; 30:3986–3991.

22. Yoshioka N, Lahti PM, Kaneko T, Kuzumaki Y, Tsuchida E, Nishide H. Semiempirical Investigation of Stilbene-Linked Diradicals and Magnetic Study of Their Bis(*N-tert*-butylnitroxide) Variants. J Org Chem. 1994; 59:4272–4280.

23. Nishide H, Kawasaki T, Morikawa R, Tsuchida E. Synthesis and Magnetic Property of Phenoxyl-Substituted Poly(phenyleneethynylene)s. Polym Preprints, Jpn. 1996; 45:451.

24. Hayashi H, Yamamoto T. Synthesis of Regioregular π-Conjugated Poly(thienyleneethynylene) with a Hindered Phenolic Substituent. Macromolecules. 1997; 30:330–332.

25. (a) Miura Y, Ushitani Y. Synthesis and Characterization of Poly(1,3-phenyleneethynylene) with Pendant Nitroxide Radicals. Macromolecules. 1993; 26: 7079–7082. (b) Miura Y, Ushitani Y, Inui K, Teki Y, Takui T, Itoh K. Synthesis

and Magnetic Characterization of Poly(1,3-phenyleneethynylene) with Pendant Nitronyl Nitroxide Radicals. Macromolecules. 1993; 26:3698–3701.

26. Kanno F, Inoue K, Koga N, Iwamura H. Persistent 1,3,5-Benzenetriyltris(*N*-*tert*-butyl nitroxide) and Its Analogs with Quartet Ground States. Intramolecular Triangular Exchange Coupling among Three Nitroxide Radical Centers. J Phys Chem. 1993; 97:13267–13272.

27. Kaneko T, Toriu S, Tsuchida E, et al. Ferromagnetic Spin Coupling of *tert*-Butylnitroxide Diradicals Through a Conjugated Oligo(1,2-phenylenevinylene)-Coupler. Chem Lett. 1995; 421–422.

28. Nishide H, Kaneko T, Gotoh R, Tsuchida E. Polyacetylene Derivatives with Chain-sided Phenoxy and Galvinoxyl Radicals. Mol Cryst Liq Cryst. 1993; 233:89–96.

29. Nishide H, Kaneko T, Toriu S, et al. π-Conjugated Polyradicals with Poly(phenylenevinylene) Skeleton and Their Through-Bond and Long-Range Interaction. Mol Cryst Liq Cryst. 1995; 272:131–138.

30. Kaneko T, Ito E, Seki K, Tsuchida E, Nishide H. UPS Study of Poly(phenylenevinylene)s Substituted with Hexyloxy, Phenoxy, and Nitroxy Residues. Polym J. 1996; 28:182–184.

31. Tyutyulkov NN, Karabunarliev SC. Structure and Properties of Nonclassical Polymers. III. Magnetic Characteristics at Finite Temperatures. Internal J Quantum Chem. 1986; 29:1325–1337.

32. Kaneko T, Toriu S, Nii T, Tsuchida E, Nishide H. Poly[(*N*-oxyamino) and (oxyphenyl)phenylenevinylene]s: Magnetically Coupled Polyradicals in the Chain. Mol Cryst Liq Cryst. 1995; 272:153–160.

33. Sato K, Shiomi D, Takui T, et al. FT Pulsed EPR/Transient Quantum Spin Nutation Spectroscopy Applied to Inorganic High-Spin Systems and a High-Spin Polymer as Models for Organic Ferromagnets. J Spectro Soc Jpn. 1994; 43:280–291.

34. (a) Rajca A, Lu K, Rajca S. High-Spin Polyarylmethyl Polyradical: Fragment of a Macrocyclic 2-Strand Based upon Calix[4]arene Rings. J Am Chem Soc. 1997; 119:10335–10345. (b) Rajca A, Wongsriratanakul J, Rajca S. Organic Spin Clusters: Ferromagnetic Spin Coupling through a Biphenyl Unit in Polyarylmethyl Tri-, Penta-, Hepta-, and Hexadecaradicals. J Am Chem Soc. 1997; 119:11674–11686.

35. Nishide H, Miyasaka M, Tsuchida E. Polyphenoxyl Based on a Star-Shaped *p*-Conjugation: A Radical Polymer with an Octet Ground State. Angew Chem Int Ed Engl. 1998; 37:2400–2402.

15

Liquid Crystalline Radicals: An Emerging Class of Organic Magnetic Materials

Piotr Kaszynski
Vanderbilt University, Nashville, Tennessee

I. INTRODUCTION

Magnetism of molecular solids that arises from cooperative spin–spin interactions requires specific 3-dimensional orientational and positional order of the spin-containing species [1–3]. The magnitude and the type of the intermolecular exchange interactions may not be the same in all three dimensions, and 2-D and 1-D magnetic materials often result [2]. The most common way to obtain molecular magnetic materials is to engineer their crystal structures by using hydrogen bonding [4–7], chelating metals [8,9], and crystallization or cocrystallization of individual components from various solvents [10,11]. The materials may exhibit several polymorphs, depending on the crystallization conditions or the temperature of the material [12,13].

An alternative approach [14] to molecular magnetic materials is via partially oriented fluids, or liquid crystals (LCs), which lack positional order but offer various degrees of orientational correlations. These materials, while being liquidlike, exhibit some characteristics of ordered crystalline solids, e.g., optical, dielectric, and magnetic anisotropy [15,16]. Figure 1 shows examples of spatial arrangements of rodlike and disclike molecules in three calamitic phases (Figure 1a–c) and two discotic phases (Figure 1d,e) [17–19].

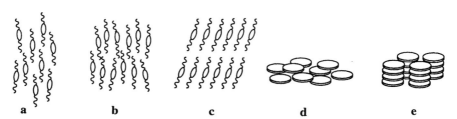

Figure 1   Molecular arrangements in different liquid crystals: (a) nematic; (b) smectic A; (c) smectic F; (d) nematic discotic; (e) columnar discotic. The ovals represent the rigid cores; the lines are the flexible tails. Discs are comprised of a rigid core and radially distributed flexible tails.

II. THE LIQUID CRYSTAL APPROACH: BENEFITS AND PROBLEMS

There are over two dozen recognized and morphologically distinct LC phases (polymorphs) and their chiral modifications, which belong to the two major families of thermotropic LCs, calamitics and discotics (Figure 1) [20]. In contrast to solid crystals, all these mesophases are thermodynamic structures. A single compound may display several mesophases, and as many as nine phases have been observed [21]. The degree of molecular order varies in different structures: nematics, having only orientational correlations, are more liquidlike (least ordered); smectics and columnar discotics are more highly ordered. The molecules in lamellar smectic phases [22] can be oriented with their long molecular axes perpendicular to the layer plane (e.g., smectic A or B) or form a tilt angle (e.g., smectic C or G), which typically changes with the temperature. The in-plane order can be nonexistent (smectics A and C), partial (smectics B, F, and I), or positional (crystal smectics B, E, G, H, J, and K). The molecular rotations in LCs are usually unrestricted in most mesophases except for crystal smectics E, H, and K [18,19]. Thus, liquid crystals offer a continuum of molecular order between an isotropic liquid and a crystal, with a wealth of possible spatial molecular arrangements and intermolecular interactions.

Intermolecular multipolar and dispersive interactions in smectic and columnar phases are usually stronger between the rigid cores than between the alkyl chains, and they become the dominant forces in higher-order mesophases and crystals. Therefore it is beneficial for materials properties to place an active element such as a spin in the rigid core and to spread it (delocalize it) over a large area in order to achieve maximum intermolecular interactions. Unfortunately, the dynamic nature of LC phases characterized by fast molecular on-axis rotations (10^{10}–10^{11} s^{-1}) precludes strong and specific interactions such as the McConnell type necessary for effective inter-molecular spin–spin exchange [23,23a]. In other words, the thermal energy kT to form the LC state effectively competes with the possible spin–spin exchange interactions, and therefore the most desirable magnetic properties (ferro- and ferrimagnetism) may not be ascertainable at temperature ranges typical for LC (>0°C) [24].

One application of LC behavior in the study of molecular magnetism is to use the mesogenic states to obtain a designer, rigid solid in which molecular motions are inhibited and specific static intermolecular interactions (spin–spin) are in place. The viscosity of LC phases increases with their increasing order, and it is commonly observed that certain highly ordered phases such as smectic E, discotic, and polymeric LCs easily super-cool, preserving the molecular arrangement of the LC phase in the glassy phase [25]. It has also been noted that the structure of the solid phase is largely defined by the structure of only the preceding LC phase and the presence of external magnetic or electric fields. Thus the formation of a glass or crystalline phase from LC phases can be used to engineer the solid structure. Consequently, the magnetic behavior of a rigid solid that retains most of the LC-phase structural features can be examined at the very low temperatures typically required for studies of molecular organic magnetism.

III. METALLOMESOGENS VS. ALL-ORGANIC LIQUID CRYSTALLINE RADICALS

The majority of LCs with permanent spins studied thus far are metallome-sogens [26] (metal-containing LC molecules), in which the spin originates from transition-metal centers. Some of them exhibit ferromagnetic order in the solid state [27], some show dramatic changes in bulk magnetization upon phase transition [28], and the most recently prepared mesogenic derivatives of Eu and Tb show very high magnetic birefringence [29–31]. Most of the spin density resides on the metal center, although some is distributed into the ligand via a spin-polarization mechanism. Typically, very little is trans-ferred to the π-framework [32–34]. This limits the extent of intermolecular interactions and spin–spin magnetic exchange.

There are very few LCs containing an organic spin, and the only π-radical compounds are derivatives of bipyran-4,4'-ylidene **1** (Scheme 1), which form charge-transfer discotic LCs upon treatment with TCNQ [35,36]. Other sources of organic radicals [37] are much less suitable for the synthesis of LCs, due mainly to inappropriate substitution patterns or the geometry and bulkiness of radical stabilizing substituents, which are detrimental to the mesophase stability. The nitroxyl group has been explored as a potential

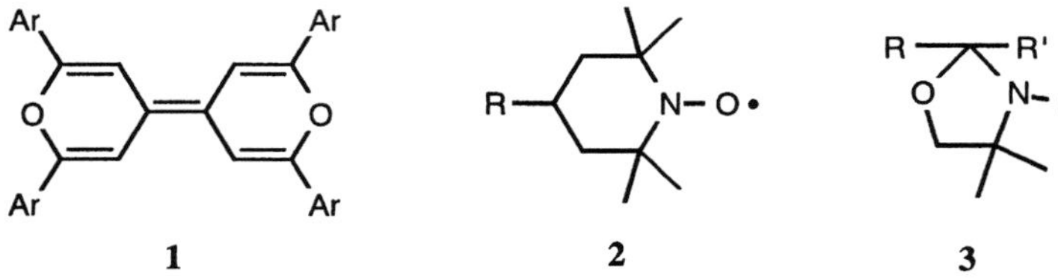

Scheme 1

spin source, but all attempts to prepare monomeric or polymeric mesogens using **2** as a part of the rigid core have failed [38–41]. Only by placing the nitroxyl group **3** within the alkyl chain, have stable smectic A, C and E phases been observed [41,42].

IV. A NEW APPROACH TO ALL-ORGANIC LIQUID CRYSTALLINE RADICALS

All molecules forming liquid crystalline phases are characterized by a high aspect ratio (either elongated or flattened molecular shapes; see Figure 1), and any broadening of the molecule has a detrimental effect on the stability of LC phases. Therefore, an ideal radical source should be a chemically stable, electrically neutral, bidentate fragment that is easy to incorporate into a six-membered ring system (to minimize the effect on the aspect ratio), and electronically conjugated so that the spin is π-delocalized over the core to maximize intermolecular spin–spin interactions. A system that is chemically versatile and satisfies these requirements is the thioaminyl radical fragment shown in Figure 2.

IV.A. Thioaminyl Radicals

The thioaminyl group (Figure 2) has been featured as a structural element in a number of chemically and thermally stable, free radicals [43–46]. Most of these acyclic [47–50] and cyclic [51–57] species are π-delocalized radicals with the unpaired electron in the π* orbital. Some of these radicals melt above 150°C without decomposition [47], some exist as liquids [54],

Figure 2 Resonance structures of the thioaminyl radical fragment.

and some have been explored as possible unidimensional conductors [58–61] and molecular magnetic materials [59,62–71]. In the solid state, most of these radicals behave as diamagnetic substances at ambient temperature [59,60] as a result of spin-paired interactions between the sulfur centers and the formation of weak out-of-plane S–S bonds ($\Delta H_{dimer} > -40$ kJ/mol) [54]. In the liquid phase, however, the magnetic moments of some of these radicals are essentially temperature independent [54,72].

IV.B.   Liquid Crystal Core Structural Elements

Although virtually none of the known heterocyclic radicals has a geometry suitable for the formation of LC phases, it is easy to envision incorporation of the N–S fragment into structures of known mesogenic molecules. For instance, a —CH=CH— fragment in the central ring of **4** might be replaced with the thioaminyl fragment —S—N—, as shown in Figure 3. The remaining two CH fragments may be halogenated or replaced with nitrogen atoms in order to stabilize the radical or otherwise modify its electronic structure. Since these structural modifications have a rather small effect on the molecular shape, one might predict that **5** will display LC behavior similar to that of **4** [73].

Following this general idea, a number of benzene, naphthalene, phenanthrene, and pyrene derivatives have been designed as rigid core structural elements for liquid crystalline radicals (Figure 4) [74]. Thus, the thioaminyl fragment has become the centerpiece of a new, broad class of potential calamitic and discotic mesogens.

Cr 89.3 S$_B$ 189.9 S$_A$ 216.4 N 217.0 I

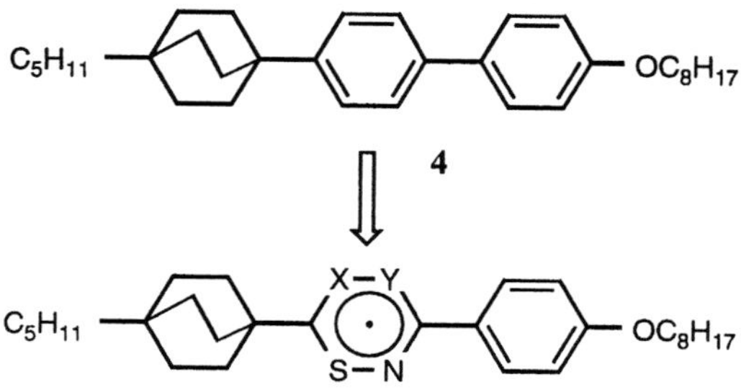

Figure 3   Conceptual transformation of the known three-ring hydrocarbon mesogen into a mesogenic radical **5**. Phase transition temperatures for **4** are shown in °C. S$_B$, smectic B; S$_A$, smectic A; N, nematic; Cr, crystalline; I, isotropic.

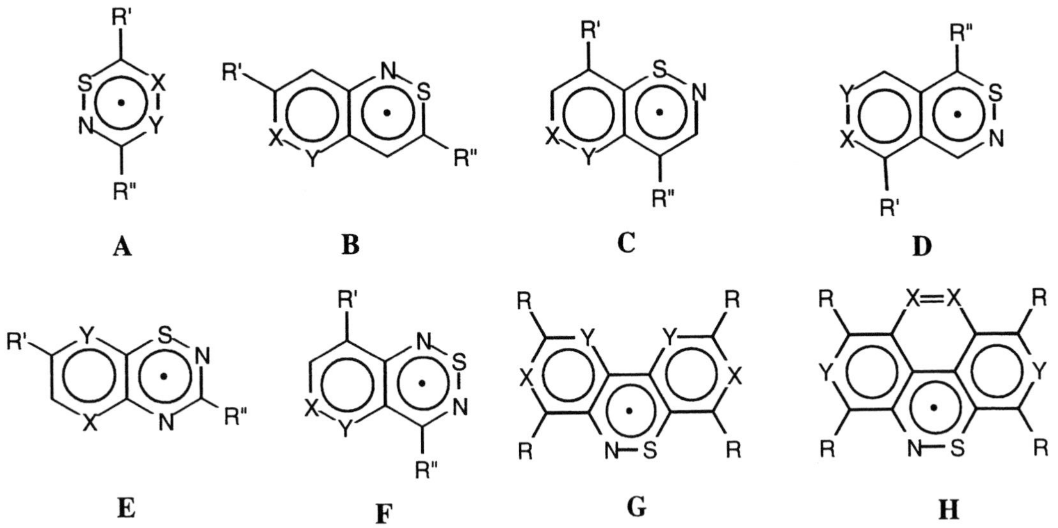

Figure 4 Potential liquid crystal core elements suitable for the preparation of calamitic liquid crystals (**A**–**F**) and discotic liquid crystals (**G** and **H**) when substituted with R, R', or R".

X, Y = N, CH, C-F, C-Cl; R = poly-alkyl(oxy)phenyl, poly-alkyl(oxy)phenylethynyl; R', R" = alkyl(oxy), cycloalkyl, aryl

The choice of substituents is typical for mesogenic materials [19] and affects the spin distribution within the heterocyclic ring. More extensive spin delocalization will be possible using aromatic substituents on the radical core, and stabilization of the reactive sites on the heterocyclic ring can be achieved by using, for example, bulky yet mesogeneity-promoting bicyclo[2.2.2]octane substituents. This offers the possibility to control the liquid crystalline behavior and the intermolecular spin–spin interactions.

The basic building blocks shown in Figure 4 can also be used, in principle, as mesogenic ligands for the construction of high-spin metallomesogenic complexes.

V. THEORETICAL TREATMENT

Prediction of parent radical properties such as spin distribution, redox potentials, molecular geometry, and also modeling of the intra- and intermolecular interactions of radicals are integral components of our strategy. Spin-distribution maps of the parent radicals indicate ring positions to be chemically modified either by extending the conjugation or by halogenation or substitution with nitrogen in order to reduce their chemical reactivity (see Figure 4). The halogenation and introduction of nitrogen atoms to the heterocycles affect the molecular dipolar properties and redox potentials, which are important for liquid crystalline behavior of pure compounds and in binary mixtures. The modeling of radical derivatives yields information about conformational preferences of the substituents and allows an estimate of the degree of intermolecular steric and electronic interactions.

The relatively small size of the parent radicals makes it possible to use fairly high levels of theory for computing their properties. A comparison of theoretical and experimental results for over two dozen known thioaminyl radicals shows that density functional theory methods, and B3LYP in particular, with 6-31G* or cc-pVDZ basis sets are particularly well suited for predicting spin density, electron spin resonance (ESR) hyperfine coupling constants, geometries, ionization, and redox potentials of thioaminyl radicals [75]. The large molecular sizes and lack of symmetry in some of the LC radical derivatives precludes the use of ab initio formalisms. Instead, semi-empirical hamiltonians, especially PM3 [74], are methods of choice in these cases.

V.A. Parent Radicals

Most of the potential LC rigid core elements shown in Figure 4 are new heterocyclic radicals. Radicals **6–18** were subjected to computational analysis, with emphasis on molecular geometry, spin distribution, and oxidation potentials. Some of the results of the B3LYP/6-31G* calculations are shown in Figure 5, and they can be summarized as follows:

1. All radicals except for **9** (**A**, X=Y=N) have planar heterocyclic rings.
2. The antibonding SOMO is largely localized on the thioaminyl fragment. In the case of **6–13**, it extends to the ring atoms in the 3 and 5 positions with respect to the nitrogen atom. In three- and four-ring radicals **14–18**, the SOMO is almost evenly distributed over the biphenyl system but not to the —CH=CH— and —N=N— bridges in **15** and **16**.
3. Spin density is concentrated on the —S—N— fragment and typically on ring atoms in the 3 and 5 positions with respect to the nitrogen atom.
4. Estimated oxidation potentials for the radicals decrease with an increasing number of rings and increase with an increasing number of nitrogen atoms. Thus, thiatriazinyl **9** is predicted to show the highest oxidation potential of 1.14 V vs. SCE, and two- and three-ring radicals without additional heteroatoms (**10**, **11**, **13–15**) show low oxidation potentials about 0 V vs. SCE.
5. The dipole moment associated with the N–S group is calculated to be about 2.5 D. The presence of other nitrogen atoms in the molecule may alter the net dipole moment, and a value as low as 0.78 D has been calculated for **8**.

Inspection of Figure 5 indicates ring systems best suited for LC rigid core elements and suggests further structural modifications. Among the monocyclic radicals, 1,2,4-thiadiazinyl (**7**) and 1,2,4,5-thiatriazinyl (**9**) are appropriate for the structural elements, since the heteroatoms are present at the high-spin-density sites and positions 3 and 6 in the ring will be used for further substitution (Figure 4, type **A**). 1,2,4-Benzothiadiazinyl (**12**) appears to be a particularly useful bicyclic system for incorporation into LC structures; most of the spin density is contained in the heterocyclic ring, and after some modification of the position 5 and disubstitution in the 3 and 7 positions one might obtain a chemically stable liquid crystalline radical (vide infra). Other bicyclic radicals (**10**, **11**, **13**) would require extensive chemical modification to stabilize the high-spin-density sites.

All three- and four-ring radicals **14–18** shown in Figure 5 show high-positive-spin densities, mainly in positions of intended substitution (see Figure 4), so no further structural modifications are necessary.

Heterocycles **17** and **18** are predicted to be ground state triplets [75], according to the CASSCF(2,2) calculations, with the singlet-triplet gaps of

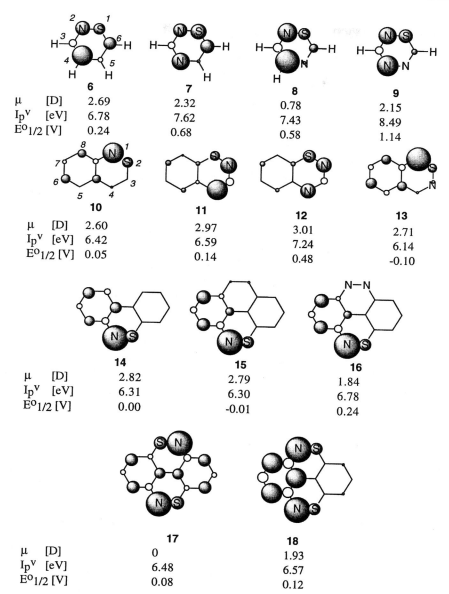

Figure 5 Spin-distribution maps of parent radicals of types **A–H** with indicated dipole moments μ, vertical ionization potentials, I_p^v and expected oxidation potentials $E_{1/2}^0$ in MeCN vs. SCE (B3LYP/6-31G*). Circles represent relative total spin densities.

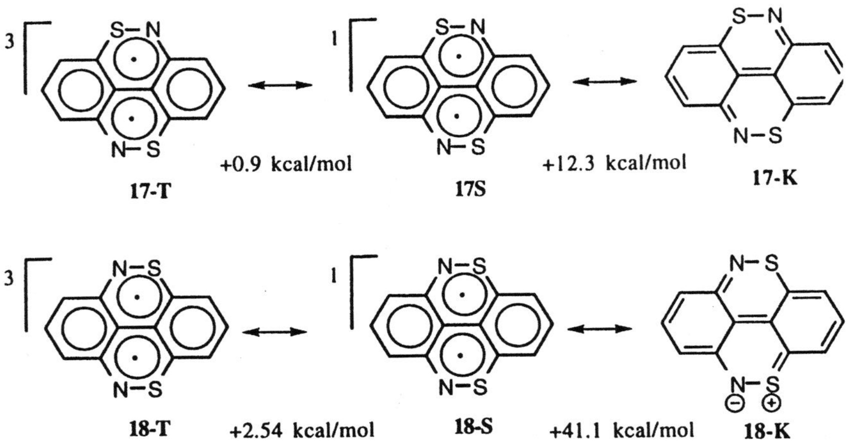

Figure 6 Closed-shell and open-shell structures for **17** and **18** and their calculated relative energies.

0.9 kcal/mol and 2.5 kcal/mol, respectively (Figure 6). The available closed-shell Kekulé singlets **17-K** and the zwitterionic **18-K** are substantially less stable than the triplets by 13.2 and 43.6 kcal/mol, respectively. While a triplet ground state would be expected for **18**, a distant relative of m-xylylene, it is rather surprising that **17** also has the calculated preference for the triplet. These computational predictions are still awaiting experimental verification.

V.B.   Intermolecular Interactions

The extent of intramolecular spin delocalization, the type and magnitude of intermolecular electronic interactions, and hence the spin–spin exchange between two radicals in the mesogenic and solid states will largely depend on the accessibility of their π-electronic systems to interact. This, in turn, depends on the distribution of conformational minima and torsional potentials between the rigid core elements.

In compounds of types **A**, **B**, and **E**, the torsional potential between the heterocycle and the phenyl substituent should be similar to that of biphenyl, which exhibits a 44° dihedral angle in the gas phase [76] and a high tendency for planarization in the solid state [77]. Placing a phenyl substituent in the positions indicated in the remaining rings shown in Figure 4 leads to a high torsional angle of 60–70° and four to five times greater barriers to planarization [75], a situation similar to that found in 1-phenylnaphthalene

Figure 7 Two possible molecular arrangements in a binary columnar discotic. The black-edge and white-edge discs represent two different isosteric molecules.

[78]. This, in turn, is expected to unfavorably affect the mesogenic properties (vide infra), spin delocalization, and packing in the solid phase.

To alleviate the torsional strain, some of the carbon atoms may have to be replaced with nitrogen atoms (e.g., X, Y=N in Figure 4), or an acetylene linking group between the phenyl and the heterocycle may be used. Thus, a judicious choice of the substituents and use of nitrogen atoms provides opportunities to control the torsional potentials and intermolecular contacts, and hence the electronic/magnetic properties.

Accurate prediction of the type and magnitude of intermolecular interactions of two disklike radicals in the bulk material is a rather difficult task. It is expected, however, that discotic mesogens of types **G** and **H** prepared from radicals **14–18** will be almost isosteric and should easily form binary mixtures. Molecules in the resulting mixtures may form alternating or segregated stacks (Figure 7), which are expected to exhibit different bulk properties. This offers the exciting possibility of engineering ferrimagnetic interactions [79] if "doublet discs" (e.g., derivatives of **14–16**) are mixed with the isosteric "triplet discs" (e.g., derivatives of **17** and **18**).

VI. PREDICTING LIQUID CRYSTALLINE BEHAVIOR

Liquid crystalline properties of the potential calamitic radicals (types **A–F** in Figure 4) can be extrapolated using Vill's LiqCryst 3.0 database [80], which contains information on over 70,000 substances and is capable of predicting mesogenic behavior of structurally similar molecules. These established structure–property relationships allow for the design of calamitic materials (Figure 4), but discotic radicals of types **G** and **H** have no structural counterparts in the literature.

A better understanding of the substituent effect on mesogenic properties of radicals of types **G** and **H** comes from studies of rings **19–21** (Scheme 2), which are virtually isosteric closed-shell analogs of radicals **14–18**. The incorporation of rings **19–21** as the core in **I**, a general structure that includes radicals **G** and **H**, allows for investigation of mesogenic prop-

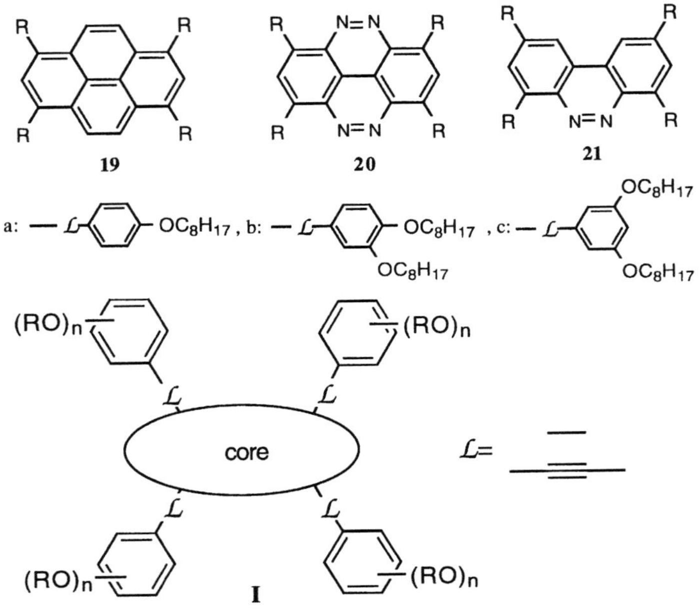

Scheme 2

erties of these tetraaryl derivatives as a function of the number and distribution of alkoxy chains on the phenyl rings.

The initial findings [81] can be summarized in the following points:

1. At least two octyloxy substituents on each phenyl ring are necessary to induce LC behavior.
2. Pyrene derivatives **19** have the least stable mesophases in the series, presumably due to high barriers to planarization and low polarity of the core as compared with tetraazapyrene derivatives **20**. Lowering the planarization energy and introduction of a permanent dipole moment in benzocinnoline derivatives **21** further stabilizes liquid crystalline phases and increases the melting points.
3. Typically, a liquid crystalline polymorphism is observed. For example, derivative **21b** exhibits two discotic mesophases and forms an optically isotropic cubic phase above 159°C, according to XRD.
4. As expected, these liquid crystalline phases easily supercool, and the resulting solid phase retains structural features of the preceding mesophase.
5. All compounds show high solubility in hydrocarbons and low melting temperatures. This implies good miscibility with other

structurally similar compounds and the possibility of formulation of binary mixtures with "doublet" and "triplet" discs (vide supra).

It is expected that when complete, these results will provide an excellent guidance for designing discogenic radicals and that many of these findings will be directly applicable for the mesogenic radicals.

VII. PREPARATION AND CHARACTERIZATION OF RADICALS AND LIQUID CRYSTALLINE RADICALS

Experimental pursuit of this novel class of materials is hindered by synthetic challenges surrounding the construction of new sulfur–nitrogen heterocycles and development of novel ways to generate and isolate the radicals in bulk. Therefore, the experimental work follows two parallel avenues: the chemistry of parent radicals and the preparation of potentially liquid crystalline radicals.

VII.A. Generation of the Radicals

Thioaminyl radicals are typically obtained by two general methods (Scheme 3): (1) hydrogen atom abstraction or deprotonation and one-electron oxidation of the corresponding sulfenamides [47,48,50,57], and (2) one-electron reduction of the corresponding sulfiliminyl chlorides with Ph_3Sb [54, 55,72,82], sodium metal [56,57], or verdazyl radical [56].

Among the one-electron oxidants are t-butyl peroxalate [83], PbO_2/K_2CO_3 [47,48,50,84], Ag_2O/K_2CO_3 [85], and thianthrenium perchlorate/pyridine systems [74].

VII.B. Progress Towards Calamitic Materials

Radicals 22a–c (Scheme 4) have been prepared using both methods shown in Scheme 3. Solutions of 22a and 22b show almost identical ESR spectra, with the characteristic five-line pattern (pseudoquintet, $a_N = 0.57$ mT). The stability of 22a toward oxygen is limited ($\sim$2 h) but is significantly increased

Scheme 3

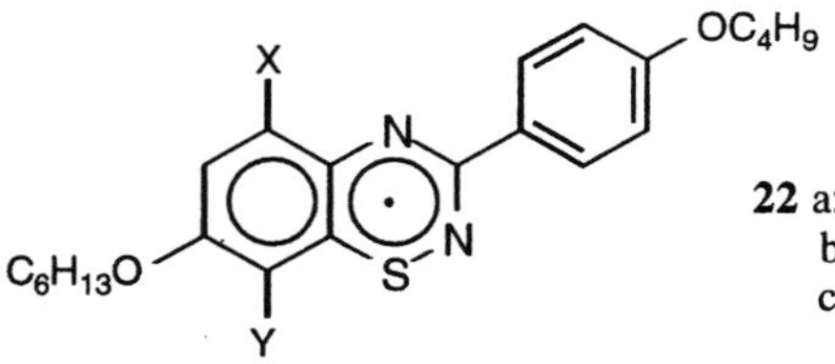

22 a: X=Y=H;
      b: X=Cl, Y=H
      c: X=Y=Cl

Scheme 4

upon introduction of a chlorine atom in **22b**, which now can be chromato-graphed and isolated as a dark green solid [86].

The incorporation of another chlorine atom into the ring and the for-mation of **22c** further stabilizes the radical, which now is stable for several days in solution [86]. It is expected, however, that the presence of two chlorine atoms will have a detrimental effect on the mesogenic properties.

The preparation of radical **23** from precursor [74] **24** is under inves-tigation (Scheme 5). Standard redox techniques have been unsuccessful in the generation of radical **23** thus far.

VII.C.   Progress Towards Discotic Materials

The preparation [74,87] of cyclic sulfenamides **25–27** has opened access to the parent radicals **14–17** (Scheme 6). Oxidation of **25–27** with thianthren-ium perchlorate in the presence of one equivalent of pyridine resulted in the immediate disappearance of the purple color of the oxidant and the formation of persistent radicals **14**, **16**, and **28**. The same reaction performed with PbO_2/K_2CO_3 failed to produce any ESR signal. The ESR spectra for the radicals are about 3.5 mT wide, with the characteristic seven groups of multiplets. Based on the total spin densities for **14** and **16** shown in Figure 5, the principal contributors to the hyperfine splitting are assigned to the nitrogen atom ($a_N \sim 0.9$ mT) and two hydrogen atoms in the ortho and para positions with respect to the N atom ($a_H \sim 0.45$ mT) [74,87]. The values of these and other hyperfine coupling constants obtained for both radicals are consistent with those values calculated with the DFT methods.

Scheme 5

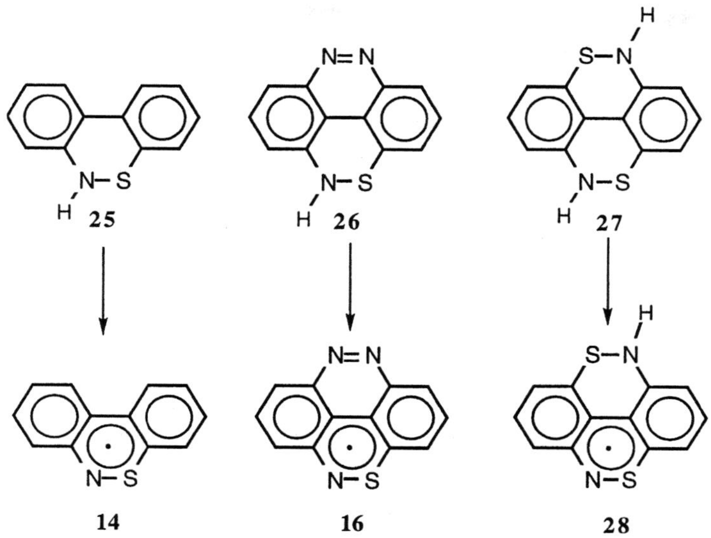

Scheme 6

The relatively large g-value of 2.0051 obtained for **14** is consistent with g-values reported for other thioaminyl radicals [43,45]. This indicates a significant delocalization of the unpaired electron onto the sulfur atom, resulting in a large spin-orbit coupling parameter. Again, this is consistent with our DFT calculations.

Solutions of radicals **14** and **28** are persistent for hours, while **16** shows very little decomposition after several days, even in the presence of pyridinium salts. They exhibit low reactivity toward oxygen, but they are susceptible to acid- or base-induced cleavage of the N—S bond. The observed higher stability of **16** as compared with **14** and **28** is presumably related to the presence of the azo bridge, which stabilizes the planar geometry of the heterocycle. The oxidation of **28** and the formation of biradical **17** is currently being studied in our laboratory.

Preparation of the potentially discotic tetraaryl derivatives of **14–18**, such as **29**, relies on synthetic methodologies developed for the parent radicals and the availability of appropriate polysubstituted biphenyls [88]. Thus our first potentially discotic radical **29a** was prepared in an analogous way to **14** (Scheme 7) [89]. ESR spectra of **29a** and **29b** [90] consist of a broad triplet with a hyperfine coupling to the nitrogen atom of 0.87 mT, which is similar to the a_N observed in **14**. Isolation of these and preparation of other

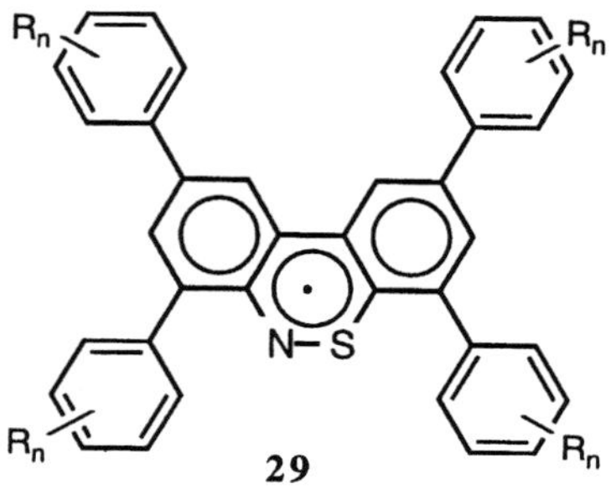

a: $R_n = 4\text{-}C_{10}H_{21}$
b: $R_n = 3,4\text{-}(C_8H_{17}O)_2$

Scheme 7

potentially mesogenic radicals and their full characterization is currently being pursued in our laboratory.

VIII. OUTLOOK

A new and potentially broad class of liquid crystalline radicals is emerging from our research. The parent radicals and their derivatives synthesized thus far exhibit remarkable chemical and thermal stability, which is expected to increase even further in the liquid crystalline derivatives. This will allow for facile isolation and study of these novel materials.

The potentially discotic radicals are particularly interesting, because the arrangement of molecules in columns may result in unusual magnetic and electric properties of the materials. Preliminary results obtained for closed-shell materials that are isosteric with the radicals suggest a rich discotic polymorphism in the target liquid crystalline radicals.

ACKNOWLEDGMENTS

This project has been supported by the National Science Foundation (CHE-9528029).

REFERENCES

1. Miller JS, Epstein AJ. Angew Chem Int Ed Engl. 1994; 33:385.
2. Kinoshita M. Jpn J Appl Phys. 1994; 33:5718.
3. Ouahab L. Chem Mater. 1997; 9:1909.
4. Veciana J, Cirujeda J, Rovira C, Vidal-Gancedo J. Adv Mater. 1995; 7:221.

5. Yoshioka N, Irisawa M, Mochizuki Y, Kato T, Inoue H, Ohba S. Chem Lett. 1997; 251.
6. Cirujeda J, Hernandez-Gasio E, Rovira C, Stanger J-L, Turek P, Veciana J. J Mater Chem. 1995; 5:243.
7. Awaga K. In: M. M. Turnbull, T. Sugimoto, L. K. Thompson, eds. Molecule-Based Magnetic Materials. Washington, DC: American Chemical Society, 1996, pp 236–246.
8. Caneschi A, Dei A, Gatteschi D. J Chem Soc, Chem Commun. 1992; 630.
9. Cogne A, Grand A, Rey P, Subra R. J Am Chem Soc. 1989; 111:3230.
10. Izuoka A, Fukada M, Kumai R, Itakura M, Hikami S, Sugawara T. J Am Chem Soc. 1994; 116:2609.
11. Akita T, Mazaki Y, Kobayashi K. J Chem Soc, Chem Commun. 1995; 1861.
12. Kahn O. In: J. Michl, ed. Modular Chemistry. Boston: Kluwer, 1997, pp 287–302.
13. Kahn O. In: D. Gatteschi, O. Kahn, J. S. Miller, F. Palacio, eds. Magnetic Molecular Materials. Boston: Kluwer, 1991, pp 35–52.
14. Haase W, Borchers B. In: D. Gatteschi, O. Kahn, J. S. Miller, F. Palacio, eds. Magnetic Molecular Materials. Boston: Kluwer, 1991, pp 245–253.
15. Gray GW, ed. Thermotropic Liquid Crystals. New York: Wiley, 1987.
16. Blinov LM. Electro-optical and Magneto-optical Properties of Liquid Crystals. New York: Wiley, 1983.
17. Demus D. Mol Cryst Liq Cryst. 1988; 165:45.
18. Demus D. In: B. Bahadur, ed. Liquid Crystals: Applications and Uses. Singapore: World Scientific, 1990, pp 1–36.
19. Collings PJ, Hird M. Introduction to Liquid Crystals, Chemistry and Physics. Bristol, England: Taylor & Francis, 1997.
20. Leadbetter AJ. In: G. W. Gray, ed. Thermotropic Liquid Crystals. New York: Wiley, 1987, pp 1–27.
21. Piecek W, Czuprynski K. Personal communication, 1998.
22. Gray GW, Goodby JWG. Smectic Liquid Crystals—Textures and Structures. Philadelphia: Leonard Hill, 1984.
23. McConnell HM. J Chem Phys. 1963; 39:1910.
23a. Yoshizawa K, Hoffmann R. J Am Chem Soc. 1995; 117:6921.
24. Kats EI, Lebedev VV. Sov Phys JETP, 1989; 69:1155.
25. Blumstein RB, Kim DY, McGowan CB. In: R. A. Weiss, C. K. Ober, eds. Liquid Crystalline Polymers. Washington, DC: American Chemical Society, 1990, pp 294–307.
26. Serrano, LJ, ed. Metallomesogens. New York: VCH, 1996.
27. Griesar K, Athanassopoulou MA, Bustamante EAS, Tomkiewicz Z, Zaleski AJ, Haase W. Adv Mater. 1997; 9:45.
28. Alonso J. In L. J. Serrano, ed. Metallomesogens. New York: VCH, 1996, pp 349–385, 385–418.
29. Ovchinnikov IV, Galyametdinov YG, Prosvirn AV. Russian Chem Bull. 1995; 44:768.
30. Galyametdinov YG, Atanassopoulo M, Khaaze V, Ovchinnikov IV. Russ J Coord Chem. 1995; 21:718.

31. Galyametdinov Y, Athanassopoulou MA, Griesar K, Kharitonova O, Busta-mante EAS, Tinchurina L, Ovchinnikov I, Hasse W. Chem Mater. 1996; 8: 922.
32. Gillon B, Journaux Y, Kahn O. Physica B. 1989; 156 & 157:373.
33. Oshio H, Ichida H. J Phys Chem. 1995; 99:3294.
34. Gillon B, Cavata C, Schweiss P, Journaux Y, Kahn O, Schneider D. J Am Chem Soc. 1989; 111:7124.
35. Saeva FD, Reynolds GA, Kaszczuk L. J Am Chem Soc. 1982; 104:3524.
36. Strzelecka H, Gionis V, Rivory J, Flandrois S. J Physique, 1983; 44 C3:1201.
37. Forrester AR, Hay JM, Thomson RH. Organic Chemistry of Stable Free Radicals. New York: Academic Press, 1968.
38. Tanaka H, Shibahara Y, Sato T, Ota T. Eur Polym J. 1993; 29:1525.
39. Haase W, Griesar K, Kinoshita M. IV International Conference on Molecule-Based Magnets, Salt Lake City, Utah, October 16–21, 1994.
40. Griesar K. PhD dissertation, Technische Universität Darmstadt, 1996.
41. Allgaier J, Finkelmann H. Macromol Chem Phys. 1994; 195:1017.
42. Dvolaitzky M, Billard J, Poldy F. Tetrahedron, 1976; 32:1835.
43. Rawson JM, Banister AJ, Lavender I. Adv Heterocyc Chem. 1995; 62:137.
44. Oakley RT. Prog Inorg Chem. 1988; 36:299.
45. Miura Y. Rev Heteroatom Chem. 1990; 3:211.
46. Preston KF, Sutcliffe LH. Magn Reson Chem. 1990; 28:189.
47. Miura Y, Tanaka A, Hirotsu K. J Org Chem. 1991; 56:6638.
48. Miura Y, Ohana T. J Org Chem. 1988; 53:5770.
49. Miura Y, Yamamoto A, Katsura Y, Kinoshita M. J Org Chem. 1982; 47:2618.
50. Miura Y, Yamano E, Tanaka A. J Org Chem. 1994; 59:3294.
51. Hofs H-U, Bats JW, Gleiter R, Hartmann G, Mews R, Eckert-Maksic M, Oberhammer H, Sheldrick GM. Chem Ber. 1985; 118:3781.
52. Boere RT, Oakley RT, Reed RW, Westwood NPC. J Am Chem Soc. 1989; 111:1180.
53. Burford N, Passmore J, Schriver MJ. J Chem Soc, Chem Commun. 1986; 140.
54. Brooks WVF, Burford N, Passmore J, Schriver MJ, Sutcliffe LH. J Chem Soc, Chem Commun. 1987; 69.
55. Hayes PJ, Oakley RT, Cordes AW, Pennington WT. J Am Chem Soc. 1985; 107:1346.
56. Shermolovich YG, Simonov YA, Dvorkin AA, Polumbrik OM, Borovikova GS, Kaminskaya EI, Levchenko ES, Markovskii LN. J Org Chem USSR, 1989; 25:550.
57. Markovskii LN, Talanov VS, Polumbrik OM, Shermolovich YG. J Org Chem. USSR. 1981; 17:2338.
58. Cordes AW, Haddon RC, Hicks RG, Kennepohl DK, Oakley RT, Schneemeyer LF, Waszczak JV. Inorg Chem. 1993; 32:1554.
59. Andrews MP, Cordes AW, Douglass DC, Fleming RM, Glarum SH, Haddon RC, Marsh P, Oakley RT, Palstra TTM, Schneemeyer LF, Trucks GW, Tycko R, Waszczak JV, Young KM, Zimmerman NM. J Am Chem Soc. 1991; 113: 3559.

60. Cordes AW, Haddon RC, Oakley RT, Schneemeyer LF, Waszczak JV, Young KM, Zimmerman NM. J Am Chem Soc. 1991; 113:582.
61. Bryan CD, Cordes AW, Fleming RM, George NA, Glarum SH, Haddon RC, Oakley RT, Palstra TTM, Perel AS, Schneemeyer LF, Waszczak JV. Nature. 1993; 365:821.
62. Teki Y, Miura Y, Tanaka A, Takui T, Itoh K. Mol Cryst Liq Cryst. 1993; 233: 119.
63. Teki Y, Itoh K. IV International Conference on Molecule-Based Magnets, Salt Lake City, Utah, October 16–21, 1994.
64. Barclay TM, Cordes AW, George NA, Haddon RC, Oakley RT, Palstra TTM, Patenaude GW, Reed RW, Richardson JF, Zhang HJ. Chem Soc, Chem Commun. 1997; 873.
65. Banister AJ, Bricklebank N, Lavender I, Rawson JM, Gregory CI, Tanner BK, Clegg W, Elsegood MRJ, Palacio F. Angew Chem Int Ed Engl. 1996; 35: 2533.
66. Miura Y, Oka H, Yamano E, Teki Y, Takui T, Itoh K. Bull Chem Soc Jpn. 1995; 68:1187.
67. Miura Y, Momoki M, Fuchikami T, Mizutani H, Teki Y, Itoh K. Mol Cryst Liq Cryst. 1997; 306:271.
68. Teki Y, Ohmura Y, Itoh K, Miura Y. Mol Cryst Liq Cryst. 1997; 306:315.
69. Teki Y, Itoh K, Miura Y, Kurokawa S, Ueno S, Okada A, Yamakage H, Kobayashi T, Amaya K. Mol Cryst Liq Cryst. 1997; 306:95.
70. Banister AJ, Bricklebank N, Clegg W, Elsegood MRJ, Gregory CI, Lavender I, Rawson JM, Tanner BK. J Chem Soc, Chem Commun. 1995; 679.
71. Barclay TM, Cordes AW, deLaat RH, Goddard JD, Haddon RC, Jeter DY, Mawhinney RC, Oakley RT, Palstra TTM, Patenaude GW, Reed RW, Westwood, NPC. J Am Chem Soc. 1997; 119:2633.
72. Awere EG, Burford N, Mailer C, Passmore J, Schriver MJ, White PS, Banister AJ, Oberhammer H, Sutcliffe LH. J Chem Soc, Chem Commun. 1987; 66.
73. Czuprynski K, Douglass AG, Kaszynski P, Drzewinski W. Liq Cryst. 1999; 26:261.
74. Patel MK, Huang J, Kaszynski P. Mol Cryst Liq Cryst. 1995; 272:87.
75. Kaszynski P. Unpublished results.
76. Almenningen A, Bastiansen O, Ferenholt L, Cyvin BN, Cyvin SJ, Samdal S. J Mol Struct. 1985; 128:59.
77. Tsuzuki S, Tanabe K. J Phys Chem. 1991; 95:139.
78. Cheng CL, Murthy DSN, Ritchie GLD. J Chem Soc, Faraday Trans. 2, 1972; 1679.
79. Shiomi D, Nishizawa M, Sato K, Takui T, Itoh K, Sakurai H, Izuoka A, Sugawara T. J Phys Chem. 1997; 101:3342.
80. Vill V. In: J. Thiem, ed. Landolt-Bornstein, Group IV, Vol 7, 1994. Vill V. LiqCryst 3.0. Hamburg: LCI, 1997.
81. Farrar JM, Kusuma S, Kaszynski P. Unpublished results.
82. Boere RT, Cordes AW, Hayes PJ, Oakley RT, Reed RW, Pennington WT. Inorg Chem. 1986; 25:2445.
83. Miura Y, Isogai M, Kinoshita M. Bull Chem Soc Jpn. 1987; 60:3065.

84. Miura Y, Kitagishi Y, Ueno S. Bull Chem Soc Jpn. 1994; 67:3282.
85. Miura Y, Kunishi T, Isogai M, Kinoshita M. J Org Chem. 1985; 50:1627.
86. Guo F, Kaszynski P. Unpublished results.
87. Benin V, Kaszynski P. Unpublished results.
88. Guo F, Farrar JM, Huang J, Yin H, Kaszynski P. Unpublished results.
89. Huang J. Masters thesis, Vanderbilt University, Nashville, Tenn., 1995.
90. Farrar JM, Kaszynski P. Unpublished results.

16

Magnetic Properties of Transition-Metal-Containing Liquid Crystals

Klaus Griesar* and Wolfgang Haase
Darmstadt University of Technology, Darmstadt, Germany

I. INTRODUCTION

Metallomesogens are very attractive candidates for novel advanced materials. Interest in the synthesis of transition-metal-containing liquid crystals and the study of their physical properties has increased considerably during the past few years [1−4]. The interest is based on new features of these materials that combine the optical and electrical properties of conventional liquid crystals with the magnetic and electronic properties of transition-metal complexes.

A great deal of current research is devoted to metallomesogens on account of their magnetic properties. The long-range order in a mesogenic phase may provide magnetically ordered structures. To some extent, the formal analogy between magnetic and electrical phenomena connected with the existence of ferroelectric liquid crystals suggests the possibility of ferromagnetic liquid crystals [5]. Theoretical considerations on the dynamics of ferromagnetic fluids and liquid crystals based on their spin wave spectrum have been presented by Kats and Lebedev [6].

Our research activities devoted to the magnetic properties of transition-metal-containing liquid crystals have focused on two aspects. We used the analysis of magnetic behavior as a valuable tool to yield information about the arrangement and molecular structure of their crystalline and liquid crystalline (LC) phases. Based on this strategy, we investigated some magnetic properties of metallomesogens.

---

*Current affiliation; SKW Trostberg AG, Trostberg, Germany.

After the discussion of some general features we present in this chapter the highly remarkable magnetic properties of transition-metal-containing liquid crystals that might have valuable possible applications, such as thermomagnetic properties, magnetic-field-induced orientation, and magnetic order in metallomesogens.

II. METALLOMESOGENS: REPRESENTATIVE EXAMPLES AND STRUCTURAL REQUIREMENTS

Metallomesogens are defined as metal-containing complexes or organometallic compounds exhibiting liquid crystalline phases. Representative examples [1–4] are given in Figure 1. The de novo design of metal-containing complexes showing mesomorphic behavior is based on the sophisticated choice of appropriate ligands, such as monodentate ligands [aryls (A), nitriles (B), stilbazoles (C)], bidentate chelate moieties [1,3-diketones or malondialdehydes (D), 1,3-dithiolenes (E), carboxylates (F), thiocarbonic acids (G), 2,2′-bipyridines (H), azomethines (I)], and macrocyclic ligands [porphyrins (J) or phthalocyanins (K)]. The metallomesogens also include organometallic compounds such as ferrocenes (L).

In order to achieve liquid crystalline phases, metal-containing complexes have to fulfill one basic requirement: the combination of polarizable groups (e.g., ester or nitrile groups) within the molecule with an appropriate form of anisotropy of the whole molecule. The type of mesophase formed must be regarded as a consequence of the overall shape of the mesogenic molecule and intermolecular interactions. Generally, rodlike ligands with alkyl chains extended parallel to each other give rise to calamitic liquid crystals, whereas disklike ligands tend to form discotic phases.

As in instructive example, we can consider β-diketones like those in Figure 1, compound D (R_1 = p-alkylphenyl or p-alkyloxyphenyl and R_2 = H) in comparison with malondialdehydes (compound D, R_1 = H and R_2 = p-alkylphenyl or p-alkyloxyphenyl). The β-diketones form discotic phases such as the copper(II) complexes reported by Giroud-Godquin and Billard [7], whereas the compounds with an inverted substitution pattern tend to form nematic phases [8]. However, a prediction of the detailed structural arrangement at a given temperature (e.g., the type of calamitic or discotic phase) is almost impossible.

It must be emphasized that any presentation restricted to selected types of ligands is far from giving a complete overview of the metallomesogens. For a more detailed description the reader is referred to some excellent reviews [1–4].

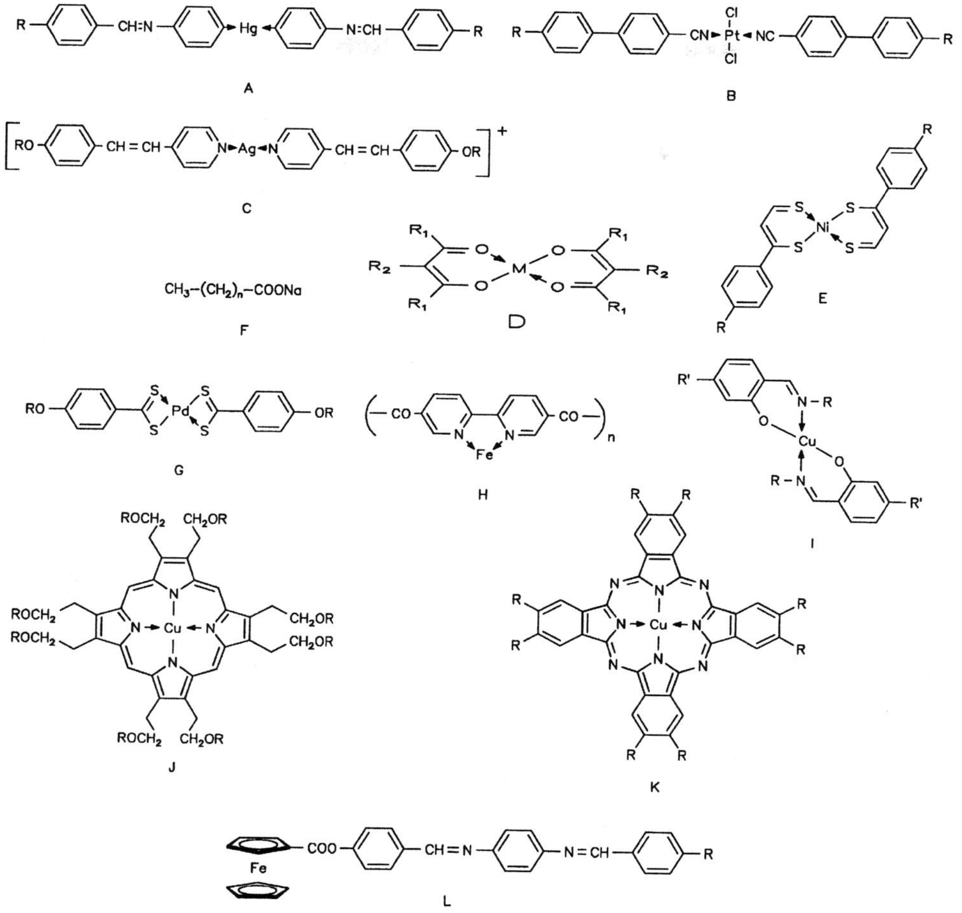

Figure 1 Representative examples of metallomesogens.

III. THERMOMAGNETIC PROPERTIES

The structural changes connected with the crystalline → mesogenic phase transition might cause variations in magnetic and optical properties. The majority of such thermomagnetic effects reported [9–11] in the past are the result of varying the exchange interaction between the metal centers. Since we will consider this type of phenomenon in our discussion of a possible magnetic order in liquid crystals (see part V), we will focus our interest here on one representative example showing a more pronounced thermomagnetic effect derived from a change in the coordination type of the metal.

In order to achieve an adequate molecular shape, rational design of metallomesogens has to take into account the anisotropy of the ligands as well as the coordination geometry around the metal. As illustrated in Figure 2, tetrahedral coordinated metal complexes (B) possess lower molecular anisotropy than square planar coordinated ones (A).

Up to 1994, there were no reports about tetrahedral transition-metal-containing liquid crystals. In order to design metallomesogens showing thermomagnetic effects, we aimed to prove the existence of tetrahedral coordinated metallomesogens [12], in spite of the general notion that this type of coordination—due to the low ansiotropy of the corresponding transition-metal-containing complexes—should not favor the formation of liquid crystalline phases.

As an appropriate class of substance we used the nickel(II) Schiff base metallomesogens **Ni1**, **Ni2** and **Ni3** with branched *N*-alkyl-substituents, which provide considerable stereochemical flexibility of the nickel(II) configuration [13]. The general structure of the investigated nickel complexes is presented in Figure 3. We concentrated our investigations on the enantiotropic compound **Ni1**. The monotropic compounds **Ni2** and **Ni3** were used as references. Unfortunately, we were unable to grow single crystals for crystal structure determination. Therefore, we used UV/Vis reflection spectra for qualitative estimation of the coordination geometry around the metal in the crystalline phase. The results of the UV/Vis reflection spectra [14] are displayed in Figure 4. While the reference compounds **Ni2** and **Ni3** show typical spectra of square-planar (P) and tetrahedral (T) coordinated nickel

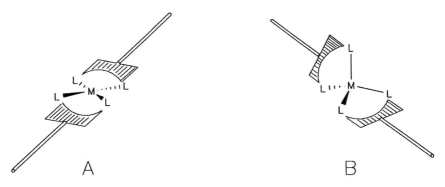

Figure 2 Form anisotropy of square planar (A) and tetrahedral (B) coordinated metal complexes.

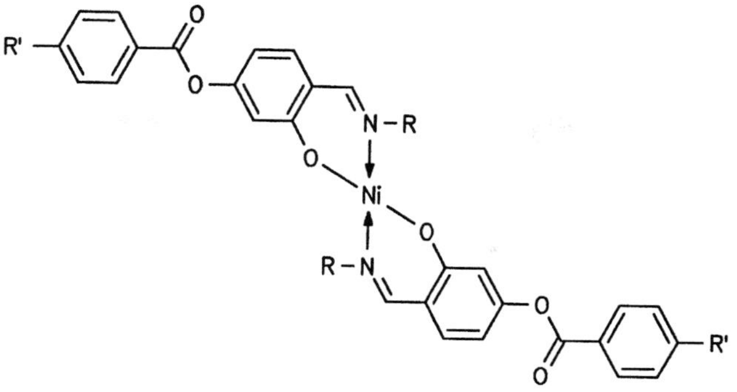

Ni1: R' = OC$_7$H$_{15}$ R = -CH$_2$C*H(CH$_3$)C$_2$H$_5$ C 431 N* 460 I
Ni2: R' = OC$_7$H$_{15}$ R = -C*H(CH$_3$)C$_{11}$H$_{23}$ C 394 I [330] N*
Ni3: R' = OC$_7$H$_{15}$ R = -C*H(CH$_3$)C$_6$H$_{13}$ C 405 I [330] N*

Figure 3 Chemical structure and phase transition temperatures (K) of the investigated nickel complexes. (C = crystalline phase; N* = chiral nematic phase; I = isotropic phase.)

ions, the spectrum of compound **Ni1** exhibits the characteristic maxima of both types of coordination geometry.

In accordance with the optical properties, the magnetic susceptibility measurements showed that compound **Ni2** is diamagnetic and compound **Ni3** paramagnetic. The magnetic behavior of compound **Ni1** (μ_{eff} = 1.73 μ_B) indicates the coexistence of diamagnetic [S = 0; 72% of all Ni(II)] and paramagnetic (S = 1) nickel centers within the powder. The experimental effective magnetic moments vs. temperature data for compound **Ni1** are shown in Figure 5. The increased effective magnetic moment in the chiral nematic phase N* (μ_{eff} = 1.96 μ_B) of compound **Ni1** reflects an increase of the relative amount of paramagnetic nickel(II) ions.

It must be mentioned that the changes in the effective magnetic moment on heating are accompanied by changes in the color of the nickel compounds: the green powder at room temperature becomes brown at the melting point. A similar mesomorphic thermochromism was already described in detail by Ohta et al. [15] for a nickel metallomesogen. Probably, such thermomagnetic effects might offer an approach for possible applications of metallomesogens—especially if they are accompanied by hysteresis effects as observed in the case of compound **Ni1** [12].

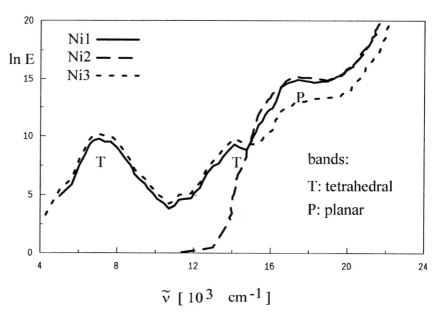

Figure 4 UV/Vis reflection spectra of the investigated nickel complexes in the crystalline phase.

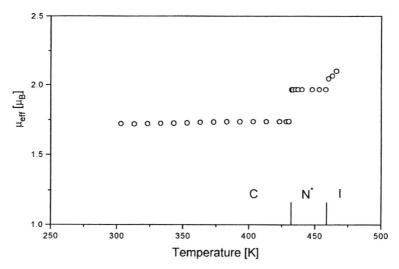

Figure 5 Effective magnetic moments μ_{eff} versus temperature for compound **Ni1**.

IV. ORIENTING METALLOMESOGENS BY MEANS OF A MAGNETIC FIELD

In liquid crystals, the anisotropic properties of crystals are combined with the low viscosity of fluids. Therefore, a mesophase can generally be oriented macroscopically by means of surface, electric, or magnetic field effects. One of our research objectives was to investigate the orientation behavior of metallomesogens in a magnetic field in order to obtain information about their magnetic anisotropy. One further aim was to search for materials that can easily be aligned by means of a magnetic field.

IV.A. Magnetic Anisotropy

In a uniaxial liquid crystalline phase the molecules tend to orient with their axes of maximum susceptibility parallel to the magnetic field, provided that the molecular units possess a magnetic anisotropy ($\Delta\chi^m \neq 0$). The magnetic anisotropy is given by $\Delta\chi = \chi_\| - \chi_\perp$, with $\chi_\|$ parallel and $\chi_\perp$ perpendicular to the director $\bar{n}$. Consequently, the increase of magnetic susceptibility associated with the phase transition isotropic $\rightarrow$ mesogenic can be observed by means of static magnetic susceptibility measurements. By comparing the isotropic value of susceptibility χ_{iso} measured in the isotropic phase with the susceptibility χ observed in the oriented liquid crystalline phase, we can obtain useful information about the magnetic anisotropy of the compounds and about the orientational order parameter P_2.

Depending on the sign of the magnetic anisotropy $\Delta\chi$ of the mesogenic material, the molecules tend to align with the director parallel ($\Delta\chi > 0$) or perpendicular ($\Delta\chi < 0$) to the direction of the magnetic field. For a quantitative interpretation, the relative orientation of the χ-tensor with respect to the molecular frame must be considered; i.e., the orientation of the molecules must be determined by means of x-ray techniques [16] or—in the case of paramagnetic compounds—by analyzing the angle dependency of the electron paramagnetic resonance (EPR) spectra [17–19]. These techniques can be used to estimate the sign of $\Delta\chi$.

In the case of calamitic organic liquid crystals [20–21], the anisotropy is produced mainly by diamagnetic phenyl groups. The axis of maximum diamagnetic susceptibility is perpendicular to the phenyl ring. The long axes of the molecules orient with the director parallel to the magnetic field. For organic liquid crystals with two phenyl groups the magnitude of the magnetic anisotropy $\Delta\chi$ is approximately 50×10^{-6} cm^3/mol.

The total magnetic anisotropy of paramagnetic metallomesogens includes the anisotropy of the diamagnetic ligand (caused mainly by the phenyl groups) and the anisotropy of the paramagnetic metal ion. Copper complexes like those given in Figure 6 may exhibit $\Delta\chi > 0$ or $\Delta\chi < 0$, depending on the nature of their ligands and the number of incorporated phenyl rings [16,22–24].

One can consider the paramagnetic ($\Delta\chi_{para}^{m} < 0$) and diamagnetic ($\Delta\chi_{dia}^{m} > 0$) molecular anisotropy that both contribute to the overall molecular anisotropy $\Delta\chi^{m}$. With the isolated copper complex as a reference system, the axis of the maximum paramagnetic susceptibility coincides with the z-axis. All tensor components of the diamagnetic susceptibility are found to be negative, with the lowest magnitudes being related to the (x,y)-plane. In uniaxial liquid crystalline phases—as a consequence of the reorientation of the molecular units around the director—we can attribute a unique susceptibility $\chi_{\perp}^{m}$ to all directions perpendicular to the director $\vec{n}$. The tensor of the molecular susceptibility in such a uniaxial liquid crystalline phase is characterized by $\chi_{\parallel\,para}^{m} < \chi_{\perp\,para}^{m}$ (i.e., $\Delta\chi_{para}^{m} < 0$) for the paramagnetic contribution and $|\chi_{\parallel\,dia}^{m}| < |\chi_{\perp\,dia}^{m}|$ (i.e., $\Delta\chi_{dia}^{m} > 0$, since $\chi_{\parallel\,dia}^{m} < 0$ and $\chi_{\perp\,dia}^{m} < 0$) for the diamagnetic one. Therefore, the sign of the overall magnetic anisotropy $\Delta\chi^{m}$

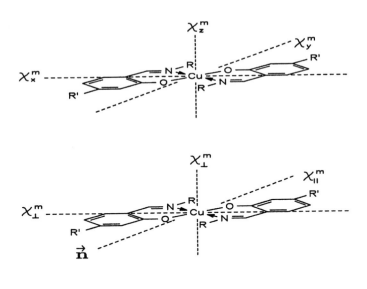

$R = $ -Ph-OC$_{10}$H$_{21}$	$R' = $ -OC$_7$H$_{15}$	$\Delta\chi < 0$
$R = $ -Ph-Ph-OC$_{12}$H$_{25}$	$R' = $ -OC*H(CH$_3$)C$_6$H$_{13}$	$\Delta\chi > 0$

Figure 6 Relationship between the molecular susceptibility tensor in the crystalline phase (top) and the molecular susceptibility tensor in uniaxial mesogenic phase (bottom) for Cu(II)-Schiff-base complexes.

of copper complexes as given in Figure 6 can be positive or negative, since the magnitude of the paramagnetic anisotropy of copper is comparable to the diamagnetic anisotropy of the ligands. The magnitude of the overall magnetic anisotropy [22] is in the range $20-130 \times 10^{-6}$ cm^3/mol.

IV.B. Alignment by Magnetic Field

In order to design metallomesogens that can be easily oriented by a magnetic field, we synthesized rare-earth-containing metallomesogens [25] with the general structure $[(L''H)_2L''M(X)_2]$ ($L''H = H_{2n+1}C_nOC_6H_3(OH)CHNC_mH_{2m+1}$; $n = 7$, 12; $m = 14$, 18; M = La, Tb, Dy, Er, Nd, Ho, Eu, Pr, Gd; X = NO_3 or Cl;). The chemical structure of the ligand $L''H$ is presented in Figure 7. The structure of all lanthanide compounds were characterized by means of elemental analysis, IR spectroscopy, and magnetic susceptibility measurements.

In these investigations, the temperature-dependent magnetic susceptibility measurement of the terbium(III) complex $[(L''H)_2L''Tb(NO_3)_2]$ was carried out in the liquid crystalline phase. The effective magnetic moments versus temperature for compound $[(L''H)_2L''Tb(NO_3)_2]$ are displayed in Figure 8. The substance was initially heated up to the isotropic phase and then cooled slowly to the smectic phase in an applied magnetic field of 1.2 T. A drastic increase of the susceptibility values on cooling compared with the initial ones recorded during the heating process were observed. This behavior indicates a magnetic-field-induced orientation of the molecules, with the axis of maximum susceptibility χ_{max} parallel to the magnetic field. Due to the extremely high paramagnetic anisotropy of the terbium ion, the experimental value of $\chi_{max} - \chi_{iso} = 22,800 \times 10^{-6}$ cm^3/mol of the terbium compound clearly exceeds the corresponding values for copper complexes. This liquid crystalline rare-earth compound displays a magnetic anisotropy that is two orders of magnitude greater than those of known liquid crystals.

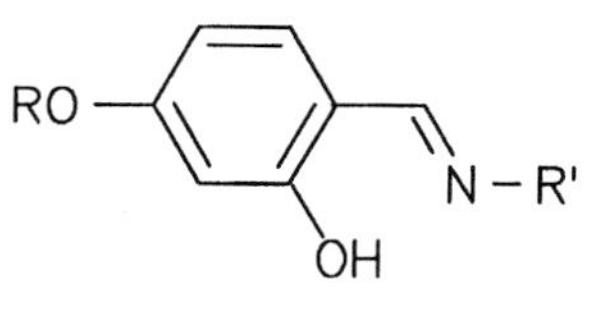

$R = C_{12}H_{25}$; $R' = C_{18}H_{37}$

Figure 7 Chemical structure of the ligand $L''H$ used for the synthesis of the compound $[(L''H)_2L''Tb(NO_3)_2]$.

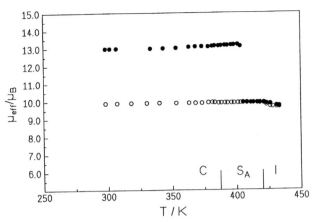

Figure 8 Magnetic moments μ_{eff} versus temperature for compound $[(L''H)_2L''Tb(NO_3)_2]$. (○: heating process; ●: cooling process; C: crystalline phase; S_A: smectic A phase; I: isotropic phase.)

Preliminary results of magnetic investigations of rare-earth containing mesogens with Schiff's base ligands [26] and β-enamino ketone ligands [27] have been presented.

V. DEVELOPMENT OF A MAGNETIC ORDER IN METALLOMESOGENS

Growing attention has been paid to the correlation between the ordering in the mesogenic phase and the related possible existence of magnetically ordered structures in liquid crystalline compounds. The challange of creating magnetic mesogens has been addressed in two different ways via transition-metal-containing liquid crystals and via liquid crystalline organic radicals [28–31]. However, the search for such compounds has been dominated by syntheses of the first type.

Moreover, liquid materials showing magnetic order can be obtained in the form of ferrofluids [32], i.e., in the form of colloidal dispersions of ferromagnetic particles such as Fe_2O_3 in polymers. Matuo et al. [33] determined the minimum concentration of ferrofluid of $CoFe_2O_4$ required to orient a liquid crystal.

It must be mentioned that it is rather difficult to realize a 3D magnetic order in common metallomesogens. Ferromagnetism and antiferromagnetism are bulk phenomena, and exchange interactions between spin centers or long-range dipolar interactions must be present and dominate throughout the material. The majority of mesogens with permanent spins studied so far include transition metals. Here, possible exchange interactions derive mainly from superexchange via bridging groups, since most of the spin density resides on the metal centers.

For example, it is well known that in mesogenic dinuclear copper(II) carboxylates, the extension of the bridging angle connected with the phase transition crystalline $\rightarrow$ discotic leads to an amplification of the antiferromagnetic exchange interaction [9,10]. In these compounds, the spin–spin interaction is restricted to two spin centers. Until a few years ago, magnetic ordered structures that include more than two spin centers in their liquid crystalline phase were completely unknown, although electron paramagnetic resonance (EPR) investigations on a copper(II) β-diketone complex [11] give evidence for the existence of a one-dimensional magnetic ordered structure in the crystalline phase. Electron paramagnetic resonance investigations on copper(II) Schiff base metallomesogens reveal an antiferromagnetic exchange interaction between the copper centers even in their liquid crystalline phase [34]. The authors interpreted the existence of this exchange interaction in terms of a correlation of the rapid rotational movements of the molecules within the mesogenic phase. Similar results have been reported for copper(II) dithiocarbamates [35].

A proper method to examine such spin–spin interactions is the measurement of the temperature-dependent magnetic susceptibility. Such measurements yield information about the exchange integral J (which describes the energy seperation between the energy levels) and the relative arrangement of spin centers. In order to illustrate this fact, one can consider the magnetic behavior of various arrangements of antiferromagnetically coupled $S = 1/2$ spins. Although the coupling strength of the antiferromagnetic interaction is equal, the $\chi(T)$ curves of dimeric (Heisenberg Hamiltonian using $[\hat{H} = -2]\hat{S}_1 \hat{S}_2$), trimeric, and infinite chain arrangements differ considerably. The temperature dependency of the magnetic susceptibility of selected arrangements of $S = 1/2$ spin centers are presented in Figure 9, using $J = -100$ cm^{-1} as an example.

We examined the magnetic properties of the nematic copper(II) malondialdhyde complex [8] **Cu1**. The chemical structure of the copper complex **Cu1** is shown in Figure 10. In the crystalline phase we found an antiferromagnetic exchange interaction between the copper ions. In order to interpret this behavior, we assume an antiferromagnetic superexchange mechanism via bridging oxygen atoms, implying the existence of axial Cu–O interactions, which can be realized in dimeric or chain-motive arrangements (see Figure 11). Of course, these axial Cu–O interactions can be antiferromagnetic or ferromagnetic, depending on magnetostructural

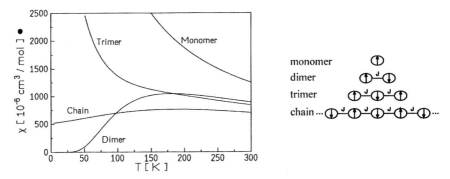

Figure 9 Temperature dependency of the magnetic susceptibility of selected arrangements of $S = 1/2$ spin centers.

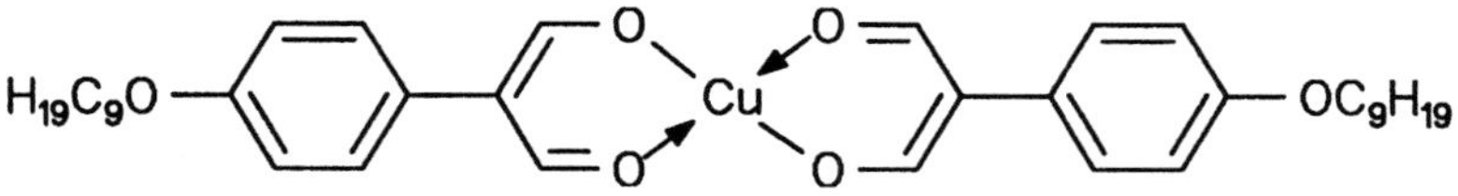

$H_{19}C_9O$ ... OC_9H_{19}

Figure 10 Chemical structure of the copper complex **Cu1**.

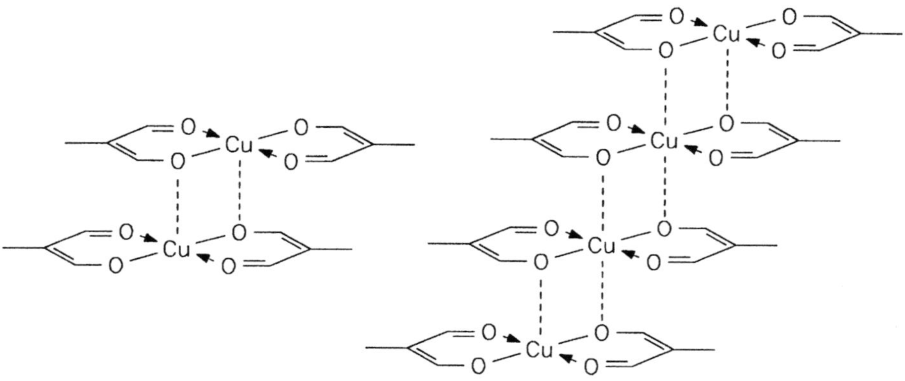

Figure 11 Dimeric and chain-motive arrangement of **Cu1**.

properties. The magnetic susceptibility vs. temperature data recorded in the temperature range 4.2–34 K are displayed in Figure 12. An antiferromagnetic exchange interaction is maintained in the nematic phase of the complex. However, since the coupling was found to be weak, it is rather difficult to estimate the type of arrangement existing in the mesogenic phase. We can assume that the rapid reorientation of the molecules along their long axis should hinder the formation of chain structures of considerable length.

In the case of the polymeric copper(II) metallopolymer **Cu2**, we found a considerably higher degree of antiferromagnetic exchange interaction [36]. The chemical structure of the copper metallopolymer is presented in Figure 13. The temperature-dependent magnetic susceptibility data of the copper metallopolymer **Cu2** are shown in Figure 14. The magnetic data in the glassy state and smectic phase are in agreement with a strong antiferromagnetic coupling between the copper centers (1.52 μ_B at 300 K). The relatively high values of μ_{eff} at 4.2 K (0.67 μ_B) are due to the remaining monomeric (i.e., noncoupled) copper sites, which follow a Curie or Curie–Weiss law behavior. In contrast to the reduced effective magnetic moment in the glassy state and smectic phase, we observed paramagnetic behavior in the isotropic phase ($\mu_{\text{eff}} = 1.85$ μ_B at 430 K).

The proposed structure displayed in Figure 13 is based on the assumption of square-planar coordinated monomeric copper ions. However, the observed antiferromagnetic exchange interaction in this copper metallopolymer can be explained if we consider more extended arrangements of

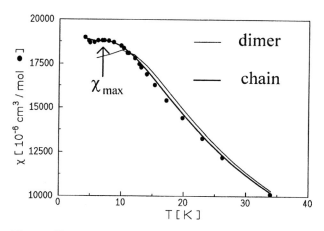

Figure 12 Magnetic susceptibilities χ versus temperature for compound **Cu1**. The thick line is based on a fit using the chain model; the thin line refers to a fit using the dimer model.

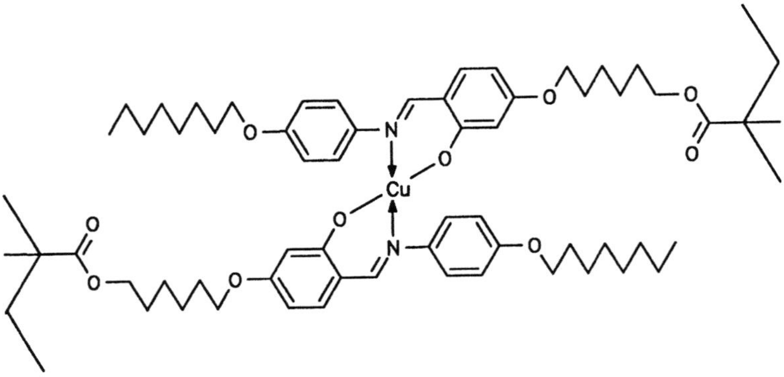

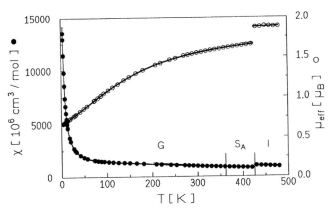

Figure 13   Chemical structure of the copper metallopolymer **Cu2**.

spin centers that may have efficient superexchange pathways, such as dimeric units and linear chain motifs. Figure 15 shows different schematic
representations of the structural motifs considered for the copper metallopolymer **Cu2**.

   Additionally, an adequate description of the magnetic behavior of an
amorphous metallopolymer has to take into account residual monomeric
copper(II) centers formed due to random polymer backbone conformation
and some degree of structural disorder. Such a model should consider the
formation of clusters with different nuclearities, such as dimers, trimers, and

Figure 14   Magnetic susceptibilities χ and effective magnetic moments μ_{eff} versus
temperature for the copper metallopolymer **Cu2**. The lines are based on the fit described on p. 339. (G: glassy state; S_A: smectic A phase; I: isotropic phase.)

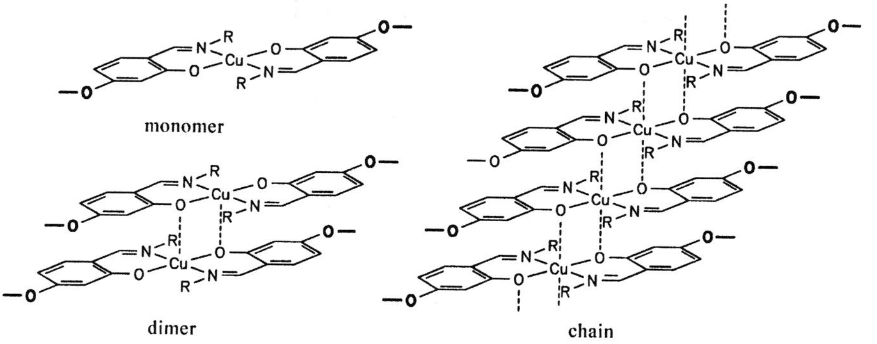

Figure 15 Schematic representation of the structural motifs considered for the copper metallopolymer **Cu2**.

tetramers, and a variety of different exchange interactions. In odd-member clusters ($n = 3, 5, 7, \ldots$), antiferromagnetic interactions will lead to some degree of noncancellation and consequently to $S \neq 0$ ground states; whereas in even-member clusters, a complete cancellation of all spins to yield an $S = 0$ ground state occurs (compare Figure 9).

Therefore, the appropriate model involves a mixture of noninteracting and antiferromagnetic coupled copper species. Accordingly, we have to fit the magnetic data using a refined procedure, which allows the coexistence of copper centers occupying monomeric (molar fraction x_P), dimeric (molar fraction x_{dim}; coupling strength J_{dim}) and chain (coupling strength J_{chain}) sites yielding $g = 2.12$; $J_{chain} = -87$ cm^{-1}, $J_{dim} = -84$ cm^{-1}, $x_{dim} = 8\%$, and $x_P = 14\%$. Figure 16 shows the relative contributions of the different structural motifs displayed in Figure 15 to the overall susceptibility χ_{total} of the copper metallopolymer **Cu2**. The specific arrangement of the copper(II) ions that leads to the 1-D-Heisenberg behavior is connected with a combination of polymeric and liquid crystalline properties. In the isotropic phase we have found a normal Curie–Weiss behavior ($g = 2.12$; $\theta = -7$ K). The structural proposal is connected with the parallel orientation of the planar copper(II) coordination planes as a result of ordering within the layered smectic A phase. This provides an efficient exchange pathway by means of axial Cu—O—Cu-bridges, which are nonexistent both in the melted paramagnetic polymer and in paramagnetic monomeric copper(II) complexes of similar structure. The network structure of the polymer as a consequence of the cross-linking of two Schiff base units by one copper(II) ion hinders the rotation of the planar chelate units along the spacer axis and gives rise to the formation of axial Cu–O interactions over a considerable correlation length.

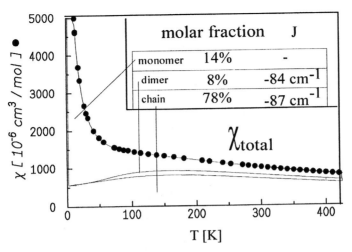

Figure 16 Relative contributions of the different structural motifs displayed in Figure 15 to the overall susceptibility χ_{total} of the copper metallopolymer **Cu2**.

The possibility of combining the bulk magnetic properties of ferri-magnetic chains such as those observed for $[Mn^{III}(TPP)](TCNE) \cdot 2$ toluene **Mn1** [37] (Mn; TPP = tetraphenylporphyrin; TCNE = tetracyanoethylene) with the liquid crystalline behavior of transition-metal-containing mesopor-phyrins prompted us to prepare the dodecyloxy-substituted compound $Mn^{III}[((OC_{12}H_{25})_4TPP)](TCNE) \cdot 2$ toluene **Mn2** [38–39] (Figure 17). The complex **Mn2** shows a discotic hexagonal columnar mesophase in the tem-

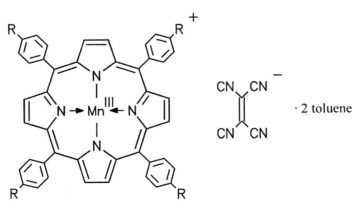

Figure 17 Chemical structure of the investigated compound **Mn2** ($R = OC_{12}H_{25}$).

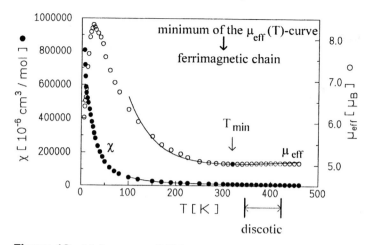

Figure 18 Molar susceptibilities χ (●) and effective magnetic moments μ_{eff} (○) versus temperature for compound **Mn2**.

perature range 108–155°C. The mesophase was characterized by x-ray diffraction experiments.

The temperature dependence of the molar susceptibility χ and effective magnetic moment μ_{eff} for compound **Mn2** is presented in Figure 18. The complex **Mn2** exhibits essentially the same temperature dependency of the effective magnetic moment μ_{eff} as that reported for compound **Mn1** [37]. The significant temperature dependence of μ_{eff} exhibiting a minimum is characteristic for 1D ferrimagnetic chains. Although it is an approximation, we used an equation derived by Drillon et al. [40] to describe the strength of the intrachain coupling in a phenomenological way. The best parameters obtained by fitting the χ(T)-data above 100 K were $g = 2.19$, $J_{intra} = -123.7$ cm^{-1}. The line with solid circles in Figure 18 originates from this fit. The strength of antiferromagnetic exchange interaction between Mn(III) and TCNE is essentially not influenced by the crystalline-to-discotic phase transition. Thus, the antiferromagnetic exchange interaction between Mn(III) and TCNE is maintained in the liquid crystalline phase as a 1D ferrimagnetic chain. AC susceptibility measurements have shown that the compound **Mn2** [37,38] exhibits a 3D ferrimagnetic order below 21.4 K.

VI. CONCLUSIONS

In spite of some theoretical predictions [5–6], the search for collective ferromagnetism in liquid crystals has not been successful up to now. Further investigations should answer the question of whether some ordered domains—leading to phenomena such as spin-glass behavior or super-paramagnetism—may exist in smectic or discotic phases. Interestingly, the manganese compound and the copper metallopolymer presented here exhibit 1D ferrimagnetic or antiferromagnetic chains in the crystalline phase as well as in the mesophase.

ACKNOWLEDGMENTS

We thank M. A. Athanassopoulou for valuable discussions. The research work was supported by the Deutsche Forschungsgemeinschaft and the Volkswagen Foundation.

REFERENCES

1. Serrano JL, ed. Metallomesogens: Synthesis, Properties, and Applications. Weinheim, Germany: VCH, 1996.
2. Giroud-Godquin AM, Maitlis PM. Metallomesogens: Metal complexes in ordered fluid phases. Angew Chem Int Ed Engl. 1991; 30:375–402.
3. Espinet P, Esteruelas MA, Oro LA, Serrano JL, Sola E. Transition metal liquid crystals: Advanced materials within the reach of the coordination chemist. Coord Chem Rev. 1992; 117:215–274.
4. Hudson SA, Maitlis PM. Calamitic metallomesogens: metal-containing liquid crystals with rodlike shapes. Chem Rev. 1993; 93:861–885.
5. Buivydas M. Some conditions of ferromagnetic liquid crystal existence. Phys Status Solidi B. 1991; 168:577–581.
6. Kats EI, Lebedev VV. Some properties of ferromagnetic liquid crystals. Mol Cryst Liq Cryst. 1991; 209:329–337.
7. Giroud-Godquin AM, Billard J. Thermotropic transition metal complexes discogens and their smectogenic intermediates. Mol Cryst Liq Cryst. 1983; 66: 287–295.
8. Salas Reyes V, Soto Garrido G, Aguilera C, Griesar K, Athanassopoulou M, Finkelmann H, Haase W. Metallo-mesogens of copper(II) with α-(*p*-alkyloxy-phenyl)malondialdehydes. Mol Mat. 1994; 3:321–327.
9. Giroud-Godquin AM, Latour JM, Marchon JC. Magnetic susceptibility as a probe of the solid–discotic phase transition in binuclear copper(II) *n*-alkano-ates. Inorg Chem. 1985; 24:4452–4454.
10. Haase W, Gehring S, Borchers B. Magnetic Properties of Liquid Crystals Incorporating Metal Centres. In: Buckley AJ, Gallagher-Daggitt G, Karasz FE, Ulrich DR, eds. Multifunctional Materials. MRS Symposium Proceedings, Vol. 175. 1990, pp 249–255.

11. Eastman MP, Horng ML, Freiha B, Sheu KW. ESR of a one-dimensional Hei-senberg antiferromagnet which undergoes a crystalline-to-discotic phase transition. Liq Cryst. 1987; 2:223–228.

12. Griesar K, Galyametdinov Y, Athanassopoulou MA, Ovchinnikov I, Haase W. Paramagnetic liquid crystalline nickel(II) compounds. Adv Mater. 1994; 6:381–384.

13. Yamada S, Takeuchi A. The conformation and interconversion of Schiff base complexes of nickel(II) and copper(II). Coord Chem Rev. 1982; 43:187–204.

14. Griesar K. Magnetische Untersuchungen an Metallomesogenen und polynuklearen Metallkomplexen der 3-d-Übergangsmetalle. PhD dissertation, Technische Hochschule Darmstadt, Darmstadt, Germany, 1996.

15. Ohta K, Hasebe H, Moriya M, Fujimoti T, Yamamoto I. Thermochromism and solvatochromism of bis[1,2-bis(3,4-di-*n*-alkoxyphenyl)ethanedionedioximato]-nickel(II) complexes. J Mater Chem. 1991; 1:831–834.

16. Borchers B, Haase W. Investigations on the liquid crystalline phase of Schiff's base complexes of copper(II) and their corresponding ligands. Mol Cryst Liq Cryst. 1991; 209:319–328.

17. Bikchantaev IG, Galyametdinov YG, Ovchinnikov IV. Molecular structure and molecular organization of the phase states of mesogenic complexes of the homologous series of copper(II) *N*-(4-alkoxyphenyl)-4-heptyloxysalicylaldimi-nates according to ESR data. J Struct Chem (USSR). 1987; 28:685–691.

18. Bikchantaev IG, Galyametdinov RM, Ovchinnikov IV. Structure and properties of the smectic phase in a mesogenic complex of copper with a Schiff base. Theoret Exper Chem. 1988; 24:360–364.

19. Galyametdinov YG, Bikchantaev IG, Ovchinnikov IV. Effect of the geometry of the chelate bond on the existence of liquid crystal properties in complexes of transition metals with Schiff bases. Zh Obshch Khim. 1988; 58:1326–1331.

20. Müller HJ, Haase W. Magnetic susceptibilities and the order parameters of some 4,4'-disubstituted biphenylcyclohexanes. J Phys. 1983; 44:1209–1213.

21. Schad H, Baur G, Meier G. Investigation of the dielectric constants and the diamagnetic anisotropies of cyanobiphenyls (CB), cyanophenylcyclohexanes (PCH), and cyanocyclohexylcyclohexanes (CCH) in the nematic phase. J Chem Phys. 1979; 71:3174–3181.

22. Bikchantaev I, Galyametdinov Y, Prosvirin A, Griesar K, Soto Bustamante EA, Haase W. Correlation between magnetic properties and molecular structure of some metallo-mesogens. Liq Cryst. 1995; 18:231–237.

23. Alonso PJ. Magnetic properties of metallomesogens. In: Serrano JL, ed. Metallomesogens: Synthesis, Properties, and Applications. Weinheim, Germany: VCH, 1996, pp 387–417.

24. Alonso PJ, Martínez JI. Relationship between the magnetic field induced orientation in the mesophase of metallomesogens derived from Schiff's bases and their mesophase structure. Liq Cryst. 1996; 21:597–601.

25. Galyametdinov Y, Athanassopoulou MA, Griesar K, Kharitonova O, Soto Bustamante EA, Tinchurina L, Ovchinnikov I, Haase W. Synthesis and magnetic investigations on rare-earth-containing liquid crystals with large magnetic anisotropy. Chem Mater. 1996; 8:922–926.

26. Galyametdinov YG, Ivanova G, Ovchinnikov I, Prosvirin A, Guillon D, Heinrich B, Dunmur DA, Bruce DW. X-ray and magnetic birefringence studies of some lanthanide metallomesogens with Schiff's base ligands. Liq Cryst. 1996; 20:831–833.

27. Bikchantaev I, Galyametdinov YG, Kharitonova O, Ovchinnikov IV, Bruce DW, Dunmur DA, Guillon D, Heinrich B. Magnetic properties of rare-earth β-enaminoketone metallomesogens. Liq Cryst. 1996; 20:489–492.

28. Dvolaitzky M, Billard J, Poldy F, Laval J. Mesomorphogenic free radicals. C R Hebd Seances Acad Sci. 1974; Ser C 279:533–535.

29. Gionis V, Fugnitto R, Meyer G, Strzelecka H, Dubois JC. Synthèse et proprietes redox des π-donneurs mesogens. Mol Cryst Liq Cryst. 1982; 90:153–162.

30. Gionis V, Fugnitto R, Strzelecka H, Dubois JC. Synthesis and properties of mesogenic π-donors: Precursors of mesomorphic organic conductors. Mol Cryst Liq Cryst. 1983; 96:215–219.

31. Veber M, Sotta P, Davidson P, Levelut AM, Jallabert C, Strzelecka H. Mesomorphic properties of short chain substituted heteroaromatic salts. J Phys France. 1990; 51:1283–1301.

32. Stierstadt K. Magnetic fluids—liquid magnets. Phys Bl. 1990; 46:377–382.

33. Matuo CY, Tourinho FA, Figueiredo Neto AM. Determination of the minimum concentration of ferrofluid of $CoFe_2O_4$ required to orient liquid crystals. J Magn Magn Mater. 1993; 122:53–56.

34. Alonso PJ, Marcos M, Martínez JI, Orera VM, Sanjuán ML, Serrano JL. Persistence of short-range order in the fluid phases of a mesogen copper complex studied by EPR. Liq Cryst. 1993; 13:585–596.

35. Martínez JI, Bruce DW, Price DJ, Alonso PJ. Mesophase order of a new smectic paramagnetic copper complex detected by EPR. Liq Cryst. 1995; 19:127–132.

36. Haase W, Griesar K, Soto Bustamante EA, Galyametdinov YG. Magnetic investigations on liquid crystalline metallopolymers. Mol Cryst Liq Cryst. 1995; 274:99–111.

37. Miller JS, Calabrese JC, Mc Lean RS, Epstein AJ. meso-(Tetraphenylporphinato)manganese(III)-tetracyanoethenide, $[Mn^{III}TPP]^{::(+)}[TCNE]^{\cdot(-)}$. A new structure-type linear-chain magnet with a T_c of 18 K. Adv Mater. 1992; 4:498–501.

38. Griesar K. Athanassopoulou MA, Soto Bustamante EA, Tomkowicz Z, Zaleski AJ, Haase W. A ferrimagnetically coupled liquid crystal. Adv Mater. 1997; 9: 45–48.

39. Griesar K, Athanassopoulou MA, Tomkowicz Z, Balanda M. Magnetic Investigations on a liquid crystalline complex showing ferromagnetic properties. Mol Cryst Liq Cryst 1997; 306:57–65.

40. Drillon M, Gianduzzo JC, Georges R. Spin alternation in one-dimensional exchange-coupled systems: Magnetic behavior of the $(1/2\text{-}1)_N$ chain. Phys Lett. 1983; 96A:413–416.

17

High-Spin Polyradicals

Andrzej Rajca
University of Nebraska, Lincoln, Nebraska

I. INTRODUCTION

Very high-spin organic polyradicals may be defined as organic molecules with ground states possessing spin values (S) greater than 7/2, exceeding the highest possible spin for a ground state of a single transition- or lanthanide-metal ion [1–7]. A characteristic feature of the electronic structure of such molecules is the high degree of near degeneracy for half-filled energy levels. In principle, this topological degeneracy can be increased without limit, provided suitable design and synthetic capabilities are in place [8]. Furthermore, as long as adequate orbital overlap (or coincidence in space) is present, the energy gaps between the high-spin ground state and the lowest electronic excited state may exceed RT at ambient temperature (strong ferromagnetic coupling), e.g., in selected π-conjugated organic di- and poly-radicals [1–7,9]. Potential for both very high-spin and strong ferromagnetic coupling (through bond) propelled the most recent efforts in the area of polyradicals, aimed at achieving the highest S and an understanding of magnetism in these unusual materials. Long-term objectives, such as an organic ferromagnet with near–ambient Curie temperature based on a 3-dimensional network and a single-molecule organic magnet (blocked superparamagnet), remain to be fulfilled.

Throughout the past few decades, the highest S for organic molecules has been obtained for hydrocarbons based on either triarylmethyls or diarylcarbenes. These two classes of molecules have the advantage of near-quantitative preparative methods [1,3,10,11].

This overview covers the rapid progress in the area of high-spin organic polyradicals during the past decade, which is hallmarked by the achievement of $S = 10$ for an organic molecule and, more importantly, an understanding of spin coupling between multiple sites in a single molecule.

II. MOLECULAR CONNECTIVITY AND HIGH SPIN

Adequate connectivity and coincidence in space between the neighboring $2p$ orbitals can be employed to attain $S > 1/2$ [12–14]. Carbene (CH_2) and trimethylenemethane (TMM) are both examples of organic molecules with $S = 1$ ground states (Figure 1) [15,16].

In CH_2, the out-of-phase (antibonding) overlap of the two orthogonal half-occupied 2p orbitals leads to the $S = 1$ ground state. Significant coincidence in space of the two orbitals is an important factor; e.g., the $S = 1$ and $S = 0$ states are nearly degenerate in the 90° twisted ethylene, where the $S = 0$ state is the ground state due to correlation effects [12]. Of course, the greater possible degeneracy of d- and f-orbitals allows S up to 5/2 and 7/2 in single metal ion complexes.

In trimethylenemethane (TMM), the four half-occupied $2p$ orbitals of the π-conjugated system are not orthogonal; their pairwise in-phase (bonding) overlap, pairing the electron spin at the adjacent $2p$ orbitals, gives rise to the net $S = 1$. It is convenient to consider TMM as composed of two spin sites with unpaired electrons 1,1-connected to ethylene moiety, a ferromagnetic coupling unit (fCU) [1,2].

In principle, ground states with $S \gg 1$ might be attained by designing systems where the spin sites alternate with fCUs [1,2,17]. This is the most successful approach in attaining organic molecules and polymers with high-spin ground states [1–3,18]. A simple example is provided by the homologue of TMM, where the insertion of three 1,3-phenylene fCUs gives the $S = 2$

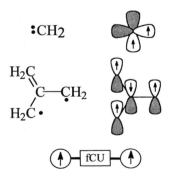

Figure 1 Carbene and tetramethylenemethane: connectivity of 2p-orbitals and ferromagnetic coupling unit (fCU).

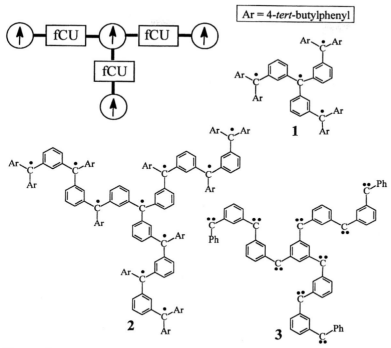

Figure 2 Selected star-branched polyradicals. Homologation of TMM with 1,3-phenylene fCUs.

ground state tetraradical **1** (Figure 2) [19]. Higher star-branched homologues, including the $S = 5$ decaradical **2**, were also prepared (Figure 2) [20]. The Iwamura and Itoh groups prepared and characterized a series of both linear and star-branched high-spin polycarbenes [3,11]. This research effort on high-spin polycarbenes, spanning three decades, culminated in the $S = 9$ nonacarbene **3** (Figure 2) [21].

Polyradicals with a greater number of spin sites for unpaired electrons, compared to **2** and **3**, were also synthesized (up to 31 arylmethyl and 12 arylcarbene sites). However, in the case of linear or dendritic connectivities, the average S-values were far below the theoretically expected values, $S < 2.5$ for polyarylmethyls and $S < 3$ for polyarylcarbenes (Figure 3) [22,23].

III. DEFECTS

The failure to generate high-spin polyradicals with more than ten interacting spin sites may be related predominantly to the chemistry of radical gener-

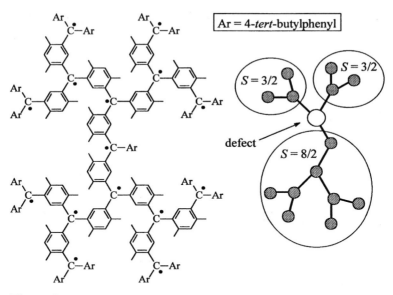

Figure 3 Dendritic polyarylmethyl pentadecaradical: interruption of spin coupling by one defect at an inner site.

ation. Near-quantitative preparative methods are available only for poly-arylmethyls and polyarylcarbenes. For polyarylmethyls, the best method re-lies on a two-step sequence: (1) treatment of the corresponding polyether with alkali metal in THF at near-ambient temperature and (2) subsequent I_2 oxidation of the resultant polyanion at low temperature in THF [19,24]. For polyarylcarbenes, irradiation of dilute frozen or solid solution of polydiazo compound at near-liquid-helium temperatures is employed. This solid-state photochemical reaction has a remarkable quantum yield; however, the re-action fails to provide near-quantitative chemical yields for higher homo-logues of nonacarbene **3** [23].

For a polyradical with n spin sites and the yield per site p for gener-ation of unpaired electrons (or carbenes) equal to 100%, the fraction of intact polyradicals is only p^n. The remainder of such a mixture contains polyrad-icals where unpaired electrons are missing from one or more spin sites; such polyradicals are said to contain defects [22]. For large n, it is not only essential to keep to a minimum the total number (or density) of defects ($p \sim 1$) but to keep the defects from significantly lowering S as well. The last problem may be illustrated by considering a dendritic polyradical with 15 sites for unpaired electrons [22]. The spin sites and fCUs are shown as full circles and bars, respectively (Figure 3) [25].

A dendritic polyradical possesses two types of spin sites, inner and terminal, in approximately equal numbers. A defect at a terminal site is relatively innocuous, i.e., S is lowered merely by 1/2. However, one defect at an inner site may interrupt π-conjugation and lead to a mixture of spin systems with rather low S. Therefore, a drastic lowering of an average S would occur compared to the theoretical value of $S = 15/2$, even at a very low density of defects. The situation is even worse in a linear oligomer/polymer, in which only two sites are terminal and the remaining ones are inner. Spin sites, for which at least n defects are needed for interruption of spin coupling, are referred to as n-defect sensitive; when $n \gg 1$, such sites may be viewed as defect insensitive. For a dendrimer or a linear chain, inner sites are 1-defect sensitive and terminal sites are defect insensitive.

The problem of defects interrupting coupling between the spin sites with unpaired electrons can be addressed by designing polyradicals with multiple coupling pathways, which circumvent the defective spin sites, and ensuring that a significant fraction of spin sites are terminal. Two connectivities are possible, designated Class I and Class II, relying on macrocycles and chain with pendants, respectively. Examples for both types of connectivities are octaradical **4** and polymeric radical **5** (Figure 4) [1,25–27].

Class II connectivities dominated early attempts to obtain high-spin organic polymers. Current limitations of Class II connectivities are relatively weak coupling, which is further weakened in the presence of defects, and lack of efficient methods for generation of suitable radicals. In accompanying chapters of this book, Nishide and Kaneko (Chapter 14) and Miura (Chapter 13) describe the current state of the art for polyradicals with Class II connectivities. The present discussion will be confined to Class I connectivities, which, in our opinion, have the greatest potential for achieving very high-spin molecules.

IV. MACROCYCLES

Several macrocyclic polyradicals were characterized (Figures 4 and 5). For polyarylmethyls, calix[4]arene was found to be the optimum ring size as far as the synthesis, spin coupling, and stability (persistence) of the polyradical are concerned. In particular, attempted generation of calix[3]arene-based triradical **6**, which is a key component of the proposed Mataga network, gave only the corresponding dimer [28].

Macrocyclic polyarylcarbenes were prepared via functionalization of the well-known calixarene derivatives [3,29]. For macrocyclic polyarylmethyls, a versatile method for construction of functionalized calixarenes was devel-

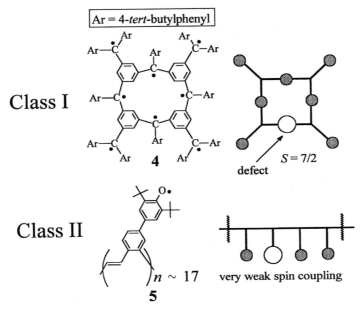

Figure 4 Class I and Class II polyradicals: macrocyclic polyarylmethyl octaradical **4** and polymeric phenoxyl **5**.

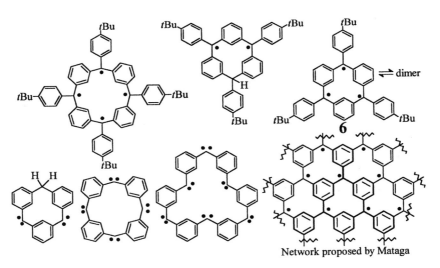

Figure 5 Macrocyclic polyarylmethyls and polyarylcarbenes, and proposed Mataga network.

oped, based upon addition of aryldilithium compounds to aryldiketones. Cis/trans isomers of various functionalized calixarenes may be separated and used in subsequent syntheses as modular building blocks [24,26].

A macrocyclic polyradical such as **4** has an equal number of 2-defect-sensitive sites and defect-insensitive sites. However, as the number of spin sites increases, the probability of having two or more defects in one polyradical will increase. Therefore, the design of a very high-spin polyradical must circumvent the disruption of spin coupling by two and more defects. Two strategies were recently reported in order to deal with this problem: (1) annelation of macrocycles and (2) organic spin clusters [24,30].

V.   ANNELATION OF MACROCYCLES

A typical lattice, such as found in most metal-containing magnetic materials, may be viewed as an annelated polycyclic system. Even 2-dimensional lattices can be quite impervious to defects; for example, the proposed Mataga network (honeycomb) has only 5-defect-sensitive sites [1]. From the point view of a classical organic synthesis, finite annelated oligomacrocycles, such as fragments of 2-strand **7**, are feasible targets (Figure 6) [30]. Such oligomacrocycles still have 2-defect-sensitive sites, similar to monomacrocycles; however, the relative weights of those 2- or 3-defect configurations, which interrupt coupling, are relatively diminished. Quantitative evaluation of impact of defects on spin coupling as a function of annelation is described elsewhere [31].

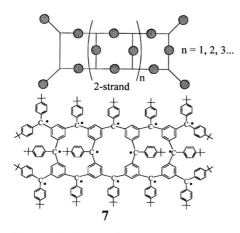

Figure 6   2-Strand and tetradecaradical **7**.

Polyradical **7** is prepared from the corresponding tetradecaether precursor. Magnetic and quenching studies of **7** in perdeuterated tetrahydrofuran are consistent with the presence of one defect ($X = D$) at one of the center triarylmethyl sites. Furthermore, S is lowered by small density of defects at the other sites. The resultant $S = 6.2$ is exceeded only by Iwamura polycarbenes and the subsequent polyarylmethyl-based organic spin clusters [30].

The design of annelated polyradicals is an important undertaking, related to the ultimate goal of an organic ferromagnet based on a network polyradical. There are two key synthetic obstacles to be overcome: (1) low yield of the macrocyclization step, and (2) the possibility of severe out-of-plane distortion of the π-conjugated system associated with annelation.

VI. ORGANIC SPIN CLUSTERS

The most recent design of organic polyradicals, referred to as organic spin clusters, encompasses synthesis, defects, and analysis of magnetic data [24]. In this approach, groups of spins 1/2 are effectively combined into larger component spins. The spins 1/2 are strongly coupled through 1,3,5-phenylene fCUs, and the component spins are weakly coupled through the biphenyl fCUs. A series of recently prepared spin dimers, trimers, and pentamers (**8–13**) is shown in Figure 7 [24,32].

Efficient modular synthesis based on the biphenyl CC-bond-forming reactions has been implemented. Polyradicals have relatively few 1-defect-sensitive sites, which are, importantly, "protected" by the 4-biphenyl substituents. The remaining sites are either 2-defect sensitive (macrocyclic) or defect insensitive (terminal). The 4-biphenyl substitution allows for more accurate titration of distinctly colored carbopolyanions in their oxidation to polyradicals, resulting in improved average yield per site for unpaired electrons (97–98%). Magnetic data can be analyzed in more detail for simple symmetric clusters, using readily available eigenvalues of the Heisenberg Hamiltonian. A simple-to-use vector method provides the eigenvalues for many symmetrical clusters [33]. Values of ferromagnetic coupling constants through biphenylene fCUs (J/k) strongly depend on connectivity; however, the energy gaps between the high-spin ground states and the lowest excited state are similar for all clusters in Figure 7 (Table 1).

Dendritic-macrocyclic spin pentamer **13** illustrates challenges provided by very high-spin polyradicals [32]. Magnetization studies reveal that only the high-spin ground state is populated at low temperatures ($T = 5$ K or below). For a sufficiently dilute solution of **13** in THF-d_8, perfect paramagnetic behavior with an average $S = 10$ is observed. This is the highest S observed for an organic molecule to date. However, $S = 12$ theoretically is

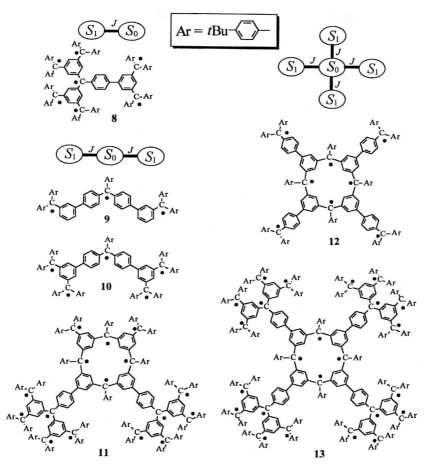

Figure 7 Organic spin clusters: polyradicals **8–13** and their representations as symmetrical spin clusters.

expected for 24 ferromagnetically coupled electron spins in the high-spin ground state. This discrepancy is associated with a small density of defects, which is estimated at 2% from a simple percolation model. At higher temperatures ($T = 5–160$ K), the high-spin ground state is partially depopulated; magnetization, calculated for a Heisenberg Hamiltonian with four $S = 5/2$ ferromagnetically coupled to one $S = 2$ with $J/k = 7$ K, shows good agreement with the experimental data [32]. Numerical tests on smaller spin clus-

Table 1   Spin Clusters **8–13**: Values of S, Equations for $\langle S \rangle$, $p100\%$, and J/k, both Experimental and from the Spin-Density Scaling

Polyradical	S	$\langle S \rangle$	$p100\%^b$	J/k(exp) [K]	J/k(scaled) [K]
8	3.28	$3.5p/(3 - 2p)$	98	13 ± 2	$90/5 = 18$
9	a			90 ± 20	90
10	2.40	$2.5p/(2 - p)$	98	90 ± 5	90
11	7.2	$8p/(5 - 4p)$	98	4 ± 1	$90/15 = 6$
12	a			35 ± 10	$90/2 = 45$
13	10.0	$12p/(9 - 8p)$	98	7 ± 1	$90/10 = 9$

(From refs. 24 and 32.)
aEstimates of $p100\%$ are not reported for polyradicals **9** and **12** because their relatively large negative values of mean field parameters (θ) introduce significant error in measured S.
bEstimated yields per site are for the best samples of polyradicals.

ters suggest that values of J/k are not significantly affected when the density of defects is in the range of a few percent [24].

The 98% average yield per site for the generation of radicals in **13** is quite remarkable for an organic polyradical. However, it corresponds to approximately 60% of **13** with all 24 unpaired electrons intact. Even though only 4 out of 24 sites in **13** are 1-defect sensitive and stabilized by the 4-biphenyl substituents, the average $S = 10$ is obtained rather than $S = 12$, which theoretically would be expected for a 100% yield.

VII.   SPIN DENSITY SCALING OF J

The values of J/k, corresponding to pairwise ferromagnetic coupling through biphenyl unit, are not identical for all spin clusters (Table 1). It was proposed that J/k should be scaled by a fraction of the spin density of the spin participating directly in the pairwise coupling. In **13**, each pairwise coupling involves the $S = 2$ and the $S = 5/2$; however, only two out of four sites in $S = 2$ and one out of five sites in $S = 5/2$ participate in pairwise coupling through each biphenyl unit. Therefore, J/k in **13** should be scaled by a factor of $(2/4)*(1/5) = 0.1$. Analogous scaling was previously proposed by Dougherty and coworkers [34]. This scaling procedure gives $J/k = 9$ K, compared to the experimental value of 7 K for **13**. The scaled values of J/k for the spin clusters in Table 1 are in approximate agreement with the experimental values; however, a general trend of overestimation of the experimental values is noticeable. Is the scaling procedure for J/k adequate?

It is instructive to consider a couple of simple clusters in which scaling of J/k can readily be verified without recourse to a major computation. In nonsymmetrical clusters, such as trimer **14** and tetramer **15**, two coupling constants J_1 and J_2 can be defined (Figure 8). The simple vector method cannot be applied to such nonsymmetrical clusters in order to obtain eigenvalues for the Heisenberg Hamiltonian. However, the analytical solution for the eigenvalues is still feasible for the clusters of $\frac{1}{2}$-spins by a relatively simple procedure, involving the use of Clebsh–Gordan coefficients and diagonalization of relatively small matrices [35]. For $J_2 > J_1 > 0$ (both couplings are ferromagnetic), the two lowest energy levels for clusters **14** and **15** are shown in Figure 8. When J_2 is at least a few times greater than J_1, these two lowest energy levels will be well separated in energy from others; i.e., for temperatures either below or not far above J_1/k K only the two lowest energy levels will dominate contribution to the magnetization. However, for $J_2 \gg J_1 > 0$, the J_2-coupled spins 1/2 may be lumped into component spins leading to readily solvable spin dimers, such as $S = 1$ and $S = 1/2$ (for **14**) and $S = 3/2$ and $S = 1/2$ (for **15**). The assumption that the dimers are good

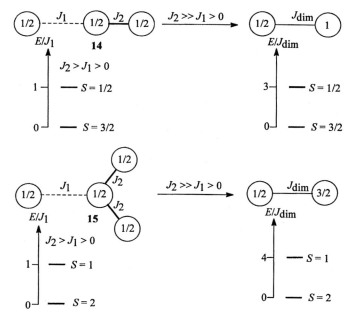

Figure 8 The two lowest-energy eigenvalues (from the Heisenberg Hamiltonian) for the ferromagnetically coupled nonsymmetrical spin clusters **14** and **15**, and the energy eigenvalues for the correseponding spin dimers.

approximation for clusters **14** and **15** requires that the energy levels for each spin dimer be exactly matched with the corresponding cluster. Consequently, ferromagnetic coupling constant J_{dim} for each dimer is related to J_1 by a scaling factor. For $J_2 \gg J_1$, the scaling factors are 1/3 and 1/4 for the dimers corresponding to **14** and **15**, respectively. However, significantly different scaling factors, 1/2 and 1/3 for **14** and **15**, respectively, are derived from spin densities.

When $J_2 > J_1$ but with comparable values, scaling factors will significantly depend on the relative magnitudes of J_2 and J_1. For example, $J_2 = 6J_1$ may be estimated for pairwise ferromagnetic couplings of spins 1/2 through the phenylene and biphenyl units, which is relevant to spin clusters in Figure 7 [24]. For $J_2 = 6J_1$, the scaling factors are 0.48 and 0.32 for the dimers corresponding to **14** and **15**, respectively, i.e., nearly identical to the respective scaling factors, which are derived from spin densities. Applicability of the spin-density scaling of J to the spin clusters in Figure 7 is based on several interdependent factors, involving values of J_2/J_1, spin-density distributions, and molecular conformations. It is rather amazing that the spin-density scaling of J only moderately overestimates the experimental J's for the spin clusters in Figure 7, thus providing an additional tool for magnetic characterization of high-spin organic polyradicals.

VIII. OUTLOOK

Since 1990, the year of the first publication on high-spin polyradicals from our laboratory, significant advances have been made in the synthesis and characterization of polyradicals based on polyarylmethyls. Recently, a poly-arylmethyl-based polyradical with an average $S = 10$, the highest spin quantum number that has been measured for an organic molecule, was prepared in our laboratory. Elaboration of the molecular design, driven primarily by the necessity to alleviate the impact of defects, led to the improvement of S from 2 to 10 in the past eight years. This research contributed to better understanding of factors affecting spin coupling between multiple sites in polyradicals. Thus, in conjunction with the refined synthetic methodology, the foundation was set for accelerated progress toward organic superparamagnets and ferromagnets. Astute design of π-conjugated networks and systems with well-defined fractal dimensionalities is especially promising strategy. For example, preliminary results from our laboratory show that an average S exceeding 30 can be obtained for networks synthesized from macrocyclic building blocks [36].

ACKNOWLEDGMENTS

I thank my coworkers, named in the references, for their contributions to this field of study. I gratefully acknowledge the National Science Foundation for the support of research on polyarylmethyl polyradicals.

REFERENCES

1. Rajca A. Organic Diradicals and Polyradicals: From Spin Coupling to Magnetism? Chem Rev. 1994; 94:871–893. Rajca A. Very High-Spin Polyradicals. In: Turnbull MM, Sugimoto T, Thompson LK, eds. ACS Symposium Series 644, Molecule-Based Magnetic Materials. Washington, DC: ACS, 1996, chap 17.

2. Dougherty DA. Spin Control in Organic Molecules. Acc Chem Res. 1991; 24: 88–94. Jacobs JS, Dougherty DA. Combination of Superexchange and Spin Polarization Mechanisms Leads to a Novel Hydrocarbon Tetraradical with a High-Spin Ground State. Angew Chem Int Ed Engl. 1994; 33:1104–1106.

3. Iwamura H, Koga N. Studies of Organic Di-, Oligo-, and Polyradicals by Means of Their Bulk Magnetic Properties. Acc Chem Res. 1993; 26:346–351. Iwamura H. High-Spin Organic Molecules and Spin Alignment in Organic Molecular Assemblies. Adv Phys Org Chem 1990; 26:179–253.

4. Berson JA. A New Class of Non-Kekulé Molecules with Tunable Singlet-Triplet Energy Spacings. Acc Chem Res. 1997; 30:238–244. Borden WT, Iwamura H, Berson JA. Violations of Hund's Rule in Non-Kekulé Hydrocarbons: Theoretical Prediction and Experimental Verification. Acc Chem Res. 1994; 27: 109–116.

5. Lahti PM. Design of Organic-Based Materials with Controlled Magnetic Properties. In: Turnbull MM, Sugimoto R, Thompson LK, eds. ACS Symposium Series 644, Molecule-Based Magnetic Materials. Washington, DC: ACS, 1996, chap 14.

6. Buchachenko AL. Organic and Molecular Ferromagnetics: Advances and Problems. Russ Chem Rev. 1990; 59:307–319.

7. Yoshizawa K. Hoffmann R. Potential Linear-Chain Organic Ferromagnets. Chem Eur J. 1995; 1:403–413.

8. Mataga N. Possible "Ferromagnetic States" of Some Hypothetical Hydrocarbons. Theor Chim Acta. 1968; 10:372–376.

9. Wenthold PG, Kim JB, Lineberger WC. Photoelectron Spectroscopy of m-Xylylene Anion. J Am Chem Soc. 1997; 119:1354.

10. Kothe G, Ohmes E, Brickman J, Zimmermann H. 1,3,5-Benzenetriyltris[di(p-biphenyl)methyl], a Radical Having a Quartet Ground State that Dimerizes by Entropy Bonding. Angew Chem Int Ed Engl. 1971; 10:938–940.

11. Itoh K. Electronic Structure of Aromatic Hydrocarbons with High-Spin Multiplicities in the Electronic Ground State. Pure Appl Chem. 1978; 50:1251–1259.

12. Borden WT, Davidson ER. Effects of Electron Repulsion in Conjugated Hydrocarbon Diradicals. J Am Chem Soc. 1977; 99:4587–4594. Borden WT. Effects of Electron Repulsion in Diradicals. In: Borden WT, ed. Diradicals. New York: Wiley, 1982, pp 1–73.

13. Ovchinnikov AA. Multiplicity of the Ground State of Large Alternant Organic Molecules with Conjugated Bonds. Theor Chem Acta. 1978; 47:297–304.

14. West AP Jr, Silverman SK, Dougherty DA. Do High-Spin Topology Rules Apply to Charged Polyradicals? Theoretical and Experimental Evaluation of Pyridiniums as Magnetic Coupling Units. J Am Chem Soc. 1996; 118:1452–1463.

15. Schaefer HF. Science. 1986; 231:1100.

16. Dowd P. J Am Chem Soc. 1966; 88:2587.

17. Nau WM. Organic Tri- and Tetraradicals with High-Spin or Low-Spin States. Angew Chem Int Ed Engl. 1997; 36:2445–2448.

18. Murray MM, Kaszynski P, Kaisaki DA, Chang W, Dugherty DA. Prototypes for the Polaronic Ferromagnet. Synthesis and Characterization of High-Spin Organic Polymers. J Am Chem Soc. 1994; 116:8152–8161.

19. Rajca A. A Polyarylmethyl Quintet Tetraradical. J Am Chem Soc. 1990; 112:5890–5892.

20. Rajca A, Utamapanya S, Thayumanavan S. Poly(arylmethyl) Octet ($S = 7/2$) Heptaradical and Undecet ($S = 5$) Decaradical. J Am Chem Soc. 1992; 114:1884–1885.

21. Nakamura N, Inoue K, Iwamura H. A Branched-Chain Nonacarbene with a Nonadecet Ground State: A Step Nearer to Superparamagnetic Polycarbenes. Angew Chem Int Ed Engl. 1993; 32:872–874.

22. Rajca A, Utamapanya S. Toward Organic Magnetic Particle: Dendritic Polyradicals with 15 and 31 Centers for Unpaired Electrons. J Am Chem Soc. 1993; 115:10688–10694. Rajca S, Rajca A. Dendritic Polyradicals. In: Salamone JC, ed. The Polymeric Materials Encyclopedia. Boca Raton; FL: CRC Press, 1996, Vol. 3, pp 1830–1834.

23. Matsuda K, Nakamura N, Inoue K, Koga N, Iwamura H. Toward Dendritic Two-Dimensional Polycarbenes: Syntheses of "Starburst"-Type Nona- and Dodecadiazo Compounds and Magnetic Study of Their Photoproducts. Bull Chem Soc Jpn. 1996; 69:1483–1494.

24. Rajca A, Wongsrirantanakul J, Rajca S. High-Spin Organic Polyradicals as Spin Clusters: Ferromagnetic Spin Coupling through Biphenyl Unit in Polyarylmethyl Tri-, Penta-, Hepta-, and Hexadecaradicals. J Am Chem Soc. 1997; 119:11674–11686.

25. Rajca A. High-Spin Organic Polyradicals. Polymer News. 1996; 21:405–409.

26. Rajca A, Rajca S, Padmakumar R. Calixarene-Based Macrocyclic Nonet ($S = 4$) Octaradical and its Acyclic Sextet ($S = 5/2$) Pentaradical Analogue. Angew Chem Int Ed Engl. 1994; 33:2091–2093.

27. Nishide H, Kaneko T, Nii T, Katoh K, Tsuchida E, Lahti PM. Poly(phenylenevinylene)-Attached Phenoxyl Radicals: Ferromagnetic Interaction through Planarized and π-Conjugated Skeletons. J Am Chem Soc. 1996; 118:9695–9704.

28. Rajca A, Rajca S, Desai SR. Macrocyclic π-Conjugated Carbopolyanions and Polyradicals Based upon Calix[4]arene and Calix[3]arene Rings. J Am Chem Soc. 1995; 117:806–816.
29. Matsuda K, Nakamura N, Takahashi K, Inoue K, Koga N, Iwamura H. Design, Synthesis, and Characterization of Three Kinds of π-Cross-Conjugated Hexacarbenes with High-Spin ($S = 6$) Ground States. J Am Chem Soc. 1995; 117: 5550–5560.
30. Rajca A, Lu K, Rajca S. High-Spin Polyarylmethyl Polyradical: Fragment of a Macrocyclic Double-Strand. J Am Chem Soc. 1997; 119:10335–10345.
31. Rajca A. Assembling Triarylmethyls into Mesoscopic-Size Polyradicals: How to Maintain Strong Interactions Between Multiple Sites in a Single Molecule. In: Michl J, ed. Modular Chemistry. Dordrecht: Kluwer, 1997, pp 193–200.
32. Rajca A, Wongsrirantanakul J, Rajca S, Cerny R. Dendritic-Macrocyclic Very-High-Spin Organic Polyradical with $S = 10$. Angew Chem Int Ed Engl. 1998; 37:1229–1232.
33. Belorizky E, Fries PH. J Chim Phys (Paris). 1993; 90:1077–1100.
34. Jacobs SJ, Schultz DA, Jain R, Novak J, Dougherty DA. Evaluation of Potential Ferromagnetic Coupling Units: The Bis(TMM) Approach to High-Spin Organic Molecules. J Am Chem Soc. 1993; 115:1744–1753.
35. Kahn O. Molecular Magnetism. New York: VCH, 1993.
36. Rajca A, Rajca S, Wongsriratanakul J. Manuscript in preparation.

18

Band Theory of Exchange Effects in Organic Open-Shell Systems

Nikolai Tyutyulkov
University of Sofia, Sofia, Bulgaria

Fritz Dietz
University of Leipzig, Leipzig, Germany

I. INTRODUCTION

Why do we use the band theory for the interpretation of exchange effects? Two basic types of magnetic ordering can be distinguished: a ferromagnetic (FM) one, when all spins have an identical orientation, and the antiferromagnetic (AFM) one, when the spins belong to two sublattices of opposite orientation with compensated magnetic moments of the two sublattices. If the latter are uncompensated the state is a ferrimagnetic one.

Depending on the number N of the unpaired electrons, molecular open-shell systems can be classified as biradicals, polyradicals, and systems for which $N \gg 1$. The interaction between two magnetic moments can be called ferromagnetic if the interaction tends to align them parallel. In accordance with this definition, the interaction between the uncompensated spins in the triplet Schlenk biradical **R-1** is ferromagnetic (Scheme 1); however, the biradical is not a ferromagnet.

There exists, however, a substantial difference between the magnetic states of a biradical and of a many-electron system ($N \gg 1$). Bulk magnetism as well as superconductivity are typical cooperative phenomena, since they result from the interaction of a large number of electrons [1]. At specific conditions, below a critical temperature T_c, the interaction gives rise to the occurrence of an internal ordering. That is why the theory used for the design of molecular materials with magnetic ordering must be a many-body (elec-

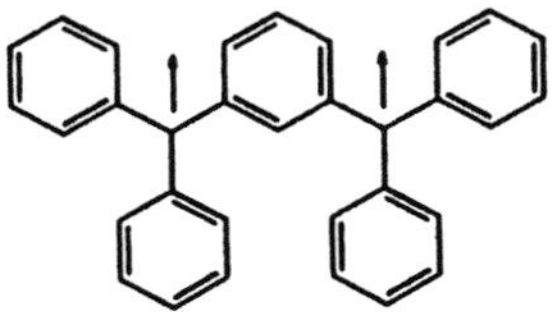

R-1

Scheme 1

tron) band theory including the electron correlation [1,2]. For other many-electron formalisms, see refs. 3–5. Therefore, we shall focus our attention on cooperative magnetism, responsible for the properties of magnetic materials with bulk magnetism, rather than on the magnetic properties of bi- and polyradicals.

II. METHODS OF INVESTIGATION

II.A. Energy Spectra of the Polymers

If the molecular orbitals (MOs) of a one-dimensional (1-D) system have the form of Bloch running waves

$$|k\rangle = N^{-1/2} \sum_{\mu} \sum_{r} C_r(k) \exp{(-ik\mu)}|r, \mu\rangle \qquad (1)$$

($k \in [-\pi, \pi]$ is the wave vector, μ denotes the number of elementary units (EUs), and $|r, \mu\rangle$ is the rth atomic orbital (AO) within the μth EU), in the Hückel–Hubbard version of the Bloch method the MO energies $e(k)$ are eigenvalues of the energy matrix:

$$\mathbf{E}(k) = \mathbf{E} + \mathbf{V} \exp{(ik)} + \mathbf{V}^+ \exp{(-ik)} \qquad (2)$$

In Eq. (2), $\mathbf{E}$ is the energy matrix of the EU, $\mathbf{V}$ is the interaction matrix between neighboring EUs [μth and ($\mu + 1$)th] and $\mathbf{V}^+$ is the tranposed matrix.

II.B. Spin-Exchange Interaction of the Electrons in the Half-Filled Band

A simple but very effective approach to determine the relative stability of a localized high-spin (HS) (ferromagnetic) state and a low-spin state (nonmagnetic) delocalized in the half-filled band (HFB) can be found in the studies of Whangbo [6] and Hubbard [7]. Let us denote the Coulomb repulsion integrals of two electrons occupying the same Wannier state and of adjacent states by U_0 and U_1, respectively, and the width of the HFB by $\Delta\varepsilon$.

Whangbo's condition reads [$U = U_0 - U_1$ is the renormalized Hubbard integral (see Eq. (7)]:

$$\Delta\varepsilon < \left(\frac{\pi}{4}\right) U \tag{3}$$

Equation (3) is similar to those derived by Hubbard:

$$\Delta\varepsilon < \left(\frac{2}{\sqrt{3}}\right) U \tag{4}$$

From the Whangbo–Hubbard conditions the following conclusions can be drawn: if $\Delta\varepsilon$ is small compared with U, then the magnetic HS state is favored, and the opposite relation should hold for conducting properties of polymers. These qualitative results for the nature of the ground state agree with the quantitative ones obtained by means of the direct spin-exchange determination (see Sections IV and V and Table 1).

A more established method used for the determination of the nature of the spin exchange within the HFB follows from Anderson's general theory of magnetism [2]. The formalism is valid for infinite π-conjugated polymers with translational symmetry, in particular for 1-D systems. For descriptions of spin correlation within a narrow HFB it is more convenient to pass from the Bloch wave functions to Wannier states, if they are well localized within the EU. Based on Anderson's theory [2] it was shown [8] that the effective exchange integral J_{eff} between Wannier functions localized upon the νth and ρth EU in the Heisenberg–Dirac–Van Vleck Hamiltonian:

$$\mathbf{H} = -2 \sum_{\nu\neq\rho} J_{eff}(\nu,\rho)\, \mathbf{S}_\nu \mathbf{S}_\rho = -2 \sum_{\nu\neq\rho} J_{eff}(\tau) \mathbf{S}_\nu \mathbf{S}_\rho \tag{5}$$

can be expressed as a sum of three contributions (for the sake of simplicity the dimensionless distance parameter $\tau = |\nu - \rho|$ is omitted):

$$J_{eff} = J + J_{kin} + J_{ind} \tag{6}$$

The terms in Eq. (6) have the following physical meaning: J is the Coulomb exchange integral between the localized Wannier states within the νth and ρth sites. The kinetic exchange parameter J_{kin} and the indirect exchange term J_{ind} arise if the electron correlation is taken into account [4,7–8]. The kinetic exchange parameter:

$$J_{kin} = \frac{-2t^2}{U_0 - U_1 - 2J} \tag{7}$$

represents the antiferromagnetic contribution to the spin exchange.

The parameter t is the transfer (hopping) parameter between adjacent Wannier functions:

$$t = \langle \mu | \boldsymbol{h}(1) | \mu + 1 \rangle \quad (\boldsymbol{h} \text{ is the one-electron periodic Hamiltonian})$$

The exchange parameter (spin polarization exchange) J_{ind} expresses the indirect exchange ("superexchange") of the electrons in the HFB via delocalized π-electrons (along the polymer chain) in the filled energy bands. This term can be calculated using a formalism described in ref. 9. The sign of J_{ind} is determined by the structure of the EU and by the interaction between the EUs.

III. SPECIFIC DISCUSSION GOALS: POLYMERS, RADICAL CRYSTALS, CHARGE TRANSFER

The stated title topic, which is quite general, will be discussed with specific reference to target systems of various composition, structure, mechanism and character of exchange interaction.

The investigations of purely organic 1-D systems are focused mainly on three classes:

  I.   Systems with intramolecular magnetic interaction in polymers with a continual π-system of conjugation [10,11].

  II. Molecular radical crystals in which the EUs are weakly interaction subsystems, namely, radicals or ionradicals with delocalized π-electrons.

  III. Systems for which the magnetic ordering is caused by a charge transfer (CT) mechanism [12] (see Kollmar and Kahn [13]). The problem of exchange interaction in infinite 1-D CT systems was studied in the papers of Soos et al. [14].

Our considerations will be limited to the most frequently investigated purely organic systems, the regular 1-D systems of the types I and II.

Two kinds of character of the degenerate MOs of the HFB will be used for the classification of the polymers: (1) *delocalized* MOs (through the whole π-system of conjugation of the polymer), and (2) *localized* MOs (within each EU). Such a MO classification allows the prediction of the type of the spin-exchange mechanism that determines the character of the magnetic coupling of the electrons in the HFB: *direct* spin exchange (Section IV) or *indirect* spin exchange (Section V).

IV. POLYMERS WITH DIRECT SPIN EXCHANGE

The presence of a degenerate or nearly degenerate MO band is neither a necessary nor a sufficient condition for the occurrence of a high-spin state in a system, but it is an essential implication for the existence of such a possibility (the presence of degenerate quantum states is a sufficient condition only within the one-electron approximation—Hund's rule). Therefore, it is reasonable to discuss the role of the structural conditions for the presence of an HFB of degenerate MOs.

IV.A. Conditions for the Occurrence of Degenerate Molecular Orbitals Determined by the Molecular Topology

The possibility of parallel alignment of electron spins in alternant nonclassical (ANC) hydrocarbons with degenerate MOs (Hund's rule) was first suggested by Longuet-Higgins [15]. Nonclassical molecular π-electron systems are those that cannot be described by means of a classical—a Kekulé-type —valence formula. That is why the terms nonclassical and non-Kekulé systems are identical. The basic idea determining the structural principle of ANC polymers (ANCPs) is connected with the Coulson–Rushbrooke–Longuet-Higgins (CRLH) theorem for the presence of degenerate nonbonding MOs (NBMOs) [15,16]. According to this theorem, an alternant hydrocarbon containing $\{S\}$ nonneighboring starred π-centers and $\{R\}$ nonneighboring nonstarred π-centers $(S > R)$ possesses $n = S - R$ NBMOs. Polymers **1** [4,17] $[n = N(2* - 1) = N$ NBMOs, N is the number of EUs] and **2** $[n = N(4* - 3) = N$ NBMOs] [18,19] are typical examples of ANCPs with an HFB of degenerate *delocalized* NBMOs (Scheme 2). The experimental and theoretical [10,11,18,19] studies show that ANCPs are some of the most promising candidates for high-spin polymers with FM coupled electrons.

The CRLH theorem has been generalized [20] for some nonalternant nonclassical polymer (NCP) and heteroatomic π-systems. This is the basis of a structural principle of a new class of NCPs denoted as quasi-ANCPs (QANCPs) [20]. In the generalized theorem, only one subset of nonbonded π-centers is responsible for the appearance of NBMOs. A conjugated system with M π-centers and a maximum set of $M*$ homonuclear disjoint π-centers must have at least $n = 2M* - M$ NBMOs. The NBMOs are present even if the π-centers not belonging to the disjoint subset are heteroatomic. Repre-

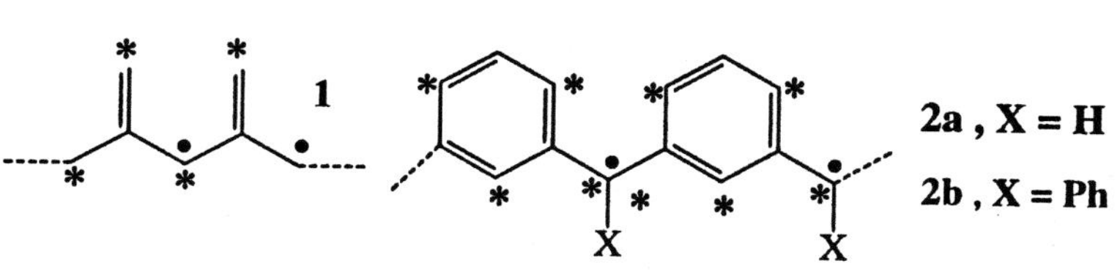

2a , X = H

2b , X = Ph

Scheme 2

sentative examples are the radicals **R-2** and **R-3** and the QANCP **3** (Scheme 3) investigated in many papers of Berson (see ref. 21 and Chapter 2 in this book).

All ANCPs and QANCPs studied so far are characterized by a wide gap in which the HFB consisting of degenerate MOs is situated. The HFB is strictly degenerate within the Hückel approximation. In order to go beyond the limits of this approximation, the band structure of the polymers can be calculated (see ref. 8) within the restricted open-shell Hartree–Fock approach [22]. If the MOs are calculated within the self-consistent-field (SCF) method with the Pariser–Parr–Pople Hamiltonian using periodic boundary conditions, the consideration of the electron–electron interaction leads to the splitting of the degenerate MOs, resulting in a small band with a width $\Delta\varepsilon$ (HFB) ~ 0.1–0.2 eV.

For all ANCPs and QANCPs the effective exchange integral $J_{eff} > 0$; i.e., all polymers exhibit a net spin exchange of ferromagnetic nature, with the dominant contribution of the potential exchange J (see Table 1).

A weak perturbation (WP) (see Figure 1) caused by the substitution of a starred π-center, e.g., the exocyclic CH group of **2a** by $-N^-$, $-HN^{+\cdot}-$ or $-HB^-$, results in the polymers **4a–c**, which have the same topology as **2** (Scheme 4). For these polymers, the NBMOs split in an HFB whose width $\Delta\varepsilon$ depends on the nature of X. However, the ground state of all these polymers is a high-spin one with FM coupled electrons within the HFB [23–25]. As in the case of ANCPs and QANCPs, the dominant contribution to J_{eff} is the potential exchange J (Table 1). These theoretical results also explain the magnetic properties of poly(*meta*-aniline) synthesized by Yoshizawa et al. [26]. Recently, Rajca and coworkers have published the synthesis of a stable triplet diradical dianion of a diborane [27].

In the case of a strongly perturbed (SP) NCP that may be anticipated by changing the topology, the NBMO band splits into a wide band. An example is the iso-π-electronic polymer of the dehydrogenated form of

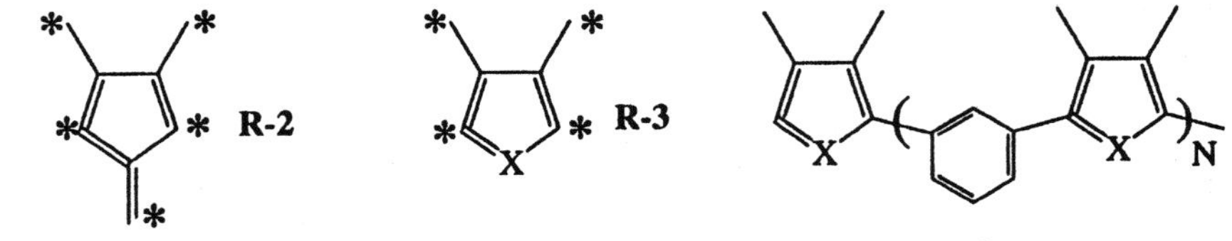

$n = 2.5* - 8 = 2$ NBMOs $n = 2.4* - 7 = 1$ NBMOs   **3**

$X = O, N, S$

Scheme 3

Table 1 Calculated Values of Different Contributions[a] to Effective Spin Exchange Between Unpaired Electrons in the HFB (Eq. 6), and Energy Characteristics of the HFB

Polymer	$\Delta\varepsilon$	$\pi U/4$	J	$-J_{kin}$	J_{ind}	J_{eff}
1	3	1.911	0.199	0	0.070	0.269
4a	0.193	1.904	0.184	0.003	0.061	0.242
4b	0.272	1.167	0.126	0.014	0.026	0.138
4c	0.227	1.273	0.089	0.004	0.015	0.100
5	0	1.317	0.053	0	0.032	0.085
11	0.292	1.597	0.432	0	0.017	0.449
12	0.091	1.616	0.435	0	0.078	0.513
15	0	1.224	0	0	−0.009	−0.009
16	0	1.355	0	0	−0.012	−0.012

$\Delta\varepsilon$ is the band width of the HFB, and U is the renormalized Hubbard integral [see the Whangbo condition, eq. (3)]. All entries in electron-volts.
[a]The values are estimated with Wannier functions calculated using an MO basis obtained within the Hückel–Hubbard approach (Eq. 1).

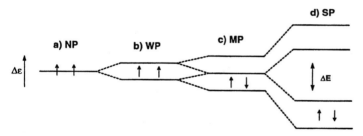

Figure 1 Energy spectra of an NCP as a function of the degree of perturbation. **NP**: nonperturbed NCP; **WP**: weakly pertubed NCP; **MP**: medium perturbed NCP; **SP**: strongly perturbed NCP.

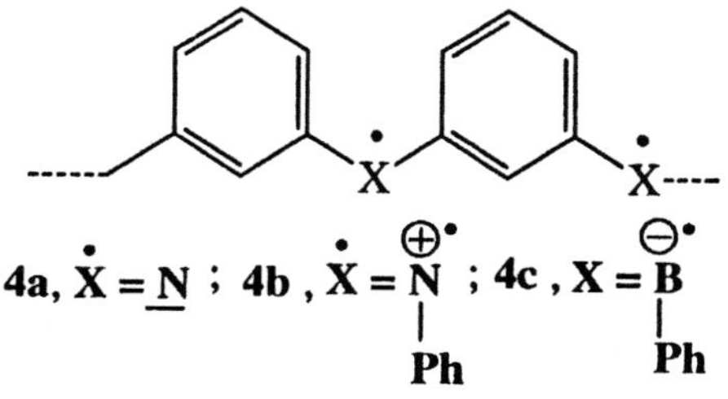

Scheme 4

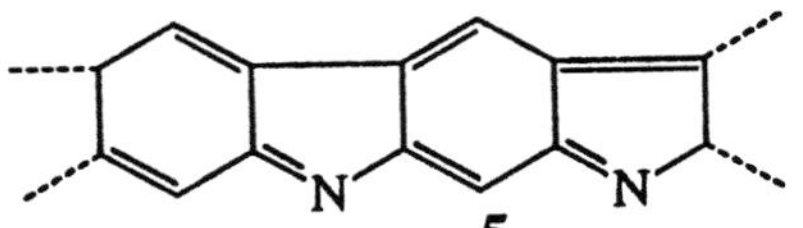

Scheme 5

poly(*meta*-aniline) **5** (Scheme 5). In this case, the ground state is nonmagnetic dielectric with a band gap $\Delta E > 1.5$ eV [24].

In Figure 1 are shown the energy spectra of a non-Kekulé 1-D polymer as a function of the degree of perturbation. Of special interest are medium perturbed (MP) polymers. Such 1-D systems proposed by Berson et al. [21,28] are a principal new class of organic conducting polymers with a metallic ground state [29].

The CRLH and the generalized CRLH theorem provide only a sufficient, but not a necessary, condition for the presence of degenerate MOs in a molecular system. There exist many nonclassical molecules and polymers having degenerate MOs, the occurrence of which can be explained with the Sachs theorem [30]. An example is the nonclassical monoradical **R-4** forming the HS polymer **6** with FM-coupled electrons in the HFB with degenerate NBMOs (Scheme 6) [31].

IV.B. Degenerate Molecular Orbitals Caused by the Steric Hindrance

The presence of a narrow band of degenerate or nearly degenerate quantum states can also be caused by steric hindrance that reduces the overlap between the interacting singly occupied orbitals to zero in the limiting case. A simple organic system that illustrates this idea is the model polymer **7** suggested by Lahti [32] (see also the references). The theoretical [33] and experimental studies [34] of the polyanion **8** show that the FM alignment is favored when $\Theta \rightarrow \pi/2$ (Scheme 7). For other polyaryl systems in which

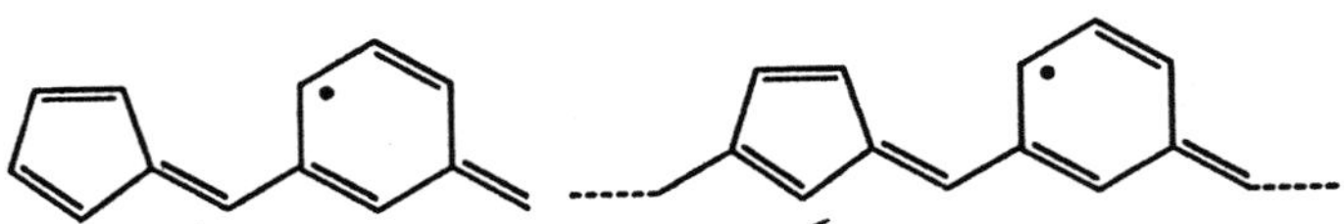

Scheme 6

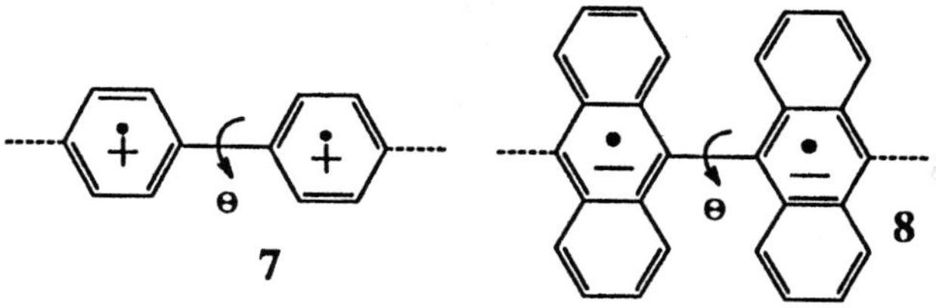

Scheme 7

the rings are orthogonal and they are oxidized to radical cations or reduced to radical anions, see Chapter 8 in this book.

IV.C.   Polaronic High-Spin Ferromagnets

Polaronic ferromagnets also belong to the classes of polymers with direct exchange. In the Fukutome model [35] a charged fragment **A** with a degenerate ground state is connected to a fragment **B** with nondegenerate ground state: ..$A^{-(+)} - B - A^{-(+)} - B$.... In the model investigated in ref. 36, a narrow HFB appears within the energy gap of a dendralene chain as a result of doping with electrons or holes. It was shown [36] that in contrast to polyacetylene **9**, the polaron (at reductively or oxidatively doping) in a dendralene chain **10** is strictly localized in one elementary unit (Scheme 8).

A structural relaxation corresponding to a polaron delocalized in more than one EU is not possible. For such a polaron there is no corresponding valence formula. This property caused by the topology of the π-network (1,1-linking of the vinylidene units) yields good localization of the Wannier functions and an HS ground state with FM coupled electrons within the narrow HFB. The dendralenes **10** are the simplest examples of a large class of π-conjugated 1-D systems [36] having this property of localization of the polaron in one fragment and FM exchange-coupled electrons. Examples are the fully reduced or oxidized polymers **11** and **12** (Scheme 9).

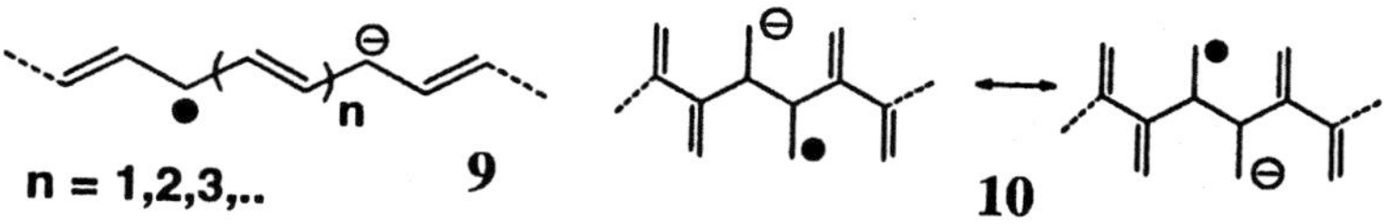

n = 1,2,3,..   **9**   **10**

Scheme 8

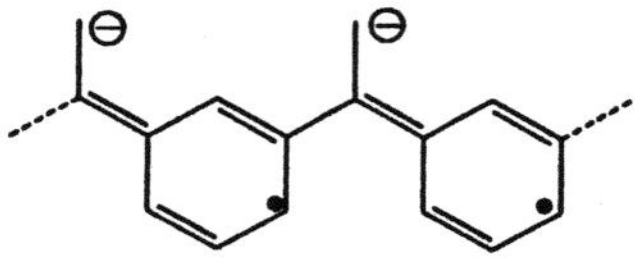

11 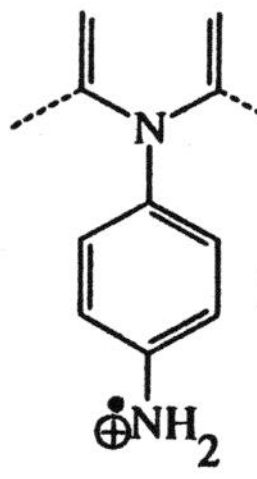 **12**

Scheme 9

V. POLYMERS WITH INDIRECT SPIN EXCHANGE

The indirect magnetic exchange interaction is responsible for the magnetic ordering in many crystals with paramagnetic metal ions. This widespread, well-investigated phenomenon within inorganic materials [37] has not been investigated extensively theoretically in the field of purely organic molecular systems. An exception is ref. 9, in which the concept of indirect exchange was extended to some classes of purely organic π-systems. The indirect communication between the conjugatively isolated unpaired electrons in quinonoid dinitrenes was discussed in a recent paper of Minato and Lahti [38].

The appearance of localized, partially occupied MOs (LPOMOs) can be caused by the topology [9], steric hindrance [9], or symmetry [39]. If at least one of these (sufficient but not necessary) conditions is fulfilled, then the magnetic coupling between the electrons occupying the LPOMOs is of indirect nature.

V.A. Degenerate Molecular Orbitals Determined by the Topology or Steric Hindrance

The model polymers with indirect exchange interaction suggested in ref. 9 consist of 1-D π-conjugated chains of subunits **L** with radical centers **R·** attached (Scheme 10). The radical **R·** should possess an LPOMO. The occurrence of LPOMOs is caused either by steric hindrance (RL-S polymers),

$$R^\bullet \quad R^\bullet \quad R^\bullet \quad R^\bullet$$
$$| \quad | \quad | \quad |$$
$$\cdots L - L - L - L \cdots$$

Scheme 10

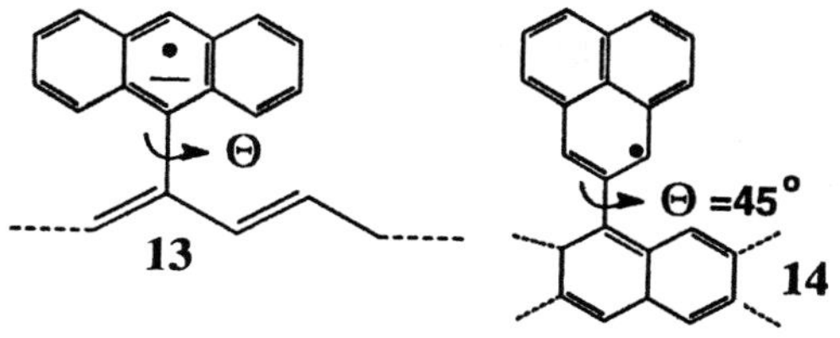

Scheme 11

which inhibits the $\pi-\pi$ interaction of LPOMOs and the delocalization of π-electrons over more than one unit, or by the topology (RL-T polymers) of the radical site $\mathbf{R}^{\cdot}$. The uncompensated spins in $\mathbf{R}^{\cdot}$ induce spin polarization of the π-electrons delocalized along the polymer chain. Examples of polymers with an LPOMO in the EU are **13** (RL-S, $\Theta = \pi/2$) and **14** (RL-T) (Scheme 11).

The exchange constants for the polymers RL-S are very small and of varying sign, depending on the structure of $\mathbf{R}^{\cdot}$ and **L**. For **13**, for instance, $J_{\text{eff}} \sim J_{\text{ind}} \sim 0.003$ eV. All studied alternant RL-T polymers [9] have an FM exchange coupling with $J_{\text{ind}} \sim 0.01-0.2$ eV.

V.B. Degenerate Molecular Orbitals Determined by the Symmetry

In another model, the localization of the orbitals giving rise to indirect exchange is determined by the symmetry of the EU. These theoretical studies [39] are connected with the experimental results [40] related to the synthesis, spectroscopic, and the magnetic properties of the ion-radicals of 2,2′-bipyrenyl (**Bi**) and *para*-terpyrenyl (**Te**) (Scheme 12).

In the oligopyrenyls, the MO coefficients of the bridgehead π-centers of the frontier MOs are zero, and consequently the pyrenyl units are electronically decoupled. The ground state of the dianion $\mathbf{Bi}^{2-}$ is not a diamagnetic one but corresponds to a singlet diradical [40].

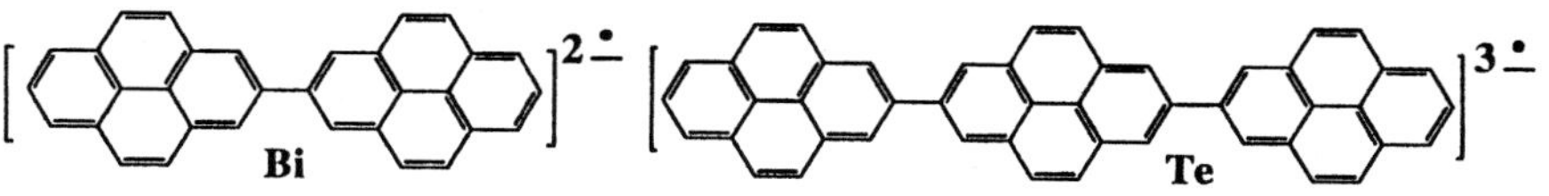

Scheme 12

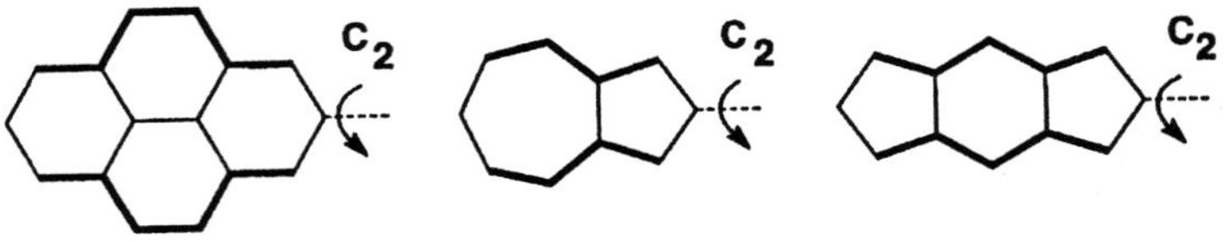

Scheme 13

The π-electron molecules, such as pyrene (**Py**), azulene (**Az**), and *s*-indacene (**I**), possess a symmetry axis C_2 crossing two π-centers (Scheme 13). For molecules of this class, the frontier MOs (HOMO and/or LUMO) are asymmetric with respect to rotation at C_2, and the MO coefficients of the AOs of the frontier MOs lying on the symmetry axis are zero. The MOs are localized within the fragments shown in Scheme 13. On this account it follows that at linking of the fragments **Q** such as **PY, AZ, I** with other π-fragments **R** such as *p*-phenylene, the corresponding frontier MOs of the polymers $(..[\mathbf{Q} - \mathbf{R}]_N^{\cdot - \text{ or } \cdot +}...)$ remain degenerate and localized within the **Q** fragment. Examples include the polymers shown in Scheme 14. The localization and nonoverlapping of the MOs result in vanishing of the direct exchange and the kinetic exchange interaction between the unpaired electrons. The indirect exchange of the unpaired electrons (spin polarization exchange) is antiferromagnetic, $J_{\text{ind}} < 0$ (Table 1).

VI. MOLECULAR RADICAL CRYSTALS

The nature of *intermolecular* magnetic interaction in molecular radical crystals (MRCs) is determined by the structure of the radicals and by the topology and the geometry of the 1-D stacks [41,42]. The monoradicals can be arranged in the MRCs in a full face-to-face stacking (**A**), a rotated and slipped face-to-face stacking (**B**), or a slipped face-to-face stacking (**C**),

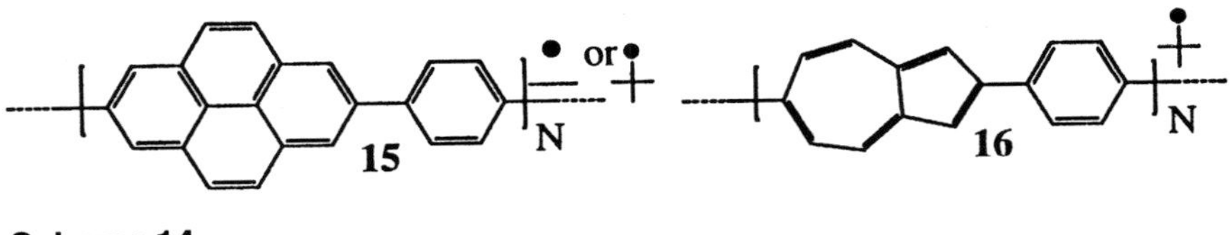

Scheme 14

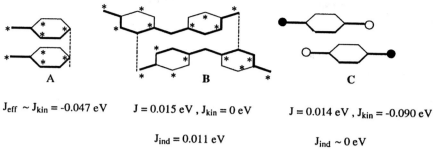

$J_{eff} \sim J_{kin} = -0.047$ eV $J = 0.015$ eV , $J_{kin} = 0$ eV $J = 0.014$ eV , $J_{kin} = -0.090$ eV

$J_{ind} = 0.011$ eV $J_{ind} \sim 0$ eV

Figure 2 (A) AFM stack consisting of benzyl radicals with full face-to-face arrangement. (B) FM rotated and slipped stack of a hydrocarbon with the same topology as the galvinoxyl radical (the exocyclic methylene groups are substituted by oxygen atoms). (C) AFM stack of Wurster's radicals in Wurster's red bromide [○ = NH_2, ● = $N(CH_3)_2$] with slipped face-to-face arrangement.

respectively [43–45] (see Figure 2). In all cases, the **A** stacking determines the AFM exchange interaction. The relatively large HFB width causes the large values of the kinetic term J_{kin}.

If the MRCs have a topology that corresponds to the CRLH (or generalized) theorem, the HFB consists of degenerate quantum states. The effective spin exchange is ferromagnetic, with a predominant contribution of the Coulomb exchange J and a partial contribution of the indirect exchange (model **B** in Figure 2) [44].

If the topology of the radical does not correspond to the CRLH (or generalized) theorem (model **C**), the AFM character of the interaction in the stacks is determined mainly by the kinetic spin exchange J_{kin}, and J and J_{ind} have small values. An example is the spin exchange within the HFB of infinite 1-D stacks of Wurster's radicals in Wurster's red bromide [45]. The value of J_{kin} is determined by the width of the HFB. The large width of the HFB ($\Delta\varepsilon \sim 1.3$ eV) is caused by the relatively short interplanar distance between the radicals in the stacks ($R_{exp} = 3.105$ Å) and the strong overlap between the radicals.

ACKNOWLEDGMENTS

This work was supported by the Volkswagen Stiftung, Projekt I/71 601 Ak, the Deutsche Forschungsgemeinschaft, and by the Alexander von Humboldt Stiftung (N.T.).

REFERENCES

1. Harrison WA. Solid State Theory. New York: McGraw-Hill, 1970.
2. Anderson PW. Theory of Magnetic Exchange Interaction: Exchange in Insulators and Semiconductors. In: Seitz F, Turnbul D, eds. Solid State Physics. New York: Academic Press, 1963, chap 14, pp 99–214.
3. Bulaewski LN. Quasihomopolar Electron States in Crystals and Molecules. Zh Eksp Theor Fis. 1966; 51:230–239. [Sov Phys-JETP. 1967; 24:154].
4. Klein DJ, Nelin CJ, Alexander S, Matsen FA. High-spin Hydrocarbons. J Chem Phys. 1982; 77:3101–3108.
5. Maynau D, Durand Ph, Daudey JP, Malreu JP. Direct Determination of Effective Hamiltonians by Wave-Operator Methods. II. Application to Effective-Spin Interaction in π-Electron Systems. Phys Rev. 1983; A28:3193–3206.
6. Whangbo M-H. Mott-Hubbard Condition for Electronic Localization in the Hartree-Fock Band Theory. J Chem Phys. 1979; 70:4963–4966.
7. Hubbard J. Electron Correlation in Narrow Energy Band. Proc Roy Soc London. 1963; A276:238–257; 1964; A277:401–419.
8. Ivanov CI, Tyutyulkov N, Karabunarliev S. On the Effective Spin Exchange in Non-Classical Polymers. J Magnet Magn Magn Mater. 1990; 92:171–179.
9. Tyutyulkov N, Karabunarliev S, Müllen K, Baumgarten M. A New Class of Narrow-band High-spin Organic Polymers. II. Polymers with Indirect Exchange Interaction. Synth Met. 1992; 52:71–85.
10. Iwamura H. High-spin Organic Molecules and Spin Alignment in Organic Molecular Assemblies. Adv Phys Org Chem. 1990; 26:179–253.
11. Dougherty DA. Progress Toward an All-organic Magnet. High-spin Molecules and Polymers. In: O'Connor CJ, ed. Research Frontiers in Magnetochemistry. Singapore, New Jersey: World Scientific, 1993, pp 327–349.
12. McConnell HM. Ferromagnetism in Solid Free Radicals. J Chem Phys. 1963; 39:1910.
13. Kollmar C, Kahn O. Is the McConnell Mechanism a Suitable Strategy for the Design of Molecular Ferromagnets? J Am Chem Soc. 1991; 113:7987–8005.
14. Strebel P, Soos ZG. Theory of Charge Transfer in Aromatic Donor-Acceptor Crystals. J Chem Phys. 1970; 53:4077–4090.
15. Longuet-Higgins HC. Resonance Structures and Molecular Orbitals in Unsaturated Hydrocarbons. J Chem Phys. 1950; 18:265–269.
16. Coulson CA, Longuet-Higgins HC. The Electronic Structures of Conjugated Systems. I. General Theory. Proc Roy Soc London. 1947; A191:39–66.
17. Tyutyulkov N, Bangov I. Electronic Structure of Some Hypothetical Polymeric Non-Classical Hydrocarbons. Compt rend Acad Bulg Sci. 1974; 27:1517–1519.
18. Tyutyulkov N, Schuster P, Polansky OE. Structure and Properties of Non-Classical Polymers. I. Band Structure of Non-Classical Polymers. Theor Chim Acta. 1983; 63:291–304.
19. Tyutyulkov N, Polansky OE, Schuster P, Karabunarliev S, Ivanov CI. Structure and Properties of Non-Classical Polymers. II. Band Structure and Spin Densities. Theor Chim Acta. 1985; 67:211–228.

20. Tyutyulkov N, Karabunarliev S, Ivanov CI. A Class of Narrow-Band Organic Polymers with a Ferromagnetic Ground State. Mol Cryst Liq Cryst. 1989; 176: 139–150.
21. Berson JA. A New Class of Non-Kekulé Molecules with Tunable Singlet-Triplet Energy Spacing. Acc Chem Res. 1997; 30:238–244.
22. McWeeny R, Sutcliffe BT. Methods of Molecular Quantum Mechanics. New York: Academic Press, 1976.
23. Baumgarten M, Müllen K, Tyutyulkov N, Madjarova G. On the Nature of the Spin Exchange Interaction in Poly(m-aniline). Chem Phys. 1993; 169:81–85.
24. Madjarova G, Baumgarten M, Müllen K, Tyutyulkov N. Structure, Energy Spectra and Magnetic Properties of Poly(m-aniline)s. Macromol Theory Simul. 1994; 3:803–815.
25. Tyutyulkov N, Dietz F, Müllen K, Kübel Ch. Structure and Properties of Non-classical Polymers. XI. Heteroatomic Analogues of Poly(1,3-phenylenemethylene)s. Bull Chem Soc Japan. 1997; 70:1517–1521.
26. Yoshisawa K, Tanaka K, Yamabe T, Yamauchi J. Ferromagnetic Interaction in poly(m-aniline): Electron Spin Resonance and Magnetic Susceptibility. J Chem Phys. 1992; 96:5516–5522.
27. Rajka A, Rajka S, Desai SR. Boron-Centered Diradical Dianion: A New Triplet State Molecule. J Chem Soc Chem Commun. 1995; 1957–1958.
28. Lu HSM, Berson JA. Catenation of Heterocyclic Non-Kekulé Biradicals to Tetraradicals. Prototypes of Conductive or Magnetic Polymers. J Am Chem Soc. 1997; 119:1428–1448.
29. Baumgarten M, Tyutyulkov N. Non-classical Conducting Polymers: New Approaches to Organic Metals. Chem Eur J. 1998; 4:987–989.
30. Sachs H. Beziehungen zwischen den in einen Graphen enthalteten Kreisen und seinem charakteristischen Polynom. Publikationes Mathematicae Debrecen. 1964; 11:119–134.
31. Tyutyulkov N, Dietz F, Müllen K, Baumgarten M, Karabunarliev S. Structure and Properties of Non-classical Polymers. IX. Non-alternant Polymers. Theor Chim Acta. 1993; 86:353–367.
32. Lahti PM. A Semiempirical Investigation of Interelectronic Exchange Coupling in Bisected Poly(1,4-phenylene) Polycation Model Systems. Int J Quantum Chem. 1992; 44:785–794.
33. Müllen K, Baumgarten M, Tyutyulkov N, Karabunarliev S. A New Class of Narrow-Band High-Spin Organic Polymers. I. Polymers with Direct Exchange Interaction Between Orthogonal π-Orbitals. Synth Met. 1991; 40:127–135.
34. Baumgarten M. Conjugation and High-Spin Formation in Radical Ions of Extended π-Systems. Acta Chem Scand. 1997; 51:193–198.
35. Fukutome H, Takahashi A, Ozaki M. Design of Conjugated Polymers with Polaronic Ferromagnetism. Chem Phys Lett. 1987; 133:34–38.
36. Tyutyulkov N, Dietz F, Müllen K, Baumgarten M. Energy Spectra and Magnetic Properties of Radialenes, Dendralenes and Their Polyradical Ions. A New Class of Polaronic High-Spin π-Systems. Chem Phys. 1994; 189:83–97.
37. Goodenough JB. Magnetism and the Chemical Bond. New York: Wiley, 1963.

38. Minato M, Lahti PM. Characterizing Triplet States of Quinoidal Dinitrenes as a Function of Conjugation Length. J Am Chem Soc. 1997; 119:2187–2195.

39. Tyutyulkov N, Madjarova G, Dietz F, Baumgarten M. Organic polymers with indirect magnetic interaction caused by the symmetry of the elementary units. Int J Quantum Chem. 1998; 66:425–434.

40. Kreynschmidt M, Baumgarten M, Tyutyulkov N, Müllen K. 2,2′-Bipyrenyl and *para*-Terpyrenyl—A New Type of Electronically Decoupled Oligoarylene. Angew Chem Int Ed. 1994; 33:1957–1959.

41. Yoshizawa K, Hoffmann R. The Role of Orbital Interactions in Determining Ferromagnetic Coupling in Organic Molecular Assemblies. J Am Chem Soc. 1995; 117:6921–6926.

42. Kollmar C, Kahn O. A Heisenberg Hamiltonian for Intermolecular Exchange Interaction: Spin Delocalization and Spin Polarization. J Chem Phys. 1993; 98:453–472.

43. Tyutyulkov N, Dietz F, Madjarova G. Nature of Magnetic Interaction in Organic Radical Crystals. I. Alternant Systems. Mol Cryst Liq Cryst. 1997; 305:249–258.

44. Dietz F, Tyutyulkov N, Baumgarten M. Nature of the Magnetic Interaction in Organic Radical Crystals. III. Galvinoxyl Radicals in 1-D Crystals. J Phys Chem B. 1998; 102:3912–3916.

45. Dietz F, Tyutyulkov N, Lüders K, Christen C. Nature of magnetic interaction of Wurster's radicals in the solid state. Chem Phys. 1997; 218:43–48.

19

Orbital Interactions Determining the Exchange Effects in Organic Molecular Crystals

Kazunari Yoshizawa
Kyoto University, Kyoto, Japan

I. INTRODUCTION

Ferromagnetic properties are traditionally exhibited by inorganic solids such as metals and alloys. The positive exchange interactions in such ferromagnetic materials can arise from the complicated nodal properties of d or f atomic orbitals at the magnetic centers. There has been much interest in the design and characterization of ferromagnetic materials based on extended arrays of pure organic and organometallic compounds [1–9]. Observations of bulk ferromagnetic properties in many kinds of nitroxide crystals [10–13] and TDAE-C_{60}, where TDAE is tetrakis(dimethylamino)ethylene [14–17], are of great interest because these compounds are viewed as molecular crystals composed of light elements such as H, C, N, and O that have only s and p atomic orbitals.

II. McCONNELL'S MODEL OF MAGNETIC INTERACTIONS

Although the relationship between through-bond magnetic interactions in organic materials and molecular topology is well understood from theoretical viewpoints [18–25], it is difficult to control magnetic interactions through the bulk. In 1963 McConnell [26] suggested that the exchange interaction in stacked aromatic radicals could be approximated by the following Heisenberg-type Hamiltonian:

$$H^{AB} = -\sum_{i,j} J_{ij}^{AB} S_i^A \cdot S_j^B \tag{1}$$

where J_{ij}^{AB} is the valence bond exchange integral between atom i of molecule A and atom j of molecule B. S_i^A is the π-electron spin density on atom i of molecule A, and S_j^B is that on atom j of molecule B. Equation (1) can be written in the form

$$H^{AB} = -S^A \cdot S^B \sum_{i,j} J_{ij}^{AB} \rho_i^A \rho_j^B \tag{2}$$

in which S^A and S^B are the total spin operators for A and B; ρ_i^A and ρ_j^B are the spin densities on atom i of molecule A and atom j of molecule B, respectively.

Since the exchange interaction between two aromatic radicals is a many-center problem, it can become positive in certain stacking modes. J_{ij}^{AB} is usually negative in a $p\sigma-p\sigma$ interaction, which would govern the intermolecular π-stacking of aromatic radicals, so that the effective exchange interaction between two aromatic radicals can be ferromagnetic if the spin-density product $\rho_i^A \rho_j^B(i \bullet j)$ is negative. Since in certain organic radicals there exist alternating large positive and negative spin densities, these radicals can in principle be ferromagnetically coupled through the exchange interaction between atoms of positive spin density and those of negative spin density.

In 1967 McConnell proposed a second model for ferromagnetic spin alignment [27]. Although our present study is concerned mainly with his first model, it is also an important strategy for intermolecular ferromagnetic coupling. McConnell's second proposal is that if an ionic charge-transfer pair (D^+A^-) could be built with a donor molecule whose neutral ground state is a triplet, then the $D^+A^-D^+A^- \bullet\bullet\bullet$ array would show ferromagnetic interaction, due to mixing of the charge-transfer state with the ground state. This model has been extended in various ways by Breslow [28,29].

The applicability of McConnell's first model of magnetic interaction through space was nicely confirmed in 1985 by Iwamura and collaborators [30,31] for the case of diphenylcarbenes with a [2.2]paracyclophane-type structure. From molecular orbital calculations, Yamaguchi et al. [32,33] showed that the effective exchange integral is positive for the *ortho-* and *para*-stacking modes of phenylcarbenes, while it is negative for the *geminal-* and *meta*-stacking modes. Moreover, Buchachenko [34] showed that the effective exchange integral is positive when the methyl radical is aligned with the central carbon of an allyl radical. Crystal orbital calculations of one-dimensional arrays of diphenylcarbene in various stacking modes were carried out by Tanaka et al. [35].

In addition to these quantum-chemical computations, important analytical studies of McConnell's first [26] and second [27] models for magnetic interaction have been published by Kollmar and Kahn [36–39]. The role of the spin-polarization effect in McConnell's models has been stressed by these authors. Moreover, Kollmar and Kahn [36] in their investigation of McConnell's second model showed that the singlet state may be stabilized, in contrast to McConnell's proposal, if higher-order charge-transfer mixing terms are taken into account.

We demonstrate that McConnell's first model of magnetic interaction through space may be easily interpreted on the basis of orbital interactions. There are important orbital interactions behind McConnell's first model. In this study we place special emphasis on a qualitative understanding of the ferromagnetic interactions in organic molecular assemblies, not on the quality of the computation.

III. INTERACTION BETWEEN TWO OPEN-SHELL MOLECULES

Let us first look at the intermolecular magnetic coupling based on McConnell's model, indicated in Eq. (2), between two open-shell molecules such as allyl and diphenylmethyl radicals. Allyl radical, an alternant hydrocarbon [40] (where the conjugated atoms are alternately labeled as "starred" and "unstarred," such that no two atoms of the same label are directly linked), is a classic spin-polarized system. As indicated in Scheme 1, in this radical there are substantial positive spin densities on the terminal starred carbons and small negative spin density on the central unstarred carbon. The length of the arrows is just an iconic representation of the magnitude of the density. This has been known for some time, and is due to the spin-polarization effect.

Now imagine a stack of allyl radicals. We assume a typical π-stacking geometry with the radicals in roughly parallel planes. The three configura-

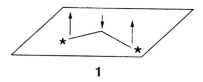

1

Scheme 1

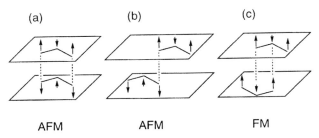

Figure 1 Ferromagnetic (FM) and antiferromagnetic (AFM) couplings of two allyl radicals based on a spin-polarization mechanism. (From ref. 51.)

tions shown in Figure 1 indicate approximately the topological range of contact geometries through singling out a pair of neighbor radicals. We see from Figure 1 that ferromagnetic coupling should appear when the large positive spin density on the atom of one molecule is coupled with the small negative spin density on the atom of another molecule, i.e., coupling mode (c). To put it another way, if starred atoms of one molecule are interacted with unstarred atoms of another molecule, the coupling between two radicals can be ferromagnetic. In the other cases, (a) and (b), McConnell's model indicates antiferromagnetic coupling.

Diphenylmethyl is another interesting radical [41], with the spin distribution shown in Scheme 2. It has large positive spin density on the central carbon and alternating positive and negative spin densities on the two phenyl rings. The real molecular structure does not have the phenyl rings coplanar with the central carbons, for steric reasons, so the planar system we discuss here serves as just a theoretical model.

Let us assume several stacking modes of two diphenylmethyl radicals. In each, one phenyl ring of one radical is in π-contact with its neighbor, but differing in the relative orientations of the remainder of the molecule. The disposition of neighbors in this model is governed by a rotational angle θ defined in Figure 2. Ferromagnetic coupling can appear if the stacking angle θ is 60° or 180°, while antiferromagnetic coupling is likely for θ = 0° and 120°. Also in these geometries, we find that two radicals can be ferromagnetically coupled when starred atoms of one molecule are interacted with unstarred atoms of another molecule.

The π-electronic structure of diphenylcarbene is very much like that of diphenylmethyl, as indicated in Scheme 3. Again, we assume the phenyl rings are coplanar with the carbene center; they are actually kept nearly so in the paracyclophanes for which this carbene is a model. The important triplet state of the carbene has one electron in the conjugated π-system, just

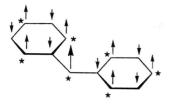

Scheme 2

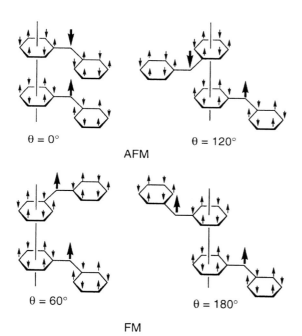

Figure 2 Ferromagnetic (FM) and antiferromagnetic (AFM) couplings of two diphenylmethyl radicals based on a spin-polarization mechanism. (From ref. 51.)

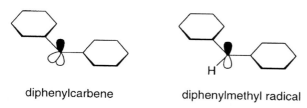

Scheme 3

as the diphenylmethyl radical. The carbenes also may be imagined to stack; the four kinds of stacking modes of two diphenylmethyl radicals shown in Figure 2 are thus also good models for the experiments of Iwamura et al. [30,31].

As already indicated, McConnell's first model is clearly based on the spin-polarization of π-electrons. The importance of spin polarization was stressed in a careful analysis of the zero-field splittings of triplet trimethyl-enemethane by McConnell himself [42]. Dowd [43] confirmed this effect and observed large electron spin resonance (ESR) proton splittings related to the negative spin density on the central carbon of trimethylenemethane. Ramasesha and Soos [44] successfully evaluated negative spin densities using diagrammatic valence bond theory. A study by Zheludev et al. [45] showed that the density functional method can predict well the experimentally determined (though small) negative spin density at the unstarred atoms. In this way the spin-polarization mechanism provides us with a useful strategy for predicting intra- and intermolecular magnetic coupling.

IV. NODAL PROPERTIES OF NONBONDING MOLECULAR ORBITALS

Before discussing the role of orbital interactions for the exchange effects in organic molecular assemblies, let us look at the interesting properties of nonbonding molecular orbitals (NBMOs) in π-conjugated systems, with energy α corresponding to the energy of the $2p$ atomic orbital (AO) of an isolated carbon atom. We must usually solve the Hückel MO equations to find orbital coefficients, but NBMOs are an exception. Here one can find the orbital coefficients by an extremely simple procedure, known as the "zero-sum rule" [40]. Consider the equations for the coefficients in an MO of an alternant hydrocarbon. In the case of an NBMO, the equations become

$$\sum_{k}^{\langle i \rangle} c_{0k} = 0 \tag{3}$$

in which $\langle i \rangle$ specifies the summation of sites k directed to site i and c_{0k} NBMO coefficients. This relation can be conveniently used to estimate NBMO coefficients, as the following example shows.

Let us determine the coefficients of the NBMO of benzyl radical. Suppose we denote the NBMO coefficient of atom 2 by c, as shown in Figure 3(1), then we take atom 3 to be atom i in the context of Eq. (3). Equation (3) tells us that the sum of NBMO coefficients of atoms adjacent to atom 3 must vanish; therefore the coefficient at position 4 must be $-c$. Repeating

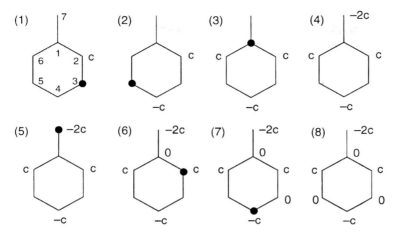

Figure 3 Calculation of NBMO coefficients of benzyl radical.

this process, we find that the coefficient of atom 6 is c, and that of the methylene group $-2c$. Next we can go through the same procedure, starting with atom 1. The sum of coefficients of atoms adjacent to atom 7 must vanish. Thus the coefficient of atom 1 is 0 because the only atom adjacent to atom 7 is atom 1. Following through the same argument as before, we find that the coefficients of atoms 3 and 5 are also 0. The final coefficients are now known in terms of c, as indicated in Figure 3(8). To determine c, we normalize the NBMO:

$$c^2 + (-c)^2 + c^2 + (-2c)^2 = 7c^2 = 1$$

Thus, c is $1/\sqrt{7}$. In this way, one can easily obtain NBMOs of odd-numbered alternant hydrocarbons.

It is important to note that the NBMO coefficients at all the unstarred positions in benzyl vanish, as indicated in Scheme 4. Thus in the NBMO there are nodes at all the unstarred positions. We will show that such nodal properties of the NBMOs play an important role in intermolecular magnetic couplings.

According to the well-known theorem of Coulson and Longuet-Higgins [40], an alternant hydrocarbon with n starred atoms and m unstarred atoms has $(n - m)$ NBMOs, which are composed entirely of AOs of starred atoms, and a neutral alternant hydrocarbon with n starred atoms and m unstarred atoms will exist as an $(n - m)$fold radical, having $(n - m)$ unpaired π electrons occupying the $(n - m)$ NBMOs. Therefore m-benzoquinodimethane is predicted to be a ground state triplet, from Hund's rule [46]. The

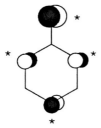

Scheme 4

two NBMOs, which are degenerate within the framework of the Hückel MO theory, are shown in Scheme 5. These NBMOs are also composed of AOs of only starred atoms. There are nodes at all the unstarred positions of these NBMOs.

The reason why *m*-benzoquinodimethane is a ground state triplet derives from the *nondisjoint* character of the π NBMOs of this molecule. As explained by Borden and Davidson [21], when NBMOs are nondisjoint, i.e., while orthogonal they are very much coextensive in space, a ground state triplet can be expected. This is because exchange interactions are large in such a case. In some other systems, the NBMOs *can be chosen* in such a way that they are *disjoint*; i.e., they do not share any atoms [21]. For such molecules, to a first approximation singlet and triplet states are degenerate, since the exchange interactions are small. The best example of a molecule with disjoint NBMOs is probably that of square cyclobutadiene, another is biallyl (tetramethyleneethane, TME).

We can construct a potential linear-chain organic ferromagnet by extending triplet *m*-benzoquinodimethane and its analogs [47–50]. The magnetic interactions in such one-dimensional ferromagnets have been studied with great interest. However, our interest in this contribution is in exchange interactions in three-dimensional organic solids, so we will not discuss at this time the interesting magnetic properties of such polymeric systems.

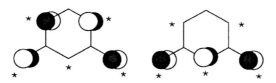

Scheme 5

V. ORBITAL INTERACTIONS IN FERROMAGNETIC COUPLING

Let us now consider intermolecular magnetic coupling in organic molecular assemblies in terms of orbital interaction [51]. Suppose one has two singly occupied molecular orbitals (SOMOs) a and b, which interact weakly. The well-known Heitler–London type of singlet and triplet wavefunctions [52] are written in the form of Eq. (4):

$$^1\Psi(1,2) = \frac{a(1)b(2) + b(1)a(2)}{\sqrt{2 + 2S_{ab}^2}} \frac{1}{\sqrt{2}} \{\alpha(1)\beta(2) - \beta(1)\alpha(2)\}$$

$$^3\Psi(1,2) = \frac{a(1)b(2) - b(1)a(2)}{\sqrt{2 - 2S_{ab}^2}} \left\{ \begin{array}{c} \alpha(1)\alpha(2) \\ \dfrac{\alpha(1)\beta(2) + \beta(1)\alpha(2)}{\sqrt{2}} \\ \beta(1)\beta(2) \end{array} \right.$$

(4)

where S_{ab} is the overlap between a and b. The corresponding total energies are

$$^1E = \frac{\{h_{aa} + h_{bb} + 2S_{ab}h_{ab} + (aa|bb) + (ab|ba)\}}{(1 + S_{ab}^2)}$$

$$^3E = \frac{\{h_{aa} + h_{bb} - 2S_{ab}h_{ab} + (aa|bb) - (ab|ba)\}}{(1 - S_{ab}^2)}$$

(5)

where

$$h_{ab} = \int a^*(1)h(1)b(1)\, d\tau(1)$$

and the Coulomb integral $(aa|bb)$ and exchange integral $(ab|ba)$ are

$$(aa|bb) = \int\int a^*(1)a(1)\frac{1}{r_{12}}b^*(2)b(2)\, d\tau(1)\, d\tau(2)$$

$$(ab|ba) = \int\int a^*(1)b(1)\frac{1}{r_{12}}b^*(2)a(2)\, d\tau(1)\, d\tau(2).$$

The triplet–singlet separation is then

$$^1E - {}^3E = \frac{1}{1 - S_{ab}^4}\{-2h_{aa}S_{ab}^2 - 2h_{bb}S_{ab}^2$$

$$+ 4h_{ab}S_{ab} - 2(aa|bb)S_{ab}^2 + 2(ab|ba)\}$$

(6)

Since the intermolecular overlap S_{ab} is small,

$$^1E - {}^3E \cong -2(h_{aa} + h_{bb})S_{ab}^2 + 4h_{ab}S_{ab} - 2(aa|bb)S_{ab}^2 + 2(ab|ba) \qquad (7)$$

This equation provides us with the exchange constant (J) between two open-shell molecules A and B in the context of a Heisenberg-type Hamiltonian; triplet coupling ($J > 0$) is the necessary precondition for ferromagnetism. The first term in Eq. (7) is positive. On the other hand, the sign of the second term is minus in general. That term is also of the order S_{ab}^2, if we assume the Wolfsberg–Helmholz relation of Eq. (8):

$$h_{ab} = KS_{ab}(h_{aa} + h_{bb}) \qquad (8)$$

K is an adjustable parameter; within the widely used extended Hückel theory [53], this expression has found some success, with K typically 1.75. Assuming the extended Hückel method's value of K, we obtain a very simple expression for the exchange constant between two open-shell molecules in terms of the SOMO–SOMO interactions:

$$J \cong 5(h_{aa} + h_{bb})S_{ab}^2 - 2(aa|bb)S_{ab}^2 + 2(ab|ba) \qquad (9)$$

This formulation is of course a result of neglecting ionic configurations, which may stabilize a singlet state in delocalized π-electronic systems, as discussed by Girerd et al. [54]. However, the contribution of ionic configurations seems small in intermolecular interactions, as suggested in several treatments of the hydrogen molecule [55].

The first and second terms of Eq. (9) are typically negative, thus favoring antiferromagnetic interaction. The third term may lead to net ferromagnetic coupling. The sign of J ultimately depends on magnitudes of h_{aa}, h_{bb}, S_{ab}^2, $(aa|bb)$, and $(ab|ba)$. We think that one may control S_{ab}^2 most easily among these parameters by considering the geometrical arrangement of molecules, and this will ultimately form the basis of our analysis.

The first consequence of the general expression we have derived is that the triplet state is stabilized when the overlap is small. If orbital overlap is large, inevitably a singlet state will be prevail. A precondition for effective ferromagnetic coupling between two open-shell molecules is to arrange the molecules in such a way that the two SOMOs are orthogonal or as nearly so as possible. In traditional inorganic magnets, the nodal properties of d or f atomic orbitals would lead to the cancellation of overlap in their crystals. Since organic molecules have at most $2p$ atomic orbitals, the detailed geometrical arrangement of the SOMOs is important for the cancellation of overlap.

Second, to achieve the situation of a triplet well below a singlet, it is essential that the interacting orbitals be noded [56], and arranged in very specific ways. It is best if the relevant orbitals are in same region of space (as in a carbene), so as to maximize the exchange integral $(ab|ba)$. To put it simply, the SOMOs of two open-shell molecules must be highly over-lapped, although their net overlap must cancel. Skillful control of geometry (assuming that control is in our hands) is needed to engineer this. 1s-type wavefunctions cannot lead to a triplet state below a singlet state, because they have no nodes [57].

In Figure 4 we show important interactions between two AOs in order to clarify what we just stated. These illustrations show several pairs of AOs, in which the orbital overlaps strictly vanish due to the nodal properties in their specific configurations. Thus, the exchange constant J in the context of Eq. (9) is clearly positive when each orbital of the pair is filled by a single electron. Although in these examples the sign of J is independent of the distance between the paired orbitals, its magnitude strongly depends on the distance. These orbitals should be highly overlapped in same region of space in order to maximize the exchange integral $(ab|ba)$, i.e., J.

The general "two-electron, two-level" problem, in the context of sta-bilizing singlet states, has been discussed by Hoffmann and collaborators [58–63]. If SOMO a of molecule A is degenerate with SOMO b of molecule B, the MOs after interaction are as defined in Eq. (10):

$$\varphi_+ = \frac{a + b}{\sqrt{2 + 2S_{ab}}}$$

$$\varphi_- = \frac{a - b}{\sqrt{2 - 2S_{ab}}} \tag{10}$$

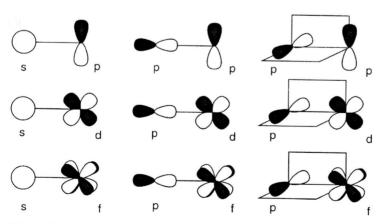

Figure 4 Pairs of atomic orbitals (AOs) leading to zero overlap.

These MOs have first-order one-electron energies given by

$$\varepsilon_+ = \varepsilon_0 + (H'_{ab} - \varepsilon_0 S_{ab})$$

$$\varepsilon_- = \varepsilon_0 - (H'_{ab} - \varepsilon_0 S_{ab}) \tag{11}$$

where H'_{ab} is the perturbation matrix element, approximately proportional to the overlap in general. H'_{ab} has the opposite sign to the overlap; that is, H'_{ab} is negative for positive S_{ab}, positive for negative S_{ab} [62]. This of course means that positive overlap implies stabilization or bonding. We can choose as a primary measure of orbital interaction the energy splitting between two orbitals after interaction compared with that before interaction. When S_{ab} is zero or nearly zero, however, ε_+ and ε_- are degenerate even after interaction; therefore the triplet state can lie below the singlet state, as Hund's rule implies [46]. A discussion of the singlet-triplet splitting on the basis of a localized MO picture has been given [64].

VI. EXCHANGE EFFECTS IN MOLECULAR ASSEMBLIES

The SOMOs of odd-numbered alternant hydrocarbons are, within the Hückel MO theory, NBMOs. The most interesting property of the NBMO is that the NBMO coefficients at all the unstarred positions are zero; thus in the NBMO there are nodes at all the unstarred positions, as mentioned earlier. To put it in another way, the signs of the coefficients at the starred positions flanking both sides of every unstarred position are always plus and minus, respectively. This leads to interesting intermolecular magnetic coupling, as described later.

Let us first look at the orbital interactions between two allyl radicals. Figure 5 shows the three types of stacking modes between the two SOMOs of allyl radicals a and b. φ_+ and φ_- signify, respectively, the bonding and antibonding combinations between the two SOMOs after interaction. In stacking modes (a) and (b), the overlap is not zero; therefore the wavefunction φ_+ is stabilized and φ_- is destabilized, following Eq. (11). We can gauge the energy splitting between the two orbitals after interaction on the basis of the orbital phases or orbital coefficients; since the two SOMOs are maximally overlapping in mode (a), whereas their overlap is small in (b), the energy splitting should be larger in (a), as shown in Figure 5. Under certain specific symmetry conditions (in a spiro system) there may also occur other small through-space interactions between allyls [58].

On the other hand, in stacking mode (c) (see Figure 5) the overlap between the two SOMOs almost vanishes, at least in the Hückel approximation. We see that the node on the central carbon in the SOMO of allyl radical plays an important role in the cancellation of the partial overlaps in

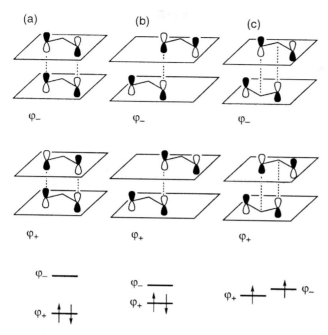

Figure 5 Orbital interactions in the magnetic coupling of two allyl radicals. (From ref. 51.)

this stacking mode. The energy splitting after interaction is therefore nearly zero. As a result, the triplet state should lie below the corresponding singlet state, following Hund's rule [46].

Fragment molecular orbital (FMO) analyses, based on the extended Hückel method [53], for stacking modes (a) and (c) of allyl dimer are shown in Figure 6. We do not always need to know the details of the electronic structures of the fragments; it is sufficient that we look at the frontier orbitals of the fragments and their reconstructions. The SOMO–SOMO overlap is almost canceled in stacking mode (c), so the HOMO–LUMO gap in the united molecule is small, in which HOMO and LUMO are highest occupied and lowest unoccupied molecular orbitals, respectively. Therefore ferromagnetic coupling is expected to appear in the stacking mode.

Next we consider the orbital interactions between two diphenylmethyl radicals, a more realistic model for the experiments of Iwamura et al. [30,31]. We can easily find the NBMO coefficients from the zero-sum rule mentioned earlier. The amplitude of the Hückel SOMO of diphenylmethyl radical is, of course, largest at the central carbon and smaller at the alternate

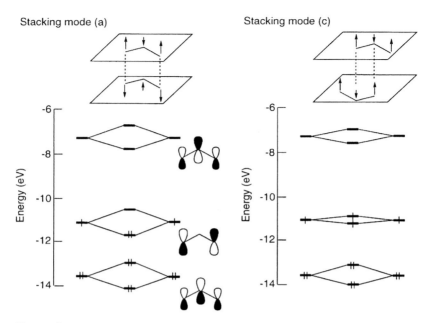

Figure 6 FMO (fragment molecular orbital) analyses for the couplings of two allyl radicals.

starred carbons on the two benzene rings. Two types of antiferromagnetic couplings between the SOMOs are illustrated in Figure 7. In these the wave functions, φ_+ and φ_- are stabilized and destabilized, respectively. For a stacking rotation angle of $\theta = 0°$ the overlap is complete, especially at the central carbons. Therefore the energy splitting at $\theta = 0°$ should be larger than that at $\theta = 120°$.

On the other hand, ferromagnetic coupling is favored at $\theta = 60°$ and 180° (as shown in Figure 8), for at these θ, φ_+ and φ_- are nearly degenerate. We see again that the partial overlaps are almost canceled in these stacking modes, due to the nodal structure of the SOMO [65]. Thus control of molecular arrangement is important for creating ferromagnetic intermolecular coupling.

The through-space interaction between the central carbons of both molecules cannot be entirely neglected at $\theta = 60°$, as indicated in Figure 8. As a result there is some small bonding in the stacking mode at $\theta = 60°$, and a consequent magnetic level splitting. The ideal ferromagnetic stacking in this case is for $\theta = 180°$. In the stacking of diphenylcarbenes, the observed quintet-triplet splitting at $\theta = 180°$ is 110 cm^{-1}, while that at $\theta = 60°$ is

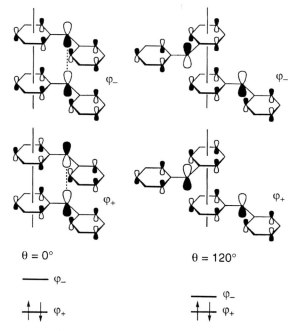

$\theta = 0°$

$\theta = 120°$

Figure 7 Orbital interactions in the magnetic coupling of diphenylmethyl radicals, for $\theta = 0°$ and $120°$. (From ref. 51.)

reduced to 60 cm^{-1} [30,31]. These experimental results can now be easily understood on the basis of our qualitative analysis of the orbital interactions.

A Walsh-type diagram for energy of the frontier orbitals of the diphenylmethyl dimer is shown in Figure 9, as a function of rotation angle θ. The distance between the stacked benzene rings is assumed as 2.5 Å in this dimer model. As clearly seen from this diagram, the energies of the HOMO and LUMO are largely dependent on θ. When $\theta = 0°$, the HOMO and LUMO are split substantially in energy because the SOMO–SOMO overlap is large, as shown in Figure 7 (left). Therefore we can expect that the singlet state is well below the corresponding triplet state in this stacking mode. On the other hand, when θ is nearly equal to 50° or 160°, the HOMO–LUMO gap is extremely small so that ferromagnetic coupling can appear in these stacking modes. When $\theta = 120°$, the HOMO–LUMO gap is not small. Thus this configuration is likely to favor an antiferromagnetic interaction.

Let us next consider the through-space magnetic coupling between two nitroxides, because several organic molecular ferromagnets composed of light elements have been realized using this moiety [10–13]. The first ex-

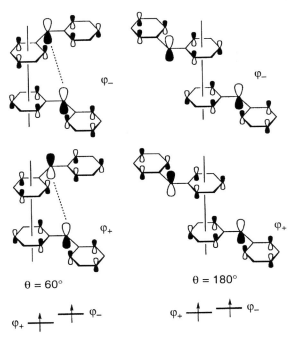

Figure 8 Orbital interactions in the magnetic coupling of diphenylmethyl radicals, for $\theta = 60°$ and $180°$. (From ref. 51.)

ample of an organic ferromagnet, discovered by Kinoshita and collaborators [10,11], is the orthorhombic β-phase crystal of p-nitrophenyl nitronyl nitroxide, **6** (Scheme 6), which exhibits a T_c of 0.6 K. A slightly higher T_c of 1.48 K was observed in the crystal structure of an adamantane-type dinitroxide, **7** [12]. Dipole–dipole interactions [2], not involving the overlap of orbitals, have been suggested as a possible mechanism to achieve the weak, cooperative magnetic behavior in the vicinity of 1 K in the nitroxide crystals. However, such dipole–dipole interactions are estimated to afford a T_c of one order of magnitude smaller than 1 K. Thus, we consider that such low T_c's should be ascribed to weak exchange interactions in the nitroxide crystals.

There is a variety of potentially coupled orientations of nitroxides in discrete molecules or extended structures; perhaps it is easiest to define the extremes once again by a rotational angle θ, as in Scheme 7. Yamaguchi et al. [66] calculated with the UHF-based Møller–Plesset perturbation method that the effective exchange integral is negative in the *syn* stacking mode whereas it is positive in a specific stacking mode.

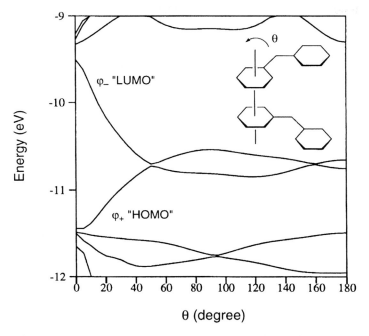

Figure 9 Walsh diagram for the frontier orbitals of diphenylmethyl dimer as a function of θ (degrees). The distance between the stacked benzene rings is 2.5 Å.

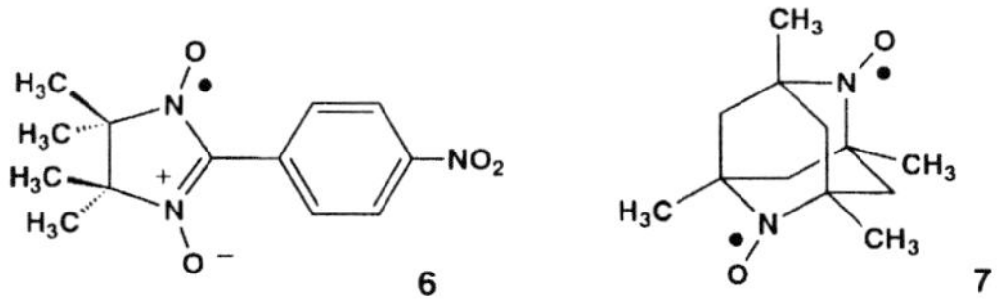

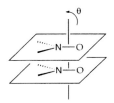

Scheme 6

Scheme 7

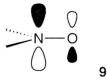

9

Scheme 8

The characteristic feature of nitroxides is that the SOMO is an anti-bonding π^*; i.e., the MO housing the odd electron has a node between nitrogen and oxygen, as indicated in Scheme 8. The SOMO of nitroxide may be written as $c_N c_N - c_O c_O$, where c_N and c_O are $2p$ atomic orbitals of nitrogen and oxygen, respectively; c_N and c_O are positive orbital coefficients. Since oxygen is more electronegative than nitrogen, the p orbital is localized more on oxygen, and the π^* here relevant more on nitrogen. Thus in the π^* SOMO $c_N > c_O$.

Consider the two extremes of nitroxide stacking shown in Figure 10. Clearly, the overlap is large in the *syn* stacking mode ($\theta = 0°$), leading to a level splitting. The singlet state will be consequently stabilized, compared to the corresponding triplet state, in this geometry. On the other hand, the radical–radical overlap in the *anti* stacking mode ($\theta = 180°$) has σ and π

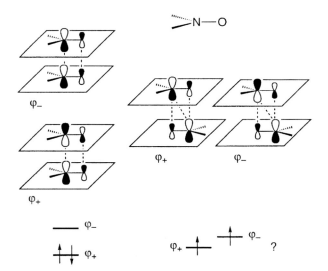

Figure 10 Orbital interactions in *syn* and *anti* stacking modes of nitroxides. (From ref. 51.)

components in it; the "cross" N–N interactions (see Figure 10) cannot be neglected, since $c_N > c_O$. The N–N ($\sigma + \pi$) interactions are of opposite sign to two N–O (both σ) interactions; the net overlaps are likely to be smaller.

We may estimate the actual SOMO–SOMO overlaps by means of an FMO analysis, based on the extended Hückel method [53]. Figure 11 shows the computed S_{ab}^2 of the SOMOs of two nitroxides as a function of stacking angle θ. In fact the overlaps are very small in actual molecular crystals, in which individual molecules are separated by at least a typical π van der Waals contact of 3.05 Å. S_{ab}^2 is maximized at $\theta = 0°$ (*syn* stacking) and minimized at $\theta = 110°$. S_{ab}^2 at $\theta = 180°$ (*anti* stacking) is about one-fourth of that in *syn* stacking, as expected from a qualitative MO analysis, described earlier. Thus, the overlap becomes small in the *anti* stacking mode of two nitroxides, due to the nodal properties of the π^* SOMO. In the *syn* stacking mode, in contrast, it is large.

Ferromagnetic coupling may have a chance of being realized in the *anti* stacking mode of nitroxides. It could be optimized still further; our considerations suggest, if one could engineer a stacking with θ near 110°.

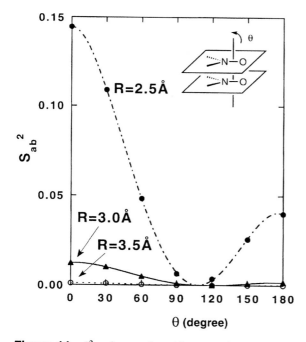

Figure 11 S_{ab}^2 of two nitroxides as a function of stacking rotation angle θ (degrees). (From ref. 51.)

One might note that this type of magnetic interaction cannot easily be rationalized by McConnell's model, because in the N—O bond there is no spin polarization, on which McConnell's model is based.

A Walsh-type diagram for energy of the frontier orbitals of the nitroxide dimer is shown in Figure 12, as a function of rotation angle θ. The distance between the stacked nitroxide radicals is assumed as 3.0 Å in this dimer model. As clearly seen from this diagram, the energies of the HOMO and LUMO are largely dependent on θ. When $\theta = 0$, the HOMO and LUMO are split substantially in energy because the SOMO–SOMO overlap is large, as mentioned earlier. Thus we can expect that the singlet state lies well below the corresponding triplet state in this stacking mode. On the other hand, when θ is nearly equal to 110°, the HOMO–LUMO gap is actually zero so that ferromagnetic coupling can appear in this stacking mode.

As already mentioned, Rassat et al. [12] prepared and characterized the adamantane-type dinitroxide N,N'-dioxy-1,3,5,7-tetramethyl-2,6-diazaadamantane **7**, which shows the highest Curie temperature (1.48 K) known

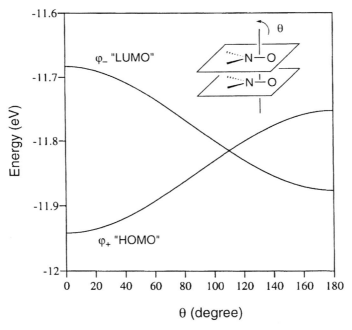

Figure 12 Walsh diagram for the frontier orbitals of nitroxide dimer as a function of θ (degrees). The distance between the stacked nitroxides is 3 Å.

to date within the ferromagnetic nitroxide family. This molecule is nicely designed because the two nitroxides are arranged in such a way that they are orthogonal within a single molecule, producing triplet coupling. In addition to this *intra*molecular ferromagnetic coupling, two nitroxides are coupled in a nearly *anti* stacking mode between neighboring molecules. The four methyl groups attached to the adamantane skeleton shown in **7** lead to this proper stacking mode in the crystal structure. Thus the interesting bulk magnetic properties of this material may be ascribed to both the *intra-* and *inter*molecular ferromagnetic couplings.

Finally let us refer to the exchange interaction in the charge-transfer TDAE-C_{60} crystal, which shows a T_c of 16 K [14–17]. Recently the J value between two $C_{60}-$ anions was computed to be strongly dependent on their relative orientations [67]. The SOMO of $C_{60}-$, which exhibits complicated nodal properties, is one of the threefold degenerate t_{1u} LUMO in the Jahn –Teller distorted D_{5d}, D_{3d}, and C_{2h} symmetry structures [68]. The orientational dependence in the exchange interaction is a direct consequence of SOMO–SOMO interactions between the two anions. The observed ferromagnetic properties in TDAE-C_{60} can be derived from the specific orbital interactions.

VII. CONCLUSIONS

This chapter has been concerned with the qualitative MO analysis of ferromagnetic coupling in molecular assemblies, especially in organic materials. We have demonstrated that McConnell's model of through-space magnetic interaction, based on spin polarization, may also be interpreted from a starting point of specific SOMO–SOMO interactions. The nodal properties of the SOMOs are essential for the cancellation of partial overlaps, creating the necessary preconditions for maximizing ferromagnetic intermolecular coupling. We think that the concept of zero or minimal overlap between the SOMOs should play a more fundamental role in the optimization of ferromagnetism. The well-known spin-polarization mechanism and our orbital interaction analysis are complementary viewpoints in the analysis of magnetic coupling of molecules. For realization of higher Curie temperature in molecular systems, the SOMOs must be highly overlapped in space to increase exchange interactions, while their net overlap must be nearly zero. To achieve and optimize interesting magnetic properties one must control the geometrical arrangement of molecules by introducing appropriate substituents, taking proper cognizance of the nodal properties of the SOMOs.

Table 1 Extended Hückel Parameters for
H, C, N, and O Atoms[a]

Atomic orbital	H_{ii} (eV)	ζ
H-1s	-13.6	1.3
C-2s	-21.4	1.625
C-2p	-11.4	1.625
N-2s	-26.0	1.950
N-2p	-13.4	1.950
O-2s	-32.3	2.275
O-2p	-14.8	2.275

[a]H_{ii} is orbital energy and ζ is the Slater exponent.

ACKNOWLEDGMENTS

The largest part of this study was performed at Cornell University, with Professor Roald Hoffmann. It is a pleasure to acknowledge helpful discussions with him concerning orbital interactions—a vantage point in qualitative molecular orbital theory.

APPENDIX

Fragment molecular orbital (FMO) and Walsh analyses, based on the extended Hückel method, were carried out using YAeHMOP developed by Dr. Greg Landrum [69]. Parameters used for H, C, N, and O atoms are listed in Table 1.

REFERENCES

1. Miller JS, Epstein AJ, Reiff WM. Ferromagnetic molecular charge-transfer complexes. Chem Rev. 1988; 88:201–220.
2. Miller JS, Epstein AJ. Organic and organometallic molecular magnetic materials—designer magnets. Angew Chem Int Ed Engl. 1994; 33:385–415.
3. Caneschi A, Gatteschi D, Sessoli R. Toward molecular magnets: The metal-radical approach. Acc Chem Res. 1989; 22:392–398.
4. Iwamura H. High-spin organic molecules and spin alignment in organic molecular assemblies. Adv Phys Org Chem. 1990; 26:179–253.
5. Dougherty DA. Spin control in organic molecules. Acc Chem Res. 1991; 24:88–94.
6. Kahn O. Molecular Magnetism. New York: VCH, 1993.

7. Borden WT, Iwamura H, Berson JA. Violation of Hund's rule in non-Kekulé hydrocarbons: Theoretical prediction and experimental verification. Acc Chem Res. 1994; 27:109–116.

8. Rajca A. Organic diradicals and polyradicals: From spin coupling to magnetism? Chem Rev. 1994; 94:871–893.

9. Berson JA. A new class of non-Kekulé molecules with tunable singlet-triplet energy spacings. Acc Chem Res. 1997; 30:238–244.

10. Kinoshita M, Turek P, Tamura M, Nozawa K, Shiomi D, Nakazawa Y, Ishikawa M, Takahashi M, Awaga K, Inabe T, Maruyama Y. An organic radical ferromagnet. Chem Lett. 1991; 1225–1228.

11. Tamura M, Nakazawa Y, Shiomi D, Nozawa K, Hosokoshi Y, Ishikawa M, Takahashi M, Kinoshita M. Bulk ferromagnetism in the β-phase crystal of the p-nitrophenyl nitronyl nitroxide radical. Chem Phys Lett. 1991; 186:401–404.

12. Chirarelli R, Novak MA, Rassat A, Tholence JL. A ferromagnetic transition at 1.48 K in an organic nitroxide. Nature. 1993; 363:147–149.

13. Tomioka K, Mitsubori S, Ishida T, Nogami T, Iwamura H. Intermolecular ferromagnetic interaction of 4-benzylideneamino-2,2,6,6-tetramethylpiperidin-1-oxyl. Chem Lett. 1993; 1239–1242.

14. Allemand P-M, Khemani KC, Koch A, Wudl F, Holczer K, Donovan S, Grüner G, Thompson JD. Organic molecular soft ferromagnetism in a fullerene C_{60}. Science. 1991; 253:301–303.

15. Stephens PW, Cox D, Lauher JW, Mihaly L, Wiley JB, Allemand P-M, Hirsch A, Holczer K, Li Q, Thompson JD, Wudl F. Lattice structure of the fullerene ferromagnet TDAE-C_{60}. Nature. 1992; 355:331–332.

16. Tanaka K, Zakhidov AA, Yoshizawa K, Okahara K, Yamabe T, Yakushi K, Kikuchi K, Suzuki S, Ikemoto I, Achiba Y. Magnetic properties of TDAE-C_{60} and TDAE-C_{70}. A comparative stdy. Phys Lett A. 1992; 164:221–226.

17. Tanaka K, Zakhidov AA, Yoshizawa K, Okahara K, Yamabe T, Yakushi K, Kikuchi K, Suzuki S, Ikemoto I, Achiba Y. Magnetic properties of TDAE-C_{60} and TDAE-C_{70}, where TDAE is tetrakis(dimethylamino)ethylene. Phys Rev B. 1993; 47:7554–7559.

18. Longuet-Higgins HC. Some studies in molecular orbital theory I. Resonance structures and molecular orbitals in unsaturated hydrocarbons. J Chem Phys. 1950; 18:265–274.

19. Higuchi J. Electron spin–spin interaction in higher molecular spin multiplets. J Chem Phys. 1963; 39:1847–1851.

20. Mataga N. Possible ferromagnetic states of some hypothetical hydrocarbons. Theor Chim Acta. 1968; 10:372–376.

21. Borden WT, Davidson ER. Effects of electron repulsion in conjugated hydrocarbon diradicals. J Am Chem Soc. 1977; 99:4587–4594.

22. Ovchinnikov AA. Multiplicity of the ground state of large alternant organic molecules with conjugated bonds. Theoret Chim Acta. 1978; 47:297–304.

23. Döhnert D, Koutecky J. Occupation numbers of natural obitals as a criterion for biradical character. Different kinds of biradicals. J Am Chem Soc. 1980; 102:1789–1796.

24. Klein DJ, Nelin CJ, Alexander S, Matsen FA. High-spin hydrocarbons. J Chem Phys. 1982; 77:3101–3108.

25. Maynau D, Said M, Malrieu JP. Looking at chemistry as a spin ordering problem. J Am Chem Soc. 1983; 105:5244–5252.

26. McConnell HM. Ferromagnetism in solid free radicals. J Chem Phys. 1963; 39:1910.

27. McConnell HM. Proc Robert A. Welch Found Conf Chem Res. 1967; 11:144.

28. Breslow R. Stable 4n pi-electron triplet molecules. Pure Appl Chem. 1982; 54: 927–938.

29. Breslow R, LePage TJ. Charge-transfer complexes as potential organic ferromagnets. J Am Chem Soc. 1987; 109:6412–6421.

30. Sugawara T, Tukada H, Izuoka A, Murata S, Iwamura H. Magnetic interaction among diphenylmethylene molecules generated in crystals of some diazodiphenylmethanes. J Am Chem Soc. 1986; 108:4272–4278.

31. Izuoka A, Murata S, Sugawara T, Iwamura H. Molecular design and model experiments of ferromagnetic intermolecular interaction in the assembly of high-spin organic molecules. Generation and characterization of the spin states of isomeric bis(phenylmethyl)[2,2]paracyclophane. J Am Chem Soc. 1987; 109:2631–2639.

32. Yamaguchi K, Toyoda Y, Fueno T. Ab initio calculations of effective exchange integrals for triplet carbene clusters. Importance of stacking modes for ferromagnetic interactions. Chem Phys Lett. 1989; 159:459–464.

33. Yamaguchi K, Fueno T. An effective spin Hamiltonian for clusters of organic radicals. Application to allyl radical clusters. Chem Phys Lett. 1989; 159: 465–471.

34. Buchachenko AL. Organic and molecular ferromagnetics: advances and problems. Russ Chem Rev. 1990; 59:307–319.

35. Tanaka K, Takeuchi T, Yoshizawa K, Toriumi M, Yamabe T. Theoretical study on ferromagnetic interaction in stacked diphenylcarbene polymer. Synth Met. 1991; 44:1–8.

36. Kollmar C, Kahn O. Is the McConnell mechanism a suitable strategy for the design of molecular ferromagnets? J Am Chem Soc. 1991; 113:7987–7994.

37. Kollmar C, Couty M, Kahn O. A mechanism for the ferromagnetic coupling in decamethylferrocenium tetracyanoethenide. J Am Chem Soc. 1991; 113: 7994–8005.

38. Kollmar C, Kahn O. A Heisenberg Hamiltonian for intermolecular exchange interaction: Spin delocalization and spin polarization. J Chem Phys. 1993; 98: 453–472.

39. Kollmar C, Kahn O. Ferromagnetic spin alignment in molecular systems: An orbital approach. Acc Chem Res. 1993; 26:259–265.

40. Dewar MJS. The Molecular Orbital Theory of Organic Chemistry. New York: McGraw-Hill, 1969.

41. Forrester AR, Thomson RH. Organic Chemistry of Stable Free Radicals. London: Academic, 1968.

42. McConnell HM. Antiparallel spin polarization in triplet states. J Chem Phys. 1961; 35:1520–1521.

43. Dowd P. Trimethylenemethane. Acc Chem Res. 1972; 5:242–248.
44. Ramasesha S, Soos ZG. Magnetic and optical properties of exact PPP states of naphthalene. Chem Phys. 1984; 91:35–42.
45. Zheludev A, Barone V, Bonnet M, Delley B, Grand A, Ressouche E, Rey P, Subra R, Schweizer J. Spin density in a nitronyl nitroxide free radical. Polarized neutron diffraction investigation and ab initio calculations. J Am Chem Soc. 1994; 116:2019–2027.
46. Kutzelnigg W. Friedrich Hund and chemistry. Angew Chem Int Ed Engl. 1996; 35:573–586.
47. Tyutyulkov N, Schuster P, Polansky O. Band structure of nonclassical polymers. Theor Chim Acta. 1983; 63:291–304.
48. Tyutyulkov N, Polansky O, Schuster P, Karabunarliev S, Ivanov CI. Structure and properties of non-classical polymers II. Band structure and spin densities. Theor Chim Acta. 1985; 67:211–228.
49. Yoshizawa K, Tanaka K, Yamabe T. Ferromagnetic coupling through m-phenylene. Molecular and crystal orbital study. J Phys Chem. 1994; 98:1851–1855.
50. Yoshizawa K, Hoffmann R. Potential linear-chain organic ferromagnets. Chem Eur J. 1995; 1:403–413.
51. Yoshizawa K, Hoffmann R. The role of orbital interactions in determining ferromagnetic coupling in organic molecular assemblies. J Am Chem Soc. 1995; 117:6921–6926.
52. Pauling L, Wilson EB. Introduction to quantum mechanics. New York: McGraw-Hill, 1935.
53. Hoffmann R. An extended Hückel theory. I. Hydrocarbons. J Chem Phys. 1963; 39:1397–1412.
54. Girerd JJ, Journaux Y, Kahn O. Natural or orthogonalized magnetic orbitals: Two alternative ways to describe the exchange interaction. Chem Phys Lett. 1981; 82:534–538.
55. Coulson CA. Valence. 2nd ed. Oxford: Clarendon, 1961.
56. Herring C. Critique of the Heitler–London method of calculating spin couplings at large distances. Rev Mod Phys. 1962; 34:631–645.
57. The triplet state of H_2 molecule has a very small potential minimum at about R = 8 a.u., due to van der Waals interactions. (a) Hirschfelder JO, Linnett JW. The energy of interaction between two hydrogen atoms. J Chem Phys. 1950, 18:130–144. (b) Kolos W, Wolniewicz L. Potential-energy curves for the $X^1\Sigma_g^+$, $b^3\Sigma_u^+$, and $C^1\Pi_u$ states of the hydrogen molecule. J Chem Phys. 1965; 43:2429–2441.
58. Hoffmann R, Imamura A, Zeiss GD. The spirarenes. J Am Chem Soc. 1967; 89:5215–5220.
59. Hoffmann R. Trimthylene and the addition of methylene to ethylene. J Am Chem Soc. 1968; 90:1475–1485.
60. Hoffmann R, Zeiss GD, Van Dine GW. The electronic structure of methylenes. J Am Chem Soc. 1968; 90:1485–1499.
61. Hoffmann R, Imamura A, Hehre WJ. Benzenes, dehydroconjugated molecules, and the interaction of orbitals separated by a number of intervening σ bonds. J Am Chem Soc. 1968; 90:1499–1509.

62. Hoffmann R. Interaction of orbitals through space and through bonds. Acc Chem Res. 1971; 4:1–9.

63. Albright TA, Burdett JK, Whangbo MH. Orbital Interactions in Chemistry. New York: Wiley, 1985.

64. Hay PJ, Thibeault JC, Hoffmann R. Orbital interactions in metal dimer complexes. J Am chem Soc. 1975; 97:4884–4899.

65. In the Hückel SOMO of diphenylmethyl $|c_c| = 2|c_p|$, where $|c_c|$ is the absolute value of orbital coefficient on the central carbon and $|c_p|$ is those on the two phenyl rings.

66. Yamaguchi K, Okumura M, Maki J, Noro T, Namimoto H, Nakano M, Fueno T, Nakasuji K. MO theoretical studies of magnetic interactions in clusters of nitronyl nitroxide species. Chem Phys Lett. 1992; 190:353–360.

67. Tanaka K, Asai Y, Sato T, Kuga T, Yamabe T, Tokumoto M. Orientation dependent magnetic interaction in TDAE-C_{60}, where TDAE is tetrakis(dimethylamino)ethylene. Chem Phys Lett. 1996; 259:574–578.

68. Koga N, Morokuma K. Ab initio MO study of the C_{60} anion radical: The Jahn–Teller distortion and electronic structure. Chem Phys Lett. 1992; 196:191–196.

69. Landrum GA. YAeHMOP (Yet Another Extended Hückel Molecular Orbital Package), version 2.0. Ithaca, New York: Cornell University, 1997.

20

MO-Theoretical Elucidation of Spin Alignments in Organic Magnetic Crystals

Kizashi Yamaguchi, Takashi Kawakami, Akifumi Oda, and Yasunori Yoshioka
Osaka University, Osaka, Japan

I. INTRODUCTION

Experimental results reported until 1986 in the field of molecular magnetism showed that antiferromagnetic exchange interactions are operative between free radicals in most cases [1–4]. Therefore, it was believed that antiferromagnetism should be expected even if free-radical solids exhibited a long-range magnetic order. This situation motivated us to perform molecular orbital (MO) calculations that would provide quantum mechanical information to clarify why free-radical pairs should generally have antiferromagnetic interactions and to elucidate possibilities of ferromagnetic interactions between these species [5–16]. The MO calculations indeed showed that diphenyl nitroxide and quinone anion radical exhibited the ferromagnetic exchange interaction, even in the face-to-face stacking mode, because of the *no-SOMO-SOMO* overlap integral and strong spin-polarization effects [6]. The computations also indicated that crystalline organic ferromagnets should be achieved if stacking modes of free radicals such as allyl [9] and benzyl [8] radicals are well controlled [10]. Thus, the no-SOMO-SOMO overlap and orientation principles were presented as useful guides (G1 and G2) for molecular design of organic ferromagnets [5–16]:

> G1. The orbital overlap (OO) term should be reduced by controlling stacking modes of free radicals to suppress the antiferromagnetic interaction.

G2. The ferromagnetic spin-density product (SDP) term should be increased by an enhancement of the spin-polarization (SP) effect.

Kanamori and Goodenough first presented the orbital symmetry rules for effective exchange interactions between spins in transition-metal complexes and related crystals [17,18]. The symmetry demand for atomic d-orbitals and ligand orbitals determines signs of J_{ab} values via superexchange mechanisms in these systems [17,18]. We have extended Kanamori–Goodenough (KG) rules for orbital interactions between organic radicals in order to determine the sign and magnitude of their effective exchange interactions in the course of radical reactions [19,20]. The extended KG (EKG) rules were successfully applied to derive selection rules for organic-radical reactions [19,20]. In order to emphasize an important role of the SOMO–SOMO interaction term in molecular magnetism, an extended McConnell model (EMM) [6–13] involving the orbital overlap (OO) term, together with the spin-density-product (SDP) term [21], was also presented for determination of effective exchange interactions between organic radicals. The extended McConnell model enabled us to derive qualitative selection rules for signs of J_{ab} and for the molecular design of organic magnetic materials. The careful examination of a number of ab initio computational results [5–16] have supported the first guiding principles (G1 and G2) for ferromagnetic effective exchange interactions. In this chapter, our no-orbital-overlap and orientation principles [5–16] are explained by considering typical important stacking models for radical pairs. Ab initio results [22–24] are briefly discussed for confirmations of predictions based on the symmetry rules and for explanations of experimental results.

II. NO-ORBITAL-OVERLAP AND ORIENTATION PRINCIPLES FOR FERROMAGNETIC EFFECTIVE EXCHANGE INTERACTIONS

II.A. Shapes and Symmetries of SOMOs

The intermolecular perturbation (IPT) and intermolecular configuration interaction (ICI) models described previously [24] have been applied to derivations of orbital symmetry rules for effective exchange interactions between organic radicals. To this end, orbital phases and/or topologies of SOMOs play important roles. Figure 1 illustrates shapes of SOMOs for typical radical species. Their orbital energy levels and electronic configurations are illustrated in Figure 2. The SOMOs **1–11** are easily characterized as symmetric (S) or antisymmetric (A) by using the symmetry plane. Their symmetry properties are therefore given in parentheses. Alkyl-substituted

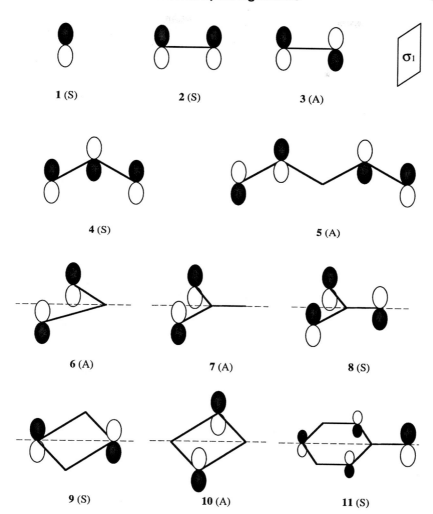

Figure 1 Shapes **1–5** of *p*-, π- and π*-type SOMOs for methyl (**I**), vinyl-cation (**II**), and vinyl-anion (**III**) radical analogs, and alkyl or amino nitroxides (**IV**) and nitronyl nitroxide derivatives (**V**), respectively. Shapes **6–11** of NBMO-type SOMOs for allyl (**VI**) radical, trimethylenemethane derivatives (**VII, VIII**), cyclobutadiene derivatives (**IX, X**), and benzyl radical derivatives (**XI**). Their symmetry properties are given in parentheses.

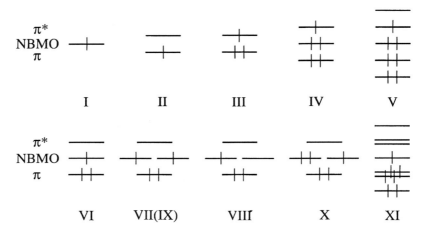

Figure 2 Electronic configurations for methyl (**I**), vinyl-cation (**II**), and vinyl-anion (**III**) radical analogs, and alkyl or amino nitroxides (**IV**) and nitronyl nitroxide derivatives (**V**), allyl (**VI**) radical, trimethylenemethane derivatives (**VII, VIII**), cyclobutadiene derivatives (**IX, X**), and benzyl radical derivatives (**XI**).

carbon radicals **Ia** ($R_1R_2R_3C$), such as ethyl radical, have the π-type SOMO **1**. Similarly, SOMO **1** is involved in anion radicals of boron compounds **Ib** ($R_1R_2R_3B^-$), cation radicals of amino compounds **Ic** ($R_1R_2R_3N^+$), and alkoxide radicals **Id** (RO). Generally speaking, these species might not have the σ-symmetry plane. However, the π-type shape **1** is taken as a representative of SOMOs of these species. Therefore, **I** is regarded essentially as a one-orbital, one-electron {1,1} system. The cation radicals of vinyl compounds and isoelectronic species **II** such as $R_1R_2C\text{-}BR_3R_4$ have SOMO **2** with the π-bonding nature. On the other hand, anion radicals of vinyl groups and isoelectronic compounds **IIIa**–**IIId** have the π^*-type SOMOs **3**. These ion-radical species are of practical utility, since they can be generated by chemical and electrochemical dopings, as illustrated in Figure 3. For qualitative discussions, **II** and **III** are regarded as two-orbital, one- and three-electron systems, {2,1} and {2,3}, respectively [22–24].

Previous theoretical calculations [6] predicted that ketyl radical **IIIb** and imino nitroxide **IIId** with the π^*-type SOMO exhibit the ferromagnetic effective exchange interactions in the appropriately stacked conformations. The importance of the π^*-type SOMOs in nitroxide **IIId** has been clarified in relation to no orbital overlap interaction [14,15]. The π^*-type SOMO also plays an important role for ferromagnetic interactions between alkyl nitroxides **IV**, for which the hyperconjugation [16] of alkyl group with the π^*-SOMO of **IIId** provides the SOMO **4**. The nitronyl nitroxide and isoelec-

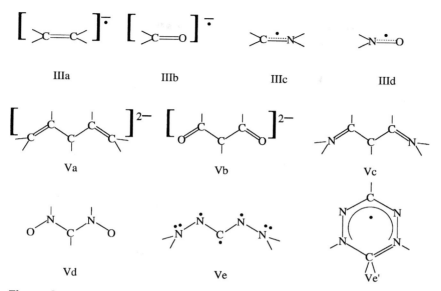

Figure 3   Molecular structures having types **III** and **V** electronic structures.

tronic compounds **Va–Vd** in Figure 3 have the π^*-type SOMO **5**, which consists of the p-type (**1**) and two π^*-type (**3**) SOMOs. Recently, nitronyl nitroxide derivatives **Vd** have been found to be stable spin sources for organic ferromagnets [24–29]. Similarly, verdazyl derivatives have the isoelectronic skeleton **Ve** and the π^*-type SOMO **5**. The electronic configurations of **IV** and **V** are regarded as the three-orbital, five-electron {3,5} and five-orbital, seven-electron {5,7} systems, respectively [22,23].

Alternant hydrocarbon radicals have the nonbinding MO (NBMO) [30] as SOMOs [31], some of which are illustrated in Figure 1. The allyl radical and its isoelectronic series **VI**, such as galvinoxyl radical [32], have SOMO **6**, where the linear combination of atomic orbital (LCAO) coefficient becomes zero on the central carbon atom and those on terminal carbon atoms have opposite signs. These topological properties of NBMOs play an important role in ferromagnetic intermolecular interactions [6]. Trimethylene methane derivatives [33] **VII** have two NBMO-type SOMOs **7** and **8**, which have different spatial symmetries. Therefore, cation and anion radicals **VIII** of these species have **7** or **8** as the SOMO. Similarly, triplet cyclobutadiene derivatives [34] **IX** and their cation and anion radical analogs **X** have NBMO-type SOMOs **9** and/or **10**. On the other hand, benzyl radical derivatives **XI** have SOMO **11**. The orbital overlaps between the NBMO-type

SOMOs often disappear because of the zero MO coefficients at the non-starred carbons, as shown previously in the cases of **VI** [9] and **XI** [8].

The electronic configurations of **VI** and **XI** are, respectively, described as the {3,3} and {7,7} systems, whereas **VII** and **IX** are the {4,4} systems, and **VIII** and **X** are the {4,3} or {4,5} system, as shown in Figure 2. Therefore, the NBMOs are the SOMOs for these species.

II.B.   Intermolecular Hund Rules

The first step for the theoretical prediction of the sign of effective exchange integrals (J_{ab}) is an application of the symmetry rules to orbital interactions [5–16,19,20]. Let us first consider simple examples of SOMO–SOMO interactions **12–18**, as illustrated in Scheme 1A. We can imagine several combinations of radical species among **12–18**. For example, **12** represents the orbital interactions between methyl radical analogs **Ia–Id**. Since the SOMO–SOMO interaction is symmetry allowed in **12**, the sign of J_{ab} should be negative (antiferromagnetic). **13** shows the SOMO–SOMO interaction between methyl (**I**) and vinyl-cation radical (**II**) analogs. **14** denotes the orbital interactions between **II**, whereas **15** expresses the orbital interactions between vinyl anion analogs **IIIa–IIId**. Ab initio calculations [35] for *syn* and *anti* stackings of iminonitroxide indicated the antiferromagnetic exchange interactions in conformity with the symmetry rule **15**. The orbital interactions between **II** plus alkyl nitroxide (**IV**)—and between **III** plus allyl radical analogs (**VI**)—are expressed by **16** and **17**, respectively. The SOMO–SOMO interactions between **IIIp** ($p = a–d$) and nitronyl nitroxide derivative **Vq** ($q = a–d$) are expressed schematically by **18**. The SOMO–SOMO interactions in Scheme 1 also occur in many other radical pairs.

The SOMO–SOMO interaction is orbital-symmetry allowed in the orientations **12–18** in Scheme 1A, since the interacting SOMOs have the same orbital symmetry with respect to symmetry plane σ. Therefore, the SOMO–SOMO overlap term $J_{ab}(OO)$, which is called as the kinetic exchange (KE) term [19–22], plays a predominant role, giving rise to the negative J_{ab} value. This indicates that the covalent bondings are feasible, leading to a greater stabilization of the singlet biradical (BR) state than the triplet BR state, as confirmed by the ab initio computations [8–9] of simple examples **12** and **15**.

On the other hand, the SOMO–SOMO overlaps are zero because of different symmetries of SOMOs for the perpendicular orientations **19–25**, as shown in Scheme 1B. Moreover, the potential exchange (PE) terms are not zero because of the orbital symmetry, together with close contact between SOMOs; $K_{ab} = \langle AS|SA \rangle > 0$. The ferromagnetic interactions are sym-

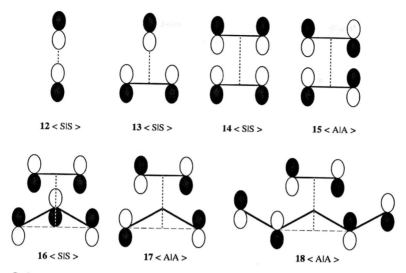

12 < SIS >   **13** < SIS >   **14** < SIS >   **15** < AIA >

16 < SIS >   **17** < AIA >   **18** < AIA >

Scheme 1A

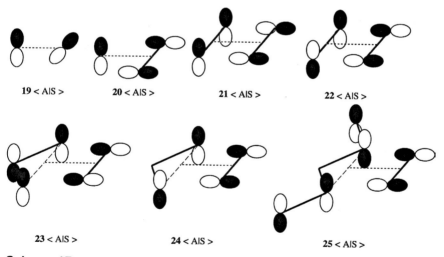

19 < AIS >   **20** < AIS >   **21** < AIS >   **22** < AIS >

23 < AIS >   **24** < AIS >   **25** < AIS >

Scheme 1B

metry allowed in the orientations **19–25**, in conformity with the intermolecular variant of the Hund rule. Many ferromagnetic radical pairs are conceivable from **19–25**. For example, **22** represents the orthogonal orbital interaction responsible for the ferromagnetic interaction between nitroxide analogs **IIIp**($p = a-d$). However, it is noteworthy that there are many secondary orbital interactions in real organic systems [24–29]. In fact, J_{ab} values determined by experiments [24–29] involve all these intermolecular interactions, indicating the necessity of ab initio computations for confirmation of the selection rules (Section III).

II.C. Importance of π^*-type SOMOs and NBMOs

Following the foregoing intermolecular Hund picture [6,14,15], the effective exchange interactions have been taken as antiferromagnetic in the case of parallel orientations or organic radicals because of the nonzero overlaps between SOMOs. However, the antiferromagnetic KE interactions often become almost zero even in the parallel orientation, since the SOMO–SOMO overlap integrals $S_{ab} = \langle S|A \rangle$ disappear because of the different orbital symmetries: symmetric (S) for SOMO and antisymmetric (A) for the other SOMO. For example, the orbital overlap between SOMOs becomes zero even at parallel conformations **26–32** in Scheme 2, for which the Coulombic exchange term K_{ab} is still nonzero because of the van der Waals contact between SOMOs, giving rise to the ferromagnetic interactions. As shown in **26–32**, the π^*-nature of SOMOs plays an important role in ferromagnetic interactions, since the orbital overlap disappears even in these parallel stacking modes.

Many radical combinations are conceivable for **26–32**. For example, **26** represents the SOMO–SOMO interaction between cation radical of amine derivatives (**I**) and vinyl-anion radical (**III**) analogs. The charge-transfer (CT) complexes between these species are possible candidates for organic ferromagnets [11]. On the other hand, SOMO–SOMO interactions between vinyl-cation (**II**) analogs and vinyl-anion radical (**III**) analogs are expressed by **27** and **29**. The CT complexes between donor (D) and acceptor (A) pairs could be ferromagnetic if these species are stacked so as to guarantee orthogonal SOMO–SOMO interactions. **28** illustrates the orthogonal orbital interactions between **III**: the anionic species can be generated by doping. Thus, systematic investigations of CT complexes are desirable from our theoretical point of view [11].

30 illustrates the orbital interactions between nitroxides (**IIId**) and/or alkyl nitroxides (**IV**). Similarly, the π^*-type SOMO or nonbonding MO (NBMO) plays an important role for ferromagnetic interactions in the parallel orientations **31–32**. Allyl radical analogs (**VI**) [9] and nitronyl nitroxide

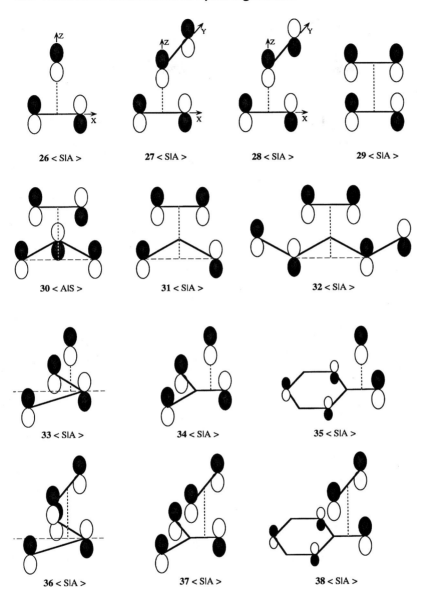

Scheme 2

analogs (**V**) [13] play central roles in such orbital interactions. The SOMO–SOMO overlaps become zero for more complex systems **33–38**, since the phase relationships or orbital topologies for interacting SOMOs are different. There are a lot of radical pairs represented by SOMO–SOMO interaction modes **33–38**.

II.D. Importance of Rhombus Conformations for Ferromagnetic Interaction

From the preceding examples [6,14,15], the ferromagnetic effective exchange interactions are feasible even in the case of parallel orientations of organic radicals, because of the nonzero overlap integrals between SOMOs. There are several important sliding conformations where the antiferromagnetic KE interactions become almost zero even in the parallel orientations, because of the zero SOMO–SOMO overlap integral. For example, the orbital overlap between SOMOs becomes almost zero even at parallel conformations **39–43** in Scheme 3, for which the Coulombic exchange term K_{ab} is still nonzero, providing ferromagnetic interactions [6,14,15]. In the sliding conformations, two symmetry planes, illustrated by the dotted lines, are necessary to characterize the orbital interactions, and two symmetry properties for the SOMO–SOMO interactions are given in Scheme 3.

As shown in **39–43**, the π^*-nature and NBMO property of a SOMO plays an important role in ferromagnetic interactions, since the orbital over-

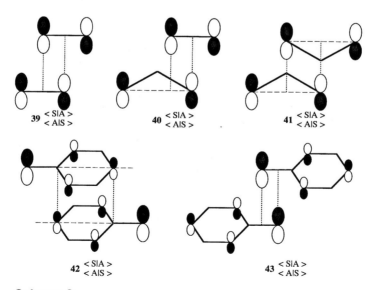

Scheme 3

laps disappear even in these parallel but sliding modes [6,14,15]. The effective exchange interactions between NBMOs in allyl [9] and benzyl radicals [8] were confirmed by ab initio computations. It was found that the effective exchange interaction between allyl radicals is ferromagnetic for the zigzag conformation **41** [9]. Ferromagnetic exchange interactions [32] between galvinoxyl radicals could be explained by this type of orbital interaction. Both the PE and spin-polarization (SP) mechanisms [7] favor ferromagnetic interactions between benzyl radicals in appropriate conformations **42** where the KE terms become zero. In fact, ab initio APUHF and CASSCF computations support these orbital symmetry rules [8]. The SOMO–SOMO interaction **42** [5] almost disappears in the case of the cyclophane-type dimers [36]. The ferromagnetic interactions observed for their *para-* and *meta-*isomers are indeed consistent with the no-SOMO–SOMO overlap (G1) and spin-polarization (G2) rules [5,6]. There are a lot of other radical pairs represented by the SOMO–SOMO interaction modes **39–43**.

III. AB INITIO CALCULATIONS OF TYPICAL EXAMPLES

In order to confirm the orbital symmetry rules in Schemes 2 and 3, ab initio calculations [35,37,38] were performed for the three different stacking modes illustrated in Figure 4.

III.A. Bridge Conformation

As illustrated in Figure 4A, let us consider the intermolecular interaction between ketyl radical (**IIIb**) and methyl radical in order to clarify the im-

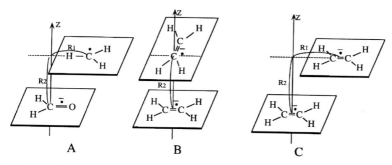

Figure 4 (A) The face-to-face stacking mode for ketyl and methyl radicals. (B) T-shape approach models for vinyl-anion radicals. (C) Sliding conformation between anion radicals of ethylene.

portance of the π^*-nature of the SOMO for ferromagnetic interaction. Since the π^*-type SOMO of **IIIb** has a node near the center of the C—O bond, the $p\pi^*$ orbital-overlap integral $S_{p\pi^*}$ should disappear at a bridge structure, as illustrated by **26** in Scheme 2. As equation (1) shows

$$S_{p\pi^*} = \langle \chi_1 | c_2\chi_2 - c_3\chi_3 \rangle = S_{12} - S_{13} = 0 \tag{1}$$

where $S_{12(3)}$ denotes the orbital overlap between atomic sites 1 and 2(3). Therefore the effective exchange integral J_{ab} should become positive (ferromagnetic) at this structure because of the nonzero potential exchange. The exchange integral is given by equation (2)

$$K_{ab} = \langle \chi_1(c_2\chi_2 - c_3\chi_3) | \chi_1(c_2\chi_2 - c_3\chi_3) \rangle = S_{12}^2 U > 0 \tag{2}$$

where the Mulliken approximation was used for the calculation of four-center atomic exchange integrals and the on-site repulsion integral $U = \langle \chi_p \chi_p | \chi_p \chi_p \rangle$. The J_{ab}-values were calculated for this parallel interplane stacking mode A in Figure 4 by APUMP2(4)/4-31G [35] in order to confirm the no-overlap and orientation principles [5–16]. Figure 5A shows variations of the calculated effective exchange integrals with R_1 at the fixed intermolecular distance $R_2 = 3.4$Å. From Figure 5A, the J_{ab}-values are positive at the bridge structure **26**, whereas they are negative at the no-bridge conformations. Since the π^*-SOMO of **IIIb** is unsymmetric, the bridge structure is also unsymmetric for the zero SOMO–SOMO overlap **26a**. The maximum positive J_{ab}-value at this conformation ($R_1 = 1.4$Å) is about 169 cm^{-1} by the APUMP4/4-31G method. The potential exchange interaction is rather strong even at the van der Waals contact between ketyl and methyl radicals.

The APUMPn and APUCC SD(T) calculations were also performed for the bridge structure ($R_1 = 1.4$Å), changing basis sets to determine their effect on results. The magnitude of calculated J_{ab} was not so dependent on the computational methods and basis sets. This indicates that the potential exchange (PE) K_{ab} plays a dominant role for the ferromagnetic exchange interaction in a SOMO–SOMO contact region. The magnitude of positive J_{ab}-value for the direct exchange of the ketyl-methyl radical pair is quite a bit larger than that for the indirect exchange in nitroxide crystals [24–29]. Ab initio computations indicate that a ferromagnetic exchange interaction between the anion radical of a carbonyl compound and an alkyl radical is feasible through the potential exchange (PE) term, even though the SOMO–SOMO direct contact is weak. The same situation is expected for

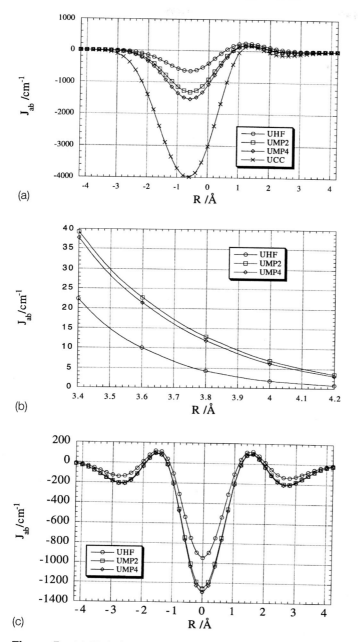

Figure 5 (a) Variations of J_{ab}-values calculated by the APUMP/4-31G method for model A in Figure 4. (b) Variations of J_{ab} values calculated for model B in Figure 4. (c) Variations of J_{ab} values calculated for model C in Figure 4.

the cation radical of amine derivatives (**I**) and anion radical (**III**) of carbonyl compound, though experimental results are lacking.

III.B.   T-Shape Conformation

Three-center interactions **28** in the T-shape conformations in Scheme 2 satisfy the no-overlap principle because of the π^*-nature of one of the SOMOs, suggesting ferromagnetic interactions. Here, the T-shape conformations between anion radicals of ethylene in Figure 4B were examined by the AP-UMP/4-31G method. Figure 5B shows variations of J_{ab} with change in the intermolecular distance R_2 for **28**. As expected from the orbital symmetry rule for **28**, the J_{ab} values calculated for model B in Figure 4 are positive (ferromagnetic), and decrease monotonically with the increase of R_2.

III.C.   Rhombus Conformation

In order to confirm the no-overlap but parallel orientation principles for ferromagnetic interactions in Scheme 3, the face-to-face stacking of anion radical ($m = -1$) of ethylene (model C in Figure 4) was examined by changing the sliding distance (R_1) at a fixed interplane distance ($R_2 = 3.4 \text{Å}$). Because of the π^*-nature of the SOMO for the species, the SOMO–SOMO overlap should disappear at the rhombus conformation, as illustrated in **39** of Scheme 3. Therefore the KE term is almost zero, but the potential exchange (PE) term remains nonzero because of the close SOMO–SOMO contact as given in equation (3):

$$K_{ab} = \frac{(S_{14}^2 + S_{23}^2 - S_{13}S_{24})U}{2} > 0 \tag{3}$$

The APUMP4(2) calculations were carried out for confirmation of this prediction. Figure 5C shows variations of the calculated J_{ab}-values with the sliding distance R_1. From Figure 5C it is seen that the J_{ab}-values become positive (ferromagnetic) near the rhombus conformation **39** in Scheme 3, whereas they are negative near the rectangular conformation **15** in Scheme 1. The ab initio results are consistent with the orbital symmetry rule. The maximum positive J_{ab}-value by the APUMP4/4-31G method is 93 cm^{-1} at the rhombus conformation with the sliding distance $R_1 = 1.6 \text{Å}$.

In order to elucidate the basis-set dependence, an APUMP/6-31G* calculation was carried out for the rhombus conformation at $R_1 = 1.6 \text{Å}$. The J_{ab}-value increases by about 4 cm^{-1} by the addition of the d-polarization at the MP4 level, showing that the ferromagnetic exchange interaction is symmetry derived. Therefore, the potential exchange (PE) term outweighs the kinetic exchange (KE) term in the rhombus conformation, showing the important role of the no-overlap but parallel orientation for ferromagnetic in-teraction between π^*-SOMOs [15]. The situation should be the same for other radical pairs in Scheme 3.

IV. INVESTIGATIONS OF ORGANIC MAGNETIC CRYSTALS

As typical examples, let us consider magnetic crystals composed of α-sub-stituted nitronyl nitroxide (NN) (4,4,5,5-tetramethyl-4,5-dihydro-1H-imida-zoyl-1-oxyl-3-oxide). Ab initio MO calculations [13] showed that the SO-MOs of these species are essentially localized on the nitronyl nitroxide skeleton (ON—C—NO) (**V** in Figure 2 and **Vd** in Figure 3), as illustrated by **5** in Figure 1. In order to elucidate the effective exchange interactions between SOMOs, the J_{ab}-values were calculated for pairs of NN skeletons, whose geometries were extracted from the x-ray structures of crystalline magnets. The calculated J_{ab}-value for the SOMO pair of α-hydro NN (HNN) was -7.25 cm^{-1} by the complete active space (CAS) SCF method using a two-orbital, two-electron {2,2} system. This value is close to the observed value (-7.65 cm^{-1}) for the α-phase crystal of HNN [39]. This in turn in-dicates that the SOMO–SOMO overlap (OO) term is predominant, as shown by **44** in Scheme 4. The situation should be similar for **12–18** in Scheme 1A.

On the other hand, the J_{ab}-values were calculated to be 8.3 and 6.4 cm^{-1}, respectively, by CASSCF {6,6} [40] for the SOMO pairs of p-car-boxyphenyl NN (p-CPNN), whose geometries were taken from the x-ray structures of p-CPNN$^-$Li$^+$MeOH and p-CPNN$^-$Li$^+$MeOD crystals [28]. They were slightly smaller than the observed values (10.2 and 7.2 cm^{-1}), but the ratio 8.3/6.4 was close to the observed one (10.2/7.2), showing the reliability of the computational results. The ferromagnetic exchange inter-action for p-CPNN can be explained by the no-OO term but the nonzero potential exchange (PE) term, which is characteristic of the T-shape confor-mation **45** in Scheme 4. These results are consistent with preceding ab initio results of bridged and T-shape conformations in Figure 5. The ab initio results do indeed support the no-orbital-overlap and orientation principles [5–16] for **26–38** in Scheme 2.

Both the OO and PE terms become zero for β- and γ-phase crystals [25–27] of p-nitrophenyl NN (p-NPNN) because of too long a distance between the NN skeletons, as illustrated in **46** in Scheme 4. The indirect effective exchange interactions via the spin-polarization (SP) effect of the nitrophenyl group play important roles, as shown in **47** and **48** [13,41,42]. In conformity with the second guiding principle (G2), the signs of the cal-

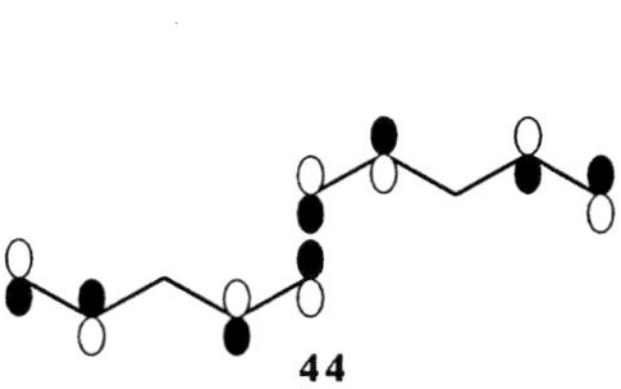

44 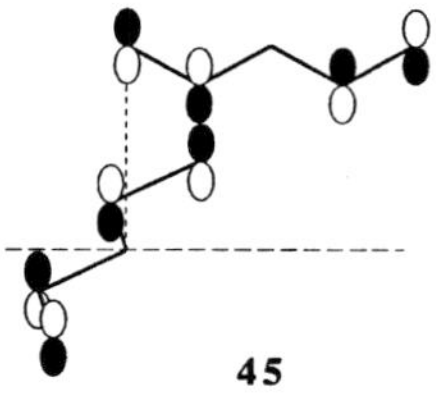   **45**

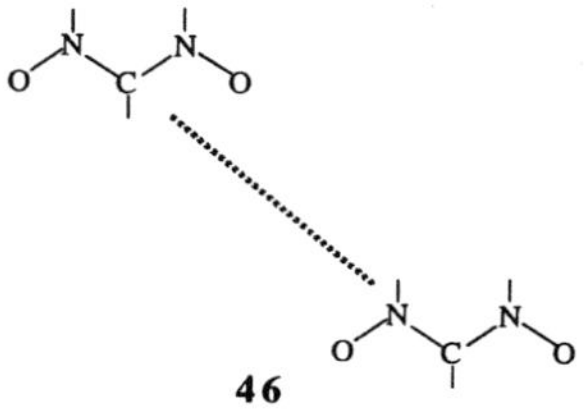

46 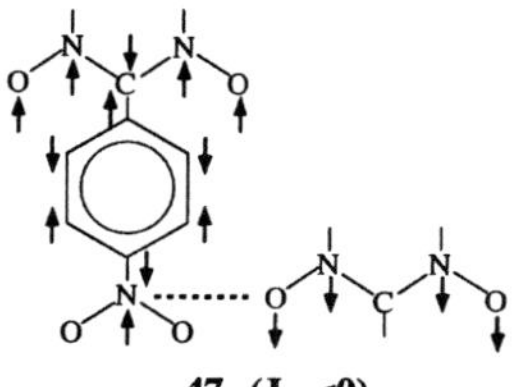   **47** ($J_{ab}<0$)   **48** ($J_{ab}>0$)

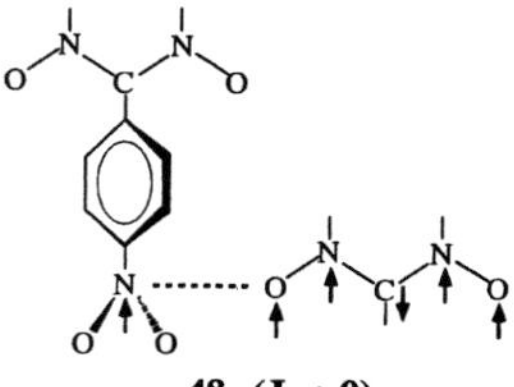

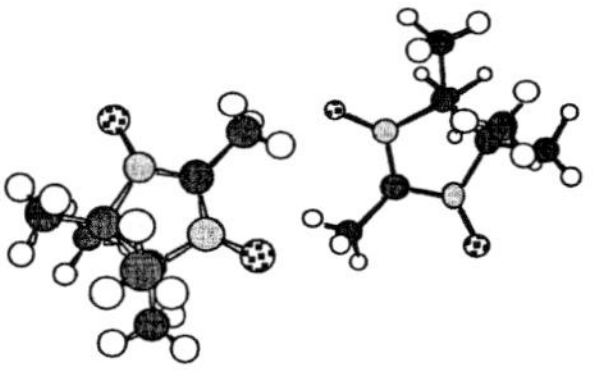

49 (MeNN) 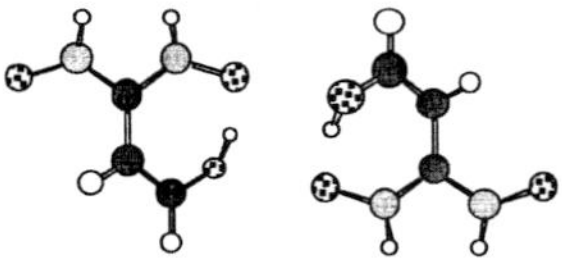   **50** (HQNN)

Scheme 4

culated J_{ab}-values are negative (-0.81 cm^{-1}) and positive (1.51 cm^{-1}) for the parallel and perpendicular orientations between the nitrophenyl and nitronyl nitroxide groups as shown in **47** and **48**. The latter conformation is realized for the β-phase crystal [27] of p-NPNN at ambient pressure, but internal rotation of the nitrophenyl group induced by high pressure may encourage spin crossover from **48** to **47**, in accord with recent experiments [43]. The SP mechanism is also operative for the ferromagnetic exchange interaction between p-NPNN in the γ-phase crystal [42].

The through-bond interactions via methyl or alkyl groups [39] and hydrogen bonds [29] have received great interest in relation to molecular magnetism of several nitronyl nitroxide crystals. For example, the calculated J_{ab}-value by CASSCF {2,2} was 3.52 cm^{-1} for the dimer **49** of 2-methyl NN (MeNN) with the experimental geometry, showing the important role of methyl bridging in explaining the observed ferromagnetic interaction [39]. The J_{ab}-value by CASSCF {6,6} was 0.11 cm^{-1} for the simplified model **50** of 2,5-dihydroxyphenyl NN (HQNN), while it was -1.23 cm^{-1} when the hydroxy groups were replaced by hydrogen atoms [44]. This clearly manifests the important role of the hydrogen bonds for the observed ferromagnetism of HQNN [29]. Thus ab initio calculations are now reliable enough for semiquantitative rationalization and explanation of weak intermolecular interactions in magnetic molecular crystals.

V. PHASE TRANSITION TEMPERATURE AND PRESSURE EFFECTS

The effective exchange integrals for three different pairs of p-NPNN were calculated theoretically in order to estimate the ferromagnetic phase transition temperature (T_c) in the case of the β-phase with Fdd2 space symmetry, as illustrated in Figure 6a. The magnitude of the ($J_{13} + J_{14}$)-value was about one-third of the J_{12}-value in the $a-c$ plane. This indicates the adequacy of the quasi-three-dimensional Heisenberg-type model. The magnetization is described by the Brillouin function, and it disappears at the critical temperature (T_c), leading to the so-called Langevin–Weiss relation given by $k_B T_c = 2S(S + 1); J_{ab}/3$, where S is 1/2 for a free radical. The relation was successfully applied to theoretical estimation of T_c for the β-phase, assuming that the Weiss effective molecular magnetic field was given approximately by taking twelve nearest-neighbor p-NPNN. T_c was calculated to be 0.64 K [41], in accord with the experiment (0.60 K) [27].

The present theoretical calculations for the β-phase of p-NPNN have revealed that the Langevin–Weiss (LW) model utilizing the calculated J_{ab}-values works well for a theoretical explanation of organic ferromagnetism. Then the Langevin–Weiss–Neel model was also used to estimate the antiferromagnetic transition temperature (T_N) for the γ-phase of p-NPNN [26], although its refinement with the inclusion of the low-dimensional effects was essential for quantitative purposes [42]. These models were used for theoretical predictions of possible high-T_c organic ferromagnets or ferromagnets on the basis of the calculated effective exchange integrals. Thus, the first principle cluster calculations are applicable to the design of molecular magnetic insulators, for which the Heisenberg model gives reliable theoretical descriptions.

Theoretical computations revealed three characteristic modes consistent with the observed pressure-induced ferromagnetic-to-antiferromagnetic transition in the β-phase: (1) the sliding of p-NPNN along the z-axis to alter the sign of the interplane exchange integral (J_{13}), as shown in Figure 6b, (2) rotation of the nitro group to provide a negative intraplane exchange integral (J_{12}) (Figure 6c), and (3) rotation of the nitrophenyl group to reverse the

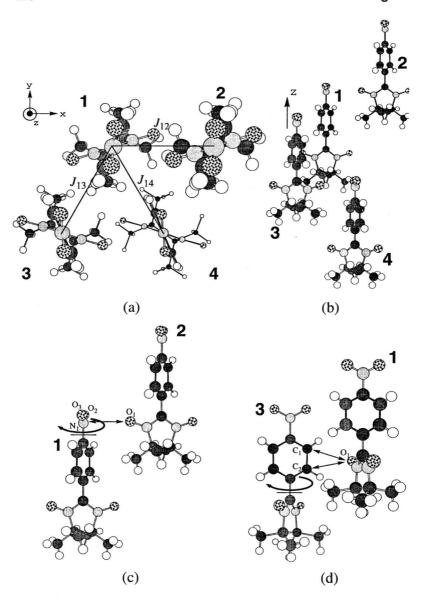

(a)

(b)

(c)

(d)

Figure 6   (a) Molecular structure of the β-phase crystal of *p*-NPNN with definition of the effective exchange integrals (J_{ab}) and three crystal deformations under high pressure. (b) Sliding mode along the *z*-axis. (c) Rotation of the nitro group. (d) Rotation of the nitrophenyl group.

sign of J_{13} (Figure 6d). Judging from experiments [43], the third mode is conceivable as the simplest model for explaining the pressure effect. Since rotations of substituents such as methyl groups are feasible under high pressure, the pressure-induced variations of $T_c(T_N)$ or ferromagnetic-to-antiferromagnetic (vice-versa) transitions would be equally anticipated for many other organic ferromagnets composed of alkylnitroxides.

VI. CONCLUSIONS

The effective exchange integral J_{ab} for a radical pair is generally expressed by five different terms, in both the spin-restricted and spin-unrestricted computational methods [5–16,24] as:

$$J_{ab}(\text{total}) = J_{ab}(\text{KE}) + J_{ab}(\text{PE}) + J_{ab}(\text{SP}) + J_{ab}(\text{EC}) + J_{ab}(\text{MDD}) \quad (4)$$

The kinetic exchange (KE) and potential exchange (PE) terms are, respectively, determined by the SOMO–SOMO overlap (S_{ab}) and Coulombic exchange integral K_{ab}. The spin-polarization (SP) term is given approximately by the product of spin densities (ρ_{ab}) induced on the π-network by the SP effect of the spin component. The electron correlation (EC) term involves all higher-order EC contributions except for anisotropic magnetic dipole–dipole (MDD) interaction. Ferromagnetic exchange interaction in molecular crystals can be classified into these terms, providing selection rules as shown in Table 1. An extended McConnell model (EMM) considering both the SOMO–SOMO interaction (G1) and potential exchange plus spin polarization (G2) terms in a well-balanced manner is a useful guide for molecular design. The ab initio computations were found to be useful for theoretical elucidation of spin alignment rules in Table 1.

Cases I and II are understood intuitively by the symmetry of SOMO–SOMO contact. The effective exchange interaction between closely located radical groups is usually antiferromagnetic ($J_{ab} < 0$), since the KE interaction stabilizes the low-spin or singlet state (Case I). However, if the mutual orientation of radical groups is controlled to reduce the KE term, a ferromagnetic interaction ($J_{ab} > 0$) is expected at a short intermolecular distance (Case II) because of the nonzero Coulombic exchange integral shown in Scheme 2. On the other hand, the SP term induced by through-bond or through-space indirect interactions becomes important when the distance between SOMOs is large (Cases III and IV). For example, the β-phase crystal of p-NPNN corresponds to Case III. On the other hand, the γ-phase of p-NPNN is regarded as Case IV. The sign of $J_{ab}(\text{SP})$ depends on the phase of spin alternation by the SP effect and mutual orientations of spins, as illustrated in Scheme 4. Magnetic dipolar interactions (MDD) are not discussed here,

Table 1   Spin-Alignment Rules Derived from Ab Initio Calculations of Magnetic Crystals

Case	KE (≤ 0)	PE (≥ 0)	SP	J_{total}	Notes
I_a	< 0	> 0	> 0	< 0	KE forbidden
I_b	< 0	> 0	< 0	< 0	
II_a	~ 0	> 0	> 0	> 0	PE allowed
II_b	~ 0	> 0	< 0	> 0	
III	~ 0	~ 0	> 0	> 0	SP allowed
IV	~ 0	~ 0	< 0	< 0	SP forbidden
V	~ 0	~ 0	Higher-order term (EC)		

though they become important in some cases. The calculated spin-density populations can be used for estimation of the MDD term.

REFERENCES

1. Yamauchi J. Linear antiferromagnetic interaction in organic free radicals. Bull. Chem. Soc. Jpn. 1971; 44:2301–2308.
2. Veyret C, Blaise A. Magnetisme dans certains composes organiques: Les radicaux libres stables nitroxydes. Mol. Phys. 1973; 25:873.
3. Kearna JF. Newer aspects of the synthesis and chemistry of nitroxide spin labels. Chem. Rev. 1978; 78:37–64.
4. Benoit A, Flouquet J, Gillon B, Schweizer J. The antiferromagnetic structure of tanol suberate. J. Magn. Magn. Mater. 1983; 3-4:1155.
5. Yamaguchi K, Fukui H, Fueno T. Molecular orbital (MO) theory for magnetically interacting organic compounds. Ab initio MO calculations of the effective exchange integrals for cyclophane-type carbene dimers. Chem. Lett. 1986; 625–628.
6. Yamaguchi K, Fueno T, Nakasuji K, Murata I. Semiempirical molecular orbital (MO) calculations of the effective exchange integrals for sandwich dimers of free radical species. Anti- and ferromagnetic spin couplings of organic free radicals. Chem. Lett. 1986; 629–632.
7. Yamaguchi K, Toyoda Y, Nakano M, Fueno T. Ab initio and semiempirical MO calculations of intermolecular effective exchange integrals between organic radicals. Designing of organic ferromagnet, ferrimagnet and ferromagnetic conductors. Synthetic Metals. 1987; 19:87–92.
8. Yamaguchi K, Namimoto H, Fueno T. Ab initio calculations of effective exchange integrals. Possibilities of superparamagnetic, mictomagnetic and

amorphous ferromagnetic states for aggregates of aromatic free radicals and polymer radicals. Mol. Cryst. Liq. Cryst. 1989; 176:151–161.

9. Yamaguchi K, Fueno T. An effective spin Hamiltonian for clusters of organic radicals. Application to allyl radical clusters. Chem. Phys. Lett. 1989; 159: 465–471.

10. Yamaguchi K, Toyoda Y, Fueno T. Ab initio calculations of effective exchange integrals for triplet carbene clusters. Importance of stacking modes for ferromagnetic interactions. Chem. Phys. Lett. 1989; 159:459–464.

11. Yamaguchi K, Namimoto H, Fueno T, Nogami T, Shirota Y. Possibilities of organic ferromagnets and ferrimagnets by the use of charge-transfer (CT) complexes with radical substituents. Ab initio MO studies. Chem. Phys. Lett. 1990; 166:408–414.

12. Yamaguchi K, Okumura M, Maki J, Noro T, Namimoto H, Nakano M, Fueno T, Nakasuji K. MO theoretical studies of magnetic interactions for clusters of nitronyl nitroxide and related species. Chem. Phys. Lett. 1992; 190:353–360.

13. Yamaguchi K, Okumura M, Nakano M. Theoretical calculations of effective exchange integrals between nitronyl nitroxides with donor and acceptor groups. Chem. Phys. Lett. 1992; 191:237–244.

14. Kawakami T, Yamanaka S, Mori W, Yamaguchi K, Kajiwara A, Kamachi M. No-overlap and orientation principles for ferromagnetic interactions between nitroxide groups. Chem. Phys. Lett. 1995; 235:257–265.

15. Kawakami T, Yamanaka S, Nagao H, Mori W, Kamachi M, Yamaguchi K. Theoretical approaches to molecular magnetism II: No-overlap and orientation principles for ferromagnetic interactions. Mol. Cryst. Liq. Cryst. 1995; 272: 117–129.

16. Kawakami T, Takeda S, Mori W, Yamaguchi K. Theoretical studies of effective exchange interactions between nitroxides via hydrogen atoms. Chem. Phys. Lett. 1996; 261:129–137.

17. Goodenough JB. An interpretation of the magnetic properties of the pervoskite-type mixed crystals. J. Phys. Chem. Solids. 1958; 6:287–297.

18. Kanamori J. Superexchange interaction and symmetry properties of electron orbitals. J. Phys. Chem. Solids. 1959; 10:87–98.

19. Yamaguchi K, Yoshioka Y, Fueno T. Heisenberg models for radical reactions: Local spin (magnetic) symmetry conservations of biradical species. Chem. Phys. 1977; 20:171–181.

20. Yoshioka Y, Yamaguchi K, Fueno T. Heisenberg models of radical reactions II: Conservation of the local spin-permutation symmetry in reactions of biradical species. Theoret. Chim. Acta 1978; 45:1–20.

21. McConnell HM. Ferromagnetism in solid free radicals. J. Chem. Phys. 1963; 39:1910.

22. Andersson K, Malqvist P, Roos BO, Sadlej AJ, Wolinski K. Second-order perturbation theory with a CASSCF reference function. J. Phys. Chem. 1990; 94: 5483–5488.

23. Yamaguchi K. Multireference (MR) configuration interaction (CI) approach for quasidegenerate systems. Int. J. Quant. Chem. 1980; S14:269–284.

24. Kawakami T, Yamanaka S, Yamaki D, Mori W, Yamaguchi K. Theoretical approaches to molecular magnetism. In: M. K. Turnbull, T. Sugimoto, L. K. Thompson, eds. Molecule-Based Magnetic Materials: Theory, Technique and Applications. ACS Symp. Series 644. Washington, DC: ACS, 1996, p 30.

25. Awaga K, Maruyama Y. Ferromagnetic and antiferromagnetic intermolecular interactions of organic radicals, α-nitronyl nitroxide II. J. Chem. Phys. 1989; 91:2743.

26. Kinoshita M, Turek P, Tamura M, Nozawa K, Shiomi D, Nakazawa Y, Ishikawa M, Takahashi M, Awaga K, Inabe T, Maruyuma Y. An organic radical ferromagnet. Chem Lett. 1991; 1225–1226.

27. Nakazawa Y, Tamura M, Shirakawa N, Shiomi D, Takahashi M, Kinoshita M, Ishikawa M. Low-temperature magnetic properties of ferromagnetic organic radical, p-NPNN. Phys. Lev. B. 1992; 46:8906.

28. Inoue K, Iwamura H. Magnetic properties of the crystals of p-(1-oxyl-3-oxido-4,4,5,5,-tetramethyl-2-imidazolin-2-yl)benzoic acid and itd alkali metal salts. Chem. Phys. Lett. 207:551–554.

29. Matsushita MM, Izuoka A, Sugawara T, Kobayashi T, Wada N, Takeda N, Ishikawa M. Hydrogen-bonded organic ferromagnet. J. Am. Chem. Soc. 1997; 119:4369–4379.

30. Longuet-Higgins HC. Some studies in molecular orbital theory I. Resonance structures and molecular orbitals in unsaturated hydrocarbons. J. Chem. Phys. 1950; 18:265.

31. Itoh K. Electronic structures of antiaromatic hydrocarbons with high spin multiplets in the electronic ground state. Pure Appl. Chem. 1978; 50:1251.

32. Awaga K, Sugano T, Kinoshita M. Ferromagnetic intermolecular interactions in a series of organic mixed crystals of galvinoxyl and its precursory closed shell compound. J. Chem. Phys. 1986; 85:2211.

33. Yamaguchi K. The electronic structures of biradicals in the unrestricted Hartree–Fock approximation. Chem. Phys. Lett. 1975; 33:330–335.

34. Yamaguchi K. Electronic structures of antiaromatic molecules. Chem. Phys. Lett. 1975; 35(2):230–235.

35. Yamaguchi K, Okumura M, Mori W, Maki J, Takada K, Noro T, Tanaka K. Comparison between spin restricted and unrestricted post Hartree–Fock calculations of effective exchange integrals in Ising and Heisenberg models. Chem. Phys. Lett. 1993; 210(1–3):201–210.

36. Izuoka A, Murata S, Sugawara T, Iwamura H. Ferro- and antiferromagnetic interaction between two diphenylcarbene units incorporated in the [2,2]paracyclophane skeleton. J. Am. Chem. Soc. 1985; 107:1786–1787.

37. Yamanaka S, Okumura M, Yamaguchi K, Hirao K. CASPT2 and MRMP2 calculations of potential curves and effective exchange integrals for the dimer of triplet methylene. Chem. Phys. Lett. 1994; 225:213–220.

38. Yamanaka S, Kawakami T, Nagao H, Yamaguchi K. Effective exchange integrals for open-shell species by density functional methods. Chem. Phys. Lett. 1994; 231:25–33.

39. Hosokoshi Y. Study on the Magnetic Interactions in Stable Organic Radical Crystals. PhD dissertation, University of Tokyo, 1995.

40. Kawakami T, Oda A, Mori W, Yamaguchi K, Inoue K, Iwamura H. Theoretical studies of the ferromagnetic intermolecular interaction of *p*-carboxylate phenyl nitronyl nitroxide. Mol. Cryst. Liq. Cryst. 1996; 279:29–38.

41. Okumura M, Yamaguchi K, Nakano M, Mori W. A theoretical explanation of the organic ferromagnetism in the *b*-phase of *para* nitrophenyl nitronyl nitroxide. Chem. Phys. Lett. 1993; 207(1):1–8.

42. Okumura M, Mori W, Yamaguchi K. An MO-theoretical calculation of the antiferromagnetism in the *g*-phase of *p*-nitrophenyl nitronyl nitroxide. Chem. Phys. Lett. 1994; 219(1):36–44.

43. Mito M, Kawae T, Takumi M, Nagata K, Tamura M, Kinoshita M, Takeda K. Pressure-induced ferro- to antiferromagnetic transition in a purely organic compound. β-phase *para*-nitrophenyl nitronyl nitroxide. Phys. Rev. B. 1997; 56: R14255–R14258.

44. Oda A, Kawakami T, Takeda S, Mori W, Matsushita MM, Izuoka A, Sugawara T, Yamaguchi K. Theoretical studies on magnetic interactions in 2′,5′-dihydroxyphenyl nitronyl nitroxide crystal. Mol. Cryst. Liq. Cryst. 1997; 306: 151–160.

21

Exchange Effects in Three Dimensions—Real Materials: Bulk Magnetic Properties of Organic Verdazyl Radical Crystals

Kazuo Mukai
Ehime University, Matsuyama, Japan

I. INTRODUCTION

The study of molecular magnetism of organic radical crystals has attracted much attention since the recent discovery of bulk ferromagnetism in the β-phase of *p*-nitrophenyl nitronyl nitroxide (*p*-NPNN) [1]. A number of organic ferromagnets have been found since then. However, the examples are mostly limited to nitroxide and nitronyl nitroxide radicals [2–8], because these radicals have high stability and the syntheses of their derivatives are comparatively easy.

The 1,3,5-triphenylverdazyl (TPV) radical (see Figure 1) is well known as one of the stable organic free radicals. The magnetic properties of TPV and its derivatives have been studied extensively, indicating the antiferromagnetic (AFM) intermolecular interaction in these radicals, as observed for most organic free-radical solids [9]. A positive Weiss constant ($\theta = +1.6$ K) was observed for one of the verdazyl radicals (3-(4-nitrophenyl)-1,5,6-triphenylverdazyl, NTV) [10]. However, the presence of AFM coupling between ferromagnetic (FM) chains has been suggested, because the magnetic susceptibility, $\chi_M T$, exhibits a maximum at 3.9 K. Furthermore, weak ferromagnetic behavior was observed for TPV radical solid at $T_N = 1.78$ K [11]. X-ray structure analyses have been performed for these verdazyl radicals, in order to find a relation between crystal structure and observed spin–

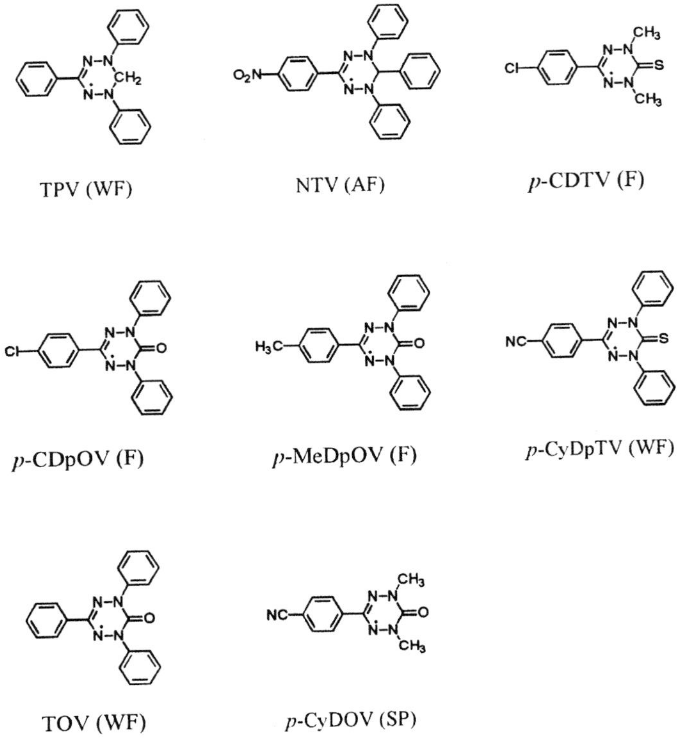

TPV (WF)   NTV (AF)   *p*-CDTV (F)

p-CDpOV (F)   *p*-MeDpOV (F)   *p*-CyDpTV (WF)

TOV (WF)   *p*-CyDOV (SP)

Figure 1 Molecular structures of verdazyl radicals. F: ferromagnetism; WF: weak ferromagnetism; AF: antiferromagnetism; SP: spin–Peierls transition.

spin exchange interaction, e.g., AFM and FM behavior at low temperature [9].

The synthesis of a series of new 1,5-dimethyl- and 1,5-diphenyl-6-oxo- and 6-thioxo-verdazyl radicals (Figure 1), in which the methylene ($>CH_2$) bridge in TPV is replaced by carbonyl ($>C=O$) or thiocarbonyl ($>C=S$), has been performed by Neugebauer et al. [12,13]. These verdazyl radicals have a delocalized *p*-electron system, differing from nitroxide and nitronyl nitroxide radicals with a π-electron system localized in the part of the NO group, and have a comparatively small molecular weight. Consequently we can expect a strong intermolecular exchange interaction ($H = -2J_{ij}S_iS_j$) between neighboring radical molecules.

Recently, we studied the magnetic properties of thirty examples of such verdazyl radical crystals and found that three kinds of verdazyls, *p*-CDTV, *p*-CDpOV, and *p*-MeDpOV (Figure 1) [6,7], are new ferromagnets with Curie transition temperatures (T_C) of 0.68 K, 0.21 K, and 0.67 K, respectively. Furthermore, several interesting magnetic properties, such as weak ferromagnetism [14–17], antiferromagnetism [18], and a spin–Peierls transition [19–21], were observed for these verdazyl radical crystals. Ferromagnetic intermolecular exchange interactions were observed with high probability for these verdazyl radical crystals, and thus the ab initio MO calculations were performed for five kinds of 6-oxoverdazyl radicals to ascertain whether the verdazyls have an electronic configuration that can yield a FM interaction or not. A very strong spin-polarization effect has been found, which is advantageous to intermolecular FM interaction.

II. GENERAL FEATURES OF THE MAGNETIC PROPERTIES OF VERDAZYL RADICAL CRYSTALS

Generally, organic free radicals have g-values very close to that of the free electron ($g = 2.0023$). In fact, the g-values of the 6-oxo- and -thioxoverdazyl radicals determined by an electron spin resonance (ESR) measurement are 2.0035–2.0037 in toluene at room temperature [12,13]. From this fact we can presume a small spin–orbit interaction and consequently an isotropic exchange interaction for the verdazyl radicals. Thus, the $S = 1/2$ isotropic Heisenberg model is most suitable for describing the present system.

Measurement of the magnetic susceptibilities of thirty kinds of verdazyl radical crystals has been performed in the temperature range of 4.2–300 K using a SQUID magnetometer. Generally, the susceptibility of these verdazyls shows a broad maximum, and the temperature dependence of magnetic susceptibility was explained by a one-dimensional (1D) AFM Heisenberg nonalternating or alternating chain model [6,9,18–20,22,23]. Comparatively large exchange interactions were observed for these verdazyls, as expected. The corresponding spin Hamiltonian is given by

$$H = -2J_1 \sum_{i=1}^{N/2} S_{2i}S_{2i+1} - 2J_2 \sum_{i=1}^{N/2} S_{2i-1}S_{2i} \qquad (1)$$

where J_1 and J_2 are the nearest-neighbor exchange integrals. Negative values of J_1 and J_2 corresponding to AFM coupling are appropriate to the free radicals considered here. $\alpha(= J_2/J_1)$ is a parameter that conveniently indicates the degree of alternation, and $\alpha = 1$ corresponds to the uniform limit, that is, a nonalternating AFM chain [24,25]. $\alpha = 0$ corresponds to an isolated dimer (spin-pair) system, and in this case the susceptibility should be interpreted by the singlet-triplet equilibrium model. A spin–Peierls transition was observed for the p-CyDOV radical crystal.

Table 1   Magnetic Properties of Verdazyl Radical Crystals

Radical	Magnetism[a]	Transition temperature (K)	θ (K)	$2J/k_B$ (K)	zJ'/k_B (K)
p-CDTV	F	$T_C = 0.68$	+3.0	+12.0	+0.21
p-CDpOV	F	$T_C = 0.21$	+2.5	+11.0	+0.03
p-MeDpOV	F	$T_C = 0.67$	+2.5	+7.0	+0.31
p-NDpOV	—	—	+1.5	+5.0	
TOV	WF	$T_N = 4.9$	−12	−9.0	
p-CyDpTV	WF	$T_N = 0.41$	+2.9	+7.0	−0.19
NTV	AF	$T_N = 1.16$	+3.1	+7.0	−1.5
p-CyDOV	SP	$T_{SP} = 15$		−84	
		$T_{max} = 54$			

[a]F: ferromagnetism; WF: weak ferromagnetism; AF: antiferromagnetism; SP: spin–Peierls transition.

On the other hand, positive Weiss constants (+1 − +3 K) were observed for six kinds of verdazyls (Figure 1 and Table 1), suggesting intermolecular FM exchange interaction between neighboring radicals [6,7, 10,17,18,22,23]. Furthermore, low-temperature transitions to ferromagnetism, weak ferromagnetism, and antiferromagnetism were observed for these verdazyl radical crystals. Above the transition temperatures, the verdazyls behave as quasi-one-dimensional (quasi-1D) Heisenberg ferromagnets [6,7,26].

III.   FERROMAGNETISM IN VERDAZYL RADICAL CRYSTALS

The ac susceptibility of 3-(4-chlorophenyl)-1,5-dimethyl-6-thioxoverdazyl (p-CDTV) was measured in the temperature region 1.7–300 K, using an ac susceptometer [6]. Above 20 K, the ac susceptibility (χ) of p-CDTV follows the Curie–Weiss law, with a positive Weiss constant of +3.0 K, indicating dominant FM exchange interaction between neighboring radical molecules. A plot of $1/\chi$ against T is no longer linear at lower temperatures (<20 K), as shown in Figure 2. Such magnetic behavior is characteristic of an FM substance.

The low-temperature ac susceptibility (χ) of the p-CDTV was also measured down to about 0.1 K at ac field $H(\nu) < 10$ Oe ($\nu = 164$ Hz), using the Hartshorn bridge method [6]. As shown in Figure 3, the susceptibility of p-CDTV increases rapidly below 1 K and reaches a maximum at $T =$

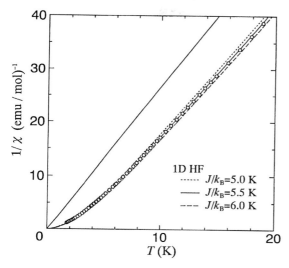

Figure 2 Inverse magnetic susceptibility ($1/\chi$) of p-CDTV at low temperatures. Above 5 K, the observed values (open circles) obey the theoretical results for a 1D isotropic Heisenberg ferromagnet (HF) with $2J/k_B = +11 \pm 1$ K. The straight line corresponds to χ of a paramagnet with $g = 2.00$ and $S = 1/2$.

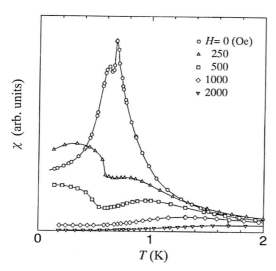

Figure 3 External field dependence of the ac magnetic susceptibility (χ) of p-CDTV.

0.68 K under zero external field. The appearance of the peak in the susceptibility may be due to time-dependent effects of the ac method. At low
temperatures, the susceptibility seems to reach a limiting value that corresponds to the inverse of the demagnetization field factor for a ferromagnet.
As is seen in Figure 3, the fluctuation of the magnetic moments is suppressed
by the external fields and the peak becomes smaller. Below 0.5 K, the susceptibility shows a plateau that seems to be characteristic of the ferromagnetically ordered state, as seen in the results for the 500-Oe field. The small
hump around 1 K is due to 1D short-range ordering effects in the paramagnetic region. The susceptibility at higher fields than 100 Oe is almost suppressed, as observed in the case of the FM *p*-NPNN (β-phase) [1]. The
results suggest that the *p*-CDTV crystal undergoes a bulk FM transition at
$T_C = 0.68$ K.

The low-temperature heat capacity, C_P, of the *p*-CDTV radical was
measured in external fields of 0–30 kOe. The magnetic heat capacity, C_m,
was obtained by subtracting the lattice contribution, which was evaluated
by reference to the lattice heat capacity of the *p*-BDTV radical crystal, with
correction for the difference in the mass of Br and Cl atoms. C_m is larger
than the lattice contribution below 5 K in zero field. As shown in Figure 4,
C_m of *p*-CDTV exhibits a sharp λ-like peak at 0.67 K in zero external field,

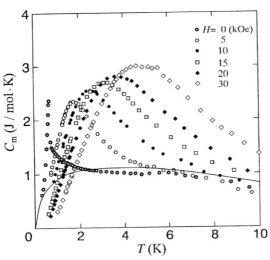

Figure 4   External field dependence of the magnetic heat capacity (C_m) of *p*-CDTV.
The solid line represents the results of theoretical calculation for an isotropic 1D
Heisenberg ferromagnet with $J/k_B = 6.0$ K.

corresponding to a three-dimensional magnetic phase transition. From these facts, we may conclude that the p-CDTV is a new ferromagnet, with a Curie temperature $T_C = 0.68$ K, analogous to those of other genuine organic ferromagnets [6].

In zero field and in the paramagnetic state, C_m of p-CDTV maintains values comparable to $C_m(\text{max}) = 0.134R = 1.12$ J/(mol K) above 2 K, which is a characteristic feature of the isotropic 1D Heisenberg ferromagnet [24,26]. This is ascertained by a theoretical calculation of the magnetic heat capacity of a 1D FM Heisenberg chain with $J/k_B = 6.0$ K. The heat capacity is a sensitive function of applied external field; even in a small field of $H = 0.25$ kOe, the sharp peak is much reduced, leaving only a vague trace of it at T_C and another broad hump of heat capacity above T_C. This hump moves toward higher temperatures as the field is increased. Such a field dependence of the heat capacity of p-CDTV is well explained as a characteristic feature of a 1D Heisenberg ferromagnet [7,26]. These results indicate that the p-CDTV behaves as a quasi-1D ferromagnet above the transition temperature T_C.

The susceptibility of a 1D ferromagnet with $S = 1/2$, χ_{1D}, is given by

$$\chi_{1D} = \frac{Ng^2\mu_B^2}{4kT}\left\{1 + \left(\frac{J}{kT}\right)^a\right\} \tag{2}$$

with $a = 1$ for $k_BT/J > 1$ [24]. For lower temperatures ($k_BT/J < 1$), however, a depends on T ($a \rightarrow 4/5$ for $T \rightarrow 0$). In fact, above 5 K the susceptibilities of p-CDTV are well reproduced by Eq. (2) with $2J/k_B = +11.0$ K and $g = 2.00$, as shown in Figure 2. At temperatures below 5 K, we have to consider the effects from the interchain interaction (J'), which induces three-dimensional ordering at $T_C = 0.68$ K. More detailed analysis of the temperature dependence of χ has been performed on the basis of both the modified spin-wave theory and the Bethe-ansatz integral-equation method by Takahashi [27], giving $2J/k_B = +12.0$ K and $zJ'/k_B = 0.21$ K or $|zJ'|/J = 3.8 \times 10^{-2}$, where z is the number of neighboring chains [6,26].

Similar measurements were performed for p-CDpOV and p-MeDpOV (see Figure 1). These verdazyl radicals have also been found to be quasi-1D ferromagnets, with T_C of 0.21 K and 0.67 K, respectively [7]. The values of T_C, $2J/k_B$, and zJ'/k_B obtained are summarized in Table 1. In the case of inorganic compounds, we can rarely get 1D quantum FM substances [6]. It is interesting that the prototypical thermodynamical behavior expected of most one-dimensional isotropic Heisenberg ferromagnets was found for these verdazyl radical crystals [6,7,26].

IV. WEAK FERROMAGNETISM IN VERDAZYL
   RADICAL CRYSTALS

Figure 5 shows the temperature dependence of the magnetic susceptibility (χ_M) of 3-(4-cyanophenyl)-1,5-diphenyl-6-thioxoverdazyl (*p*-CyDpTV) at external fields of 3, 11, 30, and 100 Oe [17]. By decreasing the temperature from 300 K, the susceptibility of *p*-CyDpTV increases gradually until about 0.5 K, independent of external fields. The susceptibility of *p*-CyDpTV shows maximuma at $T = 0.40$ K and 0.38 K under external fields of 100 and 30 Oe, respectively. A sharp increase of the susceptibility was observed near 0.41 K in the case of an external field of 3 Oe, corresponding to the onset of weak ferromagnetism associated with three-dimensional AFM ordering, as described shortly.

Figure 6 shows the magnetization curves at 0.05, 0.3, 0.4, and 1.5 K. At 1.5 K remanence was not observed, and the magnetization was almost proportional to the magnetic field up to 100 Oe. At 0.4, 0.3, and 0.05 K, the extrapolation of the linear part of the magnetization to zero field shows a finite magnetization. The low-field magnetization curve is expressed approximately by the sum of the FM moment (M_S) and the linear contribution

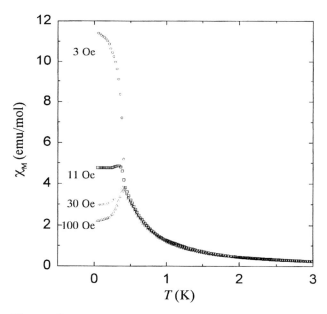

Figure 5 Temperature dependence of the magnetic susceptibility (χ_M) of *p*-CyDpTV measured at the applied magnetic fields of 3, 11, 30, and 100 Oe.

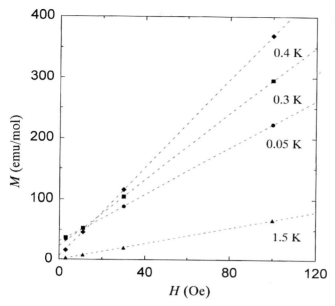

Figure 6 Magnetization (M) as a function of magnetic field (H) in the range of 0–100 Oe at 0.05, 0.3, 0.4, and 1.5 K in p-CyDpTV.

($\chi_M H$); i.e., $M(H) = M_S + \chi_M H$. From this, the magnitude of the spontaneous magnetization (M_S) was estimated to be 31.1, 22.9, and 6.4 emu/mol at 0.05, 0.3, and 0.4 K, respectively. These values correspond to the order of about 10^{-3} μ_B/molecule, which is a reasonable value for typical organic [11,14–16,28–30] and inorganic [31] weak ferromagnets. These results clearly indicate that the p-CyDpTV is a weak ferromagnet.

Similarly, the value of spontaneous magnetization (M_S) was calculated at each temperature and plotted against temperature (Figure 7). This is a first observation of temperature dependence of M_S for a weak ferromagnet, as far as we know. As shown in Figure 7, spontaneous magnetization (M_S) increases rapidly at 0.41 K, followed by a gradual increase down to 0.05 K. The result suggests that magnetic ordering occurs in the present system below a Neel temperature $T_N = 0.41$ K. This increase in M_S is attributed to the appearance of canted moments. From the value of spontaneous magnetization at 0.05 K, we can estimate the canting angle $\phi = 0.32°$ between the neighboring sublattice moments. The temperature dependence of $M_S(T)$ is very similar to that observed for usual inorganic ferromagnets, such as Ni. In fact, at temperatures near T_N, $M_S(T) \propto [1 - (T/T_N)]^\beta$ with $T_N = 0.40$ K

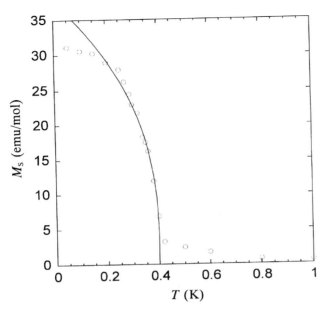

Figure 7 Temperature dependence of the spontaneous magnetization (M_S) of *p*-CyDpTV. The solid line shows a fit of $M_S(T) \propto [1 - (T/T_N)]^\beta$ with $\beta = 0.36$ and $T_N = 0.40$ K.

and critical index $\beta = 0.36$, in agreement with a value of 0.36 expected for an isotropic 3D Heisenberg spin system [32].

Kremer et al. [14] found that the TOV radical (see Figure 1) also shows the property of weak ferromagnetism below $T_N = 4.9$ K with a spin-canting angle of 0.26°. Mito et al. [15] reported that the anomalous temperature dependence of the weak ferromagnetism at low temperature is qualitatively explained by a four-sublattice model with two kinds of Dzyaloshinsky–Moriya (D–M) vectors of opposite sign in the ac-plane. Further, detailed study of the magnetism of TOV has been performed by using single crystals of TOV [16].

Weak ferromagnetism due to spin canting has recently been detected in several organic radical crystals, such as verdazyl (TOV [14–16] and TPV [11]), nitroxide (AOTMP and MATMP) [28], a dithiadiazolyl radical [29], and TCNQF$_4^-$ anion radical [30]. However, the origin of the anisotropy in these organic weak ferromagnets has not been well understood. In the high-temperature region, the susceptibility of these weak ferromagnets, except for MATMP, follows the Curie–Weiss law, with a negative Weiss constant. Above T_N, these radicals behave as quasi-1D antiferromagnets. On the other hand, p-CyDpTV shows a positive Weiss constant of $\theta = +2.9$ K, indicating a predominant FM exchange interaction ($2J/k_B = +7.0$ K) between unpaired electrons of adjacent verdazyls, and behaves as a 1D Heisenberg ferromagnet above $T_N = 0.41$ K. The weak ferromagnets are essentially antiferromagnets. Therefore, the interchain interaction in p-CyDpTV is an AFM one. Based on the mean field theory, we can estimate the effective interchain coupling J' by using the relation $kT_N \sim 2S^2(|zJ'|J)^{1/2}$, where z is the number of neighboring chains [17]. Substituting the values of $T_N = 0.41$ K and $J/k_B = 3.5$ K into this relation, we can obtain $zJ'/k_B \sim -0.19$ K, or $|zJ'|/J \sim 5.4 \times 10^{-2}$. The origin of the weak ferromagnetism observed for p-CyDpTV is a slight spin canting in the antiferromagnetically ordered interchains. An antisymmetric spin coupling of D–M type may contribute to the spin canting in p-CyDpTV, because the fine-structure energy (or the single-ion anisotropy energy in inorganic weak ferromagnets) is absent for $S = 1/2$ spin systems like organic monoradicals.

V. ANTIFERROMAGNETISM IN VERDAZYL RADICAL CRYSTALS

The ac susceptibility (χ_{ac}) of NTV follows the Curie–Weiss law, with a positive Weiss constant (θ) of $+3.1$ K, indicating a dominant FM exchange interaction between neighboring verdazyl molecules, as reported by Allemand et al. [10]. At lower temperatures (<20 K), however, the inverse susceptibility versus T is no longer linear, as shown in Figure 8. The low-temperature χ_{ac} shows a sharp peak at $T_{max} = 1.16$ K and decreases toward 0 K, with its magnitude of about two-thirds of the peak value of χ_{ac} at 1.16 K, as in ordinary powder crystal susceptibilities of antiferromagnets. The result suggests that NTV undergoes an AFM transition at $T_N = 1.16$ K [18].

The low-temperature heat capacity (C_P) of NTV radical crystals was measured in applied fields of $H = 0$, 480, and 1530 Oe. A distinct peak that corresponds to a three-dimensional magnetic phase transition is detected at $T_N = 1.10$ K, and a flat plateau follows at temperatures above T_N. The magnetic heat capacity (C_m) was obtained by subtracting the lattice contribution (C_l) from the total heat capacity (C_P). The value of C_l was estimated under the following assumptions: (1) The lattice heat capacity C_l is expressed as the Debye function, and at low temperatures it usually gives a T^3 dependence. (2) The total magnetic entropy is $Rln(2S + 1)$, where $S = 1/2$, as confirmed by the susceptibility measurement. (3) Above 10 K, the magnetic heat capacity estimate $C_m = bT^2$ is valid, as is usually reasonable in the paramagnetic region.

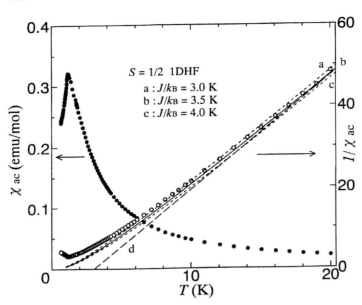

Figure 8   Molar and inverse molar magnetic susceptibilities (χ_{ac} and $1/\chi_{ac}$) of NTV at low temperatures. Above 6 K, the observed values (open circles) obey the theoretical results for the 1D isotropic Heisenberg ferromagnet (1DHF, Eq. 2) with $2J/k_B = +7.0 \pm 1.0$ K indicated by the solid line, b. The broken line d represents the Curie–Weiss law with $\theta = +3.1$ K.

The intrinsic magnetic heat capacity (C_m) of NTV is shown in Figure 9. It is noted that the plateau of the heat capacity is reproduced by the theory for a 1D Heisenberg ferromagnet [24] with $2J/k_B = 8.0$ K (Figure 9), as in the case of the magnetic susceptibility ($2J/k_B = 7.0$ K). The sharp heat capacity peak is stable even in external fields up to 1.53 kOe, as seen in the inset of Figure 9. This field dependence of the peak makes a good contrast to the sensitive field dependence in the p-NPNN (β-phase) [1], p-CDTV [6,26], and p-CDpOV [7], where the peak becomes a round maximum even in a field of 1.0 kOe. Rather, the behavior of NTV in the fields resembles the case of the AFM γ-phase of p-NPNN [1].

In zero field and in the paramagnetic state, C_m of NTV maintains values comparable to $C_m(\text{max}) = 0.134R = 1.12$ J/(mol K) above 3 K, which is a characteristic feature of an isotropic 1D Heisenberg ferromagnet [24]. Further evidence for the 1D interaction in NTV is provided by its crystal structure determination. The result of crystal structure analysis of NTV indicates that the NTV radical molecules are stacked along the a-axis, building up a

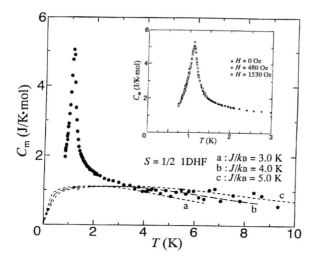

Figure 9 Temperature dependence of the magnetic heat capacity (C_m) of NTV (closed circles). Theoretical results for the isotropic 1D Heisenberg ferromagnet (1DHF) are drawn for $2J/k_B$ = +6.0, 8.0, and 10.0 K. The inset shows the field dependence of the magnetic heat capacity (C_m).

stairlike chain structure. In conclusion, NTV is an antiferromagnet with a Neel temperature of T_N = 1.16 K and behaves as a quasi-1D Heisenberg ferromagnet, with the intrachain exchange interaction of $2J/k_B$ = +7.0 K above the transition temperature T_N [18].

Recently, the magnetic properties of organic radical crystals have been studied extensively, generally indicating the AFM intermolecular interaction in these radical crystals. However, as far as we know, examples of real antiferromagnetic long-range order are very few. Furthermore, these are typically nitroxides [33] and nitronyl nitroxide [1], except for the case of the 1,3-bis-diphenylene-2-(4-chlorophenyl)allyl (*p*-Cl-BDPA) radical, which is known as the first organic antiferromagnet [34] and has the highest Neel temperature, T_N = 3.25 K, observed for a carbon-centered radical. The Neel temperature of NTV, T_N = 1.16 K, is the second highest one of genuine organic antiferromagnets [18].

VI. SPIN–PEIERLS TRANSITIONS IN VERDAZYL RADICAL CRYSTALS

The molar susceptibility (χ_M) of 3-(4-cyanophenyl)-1,5-dimethyl-6-oxoverdazyl (*p*-CyDOV) is shown in Figure 10 as a function of the temperature.

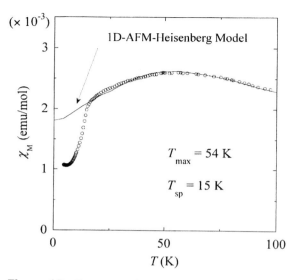

Figure 10 Paramagnetic molar susceptibility (χ_M) of p-CyDOV radical (4.2–100 K). The open circles are the experimental points. The solid line is the theoretical susceptibility calculated with the 1D AFM Heisenberg linear chain model.

The susceptibility of p-CyDOV shows a broad maximum at 54 K. However, at 15 K the susceptibility decreases suddenly, and it shows a minimum at 5.6 K. Such a magnetic behavior cannot be explained by a 1D AFM Heisenberg alternating chain model. The temperature dependence of magnetic susceptibility in p-CyDOV radical crystal may be explained by a spin–Peierls (SP) transition [19,20].

The SP transition occurs when a system of uniform AFM Heisenberg linear chains undergoes a transformation to a system of dimerized or alternating AFM linear chains [35]. This dimerization is caused mainly by spin–phonon coupling between the 1D spin system and the three-dimensional (3D) phonon system. Below the transition temperature (T_{SP}), the ground state is spin singlet (nonmagnetic), and a finite energy gap opens in the excitation spectrum. In such a case, the susceptibilities drop exponentially to small constant values below T_{SP}.

In fact, the magnetic susceptibility of p-CyDOV above 15 K can be well described by the 1D Heisenberg linear chain model ($\alpha = 1$) with AFM exchange of $2J/k_B = -84$ K between neighboring spins, as shown in Figure 10. The small increase in the susceptibility below 5.6 K was observed, which is attributable to isolated monoradicals and/or broken-chain effects. The residual paramagnetic radical concentration, calculated from the susceptibility at 4.2 K and assuming validity of the Curie law, is only 1.3%.

The field dependence of the magnetization (M) of polycrystalline p-CyDOV has been measured. The measurements were performed at 4.2 K for increasing and decreasing modes of pulsed magnetic field (H) up to 35 T. Plots of M and dM/dH vs. H are shown in Figure 11. We observed a characteristic nonlinearity of magnetization, implying a transition from dimerized to another magnetic phase [35–37]. A hysteresis of magnetization for increasing and decreasing H was also observed. The critical field (H_C) —which is defined as the field of the peak position in the dM/dH curve— is indicated by an arrow in Figure 11(b). The values of H_C^{up} and H_C^{down}, measured in increasing and decreasing field, are 25 T and 11 T, respectively. The former is 14 T larger than the latter. Thus the transition is of first order. Similar nonlinearity and hysteresis of magnetization have also been reported

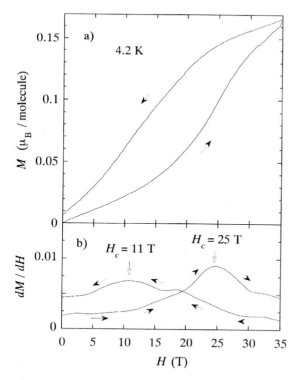

Figure 11 The magnetic-field dependence of M and dM/dH of p-CyDOV crystal at 4.2 K. Solid arrow represent the directions of the scan of the magnetic field.

in organic and inorganic spin–Peierls materials, such as TTF-CuBDT [35], MEM-(TCNQ)$_2$ [36], and CuGeO$_3$ [37]. Such magnetic behavior is considered to be a common feature of the spin–Peierls system. Magnetic phase diagram of p-CyDOV was obtained from the detailed study of both the field dependence of the transition temperature (T_{SP}) and the temperature dependence of the critical field (H_C) [21].

The susceptibility data of p-CyDOV were analyzed following the method of Jacobs et al. [35]. Above the SP transition temperature (T_{SP}), the chains are uniform, and χ_M is fit by the 1D AFM Heisenberg model. Below T_{SP}, the lattice progressively dimerizes, and two unequal and alternating J's are produced:

$$2J_{1,2}(T) = 2J[1 \pm d(T)] \tag{3}$$

The BCS theory predicts Eq. (4):

$$\frac{\Delta(0)}{T_{SP}} = 1.76 \tag{4}$$

where $\Delta(0)$ is $\Delta(T)$ at $T = 0$ and $\Delta(T)$ is the magnetic energy gap. $\Delta(T)$ follows a BCS-type temperature dependence. According to the mean-field theory, the relationship between $\delta(T)$ and the magnetic energy gap $\Delta(T)$ at temperature T is expressed as

$$\delta(T) = \frac{\Delta(T)}{2p|J|} \tag{5}$$

where the value of p is 1.637. The spin-lattice system dimerizes (in zero field) at a temperature T_{SP} given by [35]

$$T_{SP} = 4.56p|J| \exp\frac{-1}{\lambda} \tag{6}$$

where λ is the spin-lattice coupling constant. Substituting $T_{SP} = 15$ K and $2|J|/k = 84$ K from the high-temperature data into these equations, we get

$$\Delta(0) = 26 \text{ K}; \quad \delta(0) = 0.192; \quad \lambda = 0.33;$$

$$\frac{2|J_1|}{k_B} = 101 \text{ K}; \quad \alpha = 0.68 \tag{7}$$

The magnetic susceptibility of p-CyDOV radical crystals exhibits the characteristic properties of an SP transition. To our knowledge, this is the first example of an SP transition found for a genuine organic nonionic radical crystal. The preceding values of $\Delta(0)$, $\delta(0)$, and λ observed for p-CyDOV (26 K, 0.192, 0.33) closely correspond to those of ionic radical salts such as TTF-CuBDT (21 K, 0.167, 0.32) [35] and MEM-(TCNQ)$_2$ (28 K, 0.164,

0.32) [36], and inorganic compounds such as $CuGeO_3$ (24 K, 0.167, 0.20) [37]. The similarities in these values are very interesting, although the reason is not clear at present.

The magnetic susceptibilities of diluted crystals (p-CyDOV)$_{1-x}$(p-CyDOV-H)$_x$ ($x = 0-0.10$) were measured to study the effects of nonmagnetic impurities, p-CyDOV-H (amine precursor of p-CyDOV), on the spin-Peierls transition [20]. The transition temperature (T_{SP}) decreased rapidly upon p-CyDOV-H substitution, and the spin–Peierls state disappeared around $x = 0.07$. The x-dependence of T_{SP} and the magnetic behavior of diluted p-CyDOV crystal is also similar to that of Zn-doped $CuGeO_3$ crystals [38].

VII. ELECTRONIC STRUCTURE OF 6-OXOVERDAZYL RADICALS

Ab initio MO calculations were performed for 3-(4-R-phenyl)-1,5-diphenyl-6-oxoverdazyl radicals (R = H, OCH_3, CH_3, CN, and NO_2) to clarify the origin of the intermolecular FM exchange interaction that is observed with high probability for these verdazyl radicals. The π-MO energies of the next-lowest unoccupied MO (NLUMO), the singly occupied MO (SOMO), and the next-highest occupied MO (NHOMO) obtained are listed in Table 2. For instance, in p-MeDpOV radical (R = CH_3), the orbital energy of the NHOMO of the β-spin (β-NHOMO) is higher than the SOMO of the α-spin (α-SOMO). The exchange interaction within the molecule is great enough to stabilize the α-SOMO energy. This situation is one of the important requirements for intermolecular FM exchange interaction in organic radical crystals [39,40], because the electron in the β-NHOMO is more easily transferred in the charge transfer (CT) interaction than that in the α-SOMO, leaving the triplet state behind. As listed in Table 2, the energy level of the β-NHOMO is higher than that of the α-SOMO in p-MDpOV (R = CH_3O) and p-MeDpOV (R = CH_3) radicals, and the energy level of the β-NHOMO is close to that of the α-SOMO in TOV (R = H) and p-NDpOV (R = NO_2). The result indicates that the spin correlation causes a large spin-polarization effect in 6-oxoverdazyl radicals. This feature closely resembles that in the FM organic radicals galvinoxyl and β-p-NPNN [39,40]. Thus, the FM interaction of the verdazyl radicals is considered to originate mainly from the α-SOMO-β-NHOMO charge-transfer interaction.

As just described, an intermolecular FM interaction was observed for p-MeDpOV and p-NDpOV radicals. On the other hand, TOV, p-MDpOV, and p-CyDpOV show AFM interactions, although the electronic structure of the verdazyls except for p-CyDpOV satisfies one of the important conditions for the intermolecular FM exchange interaction in organic radical crystals,

Table 2 π-MO Energy Levels of the NLUMO, SOMO, and NHOMO with the α- and β-Spin of 3-(4-R-phenyl)-1,5-diphenyl-6-oxoverdazyl Radical Molecules

Radical	Orbital	α-spin E/eV	β-spin E/eV
TOV	NLUMO	2.52	3.03
	SOMO	−8.48	2.10
	NHOMO	−8.88	−8.51
p-MDpOV	NLUMO	2.69	3.11
	SOMO	−8.49	2.10
	NHOMO	−8.84	−8.28
p-MeDpOV	NLUMO	2.68	3.13
	SOMO	−8.44	2.15
	NHOMO	−8.82	−8.33
p-CyDpOV	NLUMO	2.84	2.73
	SOMO	−8.65	2.10
	NHOMO	−9.06	−8.92
p-NDpOV	NLUMO	2.07	2.53
	SOMO	−8.61	1.94
	NHOMO	−9.00	−8.60

proposed by Awaga et al. [39,40]. To discuss the details of the origin of intermolecular FM interaction observed for verdazyl radicals, further information about the crystal structures of the verdazyl radicals is necessary.

ACKNOWLEDGMENTS

The study of the magnetic properties of verdazyl radical crystals has been performed in collaboration with Profs. Kazuyoshi Takeda and Norio Achiwa. The author would like to express sincere gratitude for their kind collaboration. The author is very grateful to all collaborators for their help in accomplishing this work.

REFERENCES

1. Nakazawa Y, Tamura M, Shirakawa N, Shiomi D, Takahashi M, Kinoshita M, Ishikawa M. Low-temperature magnetic properties of the ferromagnetic organic

radical, *p*-nitrophenyl nitronyl nitroxide. Phys Rev B. 1992; 46:8906–8914, and references cited therein.

2. Chiarelli R, Novak MA, Rassat R, Tholence JL. A ferromagnetic transition at 1.48 K in an organic nitroxide. Nature. 1993; 363:147–149.

3. Nogami T, Ishida T, Yasui M, Iwasaki F, Takeda N, Ishikawa M, Kawasaki T, Yamaguchi K. Proposed mechanism of ferromagnetic interaction of organic ferromagnets: 4-(Arylmethyleneamino)-2,2,6,6-tetramethylpiperidin-1-oxyls and related compounds. Bull Chem Soc Jpn. 1996; 69:1841–1848, and references cited therein.

4. Sugawara T, Matsushita MM, Izuoka A, Wada N, Takeda N, Ishikawa M. An organic ferromagnet: *a*-Phase crystal of 2-(2′,5′-dihydroxyphenyl)-4,4,5,5-tetramethyl-4,5-hihydro-1H-imidazolyl-1-oxy-3-oxide (a-HQNN). J Chem Soc Chem Commun. 1994; 1723–1724.

5. Cirujeda J, Mas M, Molins E, de Panthou FL, Laugier J, Park JG, Paulsen C, Rey P, Rovira C, Veciana J. Control of the structural dimensionality in hydrogen-bonded self-assemblies of open-shell molecules: Extension of intermolecular ferromagnetic interactions in *a*-phenyl nitronyl nitroxide radicals into three dimensions. J Chem Soc Chem Commun. 1995; 709–710.

6. Mukai K, Konishi K, Nedachi K, Takeda K. Magnetic properties of 1,5-dimethylverdazyl radical crystals: Ferromagnetism in 3-(4-chlorophenyl)-1,5-dimethyl-6-thioxoverdazyl radical crystal. J Phys Chem. 1996; 100:9658–9663.

7. Takeda K, Hamano T, Kawae T, Hidaka M, Takahashi M, Kawasaki S, Mukai K. Experimental check of Heisenberg chain quantum statistics for a ferromagnetic organic radical crystal. J Phys Soc Jpn. 1995; 64:2343–2346.

8. Sugimoto T, Tsuji M, Suga T, Hosoito N, Ishikawa M, Takeda N, Shiro M. New charge-transfer complex-based organic ferromagnets: Pyridinium-substituted imidazolin-1-oxyl/tetrafluorotetracyanoquinodimethanide or hexacyanobutadienide salts. Mol Cryst Liq Cryst. 1995; 272:183–194.

9. Nagao A. Relation between exchange interaction and crystal structure of verdazyl radicals: As studied by means of McConnell's spin density hamiltonian. Bull Chem Soc Jpn. 1982; 55:1357–1361, and references cited therein.

10. Allemand PM, Srdanov G, Wudl F. Molecular engineering in the design of short-range ferromagnetic exchange in organic solids: The 1,3,5-triphenylverdazyl system. J Am Chem Soc. 1990; 112:9391–9392.

11. Tomiyoshi S, Yano T, Azuma N, Shoga M, Yamada K, Yamauchi J. Weak ferromagnetism and antiferromagnetic ordering of 2*p* electrons in the organic radical compound 2,4,6-triphenylverdazyl. Phys Rev B. 1994; 49:16031–16034.

12. Neugebauer FA, Fisher H, Siegel R. 6-Oxo- und 6-thioxoverdazyls. Chem Ber. 1988; 121:815–822.

13. Neugebauer FA, Fischer H, Krieger C. Verdazyls. Part 33. EPR and ENDOR studies of 6-oxo- and 6-thioxoverdazyls. X-ray molecular structures of 1,3,5-triphenyl-6-oxoverdazyl and 3-*tert*-butyl-1,5-diphenyl-6-thioxoverdazyl. J Chem Soc Perkin Trans 2. 1993; 535–544.

14. Kremer RK, Kanellakopulos B, Bele P, Brunner H, Neugebauer FA. Weak ferromagnetism and magnetically modulated microwave absorption at low

magnetic fields in 1,3,5-triphenyl-6-oxoverdazyl. Chem Phys Lett. 1994; 230: 255–259.

15. Mito M, Nakano H, Kawae T, Hitaka M, Takagi S, Deguchi H, Suzuki K, Mukai K, Takeda K. Magnetism of a two-dimensional weak-ferromagnetic organic radical crystal, 1,3,5-triphenyl-6-oxoverdazyl. J Phys Soc Jpn. 1997; 66: 2147–2156.

16. Jamali JB, Achiwa N, Mukai K, Suzuki K, Ajiro Y, Matsuda K, Iwamura H. Single-crystal weak ferromagnetism of 1,3,5-triphenyl-6-oxoverdazyl free-radical and ferromagnetic behavior of $(TOV)_{1-x}(TOV-H)_x$ diluted system. J Magn Magn Mater 1998; 177–181:789–791.

17. Mukai K, Nuwa M, Morishita T, Muramatsu T, Kobayashi TC, Amaya K. Observation of weak ferromagnetism in organic verdazyl radical crystal, p-CyDpTV. Chem Phys Lett. 1997; 272:501–505.

18. Mito M, Takeda K, Mukai K, Azuma N, Gleiter MR, Krieger C, Neugebauer FA. Magnetic properties and crystal structures of 1,5-diphenylverdazyls with electron acceptor groups in the 3-position. J Phys Chem B. 1997; 101:9517–9524, and references cited therein.

19. Mukai K, Wada N, Jamali JB, Achiwa N, Narumi Y, Kindo K, Kobayashi T, Amaya K. Magnetic properties of 1,5-dimethyl-6-oxoverdazyl radical crystals: Observation of a spin-Peierls transition in 3-(4-cyanophenyl)-1,5-dimethyl-6-oxoverdazyl radical, p-CyDOV. Chem Phys Lett. 1996; 257:538–544.

20. Mukai K, Wada N, Jamali JB, Achiwa N, Narumi Y, Kindo K, Kobayashi T, Amaya K. Magnetic properties of 1,5-dimethyl-6-oxoverdazyl radical crystals: Observation of a spin-Peierls transition in 3-(4-cyanophenyl)-1,5-dimethyl-6-oxoverdazyl radical, p-CyDOV. Mol Cryst Liq Cryst. 1997; 305:499–508.

21. Hamamoto T, Narumi Y, Kindo K, Mukai K, Shimobe Y, Kobayashi TC, Muramatsu T, Amaya K. Phase diagram of a new spin-Peierls compound: p-CyDOV. Physica B, Condensed Matter 1998; 246–247:36–39.

22. Mukai K, Kawasaki S, Jamali JB, Achiwa N. Magnetic properties of a new type of 1,5-diphenyl-6-oxo- and -thioxo-verdazyl radicals. Chem Phys Lett. 1995; 241:618–622.

23. Mukai K, Nuwa M, Suzuki K, Nagaoka S, Achiwa N, Jamali JB. Magnetic properties of 3-(4-R-phenyl)-1,5-diphenyl-6-oxo- and -thioxoverdazyl radical crystals (R = OCH₃, CH₃, CN, and NO₂). J Phys Chem B. 1998; 102:782–787.

24. Bonner JC, Fisher ME. Linear magnetic chains with anisotropic coupling. Phys Rev. 1964; A135:640–658.

25. Bonner JC, Blote HWJ, Bray JW, Jacobs ISJ. Susceptibility calculations for alternating antiferromagnetic chains. J Appl Phys. 1979; 50:1810–1812.

26. Takeda K, Konishi K, Nedachi K, Mukai K. Experimental study of quantum statistics for the $S = 1/2$ quasi-one-dimensional organic ferromagnet. Phys Rev Lett. 1995; 74:1673–1676.

27. Takahashi M. Few-dimensional Heisenberg ferromagnets at low temperature. Phys Rev Lett. 1987; 58:168–170.

28. Kobayashi TC, Takiguchi M, Hong C, Okada A, Amaya K, Kajiwara A, Harada A, Kamachi M. Magnetic properties of organic radical magnets, MOTMP, MATMP and AOTMP. J Phys Soc Jpn. 1996; 65:1427–1429.

29. Palacio F, Antorrena G, Castro M, Burriel R, Rawson J, Smith JNB, Bricklebank N, Novoa J, Ritter C. High-temperature magnetic ordering in a new organic magnet. Phys Rev Lett. 1997; 79:2336–2339.

30. Sugimoto T, Tsujii M, Matsuura H, Hosoito N. Weak ferromagnetism below 12 K in a lithium tetrafluorotetracyanoquinodimethanide salt. Chem Phys Lett. 1995; 235:183–186.

31. Moriya T. Weak ferromagnetism. In: Rado GT, Suhl H, eds. Magnetism I. New York: Academic Press, 1963, pp 85–125.

32. De Jongh LJ, Miedema AR. Experiments on simple magnetic model systems. Adv Phys. 1974; 23:1–247.

33. Togashi K, Imachi R, Tomioka K, Tsuboi H, Ishida T, Nogami T, Takeda N, Ishikawa M. Organic radicals exhibiting intermolecular ferromagnetic interactions with high probability: 4-Arylmethyleneamino-2,2,6,6-tetramethylpiperidin-1-yloxyls and related compounds. Bull Chem Soc Jpn. 1996; 69:2821–2830, and references cited therein.

34. Yamauchi J, Adachi K, Deguchi Y. Magnetic phase transition of organic free radical p-Cl-BDPA. J Phys Soc Jpn. 1973; 35:443–447.

35. Jacobs IS, Bray JW, Hart HR Jr, Interrante LV, Kasper JS, Watkins GD, Prober DE, Bonner JC. Spin-Peierls transitions in magnetic donor-acceptor compounds of tetrathiafulvalene (TTF) with bisdithiolene metal complexes. Phys Rev B. 1976; 14:3036–3051.

36. Huizinga S, Kommandeur J, Sawatzky GA, Thole BT, Kopinga K, de Jonge WJM, Roos J. Spin-Peierls transition in N-methyl-N-ethyl-morpholinium-ditetracyanoquinodimethanide [MEM-(TCNQ)$_2$]. Phys Rev B. 1979; 19:4723–4731.

37. Hase M, Terasaki I, Uchinokura K. Observation of the spin-Peierls transition in linear Cu^{2+} (spin-1/2) chains in an inorganic compound CuGeO$_3$. Phys Rev Lett. 1993; 70:3651–3654.

38. Hase M, Terasaki I, Uchinokura K, Tokunaga M, Miura N, Obara H. Magnetic phase diagram of the spin-Peierls cuprate CuGeO$_3$. Phys Rev B. 1993; 48:9616–9619.

39. Awaga K, Sugano T, Kinoshita M. Ferromagnetic intermolecular interaction in the galvinoxyl radical: Cooperation of spin polarization and charge-transfer interaction. Chem Phys Lett. 1987; 141:540–544.

40. Awaga K, Maruyama Y. Ferromagnetic and antiferromagnetic intermolecular interactions of organic radicals, a-nitronyl nitroxides. II. J Chem Phys. 1989; 91:2743–2747.

22

Neutron Diffraction Studies of Spin Densities in Magnetic Molecular Materials

Jacques Schweizer
Atomic Energy Commission, Grenoble, France

Béatrice Gillon
Léon Brillouin Laboratory, Atomic Energy Commission—CNRS, Gif-sur-Yvette, France

I. INTRODUCTION

Neutrons are microscopic probes that travel through matter. They interact with the nuclei of the atoms, which produces nuclear scattering. They interact with the magnetic clouds of unpaired electrons, which results in magnetic scattering. They have been extensively employed to investigate the magnetic properties of materials for nearly 50 years, but the studies concerning molecular magnetism are rather recent.

Since neutrons are mobile probes that see at one glance many atoms and molecules, they can detect any order of the magnetic moments and, at least in principle, resolve it. Going one step further, they are also able to describe the shape of the magnetic clouds that form the magnetic moments and that are at the origin of the magnetic scattering.

The magnetic scattering of neutrons has been extensively practiced for metals, alloys, and inorganic compounds, where unpaired electrons are d or f electrons. More recently, it has been applied to molecular magnets, where the role of p electrons become very important, even dominating.

In this last case of molecular compounds, the number of strongly magnetic atoms embedded in organic ligands is rather small, and the amount of magnetic scattering is in general low compared to the amount of nuclear scattering. As a result, the usual investigations concerning magnetic ordering are rather difficult, and the number of magnetic structures resolved by (un-polarized) neutron diffraction is limited [1–5]. However, when it is possible to make use of polarized neutrons, with their enhanced sensitivity—that is, when the magnetic moments are either ferromagnetically ordered or partially ordered by an applied magnetic field—the investigation of the spin-density distribution becomes tractable.

This chapter reviews the principal results that have been obtained in this domain of spin densities in magnetic molecular compounds. The successive sections will describe the experimental spin distributions found on different compounds, answering successively the following questions: (a) Where is the spin density located? (b) What are the signs of the spin densities? (c) What are the shapes of the spin densities? (d) How are the spin densities related to the magnetic interactions present in the compounds? But before these questions are answered, the first section will focus on the polarized neutron diffraction (PND) method and explain how it is possible to get a spin-density map from the raw data.

II. POLARIZED NEUTRON DIFFRACTION AND SPIN-DENSITY MAPS

Polarized neutron diffraction applies to molecules in crystals presenting ferromagnetic, ferrimagnetic, or paramagnetic behavior, providing that a sufficiently high magnetization can be induced in the sample by an external magnetic field.

II.A. Principles

Whereas x-rays interact with the electronic clouds of atoms, neutrons interact both with nuclei and with electronic magnetic moments. The diffraction of a neutron beam by a magnetically ordered single crystal gives rise to Bragg reflections, the intensity of which contains nuclear and magnetic contributions, expressed in term of the nuclear and magnetic structure factors F_N and F_M, and depends on the polarization of the beam [6]:

$$I_{\pm} = F_N^* F_N + \sin^2\alpha(F_M^* F_M) \pm \sin^2\alpha(F_N^* F_M + F_M^* F_N) \qquad (1)$$

where α is the angle between the electronic spin direction and the scattering vector **K** of the Bragg reflection.

The nuclear structure factors are defined by:

$$F_N(\mathbf{K}) = \sum_j b_j e^{i\mathbf{K}\mathbf{r}_j} e^{-W_j} \tag{2}$$

where j refers to the atoms contained in the elementary cell, b_j is the nuclear scattering length, expressed in 10^{-12} cm, characteristic of the chemical element, and W_j is a thermal factor.

The magnetic structure factors are the Fourier components of the magnetization density $m(\mathbf{r})$:

$$F_M(\mathbf{K}) = \int_{\text{cell}} m(\mathbf{r}) e^{i\mathbf{k}\mathbf{r}} \, d^3\mathbf{r} \tag{3}$$

The current unit for magnetic structure factors is the Bohr magneton (μ_B)* and for magnetization density is the Bohr magneton per cubic angstrom $(\mu_B/\text{Å}^3)$. The nuclear and magnetic structure factors F_N and F_M are real quantities in the case of centric structures and complex quantities otherwise.

In the general case, the magnetization density is the sum of the spin and the orbital densities. However, for the $2p$ electrons in organic radicals and for most of the transition metals, the orbital moment is almost entirely quenched and the magnetization density $m(\mathbf{r})$ is identical to the spin density $s(\mathbf{r})$ [7].

In the polarized neutron diffraction technique, a magnetic field is applied in order to align partly or totally the electronic moments in the sample. This technique takes advantage of the fact that the sign of the cross-term between the nuclear and magnetic contributions in the diffracted intensity depends on the polarization of the incident neutron beam, which is chosen to be parallel or antiparallel to the electronic spin's direction. The direction of the polarization of the incident neutron beam is vertical, and the neutron spins can be flipped from the "up" state to the "down" state with the help of a flipping device. For each Bragg reflection a flipping ratio is measured, that is, the ratio of intensities when the incident neutron spins are "up" $(+)$ or "down" $(-)$. It has the following expression for a centrosymmetric cell:

$$R(\mathbf{K}) = \frac{I_+}{I_-} = \frac{F_N^2 + 2p \sin^2 \alpha F_N F_M + \sin^2 \alpha F_M^2}{F_N^2 - 2pe \sin^2 \alpha F_N F_M + \sin^2 \alpha F_M^2} \tag{4}$$

where p is a correction for the imperfect polarization of the incident beam and e is the flipping efficiency [8]. No absorption corrections are necessary, since the corrections on I_+ or I_- are equal, their ratio being then equal to unity. In contrast, extinction corrections have to be accounted for in the case of strong extinction.

---

*The Eq. (3) has to be scaled by a constant 0.2696×10^{-12} cm/μ_B in order to express F_N and F_M in the same unit of 10^{-12} cm.

If the crystal structure is well known, the magnetic structure factors F_M can be derived directly from the experimental flipping ratios, using Eq. (4). This is why a precise structure determination is generally undertaken in complement to the polarized neutron study, in the same temperature range.

The flipping ratio technique is characterized by a high sensitivity. Induced magnetic moments as small as 10^{-3} μ_B can be observed. The order of magnitude of the required magnetic susceptibility in a general case is 0.1 $\mu em/mol$ (or cm^3/mol).

II.B. Data Analysis

There are two principal routes by which to treat the experimental data in order to reconstruct spin densities: direct methods that are model free, and methods that require modeling of the spin-density distribution.

II.B.1. Direct Methods

Because magnetic structure factors are the Fourier coefficients of the periodic spin density, the most natural way to retrieve spin-density map is the simple Fourier inversion of the series:

$$s(\mathbf{r}) = \frac{1}{V} \sum_{\mathbf{K}} F_M(\mathbf{K}) \exp(-i\mathbf{Kr}) \tag{5}$$

where V is the cell volume. However, such a transformation is exact only if the summation is performed over all of the $\mathbf{K}$ vectors of the reciprocal space, which, obviously, cannot be done. What is done practically is a truncated summation, limited to the Bragg reflections that have actually been measured. This truncation induces spurious oscillations in the map, oscillations known as *truncation errors*. These oscillations can be reduced by well-known treatments that consist essentially of averaging the spin density over a small volume around the current point.

Recently, a Bayesian approach to solving the inverse Fourier problem appeared that gave much better results than the inverse Fourier series, particularly for a limited set of noisy data [9]. It evaluates, for each possible reconstructed map, the probability of this map. Such a conditional probability, or "posterior probability," is the product of two probabilities: the "likelihood" and the "prior." The "likelihood" represents the probability for the set of experimental data to be observed if the map were the real map, and can be represented by $e^{-1/2\chi^2}$ where χ^2 is the agreement between calculated and observed data. The "prior" probability represents the intrinsic probability of the map, prior to the neutron experiment, and can be expressed in terms of the entropy of the map. The map that is chosen as the best map is that which maximizes the entropy and keeps a good agreement with the observed data. This is why it is called the maximum entropy (MAXENT) method. It has been applied to the spin-density reconstruction from magnetic structure factors [10] and, very recently, extended to the flipping ratios for acentric structures [11].

II.B.2. Methods Modeling the Spin Density

A completely different analysis method consists of modeling the spin density, to refine the parameters of the model that best fit the experimental data and to represent the map that has been determined that way [12]. According to the function that is parametrized, one can distinguish two types of modeling.

In the orbital modeling, a Hartree–Fock type of magnetic wave function $|\Psi_i\rangle$ is constructed from standard Slater orbitals at each magnetic site: $|\Psi_i\rangle = \Sigma_j \, \alpha_{ij} |\phi_j\rangle$. To allow both positive and negative spin populations, the spin density is expanded as:

$$s(\mathbf{r}) = \sum_{\text{atoms}} S_i \langle \Psi_i | \Psi_i \rangle \tag{6}$$

The individual atomic populations S_i, the coefficients α_{ij}, and the radial exponents of the Slater functions of each orbital are the parameters of the model.

A more versatile modeling is a direct multipole expansion of the spin density around the nuclei. It consists of a superposition of aspherical densities, each described by a series expansion in real spherical harmonics:

$$s(\mathbf{r}) = \sum_{\text{atoms}} \sum_{l} R_l(r) \sum_{m=-1}^{1} P_{lm} y_{lm}(\hat{\mathbf{r}}) \tag{7}$$

where P_{lm} are the population coefficients and $R_l(r)$ are the radial functions of the spin density, also of Slater type. The parameters to be refined are the populations and the radial exponents.

Both modeling methods are very convenient and versatile. In particular they apply to acentric crystal structures. However, it is clear that these methods, in contrast to direct methods, will provide only results that are already embraced by the model and in no case can go beyond the model.

III. THE DELOCALIZATION OF THE SPIN DENSITY

The main characteristics of magnetism in a molecular compound is that the magnetic cloud is not limited to magnetic centers but extends over several

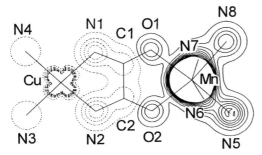

Figure 1   Induced spin-density map for [Mn(cth)Cu(oxpn)](CF$_3$SO$_3$)$_2$ at 2 K under 5 T in projection along the perpendicular to the oxamido mean plane. The solid lines represent positive levels; dashed lines represent negative levels: the equidistance between the contours is 0.01 μ_B/Å^2.

atoms of the molecule. Therefore the first and most fundamental question to ask about the spin density in such materials is "Where is it located?" or, in other words, "On which atoms is it delocalized?"

Of particular interest are the binuclear species, where two metals are bridged by organic ligands, in which the unpaired spins located in the d metal orbitals can delocalize towards the s and p orbitals of the ligands. In the molecular compound [Mn(cth)Cu(oxpn)](CF$_3$SO$_3$)$_2$,* the Mn^{2+} ion and Cu^{2+} ions are bridged by an oxamido group, which is known to favor an antiferromagnetic coupling between the metallic centers. The induced spin-density map in the quintet ground state in Figure 1 shows the negative coupling between copper and manganese as well as the spin delocalization towards their ligands [13]. The negative spin density appears to be more delocalized on the bridge than does the positive one. The copper spin population represents only 70% of the total negative population, whereas the spin population on manganese amounts to 93% of the total positive population. This reflects the ability of copper to form metal–ligand bonds with a higher covalent character than manganese.

The situation is somewhat different when there is no magnetic metal in the molecule and all the magnetism stems from $2p$ electrons: there is no magnetic center and the delocalization of the spin density over several atoms seems to be more fundamental.

One of the first studies of the spin-density distribution by polarized neutron diffraction on a molecular compound was performed on the well-known diphenylpicrylhydrazyl (DPPH) radical [14]. Here, two central ni-

---

*cth = (Me$_6$-[14]ane-N$_4$) and oxpn = N,N'-bis(3-aminopropyl)oxamido.

trogen atoms are linked to two phenyl rings on nitrogen N_α, and to one picryl ring on nitrogen N_β. The spin-density distribution is reproduced in Figure 2. It is delocalized on practically all the atoms of the molecule, but this delocalization corresponds to two levels. The major part of the density is located on the two central nitrogen atoms, and the remaining part is delocalized over the atoms of the three aromatic rings, representing positive and negative contributions (see Figure 3). The average spin population on an atom of these rings amounts to about one-tenth of the spin population on the central nitrogens.

Even less localized is the spin density on the tetracyanoethylenide ion $[TCNE]^-$, where the $C_2(CN)_4$ molecule is the acceptor of an electron coming from the nonmagnetic donor $[Bu_4N]^+$ [15]. The spin distribution is represented in Figure 4: 33% of the total density is located on each of the central carbon atoms, and 13% is carried by each of the terminal nitrogens. The *sp*

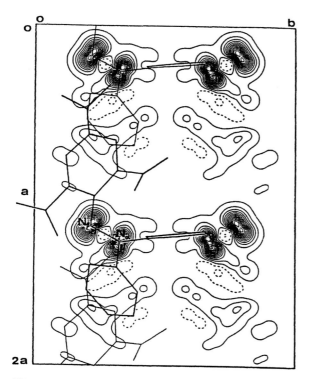

Figure 2   DPPH: spin density of the entire molecule projected along the *c*-axis of the cell. The solid lines represent positive levels; the dashed lines represent negative levels. Equidistance between the contours: 0.03 $\mu_B/\text{Å}^2$.

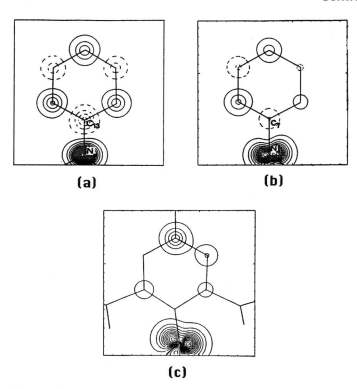

Figure 3 DPPH: spin density on each of the three rings projected onto the plane of the ring: (a) C13 phenyl, (b) C7 phenyl, (c) picryl. Equidistance between the contours: 0.03 μ_B/Å^2.

intermediate carbons present a significant but negative spin population. In such a delocalized case, it is clear that the molecular wave function of the unpaired electron concerns all the atoms of this small molecule, and that it is difficult to speak of a "magnetic center."

In alkyl nitroxide radicals, the spin density is located mainly on the NO bond [16,17], as shown in Figure 5 for the tempone radical* of formula $OC_9H_{16}NO$. The unpaired spin is equally distributed on the nitrogen and oxygen atoms, with the corresponding spin populations being equal within the experimental uncertainties: $0.41(2)\mu_B$ on N and $0.37(2)\mu_B$ on O in the tempone radical. However a non-negligible delocalization of 22% of the total spin is observed on the alkyl backbone.

---

*4-oxo-2,2,6,6-tetramethyl-1-piperidinyloxy.

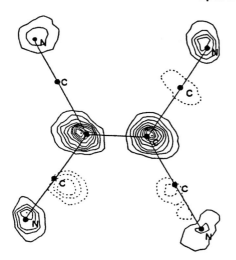

Figure 4 [TCNE]⁻: projection of the spin density onto the plane of the molecule. Step for positive contours 0.05 $\mu_B/\text{Å}^2$; for negative contours: 0.01 $\mu_B/\text{Å}^2$.

The spin delocalization is twice larger in the indolinonic nitroxide radical* [18], as shown in Figure 6. The nitrogen of the NO group is involved in conjugative interactions with the C=C double bonds of the fused benzene ring and the C=O bond of the carboxide group substituted on the five-membered ring. The spin transferred from NO to the rest of the molecule amounts to 42% of the total spin, with substantial populations on the C=O group. Positive and negative populations are carried by the carbon atoms of the benzene ring. A strong perturbation of the spin density on N is associated with this delocalization. In comparison with the nitrogen/oxygen distribution in the tempone radical, the nitrogen spin population is significantly reduced [$0.24(3)\mu_B$], whereas the oxygen population remains about the same [$0.34(92)\mu_B$].

Finally, the nitronyl nitroxide (NN) radicals, whose density has been extensively investigated during recent years, correspond to one unpaired electron, located mainly on an ONCNO group. The corresponding molecular wave function (SOMO) is doubly antibonding: it alternates signs between N and O, and also between the two NO groups, with a node on the central carbon. This radical has proven to be a versatile building block for the design of purely organic magnetic material and mixed coordination complexes as

---

*1,2-dihydro-2-methyl-2-phenyl-3*H*-indole-3-oxo-1-oxyl.

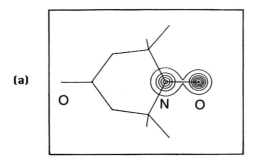

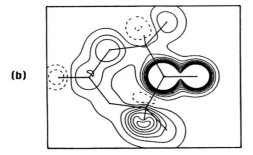

Figure 5 Tempone radical $OC_9H_{16}NO$: projection of the induced spin-density map along the π-direction at 2 K under 4.6 T. The plots are normalized to 1 μ_B per molecule: (a) high contours: step 0.20 $\mu_B/\text{Å}^2$; (b) low contours: step 0.01 $\mu_B/\text{Å}^2$.

well. There are many NN derivatives. In particular, the phenyl-substituted nitronyl nitroxide* (PNN) is the most popular representative of this family. The spin-density distribution measured on this molecule, and obtained from an orbital refinement [19], is represented in Figure 7. As predicted, it is equally shared by the two NO groups, and on each of these groups by the nitrogen and the oxygen. The part that is delocalized on the rest of the molecule is rather weak and will be discussed later. The reason for the success of this elementary brick in the synthesis of molecular magnets is rather clear. With one unpaired electron delocalized on the ONCNO fragment, it transmits the magnetic coupling over a large distance. A spin carrier that is bonded to the first oxygen will automatically be coupled with a spin carrier bonded to the last oxygen. Magnetic coupling through the phenyl ring is also possible and will be discussed later.

---

*2-phenyl-4,4,5,5-tetramethyl-4,5-dihydro-1*H*-imidazole-1-oxyl-3-oxide.

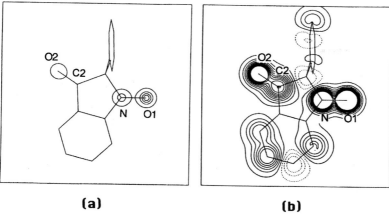

Figure 6 Indolinonic nitroxide radical $C_{15}H_{12}ONO$: projection onto the principal plane of the induced spin-density map at 9.8 K under 5 T. The plots are normalized to 1 μ_B per molecule: (a) high contours: step 0.20 $\mu_B/\text{Å}^2$; (b) low contours: step 0.01 $\mu_B/\text{Å}^2$.

IV. THE SIGN OF THE SPIN DENSITY

The second question to ask concerns the sign of the spin density. During the diffraction experiment the external applied field aligns the majority spin. Those regions where the spin density is opposed to the majority spin correspond to negative spin densities. Trivial cases are those concerning materials containing several spin carriers, with negative intramolecular spin couplings like the oxamido-bridged MnCu compound (Figure 1).

More subtle are cases of free radicals where magnetism is related to one unpaired electron only. In such conditions, we can no longer evoke the usual Hartree–Fock, or rather the restricted Hartree–Fock (RHF), scheme, where all the molecular orbitals are doubly occupied except the highest (SOMO), which is singly occupied. One is faced with a phenomenon of spin polarization: the ground state is obtained by configuration interaction (CI) from a linear combination of several RHF configurations. An example of this spin polarization is given in [TCNE]$^-$ (Figure 4), where each of the intermediate carbons carries a negative spin population that represents, in absolute values, 15% of the positive spin population on one central carbon.

Another example is the negative density that is found on the central carbon of the ONCNO fragment of the PNN radicals. The simple image of the unpaired electron molecular orbital predicts a node at this position. The spin-polarization effects produce a rather large negative contribution, of the

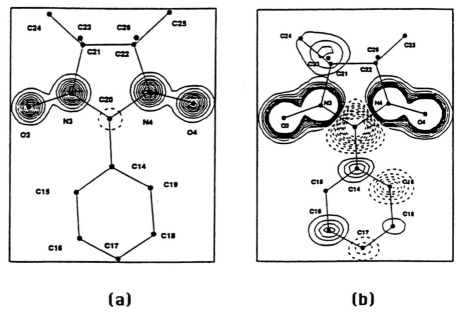

(a)   **(b)**

Figure 7 Phenyl nitronyl nitroxide (PNN): projection of the spin density onto the two rings of the molecule: (a) high contours (step 0.10 $\mu_B/\text{Å}^2$); (b) low contours (step 0.01 $\mu_B/\text{Å}^2$).

order of one-third of the positive density located on the nitrogen or the oxygen atoms (Figure 7). This negative density is systematically found on all the nitronyl nitroxides that have been studied up to now. At a lower level, one can note on the phenyl ring of the PNN an alternation of the signs. Similar alternations have been found in several cases, as, for instance, the aromatic rings of DPPH as displayed on Figure 3. On other NN radicals, the sequence of signs may be disturbed by magnetic couplings.

Spin polarization was expected to play a key role in the strong ferromagnetic coupling through the azido groups N_3^- in di-μ-azido dicopper compounds. This mechanism implies a negative spin density on the bridging nitrogen atom. The contrary is demonstrated by the induced spin-density map in $[\text{Cu}_2(t\text{-bupy})_4(\text{N}_3)_2](\text{ClO}_4)_2$* presented in Figure 8: a positive spin density is observed on the bridging atom [20]. However, a small negative spin density on the central nitrogen atom of the azido group gives evidence

---

*bupy = butylpyridine.

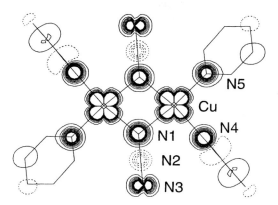

Figure 8 Induced spin-density projection for $[Cu_2(t\text{-bupy})_4(N_3)_2](ClO_4)_2$ along a perpendicular to the $(Cu-N1-Cu')$ plane at 1.6 K under 4.6 T. Low contours: step $0.01 \ \mu_B/Å^2$.

for the spin polarization of the azido bridge. The spin distribution in the triplet ground state is due mainly to the spin-delocalization phenomenon over which is superimposed spin polarization of the azido bridge.

V. THE SHAPE OF THE SPIN DENSITY

It is also very fruitful to consider the shape of the magnetic distribution around the magnetic atoms, and thus to gain insight into the nature of the spin distribution and the mechanisms of the delocalization. Clearly, for such a goal, one cannot use a parametrized model to treat the experimental data, for in such a model the shapes are already preestablished; one has to use a model-free treatment, and the maximum entropy method is best adapted for this.

A first point concerns the origin of the magnetic electrons. An evidence of the *p* shape of the magnetic orbitals on the ONCNO fragment of the nitronyl nitroxide radicals has been clearly established in the *para*-nitrophenyl nitronyl nitroxide* (*p*-NPNN), the first purely organic ferromagnet ($T_c = 0.67$ K). A MAXENT reconstruction of the spin density [11,21] is displayed in Figure 9. The projection along the ONCNO symmetry axis of the molecule clearly shows that the spin density, on each of these atoms, is carried by a *p* orbital perpendicular to the ONCNO plane. This is true also for the negative density on the central carbon.

---

*2-(4-nitrophenyl)-4,4,5,5-tetramethyl imidazoline-1-oxyl-3-oxide.

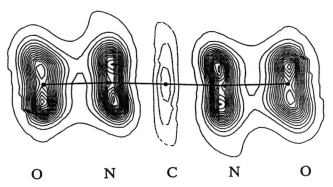

O   N   C   N   O

Figure 9   *para*-Nitrophenyl nitronyl nitroxide (p-NPNN): spin density, reconstructed by MAXENT and projected onto the π-plane of the ONCNO fragment. They clearly show the p shape of the magnetic orbitals.

Going one step further, it is also conceivable to demonstrate the bonding or antibonding character of the spin density. A careful examination of Figure 4—which represents the projection of the spin density on the plane of the molecule [TCNE]⁻—shows that on each of the central atoms the spin density is not centered at the nuclei C1 and C2, but is bent away from the midpoint of the C1–C2 bond. This shape represents the antibonding character of the molecular wave function. The reality of this bending has been very carefully checked by a detailed analysis of the multipole expansion [15] and by an extension of the maximum entropy method in order to include a nonuniform prior [22]. From both analyses it came out that the antibonding shape is really contained in the data and is not the result of an artifact.

A more complex situation occurs in the imino nitroxides. In these radicals the main part of the spin density lies on the N—C—N—O fragment, but, contrary to the case of nitronyl nitroxides, the carbon atom is not a node of the SOMO and can participate in this molecular orbital. The experimental spin densities [23] found on the two molecules (molecule A and molecule B, crystallographically independent) of a nitrophenyl imino nitroxide* (*m*-NPIN) compound, are shown in Figure 10. The central carbon carries a spin density, it is negative, and, for each of the two molecules, it is shifted from the nucleus position in the N1–N2 direction. This spin density is the sum of two contributions: the first one positive due to the delocalization of the SOMO, and a second, negative, due to the spin polarization. The SOMO has two nodes: one between O1 and N1 and one between C1

---

*2-(3-nitrophenyl)-4,4,5,5-tetramethyl-4,5,-dihydro-1H-imidazol-1-oxyle.

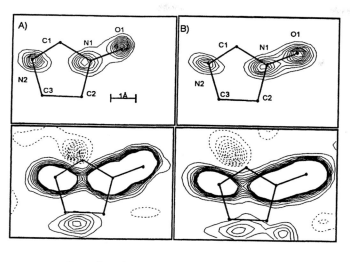

molecule A   molecule B

Figure 10 *meta*-Nitrophenyl imino nitroxide (*m*-NPIN): spin density reconstructed by MAXENT and projected onto the N—C—N—O planes of molecule A and molecule B. Contour steps 0.050 $\mu_B/\text{Å}^2$ (above) and 0.003 $\mu_B/\text{Å}^2$ (below).

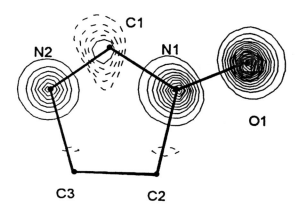

Figure 11 *meta*-Nitrophenyl imino nitroxide (*m*-NPIN): spin density calculated by the DFT method and projected onto the N—C—N—O plane. Contour step 0.020 $\mu_B/\text{Å}^2$ for positive density and 0.002 $\mu_B/\text{Å}^2$ for negative density.

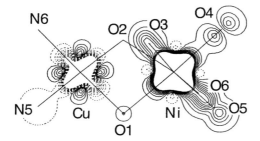

Figure 12 Induced spin-density map for Cu(salen)Ni(hfa)$_2$ at 2 K under 5 T in projection along the direction perpendicular to the CuO$_2$Ni plane. Low contours: step 0.01 μ_B/Å^2.

and N2, which implies that the orbital is antibonding on O1–N1, bonding on N1–C1, and antibonding on C1–N2. As a result, that on N1 the SOMO density is pushed in the N1–N2 direction, and on C1 it is pushed parallel to the N2–N1 direction. Because the negative spin polarization on the carbon atom is larger and centered on the nucleus, the resulting spin density is negative and clearly shifted parallel to N1–N2. This effect is reproducible. It has been retrieved [23] by first-principle calculations performed on this molecule by the density functional theory (DFT), as shown in Figure 11.

VI. SPIN DENSITY AND MAGNETIC INTERACTIONS

Finally, and probably the most productive information one can get from such investigations, is the discovery of the interplay between spin density and the magnetic interactions. For the first time, one can follow how the spin density is modified in interacting molecules, compared to isolated molecules. It is then possible, examining the spin density carefully, to find out the path of the magnetic coupling in complex systems, where it is not obvious a priori.

Heterodinuclear compounds are of fundamental interest for the study of magnetic exchange interaction. The first heterodinuclear compound to be studied by means of PND was Cu(salen)Ni(hfa)$_2$* [24]. The negative coupling between copper and nickel is shown in Figure 12 by the two spin-density regions of opposite signs associated with the metal sites on the induced spin-density map. In the bridging region no spin density appears on

---

*salen = N,N'-ethylenebis(oxosalicyldiiminato), and hfa = hexafluoroacetyl-acetonato.

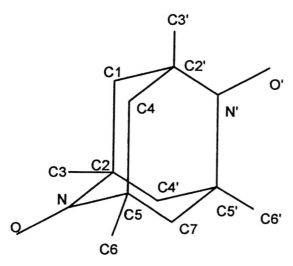

Figure 13 Structural formula of the dioxyl tetramethyl diazaadamantane (DTDA).

the oxygen atoms, while the four terminal oxygens on the nickel side carry positive spin densities and the nitrogen atoms linked to copper carry small negative spin densities. The positive spin density transferred from nickel to the phenolic oxygen atoms is almost exactly compensated by the negative spin transferred from copper, although the copper moment $(-0.22(1)\mu_B)$ is much smaller than the nickel one $(1.11(1)\mu_B)$. This reflects the more covalent character of the copper–ligand interaction, compared to nickel. The absence of spin density on the phenolic atoms demonstrates, paradoxically, their key role in the magnetic interaction between copper and nickel.

The ball-shaped radical DTDA* has raised considerable attention, for it is a purely organic ferromagnet with a Curie temperature $T_C = 1.48$ K. Its structural formula is shown in Figure 13. In this radical, the spin density stems from two unpaired electrons located on the two NO groups. The experimental density, as determined from PND [25], is presented in Figure 14. The main part corresponds, as expected, to p orbitals on the nitrogen and the oxygen of the NO groups, the orbitals of the two groups being orthogonal. A smaller part of the spin density is delocalized on the carbon atoms that form the four possible paths linking the two NO groups: C2–C1–C2′, C5–C4–C2′, C5–C7–C5′, and C2–C4′–C5′. The alternation of signs along these four paths is evidence that the ferromagnetic exchange interac-

---

*N,N′-dioxyl-1,3,5,7-tetramethyl-2,6-diazaadamantane.

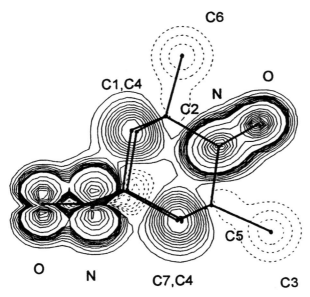

Figure 14 DTDA: spin density projected along C2–C5. The step of high positive contours is 0.100 $\mu_B/\text{Å}^2$; the step of low positive and negative contours is 0.005 $\mu_B/\text{Å}^2$.

tion between the two NO groups passes through the polarized carbon skeleton.

A case of strong modification of the spin density due to negative magnetic coupling has been met in the $CuCl_2(PNN)_2$ complex,* a compound where one copper is bonded to two oxygen atoms, each belonging to a different NN radical. The molecule carries three $S = 1/2$ spins. The coordination of the copper atom is equatorial (two O and two Cl), favoring a negative coupling between Cu and NN radicals. The spin density on half a molecule, as measured at $T = 13$ K [26], is presented in Figure 15. The ground state is a doublet $S = 1/2$, and, as expected in this state, the fraction of the spin density is twice larger on the NN than on the central Cu. On the NN part of the map, compared to the distributions usually measured on these radicals, there is a very striking difference: the spin density lies on the ONCN atoms, with the usual negative sign on carbon, but has completely disappeared from the oxygen linked to the copper. This is the result of the

---

*dichloro-bis(2-phenyl-4,4,5,5-tetramethyl-1-oxy-imidazoline 3-oxide)-copper(II).

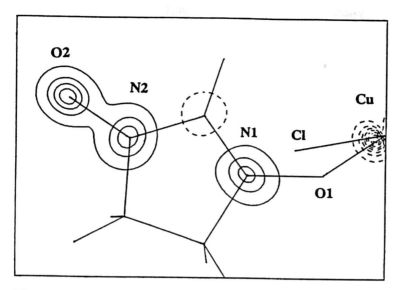

Figure 15 CuCl2(PNN)2: projection of the spin density of half the molecule onto the ONCNO plane of one of the two NITs. The equidistance between the contours is 0.005 $\mu_B/\text{Å}^2$.

strong magnetic interaction with Cu that pushes the spin density away from the connecting oxygen.

Along the same line, negative interactions can provoke a practical collapse of all the spin density on one spin carrier at low temperatures. This is the case of the enaminoketone $C_{18}H_{28}O_2N_2Cu(NO)_2$, a molecule* that carries three $S = 1/2$ spins: one on the Cu atom and one on each of the two NO groups. It crystallizes in the monoclinic system with two molecules A and B in the asymmetric unit. Its magnetic susceptibility shows a slight increase of χT between 300 K and 45 K, indicating intramolecular ferromagnetic coupling, and a steep drop below 40 K, which reveals intermolecular antiferromagnetic interactions. The spin densities [27], measured at 40 K and 4.10 K, are displayed in Figure 16. At $T = 40$ K (Figure 16a), one clearly sees localized spin densities on the copper atom and the two NO groups of each molecule. At $T = 4.10$ K (Figure 16b), the striking fact is the quasi-disappearance of the spin density on N2O2 and N6O6. These two NO units belong to the two different molecules: N2O2 to molecule A and N6O6 to molecule B. They face each other at a distance of 3.40 Å. It is obvious that

---

*Cu(II) complex of 4-(1-ethenyl-2-oxylto)-2,2,5,5-tetramethyl-imidazoline-1-oxyl.

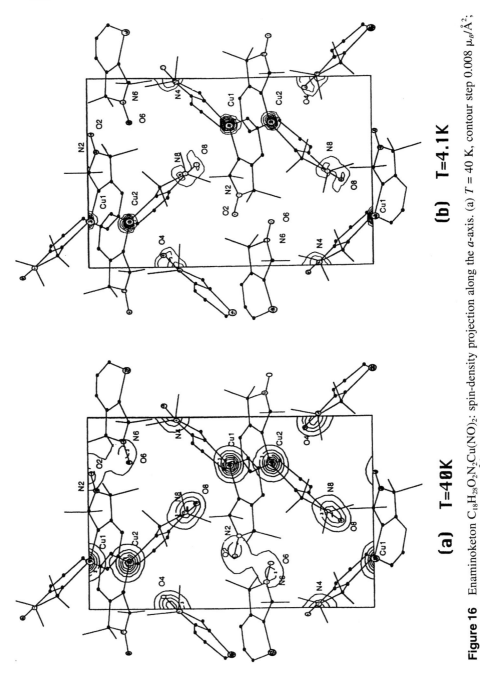

Figure 16 Enaminoketon $C_{18}H_{28}O_2N_2Cu(NO)_2$: spin-density projection along the a-axis. (a) $T = 40$ K, contour step 0.008 $\mu_B/\text{Å}^2$; (b) $T = 4.1$ K, contour step 0.090 $\mu_B/\text{Å}^2$.

(a) T=40K

(b) T=4.1K

the negative coupling between adjacent molecules, revealed by the drop of the magnetic susceptibility, actually corresponds to a negative coupling between these two spin carriers, which causes an almost complete dimerization of these two spins.

One can also focus on the appearance of unexpected but significant spin density on particular atoms, in order to detect the path of magnetic coupling. This is the case in the ferromagnetic p-NPNN [21], where the map based on orbital modeling (Figure 17) has shown the existence of a positive contribution on nitrogen N2 of the nitro group (spin population $0.020 \pm 0.005 \ \mu_B$), a contribution that corresponds to a p orbital perpendicular to the NO$_2$ plane. Figure 18 clearly shows that this density is located in the middle

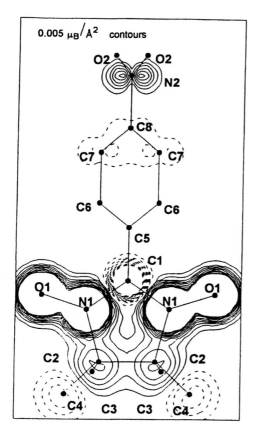

Figure 17 p-NPNN: low-contour spin-density map showing the p orbital on N2. Contour step 0.005 $\mu_B/\text{Å}^2$.

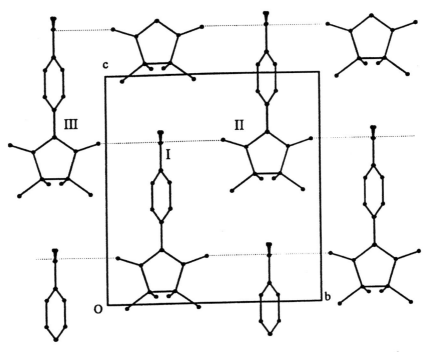

Figure 18 *p*-NPNN: relative arrangement of the molecules in the crystal.

between two oxygens O1, from the ONCNO fragments of two adjacent molecules. The distance N2–O1 is 3.22 Å, and the orthogonality of the *p* orbitals of N2 and O1 favors a ferromagnetic coupling between molecules I and II as well as between molecules II and III, leading to a two-dimensional ferromagnetic network.

Another case in which some spin density can be considered as tracing the path of the magnetic coupling has been found on hydrogen bonds. An example is the NN Py (C≡C—H).* The crystal structure consists of chains of molecules connected by the C≡C—H···O fragment, where the hydrogen bond links one carbon atom (not the usual situation) to O1, an oxygen atom of the ONCNO fragment of the next molecule. The chains are ferromagnetically coupled. The spin density is presented in Figure 19 [28]. The radical exhibits the usual features on the NN fragment and some sign alternation on the phenyl ring. However, there is a noticeable positive spin con-

---

*2-(6-ethynyl-2-pyridyl)-4,4,5,5-tetramethyl-4,5-dihydro-*1H*-imidazoline-1-oxyl-3-oxide.

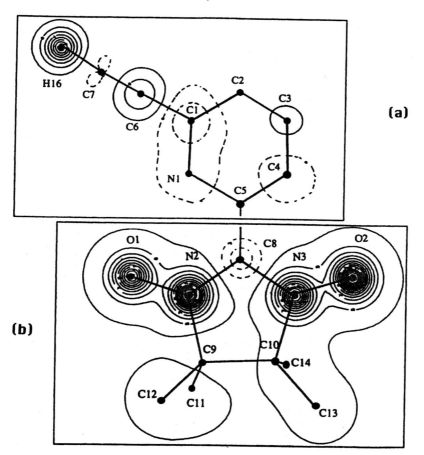

Figure 19   NIT Py(C + C—H): spin density projected: (a) onto the pyridine cycle, step 0.008 $\mu_B/\text{Å}^2$; (b) onto the NIT cycle, step 0.040 $\mu_B/\text{Å}^2$.

tribution on the H atom of the hydrogen bond: 0.042 (9) μ_B. It is significant and large enough to indicate that the hydrogen bond propagates the magnetic interaction along the chain. It is induced, in all likelihood, by the spin density of the oxygen atom O1 of the next molecule. Confirmation stems from the relative depletion of the unpaired electron on the bonded oxygen compared to the nonbonded one: 0.188 (10) μ_B instead of 0.258 (9) μ_B. Similar to the previous radical, the magnetic interaction propagates along the pyridine ring and then along the acetylenic fragment.

VII. CONCLUSIONS

Neutron diffraction, or, more precisely, polarized neutron diffraction, is certainly the best experimental technique to investigate the spin-density distribution in magnetic molecular materials. It provides maps of this distribution over the whole molecule. These maps may be compared to the distribution of spin density calculated theoretically from first principles.

There are several methods to perform these ab initio calculations. Comparisons between theory and experiment have shown that, in general, the methods based on the Hartree–Fock approximation (HF and UHF) yield spin densities that are far from what is measured experimentally, whereas those based on the density functional theory (DFT) give much closer results. Even in this last case, however, discrepancies exist that are quite noticeable. An experimental map, then, is for most situations the best solution to answer questions concerning the spin-density distribution, and polarized neutron diffraction is a unique tool to get this map.

REFERENCES

1. Benoit A, Flouquet J, Gillon B, Schweizer J. J Magn Magn Mater. 1983; 31–34:1155–1156.
2. Zheludev A, Ressouche E, Schweizer J, Turek P, Wan M, Wang H. Solid State Comm. 1994; 90:233–235.
3. Tomiyoshi S, Yano T, Azuma N, Shoga M, Yamada K, Yamauchi J. Phys Rev. B 1994; 49:16031–16034.
4. Antorrena G, LeCaër G, Malaman B, Palacio F, Ressouche E, Schweizer J. Physica B. 1997; 234–236:780–782.
5. Pellaux R, Schmalle HW, Decurtins S, Fischer P, Fauth F, Ouladdiaf B, Hauss T. Physica. B 1997; 234–236:783–784.
6. Forsyth JB. Atomic Energy Rev. 1979; 172:345–378.
7. Kittel C. Introduction to Solid State Physics. 6th ed. New York: Wiley, 1986, p 405.
8. Brown PJ, Forsyth JB. Brit J Appl Phys. 1964; 15:1529–1533.
9. Gull SF, Daniell GJ. Nature. 1978; 272:686–690.
10. Papoular RJ, Gillon B. Europhys Lett. 1990; 13:429–434.
11. Schleger P, Puig-Molina A, Ressouche E, Rutty O, Schweizer J. Acta Crystallogr. 1997; A53:426–435.
12. Gillon B, Schweizer J. In: Maruani J, ed. Molecules in Physics, Chemistry and Biology, vol II. Dordrecht: Kluwer, 1989, pp 111–147.
13. Baron V, Gillon B, Plantevin O, Cousson A, Mathonière C, Kahn O, Grand A, Öhrström L, Delley B. J Am Chem Soc. 1996; 118:11822–11830.
14. Boucherle JX, Gillon B, Maruani J, Schweizer J. Mol Phys. 1997; 60:1121–1142.

15. Zheludev A, Grand A, Ressouche E, Schweizer J, Morin BG, Epstein AJ, Dixon DA, Miller JS. J Am Chem Soc. 1994; 116:7243–7249.

16. Brown PJ, Capiomont A, Gillon B, Schweizer J. Mol Phys. 1983; 48:753–761.

17. Bordeaux D, Boucherle JX, Delley B, Gillon B, Ressouche E, Schweizer J. Z Naturforsch. 1993; 48a:117–119.

18. Caciuffo R, Francescangeli O, Greci L, Melone S, Gillon B, Brustolon M, Maniero AL, Amoretti G, Sgarabotto P. Mol Phys. 1991; 74:905–918.

19. Zheludev A, Barone V, Bonnet M, Delley B, Grand A, Ressouche E, Rey P, Subra R, Schweizer J. J Am Chem Soc. 1994; 116:2019–2027.

20. Aebersold MA, Kahn O, Bergerat P, Plantevin O, Pardi L, Gillon B, von Seggern I, Tuczek F, Öhrström L, Grand A, Lelièvre-Berna E. J Am Chem Soc. 1998; 120:5238–5245.

21. Zheludev A, Ressouche E, Schweizer J, Wan M, Wang H. J Magn Magn Mater. 1994; 135:147–160.

22. Zheludev A, Papoular RJ, Ressouche E, Schweizer J. Acta Crystallogr. 1995; A51:450–455.

23. Zheludev A, Bonnet M, Delley B, Grand A, Luneau D, Öhrström L, Ressouche E, Rey P, Schweizer J. J Magn Magn Mater. 1995; 145:293–305.

24. Gillon B, Cavata C, Schweiss P, Journaux Y, Kahn O, Schneider D. J Am Chem Soc. 1989; 111:7124–7132.

25. Zheludev A, Chiarelli R, Delley B, Gillon B, Rassat A, Ressouche E, Schweizer J. J Magn Magn Mater. 1995; 140–144:1439–1440.

26. Ressouche E, Boucherle JX, Gillon B, Rey P, Schweizer J. J Am Chem Soc. 1993; 115:3610–3617.

27. Bonnet M, Laugier J, Ovcharenko VI, Pontillon Y, Ressouche E, Rey P, Schleger P, Schweizer J. Mol Cryst Liq Cryst. 1997; 305:401–414.

28. Pontillon Y, Ressouche E, Romero F, Schweizer J, Ziessel R. Physica B. 1997; 234–236:788–789.

23

Heat Capacity Studies of Organic Radical Crystals

Michio Sorai, Yuji Miyazaki, and Takao Hashiguchi
Osaka University, Osaka, Japan

I. INTRODUCTION

Comprehensive understanding of materials will be achieved only when both their microscopic structural aspects and macroscopic energetic and/or entropic aspects are revealed. Although thermodynamic quantities principally reflect macroscopic aspects of materials, they are closely related statistically to the microscopic energy schemes of all kinds of molecular degrees of freedom. One can, therefore, gain detailed knowledge about the microscopic level on the basis of precise calorimetry. This is in particular the case for magnetic systems, in which quantum spins play fundamental roles. Among various thermodynamic measurements, heat capacity calorimetry is an extremely useful tool for investigating thermal properties arising from spin–spin interactions in magnetic materials at low temperatures [1–3]. In this chapter we shall demonstrate calorimetric evaluation of magnetic effects in molecule-based organic magnets.

II. HEAT CAPACITY CALORIMETRY

II.A. Thermodynamic Quantities Derived from Heat Capacity

Heat capacity is usually measured under constant pressure and designated as C_p. It is defined as "the enthalpy H required for raising the temperature of one mole of a given substance by 1 K." From this definition, we get C_p

$\equiv (\partial H/\partial T)_p$, where the enthalpy increment is determined by integration of C_p with respect to temperature, T:

$$H = \int C_p \, dT \qquad (1)$$

Since C_p is alternatively defined as $C_p \equiv T \, (\Delta S/\Delta T)_p$, entropy S is also obtainable by integration of C_p with respect to $\ln T$:

$$S = \int C_p \, d(\ln T) \qquad (2)$$

Since both enthalpy and entropy are derived from C_p measurements, one can estimate the Gibbs energy G as follows:

$$G = H - TS \qquad (3)$$

Heat capacity is a very useful physical quantity in that three fundamental thermodynamic quantities (H, S, and G) can be determined simultaneously solely from C_p measurements.

II.B. Phase Transitions

When magnetic interaction exists between spins of unpaired electrons, the otherwise-degenerate spin-energy levels become either less degenerate or completely nondegenerate. The heat capacity of the system clearly exhibits either cooperative phase transitions or noncooperative anomalies, depending on the magnetic interactions acting on the spins.

Heat capacity is sensitive to a change in the degree of the short-range or long-range ordering. Thus, heat-capacity calorimetry is the most reliable experimental tool for detecting the existence of phase transitions originating in the onset of long-range ordering. In the case of magnetic materials, the onset of long-range spin ordering appears as a sharp C_p peak at the Curie temperature T_C for ferromagnets and at the Néel temperature T_N for antiferromagnets. Contributions from all the degrees of freedom are in principal involved in experimental heat capacity. Therefore, for discussion of the nature of phase transitions, the contribution of the relevant degrees of freedom must accurately be separated from the observed heat capacity. In the case of a magnetic phase transition, contributions from spin–spin interactions appear as an excess heat capacity ΔC_p beyond a normal (or lattice) heat capacity. The excess enthalpy ΔH and entropy ΔS arising from the phase transition are determined by integration of ΔC_p with respect to T and $\ln T$, respectively. Based on these quantities, one can gain insight into the mechanism of the phase transition.

An example showing the sensitivity of heat capacity is seen in Figure 1, where magnetic susceptibility and heat capacity are compared for a two-dimensional spin $S = 1/2$ Heisenberg antiferromagnet, bis(5-chloro-2-aminopyridine) copper(II) tetrabromide, $(5\text{-CAP})_2\text{CuBr}_4$. As can be seen in Figure 1(a), the magnetic susceptibility data [4] seem to be well reproduced by

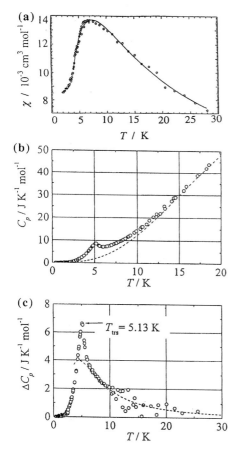

Figure 1 Two-dimensional Heisenberg antiferromagnet, $(5\text{CAP})_2\text{CuBr}_4$: (a) Magnetic susceptibility [4] (solid line is a theoretical curve determined for 2D square planar Heisenberg antiferromagnet with spin $S = 1/2$). (b) Molar heat capacity (broken curve stands for the lattice heat capacity). (c) Excess heat capacity due to three-dimensional spin ordering (the dashed curve corresponds to the theoretical curve for 2D square planar Heisenberg model with the antiferromagnetic interaction of $J/k_B = -4.0$ K [5].

a theoretical curve calculated for an S(spin) = 1/2 Heisenberg antiferromagnet in a two-dimensional (2D) square planar lattice. As far as the magnetic susceptibility is concerned, this complex seems not to exhibit a phase transition. Since it is believed that 2D Heisenberg magnets do not exhibit any phase transitions at finite temperatures [6], this substance would be regarded as an ideal 2D Heisenberg magnet. Contrary to this, the heat capacities shown in Figures 1(b) and 1(c) [7,8] clearly reveal the existence of a phase transition at 5.13 K caused by a weak 3D interaction that might exist in the actual condensed state. The large heat capacity tail above the transition temperature was well reproduced by a 2D square planar Heisenberg model with an antiferromagnetic interaction of $J/k_B = -4.0$ K. Although heat capacity and magnetic susceptibility are physical quantities similarly derived as ensemble averages in a given system, the former is often more sensitive to the onset of a long-range ordering than the latter.

II.C. Boltzmann's Principle

Although entropy is a physical quantity characteristic of macroscopic aspects of a system, the Boltzmann principle interrelates macroscopic entropy with the number of energetically equivalent microscopic states:

$$S = k_B N_A \ln W = R \ln W \tag{4}$$

where k_B and N_A are the Boltzmann and Avogadro constants, respectively, R is the gas constant, and W is the number of energetically equivalent microscopic states. If one applies this principle to the transition entropy, the following relationship is easily derived, where ΔS is the entropy difference between the high- and low-temperature phases:

$$\Delta S = R \ln \left(\frac{W_H}{W_L} \right) \tag{5}$$

where W_H and W_L mean the number of microscopic states in the high- and low-temperature phases, respectively. In many cases the low-temperature phase corresponds to an ordered state; hence, $W_L = 1$. Therefore, when a phase transition is due only to spin ordering, one can determine the spin quantum number s on the basis of the experimental ΔS value, by the relationship $W_H = 2s + 1$.

II.D. Lattice Dimensionality

Organic magnets consist of molecules as the building blocks. Since molecules are anisotropic in structure, magnetic interactions inevitably become anisotropic, leading to low-dimensional magnets characterized by one- or two-dimensional interactions. Spin ordering crucially depends on the magnetic lattice structure, resulting in various short-range and long-range order effects upon heat capacity curves. When paramagnetic species form clusters magnetically isolated from one another, the magnetic lattice should be regarded as being of zero dimension (0D), with a spin-energy scheme consisting of a bundle of levels. In such a case there exists no phase transition: heat capacity exhibits a broad anomaly characteristic of the cluster geometry. On the other hand, for a linear-chain structure (one-dimensional lattice, 1D), no phase transition is expected theoretically because fluctuation of the spin orientation is extremely large. Only a broad C_p anomaly characteristic of 1D structure is observed. For a two-dimensional (2D) structure, a phase transition showing a remarkable short-range order effect takes place when the interaction is of the Ising type, but no phase transition for the Heisenberg type. Contrary to this, three-dimensional (3D) structure gives rise to a phase transition with minor short-range order effects, independent of the type of spin–spin interaction.

In actual magnetic materials, interaction paths through which superexchange spin–spin interaction takes place are often much more complicated than can readily be classified into a single homogeneous dimension. As a result, it happens that apparent dimensionality seems to change with temperature. This behavior is called "dimensional crossover." Figure 2 shows the dimensional crossover between 1D and 2D lattices for a spin $S = 1/2$ ferromagnetic Ising model [9]. The dashed curve indicates the heat capacity

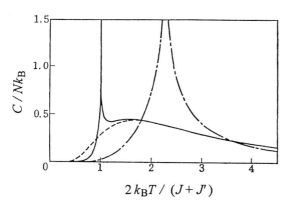

Figure 2 Dimensional crossover between 1D and 2D lattices for the spin $S = 1/2$ ferromagnetic Ising model. Dashed and dot-dashed curves correspond to 1D and 2D models, respectively; the solid curve indicates the dimensional crossover between them. J and J' mean the intra- and interchain interaction parameters, respectively. (From Ref. 9.)

anomaly arising from a linear chain with intrachain interaction J and interchain interaction $J' = 0$. The dot-dashed curve is the heat capacity of a square planar lattice ($J = J'$) showing a phase transition, corresponding to the exact solution by Onsager [10]. The solid curve shows the dimensional crossover between 1D and 2D lattices occurring in an anisotropic square lattice, in which the interchain interaction J' is assumed to be as small as 1/100 of the intrachain interaction J. At high temperatures the heat capacity curve approaches the 1D curve asymptotically, while at low temperatures the existent two-dimensional interchain interaction, though weak, diminishes the fluctuation of the system and leads to a phase transition.

II.E. Type of Spin–Spin Interaction

Fluctuation concerning the spin alignmemt strongly depends not only on the lattice dimensionality but also on the type of spin–spin interaction. The effective spin Hamiltonian accepted for the exchange interaction between spins 1 and 2 is given by the following generalized equation:

$$\mathcal{H} = -2J[\alpha(S_1^x S_2^x + S_1^y S_2^y) + \beta S_1^z S_2^z] \tag{6}$$

$\alpha = 1, \beta = 1$ Heisenberg model

$\alpha \neq \beta \ (\neq 0)$ Anisotropic Heisenberg model

$\alpha = 1, \beta = 0$ XY model

$\alpha = 0, \beta = 1$ Ising model

where S_1 and S_2 are the spin operators and J is the exchange interaction parameter.

Figure 3 presents the magnetic heat capacities for 1D $S = 1/2$ magnetic systems [1]. In the case of a magnetic system characterized by Ising- or XY-type interaction, heat capacity curves calculated for ferromagnetic and antiferromagnetic interactions are identical, unless the lattice dimensionalities are different (curves a and b in Figure 3). However, this is not the case for a system showing Heisenberg-type interactions: the heat capacities of antiferromagnetic (curve c) and ferromagnetic (curve d) Heisenberg chains. By comparing experimental heat capacity with these theoretical curves, one can easily evaluate the type of spin–spin interaction operating in the material.

II.F. Spin-Wave Excitation

Spin-wave (magnon) theory has proved to be a good approximation for describing the low-temperature properties of magnetic substances. The limiting low-temperature behavior of heat capacity due to spin-wave excitation (C_{SW}) is conveniently given by the following formula [1]:

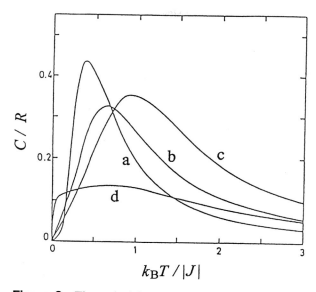

Figure 3 Theoretical heat capacities of magnetic chains with spin $S = 1/2$: (a) Ferro- or antiferromagnetic Ising model. (b) Ferro- or antiferromagnetic XY model. (c) Antiferromagnetic Heisenberg model. (d) Ferromagnetic Heisenberg model. (From Ref. 1.)

$$C_{SW} \propto T^{d/n} \tag{7}$$

where d is the dimensionality and n is the exponent in the dispersion relation. For antiferromagnetic magnons $n = 1$; for ferromagnetic magnons $n = 2$. Thus the spin-wave heat capacity of a 3D ferromagnet is proportional to $T^{3/2}$, that of a 2D antiferromagnet to T^2, and so on. However since actual low-dimensional magnets contain, more or less, weak three-dimensional interaction, the dimensionality sensed by a spin at extremely low temperatures should be $d = 3$, as exemplified later.

III. CALORIMETRIC EVALUATION OF MAGNETIC PROPERTIES OF MOLECULE-BASED MAGNETS

In this section we shall demonstrate how the tutorial just given may be applied to the characterization of actual molecular magnetism by exemplifying organic free radical crystals and mixed-metal complexes.

III.A.   Organic Free Radical Crystals

4-Methacryloyloxy-TEMPO (TEMPO: 2,2,6,6-tetramethylpiperidine-1-oxyl) (MOTMP), 4-acryloyloxy-TEMPO (AOTMP), 4-methacryloyl-amino-TEMPO (MATMP), and 4-(p-chlorophenylmethyleneamino)-TEMPO (CMTMP) are pure organic magnets having similar molecular structures based upon TEMPO (see Figure 4). The magnetic susceptibility from 2.5 to 300 K of MOTMP, for instance, shows ferromagnetic interaction between the neighboring radicals [11]. As reproduced in Figure 5, its heat capacity measured down to 0.07 K shows a λ-type transition due to the onset of long-range ordering at 0.14 K [11]. Unexpectedly, a remarkable broad anomaly was observed above the transition temperature. This anomaly was well accounted for in terms of short-range ordering in the ferromagnetic Heisenberg chains with an interaction parameter of $J/k_B = 0.48$ K [12]. As far as its crystal structure is concerned, there seems to be no indication of favorable one-dimensional exchange paths. This fact indicates how sensitive to dimensionality the heat capacity is. The finding of one-dimensional behavior was confirmed by magnetic susceptibility measurements done below 1 K [13]. As already discussed, the initial slope of the heat capacity at the lowest temperatures corresponds to the spin-wave excitation. In the case of MOTMP, the temperature dependence given by Eq. (7) was $d/n = 1.53$. This value is well approximated by 3/2, indicating that a dimensional crossover occurred from 1D to 3D and that MOTMP behaves as a three-dimensional ferromagnet below its transition temperature. The entropy gain due to the

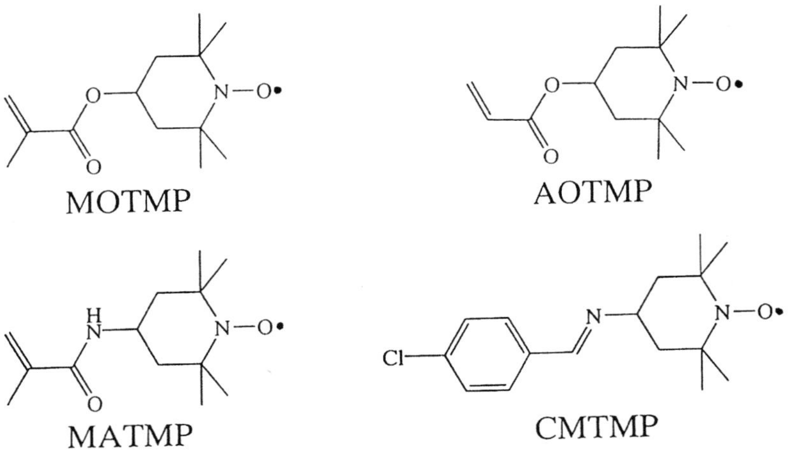

Figure 4   Molecular structure and abbreviation of the organic free radicals derived from TEMPO.

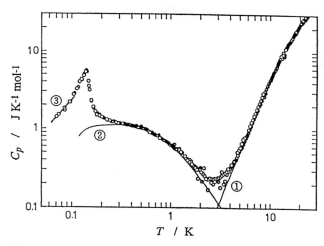

Figure 5 Molar heat capacity of MOTMP. (1) Lattice heat capacity; (2) 1D Heisenberg ferromagnet ($J/k_B = 0.45$ K); (3) spin-wave contribution.

cooperative and noncooperative anomalies was $\Delta S = 5.88$ JK^{-1} mol^{-1}. Since this value is very close to $R \ln 2 = 5.76$ JK^{-1} mol^{-1} expected for the spin $S = 1/2$ system, it turns out that 1 mol of spin is involved in the MOTMP crystal.

Heat capacities of AOTMP and MATMP [14] are compared with that of MOTMP in Figure 6. They exhibit phase transitions at 0.64 and 0.15 K, respectively. AOTMP had a broad C_p anomaly due to short-range ordering above the transition temperature, which is much higher than those of ferromagnetic MOTMP and MATMP. Since this anomaly is well reproduced by a linear antiferromagnetic Heisenberg model, AOTMP may be regarded as a 1D antiferromagnet. The spin-wave contribution to AOTMP is $d/n = 2.98$, nearly equal to the value of 3 for a 3D antiferromagnet. This fact indicates that AOTMP behaves as a 1D Heisenberg antiferromagnet at high temperatures but that weak magnetic interchain interactions lead AOTMP to become a 3D antiferromagnet below the Néel temperature.

In the case of MATMP, the temperature dependence of the heat capacity bears a close resemblance to that of MOTMP. The broad heat capacity anomaly above the Curie temperature was well reproduced by a 1D Heisenberg model with intrachain spin–spin exchange interaction of $J/k_B = 0.70$ K. The spin-wave contribution at low temperatures was $d/n = 1.57$ (approximately 3/2), indicating that MATMP is a 1D Heisenberg ferromagnet above T_C, while it behaves as a 3D Heisenberg ferromagnet below T_C. By analyzing the spin-wave heat capacity, we determined the ferromagnetic interchain

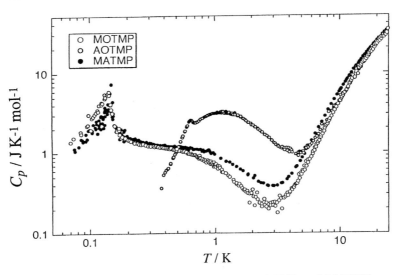

Figure 6   Molar heat capacities of MOTMP, AOTMP, and MATMP.

interaction to be $J'/k_B = 0.02$ K [14]. The entropy gain due to the spin ordering was $\Delta S = 5.49$ JK^{-1} mol^{-1}. This value agrees well with $R \ln 2 = 5.76$ JK^{-1} mol^{-1}, the value expected for a spin = 1/2 system, as expected, since the MATMP molecule possesses one unpaired electron.

On the other hand, CMTMP is known to be a 2D Heisenberg ferromagnet [15–18]. As shown in Figure 7 [19], its magnetic heat capacity exhibits a phase transition at 0.28 K. Although no phase transition would exist in a 2D Heisenberg system, the occurrence of a magnetic phase transition at 0.28 K implies the existence of weak 3D interaction. This Curie temperature is somewhat lower than the value of 0.4 K reported first on the basis of magnetic susceptibility measurements [16] but agrees well with the value of 0.28 K determined recently based on a μSR experiment [18]. The excess heat capacities above T_C correspond to a short-range spin ordering characteristic of a low-dimensional magnet. As seen in Figure 7, this broad anomaly was well reproduced by the theoretical values for a 2D square planar ferromagnetic Heisenberg model calculated by a high-temperature series expansion with the Padé approximation [5,20]. Although the spin arrangement in the actual crystal lattice is not exactly square planar [16,17], the magnetic lattice is topologically two-dimensional and is very close to being square planar. The intralayer spin–spin interaction was $J/k_B = 0.45$ K. The straight broken line drawn in Figure 7 corresponds to the spin-wave contribution given by $d/n = 1.38$ in Eq. (7). Since this value is well ap-

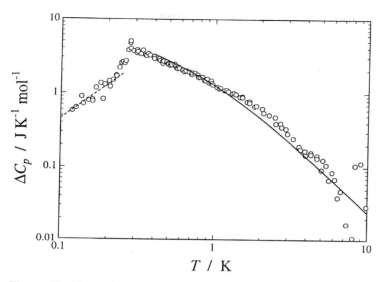

Figure 7 Magnetic heat capacities of CMTMP. The solid curve stands for the theoretical values calculated for 2D square planar $S = 1/2$ ferromagnetic Heisenberg model. The broken line approximates the spin-wave contribution to the heat capacity.

proximated by 3/2 (3D ferromagnet), the CMTMP radical crystal can be regarded as a 2D Heisenberg ferromagnet above T_C but a 3D ferromagnet below T_C. In fact, the interlayer magnetic interaction estimated on the basis of a spin-wave analysis was $J'/k_B = 0.012$ K, which supports 3D ferromagnetic ordering. The magnetic entropy gain 5.42 JK^{-1} mol^{-1} coincides well with the theoretical value expected for a spin = 1/2 system, $R \ln 2 = 5.76$ JK^{-1} mol^{-1}.

III.B. Mixed-Metal Complexes

As one of the strategies for molecule-based magnets, Okawa et al. [21] reported a series of mixed-metal assemblies $\{NBu_4[MCr(ox)_3]\}_n$ (NBu_4^+ = tetra-n-butyl-ammonium ion; ox^{2-} = oxalate ion; M = Mn^{2+}, Fe^{2+}, Co^{2+}, Ni^{2+}, Cu^{2+}, Zn^{2+}). All these complexes, except for the zinc homologue, exhibit spontaneous magnetization at low temperatures. On the basis of molecular model considerations it is suggested that these complexes form either a 2D- or a 3D network structure extended by Cr(III)-ox-M(II) bridges [21] (see Figure 8). However, since these were prepared as fine powders, single-crystal x-ray diffraction study had not been successful. As heat capacity is

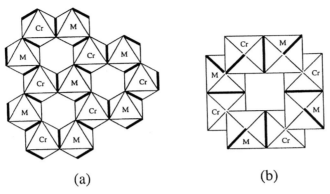

(a)   (b)

Figure 8 Schematic drawing of (a) 2D and (b) 3D network structures possible for the mixed-metal complex $\{NBu_4[MCr(ox)_3]\}_n$. The symbols M and Cr correspond to M^{2+} and Cr^{3+} ions, respectively. Thick lines stand for the edges shared by the adjacent metal octahedrals.

sensitive to lattice dimensionality, we tried heat capacity measurements of this series of complexes.

As shown in Figure 9, mixed-metal $\{NBu_4[CuCr(ox)_3]\}_n$ complex gave rise to a phase transition at 6.98 K [22,23]. This phase transition can be attributed to ferromagnetic spin ordering, because the entropy gain, $\Delta S = 16.7$ JK^{-1} mol^{-1}, coincides well with the theoretical value, $R \ln (2 \times 4) = 17.29$ JK^{-1} mol^{-1}, expected for a spin system consisting of Cu^{2+} (spin 1/2) and Cr^{3+} (spin 3/2). The transition temperature in this system agrees well with the onset temperature of spontaneous magnetization [21]. The spin–spin interaction in this complex can be regarded as being of Heisenberg type. Although no phase transition would exist in a 2D Heisenberg system, the occurrence of the magnetic phase transition does not necessarily imply that this complex should be characterized as a 3D Heisenberg ferromagnet. As in the case of CMTMP radical crystal, when 3D interaction is not neglected at low temperatures, one can expect a phase transition arising from 3D spin ordering. The heat capacities of $\{NBu_4[CuCr(ox)_3]\}_n$ in the spin-wave excitation region are well reproduced by $T^{1.51}$. Since the exponent 1.51 can be approximated by $d/n = 3/2$ in Eq. (7), the ordered state at low temperatures is actually proved to be a 3D ferromagnet.

A remarkable feature of the present phase transition is a large C_p tail above T_C. This type of short-range ordering effect is characteristic of 2D structure. In order to examine quantitatively the dimensionality of the magnetic structure, we formulated the high-temperature series expansion [24] with the Padé approximation for the 2D and 3D lattices shown in Figure 8

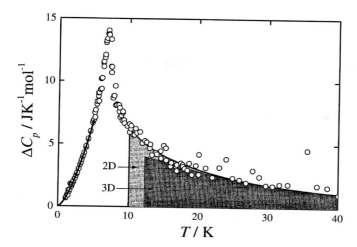

Figure 9 Excess heat capacity ΔC_p of $\{NBu_4[CuCr(ox)_3]\}_n$. Light and dark shaded areas correspond to the theoretical contribution of the short-range order effect persisting in the temperature region above T_C for 2D and 3D ferromagnetic lattices, respectively. The solid curve drawn at the low-temperature side of T_C indicates the spin-wave contribution.

and fit them to the excess heat capacities ΔC_p above T_C. As can be seen in Figure 9, good agreement is obtained for the 2D case, where the superexchange interaction parameter is $J/k_B = 5.0$ K. Although a 2D structure is favorable to this mixed-metal complex, it is of great interest that the ΔC_p estimated for 3D structure is unusually large in comparison to those for sc, bcc, or fcc lattices.* This might be caused by the fact that the number of nearest-neighbor paramagnetic ions is only three, which is extremely small compared with 6, 8, and 12 for sc, bcc, and fcc lattices, respectively. The same feature is also encountered in the heat capacity of the $\{NBu_4[FeCr(ox)_3]\}_n$ complex [23,25]. Therefore, from a thermodynamic viewpoint, we may predict that the magnetic lattice structure in this system might be two-dimensional.

It is worthwhile to remark here that Decurtins et al. [26] recently reported the single crystal x-ray study of an analogous complex $\{P(Ph)_4[MnCr(ox)_3]\}_n$ (Ph = phenyl) and concluded that it has a 2D network structure. We measured the heat capacity of this complex and found a magnetic phase transition at 5.59 K. The transition entropy $\Delta S = 26.6$ JK^{-1} mol^{-1}

---

*sc = simple cube, bcc = body-centered cubic, fcc = face-centered cubic.

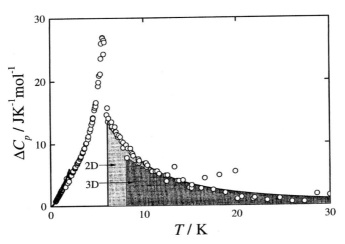

Figure 10 Excess heat capacity ΔC_p of $\{P(Ph)_4[MnCr(ox)_3]\}_n$. Light and dark shaded areas correspond to the theoretical contribution of the short-range order effect persisting in the temperature region above T_C for 2D and 3D ferromagnetic lattices, respectively. The solid curve drawn at the low-temperature side of T_C indicates the spin-wave contribution.

agreed well with the theoretical value, $R \ln (6 \times 4) = 26.42$ JK^{-1} mol^{-1}, for the spin system consisting of Mn^{2+} (spin 5/2) and Cr^{3+} (spin 3/2). As compared in Figure 10, the analysis of the heat capacity of this complex [27,28] based on the high-temperature series expansion clearly supports a 2D magnetic structure with a spin–spin interaction of $J/k_B = 0.95$ K. The coincidence of the dimensionality derived from x-ray structural analysis [26] and the present calorimetric study [27,28] confirms a 2D structure for the former two mixed-metal complexes.

IV. CONCLUSIONS

Heat capacity is a physical quantity containing contributions from all kinds of molecular degrees of freedom within the Maxwell–Boltzmann distribution. This makes a sharp contrast with various spectroscopies in which particular nuclides and/or particular spectral modes are selectively monitored. In other words, thermodynamic measurements are not restricted by selection rules or selectivity. Consequently they are applicable to a wide range of materials and phenomena. The only shortcoming of heat capacity is the ambiguity involved when the total heat capacity is separated into individual contributions. In case of molecule-based magnetic materials, the dominant contributions at low temperatures are the spin degrees of freedom and molecular vibrations (or phonons). When magnetic phenomena occur at low temperatures, separation of these two contributions can be successfully made, because the phonon contribution is extremely small. Therefore, heat capacity calorimetry is a crucially important experimental tool, one that plays a complementary role to magnetic, spectroscopic, or structural methods.

REFERENCES

1. de Jongh LJ, Miedema AR. Experiments on Simple Magnetic Model Systems. London: Taylor & Francis, 1974. Adv. Phys. 1974; 23:1.
2. Martin RL. Magnetochemistry. Berlin: Springer-Verlag, 1986.
3. Sorai M. In: Turnbull MM, Sugimoto T, Thompson LK, eds. Molecule-Based Magnetic Materials: Theory, Techniques, and Applications. Washington, DC: American Chemical Society, 1996, Chap 7, p 99.
4. Albrecht AS, Landee CP, Matsumoto T, Miyazaki Y, Sorai M. Meeting of the American Physical Society, March, 1996.
5. Baker GA Jr, Gilbert HE, Eve J, Rushbrooke GS. Phys. Lett. 1967; 25A:207.
6. Mermin ND, Wagner H. Phys. Rev. Lett. 1966; 17:1133.
7. Matsumoto T, Miyazaki Y, Hashiguchi T, Albrecht AS, Landee CP, Sorai M. 5th International Conference on Molecule-Based Magnets, Osaka, Japan, 1996.
8. Matsumoto T, Miyazaki Y, Hashiguchi T, Albrecht AS, Turnbull MM, Landee CP, Sorai M. 33rd Annual Meeting of the Japan Society of Calorimetry and Thermal Analysis, Okayama, Japan, 1997.
9. Dom C. Adv. Phys. 1960; 9:149.
10. Onsager L. Phys. Rev. 1944; 65:117.
11. Sugimoto H, Aota H, Harada A, Morishima Y, Kamachi M, Mori W, Kishit M, Ohmae N, Nakano M, Sorai M. Chem. Lett. 1991; 2095.
12. Kamachi M, Sugimoto H, Kajiwara A, Harada A, Morishima Y, Mori W, Ohmae N, Nakano M, Sorai M, Kobayashi T, Amaya K. Mol. Cryst. Liq. Cryst. 1993; 232:53.
13. Kobayashi T, Takiguchi M, Amaya K, Sugimoto H, Kajiwara A, Harada A, Kamachi M. J. Phys. Soc. Japan. 1993; 62:3239.
14. Ohmae N, Kajiwara A, Miyazaki Y, Kamachi M, Sorai M. Thermochim. Acta. 1995; 267:435.
15. Togoshi K, Imachi R, Tomioka K, Tsuboi H, Ishida T, Nogamin T, Takada N, Ishikawa M. Bull. Chem. Soc. Japan. 1996; 69:2821.
16. Nogamin T, Ishida T, Tsuboi H, Yoshikawa H, Yamamoto H, Yasui M, Iwasaki F, Iwamura H, Takada N, Ishikawa M. Chem. Lett. 1995; 635.
17. Nogamim T, Ishida T, Yasui M, Iwasaki F, Takada N, Ishikawa M, Kawakami T, Yamaguchi K. Bull. Chem. Soc. Japan. 1996; 69:1841.

18. Imachi R, Ishida T, Nogamim T, Ohira S, Nishiyama K, Nagamine K. Chem. Lett. 1997; 233.
19. Matsumoto T, Miyazaki Y, Ishida T, Nogami T, Sorai M. Annual Meeting of Molecular Sciences, Nagoya, Japan, 1997.
20. Baker GA Jr, Rushbrooke GS, Gilbert HE. Phys. Rev. 1964; 135:1272.
21. Tamaki H, Zhong ZJ, Matsumoto N, Kida S, Koikawa K, Achiwa N, Hashimoto Y, Okawa H. J. Am. Chem. Soc. 1992; 114:6974.
22. Asano K, Ohmae N, Tamaki H, Matsumoto N, Okawa H, Sorai M. 42nd Annual Meeting of Coordination Chemistry, Nara, Japan, 1992.
23. Asano K, Hashiguchi T, Miyazaki Y, Tamaki H, Matsumoto N, Okawa H, Sorai M. 44th Annual Meeting of Coordination Chemistry, Yokohama, Japan, 1994.
24. Rushbrooke GS, Wood PJ. Mol. Phys. 1958; 1:257.
25. Hashiguchi T, Asano K, Miyazaki Y, Tamaki H, Matsumoto N, Okawa H, Sorai M. 30th Annual Meeting of the Japan Society of Calorimetry and Thermal Analysis, Osaka, Japan, 1994.
26. Decurtins S, Schmalle HW, Oswald HR, Linden A, Ensling J, Gütlich P, Hauser A. Inorg. Chim. Acta 1994; 216:65.
27. Hashiguchi T, Asano K, Miyazaki Y, Tamaki H, Matsumoto N, Okawa H, Sorai M. 5th International Conference on Molecule-Based Magnets, Osaka, Japan, 1996.
28. Hashiguchi T, Miyazaki Y, Sorai M. 32nd Annual Meeting of the Japan Society of Calorimetry and Thermal Analysis, Tsukuba, Japan, 1996.

24
Electron Paramagnetic Resonance Studies of Low-Dimensional Organic Radical Magnets

Philippe Turek
Charles Sadron Institute, Louis Pasteur University, Strasbourg, France

DEFINITIONS AND NOTATIONS

Vectors, matrices, and tensors are written in bold characters.

The first derivative of the absorption signal, $I'(H)$, is usually recorded during a field-swept electron paramagnetic resonance (EPR) experiment. It is characterized by its peak-to-peak width (ΔH_{pp}), its shape (Lorentzian, Gaussian, or in between), and the resonance field, H_0, which yields the g-factor as a component of the **g** tensor with respect to the external field, H_e.

The theoretical work very often refers to the absorption line itself, $I(H)$, rather than to its first derivative. Therefore, the half-width at half-maximum ($\Delta H_{1/2}$) of the absorption line is dealt with as well. The orientation of the samples, whether single crystals or thin films, is defined with respect to H_e in Figure 1.

I. INTRODUCTION

Much of the basic study of molecular magnets involves so-called magnetostructural correlation. These aim at relating the observed magnetic behavior to a given molecular packing. The relevance of such studies actually increases when combining various experimental techniques, e.g., looking at

491

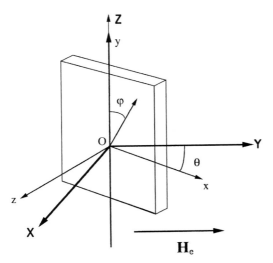

Figure 1 Diagram of a single crystal or a thin film defining its orientation with respect to the external field, with $\mathbf{H}_e = \mathbf{H}_e \mathbf{u}_y$ here. (X, Y, Z) is the laboratory frame and (x, y, z) is the sample frame. θ is the angle between $\mathbf{H}_e$ and any defined sample axis within the (XOY) plane. φ is the azimuthal angle. (From Ref. 19.)

a more or less microscopic level. The presently accepted concepts for the setup of intermolecular magnetic exchange interactions mostly consider close contacts between the molecules, hence emphasizing the spin-density distribution over the molecules. Although mostly qualitative in its conclusions, EPR study of single crystals may be of value in assessing the relevant mechanisms for the magnetic behavior of a given organic magnet. In particular, the dimensionality of the magnetic lattice (1-D, 2-D, 3-D) may be determined. This is of utmost interest, since the macroscopic studies such as, e.g., static magnetic susceptibility measurements usually suffer from a low level of distinction between the different ways/models of fitting the experimental data. A useful tool for these studies might be neutron diffraction studies. However, they require inelastic scattering experiments, which are not completely accurate in purely organic compounds. Therefore, in the absence of EPR data, most of the conclusions to be derived concerning the magnetic dimensionality of organic magnets are based solely on analysis of the crystal structure. This is very often highly subjective, since the crystal structures frequently exhibit various intermolecular contacts. The study of the anisotropy of the EPR response of single crystals may appear as a complementary tool in order to remove (at least partially) the subjectivity that a unique analysis may bring about.

The theoretical background of the EPR response of low-dimensional magnets is described in Section II. The theoretical results are then exemplified with the help of selected experimental results dealing with organic radical magnets (Section III).

II. ELECTRON PARAMAGNETIC RESONANCE IN LOW-DIMENSIONAL MAGNETS: THEORETICAL SURVEY

II.A. The Spin Hamiltonian

The following discussion is restricted to the electronic doublet state, $S = 1/2$. The EPR absorption line of an isolated, hence infinite, lifetime electron spin should ideally be infinitely narrow. However, the finite lifetime of the spin is expressed basically through the Heisenberg uncertainty principle. Any process that increases the transition rate between the two available energy levels will decrease the lifetime; hence it will broaden the absorption line. Each spin in a paramagnet is probing a variable local magnetic field. The fluctuations of the local field may be dynamic or spatial. In the former case, the line broadening is homogeneous and leads to a Lorentzian lineshape. In the second case, the inhomogeneous line broadening yields the Gaussian absorption lineshape.

However, the exceptional nature of such an isolated spin has to be stressed. The actual EPR Hamiltonian must include intermolecular spin–spin interactions of various origin [1]. We shall restrict ourselves to the Zeeman interaction $\mathbf{H}_Z$, the exchange interaction, $\mathbf{H}_{ex}$, and the dipolar interaction, $\mathbf{H}_d$, which define the EPR Hamiltonian under study:

$$\mathbf{H}_{EPR} = \mathbf{H}_Z + \mathbf{H}_{ex} + \mathbf{H}_d \tag{1}$$

$$\mathbf{H}_Z = \mu_B \sum_i \mathbf{H}_e \mathbf{g} \mathbf{S}_i \tag{1a}$$

$$\mathbf{H}_{ex} = \sum_{i<j} J_{ij} \mathbf{S}_i \cdot \mathbf{S}_j \tag{1b}$$

$$\mathbf{H}_d = \sum_{i<j} \mathbf{S}_i \mathbf{D}_{ij} \mathbf{S}_j \tag{1c}$$

where the isotropic Heisenberg exchange interaction is considered in Eq. (1b). All other processes are assumed to be of much lower order in energy. The basic effects of the exchange and dipolar interactions are dipolar broadening [2] and exchange narrowing [2,3]. The next section will summarize the results of the theory of low-dimensional spin dynamics in paramagnets. A comprehensive review may be found in review papers [4–6] or in the book of A. Bencini and D. Gatteschi [7]. We will not elaborate the mathematical details, which are well described in the cited literature, but rather outline the discussion with the basic physical key tools, e.g., the peculiar time dependence of the spin correlation function with its consequences for the EPR response [i.e., the lineshape (Section II.B), and the linewidth (Section II.C)]. Special emphasis will be devoted to the analysis of the **g** tensor (Section II.D), since its thermal behavior has recently been related to the demagnetizing field in organic radical magnets [8].

II.B.   Spin Correlations and the Electron Paramagnetic Resonance Lineshape

Within the high-temperature limit, i.e., in the paramagnetic phase, and assuming the dipolar interaction as a perturbation compared to the exchange interaction, the modulation of the dipolar interaction by the exchange interaction leads to the mechanism of spin diffusion [9].

The theory of linear response [10] sets up the EPR absorption, i.e., the imaginary part of the dynamical spin susceptibility, as the Fourier transform of a relaxation function, $r(t)$. In the case of a random Gaussian modulation of the magnetization, $r(t)$ is defined with respect to the total spin correlation function, $R(\tau)$, as follows:

$$r(t) = \exp\left[-\int_0^t (t - \tau)R(\tau)\, d\tau \right] \tag{2a}$$

$$R(\tau) = \sum_m G_m(\tau)e^{im\omega_0\tau} \tag{2b}$$

$$G_m(\tau) = \frac{\langle g_m(\tau)g_m(0)\rangle}{\langle M_+M_-\rangle} \tag{2c}$$

where the brackets denote a thermal average and $M_\pm$ represents the transverse components of the total magnetization. The m index/exponent stands for the difference in the quantum spin projection. It is worthwhile noting that $m = 0, \pm1, \pm2$, since the exchange Hamiltonian (Eq. 1b) is quadratic in the spin operators. Therefore, the EPR response is guessed at null frequency as the secular contribution ($m = 0$), and at multiples of the Larmor frequency $\omega_0(m = \pm1)$ and $2\omega_0(m = \pm2)$ as the nonsecular contributions. The dipolar correlation function $G_m(t)$ inserted in Eq. (2) depends on the secular and nonsecular geometrical contributions to the dipolar energy, F_{ij}^{m}:

$$F_{ij}^{0} = D_{ij} \frac{1}{3} (3 \cos^2\theta_{ij} - 1)$$

$$F_{ij}^{\pm 1} = D_{ij} \sin \theta_{ij} \cos \theta_{ij} \exp(-i\varphi_{ij})$$

$$F_{ij}^{\pm 2} = D_{ij} \sin \theta_{ij}^2 \exp(-2i\varphi_{ij}) \tag{3}$$

where θ_{ij} and φ_{ij} are, respectively, the polar and azimuthal angles between spins i and j, θ_{ij} being defined with respect to the Zeeman field.

The theoretical derivation of the correlation function in Eq. (2) within the limits defined at the beginning of this section yields a hydrodynamic regime for the spin motion, hence the picture of spin diffusion. For instance, given the diffusion constant of the motion D, the time dependence of the Fourier transform of the 2-spin correlation function is expressed as:

$$\langle S_{qz}(\tau)S_{-qz}(0)\rangle = \langle S_{qz}(0)S_{-qz}(0)\rangle \exp(-Dq^2\tau) \tag{4}$$

Spatially, the mechanism of spin diffusion emphasizes the role of the long-wavelength modes of the correlation function, or equivalently the uniform $\mathbf{q} = 0$ mode (i.e., the ferromagnetic mode of the spin wave). The time decay of the relaxation function is dominated for a long time by the diffusive tail of the relaxation. Whereas it may be neglected in three-dimensional lattices, the weight of the diffusive contribution increases as the dimensionality of the magnetic lattice is lowered, since the diffusion process is topologically restricted at lower dimensions, i.e., being constrained within 2-D sheets or along 1-D chains. The Lorentzian lineshape is expected at any orientation in a 3-D exchange-narrowing mechanism, where the time decay of the relaxation function follows a simple exponential behavior [3]. The long-term persistence of the relaxation function results in a departure from the Lorentzian shape. This departure will be larger in 1-D spin systems than in 2-D lattices. Concomitantly, the secular contribution will be dominating the EPR response, as described in Eq. (2b), since the Zeeman modulation will truncate the nonsecular ($m \neq 0$) contributions, whereas it has no effects on the secular ($m = 0$) contribution. This in turn points toward the peculiar orientations $\theta = 0°$ and $\theta = 54.7°$. The orientation $\theta = 0°$ corresponds to the maximum dipolar secular contribution, aligning with the magnetic axis of the spin system. It is the chain axis in a 1-D system, and it is the perpendicular to the magnetic planes in a 2-D system. A maximum departure from the Lorentzian line is expected along the magnetic axis. The orientation $\theta = 54.7°$ is the so-called magic angle, at which the secular dipolar contribution is vanishing. Accordingly, the lineshape is purely Lorentzian at this angle.

These results are summarized in the following expressions for the relaxation function in 3-D (Eq. 5a), 2-D (Eq. 5b), and 1-D (Eq. 5c):

$$r_{1D}(t) = \exp[-\eta t] \tag{5a}$$

$$r_{2D}(t) = \exp\left[-\eta t - \gamma t \ln\left(\frac{t}{t_0}\right)\right] \tag{5b}$$

$$r_{3D}(t) = \exp[-\eta t - (\gamma t)^{3/2}] \tag{5c}$$

where η depends on the dipolar correlation functions at $\omega = 0$, ω_0, and $2\omega_0$ (Eq. 3) and where the long-term tail expressed by γ depends on the secular dipolar contribution and on the exchange frequency. Keeping in mind that the EPR absorption spectra are the Fourier transforms of these expressions, it appears that check of the EPR lineshape would be of value with respect to the description of the spin dynamics in the paramagnet under study. Unfortunately, the relevant field range for such experimental studies extends over several linewidths, since the Lorentzian line decays slowly with the field, and any check of a departure from this shape must be performed in the far wings of the EPR spectra. Therefore, it has to be performed within a range of weak signal-to-noise ratio. The use of a field-frequency lock device may remove part of the experimental difficulty, since it allows precise multiple scan accumulation. However, narrow lines are required because this device is operating over a limited field range, ca. 100 gauss. Of much interest is Fourier-transformed EPR (pulsed EPR), since it probes directly the relaxation function through, e.g., a free-induction decay experiment [11]. However, its limited bandwidth again restricts work to lines narrower than 1 gauss, i.e., within the 100-ns or more relaxation time range. Whatever the experimental technique, it is clear that study of the lineshape is not the most attainable experiment. Some representative results are presented in Section III.B.

II.C. The Electron Paramagnetic Resonance Linewidth Anisotropy

The diffusive contribution to the relaxation function (Eqs. 2, 3, 5), as discussed earlier, will be responsible for a typical angular dependence of the linewidth. The typical expressions for the linewidth anisotropy are given next, for they may be used for fitting of the experimental data. The results are summarized in Figure 2 and in Eqs. (6).

The angles θ_{ij} in Eq. (3) are all equal to the angle θ between the chain axis and the Zeeman field within a 1-D chain. The resulting angular dependence of $\Delta H_{1/2}$ is given by:

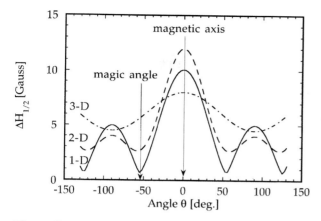

Figure 2 Angular dependence of $\Delta H_{1/2}$ in the case of spin diffusion as simulated by Eq. (6a) for the 1-D case: solid line; Eq. (6b) for the 2-D case: dashed line; and Eq. (6c) at 3-D: dash-dotted line.

$$\Delta H_{1/2} = a|3 \cos^2\theta - 1|^{4/3} + b \tag{6a}$$

It is worthwhile noting the peculiar "W-shape" of the linewidth anisotropy (Fig. 2). The minimum linewidth is observed at the magic angle, as expected from the discussion of the lineshape. The chain axis corresponds to the maximum linewidth. In such a pure 1-D model, the contribution of the nonsecular terms is neglected.

The 2-D case also emphasizes the role of the diffusive tail of the relaxation function. However, the weight of the nonsecular contributions increases as the dimensionality increases. Therefore, the resulting expression for the linewidth anisotropy includes the nonsecular dipolar contributions, e.g., for a square lattice:

$$\Delta H_{1/2} = a(3 \cos^2\theta - 1)^2 + 10b \sin^2\theta \cos^2\theta + c \sin^4\theta \tag{6b}$$

This expression is of invaluable help in fitting the experimental data, since it allows interpolation between low-D and 3-D behaviors, as described shortly. It should be noticed that the minimum linewidth may be shifted with respect to the magic angle, due to the occurrence of nonsecular contributions.

Within a 3-D lattice the secular and nonsecular contributions contribute equally to the relaxation function. This corresponds to the case $a = b = c$ in Eq. (6b):

$$\Delta H_{1/2} = a(1 + \cos^2\theta)^2 \tag{6c}$$

There is still some controversy about the temperature dependence of the EPR linewidth of low-D magnetic materials. Emphasizing theoretically the role of the uniform $\mathbf{q} = 0$ mode of the spin–spin correlation functions in the mechanism of spin diffusion [4–7], the linewidth should increase continuously for low-D ferromagnets as the temperature decreases within the high-temperature range ($kT \gg J$). A linewidth decrease should be observed for a low-D antiferromagnet within the same conditions. The increasing importance of the uniform mode is responsible for the occurrence of a minimum at the magic angle. It also emphasizes the maximum of $\Delta H_{1/2}$ along the magnetic axis. However, the experimental observations may be quite complicated, especially for 2-D ferromagnets [12]. Moreover, the effects should be proportional to the exchange interaction.

II.D. The Electron Paramagnetic Resonance g Tensor

Until very recently the study of the temperature dependence of the **g** tensor of low-dimensional magnets could be considered as a complementary approach for the study of the magnetic dimensionality of paramagnets. The basic ideas for these considerations are based on the equation of motion of the spin system. The effects of low-dimensional spin diffusion on the EPR response have previously been summarized through the lineshape and the linewidth of the absorption spectrum within the formalism of the relaxation function. It may be shown that the resonance frequency, hence the **g** tensor, is also related to the spin correlation function [13]. The short-range magnetic order is well known to have increasing effects as the dimensionality decreases. These are included in the temperature dependence of the spin correlation functions. The theories based on the effects of the short-range magnetic order could qualitatively explain numerous experimental observations of a temperature-dependent g-shift in low-D magnets, observed mostly in metal complexes [13]. These experiments were reporting a temperature-dependent and anisotropic shift of the g-factor with respect to the magnetic axis of uniaxial low-dimensional magnetic systems. This is summarized as follows:

$$\frac{\Delta g_i(T)}{g_i} = f_i \chi_c(T) n(x) \tag{7}$$

with

$$\frac{\Delta g_i(T)}{g_i} = \frac{\hbar\omega(T) - g_i\mu_B H}{g_i\mu_B H}, \quad f_i = f_\parallel = -\frac{18}{5Na^3} = -2f_\perp$$

where $\chi_c(T)$ is the Curie function, $n(x)$ is a function of the temperature derived from the temperature dependence of the 2-spin correlation functions within the classical spin approximation [13a,b], and a is the distance between neighboring spins. The indices $\perp$ and $\parallel$ mean, respectively, perpendicular to and parallel to the unique axis of a uniaxial magnetic system.

A few critical works have called attention to some possible alternate effects of the origin of the g-shift [14]. However, the wide use of the results of the former theory was obviously based on its ability to describe qualitatively the observed behavior at a more or less high level of accuracy. Given the importance of such an additional tool for the investigation of the dimensionality of the magnetic lattice, various experiments have been performed in organic radicals in that way [8,15]. The effects of a demagnetizing field were shown to be fully responsible for the observed g-shifts in organic paramagnets, extending possibly to previous observations in other so-called low-D magnets. The results are summarized next for the peculiar behavior of the **g** tensor as a function of temperature and/or angular dependence.

The internal magnetic field, $\mathbf{H}_i$, within a magnetic material does not coincide with the applied external field, $\mathbf{H}_e$. It must be corrected by the demagnetizing field, which is related to the magnetization, $\mathbf{M}$, via a demagnetizing tensor, $\mathbf{N}$. The demagnetizing tensor depends on the shape of the sample:

$$\mathbf{H}_i = \mathbf{H}_e - \mathbf{N}\mathbf{M} = \mathbf{H}_e(1 - \mathbf{N}\chi) \tag{8}$$

where χ is the volume susceptibility. The principal values of the $\mathbf{N}$ tensor may be estimated for a general ellipsoid or more complex shapes [16]. The theory of ferromagnetic resonance [17] allows a straightforward derivation of the principal values of the **g** tensor of a paramagnet. The susceptibility $\chi(T)$ will be considered as nearly isotropic in purely organic compounds. Reasonably considering that $\chi(T) << 1$, an approximated expression of the components of the **g** tensor is given by

$$g_i(T) \approx g_{\text{iref}}\left[1 + \frac{1}{2}(N_j + N_k - 2N_i)\chi(T)\right] \approx g_{\text{iref}}[1 + f_i\chi(T)] \tag{9a}$$

where $g_i(T)$ is the temperature-dependent principal value of the **g** tensor along the i-direction ($g_i = g_{ii}$) and g_{iref} corresponds to the same component measured at room temperature. Assuming no demagnetization effects, f_i is a geometrical factor determined entirely by the principal values of the $\mathbf{N}$ tensor. Defining the relative g-shift as $\Delta g/g = [g\,(300\ \text{K}) - g(T)]/g\,(300\ \text{K})$, one obtains Eq. (9b) in the particular case of a needle-shaped sample with the susceptibility $\chi(T)$ (in cgs units):

$$\left(\frac{\Delta g}{g}\right)_{\parallel} = -2\pi\chi(T) \quad \text{and} \quad \left(\frac{\Delta g}{g}\right)_{\perp} = \pi\chi(T) \tag{9b}$$

where $\parallel$ is the index for the needle axis parallel to the applied field and $\perp$ that for the perpendicular. For a disk-shaped sample, one should observe:

$$\left(\frac{\Delta g}{g}\right)_{\parallel} = 4\pi\chi(T) \quad \text{and} \quad \left(\frac{\Delta g}{g}\right)_{\perp} = -2\pi\chi(T) \tag{9c}$$

where the $\parallel$ index means that the normal to the disk is parallel to the applied field. It is worth noting the striking similarities between Eq. (7) and Eqs. (9), although the origin of the shift is basically different in both expressions.

The theory of short-range magnetic order also has considered the angular dependence of the **g** tensor through the dipolar secular contribution [13d]. No g-shift should be observed at the magic angle. This conclusion also has been shown to be valid in the case of the demagnetizing field effects, as summarized next. The anisotropy of the **g** tensor of a uniaxial system within the (i, j) plane is given by:

$$g^2(\theta) = g_{ii}^2 \cos^2\theta + 2g_{ij}^2 \sin\theta \cos\theta + g_{jj}^2 \sin^2\theta \tag{10}$$

The existence of some magic angles θ_m corresponds to the equivalence of the g-factor for two different temperatures; and combining Eq. (9) with Eq. (10) a simple application of these results is considered in the case of a rectangular parallelepiped–shaped single crystal. The **N** tensor is such that its principal axes are perpendicular to the faces of the crystal. It is assumed that the **g** tensor and the **N** tensor have the same referential. The "magic angle" is then defined

$$\theta_m = \pm\tan^{-1}\left(\frac{g_{iiref}}{g_{jjref}}\sqrt{-\frac{f_{ii}}{f_{jj}}}\right) \approx \pm\tan^{-1}\left(\sqrt{-\frac{f_{ii}}{f_{jj}}}\right)$$

$$= \pm\tan^{-1}\left(\sqrt{-\frac{1 - 3N_i}{1 - 3N_j}}\right) \tag{11}$$

Applying Eq. (11) to the case of a needle with $N_i = 0$ and $N_j = 1/2$, one gets $\theta_m = \pm 54.7°$. The same result is obtained for a disc with $N_i = 1$ and $N_j = 0$.

It is concluded that the effects of the demagnetizing field are similar to the effects of the dipolar field and/or the short-range magnetic order, i.e., θ_m does correspond to the dipolar magic angle.

The demagnetizing field effects have also explained the observed temperature-dependent rotation of the principal axes of the **g** tensor in the case of noncoincident referential for the **N** tensor and the **g** tensor [15,18].

Assuming a Curie–Weiss temperature dependence for weakly coupled $S = 1/2$ magnets, for temperatures higher than the mean-field temperature the position of the extrema of the **g** tensor may be expressed by:

$$\theta_e = \left(\frac{\pi}{4} - a + \frac{a^3}{3}\right) - b\chi(T)(1 + a^2) = A - B\chi(T) \tag{12}$$

with

$$A = \frac{\pi}{4} - a + \frac{a^3}{3} \qquad B = b(1 + a^2)$$

$$a = \frac{g_{iiref}^2 - g_{jjref}^2}{2g_{ijref}^2} \qquad b = \frac{g_{iiref}^2 f_i - g_{jjref}^2 f_j}{g_{ijref}^2}$$

Equation (12) points towards a rough hyperbolic dependence of the angular shift with temperature.

It is worthwhile noting that the foregoing theory of demagnetizing field effects on the **g** tensor of radical paramagnets succeeds in predicting both EPR line distortion and/or broadening in oriented thin films, as will be shown in Section III.A.2 [19].

III. SELECTED EXPERIMENTAL RESULTS IN ORGANIC RADICAL MAGNETS

The preceding summary of theoretical results points to various phenomena that might be observable through EPR experiments. These are exemplified in the following selection of experimental results collected in our laboratory [20].

III.A. The Lithium Phthalocyanine Radical: Single Crystals and Thin Films

The lithium phthalocyanine radical, PcLi (Figure 3), has been extensively studied due to its outstanding structural [21], magnetic [22], and electronic properties [21–23]. Various polymorphs have been reported for this molecule, namely the α-, β-, and x-phases [24]. It is worthwhile noticing that EPR studies were performed on powders, single crystals, and thin films within the various reported crystal structures. In particular the effects of dioxygen absorption on the EPR absorption line have been described in the frame of low-dimensional spin dynamics and strong spin-exchange narrowing (Section III.A.1). Moreover, the effects of the demagnetizing field on the **g** tensor are clearly shown for single crystals. Of peculiar interest is the analysis of these effects in thin films, since it allows one to draw conclusions on the arrangement of the microcrystallites within the thin films (Section III.A.2).

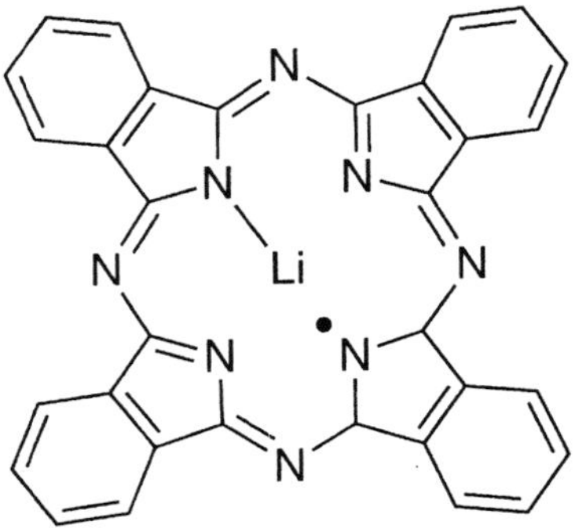

Figure 3 Scheme of the lithium phthalocyanine molecule, PcLi.

III.A.1. Extreme Spin-Exchange Narrowing in a
          One-Dimensional Semiconductor: Single Crystals of
          the *x*-Phase

The molecular packing within the tetragonal *x*-phase of PcLi allows for the diffusion of molecular oxygen, O_2. Dioxygen molecules fit within the free channels observed between the highly 1-D stacks of PcLi macrocycles. The intermolecular distance along the *c* stacking axis is less than the van der Waals distance ($d_{PcLi-PcLi} = 3.24$ Å) The strong intrastack overlap of the π-molecular orbitals has striking effects: (1) it builds up broad conduction bands at the origin of an intrinsic semiconducting behavior, and (2) extreme spin-exchange narrowing is observed along the *c*-axis for a diffusive spin system, leading to lines as narrow as a few milligauss in the absence of O_2 (Figure 4a). The exchange interaction between the triplet spin of O_2 and the surrounding radical spins has drastic effects on the linewidth, as shown in Figure 4b. Whereas the overall angular dependence of the linewidth is typical of a low-D diffusive spin system in the absence of O_2 and at room temperature (Figure 4a, Eq. 6b), a 3D-like behavior is observed at ambient atmosphere and at room temperature (Figure 4b, Eq. 6c). Thus, the spin-exchange interaction of O_2 with the radical spins sets up (1) broadening of the EPR line through spin localization, and (2) isotropic spin diffusion due to interchain bridging through O_2.

    The effects of the demagnetizing field on the **g** tensor of single crystals of the *x*-phase are well demonstrated in Figures 5a,b. The approximation of the needle-shaped crystals is valid for these very thin crystals (rectangular parallelepiped of ca. $0.1 \times 0.1 \times 1$ mm^3) elongated along the *c*-axis. The temperature dependence of the **g** tensor along the needle axis and within the orthogonal planes (Figure 5a) corresponds to the expectations from Eq. (9b);

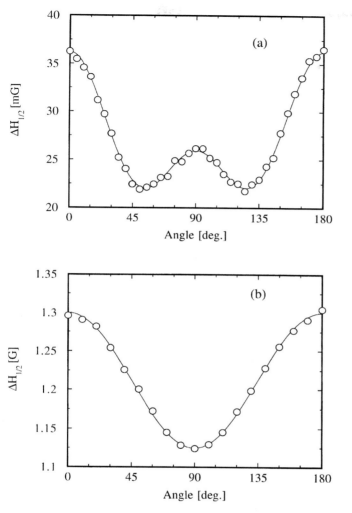

Figure 4 Linewidth anisotropy at room temperature for an x-phase PcLi single crystal held: (a) under vacuum; and (b) at the air. The sample is rotated between the stacking axis ($\theta = 0°$) and the applied magnetic field. The solid lines in Figure 4a, correspond to fits of the data to Eq. (6b) and in Figure 4b to Eq. (6c).

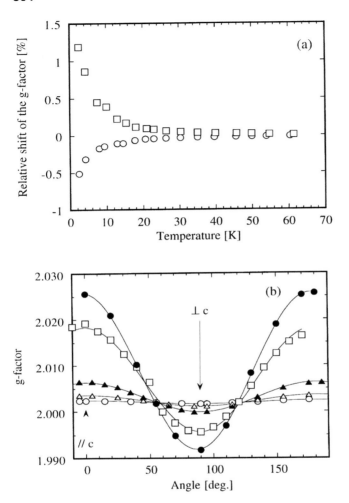

Figure 5 Evolution of the EPR resonance field for an x-phase PcLi single crystal. (a) Temperature dependence of the relative variation of the g-factor with respect to its room-temperature value, $\Delta g/g$, along the long axis of a crystal aligned with the molecular stacking axis (squares), and in a perpendicular direction (circles). (b) Anisotropy for a rotation between the chain axis ($\theta = 0°$) and the applied magnetic field of the g-factor recorded at various temperatures: $T = 50$ K, empty circles; $T = 23$ K, empty triangles; $T = 13$ K, filled triangles; $T = 4$ K, squares; $T = 2.3$ K, filled circles. The solid lines correspond to the fits of the data to Eq. (10). (From Ref. 15a.)

i.e., the g-factor along the needle-axis is positive and twice the one along the orthogonal direction. This is of great importance in studies of organic radical magnets. Since the stacking axis of a 1-D radical magnet is observed to coincide very often with the long axis of the crystalline materials, one may conclude that magnetic effects due to short-range order are fully responsible for the g-shift. Previously presented theories have found that these effects are difficult to distinguish from those of the demagnetizing field. Moreover, observation of the magic angle in the angular dependence of the g-factor recorded at various temperatures (Figure 5b) fits with our conclusions on demagnetizing field effects (Eq. 11).

III.A.2. Demagnetizing Field Effects in Thin Films

Anisotropic spin diffusion has been deduced for the thin films of PcLi, whether in the α-form or in the x-form [19]. Of peculiar interest is the previously presented analysis of the behavior of the **g** tensor, since it is related to studies of the structure of the films. It is shown later that the temperature dependence of the **g** tensor of thin films of PcLi is very likely due to the demagnetizing field according to Eqs. (9). Moreover, it suggests an in-plane distribution of the microcristallites with respect to the substrate.

Depending on the substrate temperature, T_s, the films consist of a mixing of the α-form and of the x-form of PcLi: the higher T_s, the greater the amount of the α-phase in the films. On the other hand, the spin susceptibility of the α-phase is much larger than that of the x-phase, hence the larger g-shift. Accordingly, thin films of the x-like structure exhibit a small g-factor variation (Figure 6a) as compared to α-PcLi films (Figure 6b). The analysis of the temperature dependence of the g-factor (Figures 6a,b) within the frame of the demagnetizing field effects shows that Eq. (9c) is verified in the case of PcLi thin films, regardless of the substrate temperature, i.e., whatever the crystal phase within the film. This observation indicates that the thin films consisting of needle-shaped microcrystallites in the plane of the substrate can be considered as disk-shaped samples concerning the effect of the demagnetizing field. Although the microscopic effects cannot be fully discarded, the present analysis indicates that the demagnetizing field, i.e., a purely macroscopic approach, may be considered solely in order to explain the temperature dependence of the g-factor of thin films.

The previous analyses led to a simple model based on the demagnetizing field effects in order to explain the strong angular-dependent EPR line distortion observed at low temperature in PcLi thin films (Figure 7). Each crystallite within the films is characterized by the two angles θ and φ in the plane of the substrate (see Figure 1). Accordingly, the g-factor for a given crystallite within the film is expressed as a modification of Eq. (10) to yield Eq. (13):

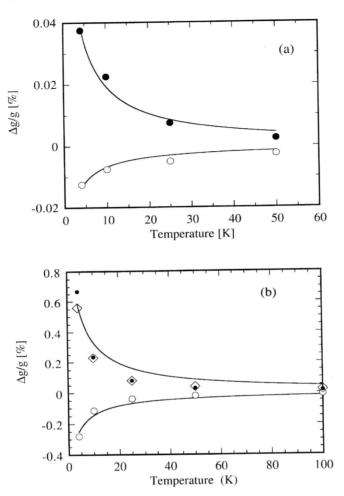

Figure 6 Temperature dependence of $\Delta g/g$ [defined in Eq. (9) and in Figure 5a] for PcLi thin films and for a direction of the applied magnetic field parallel to the normal to the film [filled circles: $(\Delta g/g)_{\parallel}$], and for a direction of the applied magnetic field within the film [open circles: $(\Delta g/g)_{\perp}$]: (a) predominant x-structure ($T_s = 24°C$); (b) predominant α-structure ($T_s = 200°C$), the diamond symbols represent $-2(\Delta g/g)_{\perp}$. The solid lines correspond to fits of the data to Eq. (9c). (From Ref. 19.)

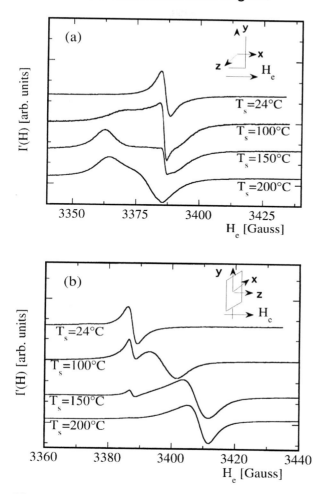

Figure 7 EPR spectra recorded at $T = 4$ K for PcLi thin films grown on glass substrate at various substrate temperatures (24°C $< T_s <$ 200°C) and of thickness in the range 40–200 nm: (a) an orientation of the normal of the films orthogonal to the applied magnetic field, and (b) the parallel orientation. The spectra are normalized to the same peak-to-peak amplitude. (From Ref. 19.)

$$g(\theta, \varphi) = \sqrt{g_{\parallel}^2 \sin^2(\theta)\sin^2(\varphi) + g_{\perp}^2[1 - \sin^2(\theta)\sin^2(\varphi)]} \qquad (13)$$

with $g_{\parallel}$ and $g_{\perp}$ being defined by Eq. (9b).

It follows from Eq. (13) that the EPR signal is the envelope of the different signals originating from the crystallites at different orientations. Simulating a "2-D powder," the EPR spectra are computed at fixed orientations θ of the film with respect to the applied field and assuming an isotropic distribution of needles in the substrate plane, i.e., summing over φ. The resulting EPR line is shown in Figure 8 at different orientations between θ = 0° and θ = 90°. It is readily concluded that the demagnetizing field is responsible for the line distortion observed in Figure 7.

III.B. Nitronyl Nitroxide Radical Derivatives

Further representative results of the analysis of the anisotropy of the EPR spectra of organic radical magnets are presented next. These have been obtained for compounds belonging to the family of organic magnets built up from nitronyl nitroxide radical derivatives.

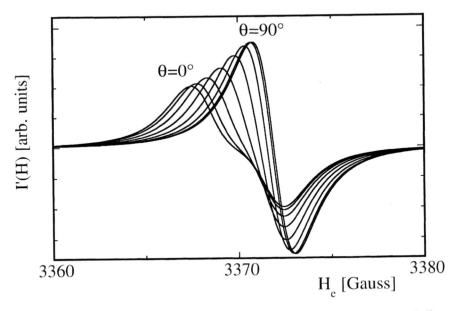

Figure 8   The effects of the demagnetizing films are simulated for the EPR line-shape of a PcLi thin film considered as a "2-D powder," i.e., consisting of needle-shaped crystallites in the plane of the substrate and exhibiting an anisotropic *g*-factor according to Eq. (13). (From Ref. 19.)

III.B.1. Linewidth and **g** Tensor Analysis Within a Two-Dimensional Organic Ferromagnet

The orthorhombic crystal structure of the compound p-hydroxyphenyl nitronylnitroxide (4-OHPNN, Figure 9) shows the existence of a network of intermolecular hydrogen bonds within the (a, b) plane [25]. Its magnetic behavior is consistent with weak 2-D ferromagnetic behavior with $J = +0.63$ K in the nearly quadratic (a, b) plane and weaker interlayer antiferromagnetic coupling [18]. This hypothesis postulates the c-axis as the magnetic axis for this 2-D magnetic system. The EPR studies reported next support this proposal by the temperature dependence of the EPR linewidth and linewidth anisotropy within the crystalline planes.

Electron Paramagnetic Resonance Linewidth. The temperature dependence of $\Delta H_{1/2}$ has been studied within the (001) and ($1\bar{1}0$) planes. The linewidth anisotropy recorded at different temperatures within the ($1\bar{1}0$) plane (Figure 10a) exhibits the peculiar features of low-D magnets as the temperature decreases. The data are well fitted to the theoretical expression for 2-D magnets (Eq. 6b). These observations suggest that c is the magnetic axis of 4-OHPNN. Moreover, observations of the magic angle as the temperature decreases demonstrates the increasing importance of the secular dipolar contribution. This is in agreement with the expected behavior for a 2-D ferromagnet, supporting the uniform $\mathbf{q} = 0$ wave vector mode of dynamical response in the diffusive regime. The overall conclusions are in agreement with the suggested behavior from the observed molecular packing and magnetic susceptibility. The overall shape of the anisotropy curves is not modified within the (a, b) plane at different temperatures (Figure 10b). The observed angular dependence cannot be related to a spin-diffusion process. It must be related to the dipolar interaction within the (a, b) plane. Although the data in Figure (10b) were fitted to Eq. (6b), it is worth noting in the present case that such a fitting procedure is simply a trigonometric analysis, including the dipolar contributions.

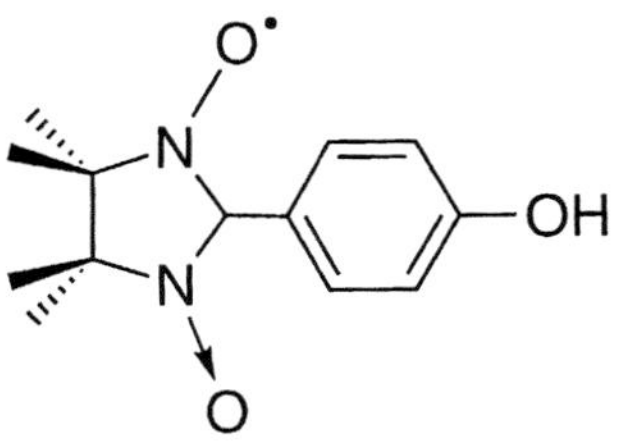

Figure 9 Scheme of a molecule of p-hydroxyphenyl nitronylnitroxide (4-OHPNN).

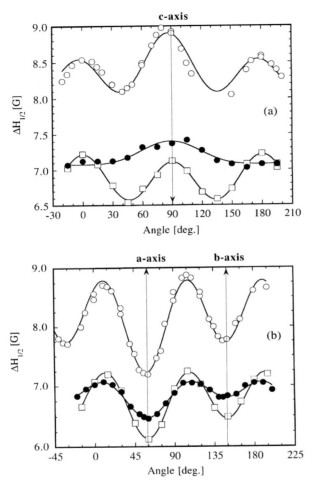

Figure 10 Angular dependence of $\Delta H_{1/2}$ for a 4-OHPNN single crystal at different temperatures within: (a) the $(1\bar{1}0)$ plane and (b) the (001) plane: $T = 4$ K, open circles; $T = 50$ K, squares; and room temperature, filled circles. The solid lines represent the fit of the data to Eq. (6b). (From Ref. 18.)

g *Tensor.* The shape of the studied sample was a rectangular parallelepiped with the c-axis perpendicular to the larger face (Figure 11). The estimation of the principal values of the **N** tensor allows one to determine the geometrical factors (Eqs. 9) along the principal axes of the **N** tensor: $f_c = -1.340$, $f_A = +0.586$, $f_B = +0.754$. The principal axes of the **g** tensor

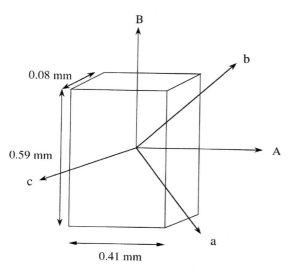

Figure 11 4-OHPNN single crystal shape and crystal dimensions used for the determination of the principal values of the tensor of the demagnetizing field. (a, b, c) is the crystallographic frame, and (A, B, c) is the frame for the tensor of the demagnetizing field. (From Ref. 8.)

coincide with the crystallographic axes. The A and B directions in Figure 11 correspond to the principal axes of the demagnetizing tensor. Because the principal axes of the **g** tensor do not coincide with the principal axes of the **N** tensor, a reorientation of the **g** tensor may occur according to Eq. (12). This phenomenon is actually observed within the (a, b) plane (Figure 12a). The relative angular shift, defined as $\delta\Omega(T) = \theta_{max}(T) - \theta_{max}(RT)$ with θ_{max} being the position of a given maximum, is plotted against temperature in Figure 12b. With knowledge of the susceptibility and of the **g** tensor, the temperature dependence of $\delta\Omega$ may be estimated from Eq. (12). The calculated curve shows a good qualitative agreement with the experimental data.

III.B.2. Lineshape Analysis Within Two-Dimensional Hexagonal Compounds

The use of the *meta*-methylpyridinium cation as an α-substituent of nitronylnitroxide radical (*m*-MPYPNN, Figure 13) led to obtaining molecular magnets having an interesting hexagonal crystal structure [26,27]. A series of related radical derivatives is closely related to the spin-frustration problem within honeycomb lattices [26b,27b]. The following results have been ob-

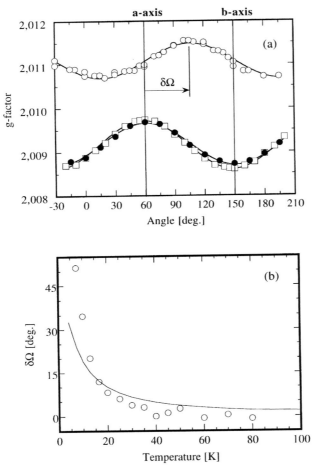

Figure 12 Evolution of the EPR resonance field for a 4-OHPNN single crystal. (a) Anisotropy of the **g** tensor within the (001) plane of the crystal described in Figure 11: $T = 4$ K, open circles; $T = 50$ K, squares; and room temperature, filled circles. The full lines represent the fits of the data to Eq. (10). (b) Temperature dependence of the angular shift, $\delta\Omega$, defined in Figure 12a. The solid line corresponds to the fit of the data to Eq. (12). (From Ref. 8.)

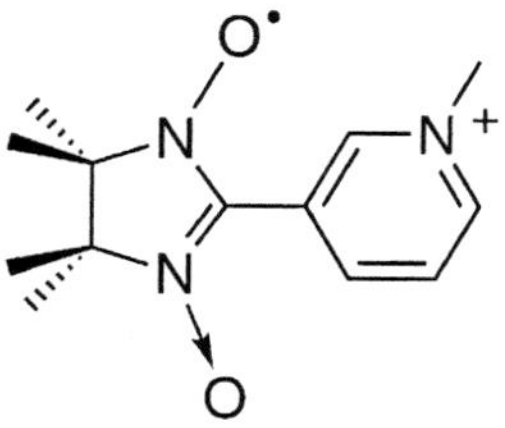

Figure 13 Scheme of the molecule of *meta*-methylpyridinium nitronylnitroxide (*m*-MPYPNN).

tained on compounds *m*-MPYPNN-BF$_3$ (**2**, [26]) and (*m*-MPYPNN)$_6$-Co(CN)$_6$I$_3 \cdot$2H$_2$O (**3**, [27a]). Both compounds show magnetic behavior resulting from competition between predominant ferromagnetic interactions and antiferromagnetic interactions. A room-temperature study [28] of the angular dependence of $\Delta H_{1/2}$ and of the lineshape along the magnetic axis and at the magic angle is described next for **2** [29] and **3**.

Due to their relatively narrow linewidths, the EPR spectra of these materials could be carefully recorded with the help of a field-frequency lock used during signal accumulation. The linewidth anisotropy of these systems behaves in a 2-D fashion at room temperature (Figures 14a,b). The data are well fitted to Eq. (6b) within crystallographic planes containing the sixfold rotation axis. A maximum in $\Delta H_{1/2}$ is observed along this well-defined axis, which is the perpendicular to the 2-D layers. It is worth noting the perfect isotropy of $\Delta H_{1/2}$ within the layers; i.e., a constant linewidth is observed within experimental accuracy. A typical analysis of the EPR lineshape is illustrated next for these two compounds. The samples are studied at room temperature along both the magnetic axis ($\theta = 0°$) and at the magic angle ($\theta = 55°$). The lineshape (Section II.B) should be purely Lorentzian at the magic angle and may depart from this shape along the magnetic axis. This departure may be small within 2-D systems, depending upon the relative weight of the nonsecular contributions to the relaxation functions. The results are shown in Figures 15 and 16 following two different representations of the experimental data. An expression for the normalized first derivative of a Lorentzian line is:

$$I'(H) = \frac{1}{2} I'_{max} \frac{16(H - H_0) \Big/ \frac{1}{2}\Delta H_{pp}}{\left\{ 3 + \left[(H - H_0) \Big/ \frac{1}{2}\Delta H_{pp} \right]^2 \right\}^2} \quad (14)$$

where I'_{max} is the peak-to-peak amplitude. Therefore, defining $\delta = (H - H_0)/$

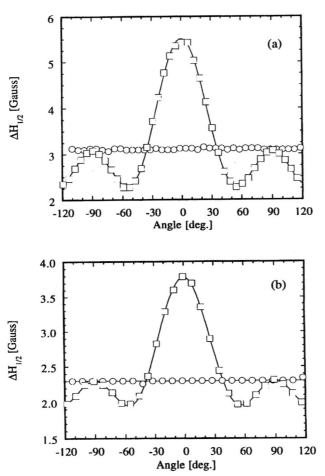

Figure 14 Linewidth anisotropy recorded at room temperature within the 2-D layers (circles) and within an orthogonal plane including the sixfold axis at null angle (squares) for single crystals of the radical derivatives. (a) **2** = m-MPYPNN-BF$_4$. (b) **3** = m-MPYPNN)$_6$Co(CN)$_6$I$_3 \cdot$ 2H$_2$O. The solid lines correspond to a fit of the data to Eq. (6b) for the squares and to a constant for the circles.

(1/2) ΔH_{pp}, a plot of $4\sqrt{\delta I'_{max}/I'(H)}$ vs. δ^2 yields a straight line with an intercept of 3 and a slope of unity for the Lorentzian shape, as shown in Figure 15. Another common representation for lineshape analysis is given by the plot of $\delta^3 I'_{max}/I'(H)$ vs. δ. Such a plot yields a constant value of 16 in the far wings of the Lorentzian line, as depicted in Figure 16. As expected,

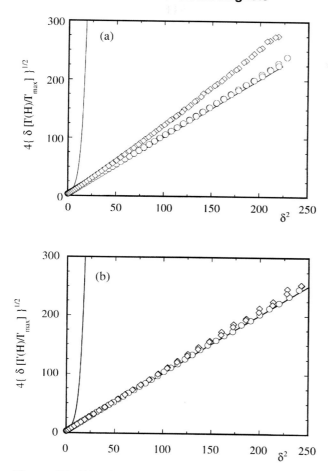

Figure 15 Lineshape analysis at room temperature along the magnetic axis (diamonds) and at the magic angle (squares) for single crystals of the derivatives: (a) m-MPYPNN-BF$_4$. (b) $(m$-MPYPNN)$_6$Co(CN)$_6$I$_3 \cdot 2$H$_2$O. The parameter $\delta = (H - H_0)/(1/2) \, \Delta H_{pp}$ is defined for the abscissas. The solid lines represent a pure Lorentzian shape for the linear plot and a pure Gaussian shape for the diverging plot.

the Lorentzian lineshape is well obeyed at the magic angle for **2** and **3**. However, whereas the EPR signal of **2** exhibits a clear departure from the Lorentzian shape at the magic angle (e.g., in Figure 15a), the lineshape of **3** remains very close to Lorentzian, whatever the orientation of the sample (e.g., in Figure 15b). Thus, although the linewidth anisotropy is similar for

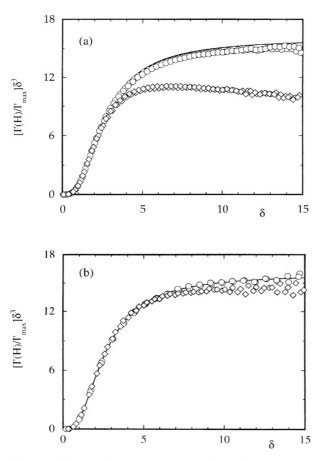

Figure 16   Another representation of the lineshape analysis. The notation is the same as in Figure 15. The solid lines represent a pure Lorentzian shape.

both compounds and is typical of 2-D spin systems, lineshape analysis points toward a more isotropic spin-diffusion regime within **3** than within **2**.

ACKNOWLEDGMENTS

I am very grateful to all of my colleagues in the EPR group [20]. It is worth mentioning the invaluable contributions of Drs. Jean-Louis Stanger and Martin Brinkmann, formerly PhD students in the group. Special thanks are due to our engineer, Maxime Bernard, to whom we owe so much experimental work and clever advice. Dr. Jean-Jacques André, although being stressed by numerous duties, was kind enough to give a careful reading to the manuscript. I gratefully acknowledge Profs. Kunio Awaga and Olivier Kahn, who have given me permission to use freely the results obtained in their beautiful materials quoted in Section III.B.2.

REFERENCES

1. (a) Weil JA, Bolton JR, Wertz JE. EPR: Elementary theory and practical applications. New York: Wiley, 1994. (b) Pake GE, Estle TL. The Physical Principles of Electron Paramagnetic Resonance. Reading, Massachusetts: Benjamin, 1973. (c) Abragam A, Bleaney B. Electron Paramagnetic Resonance of Transitions Ions. Oxford: Clarendon Press, 1970.
2. van Vleck JH. Phys. Rev. 1948; 74:1168.
3. Anderson PW, Weiss PR. Rev. Mod. Phys. 1953; 25:269.
4. Richards PM. Editrice Compositori Bologna, ed. In: Local Properties at Phase Transitions. Amsterdam: North Holland, 1976, p 539.
5. Benner H, Boucher J-P. In: L. J. De Jongh, ed. Magnetic Properties of Layered Transition Metal Compounds. Dordrecht: Kluwer, 1990, p 323.
6. Gatteschi D, Sessoli R. Mag. Res. Rev. 1990; 15:1.
7. Bencini A, Gatteschi D. EPR of Exchange Coupled Systems. Berlin: Springer-Verlag, 1990.
8. Stanger J-L, André J-J, Turek P, Hosokoshi Y, Tamura M, Kinoshita M, Rey P. Cirujeda J, Veciana J. Phys. Rev. B. 1997; 55:8398.
9. Kadanoff LP, Martin PC. Ann. Phys. 1963; 24:419.
10. Kubo R, Tomita K. J. Phys. Soc. Japan. 1954; 9:888.
11. Petit P. J. Chim. Phys. (France). 1994; 91:710.
12. Yamada I, Ikabe M. J. Phys. Soc. Japan. 1972; 33:225.
13. (a) Nagata K, Tazuke Y. J. Phys. Soc. Japan. 1972; 32:337. (b) Nagata K, Tazuke Y. J. Phys. Soc. Japan. 1972; 32:1486. (c) Nagata K. J. Phys. Soc. Japan. 1976; 40:1209. (d) Nagata K, Yamamoto I, Takano H, Yokozawa Y. J. Phys. Soc. Japan. 1977; 43:857. (e) Gatteschi D, Guillou O, Zanchini C, Sessoli R, Kahn O, Verdaguer M, Pei Y. Inorg. Chem. 1989; 28:287. (f) Karasudani T, Okamoto H. Phys. Letters A. 1976; 57:77. (g) Okamoto H, Karasudani T. J. Phys. Soc. Japan. 1977; 42:717. (h) Karasudani T, Okamoto H. J. Phys. Soc. Japan. 1977; 43:1131.
14. (a) Boucher JP. J. Magn. Magn. Mater. 1980; 15–18:687. (b) Oshima K, Kawanoue H, Haibara Y, Yamazaki H, Awaga K, Tamura M, Kinoshita M. Synth. Met. 1995; 71:1821; Mol. Cryst. Liq. Cryst. 1995; 271:29.
15. (a) Turek P. Mol. Cryst. Liq. Cryst. 1993; 233:19. (b) Stanger J-L. PhD dissertation, Université Louis Pasteur, Strasbourg, France, 1995.
16. (a) Osborn JA. Phys. Rev. 1945; 67:351. (b) Cronemeyer DC. J. Appl. Phys. 1991; 70:2911. (c) Williams CDH, Evans D, Thorp JS. J. Magn. Magn. Mater. 1989; 79:183.

17. Kittel C. Phys. Rev. 1948; 73:155.
18. Cirujeda J, Hernàndez-Gasió E, Rovira C, Stanger J-L, Turek P, Veciana J. J. Mater. Chem. 1995; 5:243.
19. Brinkmann M, Turek P, André J-J. Thin Solid Films. 1997; 303:107.
20. EPR Group for the Study of Molecular Materials, Institut Charles Sadron (CNRS).
21. (a) Brinkmann M, Chaumont C, Wachtel H, André J-J. Thin Solid Films. 1996; 283:97. (b) Brinkmann M, Wittmann JC, Chaumont C, André J-J. Thin Solid Films. 1997; 292:192.
22. (a) Turek P, André J-J, Giraudeau A, Simon J. Chem. Phys. Lett. 1987; 134: 471. (b) Turek P, André J-J, Simon J, Even R, Boudjema B, Guillaud G, Maitrot M. J. Am. Chem. Soc. 1987; 109:5119. (c) Bensebaa F, André J-J. J. Phys. Chem. 1992; 96:5739.
23. Yakushi K, Ida T, Ugawa A, Yamakado H, Ishii H, Kuroda H. J. Phys. Chem. 1991; 95:7636.
24. (a) Homborg H, Teske Chr. Z. Anorg. Allg. Chem. 1985; 527:45. (b) Brinkmann M, Turek P, Chaumont C, André J-J. J. Mat. Chem. 1998; 8:675.
25. Hernàndez E, Mas M, Molins E, Rovira C, Veciana J. Angew. Chem. Int. Ed. Engl. 1993; 22:882.
26. (a) Awaga K, Wada N. In: O. Kahn, ed. Magnetism: A Supramolecular Function. NATO ASI Ser. C 484:205. Dordrecht: Kluwer, 1996. (b) Wada N, Kobayashi T, Yano H, Okuno T, Yamaguchi A. Awaga K. J. Phys. Soc. Japan. 1997; 66:961.
27. (a) Michaut C, Ouahab L, Bergerat P, Kahn O, Bousseksou A. J. Am. Chem. Soc. 1996; 118:3610. (b) Kahn O. Chem. Phys. Lett. 1997; 265:109.
28. Gindenberger E, Turek P. Unpublished results.
29. The linewidth anisotropy of **2** has been previously reported in: Awaga K, Okuno T, Yamaguchi Y, Hasegawa M, Inabe T, Maruyama Y, Wada N. Phys. Rev. B. 1994; 49:3975.

25
Design of Solid-State Organic Ferromagnetic Materials

Kunio Awaga
University of Tokyo, Tokyo, Japan

I. INTRODUCTION

When organic radical molecules assemble, magnetic moments on neighboring molecules can have either a ferromagnetic or an antiferromagnetic correlation. The former makes the moments parallel, and the latter one antiparallel. It is known that most organic radicals possess antiferromagnetic interactions as solids. However, recently there has been extensive research about producing organic ferromagnets using ferromagnetic intermolecular interactions [1]. In order to do this, it is necessary to understand the ferromagnetic coupling mechanism. In addition, an understanding of this mechanism could lead to a clarification of the role of organic radicals in various chemical reactions and biochemical systems. In this report, we describe some mechanisms of ferromagnetic intermolecular interaction in organic radical crystals.

II. McCONNELL'S PROPOSED MECHANISMS

In the 1960s, McConnell postulated two mechanisms for the stabilization of ferromagnetic intermolecular interactions between organic radicals:

Type I: antiferromagnetic coupling between the positive spin density on one radical and the negative spin density on another [2]

Type II: admixing of a triplet charge transfer state with a ground state for a radical donor and acceptor system [3]

We will give a detailed analysis of the Type II mechanism [4]. The relationship between the two mechanisms will be discussed in a later section.

Figure 1 shows the charge-transfer (CT) excited states in a radical donor-acceptor pair, **A** and **B**, where, for the sake of simplicity, we have taken only into account the contribution of their SOMOs (singly occupied molecular orbitals) and NHOMOs (next highest-occupied MOs) to the intermolecular interaction. Based on this assumption, the CT process gives three CT configurations: S_0, S_1, and T_1. T_1 is a triplet state; S_0 and S_1 are singlet states. As McConnell noted [3], the T_1 obtained by the transfer of the NHOMO β-spin electron in molecule **A** to the SOMO in molecule **B** stabilizes the triplet ground state by the energy of $-t^2/U$, where t is a transfer

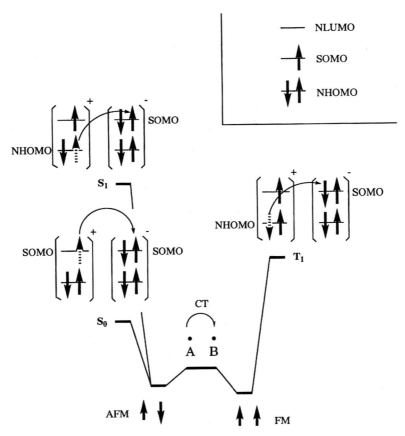

Figure 1   CT configurations in a radical dimer.

integral that is almost proportional to the overlap integral $\langle NHOMO^A/SOMO^B\rangle$ and U is the configuration energy of $\mathbf{T_1}$. Therefore, the stabilization of $\mathbf{T_1}$ (namely, a small U) and a large overlap between $NHOMO^A$ and $SOMO^B$ (namely, a large t) are advantageous for ferromagnetic coupling. However, the most stabilized excited state among $\mathbf{S_0}$, $\mathbf{S_1}$, and $\mathbf{T_1}$ is probably $\mathbf{S_0}$, which is given by the SOMO–SOMO electron transfer, and admixing of $\mathbf{S_0}$ with the ground state leads to an antiferromagnetic coupling in the radical pair. It is believed that this is the reason why most organic radicals exhibit antiferromagnetic interactions in their condensed states. The most important requirement for ferromagnetic coupling is to make overlap integral $\langle SOMO^A/SOMO^B\rangle$ as small as possible. If it is zero, i.e., if the SOMOs are exactly orthogonal, $\mathbf{S_0}$ does not contribute to the ground state magnetic coupling without being dependent on its configuration energy. $\mathbf{S_1}$ is given by the electron transfer from $NHOMO^A$ to $SOMO^B$, as well as $\mathbf{T_1}$; therefore $\mathbf{S_1}$ and $\mathbf{T_1}$ have the same transfer integral. Although the configuration energy of $\mathbf{S_1}$ is always higher than that of $\mathbf{T_1}$, in the same way that a singlet excited state in a closed-shell molecule is always more unstable than the corresponding triplet state, the contribution of $\mathbf{T_1}$ is partially canceled out by $\mathbf{S_1}$. This is another reason for the rarity of ferromagnetic interactions in organic radical solids.

From the foregoing, we can derive the following two requirements for a ferromagnetic intermolecular interaction:

a. Stable $\mathbf{T_1}$ and unstable $\mathbf{S_0}$ and $\mathbf{S_1}$
b. An orthogonal relationship between SOMOs, and a large overlap between SOMO–NHOMO

We will consider requirement (a) from a microscopic perspective for organic π-radicals. The left half of Figure 2 schematically shows what happens just before the formation of $\mathbf{S_0}$: the electron taken from the NHOMO of the neighboring molecule is entering into the SOMO. We have to make this situation unstable. The on-site Coulombic repulsion force ($U_{\pi\pi}$) is a possibility, and the Coulombic repulsions from the other electrons that occupy the other orbitals are also effective. In general, the distances between π-electrons are long, because they are delocalized and avoid each other. Therefore, repulsion $U_{\pi\pi}$ is not significant. However, nonbonding (n) electrons are localized and spread in an orthogonal direction with respect to the distributions of the π-orbitals. This usually results in a short distance between the π- and n-electrons on heteroatoms. Their interaction ($U_{n\pi}$) can effectively make $\mathbf{S_0}$ unstable. On the other hand, the electronic and/or spin states of organic radicals have been known to be characterized in terms of the spin polarization. As shown in the right half of Figure 2, this phenomenon originates in the electrostatic repulsion between the unpaired and the other elec-

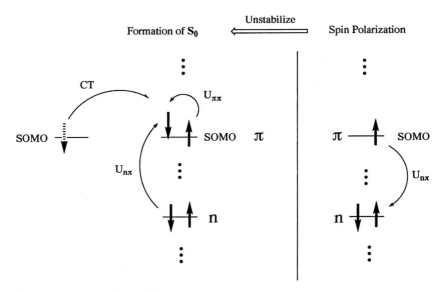

Figure 2 Formation of S_0 and spin-polarization effect.

trons in a radical molecule. Research has found that the spin polarization is enhanced by electron withdrawing groups with n electrons, such as the cyano and nitro groups, because of $U_{n\pi}$. The unstabilization of S_0 and spin polarization are different problems, but they are generally caused by the electron–electron repulsion in the organic radicals. We believe that enhancement of spin-polarization results in unstable S_0. It has been known theoretically that enhancement of spin polarization leads to a ferromagnetic intermolecular interaction [5].

On the other hand, requirement (b) is completely dependent upon the shapes of the frontier orbitals of the organic radicals. It does not appear possible to derive a universal intermolecular arrangement that satisfies requirement (b). The ferromagnetic arrangements for organic radicals will be discussed in Section IV.

III. ELECTRONIC STRUCTURE OF FERROMAGNETIC RADICALS

Although most organic radicals exhibit antiferromagnetic intermolecular interactions, reports of ferromagnetic behavior of nitronylnitroxides have appeared often in recent years [6–31]. This radical family can be regarded as a potential ferromagnetic material. The unpaired electron on the radical skeleton is localized on the two NO groups where there is a short distance between the π- and n-electrons. Researchers have speculated that there are very strong spin-polarization effects in these radical molecules.

In this section we will describe the photoelectron spectroscopy of nitronylnitroxides. Figure 3 shows the photoelectron spectrum (PES) of phenyl nitronylnitroxide (PNN) in the gas phase [32]. The first vertical ionization potential of PNN is found to be 6.8 eV, with an estimated uncertainty of ±0.05 eV. The second band is located at 7.8 eV and has almost the same intensity as that of the first. A doublet band appears at 8.9 eV, on the tail of which there seems to be a small peak at 9.7 eV. Broad and strong bands appear continuously starting at 11 eV. For comparison with the PES of PNN, the inset of Figure 3 shows part of the photoelectron spectrum of 2,2,6,6-tetramethylpiperidine-N-oxyl (TEMPO) that is discussed elsewhere [33]. The

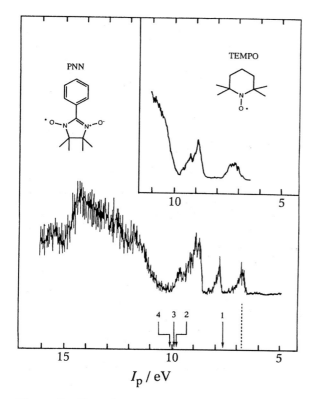

Figure 3 Photoelectron spectra of TEMPO and PNN.

first TEMPO band has been assigned to the ionization of the unpaired electron that occupied the NO π^*-orbital. The second and third bands have been assigned to the removal of the nonbonding electrons. Their doublet splitting has been interpreted as the energy difference between the triplet and the singlet states of the TEMPO cation left by the ionization of one of the oxygen lone-pair electrons.

In order to interpret the PNN spectrum, we did CNDO calculations on the PNN radical and cation, using the geometry of the PNN free radical in the solid state. The SOMO of the PNN radical is formed by the antibonding combination between the π^*-orbitals of the NO groups, making a node at the α-carbon atom. There is a small amount of delocalization of the SOMO into the phenyl ring. On the other hand, the other frontier orbital, NHOMO, is composed of the two NO π^*-orbitals, the α-carbon p_z-orbital, and the phenyl π-orbital. The NHOMO has spin population in both of the nitronyl nitroxide group, O–N–C–N–O, and the phenyl group. The ground state of the cation is found to be the singlet state where the unpaired electron of the radical is removed from the SOMO. The first ionization potential of PNN can be assigned to the energy needed for removal of the unpaired π-electron.

We calculated the excitation energies of the PNN cation that correspond to the energy differences between the first ionization potential and various others, by means of the CNDO-CI method. Table 1 shows the lowest four excitation energies of the PNN cation. The results are also shown by the arrows in Figure 3, where the energy of the ground state is assumed to be equal to the first ionization potential of 6.8 eV. Each of the calculated excitations is written as a combination of various one-electron excitations, and the largest contribution is listed in Table 1. The lowest excitation energy is 0.83 eV, primarily a NHOMO-SOMO triplet excitation. This excitation energy can explain the second PES band of PNN, i.e., that the second ionization potential could correspond to the energy needed to remove the up-spin electron in NHOMO. The largest contributions to the second and the third excitation energies are the triplet and singlet excitations of the non-

Table 1 Calculated Excitation Energies of the PNN Cation

	Excitation energy (eV)	Main contribution
1	0.83	^{3}(NHOMO→SOMO)
2	2.99	^{3}(n→SOMO)
3	3.02	^{1}(n→SOMO)
4	3.27	^{1}(NHOMO→SOMO)

bonding electrons to the SOMO, respectively. These two can be attributed to the doublet band at 8.9 eV. It is possible that the doublet bands of PNN and TEMPO look alike because they are both the results of the ionization of the oxygen lone-pair electrons of the NO group. The fourth excitation is due mostly to the NHOMO-SOMO singlet excitation, whose energy seems to explain the weak band at 9.7 eV.

The calculated excitation energies correspond closely to the observed ionization potentials of PNN. The first vertical ionization potential of 6.8 (± 0.05) eV is attributed to that needed to remove the unpaired π-electron from the SOMO, which leaves a singlet cation state. The bands at 7.8 and 9.7 eV are assigned to the triplet and singlet states of the nitroxide cation, respectively, which are formed by the photoionization of the NHOMO π-electrons. Their large energy separation reflects the strong spin-polarization effects in PNN. The strong spin polarization on nitronylnitroxide radicals has been demonstrated by magnetic resonance [34–36] and neutron scattering [37,38], and now we have observed it as the large difference in the ionization potential between the α- and β-spin electrons in NHOMO. The doublet band at 8.9 eV corresponds to the removal of the nonbonding electrons.

Based on the results of the photoelectron spectroscopy, we will next discuss the CT states of the PNN radical dimer. The lowest state among S_0, S_1, and T_1 in Figure 1 should be S_0 (which stabilizes the antiferromagnetic coupling), because the first ionization band was assigned to that of the unpaired electron. Therefore, for the realization of a ferromagnetic intermolecular interaction, the orbital overlap requirement (b) is indispensable, even in the nitronylnitroxide family. For example, PNN shows weak antiferromagnetic properties in the bulk crystal [39]. The characteristic difference in the ionization energies between the α- and β-spin electrons in NHOMO reveals in a significant energy difference between T_1 and S_1. This is very beneficial for ferromagnetic intermolecular interaction and would be a potential cause for ferromagnetic coupling in nitronylnitroxides.

IV. FERROMAGNETIC INTERMOLECULAR ARRANGEMENTS AND THE RELATIONSHIP BETWEEN McCONNELL'S TWO PROPOSALS

In the previous section, we suggested that orbital requirement (b) is important with regard to intermolecular arrangements. However, the shapes of molecular orbitals are unique to the molecules, so the intermolecular arrangement in which requirement (b) will be satisfied is definitely dependent on them. Therefore we needed to investigate the orbital interactions on a case-by-case basis, according to the features of the organic radicals. In this section we will discuss the ferromagnetic intermolecular arrangements in a model compound, the allyl radical, which is the simplest odd-alternant hydrocarbon.

Figure 4(a) shows the Hückel-level orbital energies and molecular orbitals for the allyl radical. The SOMO characteristically has a node at the center carbon. It is well known that the SOMO of an odd-alternant hydrocarbon has nodes on each of the nonstar carbons, which appear alternantly on the skeleton. Now we will analyze two kinds of intermolecular arrangement of the allyl radical, as are shown in Figure 5. In arrangement (i), one molecule is just above the other and the starred carbons on one molecule are next to the starred carbons on the other. In this arrangement, there is a notable overlap between the SOMOs, and therefore it is highly likely that a strong antiferromagnetic coupling exists. In arrangement (ii), however, the two molecules are laterally displaced by the distance of one C—C bond. The star carbons are in contact with the nonstarred ones. Here, the overlap between the SOMOs is much smaller than that between SOMO-NHOMO/NLUMO. We believe that arrangement (ii) satisfies requirement (b) (described in the previous section) and leads to a ferromagnetic coupling. The foregoing qualitative analysis has been quantitatively confirmed by MO calculations by Yamaguchi and Fueno [40]. In addition, there is a good experimental evidence in regard to the foregoing: Izuoka et al. have prepared three kinds of cyclophane-dicarbenes and studied their magnetic ground states, as is shown in Figure 6(a) [41,42]. In the cyclophane moiety of the *meta*-derivative, the starred carbons on one ring face the starred carbons on the other, and likewise for all nonstarred sites. Therefore, this is a type (i)

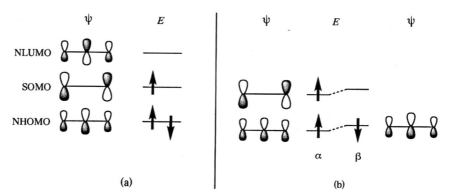

Figure 4   Hückel-level (a) and spin-unrestricted (b) molecular orbitals of the allyl radical.

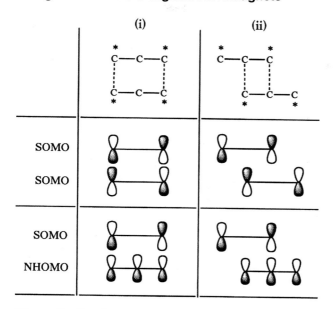

Figure 5 Two kinds of intermolecular arrangements of ally radical.

arrangement, and, in fact, an antiferromagnetic interaction between the carbene moieties was reported. The *ortho*- and *para*-derivatives, however, involve contact between the starred and nonstarred carbons. These are type (ii) arrangements, and magnetic measurements revealed their ferromagnetic ground states. Another good example is a stable phenoxyl-based radical, galvinoxyl [43]. This crystal consists of a 1D stacking chain in which a ferromagnetic intermolecular interaction operates. Figure 6(b) schematically shows a projection of two neighboring molecules in the chain. One can see contacts between the starred and nonstarred carbons in this arrangement.

Now we will return to Figure 4 and consider the higher-level molecular orbitals of the allyl radical. For the following results we used the spin-unrestricted SCF method, in which orbital energies and wave functions of the α- and β-spin electrons were calculated independently. As shown in Figure 4(b), the orbital energy of the NHOMO-α electron $E(\text{NHOMO-}\alpha)$ is always lower than $E(\text{NHOMO-}\beta)$, because the repulsion of the SOMO-α electron with the NHOMO-α electron is weaker than that with the NHOMO-β electron. This results in spin polarization, i.e., in a difference in spatial distribution between wave functions $\psi(\text{NHOMO-}\alpha)$ and $\psi(\text{NHOMO-}\beta)$. Since $|\psi(\text{NHOMO-}\beta)|^2 > |\psi(\text{NHOMO-}\alpha)|^2$ and $|\psi(\text{SOMO-}\alpha)|^2 = 0$ at the center carbon of the allyl radical, the carbon has a negative spin density.

(a)   (b)

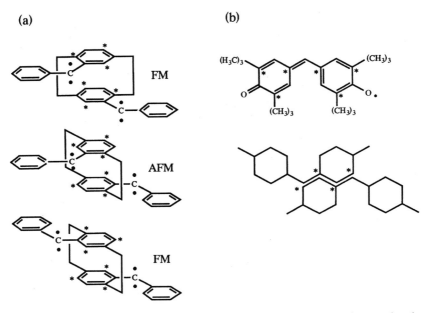

Figure 6 Three isomers of cyclophane-dicarbines (a) and the intermolecular arrangement of galvinoxyl radical (b).

Indeed, negative spin densities appear on the nonstarred carbons on which the Hückel-level SOMO has nodes. Therefore the type (ii) ferromagnetic arrangement of the allyl radical shown in Figure 5 includes overlap of the positive and negative spin densities that are required by McConnell's Type I mechanism [2].

We began by considering the Type II mechanism and came back to the Type I mechanism. McConnell's two mechanisms use the same strategy to obtain ferromagnetic intermolecular interactions; (i) enhancement of the spin polarization effect that creates the stable triplet CT state or makes the negative spin density large, and (ii) controlling the intermolecular arrangement in which SOMOs are nearly orthogonal or the positive spin densities encounter negative ones. We believe that the two mechanisms are not completely independent. However, we also would like to emphasize that we are not critical of them. It is notable that the proposals were made more than thirty years ago, when there were no experimental data about ferromagnetic magnetic intermolecular interactions.

Spin polarization can be evaluated by means of photoelectron spectroscopy, magnetic resonance, and neutron scattering. It is worth mentioning the differences in the data obtained by these techniques. As was described previously, spin polarization creates a difference between E(NHOMO-α) and E(NHOMO-β) in the allyl radical. The energy difference can be observed directly using photoelectron spectroscopy. Magnetic resonance and neutron scattering, on the other hand, can show the spin-density distribution that is obtained as a result of the difference in spatial distribution between the two electrons. Spin polarization thus can be evaluated by means of photoelectron spectroscopy, and quantitative spin densities can be evaluated by magnetic resonance or by neutron scattering. The results can be compared to those predicted by the wave functions.

V. FERROMAGNETIC INTERMOLECULAR INTERACTIONS IN THE *N*-ALKYLPYRIDINIUM NITRONYLNITROXIDE FAMILY

N-Alkylpyridinium nitronylnitroxides have been designed to produce an intermolecular arrangement that satisfies requirement (b) for ferromagnetic coupling [44]. It is known that the Hückel-level SOMO of nitronylnitroxide is localized on the two NO groups, with a node on the middle α-carbon. This SOMO has little population in the aromatic substituent on the 2-position, while the other frontier nonmagnetic orbitals are distributed on both the nitronylnitroxide group and the substituent. The positive charge on the substituent at the α-carbon where the penetration of the SOMO is very small. Since the oxygen atom in the NO group has a large negative charge resulting from an electronic polarization in the NO bond (i.e., $N^{\delta+}O^{\delta-}$), a short intermolecular distance between the NO group and the pyridinium ring is expected in the solid state, due to the electrostatic attraction force between the negative charge on the oxygen and the positive one on the pyridinium ring (see Scheme 1). In this arrangement, the distance between the SOMOs is long. This is not equal to an exactly orthogonal relationship, but can make the antiferromagnetic coupling weak. Assuming various geometric arrangements, Yamaguchi et al. calculated the magnetic interactions of *N*-alkylpyridinium nitronylnitroxides and found that the interaction was very sensitive to relative orientation, even when the head-to-tail arrangement in Scheme 1 was used [45]. However, we would like to emphasize that ferromagnetic interactions can be obtained only when the overlap between SOMOs is negligible, as was already described in Section II.

The structure of (p-*N*-butylpyridinium nitronylnitroxide$^+$ (= p-BPYNN$^+$))I$^-$ consists of a side-by-side and head-to-tail stacking chain, which is shown in Figure 7(a) [44]. This is a desirable arrangement for

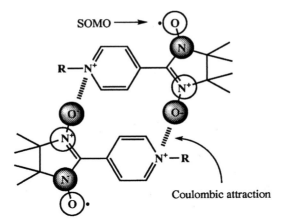

Scheme 1

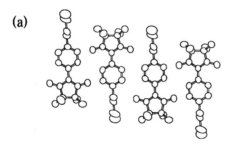

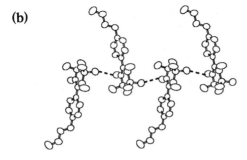

Figure 7 (a) The side-by-side and head-to-tail intermolecular arrangements of p-BPYNN$^+$ in its iodide salt. (b) Interchain arrangement of p-BPYNN$^+$.

ferromagnetic intermolecular interaction. Figure 7(b) shows a projection of the structure along the c-axis, i.e., along the 1D chain. Each 1D chain makes contact with four neighboring chains by a $CH_3 \cdots ON$ hydrogen bond, where the $C \cdots O$ distance is 3.60(1) Å. The β-hydrogens have negative spin density; Nogami et al. noted the importance of this interaction for obtaining ferromagnetic behavior in nitroxide radical crystals through their systematic research on this radical family [46]. We believe that $(p\text{-BPYNN}^+)I^-$ possesses a 3D ferromagnetic network in its crystal.

Figure 8 shows the temperature dependence of the ac susceptibility χ_{ac} for $p\text{-BPYNN}^+ \bullet I^-$ below 0.8 K. We found that χ_{ac} shows an abrupt increase below 0.3 K, suggesting a ferromagnetic transition at $T_c = 0.19$ K (peak temperature). Figure 9 shows the temperature dependence of the heat capacity C_p. Since the lattice heat capacity is negligible in this temperature range, C_p in Figure 9 is attributable to the magnetic lattice in the material. The C_p value exhibits a λ-shaped anomaly at 0.19 K, which is typical for a 3D magnetic transition; therefore, $p\text{-BPYNN}^+ \bullet I^-$ undergoes a ferromagnetic ordered state below 0.19 K. The entropy associated with the anomaly in C_p was calculated to be 88% of $R \ln 2$, using the data up to 0.8 K. This suggests the presence of high-dimensional ferromagnetic interactions in $p\text{-BPYNN}^+ \bullet I^-$.

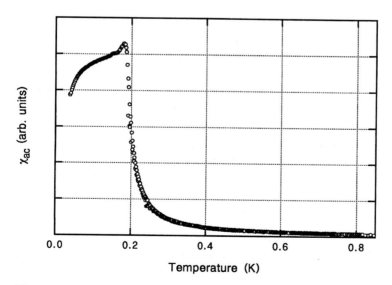

Figure 8 Temperature dependence of the paramagnetic susceptibilities of p-BPYNN$^+$I$^-$.

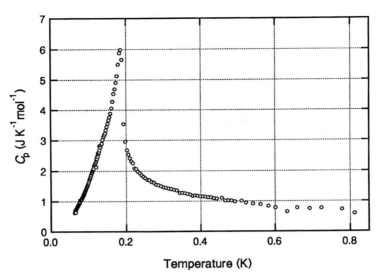

Figure 9   Temperature dependence of the heat capacity of p-BPYNN$^+$·I$^-$.

The ratio of the transition temperature T_c and the Weiss constant θ reflects the magnetic dimensionality in the crystal. The observed value for p-BPYNN$^+$·I$^-$ is 0.64, which is comparable to that of the 3D Heisenberg spin system, which is known to be 0.56–0.67. The ratio of entropy change below T_c with respect to the entire magnetic entropy ($R \ln 2$) is also a good index. The evaluated value for p-BPYNN$^+$·I$^-$ is 0.54, which is almost the same as the 0.56–0.67 of the 3D Heisenberg system. It is concluded that the magnetic dimensionality in p-BPYNN$^+$·I$^-$ is close to 3D, because of the ferromagnetic intra- and interchain interactions caused by the π-overlaps and the CH$_3$•••ON contacts, respectively.

ACKNOWLEDGMENTS

The author would sincerely like to thank Akira Yamaguchi, Tsunehisa Okuno, Nobuo Wada, Tamotsu Inabe, Yusei Maruyama, Naoki Sato, Tadashi Sugano, and Minoru Kinoshita for their kind collaboration, encouragement, and fruitful discussions. He gratefully acknowledges partial support by CREST (Core Research for Evolutional Science and Technology) of the Japan Science and Technology Corp. (JST).

REFERENCES

1. Itoh K, Miller JS, Takui T, eds. Mol. Cryst. Liq. Cryst. 1996; 305–306.
2. McConnell HM. J. Chem. Phys. 1963; 39:1910.
3. McConnell HM. Proc. R. A. Welch Found. Chem. Res. 1967; 11:144.
4. Awaga K, Sugano T, Kinoshita M. Chem. Phys. Lett. 1987; 141:540.
5. Yamaguchi K, Fueno T, Nakasuji K, Iwamura H. Chem. Lett. 1986; 629.
6. Kinoshita M, Turek P, Tamura M, Nozawa K, Shiomi D, Nakazawa Y, Ishikawa M, Takashashi M, Awaga K, Inabe T, Maruyama Y. Chem. Lett. 1991; 1225.
7. Takahashi M, Turek P, Nakazawa Y, Tamura M, Nozawa K, Shiomi D, Ishikawa M, Kinoshita M. Phys. Rev. Lett. 1991; 67:746.
8. Tamura M, Nakazawa Y, Shiomi D, Nozawa K, Hosokoshi Y, Ishikawa M, Takahashi M, Kinoshita M. Chem. Phys. Lett. 1991; 186(4,5):401.
9. Le LP, Keren A, Luke GM, Wu WD, Uemura YJ, Tamura M, Ishikawa M, Kinoshita M. Chem. Phys. Lett. 1993; 206:405.
10. Kinoshita M. Jpn. J. Appl. Phys. 1994; 33:5718.
11. Sugawara T, Matsushita MM, Izuoka A, Wada N, Takeda N, Ishikawa M. J. Chem. Soc. Chem. Commun. 1994; 1723.
12. Matsushita MM, Uzuoka A, Sugawara T, Kobayashi T, Wada N, Takeda N, Ishikawa M. J. Am. Chem. Soc. 1997; 119:4369.
13. Cirujeda J, Mas M, Molines E, Panthou FLd, Laugier J, Park JG, Paulsen C, Rey P, Rovira C, Veciana J. J. Chem. Soc. Chem. Commun. 1995; 709.
14. Cirujeda J, Hernandez E, Rovira C, Stanger JL, Turek P, Veciana J. J. Mater. Chem. 1995; 5:243.
15. Cirujeda J, Hernandez E, Panthou FL, Laugier J, Mas M, Molins E, Rovira C, Novoa JJ, Rey P, Veciana J. Mol. Cryst. Liq. Cryst. 1995; 271:1.
16. Caneschi A, Ferraro F, Gatteschi D, Lirzin Al, Novak MA, Rentschler E, Sessoli R. Adv. Mater. 1995; 7:476.
17. Caneschi A, Ferraro F, Gatteschi D, Lirzin Al, Rentschler E. Inorg. Chim. Acta. 1995; 235:159.
18. Awaga K, Inabe T, Maruyama Y. Chem. Phys. Lett. 1991; 190:349.
19. Turek P, Nozawa K, Shiomi D, Awaga K, Inabe T, Maruyama Y, Kinoshita M. Chem. Phys. Lett. 1991; 180(4):327.
20. Allemand P-M, Fite C, Canfield P, Srdanov G, Keder N, Wudl F. Synth. Metal 1991; 41–43:3291.
21. Sugano T, Tamura M, Kinoshita M, Sakai Y, Ohashi Y. Chem. Phys. Lett. 1992; 200:235.
22. Hernandez E, Mas M, Molins E, Rovira C, Veciana J. Angew. Chem. Int. Ed. Engl. 1993; 32(6):882.
23. Panthou FLd, Luneau D, Laugier J, Rey P. J. Am. Chem. Soc. 1993; 115:9095.
24. Awaga K, Inabe T, Maruyama Y, Nakamura T, Matsumoto M. Chem. Phys. Lett. 1992; 195(1):21.
25. Akita T, Mazaki Y, Kobayashi K, Koga N, Iwamura H. J. Org. Chem. 1995; 60:2092.
26. Tamura M, Shiomi D, Hosokoshi Y, Iwasawa N, Nozawa K, Kinoshita M, Sawa H, Kato R. Mol. Cryst. Liq. Cryst. 1993; 232:45.

27. Inoue K, Iwamura H. Chem. Phys. Lett. 1993; 207(4,5,6):551.
28. Okuno T, Otsuka T, Awaga K. J. Chem. Soc., Chem. Commun. 1995; 827.
29. Imai H, Inabe T, Otsuka T, Okuno T, Awaga K. Phys. Rev. B. 1996; 54:6838.
30. Lang A, Pei Y, Ouahab L, Kahn O. Adv. Mater. 1996; 8:60.
31. Ouahab L. Chem. Mater. 1996; 9:1909.
32. Awaga K, Yokoyama T, Fukuda T, Masuda S, Harada Y, Maruyama Y, Sato N. Mol. Cryst. Liq. Cryst. 1993; 232:27.
33. Morisima I, Yoshikawa K, Yonezawa T, Matsumoto H. Chem. Phys. Lett. 1972; 16:336.
34. Davis MS, Morukuma K, Kreilick RW. J. Am. Chem. Soc. 1972; 94:5588.
35. Takui T, Miura Y, Inui K, Teki Y, Inoue M, Itoh K. Mol. Cryst. Liq. Cryst. 1995; 271:55.
36. D'Anna JA, Wharton JH. J. Chem. Phys. 1970; 53:4047.
37. Ressouche E, Boucherle J, Gillon B, Rey P, Schweizer J. J. Am. Chem. Soc. 1993; 115:3610.
38. Zheludev A, Barone V, Bonnet M, Delley B, Grand A, Ressouche E, Rey P, Subra R, Schweizer J. J. Am. Chem. Soc. 1994; 116:2019.
39. Awaga K, Inabe T, Yokoyama T, Maruyama Y. Mol. Cryst. Liq. Cryst. 1993; 232:79.
40. Yamaguchi K, Fueno T. Chem. Phys. Lett. 1989; 159:465.
41. Izuoka A, Murata S, Sugawara T, Iwamura H. J. Am. Chem. Soc. 1985; 107:1786.
42. Izuoka A, Murata S, Sugawara T, Iwamura H. J. Am. Chem. Soc. 1987; 109:2631.
43. Awaga K. Ferromagnetic Intermolecular Interaction in Organic Crystals. PhD dissertation, University of Tokyo, 1988.
44. Awaga K, Yamaguchi A, Okuno T, Inabe T, Nakamura T, Matsumoto M, Maruyama Y. J. Mater. Chem. 1994; 4:1377.
45. Okumura M, Yamaguchi K, Awaga K. Chem. Phys. Lett. 1994; 228:575.
46. Togashi K, Imachi R, Tomioka K, Tsuboi H, Ishida T, Nogami T, Takeda N, Ishikawa M. Bull. Chem. Soc. Jpn. 1996; 69:2821.

26
Organic Paramagnetic Building Blocks for Ferromagnetic Materials

Tadashi Sugawara and Jotaro Nakazaki
University of Tokyo, Tokyo, Japan

Michio M. Matsushita
Tokyo Metropolitan University, Tokyo, Japan

I. INTRODUCTION

Design of organic molecular crystals is an important research target for developing novel materials having prominent physical properties. Whereas inorganic materials are composed of atoms that cannot be modified, one can create molecular architectures by assembling various types of organic molecules by virtue of their self-assembling abilities. Among physical properties of organic materials, conductivity and nonlinear optical properties have been developed extensively [1–4]. Recently, design of organic magnetic materials has drawn much attention [5–8]. Organic materials are diamagnets in general, because the electronic structure of organic molecules is closed shell. In order to construct magnetic materials, one has to prepare organic molecules of a paramagnetic open-shell structure. Such molecules are called *free radicals*, usually are highly reactive, and exist only as reactive intermediates.

This chapter describes the design of an organic radical as a building block for organic magnetic materials, and the idea of assembling those paramagnetic building blocks to exhibit prominent magnetic properties. Typical examples of organic ferromagnetic materials, which are realized according to the foregoing concept, are also introduced.

II. SELF-ASSEMBLING ORGANIC SPIN SYSTEMS

II. A. Magnetic Interactions Between Organic Radicals

Since magnetism is a bulk property, one has to take into account intermolecular magnetic interactions among organic radicals. When organic radicals assemble, the mutual magnetic interaction is usually very weak, and they behave paramagnetically. Even when they interact, unpaired electrons tend to be paired, resulting in antiferromagnetic materials. The question here is how we can align electron spins of organic radicals intermolecularly in parallel.

Organic π-radicals, in general, have characteristic electronic structures. In an allyl radical, for example, an α-spin in the SOMO (singly occupied molecular orbital) distributes on C1 and C3 carbons due to the resonance of two canonical structures (*spin delocalization*). When the molecular orbitals of an allyl radical are calculated by the UHF method, the SOMO spin distribution is found to influence the α- and β-spin distributions of the NHOMO (next-highest occupied molecular orbital) because of the spin-dependent on-site Coulombic repulsion. While the distribution of the α-spin is rather unaffected, that of the β-spin tends to avoid residing on C1 and C3, and prefers C2. As a result, a negative spin density is accumulated to some extent on C2. This modulation in the spin distribution is called *spin polarization*. The spin-dependent on-site Coulombic repulsion (ΔE_{SP}) also destabilizes the β-spin in NHOMO, as shown in Figure 1.

The next stage is to consider the intermolecular magnetic interaction of allyl radicals in various orientations. The local intermolecular interactions between spin-distributing carbon atoms are antiferromagnetic, because the carbons carrying antiparallel spins can form a covalent bond when they come close. When allyl radicals are aligned as in arrangement (a) or (c) in Figure 2, the overall intermolecular interaction becomes antiferromagnetic. On the other hand, arrangement (b) in Figure 2 leads to a net ferromagnetic interaction. Consequently, it is clear that the intermolecular magnetic interaction depends heavily on the relative orientations in a molecular assembly of π-radicals.

Magnetic interactions observed in carbene species with various mutual orientations have provided a proof for the foregoing mechanism of spin ordering. For instance, the isomeric [2.2]paracyclophane-dicarbenes have been designed and generated as model compounds to examine ferromagnetic intermolecular magnetic interactions [9] (Figure 3). The ESR spectra of the *pseudo-ortho-* and *-para*-isomers show intense ground state quintet signals due to the ferromagnetic coupling, whereas the *meta*-isomer gives a thermally populated triplet signal, indicating that this isomer has a singlet ground state due to the antiferromagnetic coupling. This is experimental

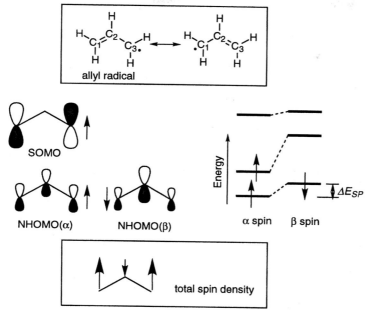

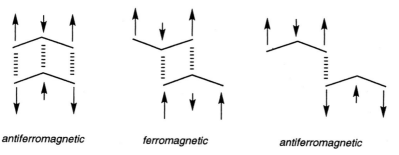

Figure 1 Spin delocalization and spin polarization in an allyl radical.

evidence for McConnell's theory concerning intermolecular spin alignment [10].

When photolysis of polycrystals of bis(p-methoxyphenyl)diazomethane is carried out at cryogenic temperatures, a ground state quintet signal is observed through the ferromagnetic intermolecular interaction [11]. In the case of bis[p-(octyloxy)phenyl]diazomethane, in particular, the photolyzed

antiferromagnetic   *ferromagnetic*   *antiferromagnetic*

Figure 2 Intermolecular interactions between allyl radicals.

ortho   meta   para

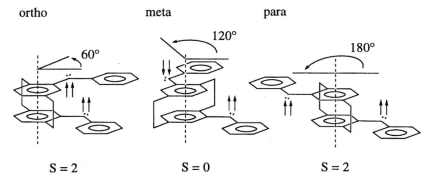

Figure 3 Magnetic interactions in cyclophane-dicarbenes.

sample shows a positive Weiss temperature of 2 K [12]. Introduction of long alkyl chains at the *para*-positions of diphenylcarbenes leads to a favorable arrangement that causes a two-dimensional ferromagnetic coupling among spin-distributing benzene rings. As a result, these carbene clusters form a two-dimensional metamagnetic system.

II. B. Chemically Modified Stable Organic Radicals

Although the essence of the magnetic interaction can be fully demonstrated by allyl radical or carbene species, the kinetic stability of these species is too low to use them as building blocks for constructing organic spin systems. This is why persistent organic radicals are required for this purpose.

Such persistent organic radicals can be obtained by extending a π-framework in order to acquire the thermodynamic stability [13]. Introduction of bulky substituents is also effective, because they block spin-distributing sites to increase kinetic stability. While the stability of the triphenylmethyl radical is not high enough to study its magnetic properties in the solid state, its *p*-phenyl, or perchloro, derivative turns out to be reasonably stable. As for the oxygen atom−centered radicals, the kinetic stability of the phenoxy radical carrying *tert*-butyl groups at 2,4,6-position is fairly high. Another class of stable radicals consist of heteroatom-centered radicals with a two-centered three-electron bond. Typical examples of this class are hydrazyls and nitroxides (Figure 4).

The bifluorenylmethyl radical [14] is an example of a chemically elaborated allyl radical. Extension of the π-electronic structure of the phenoxy radical leads to the galvinoxyl radical [15], the magnetic properties of which has been well investigated [16,17]. Verdazyls [18,19] and nitronyl nitroxides

tris(biphenylyl)methyl radical   phenoxy radical   hydrazyl   nitroxide

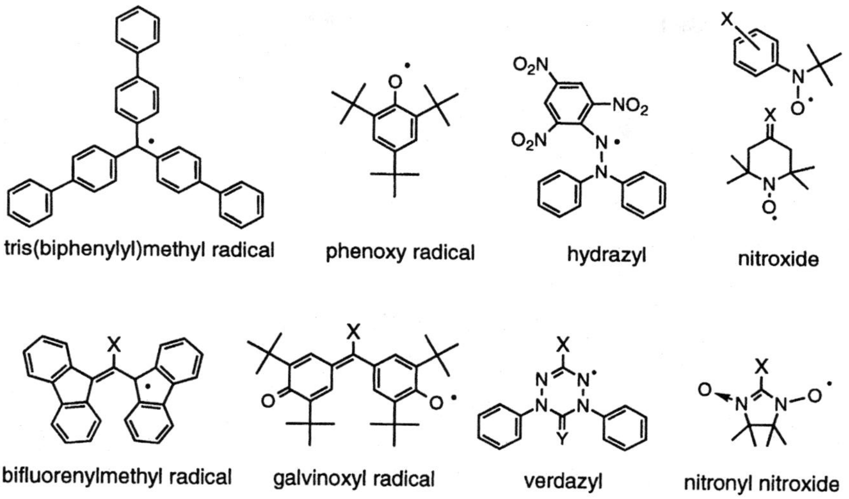

bifluorenylmethyl radical   galvinoxyl radical   verdazyl   nitronyl nitroxide

Figure 4   Stable organic radicals.

[20], which are the π-extended derivatives of the foregoing radicals, are often used as building blocks for magnetic materials.

II. C.   Self-Assembling Organic Radicals with an Orientation Controlling Site

As already described, the intermolecular magnetic interaction depends heavily on the relative orientations in a molecular assembly of π-radicals. Under such circumstances, it would be of great significance to control crystal structures of stable organic radicals so as to generate desirable magnetic properties. A method for controlling the crystal structure of organic radicals proposed here involves preparation of organic radicals equipped with an *orientation controlling site (OCS)* [21]. An OCS is a functional group that plays a significant role in the self-assembly of organic radicals by means of intermolecular interactions, such as electrostatic interactions, hydrogen bonds, or charge transfer interactions (Figure 5). Organic radicals then self-assemble by themselves as paramagnetic building blocks to create a highly ordered molecular architecture with intriguing magnetic properties.

One of the successful cases for arranging π-radicals is a crystal designing ferrimagnet [22,23]. As building blocks for a ferrimagnetic system, a ground state triplet diradical, *m*-phenylenebis(nitronyl nitroxide) (PBNN), and monoradical, *m*-dinitrophenyl nitronyl nitroxide (DNPN) were selected.

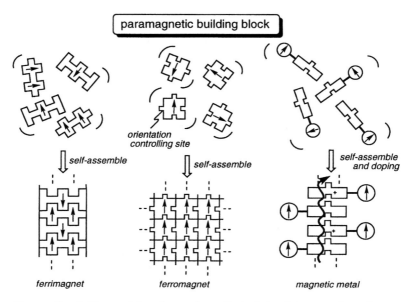

Figure 5   Self-assembly of organic radicals helped by an orientation controlling site.

The two nitro groups on the phenyl ring act as an OCS. As a result of the electrostatic interaction between the nitro groups of the monoradical and the nitronyl nitroxide groups of the diradical, they are stacked alternately to form a columnar structure (Figure 6). The organic ferrimagnetic spin system obtained is unique from the viewpoint of the competing magnetic interactions between the intramolecular ferromagnetic and the intermolecular antiferromagnetic couplings of the same order [24].

III.   HYDROGEN-BONDED FERROMAGNETS

Since the first discovery of an organic ferromagnet consisting of p-NPNN [25,26], a number of organic ferromagnets have been reported [27–34]. Among them, a hydrogen-bonded ferromagnet [27–29] has drawn much attention, because a hydrogen bond may play a role not only of organizing organic radicals but also of transmitting the spin polarization intermolecularly. As seen in the ferrimagnetic molecular assembly, nitronyl nitroxide groups are likely to interact antiferromagnetically in crystals, when they come closer due to the electrostatic interaction. In order to achieve ferro-

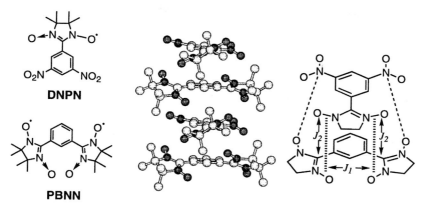

Figure 6 Self-assembled organic ferrimagnetic system composed of DNPN and PBNN.

magnetic interaction, nitronyl nitroxide groups should be located apart from each other. Since the nitronyl nitroxide group is known as a good hydrogen bond acceptor, it is interesting to connect the positively spin-polarized nitroxide group with the negatively spin-polarized aromatic carbon through a hydrogen bond. If such a hydrogen bond is formed effectively, a ferromagnetic intermolecular interaction is predicted.

III. A. Building Blocks for a Hydrogen-Bonded Magnetic System

The spin distribution of pheny nitronyl nitroxide (PhNN) has been investigated in detail by means of ESR [35], NMR [36], neutron diffraction experiments [37], and theoretical calculations [38]. The oxygen and nitrogen atoms of the nitronyl nitroxide (NN) groups have large, positive spin densities, suggesting that the unpaired electron is localized mostly at these sites, whereas positive and negative spin densities distribute alternately over the benzene ring. In contrasting, methyl hydrogens are all negatively spin polarized due to the hyperconjugative mechanism.

As building blocks for such a hydrogen-bonded assembly, phenyl nitronyl nitroxide derivatives carrying hydroxy groups as an OCS were designed as follows. Introduction of two hydroxy groups into the positively and negatively spin-polarized carbon atoms of the phenyl ring, respectively, affords a *p*-hydroquinone derivative carrying a nitronyl nitroxide group at the *ortho*-position, and the resulting radical is called HQNN. A resorcinol derivative (RSNN) carries two hydroxy groups at the *meta*-positions that are positively spin polarized. The OH proton is considered to be spin polarized in the same direction as that of the *ipso*-carbon, based on the hyperconjugative mechanism (Figure 7). Comparison between the crystal structures and the magnetic properties of crystals of HQNN and RSNN should help to elucidate the magnetostructural correlation in hydrogen-bonded crystals of nitronyl nitroxide derivatives.

III. B. Magnetostructural Correlation in HQNN and RSNN

HQNN was found to afford two phases of crystals. In the α-phase crystal, in particular, the phenolic hydroxy group at the *ortho*-position not only forms a strong intramolecular hydrogen bond with the nitroxide group (O•••O distance of 2.51 Å), but also forms an intermolecular hydrogen bond with the nitroxide oxygen of the facing HQNN. This hydrogen-bonded dimer also participates in intermolecular hydrogen bonds with the hydroxy groups at the *meta*-position of the translated molecules at a distance of 2.75 Å (Figure 8). The two parallel hydrogen-bonded chains of HQNN run along the *c*-axis, with the symmetry of inversion between facing molecules. These doubly hydrogen-bonded chains are arranged in a herringbone-type structure, as depicted in Figure 9. Furthermore, it should be noted that the oxygen atom of the NN group is located close to a hydrogen atom (2.75 Å) of the methyl group of HQNN in the neighboring chain.

The magnetic susceptibility of the polycrystalline sample of HQNN was analyzed by the S−T model ($J/k_B = 0.95$ K) with a positive Weiss temperature of $\theta = +0.44$ K. The ac susceptibility of HQNN increased rapidly at around 0.5 K, suggesting that a phase transition to the ferromagnetic phase occurred at this temperature. Since the estimated saturation value of the magnetization was very close to the theoretical one ($1 \ \mu_B \cdot mol^{-1}$), the

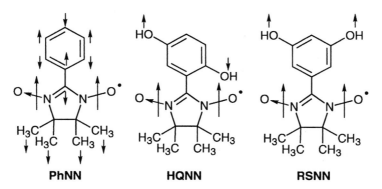

Figure 7 Spin polarization in PhNN and its dihydroxy derivatives.

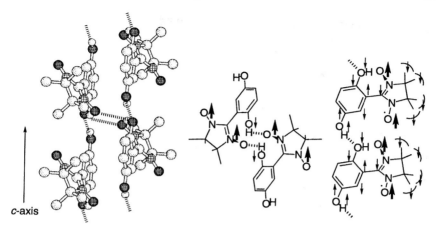

Figure 8 Doubly hydrogen-bonded chains of HQNN.

phase transition could be regarded as a bulk transition. The crystal shows a hysteretic behavior, although the coercive force was less than 20 Oe. From the heat capacity data, the spin system of HQNN turns out to be described by a three-dimensional Heisenberg model [39] with six nearest neighbors ($z = 6$, Figure 10).

The magnetic properties of HQNN were found to be consistent with the crystal structure. First, the ferromagnetic coupling can be assigned to the hydrogen-bonded face-to-face dimer between the o-hydroxy group, which is

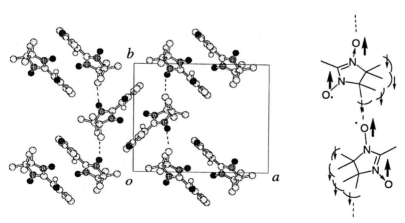

Figure 9 Interchain CH•••ON interaction in HQNN crystal.

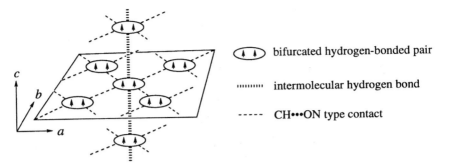

Figure 10   Three-dimensional Heisenberg spin system in HQNN.

negatively spin polarized, and the oxygen of NN, which is positively spin polarized. Second, the spin polarization should be transmitted ferromagnetically along the one-dimensional hydrogen-bonded chain, because the signs of the spin densities of these sites are opposite. The ferromagnetic interaction along this direction is enhanced, because the nitronyl nitroxide group is close to the negatively spin-polarized methyl hydrogen of the neighboring molecule. Third, a CH•••ON interaction also exists in the *ab*-plane and should increase the dimensionality of the ferromagnetic interactions in the crystal.

The crystal structure of RSNN is characterized by double hydrogen-bonded chains between the hydroxy groups and the NN groups (Figure 11). According to the spin polarization at the contacting sites, these hydrogen bonds are expected to produce the antiferromagnetic interaction. The observed magnetic properties are consistent with this prediction, suggesting that the hydrogen bond plays an influential role in transmitting spin polarization.

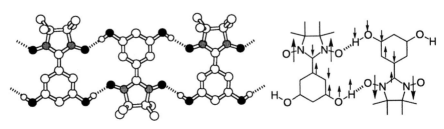

Figure 11   Hydrogen-bonded chain of RSNN.

III. C. Magnetic Coupling through Hydrogen Bonds in HQNN

Ferromagnetic coupling along the hydrogen bond in HQNN was confirmed as follows. In a crystal of HQNN, the bifurcated hydrogen bond is particularly important for the transmission of spin polarization. First, the presence of the ferromagnetic intermolecular interaction in this contact was reproduced by theoretical calculation [40]. When the *ortho*-hydroxy groups are removed, the model system exhibits a large antiferromagnetic intermolecular interaction due to the close contact of the NO groups. This result is strong evidence for the transmission of spin polarization through the hydrogen bonds.

Second, the deuterium substitution effect of HQNN on the magnetic interaction was examined [28]. If the hydrogen bond contributes to the transmission of spin polarization, the deuteration should affect the degree of the intermolecular magnetic interaction, because deuteration causes elongation of the hydrogen bond distance. In fact, the ferromagnetic interaction detected in the magnetic susceptibility of HQNN decreased upon deuteration of the hydroxy groups. The maximum of the plot of heat capacity vs. temperature was shifted to the lower-temperature side, compared with that of the hydrogen counterpart.

Third, the spin density on the *ortho*-hydroxy group is evaluated by the Knight shift of the solid-state ^{1}H- and ^{2}H-NMR signals, using a sample deuterated at the hydroxy group [41]. The spin densities determined by this method at the hydroxy groups, aromatic hydrogens, and methyl hydrogens are consistent with qualitatively predicted values. The large negative spin density at the *ortho*-hydroxy group, in particular, is strong proof for the contribution of the hydrogen bond of the OH•••NO type to the magnetic transmission of the spin polarization. All these results are consistent with the interpretation that the magnetic interaction is transmitted through the hydrogen bond.

The crystal design of organic ferromagnets is extremely difficult at the present stage. In this respect, utilization of the hydrogen bond as a ferromagnetic coupler is considered to be a promising method because the crystal structure defined by the hydrogen bond is somewhat predictable. Thus it may be applicable widely to control the molecular arrangement of paramagnetic building blocks.

IV. APPROACH TO ORGANIC FERROMAGNETIC METALS USING SPIN-POLARIZED DONORS

IV. A. Spin-Polarized Donor as a Building Block for a Conducting Ferromagnet

Although a number of organic ferromagnets have been discovered, the transition temperatures to the ferromagnetic state are still low. In order to raise the transition temperatures of organic ferromagnets significantly, it may be effective to use conduction electrons as a transmitter of the spin polarization [42–46]. For this purpose, a *spin-polarized* donor, which yields a ground state triplet cation diradical upon one-electron oxidation, may become an important building block for constructing a conducting magnetic material [47]. Namely, when the spin-polarized donor is oxidized, the resulting unpaired π-electron is delocalized over the entire molecule, coupling ferromagnetically with the local spin on the radical site. If columnar stacking of the donors is realized, the conduction electrons that are generated by hole-doping can migrate along the donor stack to align local spins on the radical sites (Figure 12).

The spin-polarized donors can be designed as follows. Trimethylene-methane (TMM) [48,49] is known to be a ground state triplet diradical, and the triplet state of TMM is significantly more stable than the singlet (ΔE_{TS} = 16 kcal/mol [50]). If a heteroatom with a lone pair of electrons (π-donor unit) is introduced to the nodal carbon of an allyl radical, the corresponding singly oxidized donor radical can be regarded as a heteroanalog of TMM. In order to enhance kinetic stability, the allyl radical part is replaced with nitronyl nitroxide (NN). According to such chemical modifications, dimethylamino nitronyl nitroxide, DMANN, and dimethylaminophenyl nitronyl nitroxide, APNN, were prepared [51,52]. The two singly occupied molecular orbitals of DMANN$^{+\bullet}$ and APNN$^{+\bullet}$ are regarded as NBMOs of

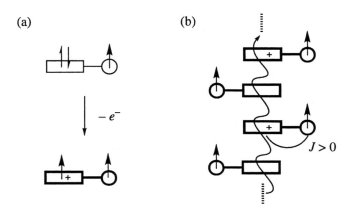

(a)

(b)

Spin-Polarized Donor   *Organic Ferromagnetic Metal*

Figure 12 Schematic drawing of: (a) spin-polarized donor; (b) electronic structure of organic ferromagnetic metal.

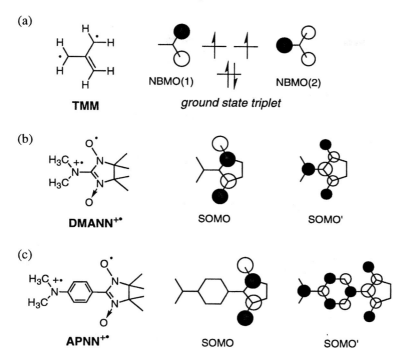

Figure 13 Electronic structures of TMM and the cation diradical of spin-polarized donors (DMANN$^{+\bullet}$ and APNN$^{+\bullet}$).

alternate hydrocarbons, although the two energy levels are not degenerate (Figure 13). As seen in Figure 13, the coefficients of the SOMO are localized on nitronyl nitroxide, whereas those of the SOMO′, which are derived from the one-electron oxidation of the HOMO, spread over the entire molecule, sharing the coefficients of atomic orbitals of the nitronyl nitroxide group

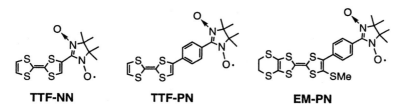

Figure 14 Spin-polarized donors as a building block for organic ferromagnetic metal.

with the SOMO. As far as the distribution of the coefficients of the SOMO and SOMO' is concerned, it resembles that of the two degenerate NBMOs of TMM closely. The cation diradicals, DMANN$^{+\cdot}$ and APNN$^{+\cdot}$, therefore, should exist as ground state triplet species.

Although DMANN and APNN are found to afford ground state triplet cation diradicals upon one-electron oxidation, the molecular assembly of these spin-polarized donors are not appropriate to realize a metallic state. Therefore, the spin-polarized donors, TTF-NN, TTF-PN, EM-PN, in which donor moieties are replaced by the TTF (tetrathia fulvalene) units, have been prepared as building blocks of organic ferromagnetic metals (Figure 14).

IV. B.   Ferromagnetic Spin Alignment in the Assembly of Spin-Polarized Donors

It is crucial to examine whether these spin-polarized donors, when they are oxidized, exhibit intra- and also intermolecular ferromagnetic coupling or not. A cation diradical derived from TTF-NN turned out to be a ground state singlet. The antiferromagnetic interaction was estimated to be $J = -100$ K, from the temperature dependence of the thermally populated triplet signal [53]. In order to surpress such an undesirable interaction, a p-phenylene group was inserted into TTF-NN between the donor site and the radical site. Now the triplet was revealed to be the ground state for the cation diradical of TTF-PN.

Based on these results, EM-PN was also prepared, in order to enhance kinetic stability and to increase the intermolecular interaction between cation diradicals. Here, a dithioethylene group was introduced at one end of the TTF moiety and a thiomethyl group was attached at the olefinic carbon. The ground state of the cation diradical of EM-PN was also proved to be triplet, and the oxidized species became reasonably stable. When excess bromine was added to a solution of EM-PN in tetrahydrofuran, a black, solid CT complex between EM-PN and bromine was obtained. The magnetization curve of the complex at 5 K showed an average spin multiplicity higher than 5/2. This result indicates that not only is the intramolecular magnetic interaction between two unpaired electrons in EM-PN$^{+\cdot}$ ferromagnetic, but EM-PN$^{+\cdot}$ also forms a ferromagnetically coupled aggregate with a neutral EM-PN through the charge-transfer interaction.

The ferromagnetically coupled cluster consisting of EM-PN$^{+\cdot}$ and the neutral EM-PN is considered to be a unit of the organic ferromagnetic metal based on the double exchange mechanism (Figure 15). The approach along this line is expected to yield a genuine organic ferromagnetic metal in the future.

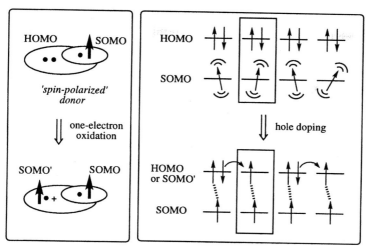

Figure 15 Electronic structure of a spin-polarized donor and of its assembly in the mixed-valence state.

REFERENCES

1. Williams JM, Ferraro JR, Thorn RJ, Carlson KD, Geiser U, Wang HH, Kini AM, Whangbo MH, eds. Organic Superconductors. Englewood Cliffs, NJ: Prentice Hall, 1992.
2. Saito G, Kagoshima S, eds. The Physics and Chemistry of Organic Superconductors. New York: Springer-Verlag, 1990.
3. Wudl F. From organic metals to superconductors: Managing conduction electrons in organic solids. Acc Chem Res. 1984; 17:227–232.
4. Kanis DR, Ratner MA, Marks TJ. Design and construction of molecular assemblies with large second-order optical nonlinearities, quantum chemical aspects. Chem Rev. 1994; 94:195–242.
5. Sugawara T. Science of organic ferromagnets. J Synth Org Chem Jpn. 1989; 47:306–320.
6. Kahn O. Molecular Magnetism. New York: VCH, 1993.
7. Iwamura H. High-spin organic molecules and spin alignment in organic molecular assemblies. Adv Phys Org Chem. 1990; 26:179–253.
8. Miller JS, Epstein AJ. Organic and organometallic molecular magnetic materials—Designer magnets. Angew Chem Int Ed Engl. 1994; 33:385–415.
9. Izuoka A, Murata S, Sugawara T, Iwamura H. Molecular design and model experiments of ferromagnetic intermolecular interaction in the assembly of high-spin organic molecules. Generation and characterization of the spin states of isomeric bis(phenylmethylenyl)[2,2]paracyclophanes. J Am Chem Soc. 1987; 109:2631–2639.

10. McConnell HM. Ferromagnetism in solid free radicals. J Chem Phys. 1963; 39:1910.

11. Sugawara T, Tukada H, Izuoka A, Murata S, Iwamura H. Magnetic interaction among diphenylmethylene molecules generated in crystals of some diazodiphenylmethanes. J Am Chem Soc. 1986; 108:4272–4278.

12. Sugawara T, Murata S, Kimura K, Iwamura H, Sugawara Y, Iwasaki H. Design of molecular assembly of diphenylcarbenes having ferromagnetic intermolecular interaction. J Am Chem Soc. 1985; 107:5293–5294.

13. Kosower EM. An Introduction to Physical Organic Chemistry. New York: Wiley, 1968.

14. Koelsch CF. Synthesis with triarylvinylmagnesium bromides. α,γ-bisdiphenylene-β-phenylallyl, a stable free radical. J Am Chem Soc. 1957; 79:4439–4441.

15. Kharasch MS, Joshi BS. Reactions of hindered phenols. I. Reactions of 4,4'-dihydroxy-3,5,3',5'-tetra-tert-butyl diphenylmethane. J Org Chem 1957; 22: 1435–1438.

16. Mukai K. Anomalous magnetic properties of stable crystalline phenoxyl radicals. Bull Chem Soc Jpn. 1969; 42:40–46.

17. Awaga K, Sugano T, Kinoshita M. Ferromagnetic intermolecular interactions in a series of organic mixed crystals of galvinoxyl radical and its precursory closed shell compound. J Chem Phys 1986; 85:2211–2218.

18. Kuhn R, Trischmann H. Über auffallend stabile N-haltige Radikale. Angew Chem 1963; 75:294–295.

19. Neugebauer FA, Fischer H, Krieger C. Verdazyls. part 33. EPR and ENDOR studies of 6-oxo- and 6-thioxoverdazyls. X-ray molecular structure of 1,3,5-triphenyl-6-oxoverdazyl and 3-tert-butyl-1,5-diphenyl-6-thioxoverdazyl. J Chem Soc Perkin Trans 2. 1993; 535–544.

20. Ullman EF, Osiecki JH, Boocock DGB, Darcy R. Studies of stable free radicals. X. Nitronyl nitroxide monoradicals and biradicals as possible small molecule spin labels. J Am Chem Soc. 1972; 94:7049–7059.

21. Sugawara T, Izuoka A. Molecular magnetism: present and future. Mol Cryst Liq Cryst. 1997; 305:41.

22. Izuoka A, Fukada M, Kumai R, Itakura M, Hikami S, Sugawara T. Magnetically coupled molecular systems composed of organic radicals with different spin multiplicities. J Am Chem Soc. 1994; 116:2609–2610.

23. Izuoka A, Kumai R, Sugawara T. Crystal designing organic ferrimagnets. Adv Mater. 1995; 7:672–674.

24. Shiomi D, Nishizawa M, Sato K, Takui T, Itoh K, Sakurai H, Izuoka A, Sugawara T. A prerequisite for purely organic molecule-based ferrimagnetics: Breakdown of simple classical pictures. J Phys Chem B. 1997; 101:3342–3348.

25. Kinoshita M, Turek P, Tamura M, Nozawa K, Shiomi D, Nakazawa Y, Ishikawa M, Takahashi M, Agawa K, Inabe T, Maruyama Y. An organic radical ferromagnet. Chem Lett. 1991; 1225–1228.

26. Tamura M, Nakazawa Y, Shiomi D, Nozawa K, Hosokoshi Y, Ishikawa M, Takahashi M, Kinoshita M. Bulk ferromagnetism in the β-phase crystal of the p-nitrophenyl nitronyl nitroxide. Chem Phys Lett. 1991; 186:401–404.

27. Sugawara T, Matsushita MM, Izuoka A, Wada N, Takeda N, Ishikawa M. An organic ferromagnet: α-Phase crystal of 2-(2′,5′-dihydroxyphenyl)-4,4,5,5-tetramethyl-4,5-dihydro-1*H*-imidazolyl-1-oxy-3-oxide (α-HQNN). J Chem Soc Chem Commun. 1994; 1723–1724.

28. Matsushita MM, Izuoka A, Sugawara T, Kobayashi T, Wada N, Takeda N, Ishikawa M. Hydrogen-bonded organic ferromagnet. J Am Chem Soc. 1997; 119:4369–4379.

29. Cirujeda J, Mas M, Molins E, Panthou FL, Laugier J, Park JG, Paulsen C, Rey P, Rovira C, Veciana J. Control of the structural dimensionality in hydrogen-bonded self-assemblies of open-shell molecules. Extension of intermolecular ferromagnetic interactions in α-phenyl nitronyl nitroxide radicals in three dimensions. J Chem Soc Chem Commun. 1995; 709–710.

30. Chiarelli R, Novak MA, Rassat A, Tholence JL. A ferromagnetic transition at 1.48 K in an organic nitroxide. Nature. 1993; 363:147–149.

31. Nogami T, Tomioka K, Ishida T, Yoshikawa H, Yasui M, Iwasaki F, Iwamura H, Takeda N, Ishikawa M. A new organic ferromagnet: 4-benzylideneamino-2,2,6,6-tetramethylpiperidin-1-oxyl. Chem Lett. 1994; 29–32.

32. Togashi K, Imachi R, Tomioka K, Tsuboi H, Ishida T, Nogami T, Takeda N, Ishikawa M. Organic radicals exhibiting intermolecular ferromagnetic interaction with high probability: 4-Arylmethyleneamino-2,2,6,6-tetramethylpiperidin-1-oxyls and rerated compounds. Bull Chem Soc Jpn. 1996; 69:2821–2830.

33. Mukai K, Konishi K, Nedachi K, Takeda K. Bulk ferro- and antiferromagnetic behavior in 1,5-dimethyl verdazyl radical crystals with similar molecular structure. J Magn Magn Mater. 1995; 140–144:1449–1450.

34. Caneschi A, Ferraro F, Gatteschi D, Lirzin A, Novak MA, Rentschler E, Sessoli R. Ferromagnetic order in the sulfur-containing nitronyl nitroxide radical, 2-(4-thiomethyl)-phenyl-4,4,5,5-tetramethylimidazoline-1-oxyl-3-oxide, NIT-(SMe)Ph. Adv Mater. 1995; 7:476–478.

35. D'Anna JA, Wharton JH. Electron spin resonance spectra of α-nitronyl nitroxide radical; solvent effects; nitrogen hyperfine tensor; *g* anisotropy. J Chem Phys 1970; 53:4047–4052.

36. Neely JW, Hatch GF, Kreilick RW. Electron-carbon coupling of aryl nitronyl nitroxide radical. J Am Chem Soc. 1974; 96:652–656.

37. Zheludev A, Barone V, Bonnet M, Delley B, Grand A, Ressouche E, Rey P, Subra R, Schweizer J. Spin density in a nitronyl nitroxide free radical. Polarized neutron diffraction investigation and ab initio calculations. J Am Chem Soc. 1994; 116:2019–2027.

38. Yamaguchi K, Okumura M, Nakano M. Theoretical calculations of effective exchange integrals between nitronyl nitroxides with donor and acceptor groups. Chem Phys Lett. 1992; 191:237–244.

39. Jongh LJ, Miedema AR. Experiments on simple magnetic model systems. Adv Phys. 1974; 23:1–260.

40. Oda A, Kawakami T, Takeda S, Mori W, Matsushita MM, Izuoka A, Sugawara T, Yamaguchi K. Theoretical studies of magnetic interactions in 2′,5′-dihy-

droxyphenyl nitronyl nitroxide crystal. Mol Cryst Liq Cryst. 1997; 306:151–160.

41. Maruta G, Takeda S, Oda A, Matsushita MM, Izuoka A, Sugawara T, Yamaguchi K. Solid-state high-resolution ^{2}H-NMR study of spin density distribution in hydrogen-bonded organic ferromagnet HQNN. Symposium on Molecular Structure, Nagoya, Japan, Oct 2–5, 1997.

42. Ruderman MA, Kittel C. Indirect exchange coupling of nuclear magnetic moments by conduction electrons. Phys Rev. 1954; 96:99–102.

43. Kasuya T. A theory of metallic ferro- and antiferromagnetism on Zener's model. Prog Theor Phys. 1956; 16:45–57.

44. Kasuya T. Electrical resistance of ferromagnetic metals. Prog Theor Phys. 1956; 16:58–63.

45. Yoshida K. Magnetic properties of Cu-Mn alloys. Phys Rev. 1957; 106:893–898.

46. Kimura T, Tomioka Y, Kuwahara H, Asamitsu A, Tamura M, Tokura Y. Interplane tunneling magnetoresistance in a layered manganite crystal. Science 1996; 274:1698–1701.

47. Sugawara T, Izuoka A, Kumai R. Approach to organic ferromagnetic metals composed of spin-polarized donors. In: Itoh K, Iwamura H, eds. Molecular Magnetism—New Magnetic Materials. Tokyo: Kodansha Gordon & Breach. In press.

48. Dowd P. Trimethylenemethane. Acc Chem Res. 1972; 5:242–248.

49. Borden WT, Iwamura H, Berson JA. Violations of Hunt's rule in non-Kekulé hydrocarbons: Theoretical prediction and experimental verification. Acc Chem Res. 1994; 27:109–116.

50. Wenthold PG, Hu J, Squires RR, Lineberger WC. Photoelectron spectroscopy of the trimethylenemethane negative ion. The singlet-triplet splitting of trimethylenemethane. J Am Chem Soc. 1996; 118:475–476.

51. Kumai R, Sakurai H, Izuoka A, Sugawara T. Ground state triplet cation diradicals having non-degenerated singly occupied molecular orbitals. Mol Cryst Liq Cryst. 1996; 279:133–138.

52. Sakurai H, Kumai R, Izuoka A, Sugawara T. Ground state triplet cation diradicals generated from N,N-dimethylamino nitronyl nitroxide and its homologues through one-electron oxidation. Chem Lett. 1996; 879–880.

53. Kumai R, Matsushita MM, Izuoka A, Sugawara T. Intramolecular exchange interaction in a novel cross-conjugated spin system composed of π-ion radical and nitronyl nitroxide. J Am Chem Soc. 1994; 116:4523–4524.

27

Crystal Control in Organic Radical Solids

Naoki Yoshioka and Hidenari Inoue
Keio University, Yokohama, Japan

I. INTRODUCTION

Recently, there has been substantial interest in the study of molecular-based magnetism. The synthesis of molecular crystals, macromolecules possessing organic radical, and/or paramagnetic metal ion as spin centers has been in progress based on various individual strategies [1–3]. More than a dozen purely organic radicals with bulk ferromagnetism caused by intermolecular magnetic coupling have been discovered by laboratories throughout the world [4–6]. Whereas these ferromagnetic orderings have been discovered fortuitously, methodology for designing intermolecular exchange coupling pathways has not been established yet. As a guiding principle of intermolecular magnetic coupling, McConnell suggested conditions for ferromagnetic interaction in stacked radical molecules that possess both positive and negative spin densities [7]. According to his mechanism, ferromagnetic interaction would be achieved when the product of the spin densities becomes negative between interacting spin sites, each belonging to the neighboring radical molecules. Figure 1 illustrates an allyl radical, which has positive spin densities on the terminal carbon atom and negative spin density on the central carbon atom. The product of spin densities at interacting sites between the two allyl radicals is positive in case (a), whereas it becomes all negative in case (b). Consequently, case (a) is predicted to be at ground singlet state, while case (b) would lead to a triplet ground state.

Organic ferromagnets so far reported have suitable arrangements to fulfill the preceding requirement; however, the control of molecular arrange-

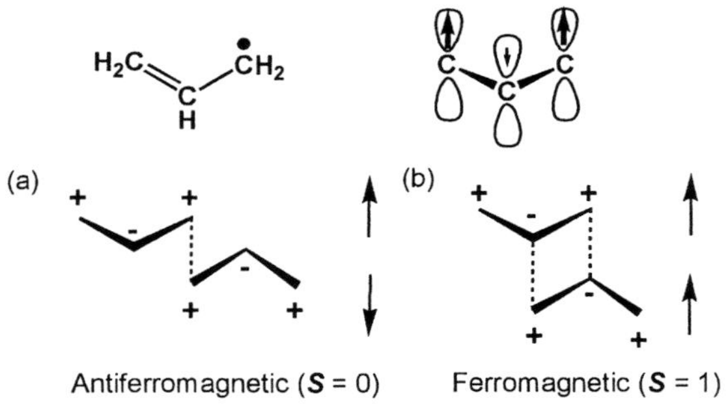

Figure 1 Interaction of two neighboring allyl radicals in parallel planes based on a spin-polarization mechanism.

ment in the crystal does not necessarily correspond to predictable strategy. For the purpose of inducing desirable stacking, van der Waals attractions [8], hydrogen bonds [9–13], elecrostatic forces [14], aromatic stackings [15], steric effects [16,17], have been introduced and their effects examined. Whereas molecular chemistry deals with the design and the synthesis of individual radical units, their assembly into a solid is governed by concepts of supramolecular chemistry. Although tremendous progress has been achieved by molecular chemistry in terms of planning the synthesis of desired open-shell molecules and the control of their electronic properties, the supramolecular design and synthesis of solid materials still remain great challenges. By using both molecular and supramolecular techniques, strictly programmed and controlled molecular systems with magnetic cooperativity can be envisaged. In principle, any type of weak, intermolecular interactions may be used to construct molecular solids. The types of bonds and interactions that have the potential to be used in the design of self-assembly of supramolecules are summarized in Table 1. Hydrogen bond, charge transfer, and weak dipole–dipole interactions are noncovalent intermolecular interactions that have been used in efforts to assemble chains, tapes, and other structures in the solid state. Electrostatic interactions and hydrogen bonds are stronger in energy than van der Waals attractions between neutral moieties. Hydrogen bonds, in particular, are well suited for building molecular assemblies with specific shapes and sizes, because their energies are roughly comparable to thermal energies and because they are fairly directional. The design of molecular assemblies using hydrogen bonds to control structure in the solid state has yielded a number of potentially useful structural motifs [18]. The criteria for identifying hydrogen bonds have been rationalized [19].

Table 1 Types of Bonds and Interactions for Designing Molecular Self-Assemblies

Bond type	Molecular system	Dependence of interaction energy on the intermolecular distance (R)
Electrostatic interactions	ionic–ionic	R^{-1}
Hydrogen bonds	neutral–neutral	Complex R^{-2} (short-range)
van der Waals interactions	neutral–neutral	R^{-6}
Charge transfer forces	neutral–neutral ionic–neutral	Complex

II. MOLECULAR DESIGN OF RADICALS WITH NH HYDROGEN BOND SITE

Our ultimate objective is to design radical molecules whose assembled structure can be predicted from their chemical structure. We have been interested in the electronic structure of diazoles, since one of the annular nitrogens has a hydrogen atom and can be regarded as a pyrrole-type nitrogen, while the other resembles the nitrogen in pyridine. For this reason it is possible to look upon imidazole as a molecule that has both a proton donor site (NH) and a proton acceptor site (N). Moreover, imidazole is characterized by rapid proton exchange between the —NH— and the =N— nitrogen atoms described by mesomeric structures (Figure 2), which is also evidenced by the magnetic equivalence of 4- and 5-hydrogen atoms in ^{1}H NMR spectra. In the solid state, an intermolecular NH···N hydrogen bond exists and forms the molecular assembly in chain shape [20].

(a)   (b)

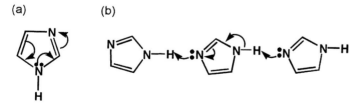

Figure 2 Mesomeric structure of imidazole (a), and illustration of rapid proton exchange between —NH— and =N— nitrogen atoms.

We combined imidazole with nitronyl nitroxide, 4,4,5,5-tetramethyl-
1*H*-imidazoline-1-oxyl-3-oxide, NN, which has two equivalent NO bonds
and whose oxygen atoms exhibit weak Lewis base character [Figure 3(a)].
Most of the spin density is distributed equally over the pair of NO groups,
whereas a large negative spin density is found on the carbon atom between
the two NO bonds. These characteristics are suitable for controlling the
molecular arrangement using hydrogen bonds. The charge distribution in Im-
NN [Figure 3(b)] [21] implies that the —NH proton is a good proton donor
(D) and the —N= and —O atom are good proton acceptors (A_1 and A_2,

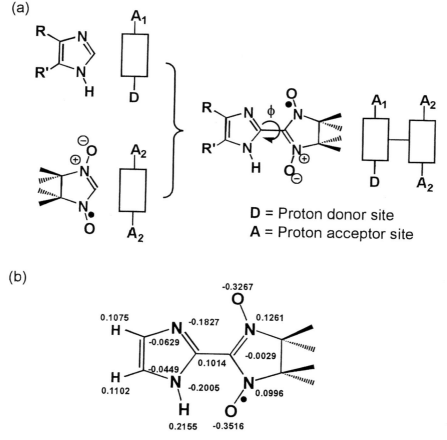

D = Proton donor site
A = Proton acceptor site

Figure 3 Molecular design of nitronyl nitroxide with imidazole ring (a), and
atomic charge of Im-NN by MNDO calculation (RHF doublet). (From Ref. 21.)

respectively). From the point of hydrogen-bond patterns, Im-NNs have four basic motifs due to the competition between the formation of $A_1 \cdots D$ versus $A_2 \cdots D$ (Table 2). In the first case, intermolecular hydrogen bonds between imidazole rings form a chain structure corresponding to the C(4) pattern of Etter's graph-set assignments [19]. The second case corresponds to a C(6) pattern, which occurs when the intermolecular NH $\cdots$ O—N hydrogen bond

Table 2 Graph-set assignments for Im-NNs

Hydrogen-bonded pattern	Graph set[a]
	C(4)
	C(6)
	S(6)
	$R_2^2(12)$

[a]Assigned based on Ref. 19. C, S, and R correspond to chain, intermolecular pattern, and ring, respectively. The number in parentheses indicates the number of atoms in the repeat unit. The number of donors and acceptors in each motif are assigned as subscripts and superscripts, respectively.

is preferred to form chains. The third example has an intramolecular NH···O—N hydrogen bond, S(6).

As a general rule for hydrogen bonds in organic compounds, six-membered-ring intramolecular hydrogen bonds form, in preference to intermolecular hydrogen bonds [19]. The last example represents intermolecular NH···O—N hydrogen bonding in a dimer leading to the formation of a ring structure. As can be seen from the table, dihedral angle ϕ between the O–N–C–N–O plane and the imidazole ring in Figure 3 is also an important structural parameter determining the hydrogen bond pattern. In the case of a three-centered bond with the (N)H atom with coexistence of both intra- and intermolecular hydrogen bonds, we can describe the overall hydrogen pattern by a combination of S(6) and other motifs.

III. MAGNETISM OF NITRONYL NITROXIDES WITH AN NH SITE

III. A. Nitronylnitroxide with an Imidazole Ring

Im-NN [22] was prepared by the coupling of 2-imidazolecarboxaldehyde and 2,3-dimethyl-2,3-bis(N-hydroxyamino)butane followed by chemical oxidation using $NaIO_4$. Im-NN crystallizes in the centrosymmetric space group $P2_1/a$. In the crystal, an NH···N hydrogen bond chain is formed with an N(H)···N distance of 2.873(5) Å, as shown in Figure 4. The crystal structure of Im-NN implies that hydrogen bonds between —NH— (D) and —N═ (A_1) are suppressed relative to those between —NH— (D) and —O (A_2), leading to a C(4) pattern. A close contact between O(1)···N'(3) equal to 3.413 Å was also found between the hydrogen bond chains.

The magnetic susceptibility of Im-NN showed a maximum at 110 K and a minimum at 20 K (Figure 5). The magnetic data were fit to a Bleaney–Bowers expression of the magnetic susceptibility for a dimer, taking into account the contribution from a small number of uncoupled isolated monoradicals:

$$\chi_m = \frac{N_A g^2 \beta^2}{3k_B T} \left\{ \left[1 + \frac{1}{3} \exp\left(\frac{-2J}{k_B T}\right) \right]^{-1} (1 - P) + \frac{3}{4} P \right\}$$

Nonlinear curve fitting yielded a singlet-triplet gap of $2J = -123$ cm^{-1} and a fraction of uncoupled monoradical of $P = 6 \times 10^{-3}$. The magnetic data can be interpreted by direct overlap between the magnetic orbitals of the two NN radicals between the hydrogen bond chains, which are related by an inversion center, as shown in Figure 6(a). The overlap of two magnetic

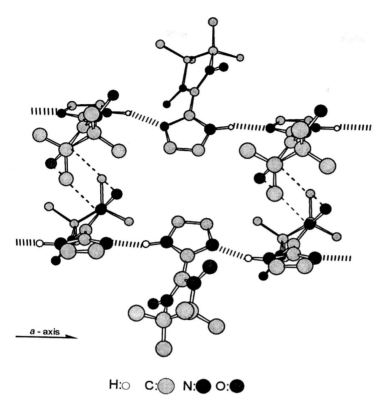

H:o   C:◐   N:●   O:●

Figure 4   View of hydrogen bond chain observed in Im-NN along the *a*-direction. Intermolecular hydrogen bonds are depicted by dotted lines (∥∥∥∥). Dashed line (- - - -) indicates the close contact between NO moieties within chains. (From Ref. 22.)

orbitals can be described by the interplane distance between the NN group (*d*), the angle $O—N\cdots O'$ (α), and the dihedral angle between the C—N—O plane and the N_2O_2 plane (β). For any given *d*, the overlap of two π^*-orbitals will be a maximum at $\alpha = \beta = 90°$, because the π^*-orbital is oriented perpendicular to the plane. Under these conditions, a strong antiferromagnetic interaction will be observed due to the σ-type overlap between the two NO bonds [Figure 6(b)], corresponding to case (a) in Figure 1. Such a strong antiferromagnetic interaction due to the head-to-head arrangement of a pair of NO bonds has often been observed in mononitroxides, such as 1,5-dimethylnortropinone nitroxide [23].

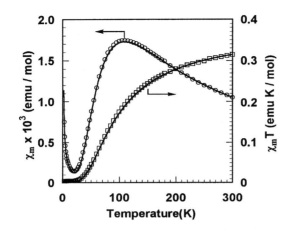

Figure 5 Temperature dependence of χ_m (○) and $\chi_m T$ (□) for Im-NN. Solid lines correspond to the least-squares fits (see text).

III. B. Nitronyl Nitroxides with Benzimidazole and Indole Rings

While imidazole itself exhibits a strong hydrogen bond corresponding to C(4) in the solid state, its hydrogen bond pattern might be modified by the steric effect of substituents R and/or R′ of the imidazole ring (Figure 7). To examine this effect we introduced a fused benzene ring to imidazole, since some benzimidazole compounds with proton acceptor groups such as carboxylic acid at the 2-position have been reported to show intramolecular hydrogen bonds [24]. 2-(Benzimidazole-2-yl)-4,4,5,5-tetramethylimidazo-line-1-oxyl-3-oxide)(BIm-NN) [22] was prepared by the reaction of benzim-idazole-2-carbaldehyde-diethylacetal and 2,3-dimethyl-2,3-bis(N-hydroxy-aminobutane) sulfate due to the low solubility of the dimerized aldehyde. BIm-NN crystallizes in the space group *Pbca*. The dihedral angle ϕ of BIm-NN is 24.3°, which is much smaller than that of Im-NN (48.4°). As seen from the molecular packing [Figure 8(a)], BIm-NN was assembled within the lattice in chains running along the c-direction. Along the chain direction, there are intermolecular hydrogen bonds in a C(6) pattern through N(5)—H(5)⋯O′(1) between the benzimidazole ring and the NN group, with intermolecular separations equal to 2.86(3) Å and of NN moiety, a close contact between O(2)⋯N′(3) equal to 3.11(3) Å, and O(2)⋯C′(7) equal to 3.16(3) Å are also observed at the nonhydrogen-bonding site [Figure 8(b)]. The longer intramolecular (N)H⋯O(2) contact corresponds to an S(6) pattern and also contributes as a three-center bond (in that case, this pattern is specified as C(6)(S(6))).

(a)

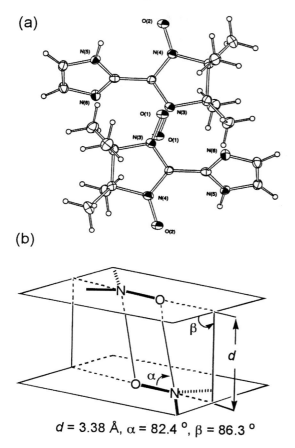

(b)

$$d = 3.38 \text{ Å}, \ \alpha = 82.4°, \ \beta = 86.3°$$

Figure 6 A dimer pair of Im-NN between hydrogen-bonded chains (a), and a conformation of two NO groups at the paired site (b).

The $\chi_m T$ value of BIm-NN at a room temperature was 0.38 emu K mol^{-1}, which corresponds to an isolated monoradical. This value increases steeply with decreasing temperature and reaches a maximum value of 1.68 emu K mol^{-1} at 3.2 K. The magnetic data above 4 K fit the Curie–Weiss law, $\chi_m = C/(T - \theta)$, with a Curie constant of 0.376 emu K mol^{-1} and a Weiss constant of $\theta = +8.2$ K (Figure 9). The temperature dependence of $\chi_m T$ was interpreted quantitatively in terms of a 1D Heisenberg ferromagnetic chain:

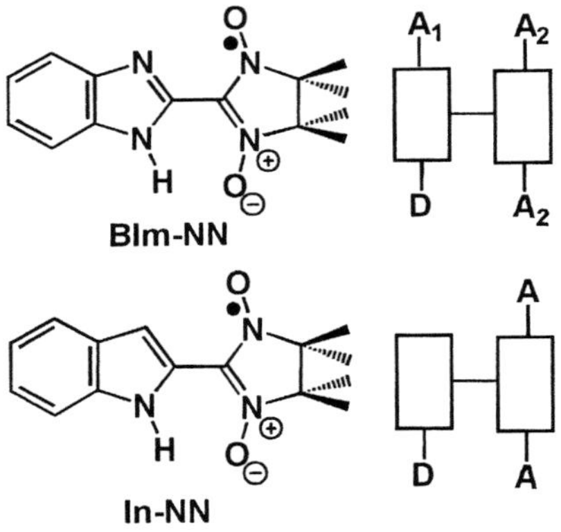

BIm-NN

In-NN

Figure 7 Chemical structures of BIm-NN and In-NN.

$$\chi_m = \frac{N_A g^2 \beta^2}{4k_B T} \left[\frac{N}{D}\right]^{2/3}$$

$$N = 1.0 + 5.7979916K + 16.902653K^2 + 29.376885K^3$$
$$+ 29.832959K^4 + 14.036918\ K^5$$
$$D = 1.0 + 2.7979916K + 7.0086780K^2 + 8.6538644K^3$$
$$+ 4.5743114\ K^4$$

with

$$K = \frac{J}{2k_B T}.$$

A good fit was obtained for a coupling constant $J = +12$ cm^{-1}, corresponding to the solid line in Figure 9. To confirm the presence of dominant ferromagnetic interaction, magnetization isotherms were measured at 2.8 and 4.0 K. At both temperatures, the magnetization curve is much above the Brillouin function for $S = 1/2$ spins, as shown in Figure 10. As the temperature is lowered further, the $\chi_m T$ value decreases to 1.49 emu K mol^{-1} at 1.8 K, which indicates that a very weak antiferromagnetic interaction operates between the BIm-NN chains.

Mechanistic consideration of strong magnetic coupling observed in BIm-NN crystal is (i) the orthogonality of SOMOs on adjacent molecules [contact between O(2) and N′(3)], (ii) intermolecular coupling following the by McConnell model (contact between O(2) and C′(7)), (iii) spin polarization through hydrogen bonds (Figure 11). Considering the structural parameters, dihedral angles between the plane containing N(4)—O(2)···N′(3) and

(a)

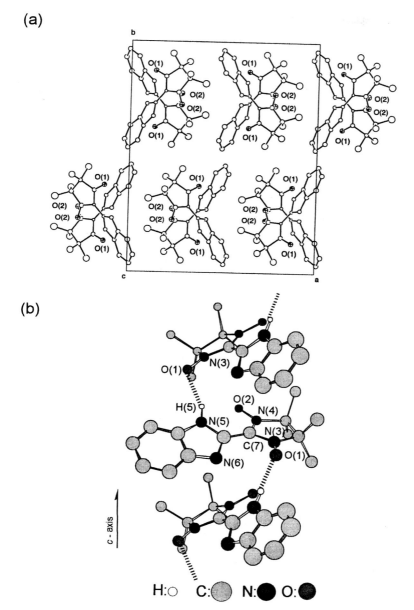

(b)

Figure 8   Crystal structure of BIm-NN: (a) Projection along the c-axis. (b) The view of a triad of molecules among the c-axis. Intermolecular hydrogen bonds are depicted by dotted lines (|||||).

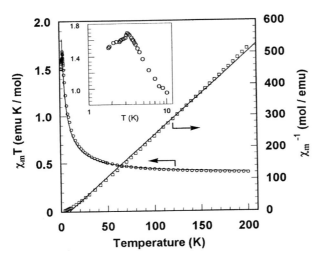

Figure 9 Temperature dependence of $\chi_m T$ ($\circ$) and χ_m^{-1} ($\square$) for BIm-NN. Solid lines correspond to the least-squares fits (see text). Inset shows the change of $\chi_m T$ at lower temperature.

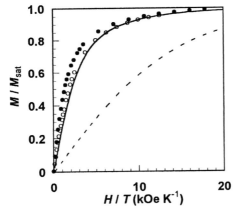

Figure 10 Magnetization isotherms of BIm-NN at 2.8 K ($\bullet$) and 4.0 K ($\circ$). Broken and solid curves are calculated using the to the Brillouin function for $S = 1/2$ and 5, respectively.

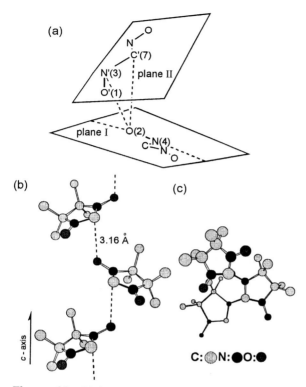

Figure 11 Molecular arrangement in the hydrogen-bonded BIm-NN chain: (a) Molecular arrangement of neighboring NN pairs (see text). (b) Side view of triad NN groups in the BIm-NN chain along the c-direction. (c) Top view.

plane I, the plane containing $O(2) \cdots N'(3)$—$O'(1)$ and plane II, and the angle $N(4)$—$O(2) \cdots N'(3)$ correspond to 70.8°, 76.1°, and 134°, respectively. We note that ferromagnetic behavior due to the orthogonality of two NO groups can be excluded due to the reduced overlap of these magnetic orbitals. The dihedral angles between the plane containing $N(4)$—$O(2) \cdots C'(7)$ and both plane I [the plane containing $O(2)$, $C'(7)$, $N'(3)$] and plane II [angle $N(4)$—$O(2) \cdots C'(7)$] correspond to 89.3°, 80.0°, and 115°, respectively, suggesting orbital overlap between $O(2) \cdots C'(7)$. Figure 11(b) shows the spatial arrangement of NN groups in the BIm-NN chain along the c-axis. Close contact between the central carbon (negative spin site) and the oxygen atom (positive spin site) of the nonhydrogen-bonded NO group favors a ferromagnetic array, according to the McConnell model. Although the spin-density distribution on the benzimidazole ring is not experimentally clear, negative spin density is polarized on the (N)—H atom, leading to the intermolecular ferromagnetic coupling.

We could switch the hydrogen-bond pattern from a C(4) to a C(6) linkage by using the steric effect of R- or R′-groups. In a parallel experiment, even a methyl group (R = CH_3, R′ = H) is enough to break the hydrogen bond between the imidazole rings. In a BIm-NN crystal the —N= site of the benzimidazole ring was not incorporated into the hydrogen bond or any intermolecular close contact. It is valuable to know the structural requirements for constructing a high-spin molecular system linked by hydrogen bond. Thus, we prepared the indole derivative (In-NN, Figure 7) in which the —N= site was replaced by a —CH= group. In-NN exhibited a hydrogen bond pattern corresponding to C(6)(S(6)) with a $d_{N(H)\cdots O}$ length of 2.944(6) Å, which is longer than that of BIm-NN. Correspondingly, close contact between the central carbon and nitroxide oxygen was extended to 3.21 Å, leading to weaker magnetic coupling, ($J = +8.6$ cm^{-1}) than that of BIm-NN. From the magnetostructural correlation, fused benzene rings in BIm-NN and In-NN separate the hydrogen-bonded chains by steric effects and reduce the antiferromagnetic interaction between the chains.

Hydrogen bond patterns could be qualitatively analyzed by IR spectra in the stretching mode of the NH site. Im-NN exhibited very broad and complex band around 2200–3200 cm^{-1}, corresponding to the hydrogen bonded N—H stretching vibration [25], while BIm-NN and In-NN showed a single strong band. Correlation between the bond distance and wave number is shown in Figure 12. Plots corresponding to BIm-NN and In-NN have the same trends to reported NH$\cdots$O pairs [26].

IV. MAGNETISM OF HYDROGEN-BONDED METAL-NITRONYL NITROXIDE COMPLEXES

In the previous section, NN derivatives with NH sites were assembled by intermolecular hydrogen bonds to form chain structures of C(4) or C(6) type. Another way of assembling these radicals is by complexation with metal ions (Figure 13). As described in Section II, =N— in imidazole resembles the nitrogen in pyridine. O (nitroxide) and N (imidazole) atoms can coordinate to a metal ion and form 6-membered chelate rings. Among the three radical molecules, Im-NN affords complexes with various metal sources, especially as perchlorate salts. The molecular structure of (Im-NN)$_2$Cu(ClO$_4$)$_2$ is shown in Figure 14 [27]. The complex was a centrosymmetric, tetragonally distorted octahedral species whose apical sites are occupied by weakly coordinated perchlorate ions with a Cu−O(perchlorate)

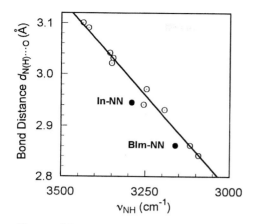

Figure 12 Relation between the hydrogenic stretching frequency and bond distances for N—H···O hydrogen bonds. (Open circles correspond to data in Ref. 26.)

distance equal to 2.523 Å. A three-center hydrogen bond existed between (N)H and the nitroxide oxygen with $d_{N(H)\cdots O}$ equal to 2.717 Å, and with a perchlorate oxygen belonging to the neighboring complex with $d_{N(H)\cdots O}$ equal to 3.136 Å. Im-NN forms an S(6) pattern in the copper complex to give a planar conformation compared to that of the pyridyl derivative [28]. Structural parameters of these two complexes are also shown in Figure 14.

The temperature dependence of $\chi_m T$ for the complex is shown in Figure 15. $\chi_m T$ at room temperature was 1.16 emu K/mol, which is almost equivalent to that expected for a noninteracting three-spin system (1.125 emu K/mol). This value increases rapidly with a lowering of the temperature to 75

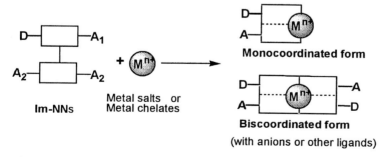

Figure 13 Schematic representation for the complexations of Im-NNs with metal ions.

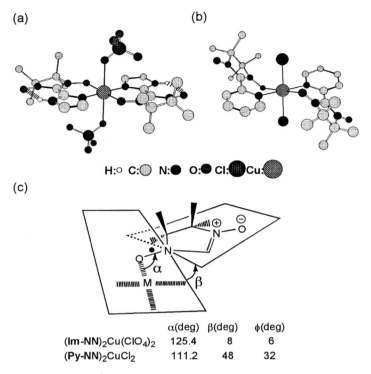

(a)   (b)

H:○ C:◯ N:● O:● Cl:◍ Cu:◍

(c)

	α(deg)	β(deg)	ϕ(deg)
(Im-NN)$_2$Cu(ClO$_4$)$_2$	125.4	8	6
(Py-NN)$_2$CuCl$_2$	111.2	48	32

Figure 14 Molecular diagrams of (a) (Im-NN)$_2$Cu(ClO$_4$)$_2$ (from Ref. 27); (b) (Py-NN)$_2$CuCl$_2$·2CH$_2$Cl$_2$ (solvated CH$_2$Cl$_2$ was omitted) (From Ref. 28); and (c) coordination geometry of NN group. Dotted lines (|||||) in (a) indicate the intramolecular hydrogen bonds.

K and shows a maximum value of 1.51 emu K/mol, indicating an intramolecular ferromagnetic interaction between the nitroxide oxygen atom and copper ions. A short bond distance of Cu—O(N) equal to 1.963 Å and an angle α close to 120° for (Im-NN)$_2$Cu(ClO$_4$)$_2$ indicate that the NO bond equatorially coordinates to copper ion through one of the two lone pair of oxygen atoms. Relevant examples of copper complexes in which an NO group is ligated equatorially have large β-angles close to 80°, corresponding to effective overlap between two magnetic orbitals, to cause a strong antiferromagnetic interaction. By comparison, (Im-NN)$_2$Cu(ClO$_4$)$_2$ has an almost coplanar geometry, requiring that there be no effective overlap. Considering these structural characteristics, ferromagnetic interaction in the complex originates from orthogonality between the two magnetic orbitals.

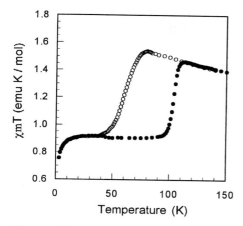

Figure 15 Temperature dependence of $\chi_m T$ for $(\text{Im-NN})_2\text{Cu(ClO}_4)_2$: (○) cooling process and (●) heating process.

The $\chi_m T$ value decreases abruptly below 75 K and reaches a plateau at 0.91 emu K/mol at 37 K. Interestingly, the change in $\chi_m T$ value shows a hysteresis between the cooling and heating cycles. In the heating process this plateau continues up to 94 K and suddenly increases, and the $\chi_m T$ values trace the cooling process above 111 K. The presence of a hysteresis loop in the plot of $\chi_m T$ vs. T indicates a crossover of ferromagnetic and antiferromagnetic interaction between the nitroxide oxygen and copper ion as a cooperative process. In the $(\text{Im-NN})_2\text{Cu(ClO}_4)_2$ complex, inter- and intramolecular hydrogen bonds produce a dense network of complex units (Figure 16). The temperature dependence of IR spectra corresponds to the magnetic behavior [29], indicating that the change in hydrogen bonding—i.e., the coordination site on the opposite side—is responsible for the "crossover" characteristics.

V. CONCLUSIONS

We designed and synthesized nitronyl nitroxide radicals having heterocycles bearing an NH site such as imidazole (Im-NN), benzimidazole (BIm-NN), and indole (In-NN). In these derivatives, we found three hydrogen bond patterns corresponding to C(4), C(6) in the pristine state, and S(6) in the state coordinated to copper ion. In the former, the hydrogen bond pattern could be controlled by the steric effect of substituents on the imidazole ring. BIm-NN and In-NN with intermolecular C(6) patterns gave 1D ferromag-

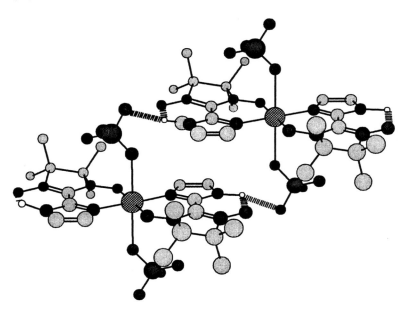

Figure 16 View of the molecular packing for $(Im-NN)_2Cu(ClO_4)_2$ crystals. Dotted lots (| | | | |) indicate hydrogen bonds.

netic chains due to the effective overlap of magnetic orbitals, following McConnell's model. A copper complex of Im-NN exhibited hysteresis in its magnetic susceptibility measurement, indicating crossover of ferromagnetic and antiferromagnetic interaction not on a molecular level but in a bulk cooperative process. The present finding of magnetic interaction and cooperativity observed in nitronyl nitroxides with NH sites will enable us to design additional new families of self-assembled radicals linked by hydrogen bonds.

ACKNOWLEDGMENTS

The authors express their sincere thanks to Professor S. Ohba (Keio University) for the collaboration on crystallographic study and to Mr. M. Irisawa for his experimental support and productive comments. This work was partially supported by a Grant-in-Aid for Scientific Research from the Ministry of Education, Science and Culture, Japan. Support from the Kanagawa Academy of Science and Technology (KAST) is also acknowledged.

REFERENCES

1. Turnbull MM, Sugimoto T, Thompson LK. Molecular-based magnetic materials. ACS Symposium Series 644. Washington DC: American Chemical Society, 1996.
2. Kahn O. Molecular Magnetism. New York: VCH, 1993.
3. Gatteschi D, Kahn O, Miller JS, Palacio F. Magnetic Molecular Materials. NATO ASI Series Vol. E-198. Dordrecht: Kluwer, 1991.
4. Kinoshita M. Ferromagnetism of Organic Radical Crystals. Jpn. J. Appl. Phys. 1994; 33:5718–5733, and references therein.
5. Togashi K, Imachi R, Tomioka K, Tsuboi H, Yasui M, Ishida T, Nogami T, Takeda N, Ishikawa M. Organic Radicals Exhibiting Intermolecular Ferromagnetic Interactions with High Probability: 4-Arylmethyleneamino-2,3,6,6-tetramethylpiperidin-1-oxyls and Related Compounds. Bull. Chem. Soc. Jpn. 1996; 69:2821–2830.
6. Chiarelli R, Novak MA, Rassat A, Tholence JL. A Ferromagnetic Transition at 1.48 K in an Organic Nitroxide. Nature. 1993; 363:147–149.
7. McConnell HM. Ferromagnetism in Solid Free Radicals. J. Chem. Phys. 1963; 39:1910.
8. Sugawara T, Murata S, Kimura K, Iwamura H, Sugawara Y, Iwasaki H. Design of Molecular Assembly of Diphenylcarbenes Having Ferromagnetic Intermolecular Interactions. J. Am. Chem. Soc. 1985; 107:5293–5294.
9. Matsushita MM, Izuoka A, Sugawara T, Kobayashi T, Wada N, Takeda N, Ishikawa M. Hydrogen-Bonded Organic Ferromagnet. J. Am. Chem. Soc. 1997; 119:4369–4379.
10. Lang A, Pei Y, Ouahub L, Kahn O. Synthesis, Crystal Structure, and Magnetic Properties of 5-Methyl-1,2,4-Triazole-Nitronyl Nitroxide: A One-Dimensional Compound with Unusually Large Ferromagnetic Inermolecular Interactions. Adv. Mater. 1996; 8:60–62.
11. Romero FM, Ziessel R, Drillon M, Tholence JL, Paulsen C, Kyritsakas N, Fisher J. Ferromagnetic Order in a Novel Imino Nitroxide (IT)Py(C≡CH) Radical Derived from 2-Ethynyl-Pyridine. Adv. Mater. 1996; 8:826–829.
12. Akita T, Mazaki Y, Kobayashi K. Ferromagnetic Spin Interaction in a Crystalline Molecular Complex Formed by Inter-heteromolecular Hydrogen Bonding: a 1:1 Complex of Phenyl Nitronyl Nitroxide Radical and Phenylboronic Acid. J. Chem. Soc. Chem. Commun. 1995; 1861–1862.
13. Veciana J, Cirujeda J, Rovira C, Molins E, Novoa JJ. Organic Ferromagnets. Hydrogen Bonded Supramolecular Magnetic Organizations Derived from Hyderoxylated Phenyl α-Nitronyl Nitroxide Radicals. J. Phys. I France. 1996; 6: 1967–1986.
14. Inoue K, Iwamura H. Magnetic Properties of the Crystals of p-(1-Oxyl-3-oxido-4,4,5,5-tetramethyl-2-imidazoline-2-yl)benzoic Acid and Its Alkali Metal Salts. Chem. Phys. Lett. 1993; 207:551–554.
15. Lanfranc de Panthou F, Luneau D, Laugier J, Rey P. Crystal Structures and Magnetic Properties of a Nitronyl Nitroxide and of Its Imino Analogue. Crystal

Packing and Spin Distribution Dependence of Ferromagnetic Intermolecular Interactions. J. Am. Chem. Soc. 1993; 115:9095–9100.

16. Rajca A, Utamapanya S, Xu J. Control of Magnetic Interactions in Polyaryl-methyl Triplet Diradicals Using Steric Hindrance. J. Am. Chem. Soc. 1991; 113:9325–9241.

17. Allemand PM, Srdanov G, Wudl F. Molecular Engineering in the Design of Short-Range Ferromagnetic Exchange in Organic Solids: The 1,3,5-Triphen-ylverdazyl System. J. Am. Chem. Soc. 1990; 112:9391–9392.

18. Desiraju GR. Supramolecular Synthons in Crystal Engineering A New Organic Synthesis. Angew. Chem. Int. Ed. Engl. 1995; 34:2311–2327.

19. Etter MC. Encoding and Decoding Hydrogen-Bond Patterns of Organic Compounds. Acc. Chem. Res. 1990; 23:120–126.

20. Craven BM. McMullan RK, Bell JD, Freeman HC, The Crystal Structure of Imidazole by Neutron Diffraction at 20°C and −150°C. Acta Cryst. 1977; B33: 2585–2589.

21. Yoshioka N, Irisawa M, Mochizuki Y, Aoki T, Inoue H. Synthesis and Magnetic Property of Stable Organic Radicals Bearing Imidazole Ring. Mol. Cryst. Liq. Cryst. 1997; 306:403–408.

22. Yoshioka N, Irisawa M, Mochizuki Y, Kato T, Inoue H, Ohba S. Unusually Large Magnetic Interactions Observed in Hydrogen-Bonded Nitronyl Nitroxides. Chem. Lett. 1997; 251–252.

23. Gloux J, Genoud F, Decorps M. Etude par R.P.E. des Excitons Triplets dans le Dimethyl-1,5-Nortropinone Nitroxyde. Mol. Phys. 1981; 42:251–266.

24. Morgan KJ. The Infrared Spectra of Some Simple Benzimidazoles. J. Chem. Soc. 1961; 2343–2347.

25. Grech E, Malarski Z, Sobczyk L. Couple of ν(NH$\cdots$N) vibration with the γ(NH$\cdots$N) overtone in solid imidazole derivatives. Spectrochim. Acta. 1992; 48A:519–523.

26. Nakamoto K, Margoshes M, Rundle RE. Stretching Frequencies as a Function of Distances in Hydrogen Bonds. J. Am. Chem. Soc. 1955; 77:6480–6486.

27. Yoshioka N, Irisawa M, Aizawa N, Aoki T, Inoue H, Ohba S. Structure and Magnetic Property of 2-Imidazolyl Nitronyl Nitroxide and Its Metal Complexes. Mol. Cryst. Liq. Cryst. 1996; 286:165–170.

28. Ohba S, Kato T, Yoshioka N, Inoue H. Dichlorobis[4,4,5,5-tetramethyl-2-(2-pyridyl)-4,5-dihydro-1H-imidazol-1-yloxy3-oxide]copper(II)bis(diloromethane) Solvate. Acta Cryst. C. 1997; IUC9700004.

29. Lahti PM. Private communication, May 1997.

28

Supramolecular Architectures and Magnetic Interactions in Crystalline α-Nitronyl Nitroxide Radicals

Jaume Veciana and Joan Cirujeda
Barcelona Institute of Material Science (CSIC), Bellaterra, Spain

Juan J. Novoa and Mercè Deumal
University of Barcelona, Barcelona, Spain

I. INTRODUCTION

The design of purely organic magnetic materials has been the goal of many research groups during the last decade [1]. As result of a systematic synthetic effort, it has been possible to obtain different kinds of crystalline organic free radicals that show relevant magnetic properties and, in very exceptional cases, even bulk ferromagnetism [2]. The number of these compounds is now large enough to make it possible to look at the general trends of the structure–magnetism relationship, even in a statistical form. That is, one can look at the connections between the crystal packing of a material and its magnetic properties. Understanding the driving forces behind already known compounds will be very useful for the design of new purely organic magnetic materials in which the magnetic properties are enhanced. Because the packing of a given molecular system can be strongly influenced by the functional groups present in the molecule, we have to look at data in families of compounds. We have chosen to focus our analysis on just one uniform family of magnetic compounds, that of the so called α-nitronyl nitroxide (or α-nitronyl aminoxyl) radicals [3], whose general formula is shown in Figure 1 and which characterized by the presence of a five-membered ring with two nitronyl groups attached in an alpha position to a $C(sp^2)$ carbon.

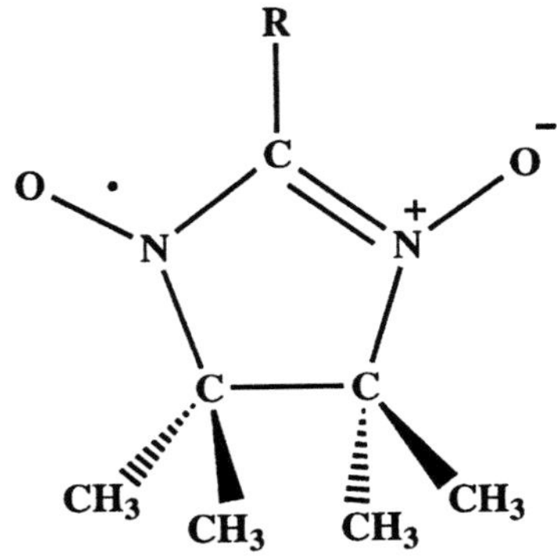

Figure 1   General structure of α-nitronyl nitroxide radicals (R is a substituent).

This is the most extensively studied family of persistent radicals among the purely organic molecular magnets. Within this family, there are many different packings and magnetic behaviors due to the effect of other functional groups present in the compounds. The latter fact reflects the strong influence induced on the packing by the functional groups attached to the five-membered ring. Sometimes even small changes on a large group, such as a functionalized phenyl, can give rise to remarkable changes in the packing. It is well established that the presence of bulk ferromagnetism in crystals of this family is strongly related to the relative disposition of the spin-containing molecules within the crystal. This result is consistent with the most commonly accepted ideas based on existing theoretical models about intermolecular magnetic interactions [1,2,4]. For instance, the very popular McConnell I model [4] is probably the one most employed today due to its simplicity, in which intermolecular ferromagnetic interactions are postulated to be possible only when short contacts are found between atoms bearing a considerable atomic spin population of opposite spin. Otherwise, the interaction is predicted to be antiferromagnetic or negligible.

The presence of ferromagnetism in a compound relies on three factors: the ability to prepare a persistent free radical, the possibility of growing a crystalline supramolecular organization with such a radical, and the ability of radicals in such crystals to interact ferromagnetically with each other. The preparation of molecular organic crystals showing a spontaneous magnetization below a certain critical temperature, T_C, is possible for persistent free radicals capable of undergoing crystalline packing that allows for the presence of intermolecular ferromagnetic interactions propagating all over the crystal [5,6]. This is possible only if each spin-containing molecule is capable of having ferromagnetic interactions with one or more of its nearest-neighbor molecules, and the strengths of the interactions overcome the thermal energy at T_C, i.e., with an energy larger than kT_C [7]. Consequently, one of the most important points in this field is to recognize among all the possible relative arrangements of two neighbor molecules those that produce intermolecular ferromagnetic interactions as well as those responsible for antiferromagnetic ones [8]. In general, the limited geometrical layouts of crystalline molecular solids received the name of *crystal packing patterns* [10], although recently these geometrical packings have been also called *synthons* [11], stressing their role as crystal-building units in the supramolecular entity constituted by the crystal.

II. PACKING PATTERNS AND CRYSTAL STRUCTURE

From the microscopic point of view, a *crystal packing pattern* is the result of the intermolecular interaction between some or all of the functional groups of the molecules that conform to the pattern. One can define many patterns within a crystal, but the most relevant are those that describe the ways in which two molecules bonded through strong intermolecular bonds are linked. These patterns describe the topology of the synthons and can be called *primary* patterns, to distinguish them from other types. Except when otherwise noted, we will concentrate on just these and drop the primary adjective.

From a microscopic perspective, a pattern is a particular geometrical distribution of a set of intermolecular interactions. Not all the interactions in the pattern are necessarily attractive in nature, but we can assume that to obtain stable crystals we need energetically stable patterns. When for a given crystal one can define more than one pattern, we will select those that are energetically stable. One can predict possible patterns looking at the forms in which one can obtain stable small aggregates (dimers, trimers, etc.). These patterns are easy to design if one knows the nature of the interactions involved in the pattern, since many interactions are highly directional, such as the hydrogen bonds, π-stacking, etc. [9]. In these cases, one only has to look at the complementarity of the functional groups of the molecule, as shown schematically in Figure 2 for a molecule with more than one complementary group for intermolecular interactions.

Not all packing patterns are going to be relevant for the analysis of the crystal packing, due to energetical and symmetrical factors. The energetical factor implies that among all possible patterns only the most stable ones are the most likely to be found in any molecular aggregate, for statistical reasons, when the energy difference between the energetically adjacent patterns is large. In this case, the most stable pattern is likely to be the

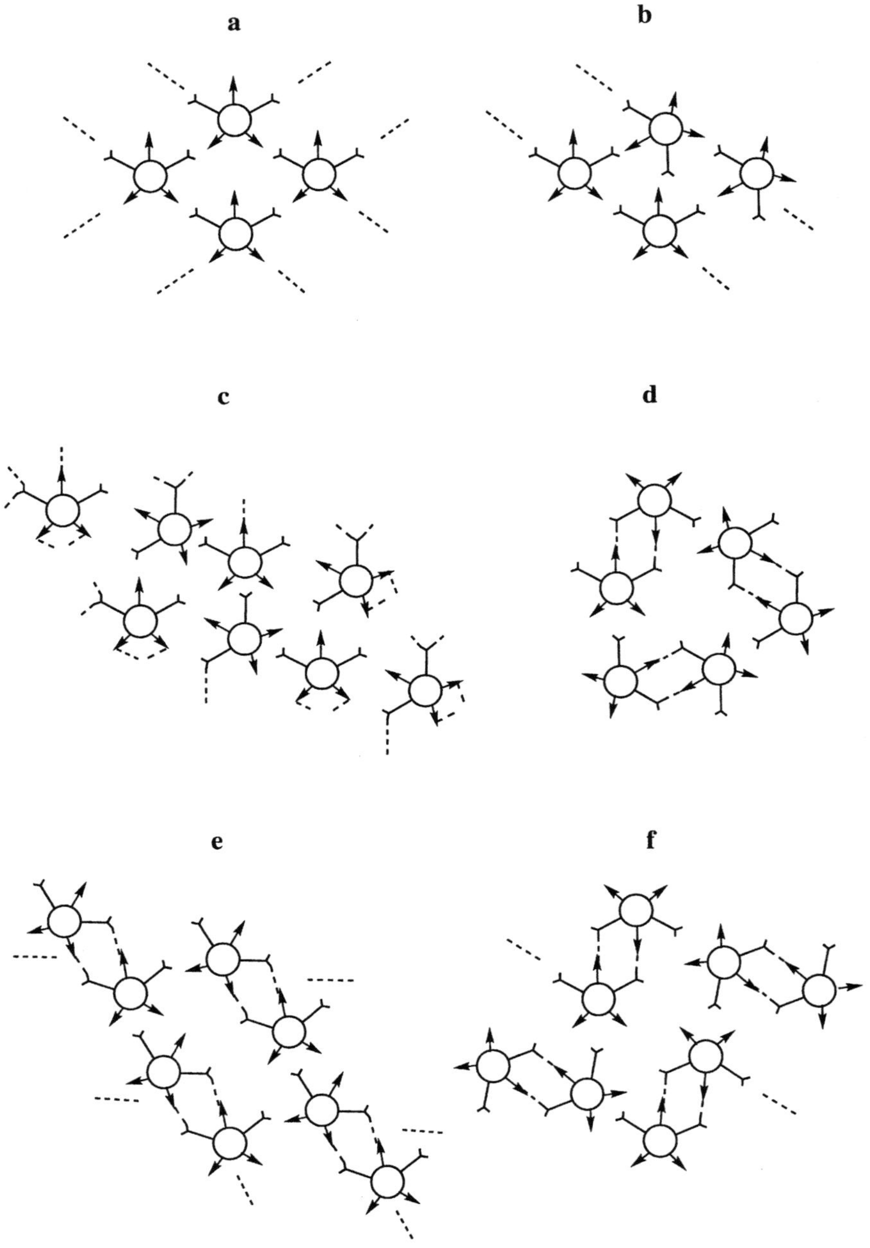

Figure 2 Examples of patterns compatible with the formation of infinite planes (a, c, e) and incompatible with that formation (b, d, f).

dominant one. Furthermore, due to the translational symmetry inherent in the nature of the crystal, some of these geometrical arrangements (that is, relative orientations) of two radicals are not possible in the crystal, because they do not allow for a propagation of the packing pattern, or this propagation is not energetically stable (see Figure 2). The combined effect of these two factors is to limit the number of relevant packing patterns present in the packing of an organic molecule. When the molecules are persistent free radicals, some of the *crystal packing patterns* could produce dominant ferromagnetic interactions, with others yielding dominant antiferromagnetic interactions, allowing these to be identified as *ferro-* and *antiferromagnetic patterns*.

Why are the primary packing patterns so critical to establish the crystal packing? The answer is that one can define the packing in a precise way by looking at all the primary packing patterns present in the crystal. The crystal is just the repetitive application of the symmetry operations of its spatial group and of the primary packing patterns along the three directions of space. This fact was first realized by Etter and coworkers [10], who introduced a graph set analysis to identify the crystal packing patterns established by hydrogen bonds. Within their method, the packing patterns are identified in terms of motifs [12]. This topological analysis of a crystal is very useful to distinguish between polymorphs of a hydrogen-bonded crystal in terms of its first- and higher-order graph sets [10]. The graph-set idea could easily be extended to other types of bonds or contacts. The main problem for complex systems lies in automatic identification of the patterns, which requires the previous identification of all the intermolecular bonds present. The latter has been done up to now using geometrical criteria based upon estimated van der Waals radii [13,14], although there have been recent suggestions of more systematic methods [15].

Can we predict the shape and stability of the relevant primary patterns formed by a molecule? We can, using theoretical ab initio methods to compute and localize the minimum-energy conformations for a dimer of the molecule of interest. Looking at the relative stabilities of these minimum-energy conformers we can get an idea of the statistical probability of each conformer. We can even study the barrier for the transformation from one conformer to another, which can give us a first approximation of the dynamics of a polymorphic transformation. This is only a first approximation to the primary packing pattern, since we have not included the effect induced by the crystal environment and the restrictions induced by the periodicity of the patterns. But we can investigate the latter by simple inspection or by improving our model using higher-order aggregates. If the molecule is too large to allow such geometrical search study even with dimers, we can get a good qualitative idea by studying the potential intermolecular contacts that each molecule could form in a dimer. Using adequate models, one can compute, for instance, the strength of the $C(sp^2)$—H$\cdots$O—N, N—O$\cdots$O—N, and $C(sp^3)$—H$\cdots$O—N contacts in the α-nitronyl nitroxides. Using this information, one can study the relative stability of the possible primary packing patterns present in a crystal of the hydro-α-nitronyl nitroxide (or HNN, for short), and rationalize the packing in this crystal [16].

A first guess of the possible orientations in which the packing patterns can be minimum-energy structures is obtained in crystals dominated by electrostatic interactions or hydrogen bonds. Looking at molecular electrostatic potential (MEP) maps [17] gives information showing where a positive charge will have a more stable electrostatic interaction. The interaction energy of a hydrogen-bonded crystal will be dominated by the electrostatic component; thus, the MEP maps give useful information on how two molecules can approach each other in stable orientations. If the two molecules overlap in regions where the MEP maps have opposite sign, the interaction is attractive; otherwise it is repulsive. We can call this qualitative approach, the *MEP overlap analysis method*. It gives a good starting point for the identification of primary packing patterns, which can be useful for the rationalization of very large molecular crystals. We have successfully used this method to rationalize the packing of many crystals [16] and large molecular clusters [18]. We will show later in this work some application examples.

Finally, one can obtain primary packing patterns by analyzing the regularities in the packing of a crystal. The structure can come from a crystal grown up experimentally. Alternatively, if no experimental crystal is available, one can use computed crystal structures obtained using any of the programs published in the literature for crystal prediction [19] or optimization (20).

Molecular electrostatic potential overlap analysis directly provides candidates for primary packing patterns. Each conformation obtained in this way is the result of overlapping complementary parts (positive with negative) of the same molecule. Thus a C—H$\cdots$O—N bond results in an α-nitronyl nitroxide radical from the combination of an acidic region (the C—H groups) and a basic region (the O—N group). We can read their acid-basic character by their positive and negative MEP potential. The MEP overlap analysis does not give energetic information. However, it is known that the interaction is most energetic if the group is more acidic or more basic. This character is manifested by more positive and negative regions in the MEP maps. Therefore, the strongest intermolecular contacts are those between the same acidic groups and the most basic groups. This gives us a qualitative ordering of the conformations obtained in an MEP overlap analysis. Better information on the energy ordering comes from ab initio computation of the interaction energy of each intermolecular contact present in the conformation or, when possible, from the interaction energy of the dimer. The accurate evaluation of the interaction energy of each interaction can be carried out by adequate ab initio methodology [21–24]. Thus, following the MEP overlap analysis we have refined our model by using a functional group analysis of the crystal packing in which we combine complementary functional groups.

What is the connection between MEP overlap analysis, or the functional group analysis, and the structure of the crystal? The answer to this question was given by Kitaigorodsky in his *Aufbau principle* [25]. This principle simply states that when organic molecules pack to form a molecular crystal, they try to minimize the total energy of the crystal. This is an energetic criterion, that is, a thermodynamic criterion. If one defines the packing energy E_p as a sum over all possible different pairs of atoms, then

$$E_p = \sum{}' E_{ij}(r_{ij}) \tag{1}$$

where $E_{ij}(r_{ij})$ is the intermolecular interaction energy between atoms i and j located at a distance r_{ij} and the prime indicates the lack of repetitions. Notice that this function has many local minima, each corresponding to a different crystal structure or polymorph that a given compound is able to show. The experimental crystal is not necessarily the most stable one or the most densely packed, as was assumed by Kitaigorodsky. In some cases, one can find many local minima in a small energy range [26]. According to this principle, the most stable crystals satisfy the strongest intermolecular contacts first and then the remaining contacts, in order of decreasing strengths. It is possible, depending on the way the crystal is grown, to obtain local minima. On the other hand, one should keep in mind that the crystal packing also depends on kinetic factors, although we will not consider them explicitly in our treatment.

Normally, a real crystal is the result of the progressive association of molecules within the solution in which they are grown. In solution, when the molecules form aggregates (molecular clusters) they do it by employing the most stable primary packing patterns in a statistical distribution. They form as many primary patterns as possible to build first-order aggregates. At this point, this kind of pattern has been saturated; if they want to aggregate further and thus build up the crystal, they have to use other patterns linking the first-order aggregates. The higher-order aggregates can be ribbons, planes, or other motifs. The aggregates can build up the crystal or further aggregate using other patterns in order to build the bulk crystal. Therefore, we have an ordering in the crystal structure that is associated with the stability of the different packing patterns. We can thus differentiate a *primary structure* of the crystal, associated with the most stable primary packing patterns, a *secondary structure* of the crystal, representing the patterns associated between clusters, and even *tertiary* or *quaternary structures*, formed by packing patterns that employ previously otherwise-functional groups.

Using the methodological approach just described, it is possible to carry out the analysis of any crystal structure by the following steps.

1. *Characterization of the functional groups.* This can be done by simple inspection, extrapolating known information on the same groups, or by studying the nature of the MEP map, which identifies regions of concentration and depletion of the electronic density and the acid-basic character of the functional groups. We thus can identify the potential nature of any intermolecular contact that can be formed by the functional groups present in the molecule.

2. *Identification of the primary packing patterns.* This can be done from an experimentally determined crystal structure by looking at the shortest intermolecular contacts, normally assumed to be attractive and responsible for the packing structure. Alternatively, we can carry out the same analysis on any structure obtained after a computational optimization of the crystal. Note here that not every short contact is always attractive, but it can be a consequence of the presence in nearby positions of two attractive interactions. We should also use relaxed cutoffs in the distances to avoid problems [13,14].

3. *MEP overlap analysis.* This helps to identify the possible relative geometric dispositions in which the functional groups can be associated in energetically stable conformations. This is sometimes done intuitively by associating complementary groups, but is more systematic when done using MEP maps obtained from ab initio computations.

4. *Computation of the strength of the primary patterns by ab initio computations.* This allows one to define an energetic order for all the patterns (that is, $E_1 < E_2 < \ldots$) required to understand the presence or absence of overall patterns in the first step.

5. *Rationalization of the structure.* Using the information obtained in the previous steps, the crystal packing can be justified in terms of a primary structure, secondary, and so on.

Examples of the application of the previous methodology to the family of the α-nitronyl nitroxides have recently been published [16,27,28] and will not be discussed here in detail. We will simply mention that these studies allow the rationalization of the packing of hydro α-nitronyl nitroxide (HNN, for short), the simplest member of the nitronyl nitroxide family, in which R

= H. This crystal has a primary structure of dimers, each involving the formation of two strong $C(sp^2)$—H$\cdots$O—N contacts (-3.71 kcal/mol each). The dimers then aggregate among themselves through weaker $C(sp^3)$—H$\cdots$O—N contacts (-0.40 kcal/mol of interaction energy each) forming planes. In this way the number of these contacts is maximized. This is the secondary structure of the HNN crystal. No short NO$\cdots$ON distances are found in this crystal because the NO groups repel each other when coplanar and in a collinear or near arrangements. Once the planes are formed, there are still some unused intermolecular associations. For instance, the methyl groups can still make contacts with the NO groups involved in the $C(sp^2)$—H$\cdots$O—N contacts, although this is not the only possibility. The planes can be linked to each other by these groups. The result is the formation of ordered stacks of planes, aligned in such a way as to maximize the interplane interaction energy. This is the tertiary structure of the crystal. Thus, the whole crystal structure can be rationalized.

III. STATISTICAL ANALYSIS OF THE PACKING PATTERNS IN FERRO AND ANTIFERRO α-NITRONYL NITROXIDES: ARE THERE FERRO AND ANTIFERRO PACKING PATTERNS?

The previous analysis can be extended to any α-nitronyl nitroxide crystal. This will allow us to obtain generalities of their crystalline packings. We must note in advance that such analysis does not provide any information about the relationship between the magnetic properties and the crystallographic structure.

The most common way of obtaining structural information about ferro- and antiferromagnetic patterns is by a detailed inspection of the molecular packing in individual crystals that show well-characterized magnetic behaviors, searching for packing patterns that are specific for ferromagnetic crystals. Usually, this kind of analysis implies that once the magnetic behavior of a crystal is known, one searches for the packing pattern responsible for it, under the light shed by the use of a theoretical model capable of associating a magnetic character to a given crystal pattern. Normally the so-called McConnell I model has been used. Most of these analyses have been performed on crystals of α-nitronyl nitroxide radicals. Although qualitative, several interesting conclusions have been achieved with this approach. For instance, close contacts between NO groups of neighboring molecules, the groups that support the largest spin density on these radicals, have been associated with antiferromagnetic patterns [1,29]. At the same time, those contacts among one NO group of one radical and certain H atoms of neighboring ones have been associated with ferromagnetic patterns, in particular when the H atoms are located on aromatic substituents or on CH_3 groups [1,30]. However, this approach has some drawbacks, because the structural information learned about the magnetic patterns is very limited, and occasionally the conclusions are contaminated by preconceptions [31]. For this reason, a more systematic and quantitative approach is required, like the one we describe here.

Magnetostructural information can be obtained by statistical analysis of a large set of crystals. We can look in these analyses at the key information defining the packing and then relate the type of packing to the ferro- or antiferromagnetic properties of the crystal. At this point, it is interesting to note that the packing of the magnetic crystals is driven only by the intermolecular interactions and not affected by the magnetic interactions, since these interactions are much weaker in magnitude than the hydrogen bonds or van der Waals interactions. According to the McConnell I model, the presence of a ferromagnetic interaction requires the overlap of significant positive and negative spin densities. It is well known, both experimentally and theoretically [1,30a,32], that the major part of the spin density in α-nitronyl nitroxide radicals is located on the ONCNO part of the five-membered ring, with the density on the central C atom being opposite to that in the N and O atoms. Thus, the nature of the magnetic interactions present in the crystal of α-nitronyl nitroxide is expected to be related directly to the spatial orientation and proximity of the atoms of the ONCNO group. Thus, statistical analysis of the geometry of the shortest ONCNO$\cdots$ONCNO contacts in crystals whose magnetic properties are well characterized can help to understand the packing–magnetism relationships in α-nitronyl nitroxides. We have performed such analysis for radicals that clearly belong to one of the following two different magnetic classes or subsets: those that show *dominant* ferromagnetic intermolecular interactions and those exhibiting *dominant* antiferromagnetic interactions. We did not want to include in our study crystals whose magnetic behavior is not clearly defined, for they would be of no help in identifying the possible patterns associated with a ferro- or antiferromagnetic packing. The analysis of the two subsets can give rise to totally different crystal packing patterns in each subset, or, instead, one can find similar patterns in both subsets. The second case would indicate the statistical presence of ferro- and antiferromagnetic patterns in both magnetic classes, combined in such a way that their number and strength result in a magnetic character for the crystal compatible with the subset in which this crystal is included.

The total number of crystals analyzed in this study (47) is large enough to make feasible a statistical analysis of the most relevant intermolecular contacts they present. The crystallographic data used in the survey were, in part, retrieved from the Cambridge Structural Database (CSD) [33], and the rest was from our own research or was found directly in the literature or supplied to us by other authors. A total of 143 crystal structures containing substituted α-nitronyl nitroxide radical units were initially used [34]. The criteria employed to select the final set of structures were as follows. (1) Structures with R factors greater than 0.10, or that were determined from very limited data, or that exhibit disorder or large molecular distortions were excluded, leaving 117 crystal structures. (2) All of the structures containing both transition-metal atoms and large closed-shell organic molecules co-crystallized (45 structures) were discarded for the present analysis, since intermolecular magnetic interactions between radicals could be complicated in these cases by the existence of other magnetic pathways through the metal atoms or between the closed-shell molecule and the radical units. This cri-terion left 72 crystal structures of purely organic compounds. (3) We dis-carded in our statistical analysis all the crystals whose magnetic interactions are not clearly dominated by ferro- or antiferromagnetic interactions. The nature of the dominant magnetic interactions is clearly manifested by the temperature dependence of the magnetic susceptibility, χ, in the temperature (T) range of 2–300 K. Thus, radicals with dominant ferromagnetic inter-actions, grouped in the subset named FM, show the characteristic signature of a continuous increase of χT when T decreases. By contrast, radicals with dominant antiferromagnetic interactions have the opposite trend and were classified in the AFM subset. It is convenient to exclude the crystals with nondominating signature because, by construction, they can show both types of crystal packing patterns; as a consequence, it would make impossible a clear identification of their ferro- and antiferromagnetic patterns. Their ex-clusion left us with only 47 purely organic crystals. Of these, the dominant intermolecular magnetic interactions were ferromagnetic in 23 cases and antiferromagnetic in the remaining 24 ones. The total number of crystals that fulfill the previous conditions and are suitable for statistical analysis belong to the following spatial groups: P-1 (3), $P2_1$ (2), Cc (2), $P2_1/c$ (25), C2/c (2), $P2_12_12_1$ (2), $Pca2_1$ (1), Ib2a (1), Pbca (4), Fdd2 (1), $I4_1/a$ (1), $P4_2bc$ (1), and P3c1 (2).

The statistical analysis of the packing of both magnetic subsets can be done by looking at the geometry of the shortest intermolecular contacts of the N—O···O—N, $C(sp^3)$—H···ON and $C(sp^2)$—H···ON types present in each crystal. We selected these contacts, first because they are the most relevant ones according to the McConnell I model and the literature [4], and, second, because they are the ones that, in light of the analysis of the previous section, are the energetically dominant ones. There are 1312 NO···ON contacts at O···O distances smaller than 10 Å. Also present are 6039 $C(sp^3)$—H···ON contacts and 2286 $C(sp^2)$—H···ON contacts at

H$\cdots$O distances smaller than 10 Å. These three sets of contacts are large enough to make feasible a statistical analysis of their geometrical parameters searching for differences that are a signature of different packing patterns.

It is important to mention here that most of the analyzed structures in both magnetic subsets were determined at room temperature, at which the thermal energy largely overcomes the strength of the intermolecular magnetic interactions. This means that we are trying to correlate a physical property whose magnitude is clearly observed only at low temperatures, with the crystal packing patterns existing at room temperature. This is a common practice in molecular magnetism that, unfortunately, cannot be avoided, since only very few crystal structures have been determined at low temperatures. At first glance the aforementioned objection may look serious, but a detailed analysis of this problem revealed that it has minor consequences for our magnetostructural correlations. Actually, except for those cases where a first-order structural phase transition occurs, the crystal packing patterns of molecular crystals show only small changes with the temperature, due to thermal contraction. These changes do not change the relative disposition of the molecules to such an extent as to reverse the nature of the dominant intermolecular magnetic interaction [35]. Therefore, it can be assumed without too much risk that similar crystal packing patterns exist at low and high temperatures in the 47 crystals selected here.

IV.   SPATIAL DISTRIBUTIONS OF N—O$\cdots$O—N CONTACTS IN THE SOLID STATE

A preliminary analysis of the geometries of the α-nitronyl nitroxide molecules in the 47 crystals analyzed—and in others with no definite or relevant or complex magnetic behaviors—showed that the spatial distribution of the atoms in the five-membered ring is nearly the same and that the five ONCNO atoms lie in the same plane. This behavior can be attributed to the delocalization of the π-electrons over the five ONCNO atoms through various resonant forms. Given this fact, we do not have to worry about small distortions in the geometry of the ONCNO group or the atoms of the imidazolidine ring, including the four methyl groups. This simplifies considerably our analysis, since we can consider the internal geometry of ONCNO groups as fixed. Consequently, given an ONCNO group in a certain crystal, whose atoms are labeled as $O_{12}N_{12}C_1N_{11}O_{11}$, the relative geometrical position of another, similar group, labeled as $O_{22}N_{22}C_2N_{21}O_{21}$ (obviously identical in geometry due to the previously mentioned behavior and the translation symmetry present in the crystal), is completely defined by six internal coordinates, as occurs for any two rigid asymmetrical objects in Cartesian space.

There are many possible choices for this six-coordinate space, all of them correlated by linear transformation among themselves. Among these choices, the one selected here is shown in Figure 3, chosen because it allows easy visualization and physical interpretation of the geometrical parameters. We first define the position of the terminal O_{21} atom relative to the $O_{12}N_{12}C_1N_{11}O_{11}$ group using three parameters, the $O_{21} \cdots O_{11}$ distance, the $O_{21} \cdots O_{11}-N_{11}$ angle, and the $O_{21} \cdots O_{11}-N_{11}-C_1$ torsion angle. Then, the position of the N_{21} atom for a fixed N—O distance is defined giving the value of the $N_{21}-O_{21} \cdots O_{11}$ angle and the $N_{21}-O_{21} \cdots O_{11}-N_{11}$ torsional angle. Finally, the plane on which the $O_{22}N_{22}C_2N_{21}O_{21}$ group lies is fully defined by the $C_2-N_{21}-O_{21} \cdots O_{11}$ torsional angle, since the C_2-N_{21} distance and the $C_2-N_{21}-O_{21}$ angle are fixed. Once this plane is known, the

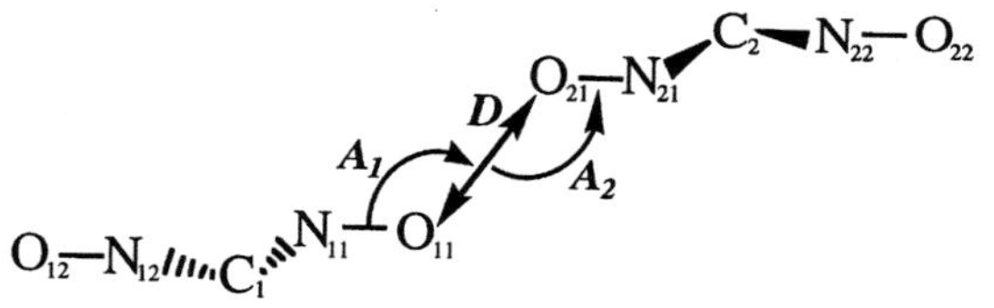

$$D = O_{21} \cdots O_{11}$$

$$A_1 = O_{21} \cdots O_{11} - N_{11}$$

$$A_2 = N_{21} - O_{21} \cdots O_{11}$$

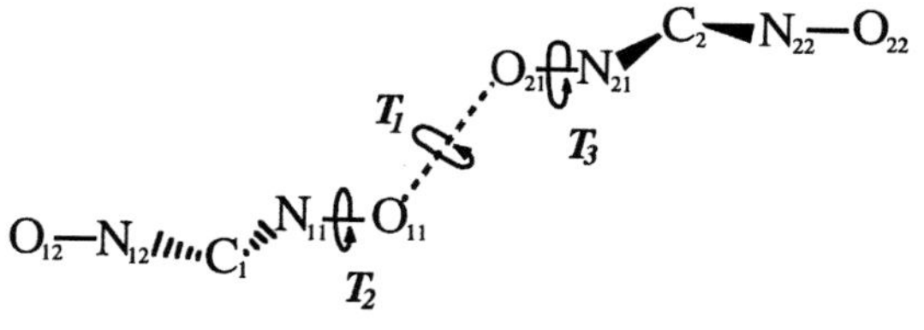

$$T_1 = N_{21} - O_{21} \cdots O_{11} - N_{11}$$

$$T_2 = O_{21} \cdots O_{11} - N_{11} - C_1$$

$$T_3 = C_2 - N_{21} - O_{21} \cdots O_{11}$$

Figure 3   Geometrical parameters employed to define the relative position of two ONCNO groups in space.

position of the N_{22} and O_{22} atoms are automatically established due to the fixed geometry of all ONCNO groups. To simplify the notation, we will identify the previous six intergroup geometrical parameters as D, A_1, A_2, T_1, T_2, and T_3 (see Figure 3 for a proper visualization).

Using the previous six parameters as a first step in our research on the spatial distribution of the ONCNO groups, we studied the number of contacts as a function of the $O \cdots O$ distance, i.e., of the D parameter. The results indicate that there are no contacts in any of the subsets at distances shorter than 3 Å. The shortest value of D (the $NO \cdots ON$ distance) is 3.158 Å within the FM subset and 3.159 Å in the AFM one. The number of contacts in the FM and AFM subsets increases as the third power of the distance in the 3–10-Å range. The proportion of contacts in the FM and AFM subsets is nearly constant for all the distance ranges tested (on average, 44% are from the FM subset and 56% from the AFM one). These values are close to the percentage of crystals in the FM and AFM subsets (49% and 51%, respectively), although the percentage of contacts in the AFM subset is slightly larger. This result suggests that the number of $ONCNO \cdots ONCNO$ contacts made by each NO group is similar in both subsets and does not depend strongly on its magnetic properties.

The previous values also indicate that the cutoff employed in analyzing the $ONCNO \cdots ONCNO$ contacts is not in principle a key factor. Therefore, given the large sizes of the R groups in many α-nitronyl nitroxide molecules, it seems adequate to begin by performing an analysis in the 0–10-Å range, to allow the possibility of unusual crystal packing patterns. Within this range, the average values of the parameters D, A_1, A_2, T_1, T_2, and T_3 for the AFM subset are 7.8 Å, 73°, 113°, 8°, −6°, 2°, while the equivalent ones for the FM subset differ by less than their standard deviation, namely, 7.9 Å, 74°, 115°, 13°, 3°, −5°. It is interesting to note that when the analysis is carried out in the 0–4-Å range—the range for which the intermolecular contacts are supposed to be stronger and more determinant of the FM or AFM character—the average values for the previous six parameters in the FM subset are 3.6 Å, 92°, 137°, 18°, −22°, and −37°, while for the AFM subset they are 3.6 Å, 88°, 93°, 32°, 44°, and 10°. Once again, the average values for the two subsets differ by less than their standard deviations. Concerning the previous values, it is interesting to note the 190° value for the average of $A_1 + A_2$, an indication of the trend in the $NO \cdots ON$ groups to being parallel. It is also worth pointing out the similarity in the average values of the six geometrical parameters for the two magnetic subsets, an indication that the FM and AFM packing patterns are similar. This can be better appreciated by plotting the position of the O_{21} atom for a representative sample of the FM and AFM subsets (see Figure 4, where the values of the parameters D, A_1, and T_2 are plotted for a random selection of 100

FM subset of crystals

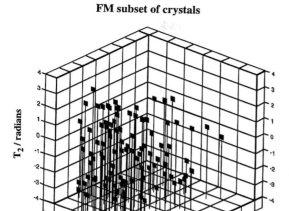

AFM subset of crystals

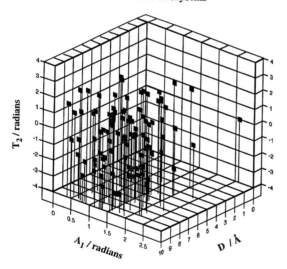

Figure 4 Tridimensional positions of the O_{21} atom relative to an $O_{12}N_{12}C_{1}N_{11}O_{11}$ group (as given by the values of D, A_1, and T_2) for a random selection of 100 contacts within the FM (upper) and AFM (lower) subsets of α-nitronyl nitroxide crystals.

contacts within each subset). We see in Figure 4 that the O_{21} atom is distributed over the whole Cartesian space in a smooth and similar manner in both magnetic subsets. The similarity in the relative-position distributions of the ONCNO groups in the FM and AFM subsets is also shown in the scattergrams of pairs of parameters for the FM and AFM subsets (Figure 5).

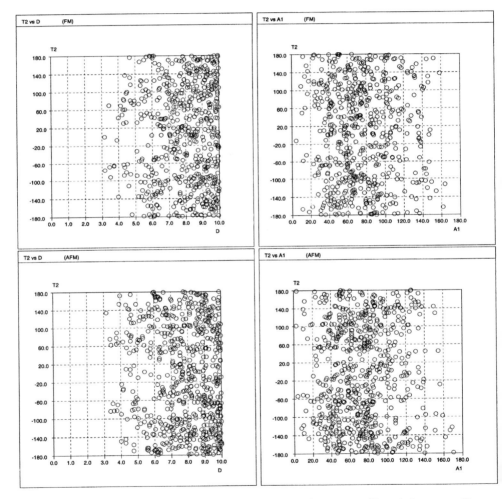

Figure 5 Scattergram showing the dependence of D versus T_2 and A_1 versus T_2 for all the contacts of the FM (upper) and AFM (lower) subsets of α-nitronyl nitroxide crystals.

Therefore, we can safely conclude that there are no characteristic patterns present in the FM subset that are excluded in the AFM subset, and vice versa. Instead, the two subsets present very similar or identical crystal packing patterns.

The scattergrams in Figure 5 also show that there is no relationship between the presence of magnetism and the values of D. This is important because it goes against some commonly accepted ideas about the relationship between magnetism and crystal packing in α-nitronyl nitroxides, specifically the idea associating the presence of short $NO \cdots ON$ contacts to crystals with dominant antiferromagnetic interactions. According to the McConnell I model [4], when two ONCNO groups approach in such a way that their closest oxygen atoms lie in the same plane with A_1 angles between 90° and −90°, they will directly overlap the positive electron spin densities [36] of the two closest O atoms and, consequently, should give rise to a strong antiferromagnetic interaction, which grows more important as the $O \cdots O$ distance gets shorter. That situation is found in our set of parameters when T_2 is equal to 0° or 180°, for values of the A_1 angles between 90° and −90°. Consequently, if the simple association between short $NO \cdots ON$ distance and dominant antiferromagnetic interactions is valid, one should find at these values a decrease in the number of $ONCNO \cdots ONCNO$ contacts in the FM subset and an increase in the contacts of the AFM subset. This is *not* found in the scattergrams in Figure 5.

We have explored whether this conclusion was a consequence of a particular distance range employed in the analysis, with no change in the conclusions. The proportion of contacts in the ferro- and antiferromagnetic subsets is nearly independent of the range of distances selected. Even at very short distance ranges, where the overlap of the spin densities is maximum and the magnetic strength associated with the $NO \cdots ON$ contact is the largest, one also finds contacts within the FM subset. It is clear from such data that similar values of all parameters for the shortest contacts are found within both the FM and AFM subsets. Therefore, *the presence of short NO···ON contacts, classically considered antiferromagnetic according to the McConnell I model, does not imply that the crystal has dominant antiferromagnetic interactions.*

Does the latter conclusion mean that the McConnell I model is not working properly for the α-nitronyl nitroxide radicals? Strictly speaking, we cannot conclude so at this stage. The failure to correlate the presence of presumably large antiferromagnetic interactions with particular crystal packing patterns in these systems could be due to the failure of the theoretical model or to the simultaneous presence of other interactions whose strengths or number can overcome the magnetic effect of the $NO \cdots ON$ contact. However, even if the second situation applies, our analysis *changes the assump-*

tions by which structure–magnetism correlation analysis of such crystals has to be carried out from now on.

We have also carried out [34] a correlation analysis [37] of the data to search for better sets of coordinates or linear dependencies. Besides a small correlation found between the A_1 and A_2 angles (not surprising, given their definition), there is no correlation for any other pair of parameters employed in this study. A factor analysis [38] concluded [34] that six is the number of parameters required to treat the geometrical data of the $ONCNO \cdots ONCNO$ contacts and that it is not possible to reduce this number [39]. Similar conclusions are found for the contacts of the AFM subset when the factor analysis is done. Finally, a cluster analysis [40] of $ONCNO \cdots ONCNO$ contact geometrical data on the set composed by the combined FM and AFM subsets—using the single-linkage method as criterion to define a cluster [40]—indicated that the FM and AFM sets of contacts are nearly identical and interpenetrated, thus being indistinguishable. Consequently, there is no identifiable statistically significant difference in the relative positions of the NO groups for the two magnetic subsets. This is equivalent to saying that there are no different crystal packing patterns in these subsets that relate to the relative positions of the NO groups. This is reasonable if one realizes that the interactions responsible for the packing of these crystals, which are rationalizable in terms of $NO \cdots ON$ and $C—H \cdots ON$ contacts, as already mentioned, are very similar in both subsets of crystals and stronger than the magnetic intermolecular interactions.

V.   SPATIAL DISTRIBUTIONS OF $C(sp^3)—H \cdots O—N$ AND $C(sp^2)—H \cdots O—N$ CONTACTS

The $C(sp^3)—H \cdots O—N$ and the $C(sp^2)—H \cdots O—N$ contacts, together with the $X—H \cdots H \cdots O—N$ ones (X = heteroatom) when they are possible, are the contacts responsible for the formation of hydrogen bonds within crystals of α-nitronyl nitroxide radicals. Therefore, if there is a general ferro- or antiferromagnetic pattern associated with these hydrogen bonds in the packing it should be reflected in the analysis of these types of bonds. Given the small amount of spin density localized in the H atoms of methyl and aromatic groups, these contacts can produce only very weak magnetic interactions, according to the McConnell I model [4]. These contacts are ferromagnetic if the spin density on such H atoms is negative and antiferromagnetic if it is positive, since the oxygen of the NO group is always positive. In principle, due to their weakness these contacts should not be the determinant ones in establishing the dominant magnetic interaction in a given crystal, although if their number is large enough, they could compensate for the stronger ones.

To carry out a systematic search on the geometry of a general C—H ⋯O—N contact, we need six independent internal coordinates: the H⋯ O distance and the H⋯O—N angle, to define the position of the hydrogen; the C—H⋯O angle and the C—H⋯O—N dihedral, to define the position of the C atom; and the R—C—H angle and R—C—H⋯O dihedral to define the position of the R group attached to the C—H. Among these parameters, the two used to define the position of the functional group R do not play a determinant role in defining the geometry of the C—H⋯O— N contacts, due to the quasi-cylindrical symmetry of the electron density on the C—H group. We will not consider them further. For simplicity, as before, we will identify the remaining four parameters as D, A_1, A_2, and T_1. Figure 6 shows these parameters for the $C(sp^3)$—H⋯O—N contacts present in the α-nitronyl nitroxide radicals; Figure 7 shows their definition for the aromatic $C(sp^2)$—H⋯O—N contacts. In the latter case, one can distinguish between *ortho, meta* and *para* aromatic carbons of $C(sp^2)$—H⋯ O—N contacts with respect to the *alpha* C atom of the five-membered ring. We will identify these by adding the suffixes o, m, and p after the name of the parameter (for instance, A_{1o}, A_{1m}, and A_{1p}). All crystals included in the

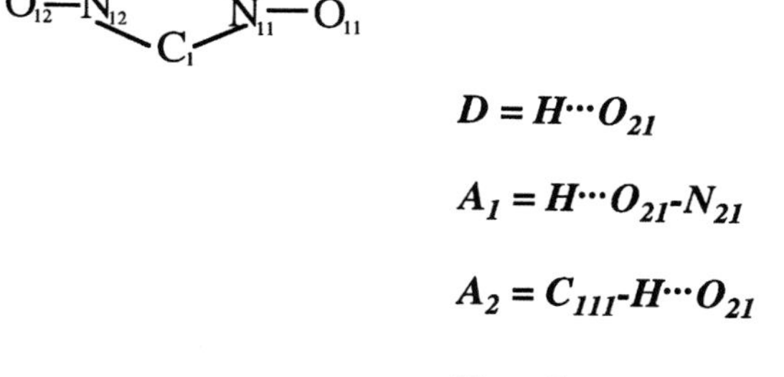

$$D = H \cdots O_{21}$$

$$A_1 = H \cdots O_{21}\text{-}N_{21}$$

$$A_2 = C_{111}\text{-}H \cdots O_{21}$$

$$T_1 = C_{111}\text{-}H \cdots O_{21}\text{-}N_{21}$$

Figure 6 Geometrical parameters employed to define the relative position of one ONCNO group relative to the $C(sp^3)$—H group of the methyl groups attached to the five-membered rings of α-nitronyl nitroxide radicals.

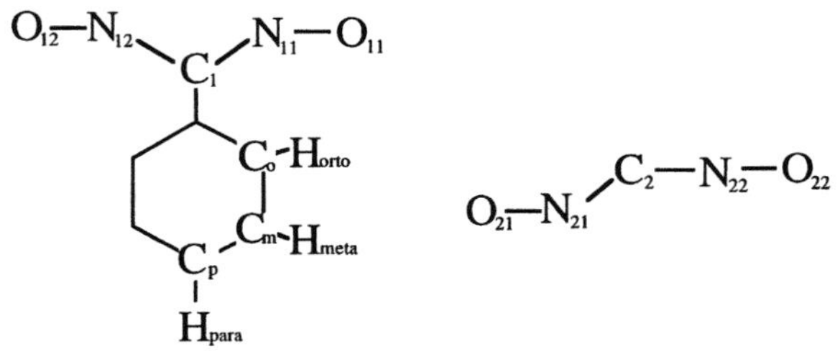

$$D_o = H_{orto} \cdots O_{21} \qquad\qquad D_m = H_{meta} \cdots O_{21}$$

$$A_{1o} = H_{orto} \cdots O_{21} \text{-} N_{21} \qquad A_{1m} = H_{meta} \cdots O_{21} \text{-} N_{21}$$

$$A_{2o} = C_o\text{-}H_{orto} \cdots O_{21} \qquad A_{2m} = C_m\text{-}H_{meta} \cdots O_{21}$$

$$T_{1o} = C_o\text{-}H_{orto} \cdots O_{21} \text{-} N_{21} \qquad T_{1m} = C_m\text{-}H_{meta} \cdots O_{21} \text{-} N_{21}$$

$$D_p = H_{para} \cdots O_{21}$$

$$A_{1p} = H_{para} \cdots O_{21} \text{-} N_{21}$$

$$A_{2p} = C_p\text{-}H_{para} \cdots O_{21}$$

$$T_{1p} = C_p\text{-}H_{para} \cdots O_{21} \text{-} N_{21}$$

Figure 7 Geometrical parameters employed to define the relative position of one ONCNO group relative to a $C(sp^2)$—H group located in *ortho, meta,* or *para* position of an aromatic ring attached to the five-membered ring of an α-nitronyl nitroxide radical.

two magnetic subsets have 12 hydrogens bonded to $C(sp^3)$ atoms that have negative spin sitting on them [41]. Therefore, a proportion of 49% ferro- vs. 51% antiferromagnetic $C(sp^3)$—H$\cdots$O—N contacts should be expected according to the number of crystals in the FM and AFM subsets. However, not all the analyzed crystals have functionalized aromatic groups as the R substituent, and the proportion of ferro- vs. antiferromagnetic $C(sp^2)$—H$\cdots$O—N contacts will depend not only on the number of analyzed crystals belonging to each magnetic subset but also on the available number of hydrogens in *ortho, meta* or *para* positions.

A search within the FM and AFM subsets for $H \cdots O$ distances smaller or equal to 3.8 Å gives 364 $C(sp^3)$—$H \cdots O$—N contacts and 102 $C(sp^2)$ —$H \cdots O$—N contacts. Of the 364 $C(sp^3)$—$H \cdots O$—N contacts, 43% belong to the FM subset and 57% to the AFM one, a proportion similar to the number of crystals in each subset. Similarly, the proportion of $C(sp^2)$—$H \cdots O$—N contacts in the FM and AFM subsets for hydrogens in the *ortho, meta*, and *para* positions are in clear accordance with the number of hydrogens in each of those positions. In conclusion, the number of C—H$\cdots$O contacts does not depend on the dominant magnetic interactions present in these crystals.

One can have an idea of the distribution of these contacts by looking at Figures 8 and 9, which present the values of the D, A_1, and T_1 parameters for the C—H$\cdots$O—N contacts found in the two magnetic subsets. In the C—H$\cdots$O—N case, the points are spread in a smooth and similar way over the whole range of values represented. Consequently, there seem to be no excluded regions, for the crystals belonging to FM and AFM subsets. Instead, the similarity of geometries in the intermolecular contacts of the two subsets is manifested, making it impossible to distinguish between the crystals showing dominant ferro- and antiferromagnetic interactions just by a direct inspection of the geometry of a particular intermolecular contact.

VI. CONCLUSIONS

We have shown how the packing of molecular crystals can be analyzed in terms of primary packing patterns by using geometrical constructions equivalent to the synthons introduced by Desiraju [11]. Using these patterns with the help of MEP overlap analysis, it is possible to rationalize the crystal packing of any molecular crystal. In particular, the packing of the HNN crystal can be rationalized in terms of $C(sp^2)$—$H \cdots O$—N and $C(sp^3)$— $H \cdots O$—N contacts, with the first being stronger [16,27,28]. The N—O$\cdots$ O—N contacts present in the HNN and other α-nitronyl nitroxide crystals are repulsive [16,27,28].

We have carried out a statistical analysis of the crystal packing of α-nitronyl nitroxide radicals. For such a purpose, we selected only crystals having dominant ferro- or antiferromagnetic intermolecular interactions, and grouped them into an FM subset and an AFM one. We found that these two subsets do not present different crystal packing patterns; that is, it is not possible to identify specific packing patterns for the ferro- and antiferromagnetic interactions. This experimental observation agrees with the fact

FM subset of crystals

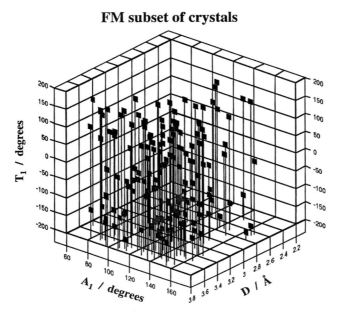

AFM subset of crystals

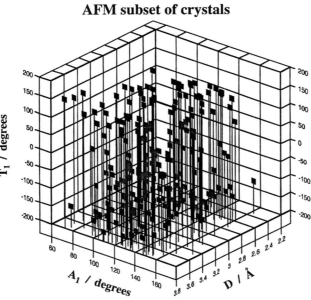

Figure 8   Tridimensional positions of the H atom relative to an $O_{12}N_{12}C_1N_{11}O_{11}$ group (as given by the values of D, A_1, and T_1) for all the $C(sp^3)$—H$\cdots$O—N contacts within the FM (upper) and AFM (lower) subsets of α-nitronyl nitroxide crystals.

FM subset of crystals

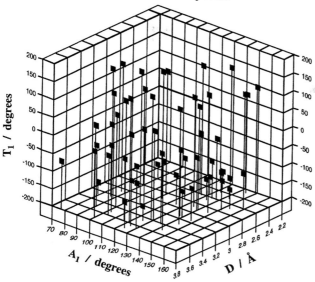

AFM subset of crystals

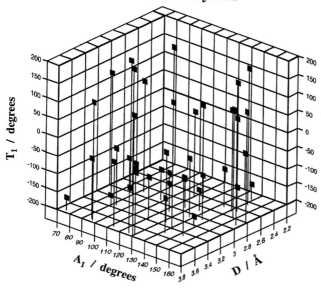

Figure 9 Tridimensional positions of the H atom relative to an $O_{12}N_{12}C_1N_{11}O_{11}$ group (as given by the values of D, A_1, and T_1) for all the $C(sp^2)$—H$\cdots$O—N contacts within the FM (upper) and AFM (lower) subsets of α-nitronyl nitroxide crystals.

that the packing of the α-nitronyl nitroxide radicals is driven by intermolecular forces that are identical in both FM and AFM subsets of crystals.

Given the fact that all the crystals of a magnetic subset must have dominant interactions of the subset type and that these interactions clearly depend on the geometry of the relative arrangements of radicals, the obvious question concerns why these differences are not seen in the statistical analysis. The only possible explanation, after all the statistical tests carried out on the data, is the simultaneous presence of ferro- and antiferromagnetic patterns in each magnetic subset. Consequently, it is impossible to guess the dominant intermolecular magnetic interactions of a given molecular crystal just by looking at some particular detail of its crystal packing pattern such as the relative orientation of the N—O$\cdots$O—N, $C(sp^3)$—H$\cdots$O—N, or $C(sp^2)$—H$\cdots$O—N contacts. It is, instead, necessary to look at the whole packing pattern of the neighboring molecules to guess the dominant magnetic character associated with such a packing pattern. Experimentally this is not an easy task, since it requires one to generate isolated clusters of molecules in solution (dimer, trimers, regular chains, etc.) having a fixed and known geometry and then to be able to determine the magnetic interactions involved. Otherwise, one can use solid samples to extract this information, although it requires the use of more elaborate techniques with oriented single crystals (magnetic and heat capacity measurements, EPR spectroscopic measurements, muon spin rotation experiments, etc.) in order to identify and anticipate the nature and directionality of the intermolecular magnetic interactions. Alternatively, high level ab-initio computations on model clusters could also give the same information, although given their cost they have not been carried out, for the moment, on realistic α-nitronyl nitroxide molecules at a computational level adequate enough to provide accurate results.

Finally, there are no simple structural features characteristic of a ferro- or antiferromagnetic pattern. Along this line, we have found that some commonly accepted ideas on the magnetic character of the NO$\cdots$ON contacts are not completely correct. For instance, it is not true that short in-plane NO$\cdots$ON contacts are always a signature for dominant AFM interactions in the crystal, as one would expect just by using the McConnell I model. All this information indicates that one has to analyze each crystal in detail to deduce the nature and strength of the magnetic interactions associated with all the crystal packing patterns present.

REFERENCES

1. For a recent overview see: (a) Gatteschi D, Kahn O, Miller JS, Palacio F, eds. Molecular Magnetic Materials. Dordretch: Kluwer, 1991. (b) Iwamura H, Mil-

ler JS, eds. Mol. Cryst. Liq. Cryst. 1993; 232/233:1–3601/1–366. (c) Miller JS, Epstein AJ. Angew. Chem. Int. Ed. Engl. 1994; 33:385–415. (d) Kahn O. Molecular Magnetism. New York: VCH, 1993. (e) Rajca A. Chem. Rev. 1994; 94:871–893. (f) Miller JS, Epstein AJ, eds. Mol. Cryst. Liq. Cryst. 1995; 272–274. (g) Kinoshita M. Jpn. J. Appl. Phys. 1994; 33:5718. (h) Coronado E, Delhaes P, Gatteschi D, Miller JS, eds. Molecular Magnetism: From Molecular Assemblies to Devices. Dordretch: Kluwer, 1996. (i) Kahn O, ed. Magnetism: A Supramolecular Function. Dordretch: Kluwer, 1996.

2. (a) Tamura M, Nakazawa Y, Shiomi D, Nozawa K, Hosokoshi Y, Isjikawa M, Takahashi M, Kinoshita M. Chem. Phys. Lett. 1991; 186:401. (b) Allemand PM, Khemani KC, Koch A, Wudl F, Holczer K, Donovan S, Grüner G, Thompson JD. Science. 1991; 253:301. (c) Chiarelli R, Novak MA, Rassat A, Tholance JL. Nature. 1993; 363:147. (d) Sugawara T, Matsushita MM, Izuoaka A, Wada N, Takeda N, Ishikawa M. J. Chem. Soc. Chem. Commun. 1994; 1081. (e) Cirujeda J, Mas M, Molins E, Lanfranc de Panthou F, Laugier J, Geun Park J, Paulsen C, Rey P, Rovira C, Veciana J. J. Chem. Soc. Chem. Comm. 1995; 709–710. (f) Caneschi A, Ferraro F, Gatteschi D, leLirzin A, Novak MA, Rentsecheler E, Sessoli R. Adv. Mater. 1995; 7:476. (g) Pei Y, Kahn O, Aebersold MA, Ouahab L, Le Berre F, Pardi L, Tholance JL. Adv. Mater. 1994; 6:681. (h) Togashi K, Imachi K, Tomioka K, Tsuboi H, Ishida T, Nogami T, Takeda N, Ishikawa M. Bull. Chem. Soc. Jpn. 1996; 69:2821, and papers cited therein.

3. α-Nitronyl nitroxide is the abbreviated name most commonly employed and it is here used to indicate 4,5-dihydro-4,4,5,5-tetramethyl-3-oxidio-1*H*-imidazol-3-ium-1-oxyl, the correct IUPAC name.

4. McConnell HM. J. Chem. Phys. 1063; 39:1910.

5. In general, bulk ferromagnetism requires the presence of ferromagnetic interactions along three (or two) directions of the solid, depending on whether there is a low (or high) magnetic anisotropy. Since organic free radicals have in general very small magnetic anisotropy, the requirement of the presence of ferromagnetic interactions in three dimensions is completely necessary for achieving such a macroscopic property. See: Palacio F. From Ferromagnetic Interactions to Molecular Ferromagnets: An Overview of Models and Materials. In: Gatteschi D, Kahn O, Miller JS, Palacio F, eds. Magnetic Molecular Materials. Dorderecht: Kluwer, 1991, pp 1–40.

6. A lack of compensation of the magnetic moments of spins coupled antiferromagnetically can also produce a net magnetic moment—spin canting—that, when propagated over the solid, produces a spontaneous magnetization. For a recent organic example of this situation, see: Banister AJ, Bricklebank N, Lavender I, Rawson JM, Gregory CI, Tanner BK, Clegg W, Elsegood RRJ, Palacio F. Angew. Chem. Int. Ed. Engl. 1996; 35:2533.

7. For molecular organic crystals with high symmetry, namely, rhombic or cubic, the presence of just one intermolecular ferromagnetic interaction among the neighbor units is enough to guarantee the propagation of magnetic interactions along two or three spatial directions. By contrast, for crystals with lower symmetries the existence of more than one intermolecular ferromagnetic interaction

for each crystallographically independent radical molecule is necessary in order to achieve a bulk ferromagnetism.

8. Other important facets to be developed in this field are the different ways to enhance the strengths of intermolecular magnetic interactions. Only by increasing these strengths will it be possible to increase the T_C of purely organic ferromagnets that nowadays are still very low, most of them in the mK region.

9. (a) Mighell AD, Himes VL, Rodgers JR. Acta Cryst. 1983; A39:737–740. (b) Wilson AJC. Acta Cryst. 1988; A44:715–724. (c) Baur WH, Kassner D. Acta Cryst. 1992; A48:356–369.

10. Etter MC. Acc. Chem. Res. 1990; 23:120–126. Bernstein J, Davis RE, Shimoni L, Chang N-L. Angew. Chem. Int. Ed. Engl. 1995; 34:1555–1573.

11. Desiraju G. Angew. Chem. Int. Ed. Engl. 1995; 34:2311–2327.

12. A motif is a pattern containing only one type of hydrogen bond. Note that some motifs can extend over more than one molecule.

13. This has been a controversial criteron, because, depending on which radii are taken, some interactions may be visualized as hydrogens bonds (see Ref. 14 for a detailed discussion).

14. Jeffrey GA, Saenger W. Hydrogen Bonding in Biological Structures. Berlin: Springer Verlag, 1991.

15. Novoa JJ, Lafuente P, Mota F. Submitted for publication.

16. See, for instance: Deumal M, Cirujeda J, Veciana J, Kinoshita M, Hosokoshi Y, Novoa JJ. Chem. Phys. Lett. 1997; 265:190.

17. (a) Scrocco E, Tomasi J. Adv. Quant. Chem. 1978; 11:115. (b) Politzer P, Murray JS. Rev. Comp. Chem. 1991; 2:273.

18. Novoa JJ, Mota F, Perez del Valle C, Planas M. J. Phys. Chem. A. 1997; 101: 7842.

19. For a review, see: Ckaka AM, Zaniewski R, Youngs W, Tessier C, Klopman G. Acta Cryst. 1996; B52:165. See also: (a) Gavezzotti A. J. Am. Chem. Soc. 1991; 113:4622. (b) Gdanitz RJ, Chem. Phys. Lett. 1992; 190:391. (c) Karfunkel HR, Gdanitz RJ. J. Comp. Chem. 1992; 13:1170.

20. Williams DE. PCK83, QCPE #41, 1983.

21. Such methodology requires the use of methods capable of including the electron correlation (as the second-order or fourth-order Moller–Plesset methods, better known by their acronyms, MP2 and MP4, respectively), the use of very extended basis sets of the adequate class (see Ref. 22), and correcting the basis-set superposition error (BSSE) using the counterpoise method (Ref. 23). With this methodology it is possible to reproduce the interaction energy and main characteristics of the potential energy surface of hydrogen-bonded and van der Waals dimers very accurately (see Ref. 24).

22. van Duijneveldt FB, van Duijneveldt, van de Rijdt JGCM, van Lenthe JH. Chem. Rev. 1994; 94:1873.

23. Boys SF, Bernardi F. Mol. Phys. 1970; 19:553.

24. (a) Novoa JJ, Planas M, Whangbo M-H. Chem. Phys. Lett. 1994; 225:240. (b) Novoa JJ, Planas M, Rovira MC. Chem. Phys. Lett. 1996; 251:33. (c) Novoa JJ, Planas M. Chem. Phys. Lett. 1998; 285:186.

25. Kitaigorodsky AI. Molecular Crystals and Molecules. London: Academic Press, 1973.

26. See, for instance: Gavezzotti A. Acc. Chem. Res. 1994; 27:309. See also Ref. 19.

27. Novoa JJ, Deumal M. Mol. Cryst. Liq. Cryst. 1997; 305:143.

28. Novoa JJ, Deumal M, Kinoshita M, Hosokoshi Y, Veciana J, Cirujeda J. Mol. Cryst. Liq. Cryst. 1997; 305:129.

29. For examples, see: (a) Awaga K, Inabe T, Maruyama Y, Nakamura T, Matsumoto M. Chem. Phys. Lett. 1992; 195:21–24. (b) Awaga K, Inabe T, Nakamura T, Matsumoto M, Maruyama Y. Mol. Cryst. Liq. Cryst. 1993; 232:69–78. (c) Awaga K, Okuno T, Yamaguchi A, Hasegawa M, Inabe T, Maruyama Y, Wada N. Phys. Rev. B. 1994; 49:3975–3981. (d) Awaga K, Yamaguchi A, Okuno T, Inabe T, Nakamura T, Matsumoto M, Maruyama Y. J. Mater. Chem. 1994; 4: 1377–1385. (e) Awaga K, Okuno T, Yamaguchi A, Hasegawa M, Inabe T, Maruyama Y, Wada N. Synth. Met. 1995; 71:1807–1808.

30. For some examples, see: (a) Veciana J, Cirujeda J, Rovira C, Vidal-Gancedo J. Adv. Mater. 1995; 7:221. (b) Cirujeda J, Hernàndez E, Rovira C, Turek P, Veciana, J. New Organic Magnetic Materials. The Use of Hydrogen Bonds as a Crystalline Design Element of Organic Molecular Solids with Intermolecular Ferromagnetic Interactions. In: Seoane C, Martin N, eds. New Organic Materials. Madrid: Universidad Complutense de Madrid, 1994, pp 262–272. (c) Cirujeda J, Hernàndez E, Rovira C, Stanger JL, Turek P, Veciana J. J. Mater. Chem. 1995; 5:243–252. (d) Cirujeda J, Hernàndez E, Rovira C, Stanger JL, Turek P, Veciana J. J. Mater. Chem. 1995; 5:243–252. (e) Cirujeda J, Rovira C, Stanger JL, Turek P, Veciana J. The Self-Assembly of Hydroxylated Phenyl a-Phenyl Nitronyl Nitroxide Radicals. In: Kahn O, ed. Magnetism. A Supramolecular Function. Amsterdam: Kluwer, 1996, pp 219–248. (f) Cirujeda J, Hernàndez E, Lanfranc de Panthou F, Laugier J, Mas M, Molins E, Rovira C, Novoa JJ, Rey P, Veciana J. Mol. Cryst. Liq. Cryst. 1995; 271:1–12.

31. When no angular data are taken into account and only intermolecular distances are considered in a given magnetostructural correlation, one is always tempted to ascribe the observed ferromagnetic intermolecular interaction to the contacts between atoms of the two interacting molecules that have opposite spin densities and are at the closest distances. However, in some molecular layouts there could be other atoms at longer distances that are more favorable to interact ferromagnetically due to angular reasons.

32. See, for instance: (a) Zheludev A, Barone V, Bonnet M, Delley B, Grand A, Ressouche E, Rey P, Subra R, Schweizer J. J. Am. Chem. Soc. 1994; 116: 2019–2027. (b) Novoa JJ, Mota F, Veciana J, Cirujeda J. Mol. Cryst. Liq. Cryst. 1995; 271:79–90.

33. Allen FH, Bellard S, Brice MD, Cartwright BA, Doubleday A, Higgs H, Hummelink T, Hummelink-Peters BG, Kennard O, Motherwell WDS, Rodgers JR, Watson DG. Acta Crystallogr. 1979; B35:2331.

34. Deumal M, Cirujeda J, Veciana J, Novoa JJ. Adv. Mater. 1998; 10:1461.

35. This fact has been well documented for organic superconducting crystals, whose critical temperatures are also much lower than room temperature. See,

for instance: Williams JM, Ferraro JR, Thorn RJ, Carlson KD, Geiser U, Wang HH, Kini AM, Whangbo M-H. Organic Superconductors. Englewood Cliffs, NJ: Prentice Hall, 1992. Magnetism is a different physical property than superconductivity, although both are intimately related to the relative arrangement of molecules in the crystal. The packing properties of molecular solids and, in particular, its variation with temperature, are determined by the intermolecular contacts, characteristic of the intermolecular contact and independent of the electronic physical property present in the crystal.

36. The presence of a positive density on the oxygen atom has been shown by experimental and theoretical methods. See, for instance, Ref. 32.

37. Barlow R. Statistics. Chichester, Eng.: Wiley, 1989.

38. Malinowski ER, Howery DG. Factor Analysis in Chemistry. New York: Wiley Interscience, 1980.

39. This conclusion shows that the use of three or even two geometrical parameters, a practice performed by some authors for analyzing series of α-nitronyl nitroxide radicals, is not correct.

40. Everitt BS. Cluster Analysis. 3rd ed. London: Edward Arnold, 1993.

41. The spin density on the hydrogen atom of the four CH_3 groups is negative, whereas those on the hydrogen atoms of a phenyl ring linked to the α-carbon of the five-membered ring alternate in sign, being positive for *ortho* and *para* H atoms and negative for the *meta* ones. See: (a) Davis MS, Morokuma K, Kreilich RW. J. Am. Chem. Soc. 1972; 94:5588–5592. (b) Neely JW, Hatch GF, Kreilich RW. J. Am. Chem. 1974; 96:652–656, and Ref. 32.

29

Design, Synthesis, and Properties of Nitroxide Networked Materials

Dante Gatteschi
University of Florence, Florence, Italy

Paul Rey
CEA-DRMFC, Grenoble, France

I. INTRODUCTION

The main difficulty encountered in the design of purely organic magnetic materials is that of establishing suitable strategies for avoiding the direct overlap of the magnetic orbitals of the individual magnetic centers, which inevitably leads to antiferromagnetic interactions at best, or more often to complete pairing of the spins. Many ingenious strategies have now been developed, leading to bulk ferromagnetism at very low temperatures, or to strong ferromagnetic interactions in discrete molecules [1–7]. However, the dream of a room-temperature organic ferro- or ferrimagnet has not yet been achieved.

In our early attempt to synthesize molecular-based magnetic materials we independently arrived at the conclusion that by using stable organic radicals, like the nitroxides, capable of acting as ligands toward transition-metal ions it might have been possible to reach the goal of a high T_c magnet. The expected advantages of this approach [8] are the following:

1. Using stable organic radicals, avoids all the complications of the very reactive systems often used in organic magnetism.
2. The possibility of strong interactions between the metal ions and the radicals, due to the fact that the magnetic orbitals of the two moieties may come in close contact, should provide extensive spin

correlation at relatively high temperatures, thus favoring the transition to magnetic order. Even if the two magnetic orbitals have a large antiferromagnetic coupling, the compounds may have a permanent magnetization due to noncompensation of the individual moments.

3.   Ample possibilities open up of choices of metal ions, of individual spins, and coordination geometries.

Even at that time a few reports were available of complexes between transition-metal ions and organic radicals, mainly of the nitroxide and the semiquinone types [9–11]. Given the long-standing tradition of the Grenoble group in the synthesis of nitroxide radicals, the choice for the development of the metal-radical approach to molecular magnetism was rather obviously made to use nitronyl nitroxide radicals: 2-(R)-4,4,5,5-tetramethyl-4,5-dihydro-1H-imidazol-1-oxy-3-oxide, NITR (Scheme 1). In fact, these systems are as stable as the nitroxides but have distinct advantages over the latter in that they have two equivalent donor oxygen atoms on which the unpaired electron is evenly spread, which may permit one to connect metal ions in infinite arrays, with full transmission of the magnetic interactions from one center to the other. The synthesis of these radicals is rather straightforward, and an almost infinite series of derivatives can be produced by changing the R groups. In this way it is also possible to introduce charged moieties, forming both radical cations and anions.

The main draw back of the NITR derivatives is that they are very poor ligands, or weak Lewis bases, so in order to form complexes relatively strong Lewis acids must be employed. For instance, it has proved to be easy to form complexes with metal hexafluoroacetylacetonates, either $M^{II}(hfac)_2$ of $M^{III}(hfac)_3$, but in this way only two additional coordination sites are available for the NITR ligands, thus making it almost impossible to form the three-dimensional arrays needed for giving rise to magnetically ordered

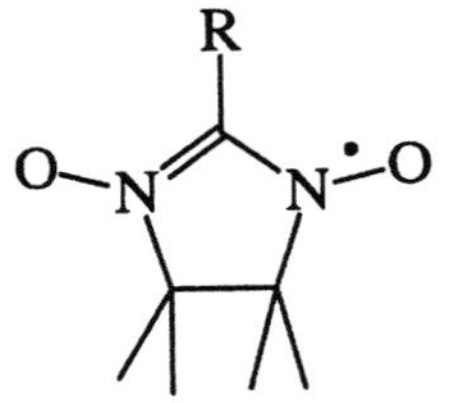

NITR

Scheme 1

structures [8]. Further, the hfac groups are very insulating, from the magnetic point of view, and thus they contribute to the effective shielding of the magnetic centers one from the other.

Notwithstanding all these flaws, the chemistry and the physics of metal nitronyl nitroxides derivatives has proved to be rather rich and is still developing after almost 15 years of intensive research. An early review of the magnetic properties of metal nitronyl nitroxide complexes was published [8] in 1991. In this chapter we do not aim to present a comprehensive review, but rather to highlight which compounds have been the most interesting reported so far and which ones may be susceptible to future developments. Further, we will focus essentially on NITR radicals, but we will briefly take into consideration other nitroxides as well, in particular those that can form extended magnetic structures.

The organization of this chapter will be as follows: (1) First we will provide some general rules for synthesizing nitronyl nitroxides and iminonitroxides; (2) the types of interactions between nitroxides and transition-metal and lanthanide ions will be briefly reviewed in order to provide suitable tools to understand the observed magnetic couplings; (3) the subsequent sections will be devoted to the description of selected examples of zero-, one-, and two-dimensional magnetic materials formed by the nitroxides, and another section will cover the interactions responsible for the transition to three-dimensional magnetic order; (4) the last section will be devoted to indications for further investigations using nitroxide-type radicals.

II. SYNTHESIS OF NITRONYL NITROXIDES AND DERIVATIVES

Several types of stable nitroxides have been reported in the literature (Figure 1): piperidinyl (a), pyrrolidinyl (b), imidazolidinyl (c), nitronyl (d), iminyl (e), and several subclasses depending on various peculiar groups present in the molecule have been tentatively described [12–23]. All these types differ in the organic backbone and particularly in the possibility of delocalization of the unpaired electron. Since this account is devoted to coordination derivatives, only nitroxides used as ligands will be considered. One is interested in extended structures; thus, particular attention will be paid to bridging nitroxides such as nitronyl and imino nitroxides and also to polynitroxides, which are m-substituents of a phenyl ring (e.g., f, g, h in Figure 1) [12–14]. In the latter, large intramolecular interactions produce (from the magnetic point of view) polydentate nitroxide ligands, where all oxygen atoms carry spin density of the same sign as in nitronyl analogs.

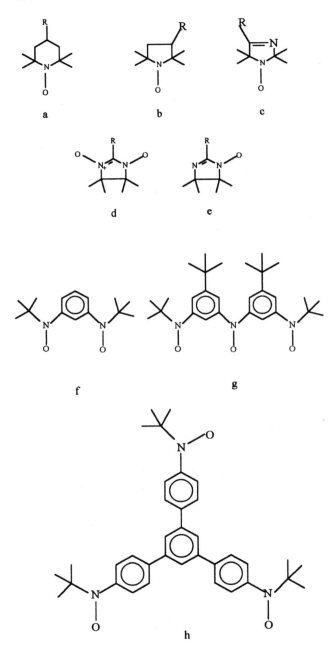

Figure 1   Some examples of nitroxide-type radicals investigated for their magnetic properties.

The synthesis of these ligands is well documented [15–22], except for the imino nitroxides that have been obtained in low yield until recently as by-products in the preparation of nitronyl nitroxides or by tedious procedures [20]. The first reliable synthesis of these ligands was reported in 1994 by Lanfranc de Panthou [23]. Since there is an increasing interest in the coordination chemistry of these free radicals [24–26], we describe next modifications of the previous synthesis that produce imino nitroxides in high yields.

Synthesis of Imino Nitroxides.

a. *Genuine Lanfranc de Panthou's synthesis.* Five g of sodium nitrite and 1 g of a nitronyl nitroxide were dissolved in 100 mL of water and 100 mL of methylene chloride. This two-phase mixture was cooled in an ice bath and diluted (N/10) hydrochloric acid was added drop to drop until the organic layer had changed color. Completion of the reaction was confirmed by thin-layer chromatography. The organic phase was dried and evaporated, and the crude imino nitroxide was chromatographed if necessary. Usual yields were 80–100%.

b. *Synthesis from acid-sensitive nitronyl nitroxides.* Dilute sulfuric acid (N/2) was added to a solution of sodium nitrite (5 g in 100 mL). A stream of nitrogen passing through the Erlenmeyer flask was then bubbled into a methylene chloride solution of a nitronyl nitroxide (1 g in 50 mL). The reaction was followed by thin-layer chromatography and stopped when complete consumption of the starting material was observed. Then the solvent was evaporated under vacuum. For alkyl- and aromatic-substituted nitroxides the reaction proceeded quantitatively.

c. *One-pot synthesis from the hydroxylamino precursor.* The hydroxylamino compound resulting from the condensation of an aldehyde with 2,3-bis(hydroxylamino)-2,3-dimethylbutane (3 g) was suspended or dissolved in 200 mL of methylene chloride. A solution containing a slight excess of sodium periodate in the minimum amount of water was added at 0°C. The mixture was stirred for one-half hour, and a solution of sodium nitrite (5 g in 100 mL of water) was added. Work up and purification were performed as described before.

Note that a modification of the genuine Lanfranc de Panthou's synthesis was recently described [26] wherein acetic acid was used in place of an inorganic acid. Since organic acids are soluble in the organic phase and thus can reduce the free radical, this modification is not reliable in all cases.

III. ELECTRONIC STRUCTURE OF THE NITRONYL NITROXIDES AND MAGNETIC INTERACTIONS WITH METAL IONS

The electronic structure and the nature of the magnetic orbital(s) play a crucial role in understanding the mechanism of magnetic coupling in a material. Nitroxide free radicals have an unpaired electron in a π^*-orbital that is localized mainly on the N and O atoms in nondelocalized species (a, b, c in Figure 29.1) and distributed over the two nitrogen-containing sites in nitronyl (d) and imino (e) nitroxides. In polynitroxides derived from m-phenylene (f, g), spin correlation results from topology and thus they are themselves high-spin species [12–14].

From a qualitative point of view it is rather easy to understand the nature of the magnetic interaction of a nitroxide radical and a metal ion. It may be convenient to assume a direct interaction between the two magnetic orbitals, even if it has been proved that spin-polarization effects may be operative [27]. If two magnetic orbitals are orthogonal to each other, the coupling is ferromagnetic; if they have a nonzero overlap, the coupling is antiferromagnetic. This is most clearly shown in the case of copper(II)-NITR complexes, in which the coupling is ferromagnetic when the radical binds in an axial position of a distorted octahedron, while it is strongly antiferromagnetic when it binds in an equatorial position [8]. In both cases the magnetic orbital on the copper(II) ion can be described as $x^2 - y^2$ and is orthogonal to the π^*-radical orbital in the former, while in the latter the two magnetic orbitals have large overlap. Recently, DFT calculations on copper-nitroxide complexes further provided a quantitative interpretation of the observed coupling constants [28].

In general, for metal ions with more than one unpaired electron, at least one magnetic orbital has some overlap with the magnetic orbital on the radical, thus yielding a relatively strong antiferromagnetic coupling. For instance, manganese(II), which has five unpaired electrons in the five d orbitals, has an antiferromagnetic coupling of $200-300$ cm^{-1} with NITR radicals [8].

A striking exception to this simple rule is provided by gadolinium(III), which has seven unpaired electrons in the seven $4f$ orbitals and is now well established to have a weak ferromagnetic coupling with NITR radicals [29]. in a qualitative way this has been explained, considering that the direct interaction between the π^*-orbital of the radical and the internal $4f$ orbitals of gadolinium is exceedingly small and that the observed coupling is determined by a spin-polarization mechanism [30,31]. The π^*-orbital has a nonzero overlap with an empty $5d$ orbital, transferring a fraction of unpaired electron to there, say with spin up. This electron polarizes the spins of the electrons in the $4f$ orbitals parallel to itself, according to Hund's rule, so yielding a weak ferromagnetic coupling of the order of 1 cm^{-1}.

In order to understand in detail the mechanism of the magnetic interactions involving organic radicals and metal ions, it would be desirable to acquire a good knowledge of unpaired spin-density maps, because this can provide a pictorial view of the couplings. In principle the unpaired spin density can be obtained by magnetic resonance (EPR, NMR, ENDOR, ESEEM, etc.) techniques and by polarized neutron diffraction experiments. The former techniques are hardly applicable to magnetically concentrated systems, but they can provide information on the isolated building blocks. Solution studies have, for instance, provided detailed information on the hyperfine coupling in NITR radicals [21].

The polarized neutron diffraction technique is much better suited for investigations on the magnetically nondiluted solid samples, as described in an earlier article [32]. In fact, the most noticeable advance in the knowledge of the electronic structure of the nitroxides and of some of their derivatives has been provided by polarized neutron diffraction experiments. This technique, which can be applied to single crystals of paramagnetic species, records the difference in absorption of left and right circularly polarized neutron beams in the presence of an external magnetic field. Under these conditions it is possible to reconstruct the map of the unpaired spin density of simple paramagnets in all space. This is the main difference from the information that can be obtained from magnetic resonance experiments by monitoring the spin densities at the nuclei. It is interesting to note that deuteration is *not* absolutely needed in these experiments. This is of course a very desirable feature for the investigation of molecular magnets in which the organic contribution is large.

The procedure of interpreting such experimental data requires the development of complex models that inevitably have some degree of arbitrariness. In the end, one must determine (a) the positions where the unpaired electrons are preferentially located and (b) the sign of the spin density on an atomic site. In a simple molecular orbital (MO) picture, we may expect to find negative spin density on some atoms due to spin-polarization effects. Therefore, polarized neutron absorptions provide useful testing grounds for the MO treatment of open-shell molecules.

Spin-density studies were devoted to: (1) nitroxides exhibiting Curie behavior in the solid state, in order to get a picture of the magnetic orbital; (2) understanding the mechanism of intermolecular ferromagnetic interactions in pure nitroxides; (3) examination of copper(II)-nitroxide complexes with the aim of characterizing their ground spin states and understanding their metal–radical coupling mechanisms.

1. *Isolated nitroxides.* The distribution of the spin density in some of these spin carriers shown in Figure 1 (a, d, e) has been determined [33–35]. These studies have confirmed the simple picture of an unpaired electron residing in a π^*-orbital and of an almost equal distribution of the spin density on the N and O atoms of the nitroxyl group. Modification of this distribution upon hydrogen bonding involving the oxyl oxygen has also been studied. Concerning nitronyl nitroxides (d), the spin density is, as predicted by elementary molecular orbital theory, equally distributed on the two NO groups; in imino nitroxides, the imino group carries, as expected, a smaller spin density than the NO group. In both types of spin carriers, a fairly large negative density is found on the sp^2 hybridized carbon atom and on remote sites of the substituent in the 2-position (attachment of —R). Thus, ferromagnetic intermolecular interactions may be favored in the solid state. This peculiar aspect of the electronic structure of nitroxide free radicals is especially relevant to the design of purely organic molecular magnetic materials.

2. *Intermolecular ferromagnetic interactions in pure nitroxides.* In contrast to the former examples, some nitroxides in the solid state exhibit a magnetic behavior that can be explained only on the basis of intermolecular ferromagnetic interactions. A few of these free radicals even order as ferromagnets and have received much attention. Recent examples of such compounds are displayed in Figure 2.

The following features are characteristic of intermolecular coupling: the oxyl oxygen atom is involved in close contacts with parts of another molecule, resulting in a lowering of the spin density on this atom and an increase of the spin density on the nitrogen atom. This modification of the distribution also results from delocalization onto the other molecule, as shown in example **4**, where the diamagnetic phenyl boric acid fragment links the free radicals in a chain structure and carries sizable spin density [36]. From this analysis, two different mechanisms of ferromagnetic coupling were observed. In the first, which corresponds to the McConnell model [37], zones of positive spin density on one center overlap zones of negative spin density on the other; in the second, the magnetic orbitals of the two centers are orthogonal to each other.

3. *Copper(II) complexes.* Earlier studies have shown the dependence of the metal–radical interaction upon the geometry of the metal coordination sphere in copper(II) complexes [8]. This behavior was qualitatively well understood by considering the symmetry of

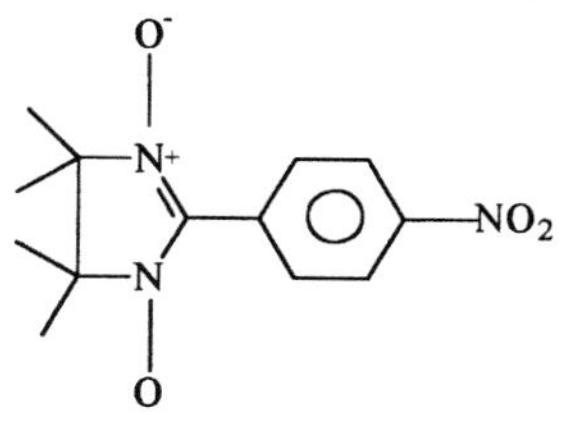

Figure 2 Some examples of nitroxide-type radicals that give rise to intermolecular ferromagnetic interactions.

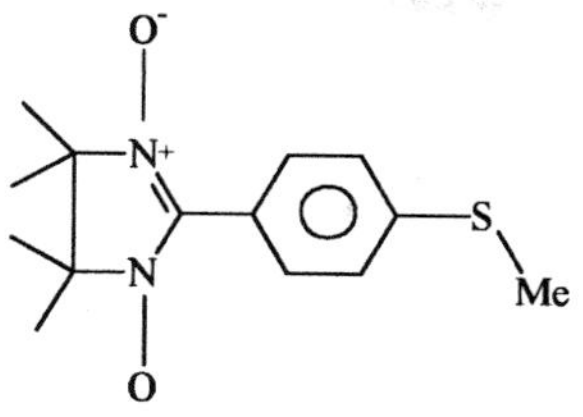

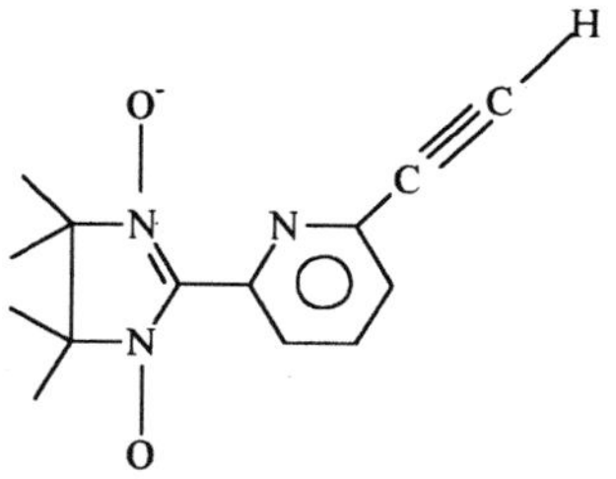

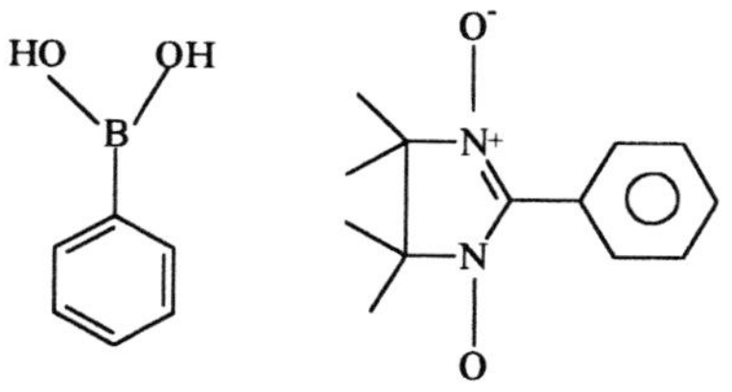

the magnetic orbitals. Polarized-neutron-diffraction data have determined the ground spin state of these complexes, and they have brought further information concerning the spin distribution after complexation. The two extreme situations of antiferro- and ferromagnetic metal–radical coupling have been studied [38,39]. As found in pure radicals, in both situations a decrease of the spin density on the coordinated oxyl oxygen is observed. In the case of large antiferromagnetic coupling this spin density may even be zero. The signs of these spin densities are in agreement with the observed ground state of the complexes.

In conclusion, thanks to neutron data the electronic structure of the nitroxides is well known. The studies involving coupled nitroxides in the solid state or some of their complexes have brought information about coupling mechanisms and clues for the design of molecular species possessing a predetermined spin ground state.

IV. ZERO-DIMENSIONAL MAGNETS

Numerous simple complexes, containing a number of metal ions ranging from one to six, have been reported [8,40–43]. These systems have been of extreme importance for the understanding of the nature of the exchange interactions between the metal ions and the radicals. However, some compounds have shown other interesting properties that make them possible candidates for use as molecular materials in the storage of information [44]. These are copper complexes that show abrupt changes in their magnetic properties associated with rearrangements in the coordination sphere, which make them similar to molecular spin-crossover systems.

Owing to the plasticity of the metal coordination sphere and to the different natures and magnitudes of the related magnetic interactions, the coordination chemistry of nitroxide free radicals with copper(II) is very rich. Thus, depending on the geometry of coordination, exchange interactions are spread over a large energy range ($-500 > J > +500$ cm^{-1}) [45–47].

For the design of magnetic materials, control of the coordination geometry is needed. It has been observed that the presence of bulky substituents induces axial coordination of the nitroxyl group in octahedral complexes because steric requirements do not offer other binding possibilities. In addition, when the free radical ligand carries another donor group so as to bridge two metal ions, the binding is also generally axial for steric crowding reasons, at least when the metal center is coordinated to hexafluoroacetylacetonato (hfac) groups. Thus, for nitronyl and imino nitroxides where there are two coordination sites, the expected (and generally observed at room temperature) binding geometry in polynuclear derivatives is axial. In contrast, in complexes of nitroxides that do not carry extra coordination sites and are not sterically demanding, the observed coordination geometry is equatorial. Therefore, the actual binding geometry is the result of a delicate balance between electronic and steric factors. In particular, it has been proposed that partial spin pairing is a driving force for equatorial binding [48]. Therefore, it is not unexpected that in complexes possessing a peculiar structure that the energy gap between axial and equatorial coordination is very small. As a consequence, these complexes may undergo conversion between these two forms and between two magnetic behaviors, under an appropriate perturbation.

Such behavior has indeed been observed [44] in copper(II) complexes derived from 3-pyridyl-substituted nitronyl and imino nitroxides (Figure 3 and Scheme 2). These complexes have very similar structures involving four metal ions and two free radical ligands. Two of the copper ions are bridged by the ligands so as to form a cyclic structure; the two other are extracyclic and pentacoordinated. The main structural differences are observed in the

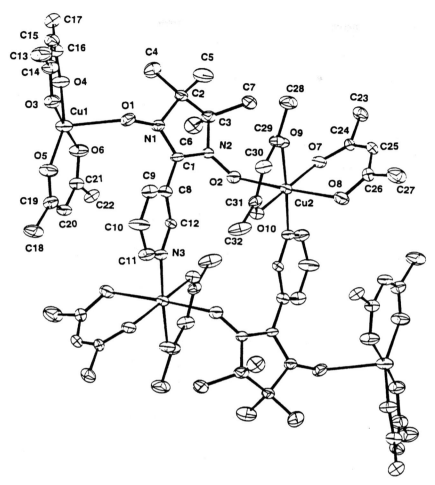

Figure 3 Structure of {[Cu(hfac)₂]₂(NIT-3Py)}₂. (From Ref. 44.)

coordination sphere of the latter; the metal environment is square pyramidal in the derivative of the nitronyl nitroxide, whereas it is trigonal bipyramidal in the complex of the imino nitroxide ligand.

At room temperature the nitronyl nitroxide complex exhibits a magnetic moment equal to the sum of the six $S = \frac{1}{2}$ independent spins and follows Curie law behavior down to 120 K, as shown in Figure 4. Below 90 K, another Curie law is observed that corresponds to only two independent $S = \frac{1}{2}$ spins. This behavior is rationalized on the basis of known struc-

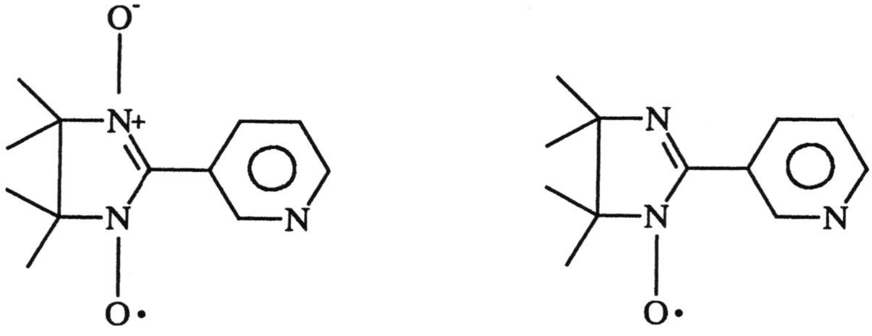

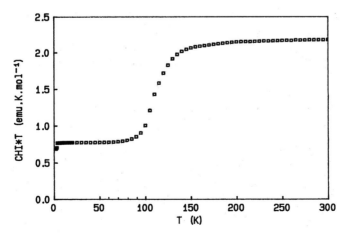

Scheme 2

tural-magnetic correlations in copper(II)-nitroxide complexes: since the observed coordination of the nitroxide ligand to the intracyclic metal ion at room temperature is axial, the metal–radical interaction is weakly ferromagnetic so that, in this temperature range, the different spins behave as if they were independent. A structural determination performed at 50 K shows that the coordination of the nitroxide ligand has switched to equatorial, resulting in a large antiferromagnetic coupling between the two nitroxides and two copper ions. Therefore, at low temperature only the two extracyclic copper(II) ions are observed. This change in spin multiplicity observed at 110 K is reversible, and no hysteresis is observed.

Figure 4 Temperature dependence of the magnetic susceptibility of $\{[Cu(hfac)_2]_2(NIT\text{-}3Py)\}_2$. (From Ref. 44.)

The magnetic behavior of the imino nitroxide complex is similar, with a transition occurring at 70 K. However, there are two differences: the transition occurs with hysteresis, and another transition is observed at higher temperature. The latter is thought to involve a change of environment of the extracyclic copper centers from a trigonal bipyramid to a square pyramid. The observation of a hysteresis loop is very important for prospects to use these materials for memory storage, as already established with spin cross-over systems [49]. In fact, the materials showing hysteresis are bistable; i.e., they can occur in two different states, depending on their history. Like spin cross-over systems, the nitroxide-based materials are characterized by the fact that the two forms are related by an intramolecular rearrangement, which means that they should not show any fatigability. Thus, they can in principle be used for an infinite number of cycles. Similar behavior was previously observed for metal dioxolene complexes [50–52].

Other systems based on pyrimidyl- and pyrazinyl-substituted nitroxides do not exhibit this molecular transition behavior, even if the same cyclic structure is observed.

V. ONE-DIMENSIONAL MAGNETS

One-dimensional magnets are easily formed by NITR radicals with transition-metal ions [8]. The most common geometry is depicted in Figure 5a, where an NITR radical connects two metal ions through its two oxygen atoms. However, several other types of chains are also possible, as shown in Figure 5b–d. It is clear that the tendency to form one-dimensional materials is associated with the fact that the NITR radicals can readily form μ-1,4 bridges, and that the metal ions, which in principle might provide a sixfold multiplicity, often actually have four coordination sites blocked by the ancillary ligands, like hfac$^-$ (which are needed in order to make the metals acid enough to coordinate the weak NITR ligands). In the geometry of Figure 5b the bridge connecting two metal ions is formed not by one NITR radical but by two, connected through a weak interaction.

In the geometry of Figure 5c the NITR radicals actually bind as μ-1,1 bridges with one oxygen atom, but then the second oxygen does not bind to a second pair of metal ions; rather, they give rise to only a weak interaction with a second NITR radical. Finally, in the geometry of Figure 5d a NITR radical is bound to one metal ion and hydrogen bridged to a water molecule that is in turn coordinated to another metal ion.

The simplest systems to analyze for their magnetic properties are those that are comprised of only diamagnetic metal ions, such as zinc(II) and yttrium(III) [29,53]. In this category we may include also the europium(III)

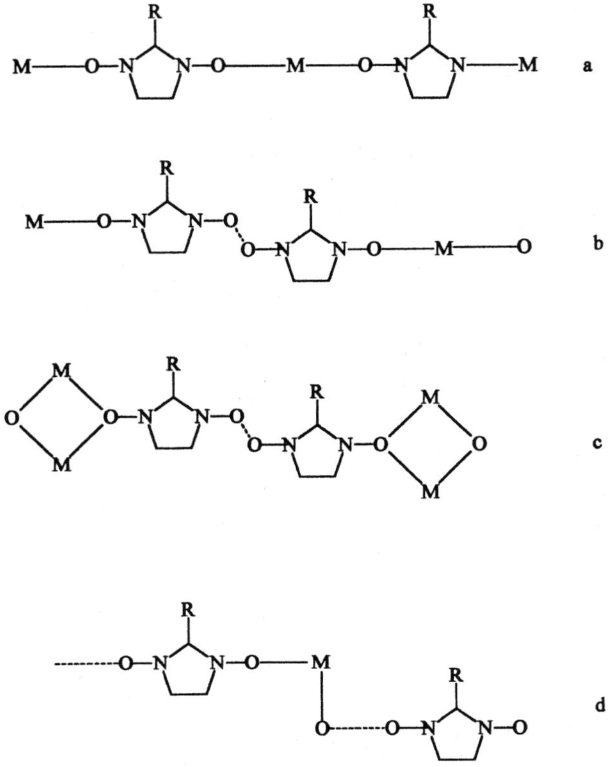

Figure 5 Types of one-dimensional metal–nitronyl nitroxide chains.

derivatives, because the metal ion has a $^7F_0$ nonmagnetic ground state, with a thermally populated magnetic state at ca. 300 K. At low temperature, the metal ion is in the nonmagnetic ground state and the magnetic properties are determined only by the radicals. Both the yttrium(III) and europium(III) derivatives have the structures of Figure 5a. They behave as one-dimensional antiferromagnets, due to the coupling transmitted by a superexchange interaction through the metal ions. The yttrium(III) ion is isoelectronic with bromide; therefore it is not difficult to figure out a possible mechanism of interaction, through superexchange mediated by the metal orbitals. In the case of zinc, the chains are of the type of Figure 5d, and the coupling between the radicals is made possible by the hydrogen bond with the water molecules bound to zinc(II).

An interesting feature of these materials is that they are excellent examples of one-dimensional Heisenberg antiferromagnets. In fact, the radicals are very isotropic (typical g-values are $g_x = 2.009$; $g_y = 2.007$; $g_z = 2.003$), which makes sure that the Heisenberg description that requires isotropic exchange is correct. In addition, the chains are very well shielded from each other, ensuring genuine one-dimensional behavior. This qualitative prediction has been confirmed by frequency-dependent NMR studies performed at room temperature [54]. In fact, in a one-dimensional Heisenberg antiferromagnet, the proton relaxation rate is expected to have a $v^{-1/2}$ dependence on the NMR frequency [55]. Deviations from this can be observed at frequencies corresponding to the interchain interactions. Experiments performed in the range 7–60 MHz did not show any deviation from ideal behavior, suggesting that the interchain interaction, J', is smaller than 4×10^{-3} cm^{-1}. Since the intrachain interactions are of the order of 10 cm^{-1}, the materials are textbook examples of one-dimensional magnetic materials, with an interchain-to-intrachain exchange ratio smaller than 10^{-4}.

When the metal ions are magnetic, two limiting behaviors can be anticipated, corresponding to nearest-neighbor ferro- and antiferromagnetic interactions, respectively. The former give rise to one-dimensional ferromagnets; the latter give rise to one-dimensional ferrimagnets, because the individual moments of the radicals and the metal ions will not compensate.

From a theoretical point of view, the magnetic properties of the ferro- and ferrimagnetic chains can easily be recognized by measuring the temperature dependence of the magnetic susceptibility. The χT product increases monotonically with decreasing temperature for a ferromagnet; it goes through a minimum and at lower temperature diverges for a ferrimagnet. In fact, at high temperature antiferromagnetic coupling dominates for the latter while χT initially decreases with temperature; but at lower temperature the product must diverge as in a ferromagnet. One must always recall that in a one-dimensional material χT diverges only at 0 K, because a one-dimensional material cannot sustain long-range order [56].

Ferromagnetic chains are formed by copper(II) ions, Cu(hfac)$_2$NITR, with the radicals occupying axial coordination sites around the metal ion. This geometry requires essentially orthogonal metal $(x^2 - y^2)$ and radical (π^*) magnetic orbitals and is consistent with ferromagnetic coupling, as observed. The temperature dependence of χT can be satisfactorily analyzed using the Bonner–Fisher approach. The largest coupling constants were found to be ca. 60 cm^{-1}. No system was observed to order magnetically above 4 K. In some cases, evidence of incipient antiferromagnetic transition was found below this temperature [8].

Ferrimagnetic chains, M(hfac)$_2$NITR, are formed by both manganese(II) and nickel(II), with coupling constants that are fairly large (200–300 cm^{-1}). The two NITR radicals that are in the octahedral coordination polyhedron of the metal ions can be either cis or trans to each other. In

$Mn(hfac)_2(NITPhOMe)$, where NITPhOMe is shown in Scheme 3, the radicals are in a cis geometric relationship, and the chains have a zigzag structure [57].

In the chains there are sizable stacking interactions between the phenyl rings and the π-orbitals of the hfac$^-$ ligands: these yield a helix structure, as shown in Figure 6. The compound crystallizes in the noncentric $P3_1$ space group, and it is interesting to notice that there is spontaneous resolution of the polar structure. The crystals show sizable second-order nonlinear optical properties, which are associated with the radicals [58]. Since these crystals order magnetically (as explained later), it would be interesting to investigate the interplay of natural optical activity and Faraday rotation at low temperatures. Indeed, this is one of the appealing features of molecular magnetism that is not easily observed in classic magnets.

The large antiferromagnetic coupling constant between manganese(II) and the NITR radicals determines a strong correlation of the spins in the chains at relatively high temperatures. In fact, at ca. 20 K the spins are on average correlated in segments comprising a few hundred spins, but the transition to three-dimensional magnetic order occurs only below 9 K. The driving force for the magnetic transition is provided by dipolar interactions [59–62]. Much of the understanding of the mechanism of crossover to three-dimensional order came from magnetic resonance experiments. In fact, EPR showed that on decreasing temperature the correlated spins tend to orient perpendicular to the chain, while NMR experiments [63] analogous to those described earlier for $Y(hfac)_3NITR$ provided evidence that the ratio of in-

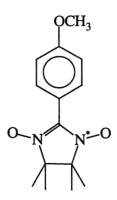

NITPhOMe

Scheme 3

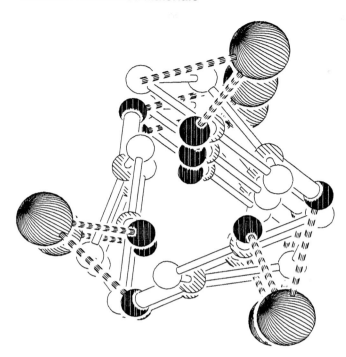

Figure 6 Scheme of the structure of Mn(hfac)$_2$(NITPhOMe), showing the trigonal helix. The large circles are manganese(II) ions; the small dark circles are oxygens; the small striped circles are nitrogens; the small open circles are carbons.

terchain to intrachain exchange interaction must be smaller than 2×10^{-6}. With this low ratio, the dipolar interactions, which can be estimated with some reasonable approximation, become important. The strong intrachain spin correlation at low temperature in these systems may trigger the magnetic transition.

Nuclear magnetic resonance spectroscopy also provided evidence of interesting spin dynamics, which had been previously observed for both one-dimensional ferro- and antiferromagnets, but not for ferrimagnets [64]. When the spins become correlated, they orient preferentially outside the chain, and there are segments of iso-oriented spins separated by domain walls, as shown in Figure 7. These domain walls can move along the chains like solitary waves, and are called solitons. The motion of these waves requires an excitation energy that depends on the exchange interaction. The frequency of the waves is therefore temperature dependent, yielding a maximum in the temperature dependence of the nuclear relaxation rate. A detailed theoretical

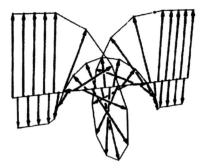

Figure 7 Solitons in one-dimensional ferrimagnets.

model was developed that satisfactorily reproduced the experimental data. Since the observation of solitary waves is associated with one-dimensional behavior, this is another proof of the good one-dimensional nature of the Mn(hfac)$_2$NITR complexes.

A surprising magnetic behavior is displayed by the Gd(hfac)$_3$NITR chains [65–68], which have structures similar to those of Y(hfac)$_2$NITEt. In fact, in many of these compounds χT decreases upon decreasing T, as in antiferromagnet, while in some it increases as in a ferromagnet. The latter qualitative behavior can be understood by remembering that weak ferromagnetic coupling is observed in simple Gd-NITR complexes, as explained in Section III. The antiferromagnetic behavior, on the other hand, cannot be reconciled with a magnetic chain in which only nearest-neighbor (nn) interactions are operative. In fact, in this case either ferro- or ferrimagnetic behaviors can be anticipated, or, in the limit of vanishingly small coupling, the Curie law should be followed.

The key to understanding of the properties of these chains comes from the observed antiferromagnetic coupling in Y(hfac)$_3$NITR. If the two radicals can interact through the bridging yttrium(III) ion, it may be reasonably postulated that a similar antiferromagnetic interaction is transmitted by the gadolinium(III) ions. Using this hypothesis and assuming a weak antiferromagnetic coupling between the gadolinium(III) ions, the systems become frustrated, because there is no possibility of orienting the spins in such a way as to respect all the interactions. Spin frustration must lead to a highly degenerate ground state, with properties that change abruptly for small perturbations. A pictorial description of the ground antiferromagnetic state can be achieved within an Ising formalism. This, of course, is not appropriate for a very isotropic system like the present one, which has radicals and gadolinium(III) ions, but it has the advantage of providing a simple starting model whose properties can be calculated with comparatively little effort. Assuming a weak ferromagnetic nn interaction, and larger next-nearest-neighbor (nnn) antiferromagnetic interactions, the ground state corresponds to a two-spins-up-two-spins-down configuration, as shown in Figure 8. It is apparent that in this way all the relatively strong antiferromagnetic interactions are respected, while only one out of two, relatively weak, ferromagnetic interactions are frustrated.

The problem of calculating the magnetic properties of these frustrated chains is still an open one for the Heisenberg case, because it requires lengthy and complex calculations. An intermediate approach, between the oversimplified Ising [69] and the too complex Heisenberg one, is that of using an XY hamiltonian [70] (Eq. 1):

$$H = \sum_{nn} J_{Gd-\text{rad}}(S_{Gd,x}S_{\text{rad},x} + S_{Gd,y}S_{\text{rad},y}) + \sum_{nnn} J_{Gd-Gd}(S_{Gd,x}S_{Gd,x}$$
$$+ S_{Gd,y}S_{gd,y}) + \sum_{nn} J_{\text{rad}-\text{rad}}(S_{\text{rad},x}S_{\text{rad},x} + S_{\text{rad},y}S_{\text{rad},y}) \quad (1)$$

Numerical solutions for the thermodynamic properties (magnetic susceptibility, specific heat, etc.) of this chain were found, which allowed an acceptable fitting of the magnetic data, showing how relatively small variations in the J parameters can explain the transition from overall ferro- to antiferromagnetic behavior. An important physical consequence of the model is the prediction that the preferred spin arrangement will not be "two-spins-up-two-spins-down," as in the Ising model, but a helical one. Of course the helices can be either right- or left-handed in nature, and the theory predicts that at low temperature a chiral structure should be observed. A confirmation of this came from specific heat measurements that showed an order–disorder transition at 2 K.

In addition to chains (in which the interest is mainly in the magnetic properties), a unique system in which self-complementary chains are formed was obtained [71], through chance, by reacting Mn(carboxylate)$_2$ with NITR

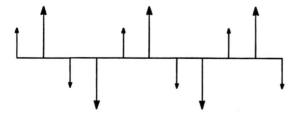

Figure 8   Preferred spin orientation in an Ising-type one-dimensional magnet with dominant next-nearest-neighbor interactions.

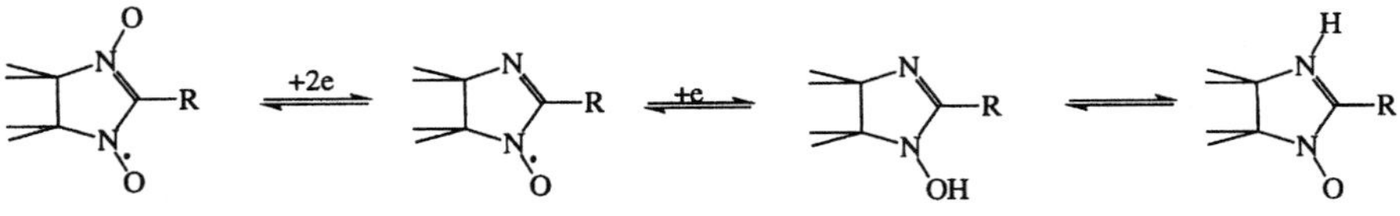

| NITR | IMR | IMHR |

Scheme 4

radicals. When Mn(pfpr = pentafluoropropionate)$_2$ is reacted with NITMe in chloroform solution, two different solids can be isolated, depending on the concentration of the reagents, the reaction temperature, and the presence of impurities in the solvent.

When the solids are allowed to precipitate slowly, a crystalline compound of formula [Mn(pfpr)$_2$]$_2$(NITMe)(IMHMe) is obtained, where IMHMe = 2,4,4,5,5-pentamethyl-4,5-dihydroimidazole-3-oxide. The process of formation of this compound is an interesting example of self-assembly, with concomitant reduction of the NITR radical according to Scheme 4.

The structure consists of infinite chains of connected octahedra in which the neighboring manganese ions are bridged by two carboxylates in μ-1,3 fashion. Scheme 5 shows two bridged metal ions, where L can be either NITMe or IMHMe. In fact, the two ligands are randomly distributed along the chains. The NITMe radical of one chain is hydrogen bonded to a IMHMe ligand of a neighboring chain. Each given chain is disordered, but the two neighboring chains are partial negative replicas of the first according to Scheme 6, where I and N stand for IMHMe and NITMe, respectively. In Scheme 6 we start from a random distribution on the sites of the central chain. The sites labeled X are not determined by the initial assignment to the central chain.

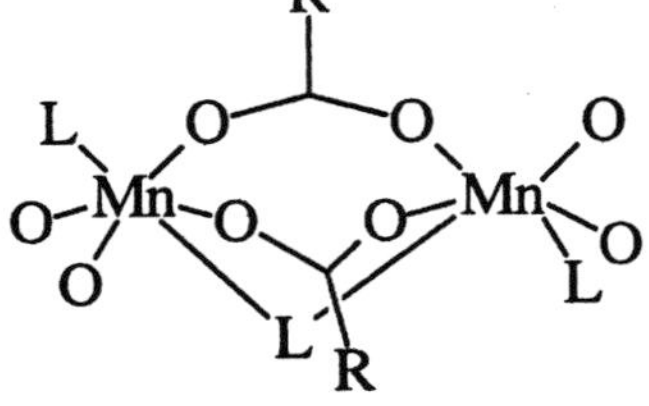

Scheme 5

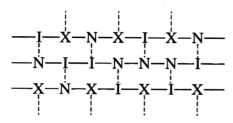

Scheme 6

VI. TWO-DIMENSIONAL MAGNETS

Most of the reported metal–radical complexes possessing an extended structure are only unidimensional because the nitroxy group is so weakly basic that it binds only to sterically congested metal centers carrying electron-withdrawing groups. The only exceptions have been reported by Iwamura's group, which, using trinitroxide ligands possessing a quartet ground spin state, succeeded in preparing manganese(II)-hexafluoroacetylacetonate derivatives that are bi- or three-dimensional and behave as a ferrimagnets [72].

Another successful strategy is based on bis-chelating nitronyl nitroxide ligands. Their manganese(II) complexes are bidimensional species in which the metal coordination sphere is free from ancillary electron-withdrawing groups and is fully occupied by paramagnetic bridging ligands only. These species, whose stoichiometry is $Mn_2(NITR)_3X^-$, are manganese(II) complexes of 2-(2-imidazolyl)-4,4,5,5-tetramethyl-4,5-dihydro-*1H*-imidazolyl-3-oxide-1-oxy (NITImH) and 2-(2-benzimidazolyl)-4,4,5,5-tetramethyl-4,5-dihydro-*1H*-imidazolyl-3-oxy (NITBzImH) [73].

A structural determination of $Mn_2(NITIm)_3ClO_4$ showed a planar honeycomb-like structure where metal ions are regularly bridged by bis-chelating nitroxide ligands, as shown in Figure 9. The interlayer space is occupied by the perchlorate anions. In this planar compound, only a OC-6-21 (*mer*) arrangement of the coordination spheres is present and Δ and Λ configurations alternate regularly. Since each crystallographically independent metal center keeps a specific chirality within a plane, the crystal as a whole has a definite chirality corresponding to one of the four diastereoisomers that could be built with alternating Δ and Λ building blocks. The two other possibilities involving metal centers of the same chirality would lead to a three-dimensional network.

The magnetic behavior of this compound at room temperature is characterized by a value of the magnetic susceptibility that is smaller than expected for independent spins, and is consistent with fairly large antiferro-

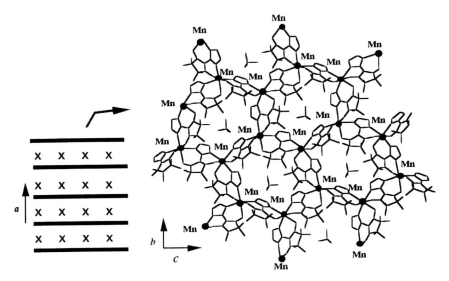

Figure 9   Structure of $Mn_2(NITIm)_3ClO_4$. (From Ref. 73.)

magnetic interactions between the manganese ions and the nitroxide radicals. As the temperature decreases, χ increases slightly and then decreases to reach a broad minimum at 260 K. On decreasing the temperature further, a sharp increase is observed that levels out gradually at low temperature, the values being field dependent. This magnetic behavior is better described by the variation of $1/\chi$ vs. T, which shows two different Curie–Weiss regimes, with θ values of 101 and 23 K at high and at low temperature, respectively. Among the series, only $Mn_2(NITIm)_3ClO_4$ exhibits such behavior. The closely related species $Mn_2(NITBzIm)_3ClO_4$ and $Mn_2(NITIm)_3BPh_4$ both exhibit only one type of Curie–Weiss behavior, with θ values of 105 and 71 K, respectively. In all compounds the magnetization saturates at values close to the theoretical value of 7 μ_B, corresponding to metal–nitroxide antiferromagnetic interactions within the layers.

Zero-field-cooled magnetization and field-cooled magnetization curves show the onset of spontaneous magnetization at low temperature. Confirmation of magnetic ordering below 1.4 K for $Mn_2(NITIm)_3ClO_4$ is indicated by the field dependence of the magnetization at 85 mK that exhibits a weak hysteresis loop whose coercive field and remnant magnetization are 270 Oe and 0.22 μ_B, respectively. Similar behavior is observed for $Mn_2(NITBzIm)_3$-ClO_4 and $Mn_2(NITIm)_3BPh_4$, for which remnant magnetization is observed below 40 K and 3.6 K, respectively. These results clearly document the presence of a weak ferromagnetic ground state resulting from spin canting.

Another important example of two-dimensional magnetic material containing a nitronyl nitroxide radical is $(NITMepy)_2^+Mn_2[Cu(opba)]_3(DMSO)_2 \cdot 2H_2O$, where $(NITMepy)^+$ is 2-(1-methylpyridinium-4-yl)-4,4,5,5-tetramethylimidazoline-1-oxyl-3-oxide; opba is *ortho*-phenylene bis(oxamato) [74]. There are two equivalent two-dimensional networks, each consisting of honeycomb layers that stack above each other in a graphitelike fashion. The two types of layers are practically orthogonal to each other and interpenetrate each other with a full interlocking of Mn_6Cu_6 hexagons. In this case the nitronyl nitroxide is a radical cation, which is not strongly involved in the exchange pathways that determine the magnetic properties of the compound. However, it may have a role in the determination of the structure through a template effect.

VII. THREE-DIMENSIONAL MAGNETS

In the previous sections we briefly mentioned some examples of nitroxide-based materials that order as ferro- or ferrimagnets. The compound with the highest critical temperature so far reported is $[\{Mn(hfac)_2\}_3 \cdot rad_2]$, where rad is shown in Scheme 7 [75]. The radical has a quartet ground state, with an exchange constant between the nearest-neighbor NO groups of $J = 167 \text{ cm}^{-1}$. The compound has a three-dimensional polymeric structure, formed by Mn–rad–Mn chains, in which the terminal NO groups of one radical bind two different manganese ions. The central NO group binds to a third manganese ion of a different chain, thus determining the three-dimensional polymeric structure. The material undergoes a transition to three-dimensional magnetic order at $T_c = 46$ K. The saturation value of the magnetization is $9\mu_B$, which can be justified by assuming that the spins of the manganese ions are up (3 × 5/2) and those of the radicals are down (2 × 3/2), to give a resultant spin $S = 9/2$ per formula unit. Therefore the compound can be described as a ferrimagnet. The critical temperature is relatively high, and it is certainly due to the fact that a three-dimensional network of relatively strong exchange interactions is built up.

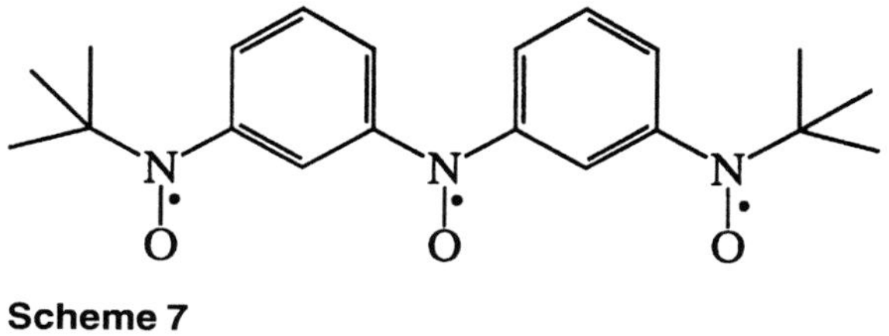

Scheme 7

The dramatic effects associated with low dimensionality is shown by the low critical temperatures observed in the $Mn(hfac)_2(NITR)$ series of compounds. These were mentioned earlier as examples of one-dimensional magnetic materials [57–62]. At low temperature they order as bulk ferrimagnets in a temperature range between 4 and 9 K. The correlations along the chains are very strong, but the interactions between chains are exceedingly weak, and the crossover to three-dimensional magnetic behavior is triggered by dipolar interactions. In fact, successful predictions of the critical temperatures were obtained by calculating the spin correlation within the chains, ξ, and the dipolar energies, using the following equation:

$$T_c \approx \xi(T_C)|E_{\text{dip}}| \tag{2}$$

The correlation length was estimated from the value of the effective spin of a manganese-radical pair:

$$\xi(T) = \frac{\chi(T)}{\chi_C(T)} \frac{S+1}{S} - \frac{1}{S} \tag{3}$$

where χ_C is the Curie susceptibility of a system with $S = 2$. The dipolar energies can be calculated by assuming a preferred spin orientation and calculating the interaction of a given spin with as many spins in the lattice as possible, given the long-range nature of the magnetic dipolar interaction. In practice, satisfactory results can be obtained by considering a sphere of ca. 7 nm around a given spin. With this approach, critical temperatures close to 10 K are calculated, in good agreement with experiment.

Another factor that can dramatically affect the critical temperatures of three-dimensional magnetic order is the magnetic anisotropy, as shown by the magnetic properties of $RE(hfac)_3(NITR)$ compounds, where RE is a rare-earth ion. Given the one-dimensional nature of these compounds, and given the weak coupling between the RE ions and the radicals, it would be reasonable to expect that transitions to three-dimensional magnetic order could be observed only at very low temperature. In fact, this is the behavior observed in the gadolinium derivatives. However, $Dy(hfac)_3(NITEt)$ orders as a weak ferromagnet at 4.3 K, a very high temperature when compared to the critical temperatures observed in ionic lattices of Dy^{3+}, where the metal–metal distances are much shorter than observed in molecular systems [76]. Our qualitative interpretation is that the large anisotropy of the ions is sufficient to give rise to relatively long-range correlations in the chains at relatively high temperatures, thus giving rise to magnetic order triggered by the interchain dipolar interactions. Relatively high critical temperatures were observed also for the Tb (1.7 K), Ho (3.2 K), and Er (1.2 K) derivatives [77].

ACKNOWLEDGMENTS

The financial support of the TMR network 3MD is gratefully acknowledged.

REFERENCES

1. Itoh K. In: Gatteschi, D, Kahn, O, Miller, JS, Palacio, F. eds. Magnetic Molecular Materials. NATO ASI Series E 198. Dordrecht: Kluwer, 1991, p. 67.
2. Gatteschi D. Adv. Mater. 1994; 6:635.
3. Miller JS, Epstein AJ. Angew. Chem. Int. Ed. Engl. 1994; 33:385.
4. Iwamura H. Pure Appl. Chem. 1993; 65:57.
5. Chiarelli R, Novak MA, Rassat A, Tholence JL. Nature. 1993; 363:147.
6. Allemand P, Khemani K, Koch A, Wudl F, Holczer K, Donovan S, Grüner G, Thompson JD. Science. 1991; 253:301.
7. Nakazawa Y, Tamura M, Shirakawa N, Shiomi D, Takahashi M, Kinoshita M, Ishikawa M. Phys. Rev. B. 1992; 4:8906.
8. Caneschi A, Gatteschi D, Rey P. Progr. Inorg. Chem. 1991; 39:331.
9. Pierpont CG, Buchanan RM. Coord. Chem. Rev. 1981; 38:45.
10. Eaton SS, Eaton GR. Coord. Chem. Rev. 1978; 26:807.
11. Eaton SS, Eaton GR. Coord. Chem. Rev. 1988; 8:29.
12. Inoue K, Iwamura H. J. Chem. Soc. Chem. Commun. 1994:2273.
13. Inoue K, Iwamura H. J. Am. Chem. Soc. 1994; 116:3173.
14. Inoue K, Hayamizu T, Iwamura H, Hashizume D, Ohashi Y. J. Am. Chem. Soc. 1996; 118:1803.
15. Forrester AR, Hay JM, Thomson RH. Organic Chemistry of Stable Free Radicals. New York: Academic Press, 1968, pp 180–246.
16. Rozantsev EG. Free nitroxyl radicals. New York: Plenum, 1970.
17. Keana JFW. Chem. Rev. 1978; 78:37.
18. Volodarsky LB, Reznikov VA, Ovcharenko VI, Synthetic Chemistry of Stable Nitroxides. Boca Raton, FL; CRC Press, 1994.
19. Breuer E, Aurich HG, Nielsen A. Nitrones, Nitronates and Nitroxides. New York: Wiley, 1989.
20. Ullman EF, Call L, Osiecki JH. J. Org. Chem. 1970; 35:3623.
21. Ullman EF, Osiecki JH, Boocock DGB, Darcy R. J. Am. Chem. Soc. 1972; 94:7049.
22. Ovcharenko VI, Lanfranc de Pauthou F, Pervukhina NV, Reznikov VA, Rey P, Sagdeev RZ. Inorg. Chem. 1995; 34:343.
23. Lanfranc de Panthou F. Thesis, Grenoble, France, 1994.
24. Luneau D, Rey P, Laugier J, Caneschi A, Gatteschi D, Sessoli R. J. Am. Chem. Soc. 1991; 113:1245.
25. Oshio H, Watanabe T, Ohto A, Ito T, Nagashima U. Angew. Chem. Int. Ed. Engl. 1994; 33:670.
26. Sutter JP, Fettouhi M, Michaut C, Ouahab L, Kahn O. Angew. Chem. Int. Ed. Engl. 1996; 35:2113.

27. Musin RN, Schastnev PV, Malinovskaya SA. Inorg. Chem. 1992; 31:4118.
28. Bencini A, Totti F, Daul CA, Doclo K, Fantucci P, Barone V. Inorg. Chem. 1997; 36:5022.
29. Benelli C, Caneschi A, Fabretti AC, Gatteschi D, Pardi L. Inorg. Chem. 1990; 29:4150.
30. Benelli C, Caneschi A, Gatteschi D, Pardi L, Rey P. Inorg. Chem. 1989; 28: 3230.
31. Andruh M, Ramade I, Codjovi E, Guillou O, Kahn O, Trombe JC. J. Am. Chem. Soc. 1993; 115:1822.
32. Lovesey SW. Theory of Neutron Scattering for Condensed Matter. Oxford: Clarendon Press, 1984.
33. Ressouche E. Thesis, Grenoble, France, 1991.
34. Zheludev A, Barone V, Bonnet M, Delley B, Grand A, Ressouche E, Rey P, Subra R, Schweizer J. J. Am. Chem. Soc. 1994; 116:2019.
35. Zheludev A, Bonnet M, Delley B, Grand A, Luneau D, Örström L, Ressouche E, Rey P, Schweizer J. J. Magnetism Magn. Mater. 1995; 145:293.
36. Akita T, Masaki Y, Kobayashi K. J. Chem. Soc. Chem. Commun. 1995:1861.
37. McConnell HM. Proc. R. A. Welch. Found. Conf. 1967; 11:144.
38. Ressouche E, Boucherle J-X, Gillon B, Rey P, Schweizer J. J. Am. Chem. Soc. 1993; 115:3610.
39. Boucherle J-X, Ressouche E, Schweizer J. Gillon B, Rey P. Z. Naturforsh. 1993; 48:120.
40. Caneschi A, Gatteschi D, Laugier J, Rey P, Sessoli R, Zanchini C. J. Am. Chem. Soc. 1988; 110:2795.
41. Luneau D, Risoan G, Rey P, Grand P, Caneschi A, Gatteschi D, Laugier J. Inorg. Chem. 1993; 32:5616.
42. Caneschi A, Chiesi P, David L, Ferraro F, Gatteschi D, Sessoli R. Inorg. Chem. 1993; 32:1447.
43. Caneschi A, David L, Ferraro F, Gatteschi D, Fabretti AC. Inorg. Chim. Acta. 1994; 217:7.
44. Lanfranc de Panthou F, Belorizky E, Calemzuc R, Luneau D, Marcenat C, Ressouche E, Turek P, Rey P. J. Am. Chem. Soc. 1995; 117:11247.
45. Laugier J, Rey P, Benelli C, Gatteschi D, Zanchini C. J. Am. Chem. Soc. 1986; 108:6931.
46. Caneschi A, Gatteschi D, Grand A, Laugier J, Pardi L, Rey P. Inorg. Chem. 1988; 27:1031.
47. Luneau D, Rey P, Laugier J, Friess P, Caneschi A, Gatteschi D, Sessoli R. J. Am. Chem. Soc. 1991; 113:1245.
48. Rey P, Luneau D. In: Kahn, O, ed. Magnetism: A supramolecular function. NATO ASI Series. Dordrecht: Kluwer 1996, pp 484, 431.
49. Kahn O, Martinez CJ. Science. 1998; 279:44.
50. Buchanan RM, Pierpont CG. J. Am. Chem. Soc. 1980; 102:4951.
51. Adams DM, Dei A, Rheingold AL, Hendrickson DN. J. Am. Chem. Soc. 1993; 115:8221.
52. Gütlich P, Dei A. Angew. Chem. Int. Ed. Engl. 1997; 36:2734.

53. Caneschi A, Gatteschi D, Sessoli R, Cabello CI, Rey P, Barra A-L, Brunel LC. Inorg. Chem. 1991; 30:1982.
54. Ferraro F, Gatteschi D. Mol. Phys. 1994; 83:933.
55. Boucher JP, Bakheit MA, Nechtschein M, Villa M, Bonera G, Borsa F. Phys. Rev. B. 1976; 13:4098.
56. Kahn O. Molecular Magnetism. Weinheim: VCH, 1993.
57. Caneschi A, Gatteschi D, Rey P, Sessoli R. Inorg. Chem. 1991; 30:3936.
58. Angeloni L, Caneschi A, David L, Fabretti AC, Gatteschi D, le Lirzin A, Sessoli R. J. Mater. Chem. 1994; 4:1047.
59. Caneschi A, Gatteschi D, Rey P, Sessoli R. Inorg. Chem. 1988; 27:1756.
60. Caneschi A, Gatteschi D, Renard J-P, Rey P, Sessoli R. Inorg. Chem. 1989; 28:1976.
61. Caneschi A, Gatteschi D, Renard J-P, Rey P, Sessoli R. Inorg. Chem. 1989; 28:3314.
62. Caneschi A, Gatteschi D, Renard J-P, Rey P, Sessoli R. Inorg. Chem. 1989; 28:2940.
63. Ferraro F, Gatteschi D, Sessoli R, Corti M. J. Am. Chem. Soc. 1991; 113:8410.
64. Ferraro F, Gatteschi D, Rettori A, Corti M, Mol. Phys. 1995; 85:1073.
65. Benelli C, Caneschi A, Gatteschi D, Laugier J, Rey P. Angew. Chem. Int. Ed. Engl. 1987; 26:913.
66. Benelli C, Caneschi A, Gatteschi D, Pardi L, Rey P. Inorg. Chem. 1989; 28: 275.
67. Benelli C, Caneschi A, Gatteschi D, Pardi L, Rey P. Inorg. Chem. 1990; 29: 4223.
68. Benelli C, Caneschi A, Fabretti AC, Gatteschi D, Pardi L. Inorg. Chem. 1990; 29:4153.
69. Pini MG, Rettori A. Phys. Rev. B. 1993; 48:3240.
70. Bartolomé, F, Bartolomé J, Benelli C, Caneschi A, Gatteschi D, Paulsen C, Pini MG, Rettori A, Sessoli R, Volokitin Y. Phys. Rev. Lett. 1996; 77:382.
71. Caneschi A, Gatteschi D, Melandri MC, Rey P, Sessoli R. Inorg. Chem. 1990; 29:4228.
72. Inoue K, Iwamura H. J. Am. Chem. Soc. 1994; 116:3173.
73. Fegy K, Luneau D, Ohm T, Paulsen C, Rey P. Angew. Chem. Int. Ed. Engl. 1998; 37:1270.
74. Stumpf HO, Ouahab L, Pei Y, Bergerat P, Kahn O. J. Am. Chem. Soc. 1994; 116:3866.
75. Inoue K, Hayamizu T, Iwamura H, Hashizume D, Ohashi Y. J. Am. Chem. Soc. 1996; 118:1803.
76. Benelli C, Caneschi A, Gatteschi D, Sessoli R. Adv. Mater. 1992; 4:504.
77. Benelli C, Caneschi A, Gatteschi D, Sessoli R. Inorg. Chem. 1993; 32:4797.

30

Metal-Dependent Regiospecificity in the Exchange Coupling of the Magnetic Metal Ion with the Free Radical Substituent on Pyridine Base Ligands: Approaches to Photomagnetic Molecular Devices

Noboru Koga
Kyushu University, Fukuoka, Japan

Hiizu Iwamura
National Institution for Academic Degrees, Nagatsuta, Japan

I. INTRODUCTION

Whereas *o*- and *p*-quinodimethanes and their analogs, e.g., the Thiele hydrocarbon, have closed-shell electronic structures, *m*-quinodimethane and the Schlenk hydrocarbon have triplet ground states [1]. The same is true for the corresponding dicarbenes; whereas the "*p*-phenylenebis(phenylcarbene)" has a low-spin ground state [2a], the *m*-isomer is a high-spin quintet species in the ground state [2,3]. In the heteroatom-centered diradicals, *p*-benzo-quinonediimine *N,N'*-dioxide (*p*) [4a] and *m*-phenylenebis(*N-tert*-butylaminoxyl) (*m*) are obtained as stable solids with singlet and triplet ground states, respectively [4]. This regiospecificity in the exchange interaction between the two radical centers at the benzylic positions is derived from differences in the phase of the spin polarization of the π-electrons on the benzene ring and has served as an important guiding principle for designing super-high-spin organic molecules as prototypes for purely organic ferromagnets [4,5].

Connection of n carbene units through the m-phenylene, 1,3,5-benzenetriyl, and other ferromagnetic coupling units gives rise to various polycarbenes with $S = n$ ground states [3] in which the highest record of the series reported so far is $S = 9$ [6]. Efforts to increase numbers of the aligned spins have been hampered by the development of antiferromagnetic intra- and/or inter-chain interactions between the carbene centers assembled in high local concentration; chemical bonds appear to be formed in the extreme case [7].

Since metal ions serve as connectors for high-dimensional molecular architecture by bridging the lower-dimensional organic ligand molecules used as building blocks [8], construction of heterospin systems [9] from organic free radicals and metal ions by self-assembly appears to be one of the most promising design strategies for real molecular-based magnets [10]. Recently we reported successful synthesis of ferri/ferromagnets by using two and three *tert*-butylaminoxyl radical units ligated to bis(hexafluoroacetylacetonato)manganese [Mn(hfac)$_2$] [11]. In order to obtain further insight into the sign and magnitude of the exchange coupling between the metal ion and the radical centers, it is of particular interest to see if the regiospecificity found in organic π-diradicals like quinodimethanes (p vs. m) is also applicable to the interaction between free radical centers (S) and coordinated magnetic metal ions (M) that are separated by π-conjugated ligands as in S$-\pi-$L--M (p' vs. m' in Scheme 1). Furthermore, the sign and magnitude of the exchange interaction between the two components in metal-radical heterospin systems are expected to depend not only on the periodicity of the ligand π-orbitals, but also strongly on the orbitals occupied by the unpaired d electrons of the metal ion. For example, manganese(II)

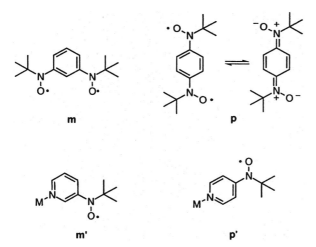

Scheme 1   Comparison of π-diradicals to radical-metal ion systems.

ions have unpaired electrons in the π-magnetic orbitals that can overlap with those of the ligand. On the contrary, the copper(II) ion in the octahedral ligand field has one unpaired electron in the $d_{x^2-y^2}$ orbital that has σ character [12,13], and therefore its complexes are expected to show different magnetic behavior from that of Mn(II) ion [13].

In the first part of this Chapter (Section II), we describe the structural and magnetic characterization of 1:2 complexes (S−π−L--M--L−π−S) of copper and manganese ions with pyridine or N-phenylimidazole bases each having a *tert*-butylnitroxide radical or a diazo group (a precursor to a triplet carbene) as a substituent on the ring and discuss what kind of regiospecificity is operative in determining the sign of the exchange coupling between the magnetic metal ions and the free radical center and whether or not the specificity will depend on the magnetic metal ions. We then develop a strategy for assembling the carbene centers into a rigid polymeric metal complex (-L−π−S−π−L--M-)$_n$ by using coordinatively doubly unsaturated paramagnetic metal ions with diazodi(4-pyridyl)methane (**1**) [14] as a bridging ligand, and analyze, on the basis of the foregoing knowledge, the formation of ferro- and ferrimagnetic super-high-spin chains after photolysis in Section III. The role of the diazo/carbene combination as photoresponsive magnetic coupler in our heterospin systems is summarized in Scheme 2.

II. 1:2 MODEL COMPLEXES, S−π−L--M--L−π−S

Complexes [15,16] of copper and manganese bis(hexafluoroacetylacetonates) {[M(hfac)$_2$], M = Cu(II) and Mn(II)}, ligated either with pyridines or N-phenylimidazoles having N-*tert*-butylaminoxyl radical substituents (NOPy and NOIm, respectively), were deemed to be suitable models and duly prepared. For investigating metal−carbene interactions through π-conjugated ligands, complexes of Cu(hfac)$_2$ and Mn(hfac)$_2$ with diazophenyl(4-pyridyl)methane (4DPy) in 1:2 molar ratios were also prepared (Scheme 3) [14a,b].

II.A. Electron Paramagnetic Resonance Studies of Free Ligands

Free ligands, 3- and 4NOPy and 4DPy, were prepared by the procedure reported previously [15a]. In electron paramagnetic resonance (EPR) spectra of the free ligands, 3NOPy and 4NOPy, the nitrogen hyperfine coupling with the aminoxyl nitrogen is reduced from a typical value of 15 G to 11.0 G and that with the pyridyl nitrogen is meaningful (2.1 G) in 4NOPy; those in 3NOPy are greater for the aminoxyl nitrogen (12.6 G) and smaller for the pyridine nitrogen (0.97 G). These values indicate that the unpaired electron at the aminoxyl radical center is effectively delocalized onto the pyri-

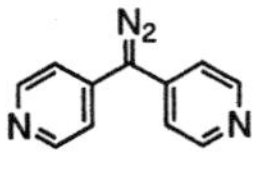

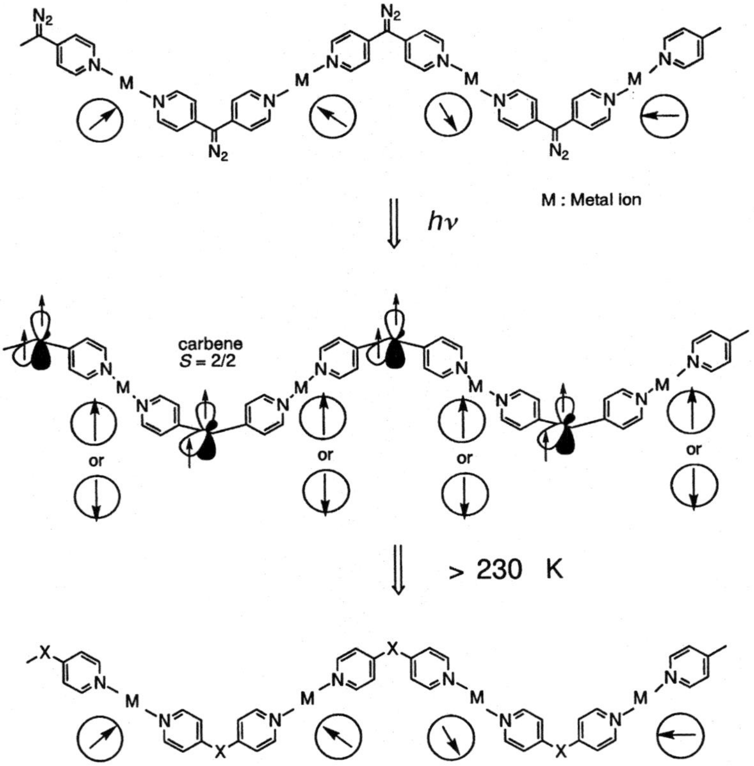

Scheme 2 Photochemical formation and thermal destruction of a ferrimagnetic chain.

dine ring. Without any ENDOR experiment, it must be logically assumed that the $p\pi$ orbitals at the ring nitrogens carry positive and negative spin densities in 4NOPy and 3NOPy, respectively. In an EPR spectrum after photolysis of 4DPy in frozen MTHF solution at cryogenic temperature, a set of signals ($|D/hc| = 0.418$ and $|E/hc| = 0.021$ cm^{-1}) characteristic of a triplet carbene and consistent with phenyl(4-pyridyl)carbene (4CPy) were observed [17].

Formulas

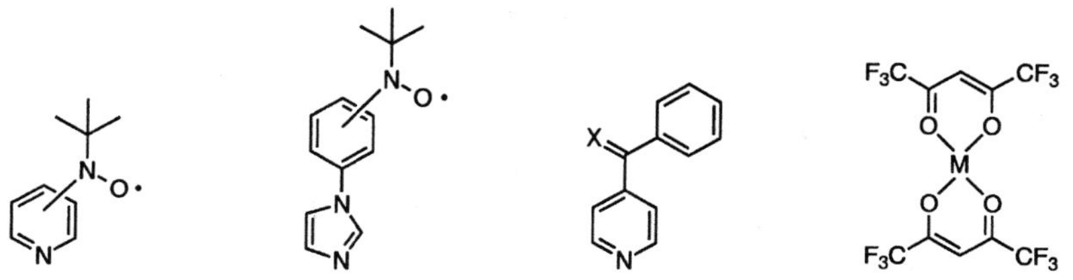

3- position : 3NOPy	3- position : 3NOIm	X = N$_2$: 4DPy	M(hfac)$_2$
4- position : 4NOPy	4- position : 4NOIm	X = •• : 4CPy	M = Cu(II) and Mn(II)

Scheme 3 Metal complexes and ligands and their abbreviations employed in this chapter.

II.B. Preparations and X-Ray Crystal and Molecular Structures of the 1:2 Complexes

1:2 Complexes, [M(hfac)$_2$(3- and 4NOPy)$_2$] and [M(hfac)$_2$(4DPy)$_2$], where M = Cu(II) and Mn(II), were prepared by mixing M(hfac)$_2$ in heptane solution and free ligands in CH$_2$Cl$_2$ solution in a molar ratio of one to two. The resulting precipitates were recrystallized from appropriate solvents to produce 1:2 complexes as greenish and orange crystals for copper and manganese complexes, respectively.

Five metal complexes [Cu(hfac)$_2$(3- and 4NOPy)$_2$], [Mn(hfac)$_2$-(4NOPy)$_2$], and [Cu- and Mn(hfac)$_2$(4DPy)$_2$] gave single crystals amenable to x-ray crystal structure analysis. As demonstrated in the ORTEP drawings of [Cu- and Mn(hfac)$_2$(4NOPy)$_2$] in Figures 1a and 1b, respectively, the molecular structures of the five complexes are similar; coordination geometries are elongated and/or distorted octahedra, and two pyridyl nitrogen atoms are coordinated to the metal ion in the trans configuration. Selected dihedral angles and bond lengths are summarized in Table 1.

As seen in the bond lengths around the copper ions in Table 1, the elongation axes lie along O(1)-Cu-O(1′)'s in three copper complexes. These indicate that the lobes of the magnetic orbital $d_{x^2-y^2}$, which is orthogonal to the d_z orbital of the Cu(II) ion, are directed toward the O(2)'s of the hfac units and N(1)'s of the pyridine units. In the manganese complexes, N(1)-Mn-N(1′)'s are the elongation axes d_z. A smaller dihedral angle between the planes of the N-tert-butylaminoxyl moiety and the pyridine ring would allow greater spin delocalization onto the pyridine ring. They are 10, 1.7, and 39° for [Cu- and Mn(hfac)$_2$(4NOPy)$_2$], and [Cu(hfac)$_2$(3NOPy)$_2$], respectively. The observed near coplanarity in the former two complexes suggests that

a)   b)

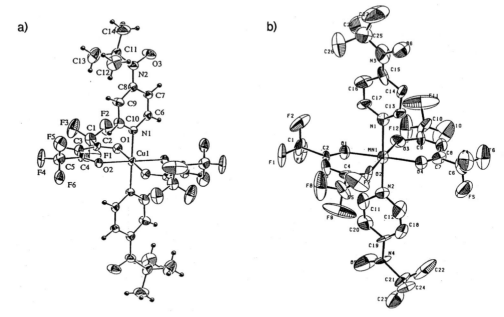

Figure 1   ORTEP drawing, (a) and (b), of the molecular structure of [Cu- and Mn(hfac)$_2$(4NOPy)$_2$], respectively.

Table 1   Selected Bond Lengths, Dihedral Angles, and the Nearest Distances from Neighboring Molecules

	Cu Complexes			Mn Complexes	
	4NOPy	3NOPy	4DPy	4NOPy	4DPy
M-O(1)(Å)	2.294	2.327	2.281	2.171	2.157
M-O(2)	2.037	2.019	2.008	2.143	2.154
M-N(1)	2.045	2.07	1.993	2.268	2.280
Dihedral angle (°)	10[a]	39[a]	58.9[b]	1.7[a]	0[b]
Distance (Å)[c]	5.85	2.45	3.86	3.31, 3.46	>5

[a]Between aminoxyl plane and pyridine ring.
[b]Between pyridine rings.
[c]The nearest distance from a neighboring molecule.

the electron spin of the aminoxyl radical center would be effectively delo-calized onto the pyridine ring.

The average distances between the radical centers of the neighboring molecules are 5.85 and >5 Å in the crystal structure of $[Cu(hfac)_2(4NOPy)_2]$ and $[Cu(hfac)_2(4DPy)_2]$, respectively, and as such the free radical centers may be regarded as well separated. In the crystal structure of $[Cu(hfac)_2$-$(4NOPy)_2]$, $[Mn(hfac)_2(4NOPy)_2]$, and $[Cu(hfac)_2(4DPy)_2]$, short distances from the adjacent complex molecules were observed. Especially, in the crystal structures of $[Cu(hfac)_2(3NOPy)_2]$ as represented in Figure 2, both aminoxyl groups in $[Cu(hfac)_2(3NOPy)_2]$ pack in close proximity to those of the neighboring molecules with a remarkably short distance of 2.45 and 2.74 Å for $N \cdots O$ and $O \cdots O$ (average distance between the two radical centers is 2.45 Å). This leads to the formation of a magnetically linear chain along the b-axis. As shown in Figure 2b, the N—O π orbitals of the adjacent aminoxyl radicals overlap with each other almost perfectly in a head-to-tail configuration. The observed short intermolecular distance between the aminoxyl groups and their relative geometry led us to predict that the crystalline compound would exhibit strongly biased magnetic properties [18], as discussed further in Section II.C.

II.C. Magnetic Properties for 1:2 Model Complexes

Magnetic properties of the model complexes in solid state were investigated by SQUID magnetrometry/susceptometry.

II.C.1. Copper and Manganese Complexes with 4NOPy and 3NOPy

The magnetic susceptibility data for copper and manganese complexes were obtained in the temperature range 2–300 K at a constant field of 100 or 800 mT and 100 mT, respectively. Temperature dependences of the molar paramagnetic susceptibility times absolute temperature $\chi_{mol}T$ for $[Cu(hfac)_2(4NOPy$ and $3NOPy)_2]$ and $[Mn(hfac)_2(4NOPy$ and $3NOPy)_2]$ are given in Figures 3a and 3b, respectively.

$[Cu(hfac)_2(4NOPy)_2]$. The $\chi_{mol}T$ value of 1.29 emu·K·mol^{-1} obtained at 300 K for $[Cu(hfac)_2(4NOPy)_2]$ is close to the theoretical one $\{\chi_{mol}T = 0.125 \times g \times 2 [3S(S + 1)] = 0.125 \times 2 \times 2 [3/2(1/2 + 1)] = 1.13$ emu·K·mol$^{-1}\}$ calculated for three isolated $S = 1/2$ spins in terms of the spin-only equation. As the temperature was decreased, the $\chi_{mol}T$ values (open circles in Figure 3a) gradually increased, reached a maximum at 36 K, and

(a)

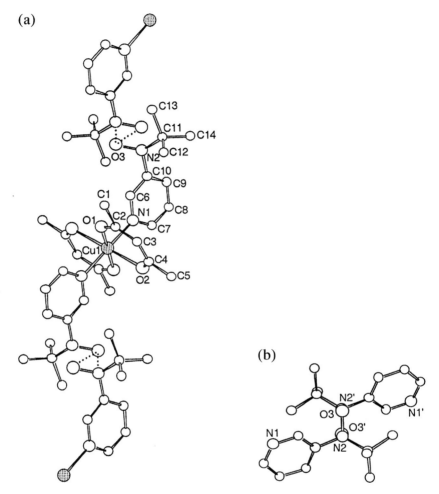

(b)

Figure 2 A stick-and-ball model for the crystal structure of [Cu(hfac)$_2$(3NOPy)] in which the nearest molecules along the *b*-axis are presented. Dotted balls and broken lines indicate copper atoms and the distances between the atoms: $r_{O-O} = 2.74$ and $r_{N-O} = 2.45$ Å, respectively.

rapidly decreased below 10 K. The maximum $\chi_{mol}T$ value of 1.69 emu·K· mol^{-1} is slightly smaller than the theoretical one (1.87) calculated for $S = 3/2$. The obtained $\chi_{mol}T$ vs. T curve for [Cu(hfac)$_2$(4NOPy)$_2$] was analyzed quantitatively on the basis of a linear three-spin model {S_1--S_M--S_2; $H = -2J(S_1S_M + S_MS_2)$} that was deemed appropriate from the x-ray molecular

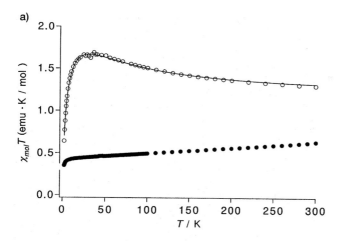

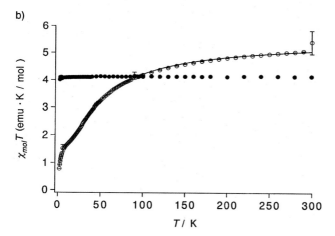

Figure 3 The plots of $\chi_{mol}T$ vs. T for crystalline samples of (a, top) [Cu(hfac)$_2$(4NOPy)$_2$] (○) and [Cu(hfac)$_2$(3NOPy)$_2$] (●) and (b), bottom) [Mn(hfac)$_2$(4NOPy)$_2$] (○) and [Mn(hfac)$_2$(3NOPy)$_2$] (●). The solid curves are theoretical ones.

and crystal structure analysis. Equation 1 for three spins, with $S_1 = S_2 = S_M = 1/2$, was applied and fitted to the observed $\chi_{mol}T$ vs. T plot for [Cu(hfac)$_2$(4NOPy)$_2$] by means of a least-squares method to give best-fit parameters: $J/k_B = 60.4 \pm 3.3$ K, $g = 2.048 \pm 0.0091$, and $\theta = -3.58 \pm 0.09$ K, where all symbols have their usual meaning. The fitted theoretical curve is represented by a solid curve in Figure 3a.

$$\chi_{mol}T = \frac{N\mu_B^2 g^2 T}{3k_B(T-\theta)} \cdot \frac{60 \exp{(3J/T)} + 6 \exp{(2J/T)} + 6}{4(4 \exp{(3J/T)} + 2 \exp{(2J/T)} + 2)} \quad (1)$$

[Cu(hfac)$_2$(3NOPy)$_2$]. The $\chi_{mol}T$ vs. T plots for [Cu(hfac)$_2$(3NOPy)$_2$] (filled circles in Figure 3a) showed quite a different temperature dependence. The $\chi_{mol}T$ value was ca. 0.6 emu·K·mol^{-1} at 300 K and gradually decreased to 0.43 at 10 K as the temperature was lowered. Since spin-only theoretical $\chi_{mol}T$ values calculated for $S = 1/2$, $3/2$, and a degenerate state among two doublets and one quartet are 0.37, 1.88, and 1.13 emu·K·mol^{-1}, respectively, the observed values in the range 0.4–0.6 emu·K·mol^{-1} are close to a theoretical one for $S = 1/2$. Whereas it is possible that the two aminoxyl radicals recombined chemically to a diamagnetic dimer [19] during the crystallization, a strong antiferromagnetic interaction between the two spins of the aminoxyl radicals canceling each other out is more likely, in good agreement with the x-ray crystal structure of [Cu(hfac)$_2$(3NOPy)$_2$].

The relatively large dihedral angle (39°) between the plane of *N-tert*-butylaminoxyl moiety and the pyridine ring is considered to weaken the interaction between the aminoxyl and the copper ion, whereas the short distance of 2.45 Å between one aminoxyl group and another one belonging to its nearest neighbor molecule strengthens the throughspace antiferromagnetic interaction between the two aminoxyl groups. Recently, we found strong intermolecular antiferromagnetic interaction of $J/k_B = -158$ and -239 K in nitronylnitroxide- and iminylnitroxide-substituted imidazole derivatives, respectively, in which the relative geometry of the π orbitals between two aminoxyl radical centers were ideal for overlap; these centers were separated by short distances of 3.49 and 3.54 Å [20]. As presented in Figure 2b, the observed geometry of the two aminoxyl radical centers in this work is closer to an ideal antiferromagnetic coupling than for the imidazole derivatives. Therefore, their interaction in [Cu(hfac)$_2$(3NOPy)$_2$] is predicted to be extremely strongly antiferromagnetic. Taking the dihedral angle of 39°, the relative geometry of the π orbitals and the distance of 2.45 Å into account, $-J_{inter}$ should be more negative than -239 K ($-J_{intra} \gg -J_{inter}$, where J_{intra} and J_{inter} are the intramolecular exchange coupling between the organic radical and copper ion and the intermolecular exchange coupling between the neighboring radicals, respectively. J_{intra} is expected to be negative, as discussed in Section II.D.2). Therefore, it appears in the $\chi_{mol}T$ vs. T plots as if the copper complex has no aminoxyl radical units.

[Mn(hfac)$_2$(4NOPy and 3NOPy)$_2$]. $\chi_{mol}T$ values of [Mn(hfac)$_2$(3NO-Py)$_2$] and [Mn(hfac)$_2$(4NOPy)$_2$] at 290 K were 4.12 and 5.04 emu·K·mol^{-1}, respectively. The latter value is slightly smaller than but nearly equal to a theoretical spin-only value of 5.12 for a degenerate state among four $S = 3/2$, $5/2$, $5/2$, and $7/2$ spins with $g = 2$. As the temperature was lowered, the

$\chi_{mol}T$ values of $[Mn(4NOPy)_2(hfac)_2]$ (open circles in Figure 3b) decreased gradually, reached an S-shaped plateau at ca. 20 K, and then decreased steeply below 10 K. A $\chi_{mol}T$ value approaching 1.90 emu·K·mol^{-1} at ca. 10–20 K is interpreted qualitatively in terms of the spin-only value for $S = 3/2$. In order to analyze the temperature dependence of the observed $\chi_{mol}T$ values more quantitatively, a linear three-spin model suggested by the x-ray crystal and molecular structure was assumed. Its spin Hamiltonian is expressed as $H = -2J(S_1S_M + S_MS_2)$. The theoretical equation was fitted to the experimental data by means of a least-squares method. The Weiss constant θ represented the intermolecular interaction expected from the short contact of 3.31 Å (Table 1) in the crystal packing of the complex. The best-fit parameters were $J/k_B = -12.4 \pm 0.1$ K, $\theta = -2.58 \pm 0.05$ K, and $g = 2.059$. The theoretical curve is included in Figure 3b.

The temperature dependence of the χ_{mol} values of $[Mn(hfac)_2(3NO-Py)_2]$ (filled circles in Figure 3b) is quite different from that of the 4NOPy complex just analyzed. The $\chi_{mol}T$ value of 4.12 at 300 K is close to 4.06 for $[Mn(hfac)_2Py_2]$ obtained under similar conditions, and remained constant in the temperature range 300–2 K. Although a bit smaller than a theoretical $\chi_{mol}T$ value of 4.38, it suggests $S = 5/2$. It appears as if the two 3NOPy ligands did not carry any spin; cancellation of the spins between the two 3NOPy's is suggested, as in $[Cu(hfac)_2(3NOPy)_2]$, although a conclusive interpretation of the $\chi_{mol}T$ value and its temperature dependence cannot be given until its molecular and crystal structure is determined to be similar to that of $[Cu(hfac)_2(3NOPy)_2]$.

II.C.2. Metal–Carbene Interactions

Magnetic measurements of ca. 0.5 mg of fine crystalline samples of $[Cu(hfac)_2(4DPy)_2]$ and $[Mn(hfac)_2(4DPy)_2]$ were carried out on a SQUID susceptometer/magnetometer before and after irradiation with light ($\lambda = 532$ and >400 nm from YAG laser and Xe lamp, respectively) through an optical fiber at ca. 5 K. The completion of the photolysis was estimated on average to be 80% and 25% for [Cu- and Mn(hfac)$_2$(4DPy)$_2$], respectively, by comparing the absorptivity at 2064 and 2060 cm^{-1} due to the diazo groups of the complexes before and after the photolysis. Temperature dependences of molar magnetic susceptibilities before and after irradiation of [Cu- and Mn(hfac)$_2$(4DPy)$_2$] are shown as $\chi_{mol}T$ vs. T plots in Figure 4.

[Cu(hfac)$_2$(4DPy)$_2$]. Before irradiation, $\chi_{mol}T$ values of a finely crystalline sample of $[Cu(hfac)_2(4DPy)_2]$ were nearly constant at 2–300 K, and the observed values of 0.38–0.40 emu·K·mol^{-1} in the whole temperature range are close to a theoretical spin-only value of 0.38 emu·K·mol^{-1} expected for paramagnetic samples of $S = 1/2$. When irradiated at 532 nm, the

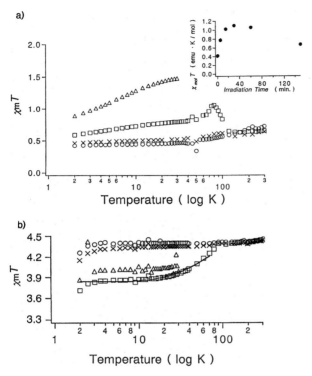

Figure 4   The temperature dependencies of products $\chi_{mol}T$ of "molar" magnetic susceptibilities times temperature per formula [Cu- and Mn(hfac)$_2$(4DPy)$_2$] of the crystalline samples: (a, top) Cu(hfac)$_2$(4DPy)$_2$] before (○) and after the irradiation of 0.5 hr (△) and 2.5 hr (□), and then after keeping at 300 K for one hour (X). The developments of the $\chi_{mol}T$ values at 5 K with irradiation time are given in the inset. (b, bottom) Mn(hfac)$_2$(4DPy)$_2$] before (○) and after the irradiation for 2 hr (△) and 18 hr (□), and then after keeping at 300 K for one hour (X). The solid curve is a theoretical one calculated on the basis of the linear three-spin system with optimized parameters given in the text.

$\chi_{mol}T$ values developed with irradiation time; values at 5 K increased for the first 0.5 hr and then decreased by further irradiation as indicated in the inset of Figure 4a. As the temperature was decreased from 100 K in the dark after irradiation for 0.5 hr, $\chi_{mol}T$ values gradually increased, reached a maximum of 1.3 emu·K·mol^{-1} at 60 K, and then started to decrease at 30 K (Figure 4a). The increase in the $\chi_{mol}T$ values as the temperature was decreased from 100 to 60 K must most probably be interpreted in terms of the ferromagnetic interaction between the photogenerated carbenes and the copper ions. The decrease of $\chi_{mol}T$ values observed in the temperature range 30–2 K and by a prolonged irradiation (>0.5 hr) is ascribed to intermolecular antiferromagnetic interaction between the photogenerated carbene centers. This antiferromagnetic interaction restricted us from estimating the magnitude of the ferromagnetic coupling by applying a theoretical linear three-spin model to the observed narrow $\chi_{mol}T$ vs. T profile.

[Mn(hfac)₂(4DPy)₂]. As the temperature was increased from 2 to 300 K in the dark after the photolysis, $\chi_{mol}T$ values of [Mn(hfac)$_2$(4CPy)$_2$] remained nearly constant at 3–20 K, gradually increased until ca. 90 K, and then exactly traced those before irradiation (Figure 4b). The behavior is characteristic of a three-spin system having antiferromagnetic interaction between the generated carbene centers and the manganese ion. The observed $\chi_{mol}T$ vs. T plots obtained between 3 and 60 K after irradiation for 18 hr were analyzed on the basis of spin Hamiltonian: $H = -2J(S_1S_M + S_MS_2)$ for the linear three-spin system S_1-S_M-S_2 revealed by the x-ray crystal and molecular structure. Photolysis factor F, Weiss temperature θ, and Lande factor g, were obtained by means of a least-squares method to give the best-fitting parameters: $J/k_B = -17.8 \pm 0.4$ K, $\theta = -0.011 \pm 0.007$ K, $g = 2.06 \pm 0.01$, and $F = 18.0 \pm 0.1\%$. The theoretical curve is included in Figure 4b.

II.D. Regiospecificity in the Exchange Coupling

Magnetic properties of related complexes [Cu(hfac)$_2$(3- and 4NOIm)$_2$] and CrTPP(3NOPy and 4NOPy) were investigated in a similar way, and the exchange-coupling parameters (J) between the manganese ions and the aminoxyl radicals were estimated from their $\chi_{mol}T$–T plots. Those values—in addition to those for [Cu(hfac)$_2$(4NOPy and 3NOPy)$_2$]—are summarized in Table 2.

II.D.1. Comparison Between the Copper and Manganese Complexes

The magnetic interactions between copper(II) and *N*-(4-pyridyl)-*N-tert*-butylaminoxyl radical are unique in that they are ferromagnetic and large in magnitude: $J/k_B = 60.4$ K. The couplings of the copper ions attached directly to the aminoxyl radicals via their oxygen atoms are well documented and typically antiferromagnetic, due to the overlap of the singly occupied orbitals of the metal ion and the free radical. Only when the oxygen atom of an aminoxyl radical is axially bound to a tetragonal copper(II) ion, a weak ferromagnetic coupling (~20 K) develops [10d,13]. The relative geometry between the magnetic orbital of the copper(II) ion and the π-orbital of the

Table 2 Exchange-Coupling Parameters (J) in Various Complexes

M	= M(4NOPy)	= M(3NOPy)	= M(4CPy)
Mn(II) $S = 5/2$	-12.4 K in Mn(4NOPy)$_2$(hfac)$_2$ -10.2 K in Mn(4NOPy)$_2$(acac)$_2$	? in Mn(3NOPy)$_2$(hfac)$_2$	-17.8K
Cu(II) $S = 1/2$	60.4 K in Cu(4NOPy)$_2$(hfac)$_2$ 58.6 K in [Cu(4NOPy)(hfac)$_2$]$_2$ 4.3 K in Cu(4NOIm)$_2$(hfac)$_2$	<-160 K in Cu(3NOPy)$_2$(hfac)$_2$ -2.7 K in Cu(3NOIm)$_2$(hfac)$_2$	$+$
Cr(III) $S = 3/2$	-77 K in Cr(TPP)(4NOPy)Cl -86 K in Cr(TAP)(4NOPy)Cl	12.3 K in Cr(TPP)(3NOPy)Cl 16.0 K in Cr(TAP)(3NOPy)Cl	

Plus and minus signs correspond to ↑ and ↓ spins at M, respectively. TPP and TAP are abbreviations for *meso*-tetraphenylporphyrin and *meso*-tetraanisylporphyrin, respectively.

pyridine ring as revealed by x-ray molecular structure analyses is illustrated in Figure 5a.

The magnetic orbital $d_{x^2-y^2}$, orthogonal to the elongated axis of octahedral Cu(II), is directed toward the two nitrogen atoms of the pyridine ligands and the oxygen atoms of the two hfac ligands in the 1:2 complexes (Figure 5a). As seen clearly in Figure 5, the magnetic orbital $d_{x^2-y^2}$ of Cu(II) and the p_π orbital at the nitrogen atom of the pyridine unit to which the $2p$ spin is polarized via the π-electrons on the pyridine ring from the aminoxyl radical center should be orthogonal. This orthogonality should be responsible for the ferromagnetic interaction. For the construction of superhigh-spin molecules in heterospin systems, these considerations lead to an interesting and useful conclusion that copper(II) ion in normal 5- and 6-coordination induces parallel spin on the ligating atom of conjugated π-ligands such as pyridine. There is one important precedent for such ferromagnetic coupling in the copper (II) complex with 2-(2-pyridyl)-4,4,5,5-tetramethyl-4,5-dihydro-1H-imidazolyl-1-oxyl in which both the imino and pyridyl nitrogens are attached to the copper ion as a bidentate ligand and the ferromagnetic coupling is a strong as >215 K [13]. In this case, delocalization of the spin by the contribution of an aminyl-nitrone resonance hybrid and proximity of the imino nitrogen atom with respect to the aminoxyl radical center, which facilitate the spin polarization at the imino nitrogen, appear to be responsible for the stronger ferromagnetic coupling.

a)

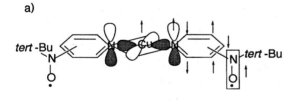

b)

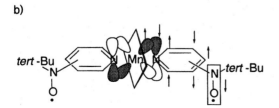

Figure 5 Scheme illustrating the intraction of the magnetic orbital of (a) the copper(II) and (b) the manganese(II) ions with the π orbitals at the nitrogen of pyridine ring in the 1:2 complexes.

When the molecular structure (Figure 1) of $[Cu(hfac)_2(4NOPy)_2]$ is compared with those of the corresponding manganese complexes, e.g., $[Mn(hfac)_2(4NOPy)_2]$, we note that both of the complexes have similar molecular structures, in which nitrogen atoms of two 4NOPys are coordinated to the metal ion in the trans configuration. However, when the J/k_B values of both the complexes are compared, the signs are opposite: positive for the former and negative for the latter with $J/k_B = 60.4$ and -12.4 K for $[Cu(hfac)_2(4NOPy)_2]$ and $[Mn(hfac)_2(4NOPy)_2]$, respectively. Furthermore, we note that the magnitude for the copper(II) ion is about five times larger than the one for the manganese complex. The torsion angles between the *N-tert*-butylaminoxyl and the pyridine ring are 10° and 1.7° for $[Cu(hfac)_2(4NOPy)_2]$ and $[Mn(hfac)_2(4NOPy)_2]$, respectively, and thus this does not explain the difference. Therefore the difference in the J/k_B values between the copper and the manganese complexes is considered to be derived from the magnetic d orbitals occupied by the unpaired electrons in the metal ions, $d_{x^2-y^2}$ for copper(II) and d_{xy}, d_{yz}, d_{xz}, $d_{x^2-y^2}$ and d_{z^2} for manganese(II) ions. The coordination geometry of the $[Mn(hfac)_2(4NOPy)_2]$ complex is a slightly elongated octahedron in which the axial ligands are the nitrogen atoms of pyridine groups. Therefore, the magnetic orbitals of manganese(II) ion can overlap with the p orbital at the nitrogen on the pyridine ring (Figure 5b) to produce an antiferromagnetic interaction, with the spin on nitrogen transmitted in the same way as in the case of the copper complex.

II.D.2.   Regiospecificity in Exchange Interaction with Respect to the Aminoxyl Radical Site on the Pyridine Ring

Now that the magnetic interaction between the organic $2p$ spins of the aminoxyl or carbene centers on the one hand and $3d$ spins of metal ions on the other in $M(hfac)_2(NOPy$ or $4CPy)_2$ complexes is explained by spin polarization of the π-electrons on the pyridine ring, as shown in Figure 5, such interactions are expected to depend on the site of the aminoxyl or carbene centers on the pyridine ring. Whereas the sign of the intramolecular exchange parameter for both Cu- and $Mn(hfac)_2$ complexes ligated with 4NOPy were determined to be positive and negative, respectively, conclusive results were not obtained for isomeric 3NOPy complexes because of interference with strong intermolecular antiferromagnetic interaction. The expected regiospecificity in the exchange coupling is clearly found in reported pairs of *meso*-tetraphenylporphyrinatochromium(III) complexes with the isomeric pyridylaminoxyls, Cr(III)TPP(3- and 4NOPy)Cl, and $Cu(hfac)_2(3$- and 4NOIm)_2$, although their molecular structure could not yet be determined by x-ray analysis. The chromium(III) ion in the porphyrin complexes has three unpaired electrons on the d_{xy}, d_{xz}, and d_{yz} orbitals, each of which is a π-magnetic orbital, and two (d_{xz} and d_{yz}) of which will be able to overlap with the π-orbital of the apical pyridine ring. Therefore, its magnetic coupling can be considered to be similar to that of the manganese(II) ion. The isomeric complexes Cr(TPP)(3- and 4NOPy)Cl show a striking contrast in their intramolecular magnetic interactions; the d^3 electrons of chromium(III)porphyrins and the unpaired $2p$ electrons on the ligating 3- and 4NOPy have been found by EPR and SQUID measurements (Figure 6a) to interact ferro- and antiferromagnetically to give quintet ($S = 3/2 + 1/2$) and triplet ($S = 3/2 - 1/2$) ground states, respectively. In the NOIm-Cu systems, similar magnetic behavior was observed; weak ferro- and presumably anti-ferromagnetic interactions (Figure 6b) between the copper ion and the aminoxyl radical in [Cu(hfac)$_2$(4NOIm and 3NOIm)$_2$], respectively; the attenuated interactions are ascribed to the weakened throughbond exchange coupling due to the presence of too many intervening π-bonds.

Thus their magnetic properties revealed that the sign of the exchange interactions between the radical center and the metal ion is controlled by the regiochemistry of the radical center on the pyridine [15] or N-phenylimidazole rings [16], which in turn is explained in terms of the spin-polarization mechanism [5] of the π-electrons (Figure 5).

III. FORMATION OF FERRO- AND FERRIMAGNETIC CHAINS

A new bridging ligand, diazodi(4-pyridyl)methane **1**, that has two kinds of functional groups, two pyridyl nitrogens and a diazo group, was designed and prepared [14]. The former would form a polymeric chain structure by the ligation with metal ions, and the latter would be photolyzed to produce a carbene unit that plays a dual role of a triplet spin source and a magnetic coupling unit (Scheme 2). The chain polymer complexes made of Cu(hfac)$_2$ and Mn(hfac)$_2$ with **1** are expected to give dilute paramagnets due to Mn(II) and Cu(II) ions, respectively, before photolysis of the diazo moieties and then to become ferro- and ferrimagnetic chains of considerable correlation length after photolysis, respectively. In the EPR spectrum after photolysis of diazo compound **1** under similar conditions, a set of triplet carbene signals with $|D/hc| = 0.434$ and $|E/hc| = 0.020$ cm^{-1} were observed [17]. The $|D/hc|$ value suggests that the delocalization of spin at the carbene center decreases as the number of pyridine rings increases: $|D/hc| = 0.404$ and 0.418 cm^{-1} for diphenylcarbene and 4CPy, respectively.

a)

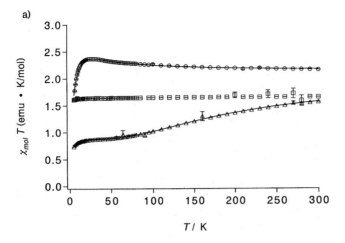

b)

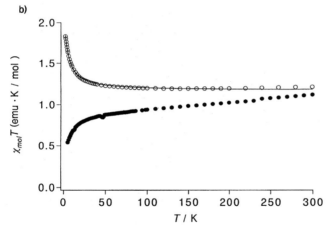

Figure 6 The plots of $\chi_{mol}T$ vs. T for microcrystalline samples of (a, top) CrTPP(4NOPy)$_2$ ($\triangle$), CrTPP(3NOPy)$_2$ ($\bigcirc$), and CrTPP ($\square$) and (b, bottom) [Cu(hfac)$_2$(4NOIm)$_2$] ($\bigcirc$) and [Cu(hfac)$_2$(3NOIm)$_2$] ($\bullet$). The solid curves are theoretical ones.

III.A.   Preparation and X-Ray Structures of the 1:1 Complexes of Copper(II) and Manganese(II) Ions

Solutions of the diazo compound **1** in CH$_2$Cl$_2$ and Cu(hfac)$_2$ in heptane— MeOH were mixed in a 1:1 molar ratio at room temperature. The resulting yellow-green precipitates were recrystallized from MeOH and MeOH—

CH$_2$Cl$_2$—benzene to give crystals **a** and **b** of [Cu(hfac)$_2$**1**] as yellow-green plates and dark green bricks, respectively [14a]. Similarly, manganese complex [Mn(hfac)$_2$**1**] [14b,21] was obtained by using anhydrous Mn(hfac)$_2$.

Selected bond lengths, bond angles, and dihedral angles between two pyridine rings for complex **a** of [Cu(hfac)$_2$**1**] as well as diazo compound **1** are listed in Table 3. The two pyridyl nitrogens of two different molecules of **1** are coordinated with a Cu(II) ion in a trans configuration to produce a one-dimensional chain structure parallel to the *ac* plane (Figure 7). Two trifluoromethyl groups in one of the two hfac ligands in Cu(hfac)$_2$ are disordered. The coordination geometry is a distorted octahedron; the bond distances are 2.031, 2.232, 2.178, and 2.020 Å for Cu-O and 2.012 and 2.063 Å for Cu-N, and the dihedral angles between the two pyridyl units through the diazomethyl unit and through the copper ion are 70.39 and 59.02°, respectively. Complex **a** has various distorted bond lengths and angles in the molecular structure, but the bond angle C3-C6-C9 of 126.4° (see Table 3) is close to 126.8° in reference compound [Cu(hfac)$_2$(4DPy)$_2$]. The distance between the chains is relatively short, and one of the nearest contacts related to any magnetic interaction is 3.94 Å, as indicated in Figure 7b. An x-ray crystal structure analysis for complex **b** of [Cu(hfac)$_2$**1**] was unsuccessful due to the limited quality of the crystals and the lability of the crystal due to included solvents of crystallization.

The x-ray crystal and molecular structure data of [Mn(hfac)$_2$**1**] [21] similarly obtained suggested that the manganese(II) ion is hexacoordinated with two pyridyl nitrogens from two different molecules of **1** in cis configuration and four hfac oxygen atoms (Figure 8). A helical one-dimensional chain was also established.

III.B. Magnetic Properties Before and After Irradiation of 1:1 Copper and Manganese Complexes with 1

Magnetic measurements of crystalline samples of [Cu(hfac)$_2$**1**] and [Mn(hfac)$_2$**1**] were carried out in a manner similar to those for model 1:2 complexes. The photolyses were estimated on average to be quantitative for both samples of [Cu(hfac)$_2$**1**]s and to be 68% complete for [Mn(hfac)$_2$**1**].

Temperature dependencies of the molar magnetic susceptibility χ_{mol} (gram susceptibility times the formula weight of the complexes) for [Cu- and Mn(hfac)$_2$**1**] in the range 2–300 K were measured at constant fields of 50 and 500 mT below and above 70 K, respectively. For both complexes, [Cu- and Mn(hfac)$_2$**1**], not only the temperature dependence of χ_{mol} but also the field dependence of the magnetization (*M*) at 3 and 5 K in the field range 0–1 T were investigated alternately for a given sample. The temperature dependencies of χ_{mol} are shown in the form of $\chi_{mol}T$ vs. *T* plots in

Table 3 Selected Bond Lengths, Bond Angles, and Dihedral Angles for **1** and [Cu(hfac)$_2$]$_2$·**1]a**

	Bond lengths (Å)		Bond angles (°)		Dihedral angles (°)	
1						
[Cu(hfac)$_2$]$_2$·**1]a**						
	Cu1-O1	2.031(8)	C3-C6-C3*	128.5(8)	C1C3C5-C1*C3*C5*	42.4(1)[a]
	Cu1-O2	2.232(7)	N1-Cu1-N4	178.5(1)	C1C3C5-C7C9C11	70.4(7)[a]
	Cu1-O3	2.178(8)	Cu1-N4-C9	171.9(6)	N4C10C8-N1C2C4	59.0(2)[b]
	Cu1-O4	2.020(8)	Cu1-N1-C3	177.0(8)		
	Cu1-N1	2.012(8)	C3-C6-C9	126.4(1)		
	Cu1-N4	2.063(8)				

[a]Dihedral angle between two pyridyl rings through diazo moiety.
[b]Dihedral angle between two pyridyl rings through copper ion.

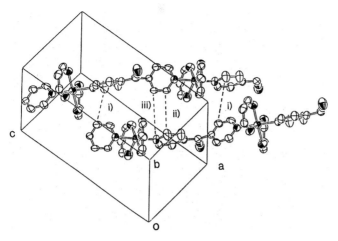

Figure 7 ORTEP drawing of the crystal structure of complex **a** of [Cu(hfac)₂**1**] in which the nearest two chains are presented. Broken lines indicate the distances between the atoms: (i) 3.94, (ii) 4.36, and (iii) 4.84 Å.

a)

b)

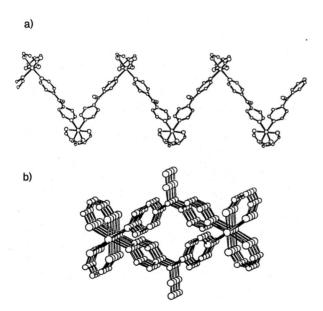

Figure 8 A stick-and-ball model for the crystal structure of [Mn(hfac)₂**1**]; (a) and (b) are the projection from the *a*- and *b*-axes, respectively.

Figures 9 and 10b for complexes **a** and **b** of [Cu(hfac)$_2$**1**], respectively. For complex **b** of [Cu(hfac)$_2$**1**], the field dependence is also shown in the form of M vs. H/T plots in Figure 11. The temperature dependencies of χ_{mol} are shown in Figure 12 for [Mn(hfac)$_2$**1**] before and after irradiation, and with subsequent annealing at 300 K.

III.B.1. $\chi_{mol}T$ vs. T Plot for Complexes **a** and **b** of [Cu(hfac)$_2$**1**]

Before irradiation, $\chi_{mol}T$ values for crystalline samples are nearly constant at 0.42 ± 0.08 and 0.38 ± 0.10 emu·K·mol^{-1} for **a** and **b**, respectively, in the whole temperature range. These values are consistent with $\chi_{mol}T = 0.375$ emu·K·mol^{-1} calculated for a dilute paramagnet of $S = 1/2$, indicating that the d electron spin of copper(II) ion is magnetically isolated.

As the photolysis of the diazo group of the samples proceeded, the $\chi_{mol}T$ values started to grow and gradually leveled off. For complex **b**, the development of the $\chi_{mol}T$ values at 5 K with irradiation time is demonstrated in Figure 10a. As the temperature was decreased, the $\chi_{mol}T$ values for complexes **a** and **b** after completion of the photolysis increased continuously, reached a maximum value of 7.43 and 18.2 emu·K·mol^{-1} at 14 and 3.0 K,

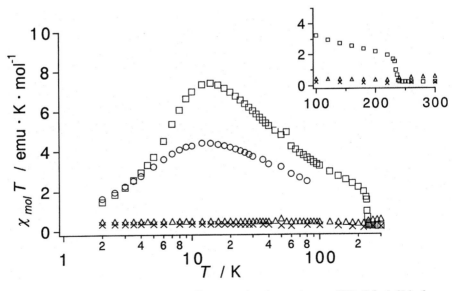

Figure 9 $\chi_{mol}T$ vs. T for the crystalline sample of complex **a** of [Cu(hfac)$_2$**1**] before ($\triangle$) and after the irradiation for 2 hr ($\bigcirc$) and 4 hr ($\square$), and then after keeping at 300 K for 15 min ($\times$). The inset shows the high-temperature region 100–300 K.

a)

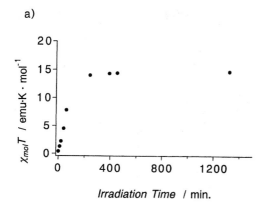

Irradiation Time / min.

b)

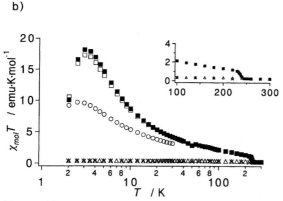

T / K

Figure 10 (a, top) Developments of the $\chi_{mol}T$ values for the crystalline sample of complex **b** of [Cu(hfac)$_2$**1**] at 5 K with irradiation. (b, bottom) $\chi_{mol}T$ vs. T of complex **b** of [Cu(hfac)·**1**] before (△) and after the irradiation for 2 hr (○), 4 hr (□), and 22 hr (■) and then after keeping at 300 K for 15 min (X). The inset shows the high-temperature region 100–300 K.

respectively, and started to decrease somewhat toward 2 K. The behavior is reversible up until 230 K, after which the $\chi_{mol}T$ values decreased sharply, as shown in the insets of Figures 9 and 10, respectively. In consecutive measurements on the same samples left at 300 K, the $\chi_{mol}T$ values again traced the horizontal lines obtained before irradiation. The data (4–200 K) of complex **b** after irradiation for 22 hr were analyzed by the method based on a model of the $S = 1/2$ and $S = 2/2$ Heisenberg ferromagnetic chain ($g = 2$)

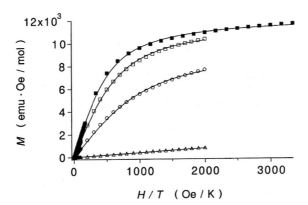

Figure 11 The plot of *M* vs. *H/T* for a crystalline sample of complex **b** of [Cu(hfac)$_2$**1**] at 5.0 K before ($\triangle$) and after irradiation for 1 hr ($\bigcirc$) and 22 hr ($\square$) and at 3.0 K after irradiation for 22 hr ($\blacksquare$). The solid curves are theoretical ones calculated on the basis of the Brillouin function with $S = 1/2$, 16.0, 26.0, and 33.6.

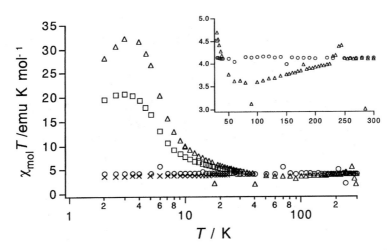

Figure 12 $\chi_{mol}T$ vs. *T* for the crystalline sample of [Mn(hfac)$_2$**1**] before ($\bigcirc$) and after the irradiation for 2 hr ($\square$) and 18 hr ($\triangle$), and then after keeping at 300 K for 1 hr ($\times$). The inset shows the high-temperature region 30–300 K.

for [Cu(hfac)$_2$**1**] to give $J/k_B = +66.8$ K [14a]. The maximum $\chi_{mol}T$ values for **a** and **b** after irradiation far surpassed the value of 11.88 emu·K·mol^{-1} for a $S = 3/2$ species that is expected for the ferromagnetically coupled set of Cu(II) ion ($S = 1/2$) and triplet carbene ($S = 1$) unit. The maximum $\chi_{mol}T$ values correspond to average correlation lengths of ca. 6 units at 15 K for **a** and ca. 15 units at 3 K for **b** along the chains. These correlation lengths should be taken as minimum estimates since $\chi_{mol}T$ values are not free from saturation at the 50-mT field employed for the measurement (see Figure 11), as well as antiferromagnetic interchain interaction (see Figures 9 and 10b).

III.B.2. *M-H/T* Plot for Complex **b** of [Cu(hfac)$_2$**1**]

Field dependencies of magnetization M for the same sample of complex **b** of [Cu(hfac)$_2$**1**] at 5 K before and after irradiation for 1 and 22 hr and at 3 K for 33 hr are reproduced in the form of M vs. H/T in Figure 11. Before irradiation the M values were nearly proportional to the applied field, in good agreement with an $S = 1/2$ paramagnet. When irradiation was started, the apparent M values increased and their curvature in the M vs. H plot depended on the irradiation time, supporting the idea that the average S values increased with increasing amounts of the generated carbene centers by the photolysis. In order to estimate the average number of coupled spins, the Brillouin function was applied to the magnetization data after completion of the photolysis at 3 K. The average spin quantum number S of the ferromagnetic chain under consideration is dependent on the correlation length, which in turn increases as the temperature is lowered and the applied field is increased. However, since the strength of the exchange coupling between the $3d$ spin of Cu(II) and the $2p$ spin of the carbene center in the present complex is estimated to be as strong as $J/k_B = 67$ K, an estimate of the apparent S values at 3.0 K was made by applying the Brillouin function describing the magnetization of a paramagnetic sample consisting of assemblies of fixed spin S as a function of the applied field strength. As in Figure 11, the best fits were obtained for a Brillouin function in which $S = 33.6 \pm 0.2$. These results correspond to the ferromagnetic coupling over ca. 23 units at 3 K along the chains.

III.B.3. Comparison Between Complexes **a** and **b**

The difference in the temperature profile between complexes **a** and **b** after complete photolysis of the diazo units might be caused by a stronger interchain antiferromagnetic interaction in complex **a** than in **b**. The complex **a** before photolysis has a short distance between the chains of 3.97 Å as revealed by x-ray analysis. Although no information of a crystal structure for complex **b** is available at the present stage, complex **b** has presumably longer interchain distances by containing the solvent molecules in the crystal and/or by cis configuration around the metal ions, as observed in [Mn(II)(hfac)$_2$**1**].

III.B.4. $\chi_{mol}T$ vs. T Plots for [Mn(hfac)$_2$**1**] [22]

Before irradiation, the $\chi_{mol}T$ values of a finely crystalline sample of [Mn(hfac)$_2$**1**] were nearly constant at 2–300 K and a value of 4.14 emu·K·mol^{-1} at 300 K, close to the theoretical spin-only value of 4.37 emu·K·mol^{-1} expected for paramagnetic samples of $S = 5/2$. When irradiated with an Xe lamp ($\lambda > 400$ nm), the $\chi_{mol}T$ values for both samples changed with irradiation time, with the gradual changes continuing even after 18 hr (Figure 12). Plots of $\chi_{mol}T$ vs. T for [Mn(hfac)$_2$**1**] showed a stark contrast to those for 1:2 model complex [Mn(hfac)$_2$(4CPy)$_2$] (Figure 4). As the temperature was decreased from 240 K, $\chi_{mol}T$ values gradually decreased to a shallow minimum of 3.58 emu·K·mol^{-1} at ca. 80 K, rapidly increased, reached a maximum of 32.2 emu·K·mol^{-1} at 3 K, and then decreased until 2 K. Abrupt change in the $\chi_{mol}T$ values was observed at 230 K, above which they traced a horizontal line before irradiation. Subsequent measurement of the $\chi_{mol}T$ values at 2–300 K of the same sample left at 300 K for an hour showed that the temperature-independent horizontal lines overlapped with those before irradiation. The data after irradiation for [Mn(hfac)$_2$**1**] was analyzed theoretically on the basis of a ferrimagnetic chain model consisting of $S = 5/2$ (classical) and $S = 2/2$ (quantum) spins to give a preliminary result of $J/k_B = -25.6$ K and $g = 2$ [22].

III.C.   Discussion

Taking into account the temperature independence of $\chi_{mol}T$ above 240 K and in consecutive measurements after leaving the sample at 300 K, we interpret the sharp change of the $\chi_{mol}T$ values at 240 K (insets of Figures 9, 10, and 12) as indicating the disappearance of the generated carbene centers by chemical reactions. It is worth noting that, whereas the carbene species generated in model complexes [M(hfac)$_2$(4DPy)$_2$] disappeared at 90 K, those in the crystal of [M(hfac)$_2$**1**] survived temperatures as high as 230 K. The observed stability of the carbene centers is novel and is most probably due to kinetic protection in the rigid crystal lattice of the metal complex.

The temperature-independent $\chi_{mol}T$ values before irradiation and after irradiation followed by annealing at >230 K show that the diazo ligand **1** and the chemically quenched carbene ligand are insulating magnetic couplers (the first and third rows in Scheme 2). The temperature dependence of the $\chi_{mol}T$ values at $T < 230$ K after photolysis of [Cu(hfac)$_2$**1**] and Mn(hfac)$_2$**1**] is contrasting and is best interpreted in terms of the formation of ferro- and ferrimagnetic chains made by alternating units of triplet di(4-pyridyl)carbenes on the one hand and of d^9 spin copper(II) and d^5 spin manganese(II) ion on the other, respectively. The increase in the $\chi_{mol}T$ values of [Cu(hfac)$_2$**1**] clearly demonstrates an increase in the correlation length of the ferromagnetic coupling along the one-dimensional heterospin chains with decreasing temperature. On the other hand, $\chi_{mol}T$ vs. T plots for [Mn(hfac)$_2$**1**] showed typical ferrimagnetic behavior. As the temperature is lowered, the antiferromagnetic coupling of the unpaired electrons of the manganese ion ($S = 5/2$) and the carbene center ($S = 2/2$) dominates thermal fluctuation of the spins and produces a minimum in the $\chi_{mol}T$ vs. T plot (as in a number of heterospin ferrimagnetic systems). The correlation of increasing length along the ferrimagnetic chain leads to an increase in $\chi_{mol}T$ values. The correlation length at 3 K is estimated to be on the order of over ca. 40 units in [Mn(hfac)$_2$**1**] from 32.2 emu·K·mol^{-1} of the $\chi_{mol}T$ value. The arrows at M in the second row of Scheme 2 should be up for copper(II) ion with $S = 1/2$ and down for manganese(II) ion with $S = 5/2$.

The observed signs of exchange coupling between the carbenes and the metal ions are consistent with those prescribed by model complexes.

IV. CONCLUSIONS

In heterospin systems containing isotropic copper(II) and manganese(II) ions, regiospecificity in the exchange coupling of p' vs. m' discussed in Section I is found to be operative and can be explained in terms of polarization of the π-electrons on the aromatic ring, as in p vs. m (Scheme 1). Additionally, whether the spin-containing π-orbital at the ring nitrogen atom of the conjugated ligand and the d orbitals containing unpaired electrons of metal ions at the coordination center have overlap or are orthogonal to each other is important in dictating the sign of the magnetic coupling (Figure 5). This metal-dependent regiospecificity in the exchange coupling of the magnetic metal ion with the free radical substituent on pyridine base ligands should serve as a powerful design strategy for constructing higher-dimensional extended heterospin systems.

It is concluded that the photochemical generation of triplet carbene centers induces polarization of the π-electrons on the pyridine rings, which in turn affects the coupling of magnetically isolated $3d$ spins of the Mn and Cu ions along the one-dimensional chains (Scheme 2). These results provide

Photochemical Production of Molecule-based Ferrimagnets

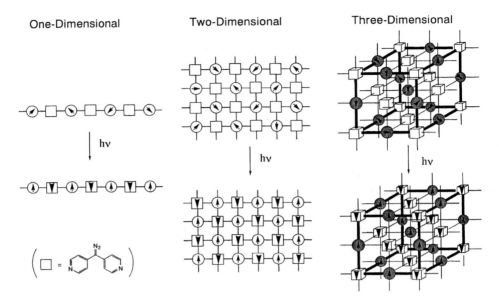

Figure 13   Schematic drawings of the one-, two- and three-dimensionally aligned spins consisting of magnetic metal ions and photoresponsive spins and/or magnetic coupling units.

the first example of the formation of extended *ferro-* and *ferri*magnetic chains containing photochemically generated $2p$ spins: a prototype of a molecular photomagnetic recording device in which only the irradiated microdomain of the nonmagnetic or weakly magnetic materials becomes strongly magnetic. Once the ordering of the spins is established for samples having higher-dimensional structure (Figure 13), photochemically generated heterospin systems will serve as molecular photomagnetic recording devices, just as photoresists become functional materials for printing circuit elements.

ACKNOWLEDGMENTS

This work was supported by Grant-in-Aid for COE Research "Design and Control of Advanced Molecular Assembly Systems'' (#08CE2005) from the Ministry of Education, Science, Sports, and Culture, Japan.

REFERENCES

1. Platz MS. In: Borden WT, ed. Diradicals. New York: Wiley, 1982, chap 5.
2. (a) Trozzolo AM, Murray RW, Smolinsky G, Yager WA, Wasserman E. J Am Chem Soc. 1963; 85:2526–2527. (b) Waserman E, Murray RW, Yager WA, Trozzolo AM, Smolinsky G. J Am Chem Soc. 1967; 89:5076–5078. (c) Mataga N. Theor Chem Acta. 1968; 10:372–376. (d) Itoh K. Chem Phys Lett. 1967; 1:235–238.
3. (a) Teki Y, Takui T, Itoh K, Iwamura H, Kobayashi K. J Am Chem Soc. 1986; 108:2147–2156. (b) Izuoka A, Murata S, Sugawara T, Iwamura H. J Am Chem Soc. 1987; 109:2631–2639. (c) Koga N, Iwamura H. Nippon Kagaku Kaishi. 1989; 1456–1462. (d) Fujita I, Teki Y, Takui T, Kinoshita T, Itoh K, Miko F, Sawaki Y, Iwamura H, Izuoka A, Sugawara T. J Am Chem Soc. 1990; 112: 4074–4075. (e) Nakamura N, Inoue K, Iwamura H, Fujioka T, Sawaki Y. J Am Chem Soc. 1992; 114:1484–1485. (f) Matsuda K, Nakamura N, Takahashi K, Inoue K, Koga N, Iwamura H. J Am Chem Soc. 1995; 117:5550–5560.
4. (a) Nakazono S, Koga N, Iwamura H. Submitted for publication in Angew Chem. (b) Chiarelli R, Ganbarelli S, Rassat A. Mol Cryst Liq Cryst. 1997; 108:455–478. (c) Calder A, Forrester AR, James PG, Luckhurst GR. J Am Chem Soc. 1969; 91:3724–3727. (d) Mukai K, Nagai H, Ishizu K. Bull Chem Soc Jpn. 1975; 48:2381–2382. (e) Ishida T, Iwamura H. J Am Chem Soc. 1991; 113:4238–4241. (f) Kanno F, Inoue K, Koga N, Iwamura H. J Phys Chem. 1993; 97:13267–13272.
5. (a) McConnell HM. J Chem Phys. 1963; 39:1910. (b) Iwamura H. Adv Phys Org Chem. 1990; 26:179–253. (c) Dougherty DA. Acc Chem Res. 1991; 24: 88–94. (d) Rajca A. Chem Rev. 1994; 94:871–893.
6. (a) Nakamura N, Inoue K, Iwamura H. Angew Chem. 1993; 105:900–901; Angew Chem Int Ed Engl. 1993; 32:872–874. (b) Matsuda K, Nakamura N, Inoue K, Koga N, Iwamura H. Bull Chem Soc. 1996; 69:1483–1494.
7. Matsuda K, Nakamura N, Inoue K, Koga N, Iwamura H. Chem Eur J. 1996; 2:259–264.
8. (a) Nierengarten J-F, Dietrich-Buchecker CO, Sauvage J-P. J Am Chem Soc. 1994; 116:375–376. (b) Lehn J-M. Supramolecular Chemistry. Weinheim: VCH, 1995. (c) Fujita M, Ibukuro F, Yamaguchi K, Ogura K. J Am Chem Soc. 1995; 117:4175–4176. (d) Nierengarten J-F, Dietrich-Buchecker CO, Amabilino DB, Stoddart JF. Chem Rev. 1995; 95:2725–2828. (e) Stang P, Olenyuk B. Angew Chem Int Ed Engl. 1996; 35:732–736.
9. (a) Kahn O. Molecular Magnetism. Weinheim: VCH, 1993. (b) Gatteschi D. Adv Mater. 1994; 6:635–645. (c) Miller JS, Epstein AJ. Angew Chem Int Ed Engl. 1994; 33:385–415. (d) Miller JS, Epstein A. J Chem Eng News. 1995; October 2:30–41. (e) Kahn O, ed. Magnetism: A Supramolecular Function. NATO ASI Series C. Dordrecht: Kluwer, 1996. (f)Turnbull MM, Sugimoto T, Thompson LK, eds. Molecule-Based Magnetic Materials. In: ACS Symposium

Series 644. Washington: ACS, 1996. (g) Gatteschi D. Curr Opin Solid State Mater Sci. 1996; 1:192–198.

10. (a) Eaton GR, Eaton SS. Acc Chem Res. 1988; 21:107–113. (b) Caneschi A, Gatteschi D, Laugier J, Rey P, Sessoli R. Inorg Chem. 1988; 27:1553–1557. (c) Caneschi A, Gatteschi D, Renard JP, Rey P, Sessoli R. Inorg Chem. 1989; 28:1976–1980. (d) Caneschi A, Gatteschi D, Sessoli R, Rey P. Acc Chem Res. 1989; 22:392–398. (e) Caneschi A, Gatteschi D, Rey P. Prog Inorg Chem. 1991; 39:3936–3941. (f) Burdukov AB, Ovcharenko VI, Ikorski VN, Pervukhina NV, Podberezskaya NV, Grigor'ev IA, Larionov SV, Volodarsky LB. Inorg Chem. 1991; 30:972–976. (g) Caneschi A, Dei A, Gatteschi D, J Chem Soc Chem Commun. 1992; 630–631. (h) Caneschi A, Chiesi P, David L, Ferraro F, Gatteschi D, Sessoli R. Inorg Chem. 1993; 32:1445–1453. (i) Stumpf HO, Ouahab L, Pei Y, Grandjean D, Kahn O. Science. 1993; 261:447–449. (j) Volodarsky LB, Reznikov VA, Ovcharenko VI. Synthetic Chemistry of Stable Nitroxides. Boca Raton, FL: CRC Press, 1994, chap 4.

11. (a) Inoue K, Iwamura H. J Am Chem Soc. 1994; 116:3173–3174. (b) Inoue K, Iwamura H. J Chem Soc Chem Commun. 1994; 2273–2274. (c) Inoue K, Iwamura H. Synth Met. 1995; 71:1791–1794. (d) Inoue K, Hayamizu T, Iwamura H. Mol Cryst Liq Cryst. 1995; 273:67–80. (e) Inoue K, Hayamizu T, Iwamura H. Chem Lett. 1995; 745–746. (f) Mitsumori T, Inoue K, Koga N, Iwamura H. J Am Chem Soc. 1995; 117:2467–2478. (g) Inoue K, Hayamizu T, Iwamura H., Hashizume D, Ohashi Y. J Am Chem Soc. 1996; 188:1803–1804. (h) Inoue K, Iwamura H. Adv Mater 1996; 8:73–76. (i) Oniciu DC, Matsuda K, Iwamura H. J Chem Soc Perkin II. 1996; 907–913. (j) Iwamura H, Inoue K, Hayamizu T. Pure Appl Chem. 1996; 68:243–252. (k) Inoue K, Iwamura H. Mat Res Soc Symp Proc. 1996; 413:313–320.

12. (a) Ishimaru Y, Kitano M, Kumada H, Koga N, Iwamura H. Inorg Chem. 1998; 37:2273–2280.

13. Luneau D, Rey P, Laugier J, Fries P, Caneschi A, Gatteschi D, Sessoli R. J Am Chem Soc. 1991; 113:1245–1251.

14. (a) Sano Y, Tanaka M, Koga N, Matsuda K, Iwamura H, Rabu P, Drillon M. J Am Chem Soc. 1997; 119:8246–8252. (b) Koga N, Ishimaru Y, Iwamura H. Angew Chem Int Ed Engl. 1996; 35:755–757. (c) Koga N, Iwamura H. Mol Cryst Liq Cryst. 1997; 305:415–424.

15. (a) Kitano M, Ishimaru Y, Inoue K, Koga N, Iwamura H. Inorg Chem. 1994; 33:6012–6019. (b) Kitano M, Koga N, Iwamura H. J Chem Soc Chem Comm. 1994; 447–448.

16. Ishimaru Y, Inoue K, Koga N, Iwamura H. Chem Lett. 1994; 1693–1696.

17. (a) Murray C, Wentrup C. J Am Chem Soc. 1975; 97:7467–7480. (b) Ono M. MS thesis, University of Tokyo, Tokyo, 1991.

18. Yamaguchi K, Okumura M, Maki J, Noro T, Namimoto H, Nakano M, Fueno T, Nakasuji K. Chem Phys Lett. 1992; 190:353–358.

19. Patai S, Rappoport Z, eds. Nitrones, Nitronates, and Nitroxides. Chichester, Eng.: Wiley, 1989, chap 4, p 346.

20. Akabane R, Tanaka M, Matsuo K, Koga N, Matsuda K, Iwamura H. J Org Chem. 1997; 62:8854–8861.

21. Recently, a new [Mn(hfac)₂·**1**] having trans coordination geometry was also prepared. Crystal structures for both manganese complexes were revealed by x-ray analysis at low temperature ($-100°C$). Furthermore, data of the magnetic measurements were improved by the replacement with argon ion laser and the prolonged irradiation for 100 hr. A part of these data are published in: Karasawa S, Sano Y, Akita T, Koga N, Itoh T, Iwamura H, Rabu P, Drillon M. J Am Chem Soc. 1998; 120:10080–10087.

22. Private communication from M. Dillion and P. Rabu of IPCMS. To be published elsewhere.

31

An Integrated Approach to Organic-Based Molecular Magnetic Materials

Paul M. Lahti
University of Massachusetts, Amherst, Massachusetts

I. INTRODUCTION

During the past thirty years, considerable effort has been expended in efforts to design new types of molecular-based magnetism. In order to achieve bulk magnetism, it is first necessary to design and then to synthesize appropriate spin-carrying units, whether they are simple paramagnetic atoms or more complex organic radicals. As described in previous chapters of this book, a number of strategies have been pursued by different research groups, including the syntheses of pure stable radicals, charge-transfer complexes, polymeric polyradicals, and chelated metal/radical complexes.

Among the strategies attempted, the most successful—at least in terms of achieving bulk magnetic effects—have been the charge-transfer and metal/radical chelate methods. Models of charge transfer and radical ion stacking have been described by Breslow [1–2], Torrance [3], Wudl [4], and Miller [5]. Miller and his collaborators have shown extremely promising magnetic properties in charge-transfer (CT) molecular solids such as decamethylferrocene/tetracyanoethylene (**1**) [5]. Metal–radical complexes have been explored by Gatteschi [6], Iwamura [7], and their respective coworkers, among others. A key principle of many such metal–radical complexes is to achieve antiferromagnetic (AFM) coupling between metal and ligand to obtain a ferrimagnetic array of spins (Scheme 1). For the purposes of this book, an arbitrary distinction has been drawn between metal–radical complexes,

661

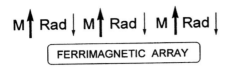

Scheme 1

where a metal ion (M^{n+}) is chelated by a stable spin-carrying radical, and magnetic molecular materials based upon metal-ligand chelates, where only the metal ion bears unpaired electron spin.

The bulk magnetic properties of molecular solids are dependent upon crystal structure and other forms of three-dimensional order, a difficult variable to control. In order to have a bulk magnetic moment, one requires ferro- or ferrimagnetic coupling in all three dimensions. In simple terms, the design of a magnetic molecular material can be no better than its "weakest link," i.e., its least favorable exchange interaction. The critical temperature of ferromagnetic order $T_c \sim S(S + 1)(J_{intra} \cdot J_{inter})^{1/2}$ for a material [8], where S is the spin quantum number and J are the intramolecular and intermolecular exchange (within molecules and between molecules). Due to this relationship, it is not unusual for a molecular solid to have strongly ferromagnetic (FM) or ferrimagnetic exchange coupling in one or two dimensions, but to have little or AFM exchange in another, leading to a net loss of moment in the bulk solid. As a result of this complexity, a number of features are desirable in an overall plan to discover new molecular magnetic materials:

> Magnetic exchange interactions should be maximized through the use of smaller molecules and/or multiple spin systems ($S > 1/2$) with high stability.
>
> Intermolecular attractions should be made as strong as possible, to encourage close contact between molecules in the solid and so strengthen the exchange interaction.
>
> Sites with high-spin density should be brought into close proximity through the use of specific intermolecular interactions such as hydrogen bonding, chelation, and strong dipolar attraction.
>
> Spin sites must exhibit good stability and resist the usual tendencies of organic radicals to dimerize or to react swiftly with atmospheric oxygen.

Finding the best combination of such effects with compatibility for magnetic properties is not straightforward, and this requires considerable "discovery-type" chemistry, in which numerous candidates of similar structure are screened for optimization through minor variations. *As a result, it is always desirable to test new candidate spin-carrier molecules and structure–property relationships for possible magnetic utility.* This chapter summarizes an integrated approach toward the discovery of desirable structure–property features for organic-based magnetic materials, going from isolated spin-bearing molecules to polyradicals and crystalline arrays.

II. THEORY: THE ROLE OF COMPUTATIONAL CHEMISTRY IN DESIGN

II.A. Methodology

Previous chapters in this book have described the qualitative parity arguments commonly being used to predict the ground state multiplicities of organic open-shell systems, such as those of Longuet-Higgins [9], Klein [10], Ovchinnikov [11–12], Sinanoglu [13], Borden and Davidson [14], and Tyutyulkov [15]. These have been the basis for much of the work in the area of non-Kekulé molecules, whether or not these systems were studied with an eye toward magnetic materials. Useful as these are, they give no quantitative insight as to the strength or weakness of exchange in one open-shell system versus another, or the variation as a function of conformation or geometric arrangement. Quantitative prediction requires computational quantum chemistry. The first section of this chapter describes much of the present state of the art for computational methodologies, including approaches for ab initio and density functional methods. The following subsection of this chapter shows that even a limited computational quantum chemical method can be of great use for semiquantitative estimation of structure–property exchange relationships.

Semiempirical quantum mechanical (QM) methods approximate a majority of the interactions in the molecular Hamiltonian expression, by substituting simple functions that are empirically parameterized to fit a set of known experimental properties. A subset of the QM interactions is retained to give some degree of approximation to the Schroedinger Hamiltonian. A variety of such approaches have been described during past thirty years. A very useful set of methods for evaluating exchange interactions are molecular orbital plus configuration interaction (MO–CI) formulations. Configuration interaction involves the mixing of various orbital occupancies within a set of MOs, in order to get a better description of the molecular electronic states. In many cases, the ground state (at least) for a molecule is quite well described by just one configuration. However, for many non-Kekulé molecules, the situation is more complex, and a state is not well described by a single orbital occupancy diagram. In this situation, CI (or some other post-

Hartree–Fock treatment) is required to get a minimally correct description of the state. For example, Scheme 2 shows that the configurational mixing in a state can increase greatly when the HOMO-LUMO gap is small, so long as symmetry constraints do not prevent mixing of configurations.

The ZINDO/S [16–17] and AMl [18] methods were shown during the 1980s to be highly effective [19–22] for computational treatment of diradicals and other non-Kekulé molecules. The former is specifically tailored for electronic spectral-state treatments, while the latter has a more generally applicable parameter set. The ZINDO/S methods has parameterization for use with single-excitation CI, while the AMl method was originally devised for use without CI. The actual "recipes" for using ZINDO/S-CI and AMl-CI methods for open-shell computations have been described elsewhere [19–23] and will not be given in detail here. AMl-CI has been used much more than the ZINDO-based method for non-Kekulé molecules, because numerical results for state energy gaps—e.g., singlet-triplet gaps—are quite similar for the two methods and because the AMl method has an internally consistent parameterization for both geometry optimization and CI state energy computation. It should be pointed out, however, that the ZINDO program [24] has a more robust set of options for the control of CI computations, including a general Rumer algorithm for generation of configurations.

A typical AMl-CI recipe for determining a state energy gap first requires geometry optimization for the separate states. The programs AMPAC

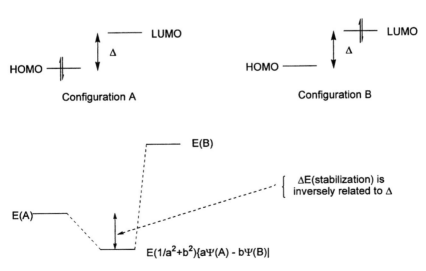

Scheme 2

[25] and MOPAC [26] can carry out a Hartree–Fock or SCF-MO-CI optimization on any given state. Sometimes, only one state geometry is determined, and is frozen for computation of other state energies. This approach can be dangerous in cases where there is considerable electronic difference between the states. For example, the geometries of the $^3B_2$ and $^1A_1$ states of oxyallyl (Table 1, **1**) are quite different [27], due to strong $C{=}O$ bond formation in the polar singlet state. Use of the triplet geometry for both states leads to overestimation of the stability of the $^3B_2$ state relative to the $^1A_1$ state—the best computations show these states to be nearly degenerate [27–28].

Having determined the state geometries, one next obtains the SCF-MO-CI state energies. Various schemes may be imagined to select active orbitals and levels of excitation (singles, doubles, etc.). Practical considerations for the programs used will in many cases limit the choice of these variables for larger molecules. This leads to considerable danger when comparing such limited CI results for a small molecule to those for a large molecule. The results for the larger molecule are inherently less trustworthy, because a larger active space and set of configurations is typically required to get as good a description as one would get with a smaller molecule. In other words, limited CI treatments are "size inconsistent." One can estimate the inconsistency by carrying out multiple MO-CI computations with different degrees of CI, to see how much the results vary. If variation is limited, the results can be used semiquantitatively, at least.

II.B. Selected Results for Model Systems

In order to calibrate semiempirical models, comparisons must be made to the limited number of singlet-triplet energy gaps and/or to related ab initio results. During the 1980s, limitations of computer capacity restricted ab initio computations to small diradicals such as oxyallyl (OA, **1**) [27], trimethylenemethane (TMM, **2**) [29–31] and/or relatively symmetrical species such as *meta*-benzoquinodimethane [28,32] (MBQM, **3**). The semiempirical MO-CI methodology proved to agree well with the few experimental and ab initio ground state multiplicities known at the time. One of the more impressive results was that for diradical **4**, which was known experimentally to have a triplet ground state, despite the parity-based expectations that it would be a ground state singlet [33,34]. The ZINDO/S-CI method gave a ground state triplet preference of 4.3 kcal/mol over the singlet state [19,33b], in qualitative accord with experiment. Table 1 summarizes crucial semiempirical benchmarks for **1–7** that gave confidence to extension of the methodology to unknown systems for predictive use. Overall, the predictive rec-

Table 1

	Semiempirical $\Delta E(T - S)$	Ab initio $\Delta E(T - S)$	Experimental $\Delta E(T - S)$
1	10.3^a (-0.3^b)	$1-2^d$	<0? (singlet ground state?)
2	19.0^a	17.5^e	16 kcal/mol^i
3	11.9^c	11.7^f	9.6^j
4	4.0^a	—	(triplet, $\gg +1.0$)^k
5	15.0^c	11.9^f	(triplet, $\gg +1.0$)^l
6	0.3^c	0.5^g	(triplet, near-degenerate?)^m
7	-10.1^c	-6.6^h	(singlet, $\ll -1.0$)^a

^{a}Ref. 19. ^{b}Ref. P. M. Lahti, unpublished results using separate AM1-CI optimization of $^1A_1$ and $^3B_2$ states. ^{c}Ref. 20. ^{d}Ref. 27. ^{e}Ref. 107. ^{f}Ref. 28. ^{g}Ref. 108. ^{h}Ref. 44. ^{i}Ref. 109. ^{j}Ref. 110. ^{k}Ref. 33. ^{l}Ref. 97. ^{m}Refs. 111 and 112. ^{n}Refs. 46 and 113.

ord for AMl-CI has been quite good since its original application to non-Kekulé molecules.

As part of our work on exchange in model systems, we studied a variety of dinitrene systems using AMl-CI. A number of benchmarks were evaluated, and several predictions were made [20]. The majority of the predictions were borne out, save that disjoint dinitrene systems such as **8** (Scheme 3) tended to be incorrectly predicted as having small high-spin preferences, whereas experiment eventually showed small low-spin preferences (see Section III.B.2). The latter result shows the semiquantitative nature of semiempirical methods. It is fairly straightforward to rank high-spin and low-spin energy gaps into strong low-spin preference, strong high-spin preference, and nearly degenerate. It can be much harder to differentiate more finely among systems with very small state energy gaps. *Trends* in semiempirical results should be viewed as more trustworthy than any isolated numerical result.

A similar finding was made for quinonoidal dinitrenes such as **9**, namely, that AMl-CI sometimes predicted a small triplet preference for these weakly coupled biradicals, whereas experiment eventually showed [34] a small singlet preference. Ichimura et al. [23] described a minor modification of the original AMl-CI computational recipe that rectified this inadequacy and that leads to proper state ordering for the quinonoidal systems as well as other systems. This example shows that improvement of semiempirical limited CI methodology occurs as new experimental results allow better calibration.

The successes of the AMl-CI methodology are much more numerous and compelling than its limitations. Predictions of the effects of heteroatom substitution upon simple connectivity-based analyses of open-shell systems have been of particular interest. The carbonyl group has been shown to be quite weak as an exchange linking unit, thanks in part to AMl-CI computations upon oxyallyl and parity-analogs such as **10** and **11** in Scheme 3 [22]. Ab initio computations confirmed that oxyallyl is a near-degenerate system, which requires little perturbation to cause its singlet state to drop below the triplet [27,35,36]. Dinitrene **11**, which should have a high-spin

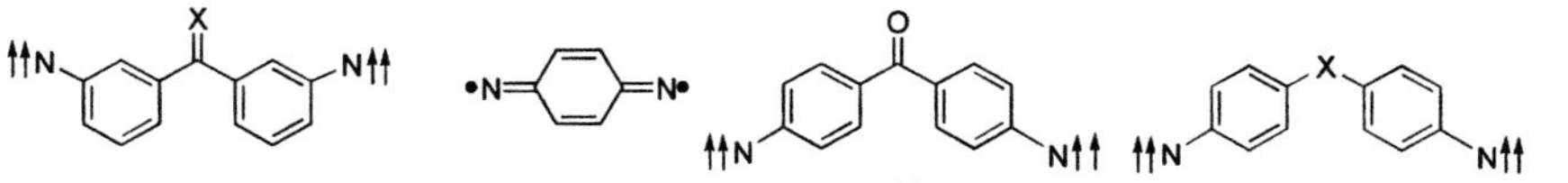

8	X=CH$_2$	**9**	**11**
10	X=O		
doubly-disjoint dinitrene			

12	X=O
13	X=S

Scheme 3

ground state by parity, appears to be a near-degenerate singlet in a crystal matrix, in accord with the picture of carbonyl as a poor exchange linker [37]. Two-electron/one-center linkers such as —O— and —S— (see **12–13** in Scheme 3) have been shown to give weak superexchange coupling that should be highly dependent upon relative torsion between the spin sites [38]. The experimental observation by Nimura et al. of a quintet state for sulfur-linked dinitrene **13** is supported by PM3-CI computations at conformationally twisted geometries (see Chapter 7, by Nimura and Yabe, in this book) [39]. Recently, ab initio multiconfiguration SCF computations have confirmed that *meta*-benzoquinomethane, **5** (Table 1), has nearly the same preference for triplet over singlet ($\sim$10 kcal/mol) as does the hydrocarbon analog **3** [28], in reasonable accord with early ZINDO/S-CI and AM1-CI computations [20,22]. Simple frontier MO arguments suggest **5** to have a smaller triplet-singlet gap than **3** due to an increase in HOMO-LUMO gap brought about asymmetrical heteroatom substitution, but the semiempirical and ab initio work agree in predicting the two to have similar gaps. Overall, AM1-CI and related semiempirical methods are excellent for finding systems in which parity effects dominate over heteroatom substitution effects, and vice versa.

Other themes more recently studied by semiempirical MO-CI methods involve the roles of hydrogen bonding for exchange coupling [40] and interesting or unusual effects upon "typical" parity expectations of exchange-coupling radical-ion spin-bearing sites that have greater or lesser degrees of torsion [41–42]. The role of heteroatoms in open-shell systems is not as straightforward as is suggested by simple connectivity arguments, so semiempirical MO-CI computations contribute considerably to the understanding and discovery of new structure–property themes [43]. A number of these results are described or referenced in Chapter 8 and will not be further described here.

Even simple hydrocarbon systems have yielded fascinating surprises upon inspection by semiempirical MO-CI methods. I recall sitting in Jerry Berson's office on the day in 1983 when he first proposed to me that 1,2,4,5-tetramethylenebenzene (Table 1, **7**, TMB) should be an excellent test of the Borden–Davidson disjointness criterion, since it is definitely a non-Kekulé system, but should be a ground state singlet according to parity treatments. ZINDO/S-CI computations quickly confirmed a robust 10-kcal/mol preference for the ^1A_g state over the $^3B_{2u}$ state in this planar system [19]. AM1-CI computations give a very similar result. Ab initio post-Hartree–Fock treatment by Borden and coworkers further supported the prediction, predicting a 5–7-kcal/mol gap [44]. Initial experimental ESR and UV-Vis studies were interpreted as suggesting TMB to have a triplet ground state [45], but by 1992 the experimental observation of the NMR spectrum of TMB in tandem with better computations of its expected UV-Vis spectrum proved TMB's singlet ground state nature [46–47]. Chapter 2 outlines the details of these studies. This is perhaps one of the best examples of the utility of a "survey"-type approximate method in selecting important target molecules for synthesis and study. Because of the speed of the semiempirical MO-CI methods, combined with their proven benchmarks relative to experimental and ab initio results, large numbers of diradicals can be rapidly evaluated for choice of the most interesting systems. The utility of the semiempirical MO-CI methods for rapid evaluation of potentially novel open-shell molecules remains very high, even as fast computers and faster algorithms make higher-level evaluation more time-efficient and more computationally precise.

II.C.   Selected Results for Polyradical Models

As already noted, limited CI methods are not size-extensive. For this reason, computations on larger polyradical systems can be problematic. Still, a limited discussion of the use of semiempirical MO-CI methods on polyradicals is in order, since very little has been done at higher levels of theory. This discussion is limited to polyradicals with pendant spin units; discussion of polyradicals with spins incorporated within the main backbone chain is given in Chapter 17.

A number of computations were carried out on oligophenoxyl-type systems [21–22]. These verified the parity-based expectation that conjugated oligoradicals can have ferromagnetically exchange-coupled high-spin states with spin quantum number $S \sim n$, where n is the degree of oligomerization. The magnitude of the high-spin to low-spin state preference decreases somewhat as a function of n, but this may be an artifact of the limited CI methodology. Results for some of these systems are reproduced in Figure 1.

In addition to the reasonably unsurprising verification of polymer parity effects, crucial findings were made with regard to the effect of conformation on exchange coupling. Pendant polyradicals—i.e., those having a continuous conjugated backbone to which radical sites are pendantly attached—are found to be very sensitive to backbone torsion. During the late 1980s and early 1990s, a number of polyacetylene systems with pendant polyradicals were synthesized (e.g., **14–17** in Scheme 4), some with spin yields of over 90% relative to the number of expected spins per monomer [48–54]. Not one of these systems showed intramolecular ferromagnetic exchange, despite parity expectations to the contrary. Our paper in 1991 noted that torsional deconjugation of spin sites can occur when backbone twisting was present, leading to effective isolation of spine sites [22]. This theme was further expanded in a 1993 summary noting that the helical nature of polyphenylacetylenes should preclude strong intramolecular exchange of

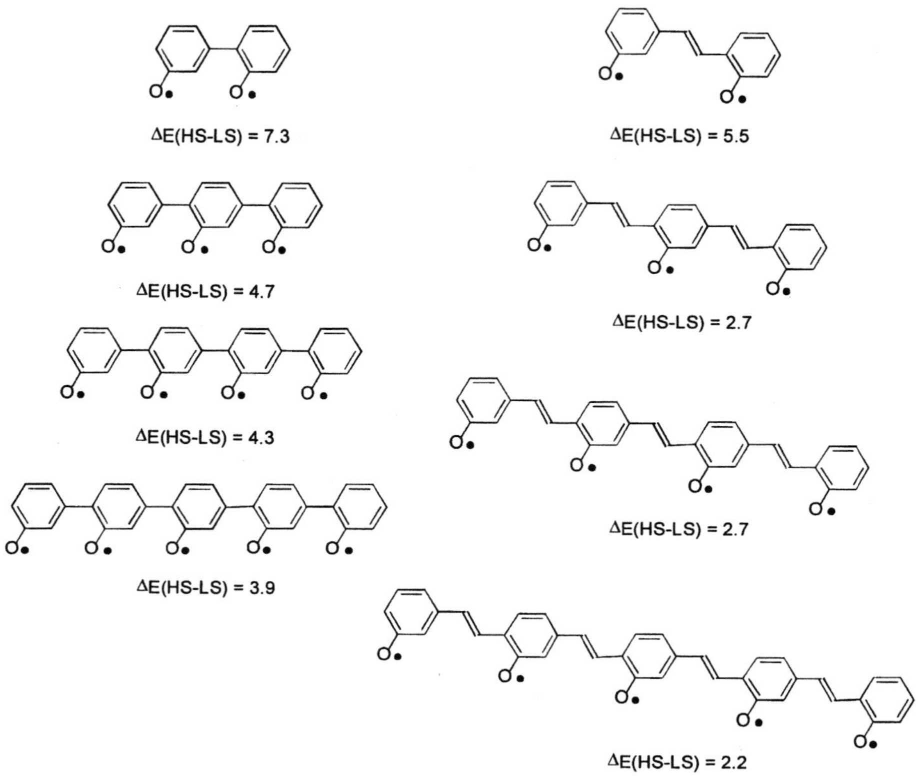

ΔE(HS-LS) = 7.3

ΔE(HS-LS) = 5.5

ΔE(HS-LS) = 4.7

ΔE(HS-LS) = 2.7

ΔE(HS-LS) = 4.3

ΔE(HS-LS) = 2.7

ΔE(HS-LS) = 3.9

ΔE(HS-LS) = 2.2

Figure 1 High-spin to low-spin gaps ($\Delta E_{HS\text{-}LS}$) for poly-p-phenylene and poly-p-phenylenevinylene based polyradicals, in kcal/mol, computed using the semiempirical SCF-MO-CI method of Ref. 22.

pendant spin sites [55]. Figure 2 shows how the AMl-CI high-spin to low-spin gap varies as a function of backbone torsion φ and pendant group torsion ψ for a dimer of a polyphenylacetylene model system. In the figure, the triplet-singlet gap decreases quite a bit faster for backbone torsion φ than for side-chain torsion ψ. These papers further noted the desirability of making systems with naturally planar backbones, such as poly(1,4-phenylene) or poly(1,4-phenylenevinylene). Shortly thereafter, this computationally based recommendation was confirmed by the first finding of a ferromagnetically exchange-coupled *pendant-spin unit* polyradical by Nishide's group, poly(2-(3′,5′-di-t-butyl-4′-oxyphenyl)-1,2-phenylenevinylene), **18** (Scheme 4) [56].

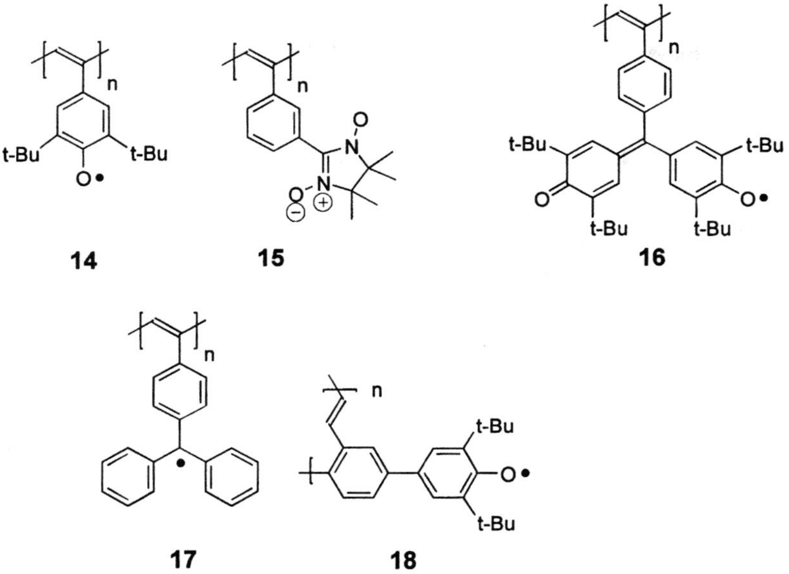

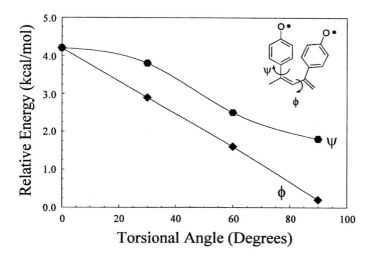

14 **15** **16**

17 **18**

Scheme 4

Figure 2 Relative energy gap from triplet ground state to singlet excited state, in kcal/mol, for a model polyphenoxylacetylene-based dimer, as a function of backbone torsion (ϕ) and pendant radical torsion (ψ). Computations used the AM1-CI OPEN(2,2) C.I. = 6 method of Ref. 22.

Other workers have also undertaken the use of computational methods for semiquantitative prediction of polyradical high-spin to low-spin state energy gaps. For example, Li et al. [57–58] have used effective valence-bond and unrestricted Hubbard models to investigate model non-Kekulé molecules and polyradicals. The method seems to show good promise for predicting properties of open-shell molecules without large computational expense. These workers have also looked at Rajca-type cyclic and backbone conjugated polyradicals, as well as starburst polyradicals. But, overall, computational methodology has not been extensively applied to oligoradicals with repeat symmetry. Three-dimensional exchange-packing effects have been treated, but for crystal motifs (see Chapter 20) rather than polyradical packing. A considerable body of investigation remains for polyradical computation, if appropriate methodology can be benchmarked by comparison to experiment.

III. PRACTICE: EXCHANGE IN MODEL ORGANIC SYSTEMS

III.A. Choice of Model Motifs

Earlier chapters of this book give considerable detail about experimental work on non-Kekulé molecules, the philosophical precursors to organic-based magnetic materials. Those chapters describe a number of model-system studies by others. Our group chose to study models made up of aryl-nitrene units connected by various conjugating linker groups —X—(:N—Ar—X—Ar—N:) for reasons that were detailed by a 1992 review [59]. These are briefly summarized as follows:

> Arylnitrene sites are readily generated by photolysis of aryl azides, which are readily synthesized by well-known methods.
> Arylnitrene electron spin resonance spectroscopy has been well understood since the 1960s, and a few dinitrene ESR studies had also been described by 1992.
> Arylnitrene sites are monovalent, and (despite possibilities for molecular rearrangement in the singlet state) are geometrically well defined by comparison to structurally more complex spin-bearing sites such as the parity-related aryl carbenes.

Chapter 7 describes much of the general background for studies of dinitrenes by us and by others; therefore the present chapter shall concentrate only on aspects of dinitrene studies that are specific to the our group. A previous review of much of our work on dinitrenes may also be found elsewhere [60]. For all of the work described in this subsection, ESR results derive from conditions where an appropriate diazide precursor molecule is dissolved in degassed 2-methyltetrahydrofuran solution and irradiated for <5 min with a Pyrex-filtered 1000-W xenon arc lamp at 77 K or below, unless otherwise stated. All ESR spectra were obtained with an X-band Bruker ESP-3000 spectrometer.

III.B. Nondisjoint Dinitrenes

III.B.1. Effects of Molecular Geometry on Zero-Field Splitting

Murata and Iwamura first described the construction of both disjoint and nondisjoint bis(arylnitrenes) **19–22** by linkage with acetylene and diacetylene [61–62], although the nondisjoint system *meta*-phenylenedinitrene **23** had been known [63] since 1967 (Scheme 5). Despite the elongation of conjugation going from systems **23** to **19** and **21**, all showed essentially the same ESR spectral features attributed to the quintet state. The zero-field splitting (zfs) of these systems appeared to depend solely upon the vector angle θ formed between the C—N bonds and not upon the conjugation length. Theoretical justification can be found for this inference in the case of the weakly coupled disjoint systems by using Itoh's previously described model for interacting triplets [64]. Briefly, Itoh's model postulates that, for a quintet system made up of interacting triplet spin sites, the quintet zfs $|D_q|$ is related to the triplet unit zfs $|D_t|$ by the following equation:

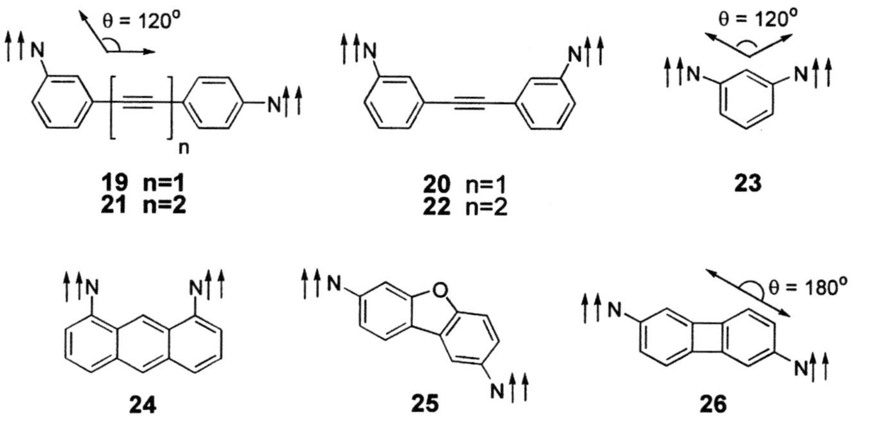

19 n=1
21 n=2

20 n=1
22 n=2

23

24

25

26

Scheme 5

$$\mathbf{D}_q = \frac{1}{6} (\mathbf{D}_t^a + \mathbf{D}_t^b) + \frac{1}{3} \mathbf{D}_t^{ab} \tag{1}$$

where the assumption is usually made that the exchange between electrons *within* each spin site is much stronger than the exchange between the spin sites (hence $\mathbf{D}^a$, $\mathbf{D}^b \gg \mathbf{D}^{ab}$). Since the one-center interactions $\mathbf{D}^a$, $\mathbf{D}^b$ for arylnitrene sites are very strong—leading to typical [65] zfs of 0.8–1.0 cm^{-1}—Itoh's model should be reasonable even in the nondisjoint case. The Murata–Iwamura results support this assertion, as do a large number of very similar nondisjoint ESR spectra obtained by us and others since that time [37]. For some time, it appeared that virtually all nondisjoint dinitrene spectra gave only minor variations in the ESR spectroscopy. In addition, all nondisjoint dinitrenes gave linear Curie law behavior in their ESR signal intensities as functions of reciprocal absolute temperature, suggesting that all were ground state quintet species in accordance with parity expectations described in Chapters 2 and 3 of this book.

In 1997, we published further experiments [66] confirming the critical role of the relative C—N vector angle θ in determining the zfs for dinitrenes. In this study it was noted that nondisjoint dinitrenes had angles θ of approximately 120°. If we assume that Itoh's model holds, it is not surprising that all such quintet ESR spectra were essentially the same, exhibiting major intensity peaks only at about 2900–3000 G at 9.3 GHz. Systems with vector angles θ ~ 120° have zfs of about 0.23 cm^{-1} [67], in good accord with the expectations of Itoh's model; this appears to be approximately the zfs for all systems with nondisjoint dinitrene spectra of this type, an example of which (for **8**) is shown in Figure 3.

We showed that the ESR spectra of nondisjoint systems **24–25** (Scheme 5), which have rigidly fixed angles θ of approximately 0° and 153°, respectively, have considerably different spectra from those of previously studied nondisjoint systems. Figure 3 shows the spectra for these dinitrenes [66] and for the related [68] system **26** (θ = 180°): all of the spectra are very different from that of a system with θ ~ 120°. Eigenfield lineshape fitting [69–71] of the quintet spectral lines for **24–26** yield experimental zfs parameters of $|D_q/hc| = 0.221$ cm^{-1} ($|E_q/hc| = 0.002$ cm^{-1}), $|D_q/hc| = 0.296$ cm^{-1} ($|E_q/hc| = 0.008$ cm^{-1}), and $|D_q/hc| = 0.233$ cm^{-1} ($|E_q/hc| = 0.002$ cm^{-1}), respectively. All of **24–26** showed ESR signal-intensity Curie law behavior consistent with the quintets' either being the ground states by >1000 cal/mol or being nearly degenerate. Using the mononitrene zfs $|D_t|$ obtained from the experimental spectra in each photolysis—0.66, 0.91 (1.00), and 0.71 for **24–26**, respectively—and applying Eq. (1) with angles

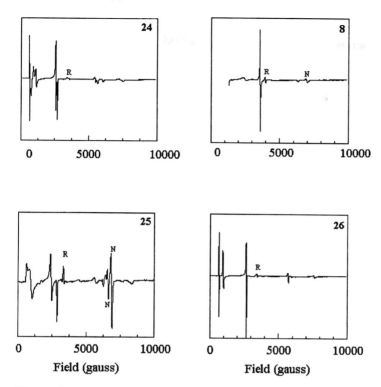

Figure 3 9.6 GHz ESR spectra for quintet dinitrenes **24, 8, 25, 26** in 2-methyltetrahydrofuran at 77 K, derived by >300-nm photolysis of appropriate diazide precursors. R = radical impurity, N = mononitrene from partial photolysis.

of $\theta = 0$, 153, and 180°, respectively, the predicted quintets $|D_q/hc|$ are 0.220, 0.293, and 0.237 cm^{-1}. The agreement with the experimental results is truly impressive. It is perhaps most intriguing to note the similarity of the spectra for **24** and **26**, which are structurally quite dissimilar but which are *expected* by Itoh's equation (1) to have very similar zfs $|D_q|$. This very interesting result demonstrates that, in general, the zfs of dinitrenes is not a straightforwardly useful yardstick for the estimation of average distance between unpaired electrons in dinitrenes, even though the zfs D-value is often used for this purpose in open-shell systems. *Because nondisjoint and disjoint dinitrenes are dominated by one-center interactions, their zfs tensors are* *generally dominated by dipolar angular interaction terms, rather than by distance terms.*

III.B.2. Effects of Conformation on Disjoint Dinitrene Zero-Field Splitting and Ground State Preference

Most of the conformational effects observed in quintet dinitrene ESR spectroscopy have been found for disjoint systems; all of the quintet spectral features found to date for disjoint systems have shown low-spin ground state Curie law behavior. Murata and Iwamura applied Eq. (1) to show that their spectra for dinitrenes **20** and **22** (Section III.B.1) were most consistent with a syn conformation of the nitrene units [62]. Nimura et al. [72] later showed what apparently was the first clearly defined example of geometric isomerism in the dinitrenes for system **27** (Scheme 6). Photolysis of the corresponding diazide **28** gave a relatively complex ESR spectrum with multiple peaks in the 1000–3000-G region. Different peaks showed different intensity versus temperature behavior, requiring the presence of at least two different putative spectral carriers having ESR-silent singlet ground states with thermally populated, ESR-visible quintet states. Given the positions of the spectral peaks, the two carriers were assumed to be quintet conformational isomers of **27**, examples of which are shown in Scheme 6. Peak assignments to specific conformers in principle is possible using Eq. (1) and the known mononitrene zfs obtained by partial photolysis of **28**. However, this analysis has not yet been reported for these systems.

   We have also observed [73] the production of (presumed) conformational isomers of pseudodisjoint dinitrenes **29–32** (Scheme 7). All show multiple ESR peaks in the 800–3000-G region, which like **27** show different Curie behavior of ESR peak intensities as functions of temperature. Figure 4 shows an example for dinitrene **32**, at 77 K in 2-methyltetrahydrofuran [74]. The quintet spectra of **29–32** show temperature behavior consistent with thermally populated excited states, as expected by their pseudodisjoint natures. As a result, the different peaks appear to be due to different conformers. We have not completed analysis of the peak positions for **29–32** by use of Eq. (1) but hope in the near future to be able to do so in order to

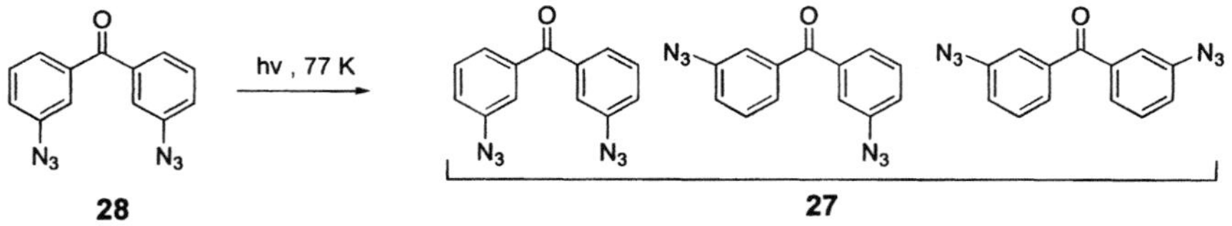

Scheme 6

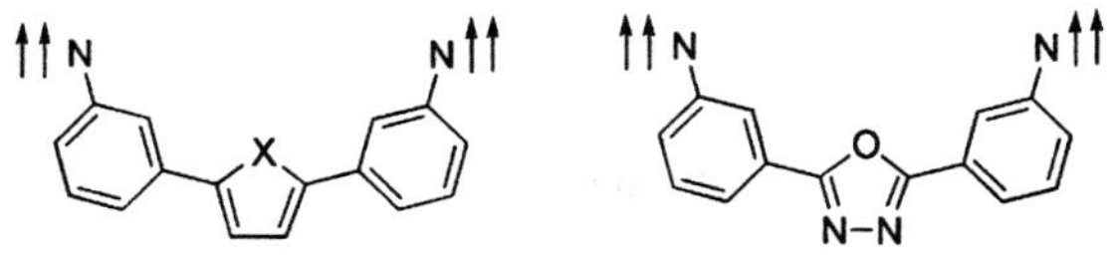

29 X=O
30 X= NH
31 X=S

32

Scheme 7

attempt the correlation of conformation with the singlet → quintet state energy $\Delta E(S - Q)$. This experimental correlation could be quite interesting, since computations to date are insufficiently precise to predict $\Delta E(S - Q)$ confidently as a function of different planar conformations for disjoint dinitrenes.

III.B.3. Effects of Heteroatom Substitution on Dinitrene Zero-Field Splitting and Ground State Preference

Until recently, no clear effect has been noted of pi-isoelectronic heteroatom substitution upon the quintet zfs of dinitrenes. As explained in the previous two subsections, molecular geometry and conformation are major determinants of dinitrene zfs. We recently found in collaboration with S. V. Chapyshev that photolysis of various di- and triazides of pyridine gave ESR spectra showing considerable effects of the pyridine nitrogen atom upon zfs of the corresponding dinitrene high-spin states. The following discussion is based upon results from 77 K frozen-phase photolysis of triazide **33** in 2-methyltetrahydrofuran (Scheme 8) [75].

The photolysis of **33** yields not only 2,4-dinitrenes of type **34**, but also 2,6-dinitrenes of type **35**. Dinitrene type **34** shows only minor effects of heteroatom substitution relative to a typical *meta*-phenylenedinitrene, with major quintet ESR peak at the typical position of about 2990 G, corresponding to $|D_q/hc| = 0.237$ cm^{-1}. However, isomeric dinitrene **35** has a major quintet peak at about 3300 G, corresponding to $|D_q/hc| = 0.284$ cm^{-1}, a major shift relative to other spectra for dinitrenes with vector angle $\theta = 120°$ (see Section III.B.1). The spectral features for **34** and **35** are shown in Figure 5. The quintet spectral change going from **34** to **35** cannot be due to a change in angle; hence it must be due to a change in the electronic spin distribution for the quintet state by heteroatom perturbation. So far as we know, this is the first evidence of an effect on dinitrene zfs that is due to an alteration in electron distribution on the aryl nitrene sites by the exchange linker group (in this case 2,6-pyridindiyl), rather than by geometric or conformational

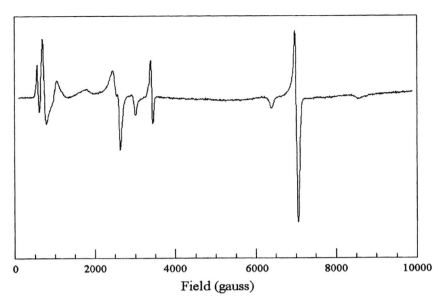

Figure 4   9.6 GHz ESR spectra for quintet dinitrene **32** in 2-methyltetrahydrofuran at 77 K, derived by >300-nm photolysis of appropriate diazide precursors. R = radical impurity, N = mononitrene from partial photolysis.

effects. The spectral assignments for these dinitrenes are supported by comparison with the spectra derived by photolysis of diazide model systems [75]. A number of "extra" peaks shown in Figure 5 are assigned to septet trinitrene **36**. All peaks for **34–36** show linear Curie law ESR-peak-intensity behavior consistent with a ground state or near perfect degeneracy. Similar peaks are also formed in photolysis of other 2,4,6-triazidopyridines [75]. The spectral analysis of **36** and related trinitrenes is ongoing.

Heteroatom effects on dinitrene ground state preference can be pronounced, even in pi-isoelectronic systems. Oxygen-linked dinitrene **37**

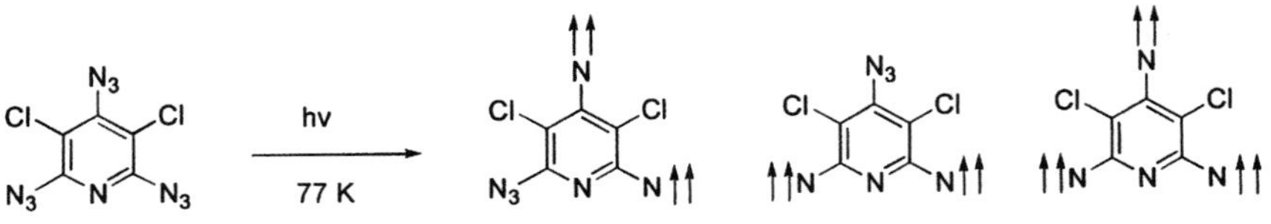

Scheme 8

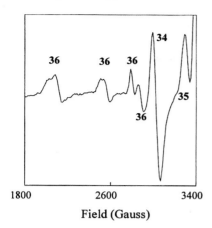

Figure 5 9.6-GHz ESR spectrum at low field from photolysis of triazide **33** in 2-methyltetrahydrofuran at 77 K. Peak assignments refer to structures in Scheme 8.

shows only weak linkage between nitrene sites [38], while sulfur-linked analog **38** shows [39] a quintet state attributed to torsion about the phenyl–sulfur bond (Scheme 9; see also Chapter 7). A notable example of hetero-atom effects is the comparison between nondisjoint 1,1-ethenediyl-linked system **39** and carbonyl-linked analog **10**. AMl-CI computations found **39** to have a much more robust $\Delta E(Q - S)$ gap of 8.5 kcal/mol, favoring the quintet, than did **10**, which had near degeneracy of states (quintet favored over the singlet by 0.4 kcal/mol) [37]. The ineffectual exchange-linkage strength of the carbonyl predicted for these systems echoes that computa-tionally predicted in the comparison of TMM and OA (Section II.A) Ex-periment appears to confirm the AMl-CI prediction. In crystalline matrix, **10** has a singlet ground state with a thermally populated quintet state [76], whereas in frozen matrix **39** appears to have a high-spin quintet ground state, based on its linear Curie law behavior. It is formally true that **39** might have near-degenerate states and still exhibit the experimentally observed

37 X=O
38 X=S

39 X=CH$_2$
10 X=O

Scheme 9

linear Curie law behavior, but the computational combined with the experimental results make this unlikely. It is worth noting that the ESR spectrum of **10** in crystalline matrix is rather different from that in 2-methyltetrahydrofuran glassy matrix, suggesting that conformational dependence on environment may be important here. *Overall, heteroatom substitution can have profound effects on intramolecular exchange even in systems that have such strong one-center interactions as the dinitrenes.*

III.B.4. Dinitrene Summary

Dinitrenes are too unstable to be useful for the construction of magnetic materials, although they can be isolated in dilute form at room temperature in solid matrices [76] under some circumstances. However, the geometric simplicity and ease of spectral interpretation for these systems has made them useful models for investigating intramolecular exchange effects, allowing extrapolation to more stable systems.

IV. EXTRAPOLATION: POLYMERIC POLYRADICALS

IV.A. Dangers of Torsional Deconjugation: Polyphenylacetylenes

Section II.C of this chapter described computational studies of oligoradicals. This section summarizes some experimental studies of such oligoradicals. One highly successful experimental approach has been the synthesis of multispin *backbone-conjugated* polyradicals (see Chapter 17) [77]. We and a number of others chose to develop *pendant* polyradicals, since incomplete generation of spin sites would not necessarily preclude exchange interaction across the defect (Scheme 10) [48–51,54,78].

Our first effort was made on derivatives of poly(4-oxyphenylacetylene), with appropriate substitution to stabilize the radicals. During this

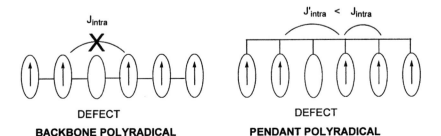

Scheme 10

work, the Nishide group published a synthesis of poly(3,5-di-*t*-butyl-4-oxyphenylacetylene), **14** (Scheme 4), which showed it to have limited spin yields and a lack of intramolecular exchange coupling between spins [48]. Our computational studies of **14** showed that its essentially helical nature should preclude exchange between pendant spins of any sort due to torsional deconjugation, unless geometric flattening of the polyacetylene backbone could be achieved (Section II.C) [55].

We attempted to extrapolate the work of Nishide by making a photochemically activated derivative of **14**, system **40**, which has aryl oxalate groups attached in place of the phenolic —OH groups shown in precursor **41** (Figure 6). Whereas Nishide's group carried out solution-phase oxidation of **41**, allowing polyradical solubility and flexibility that could lead to radical recombination reactions, we attempted to generate polyradical **14** in the solid state from precursor **40**, whose conformation we hoped to be more planarized than that of **41**. The use of aryl oxalate photolysis to generate phenoxyl radicals in the solid state had previously been investigated [79] by us to some extent, and proved to be a generally useful method, although its quantum efficiency is not as high as for other photochemical phenoxyl-generation methods developed by us (see Section V.B).

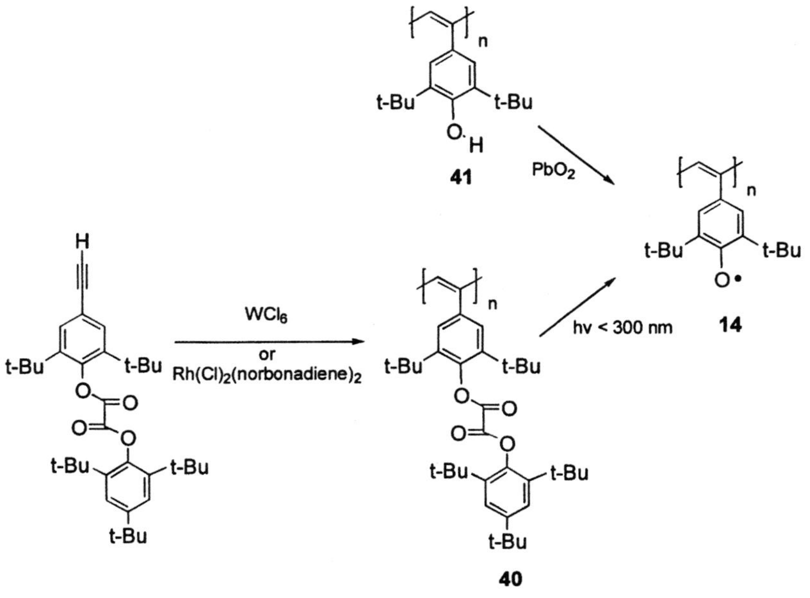

Figure 6 Synthesis of a pendant polyradical from either a solution oxidation method (via **41**) or from a photochemical precursor (via **40**). See Ref. 80.

Polymer **40** proved to be thermally stable and processible [54,80]. Variation of the metathesis catalyst used to carry out the critical polymerization resulted in variation of the product IR spectra and molecular weight, consistent with different conformational mixtures of cis-cisoid, trans-transoid, cis-transoid, and trans-cisoid linkages in the product polymer. This raised our hope that at least one of the samples might be planar enough to allow some degree of intrachain exchange, although observable [80] defects in the backbone structure raised the possibility that exchange might be interrupted. Despite ready production of considerable spin yields of room-temperature persistent phenoxyl groups upon solid-state photolysis of **40**, none of our experiments showed other than linear Curie behavior of ESR spin intensity of **14** with temperature. As mentioned in Section II.C, computational study of **14** showed it to be helical to an extent that made its backbone a very weak exchange-linking pathway. As a result of our experiments and the pessimistic computational studies, we abandoned all work on polyphenylacetylene-based pendant polyradicals. We are not aware that any such system to date has experimentally shown evidence of intrachain ferromagnetic exchange, even for systems where the nominal parity seems favorable.

IV.B. Rigidifying Polymer Backbones: Polythiophenes

The first subsections of this chapter described the importance of good backbone conjugation to produce exchange in pendant polyradicals. Chapter 14 describes the first experimental realization [56] of computational predictions [22] that backbone rigidification is required to achieve this feature, by his group's synthesis of phenoxyl- and nitroxide-pendant poly(phenylene vinylenes). This section describes work on other backbone-rigidified pendant polyradicals that should favor intrachain exchange.

Poly(2,5-thienylenes) (polythiophenes, PTs) have been targets of considerable interest for electronic purposes. We aimed to make PTs with regiospecifically attached stable radicals, such as poly(3-(3′,5′-di-t-butyl-4′-oxyphenyl)-2,5-thienylene), **42**. Figure 7 shows two of the approaches used by us [81]. We initially attempted to make precursor polyphenol **43** via $FeCl_3$-induced oxidative coupling of monomer **44** (route a). Work by Andersson et al. [82] reported that polymerization of 3(4′-octyl-phenyl)thiophene under similar conditions gave highly regiospecific poly(3-(4′-octylphenyl)-2,5-thienylene). However, polymerization of **44** by this route gave nonregiospecific polymer **45**, based on ^{1}H-NMR results. BBr_3-induced deblocking of **45** yielded **43**, which could be oxidized with aq $K_3Fe(CN)_6$/KOH to give deeply brown polyradical **42-a**. This nonregiospecific polyradical gave a strong, persistent ESR signal at $g = 2.00455$, con-

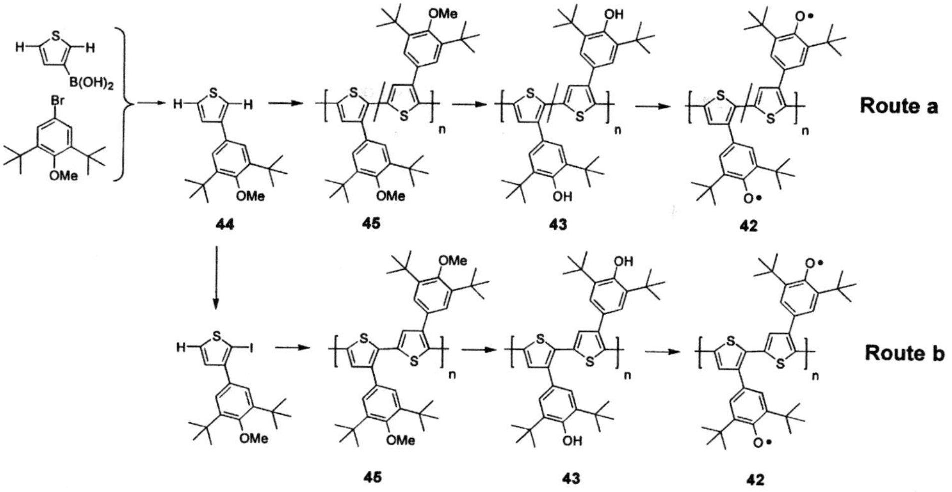

Figure 7 Regiorandom and regiospecific syntheses of a polythiophene-based poly-radical, **42**. See Ref. 81 for details.

sistent with the phenoxyl-based nature of **42-a**. At the same time that we found oxidative coupling of **44** to be nonregiospecific, Yamamoto and Hayashi reported [83] similarly motivated syntheses of **42**. The latter workers obtained higher molecular weights than we did, and like us obtained persistent radical signals upon oxidation of the appropriate polyphenol **43**, but still suffered from obtaining nonregiospecific products.

Figure 7, route b, shows our eventual solution to the problem of regiorandomization [81]. Halogenation of protected monomer **44**, followed by copper(I)-induced carbon–carbon bond polymerization, led to regiospecific formation of **45**, as shown by the presence of single, sharp peaks in the ^{1}H-NMR spectra of this polymer. The gel permeation chromatographic number average molecular weight ($\bar{M}_n$) of **45-b** relative to polystyrene standards corresponded to a degree of polymerization of only 6, so there is room for improvement. BBr$_3$-induced deblocking of **45-b** led to formation of poly-phenol **43-b**, which could be oxidized as in the nonregiospecific case to give deep brown polyradical **42-b**, with a strong, broadened, persistent radical ESR signal at g = 2.00462 and UV-Vis peaks at 285, 412, 421, 572 nm, both consistent with expected results for the phenoxyl pendant sites.

Despite its low degree of polymerization, polymer **42-b** has the regiospecific connectivity needed to yield high-spin intrachain exchange of its pendant spins. Force-field modeling of **43-b** shows it to have a highly plan-arized PT backbone, induced by the desire of the bulky t-butylated phenoxyl groups to adopt an alternating all-anti conformation (Scheme 11). Any head-to-head linkages will break the polymer backbone planarity and the train of parity through the system, which requires that $N^* > N°$. The force field-computed backbone torsion of the PT units in the regiospecific polymer is about 35°, small enough to allow reasonable intrachain exchange according to AM1-CI computations. However, force-field computations for regiorandomized **43-a** show the PT backbone torsion at the head-to-head linkages to be over 55°, large enough to insulate coupling effectively, even if the connectivity across the head-to-head linkage were correct (which it is not) to allow FM exchange.

Overall, steric forces in regiospecific **42-b** are expected to favor backbone conjugation, while analogous forces in regiorandom **42-a** favor backbone torsional deconjugation. This expectation is supported by comparative experimental UV-Vis and photoluminescence results for regiorandom polyether **36-a** versus regiospecific **36-b**. Both types of electronic spectra show red shifts in the regiospecific polymer compared to the regiorandom poly-

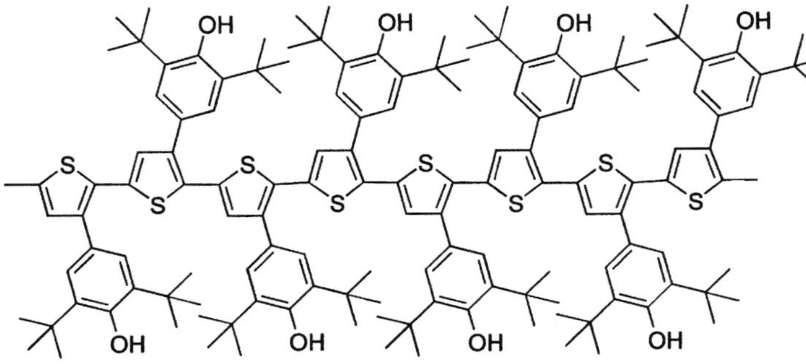

43-b

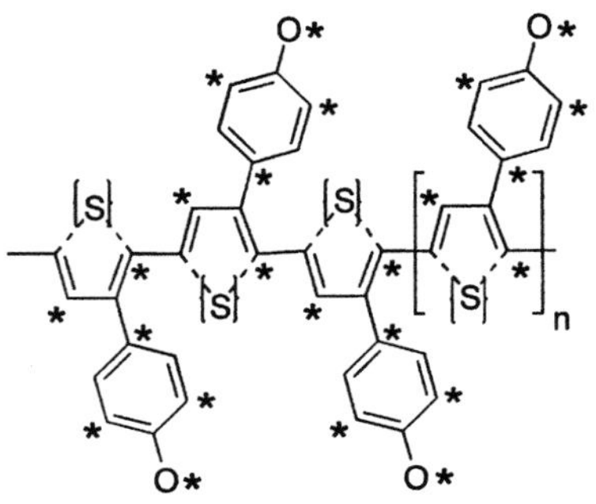

MULTIPLICITY ~ n

Scheme 11

mer, consistent with greater conjugation in the former. Since the only difference between the two systems lies in different regioconnectivity, this result supports the design expectation that **36-b** is constrained into a more planar, conjugated geometry than is **36-a**. We have yet to complete studies of the regiorandom versus regiospecific polyradicals **33**, but we hope that magnetic susceptibility and/or electron spin nutation experiments (see Chapter 11) will find intrachain FM coupling in the latter.

V. TOWARD REAL MATERIALS: SOLID-STATE RADICALS

V.A. Phenoxyl Radicals in the Solid State

The linear pendant polymers described in the previous subsection represent a step toward true bulk molecular magnetic materials, since they are "ferromagnets" in one dimension. True materials aim to achieve magnetism in three dimensions. Our first efforts at achieving 3D effects in organic systems involved the use of phenoxyl radicals. Stabilized phenoxyl radicals have been well known for decades [84], as has the fact that some can be isolated as neat solids. In addition, a number of magnetic-susceptibility studies have been carried out on the related galvinoxyl radical [85–87].

We attempted to vary the bulk exchange behavior in neat phenoxyl radicals by altering the size of the 4-substituent. This strategy was encouraged by Wudl's success in converting the antiferromagnetic behavior of verdazyl radical **45** to ferromagnetic behavior in **46** by substituent manipulation (Scheme 12) [88]. 2,4,6-Tri-*t*-butylphenoxyl and 2,4-di-*t*-butyl-6-tritylphenoxyl radicals (**47–48**, respectively) were made by solution-phase oxidation of the appropriate phenolic precursors [89]. Elemental analysis showed good purity of the products. The neat solid radicals were most persistent when sealed under vacuum in the absence of strong direct lighting,

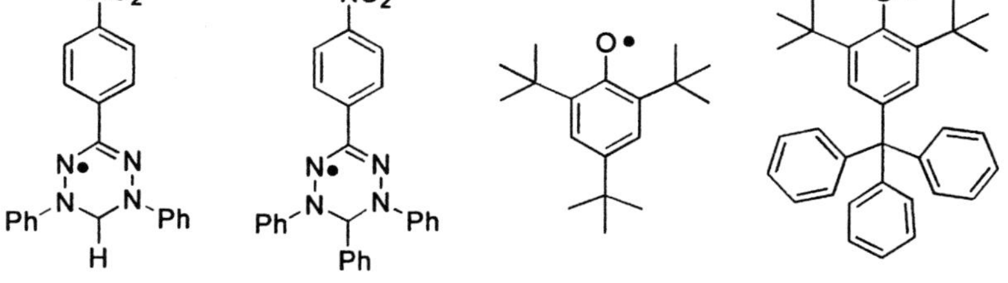

Scheme 12

under which conditions the tri-*t*-butyl derivative lasted for many months. However, we were not successful in obtaining single crystals of the radicals, and had to be satisfied with using microcrystalline powders for magnetism experiments. In collaboration with the Landee group, we obtained static magnetic-susceptibility measurements for these systems [89].

Plots of susceptibility times temperature versus temperature (χT vs. T) for both radicals showed downward curvature without a maximum, indicating antiferromagnetic behavior and/or reversible dimerization in the solid state. Considerable spin-moment loss occurred in any case, since the maximum spin count typically was 34–47% of the expected value for the samples. Curie–Weiss curve fitting of the high-temperature regions for **47–48** yielded Weiss constants of $\theta = -2.7$ K and -0.85 K, respectively, showing antiferromagnetic behavior in both cases. The lower-temperature regions could be fit to both $S = 1/2$ linear-chain or dimeric models with similar degrees of statistical confidence; but given the structures of the radicals, it seems far more plausible that a dimeric interaction mechanism is involved [89].

Overall, the 4-trityl-substituted system did show weaker interactions that the 4-*t*-butyl system, attributable to the greater steric bulk of the trityl substituent, but both systems showed AFM behavior rather than more desirable FM behavior. The exchange constants for **47** and **48** were $J/k_B \cong -6$ K and $J/k_B \cong -1$ K, respectively [89]. In addition, we were hampered by an inability to deduce the crystal structure of the samples on even a qualitative basis, as well as by the relative instability of these systems.

Alternative strategies for stabilizing phenoxyl radicals in neat crystalline lattice include the introduction of hydrogen bonds to rigidify the crystalline lattice. At present, we are working on phenoxyl radicals **49–51** (Scheme 13) that show strong, persistent ESR spectra, strong hydrogen bonding in the infrared spectra and x-ray crystallography of their precursors,

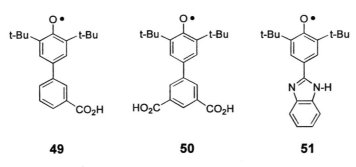

49   **50**   **51**

Scheme 13

and the high melting points associated with strong intermolecular bonding [90]. Work is ongoing to see whether the hydrogen bonding of the precursor phenols can be realized in the phenoxyl radicals, possibly leading to greater solid-state stability as well as strengthening intermolecular exchange.

V.B. Photochemical Generation of Phenoxyl Radicals in the Solid State

It is well known that the crystalline state can act like a vise to hold molecule fragments in a fixed relative geometry. We aimed to probe the exchange interaction between radical fragments by using this fact. Toward this end, we developed methods of generating phenoxyl radicals by photochemical fragmentation of appropriate crystalline precursors.

Diaryl oxalates (DAOs) are readily synthesized for a wide variety of phenyl substituent patterns, and can even be made with asymmetric substitution [79]. They are highly crystalline, thermally stable solids. Their main drawback is that their photochemical quantum efficiency appears to be about 10%, so it is difficult to get high conversion to make phenoxyl radicals, except at the surface of a crystal. This type of functionality was mentioned in Section IV.A regarding its use in photochemical generation of phenoxyl-based polyradicals [54,80].

Ultraviolet photolysis of a number of DAOs yields not only strong phenoxyl monoradical signals in the ESR spectrum, but also triplet radical-pair signals (Scheme 14). Remarkably, many of the triplet signals are quite persistent, even at room temperature (hours to days). Stabilization of photochemically generated reactive species in crystalline matrix is a known phenomenon [91] and has been observed [67] by us even for much more reactive dinitrene species. Extended irradiation causes breakdown of the lo-

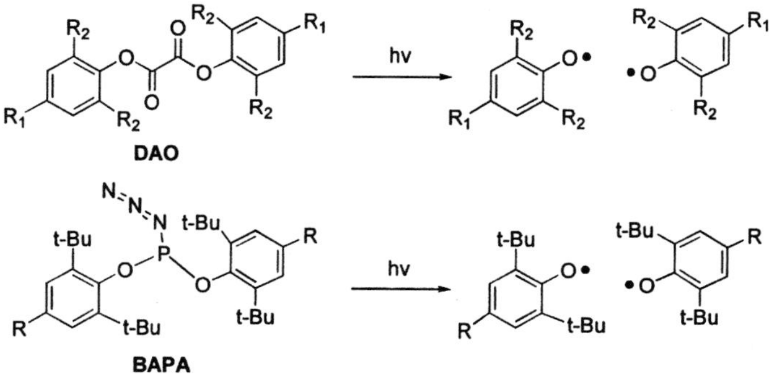

Scheme 14

cal crystal lattice, causing changes in the triplet ESR spectra that usually culminate in the observation of weak triplets with small zero-field splittings (zfs). However, briefer periods of irradiation reproducibly yield spectra such as that shown in Figure 8 from irradiation of **52** (Scheme 14). Much larger zfs are observed than for the weak signals produced upon longer irradiation. Table 2 shows a number of DAOs and zfs of the associated triplet spectra, where all zfs were obtained by simulation of the triplet portions of the appropriate ESR spectra [92].

The wide zfs triplet spectra are attributable to pairs of phenoxyl radicals held approximately at their original positions after photochemical scission of the oxalate moiety. The observed zfs can be well modeled by a combination of information. First, one uses the crystal structure of the DAO to determine the relative geometries of the component phenoxyl fragments, assuming that the crystal lattice does not permit movement [93]. The phen-

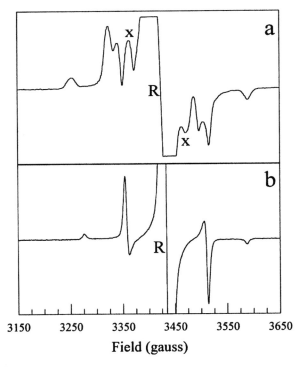

Figure 8 Comparison of radical-pair ESR spectra derived by solid-state photolysis of DAO **52** and BAPA **53** (Scheme 14), spectra a and b, respectively. R = monoradical peak, x = unknown radical pair peak.

Table 2   zfs in Solid State Radical Pairs

Radical-pair precursor	Biradical zfs parameters observed (cm^{-1})
	$D = 0.0102$, $E = 0.0001$[a]
	$D = 0.0099$, $E = 0.0004$[a]
	$D = 0.0124$, $E = 0.0006$ (initially $D = 0.0108$, $E = 0.0004$)[a]
	$D = 0.0146$, $E < 0.0003$ (initially $D = 0.0139$, $E < 0.0003$)[b]
	$D = 0.0065$, $E < 0.0003$[b]

[a]Ref. 92.
[b]Ref. 68.
Inital values refer to zfs observed before annealing of photolyzed crystals. D is abbreviation for $|D/hc|$, E is abbreviation for $|E/hc|$.

oxyl C—O bond should be shortened in accord with computational findings [94]. Next, the spin-density distribution for the phenoxyl fragments must be assigned. In cases such as 2,4,6-tri-t-butylphenoxyl radical, the distribution is experimentally [95] known; in other cases, it must be estimated by comparison to related radicals or to density functional [94] computations. Finally, a spin-dipolar approximation is used to sum over all interactions between the spin-density sites in the two radical fragments and then to find the overall zfs (Figure 9) [96,97]. A variety of models can be used to do this. All of the methods tested by us gave similar results, with the predicted zfs being in good accord with the observed zfs [98]. By implication, the least-motion model for fragmentation of the DAOs therefore is reasonable, so we can use the DAO crystal lattice to hold the phenoxyl fragments in an approximately known relative geometry.

Another photochemical source of phenoxyl radical pairs are the bis(aryloxyphosphine) azides, or BAPAs (Scheme 14) [99]. These molecules are of less general utility than the DAOs, since their efficient synthesis requires steric blockade of all *ortho* positions of the aryl groups, typically by *t*-butyl groups. However, the BAPAs have a much higher photocleavage quantum efficiency than the DAOs, even in the solid state. Irradiation of a typical colorless, transparent BAPA crystal gives sufficient amounts of stable radicals that the crystal turns faintly the color of the radical. Electron spin resonance spectra of the irradiated crystals show strong triplet biradical signals that are persistent even at room temperature. An example is shown in Figure 8 derived from photolysis of **53** (Scheme 14).

Table 2 also summarizes the zfs findings for radical pairs derived from a number of BAPAs [68]. The same procedure [97] can be used to model the zfs as was used for the DAOs, except for the case of **54**, for which we have been unable to obtain a crystal structure to date. As with the DAOs, there is reasonably good agreement between the observed and the theoretical zfs, except in the case of **54**. At present, we are uncertain as to the cause of this one discrepancy, although it seems not to be a coincidence that this is the one BAPA that gives crystals of a type that decompose without giving an interpretable x-ray scattering pattern. It is possible that this system incorporates solvent molecules in a manner to distort the typical crystal lattice

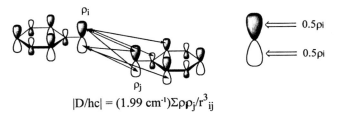

$$|D/hc| = (1.99 \text{ cm}^{-1})\Sigma\rho_i\rho_j/r^3_{ij}$$

Figure 9 Scheme for computation of triplet zero-field splitting (D/hc) in pairs of conjugated phenoxyl radicals by a spin-dipole model, with spin density (ρ_i, ρ_j) on each pi-site split above and below the plane of each ring. Distance between spin density sites is r_{ij}. See Ref. 98 for details.

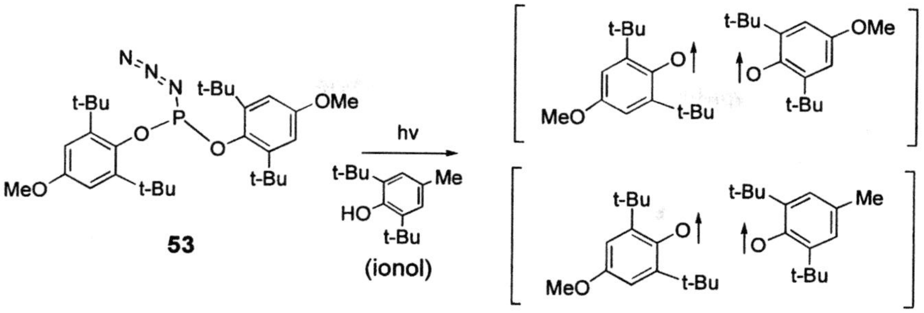

Scheme 15

and molecular conformation, and so gives a different arrangement of phenoxyl fragments than expected after photolysis.

Finally, BAPAs can be doped into the crystal lattices of phenols such as ionol (**55**, Scheme 15). In collaboration with G. Lazarev [100], we found that photolysis of pure ionol gives only weak radical signals derived from O—H bond scission, whereas photolysis of ionol crystals doped with **53** gives strong and persistent radical-pair signals [100a]. One can observe hyperfine coupling in some of the radical-pair lines that is consistent with one of the constituent phenoxyl radicals being derived from ionol! Apparently, initial scission of **53** can lead first to a pair of 2,6-di-*t*-butyl-4-methoxy-phenoxyl radicals, one of which can react with ionol in the surrounding crystal lattice. This is an interesting variation of other radical-pair-forming reactions that are observed [100b–d] when ionol crystals are doped with photoactive molecules.

Overall, DAO and BAPA photolyses allow useful insight into the interactions between delocalized phenoxyl radicals in well-defined solid-state geometries that are otherwise difficult to achieve. In addition, DAO and BAPA photochemistry allows the solid-state production of phenoxyl radicals, a technique that we have used to make polyradicals such as **14** [54] and **18** [101] (Scheme 4). The latter system was made by photolysis of **56** (Scheme 16) in collaboration with the Nishide group [101] and was the first FM exchange-coupled organic polyradical to be made by a solid-state precursor technique.

V.C.   Hydrogen-Bonded Radicals in the Solid State

Other chapters in this book detail studies of hydrogen-bonded radicals in the solid state. Work such as that of Veciana [102], Sugawara [103], Yoshioka [104], and their respective coworkers have induced magnetic ordering

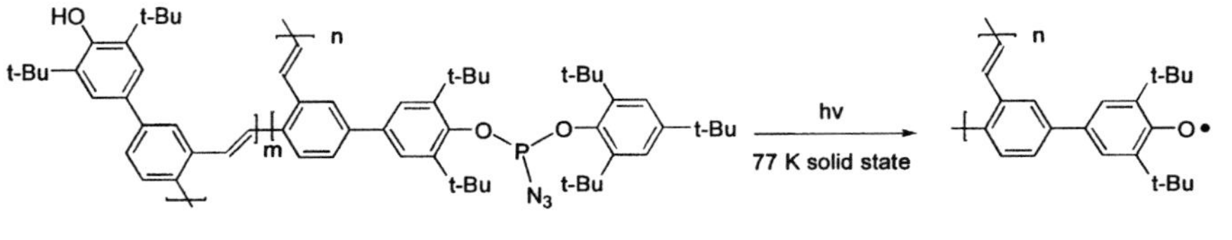

56 (70% BAPA, 30% phenol)   18

Scheme 16

in stable radicals whose crystal lattices are held together by hydrogen bonding. Semiempirical MO-CI methods [40] have also been used to investigate the hydrogen-bonding motif for crystal assembly of molecular magnetic materials. These studies are all detailed elsewhere in this book; Chapter 20, for example, explores in a more general sense some of the computational work done on exchange through crystal packing.

Recently we, also, have attempted to follow this motif toward strong intermolecular bonding for stable radicals. Scheme 13 showed some hydrogen-bonding phenoxyl radicals whose synthesis we are presently attempting. Scheme 17 shows chelatable and/or hydrogen-bonding radicals **57–63** that have recently been made by us. Most of these remain to be purified as radicals in sufficient quantities for magnetic studies, although all have been studied by ESR spectroscopy and shown to have high persistence and useful degrees of spin delocalization (based on hyperfine coupling analysis). The discussion of these molecules will therefore be limited to consideration of **57**, for which we have the most information at present. This will serve as an example of the final step in designing molecular magnets, namely, the assembly of the stable spin units into a supramolecular array of spins.

Radical **57** was synthesized by standard methods [106]. Its precursor hydroxylamine can be kept for some time in the absence of oxygen but turns red easily in solution when exposed to air. Oxidation of the hydroxylamine with lead dioxide in benzene readily yields radical **57**, which can be isolated as burgundy prisms after some effort. The radical is highly stable in solution and appears to be stable as the solid for months without major precautions.

X-ray crystallographic analysis of **57** shows it to have an orthorhombic Pbca lattice with eight molecules in the unit cell ($Z = 8$). Figure 10 shows both a view of the lattice and a scheme displaying the relative orientations of important structural features of the molecules in the lattice. The basic arrangement of molecules in the lattice has stacks of radicals along the a-xis, with hydrogen bonds between the stacks along the b-axis. Each layer is duplicated by a mirror plane relative to the benzo-moiety, to form a "bilayer" with all N—O bonds pointing in the same direction. Due to a center

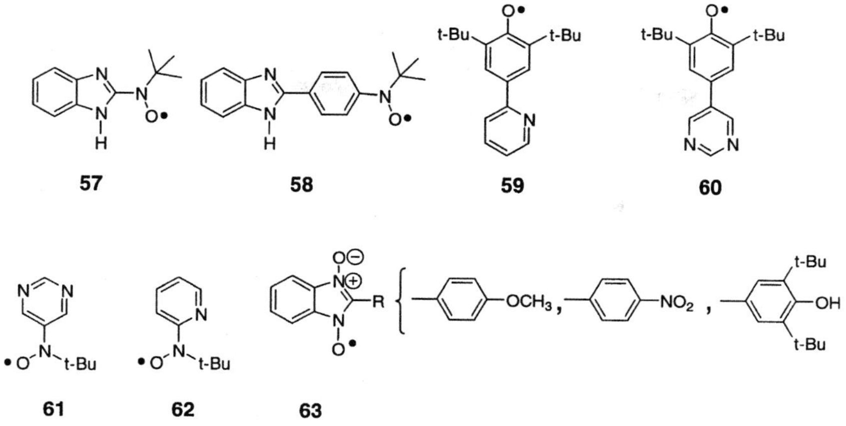

Scheme 17

of symmetry, an inverted bilayer is found relative to the initial bilayer with all N—O bonds pointing in the opposite direction relative to the first bilayer—this completes the description of the unit cell. For each molecule of **57**, the N—O bond is found to be syn to the N—H bond, presumably due to a favorable electrostatic interaction in this orientation. Within a hydrogen-bonded layer, the stacks are oriented in alternating fashion, each with molecules aligned perpendicular to one another in a herringbone arrangement, but having hydrogen bonding between the stacks at a N—H---N distance of 2.14 Å, with the N—H---N angle being somewhat bent. The closest distance between the N—O spin sites is 5.84 Å between the N—O oxygens going up a stack (along the b-axis).

Solution ESR hyperfine analysis of **57** shows the following data: $a_{N(-O)}$ = 10.1 G, $a_{=N-}$ = 2.2, $a_{N(-H)}$ = 0.7, a_1 = 0.6, a_2 = 0.6 (2H), a_3 = 0.4 (2H). Agreement between the observed hyperfine coupling constants and those predicted by UBLYP density functional theory (DFT [105]) is reasonable for all a_N, as shown in Table 3. Approximately 20% of the spin density is delocalized from the nitroxide group onto the imidazole portion of the benzimidazole group. The crystal arrangement between stacks along the b-axis allows a favorable spin arrangement between an N—H unit and the hydrogen-bond acceptor N atom on the next molecule, since both imidazole nitrogens have positive spin densities and the hydrogen-bonding H has a spin-polarized negative spin density. As a result, exchange between the stacks along the b-axis should be ferromagnetic.

Polycrystalline samples of **57** were analyzed by magnetic susceptibility analysis down to 1.8 K with the collaboration of the Palacio group [106].

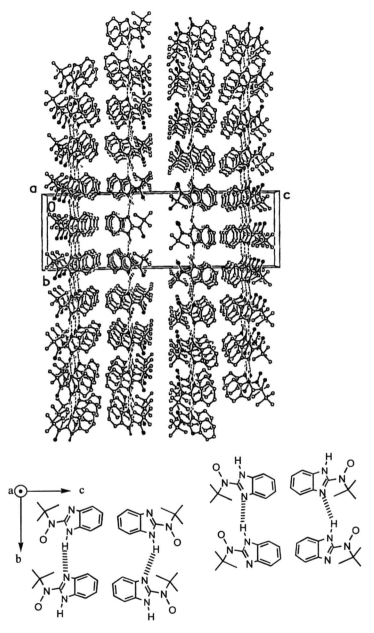

Figure 10 Crystal structure for stable radical **57**. See Ref. 106.

Table 3

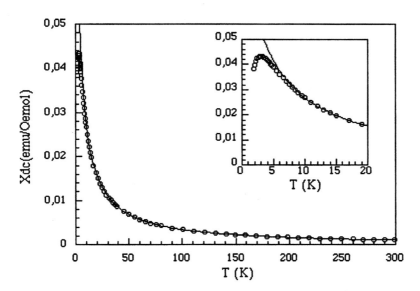

	6-31G**/UBLYP ρ_{spin}	hfc (a_x calc'd, G)^a	|hfc| (a_x expt'l, G)
N—(O)	0.333	10.3	10.2
(N)—O	0.468		
N1—(H)	0.023	0.7	0.7
=N3—	0.125	3.9	2.2
H—(N1)	−0.002(3)		0.6
H4, H7	−0.001(7), 0.000(3)		0.6
H5, H6	0.000(0), −0.000(2)		0.4
C2	−0.047		
C8, C9	0.041, −0.013		
C4, C7	0.037, 0.000(4)		
C5, C6	−0.005, 0.053		

^{a}For a_x estimates, $Q = 31.0$ G in $a_x = Q \times \rho_{spin}$.

Figure 11 Dc Magnetic susceptibility plot for radical **57** as a function of absolute temperature. Circles are data, solid line is a Curie–Weiss law fit to the higher-temperature data. See Ref. 106.

The sample was found to follow a Curie–Weiss susceptibility law at higher temperatures (Figure 11) but showed a downturn at about 3 K. The Curie–Weiss fit yielded $\theta = -4.24$ K and $g = 2.029(3)$, assuming $S = 1/2$ spin units. An antiferromagnetic magnetic-phase transition appears to be present at just below the 1.8 K limits of measurement; however, this is not directly observable under the condition possible for the experiment. Fitting of the low-temperature data to a variety of models shows a very good fit to a simple 2D Heisenberg system with $J = -1.60$ K, if $g = 2$ is assumed in accord with the organic nature of the $S = 1/2$ spin units. Without further measurements, it is not yet clear whether the putative phase transition at or below 1.8 K is due to magnetic anisotropy or interplane magnetic interactions or a combination of the two. Further work is planned to attempt to clarify this issue. The behavior of this system is quite interesting in comparison to the ferromagnetic behavior of the nitronylnitroxide analog of **57** [104], which is described in detail in Chapter 27.

VI. CONCLUSIONS

The design of molecular magnetic materials is a fascinating but quite complex business. It combines aspects of physics, chemistry, and materials science. We have attempted to develop an integrated design approach, using theory to select appropriate spin building blocks, classic organic synthetic techniques to make the building blocks, and crystal design techniques to create the supramolecular assemblies of building blocks needed to obtain multispin lattices. Much work remains to be done but this book shows how very much has been accomplished in the past decade. The most promising pathways for success seem to have been identified, and the next decade of work in this area should show whether molecular magnets become a practically utilizable set of materials, as well as a beautiful and challenging one.

ACKNOWLEDGMENTS

I gratefully acknowledge the many undergraduate and graduate students, postdoctoral associates, and collaborators whose work is summarized and variously cited in this chapter. The work would not have been possible without the support of the National Science Foundation (CHE-9204695, CHE-9521594, and CHE-9809548). The opinions expressed in this chapter are those of the author and are not necessarily those of the NSF.

REFERENCES

1. (a) Breslow R. Pure Appl. Chem. 1982; 54:927. (b) Breslow R, Jaun B, Klutz R, Xia C-Z. Tetrahedron 1982; 38:863.
2. Breslow R. Mol. Cryst. Liq. Cryst. 1985; 125:261.
3. Torrance JB, Oostra S, Nazzal A. Synth. Met. 1987; 19:709.
4. (a) Dormann E, Nowak MJ, Williams KA, Angus RO Jr, Wudl F. J. Am. Chem. Soc. 1987; 109:2594. (b) Wudl F, Closs F, Allemand PM, Cox S, Hinkelmann K, Srdanov G, Fite C. Mol. Cryst. Liq. Cryst. 1989; 176: 249.
5. (a) Miller JS, Epstein AJ, Reiff WM. Chem. Rev. 1988; 88:201. (b) Candela GA, Swarzendruber L, Miller JS, Rice MJ. J. Am. Chem. Soc. 1979; 101: 2755.
6. Caneschi A, Gatteschi D, Sessoli R, Rey P. Accts. Chem. Res. 1989; 22: 392.
7. Iwamura H, Inoue K, Koga N, Hayamizu. NATO ASI Ser., Ser. C. 1996; 484: 157.
8. (a) Villain J, Loveluck J. J. Phys. Lett. 1977; 38:L77. (b) Palacio F. In: Gatteschi D, Kahn O, Miller JS, Palacio F, eds. Magnetic Molecular Materials. Dordrecht, The Netherlands: Kluwer, 1991, pp 1–34.
9. (a) Longuet-Higgins HC. J. Chem. Phys. 1950; 18:265. (b) Longuet-Higgins HC. Theoretical Organic Chemistry, Kekulé Symposium. London: Butterwork, p 9.
10. (a) Klein DJ, Nelin CJ, Alexander S, Matsen FE. J. Chem. Phys. 1982; 77: 3101. (b) Klein DJ. Pure Appl. Chem. 1983; 55:299.
11. Misurkin IA, Ovchinnikov AA. Russ. Chem. Res. (Engl). 1977; 46:967.
12. Ovchinnnikov AA. Theor. Chim. Acta. 1978; 47:297.
13. Shen M, Sinanoglu O. In: King RB, Rouvay DH, eds. Graph Theory and Topology in Chemistry, vol 51. Amsterdam: Elsevier, 1987, p 373.
14. Borden WT, Davidson ER. J. Am. Chem. Soc. 1977; 99:4587.
15. Tyutyulkov N. Theor. Chim. Acta. 1983; 63:291.
16. Zerner MC, Ridley JE. Theor. Chim. Acta. 1973; 32:11.
17. Zerner MC, Ridley JE. Theor. Chim. Acta. 1976; 42:233.
18. Dewar MJS, Zoebisch EG, Healey EF, Stewart JJP. J. Am. Chem. Soc. 1987; 107:3902.
19. Lahti PM, Rossi AR, Berson JA. J. Am. Chem. Soc. 1985; 107:2273.
20. Lahti PM, Ichimura AS, Berson JA. J. Org. Chem. 1989; 54:958.
21. Lahti PM, Ichimura AS. Mol. Cryst. Liq. Cryst. 1989; 176:125.
22. Lahti PM, Ichimura AS. J. Org. Chem. 1991; 56:3030.
23. Ichimura AS, Koga N, Iwamura H. J. Phys. Org. Chem. 1994; 7:207.
24. Zerner M. Quantum Theory Project, University of Florida, Gainesville, FL.
25. Dewar MJS, Stewart JJP. Q.C.P.E. 1987; 506.
26. Stewart JJP. J. Computer-Aided Mol. Design. 1990; 4:1.
27. Coolidge MB, Yamashita K, Morokuma K, Borden WT. J. Am. Chem. Soc. 1990; 112:1751.

28. Horvat DA, Murcko MA, Lahti PM, Borden WT. J. Chem. Soc., Perkin 2. 1998; 1037.
29. Davidson ER, Borden WT. J. Chem. Phys. 1976; 64:663.
30. Davidson ER, Borden WT. J. Am. Chem. Soc. 1977; 99:2053.
31. Hood DM, Schaefer HF. J. Am. Chem. Soc. 1978; 100:8009.
32. Kato S, Morokuma K, Feller D, Davidson ER, Borden WT. J. Am. Chem. Soc. 1983; 105:1791.
33. (a) Seeger DE, Berson JA. J. Am. Chem. Soc. 1983; 105:5146. (b) Seeger DE, Lahti PM, Rossi AR, Berson JA. J. Am. Chem. Soc. 1986; 108:1251.
34. (a) Minato M, Lahti PM. J. Phys. Org. Chem. 1993; 6:483. (b) Minato M, Lahti PM. J. Am. Chem. Soc. 1997; 119:2187.
35. Ichimura AS, Matlin AR, Lahti PM. J. Am. Chem. Soc. 1990; 112:2868.
36. Lin D, Hrovat DA, Borden WT, Jorgensen WL. J. Am. Chem. Soc. 1994; 116:3495.
37. Ling C, Minato M, Lahti PM, van Willigen H. J. Am. Chem. Soc. 1992; 114:9959.
38. Minato M, Lahti PM. J. Phys. Org. Chem. 1994; 5:495.
39. Nimura S, Kikuchi O, Ohana T, Yabe A, Kaise M. Chem. Lett. 1994; 1679.
40. Zhang J, Baumgarten M. Chem. Phys. 1997; 222:1.
41. Okada K, Imakura T, Oda M, Kajiwara A, Kamachi M, Sato K, Shiomi D, Takui T, Itoh K, Gherghel L, Baumgarten M. Mol. Cryst. Liq. Cryst. 1997; 306:301.
42. Zhang J, Lahti PM, Wang R, Baumgarten M. Heteroatom Chem. 1998; 9:161.
43. Okada K, Imarkura T, Oda M, Kajiwara A, Mamachi M. Sato K, Shiomi D, Takui T, Itoh K, Gherghel L, Baumgarten M. Mol. Cryst. Liq. Cryst. 1997; 306:301.
44. Du P, Hrovat DA, Borden WT, Lahti PM, Rossi AR, Berson JA. J. Am. Chem. Soc. 1986; 108:5072.
45. Roth WR, Kowalczik U, Maier G, Reisenauer HP, Sustmann R, Müller W. Angew. Chem. Intl. Ed. Engl. 1987; 26:1285.
46. Reynolds JH, Berson JA, Kumashior KK, Duchamp JC, Zilm KW, Rubello A, Vogel P. J. Am. Chem. Soc. 1992; 114:763.
47. Reynolds JH, Berson JA, Scaiano JC, Berinstain AB. J. Am. Chem. Soc. 1992; 114:5866.
48. Nishide H, Yoshioka N, Inagaki K, Tsuchida E. Macromolecules. 1988; 21:3119.
49. Fujii A, Ishida T, Koga N, Iwamura H. Macromolecules. 1991; 24:1077.
50. Miura Y, Inui K, Yamaguchi F, Inoue M, Teki Y, Takui T, Itoh K. J. Polym. Sci. Polym. Chem. 1992; 30:959.
51. Nishide H, Kaneko T, Gotoh R, Tsuchida E. Mol. Cryst. Liq. Cryst. Sci. Technol., Sect. A. 1993; 233:89.
52. Nishide H, Yoshioka N, Saitoh Y, Gotoh J. Macromol. Sci., Pure Appl Chem. 1992; A29:775.
53. Nishide H, Kaneko T, Yoshioka N. Macromolecules. 1993; 26:4567.
54. Lahti PM, Inceli AL, Rossitto FC. J. Polym. Sci. Part A, Polym. Chem. 1997; 35:2167.

55. Lahti PM, Ling C, Yoshioka N, Rossitto FC, van Willigen H. Mol. Cryst. Liq. Cryst. 1993; 233:17.
56. Nishide H, Kaneko T, Nii T, Katoh K, Tsuchida E, Yamaguchi K. J. Am. Chem. Soc. 1995; 117:548.
57. Li S, Ma J, Jiang Y. J. Phys. Chem. 1997; 101:5567.
58. Li S, Ma J, Jiang Y. J. Phys. Chem. 1997; 101:5587.
59. Lahti PM. In: Menton J, ed. Research Trends, vol. 3. Council of Scientific Research. Trivandrum, India: Integration, 1992, p 179.
60. Lahti PM, Minato M, Ling C. Mol. Cryst. Liq. Cryst. 1995; 271:147. Note that all zero-field splitting estimates given for quintet dinitrenes in this review have been found incorrect by more recent discoveries; see Ref. 67.
61. Iwamura H, Murata S. Mol. Cryst. Liq. Cryst. 1989; 176:33.
62. Murata S, Iwamura H. J. Am. Chem. Soc. 1991; 113:5547.
63. Wasserman E, Murray RW, Yager WA, Trozzolo AM, Smolinsky G. J. Am. Chem. Soc. 1967; 89:5076.
64. Itoh K. Pure Appl. Chem. 1978; 50:1251.
65. Wasserman E. Prog. Phys. Org. Chem. 1971; 8:319.
66. Kalgutkar RS, Lahti PM. J. Am. Chem. Soc. 1997; 119:4771.
67. Fukuzawa TA, Sato K, Ichimura AS, Kinoshita T, Takui T, Itoh K, Lahti PM. Mol. Cryst. Liq. Cryst. Sci. Technol., Sect. A. 1996; 278:253.
68. Kalgutkar RS. PhD dissertation, University of Massachusetts, Amherst MA, 1998.
69. Teki Y. PhD dissertation, Osaka City University, Osaka, 1985.
70. Teki Y, Takui T, Yagi H, Itoh K, Iwamura H. J. Chem. Phys. 1985; 83:539.
71. Sato K. PhD dissertation, Osaka City University, Osaka, 1994.
72. Nimura S, Kikuchi O, Ohana T, Yabe A, Maise M. Chem. Lett. 1993; 837.
73. Ling C, Lahti PM. J. Am. Chem. Soc. 1994; 116:8784.
74. Ling C. PhD dissertation, University of Massachusetts, Amherst, MA, 1993.
75. Chapyshev SV, Walton R, Sanborn JA, Lahti PM. To be published.
76. Ichimura AS, Sato K, Kinoshita T, Takui T, Itoh K, Lahti PM. Mol. Cryst. Liq. Cryst. Sci. Technol., Sect A. 1995; 272:279.
77. Rajca A. Chem. Rev. 1994; 94:871.
78. Iwamura H, Koga N. Acc. Chem. Res. 1993; 26:346.
79. (a) Modarelli DA, Lahti PM. J. Chem. Soc. Chem. Comm. 1990; 1167. (b) Lahti PM, Modarelli DA, Rossitto FC, Inceli AL, Ichimura AS, Ivatury S. J. Org. Chem 1996; 61:1730.
80. Rossitto F, Lahti PM. Macromolecules. 1993; 26:6308.
81. Xie C, Lahti PM. J. Polym. Sci., Part A: Polym. Chem 1998 (in press).
82. Andersson MR, Selse P, Berggren M, Järvinen H, Hertberg T, Inganäs O, Wenneeström O, Österholm JE. Macromolecules. 1994; 27:6503.
83. Yamamoto T, Hideki Hayashi. J. Polym. Sci., Part A: Polym. Chem. 1997; 35:463.
84. Altwicker ER. Chem. Rev. 1967; 67:475.
85. Mukai K. Bull. Chem. Soc. Japan 1969; 42:40. Cf. a related study on Yang's biradical in Mukai K, Ishizu K, Deguchi Y. J. Phys. Chem. 1969; 27:783.
86. Awaga K, Sugano T, Kinoshita M. J. Chem. Phys. 1986; 85:2211.

87. Chiang LY, Upasani RB, Sheu HS, Goshorn DP, Lee CH. J. Chem. Soc., Chem. Comm. 1992; 959.
88. Allemand PM, Srdanov G, Wudl F. J. Am. Chem. Soc. 1990; 112:9391.
89. Jung K, Lahti PM, Sheridan P, Britt S, Zhang W-R, Landee C. J. Mater. Chem. 1994; 4:161.
90. Lahti PM, George C, Liu Y, Xie C. 1999. Radical **51** is solid state stable for weeks in air. Unpublished results.
91. Cf., for example, McBride JM. Acc. Chem. Res. 1983; 16:304, and references therein.
92. Modarelli DA. PhD dissertation, University of Massachusetts, Amherst, MA, 1989.
93. George C, Lahti PM, Modarelli DA, Inceli A. Acta Cryst. 1994; C50:1308.
94. Quin Y, Wheeler RA. J. Chem. Phys. 1995; 102:1689.
95. Prabhanada BS. J. Chem. Phys. 1983; 79:5752.
96. McWeeney R. J. Chem. Phys. 1961; 34:399.
97. Rule M, Matlin AR, Seeger DE, Hilinshi EF, Dougherty DA, Berson JA. Tetrahedron. 1982; 38:787.
98. Modarelli DA, Lahti PM, George C. J. Am. Chem. Soc. 1991; 113:6329. A computer program for prediction of zfs is included in the supplementary material for this reference; it requires input geometry and SOMO data for a biradical state.
99. Kalgutkar R, Ionkin A, Quin LD, Lahti PM. Tetrahedron Lett. 1994; 35:3889.
100. (a) Lazarev GG, Pilbrow JR, Kuskov VL, Lahti PM, Kalgutkar R. Appl. Magn. Res. 1998; 15:173–180. (b) Lazarev GG, Kuskov VL, Lebedev YS. Chem. Phys. Lett. 1990; 170:94. (c) Lazarev GG, Kuskov VL, Lebedev YS, Rieker A. Chem. Phys. Lett. 1991; 181:512. (d) Lazarev GG, Kuskov VL, Lara F, García F, Rieker A, Tordo P. Chem. Phys. Lett. 1993; 212:319.
101. Takahashi M, Nishide H, Tsuchida E, Lahti PM. Chem. Mater. 1997; 9:11.
102. (a) Cirujeda J, Hernàndez-Gasió E, Panthou FL-F, Laugier J, Mas M, Molins E, Rovira C, Novoa JJ, Rey P, Veciana J. Mol. Cryst. Liq. Cryst. 1995; 271:1. (b) Cirujeda J, Hernàndez-Gasió E, Rovira C, Stanger J-L, Turek P, Veciana J. J. Mater. Chem. 1995; 5:243.
103. Sugawara T, Matsushita MM, Izuoka A, Wada N, Takeda N, Ishikawa M. J. Chem. Soc., Chem. Comm. 1994; 1723.
104. Yoshioka N, Irasawa M, Mochizuki Y, Kaot T, Inoue H, Ohba S. Chem. Lett. 1997:251.
105. The efficacy of BLYP-DFT methods for hyperfine analysis of nitroxide-based radicals has been demonstrated in (for example) Novoa JJ, Mota F, Veciana J, Cirujeda. J. Mol. Cryst. Liq. Cryst. 1995; 271:79. Similar DFT methods have been employed to model solid-state exchange, as shown (for example) in Novoa JJ, Deumal M, Kinoshita M, Hosokishi Y, Veciana J, Cirujeda. J. Mol. Cryst. Liq. Cryst. 1997; 305:129, and Novoa JJ, Deuma M. Mol. Cryst. Liq. Cryst. 1997; 305:143.
106. Lahti PM, Esat B, Ferrer JR, Liu Y, Marby KA, Xie C, George C, Antorrena G, Palacio F. Mol. Cryst. Liq. Cryst. 1999. In press.
107. Cramer CJ, Smath BA. J. Phys. Chem. 1996; 100:9664.

108. Nachtigall P, Jordan KD. J. Am. Chem. Soc. 1993; 115:270.
109. Wenthold PG, Hu J, Squires RR, Lineberger WC. J. Am. Chem. Soc. 1996; 118:475.
110. Wenthold PG, Kim JB, Lineberger WC. 1996.
111. Dowd P, Chang W, Paik YH. J. Am. Chem. Soc. 1986; 108:7416.
112. Matsuda K, Iwamura H. J. Am. Chem. Soc. 1997; 119:7412.
113. Reynolds JH, Berson JA, Kumashiro K, Duchamp JC, Zilm KW, Rubello A, Vogel P. J. Am. Chem. Soc. 1992; 114:763.

32

Future Prospects for Organic Magnetic Materials

Fernando Palacio
C.S.I.C.—University of Zaragoza, Zaragoza, Spain

Jeremy M. Rawson
University of Cambridge, Cambridge, England

I. INTRODUCTION

In the last decade we have witnessed impressive developments in the area of organic magnetic materials. The primary motivating forces in this area have been both economic and scientific. The electronics industries require increasingly demanding technical specifications for their materials. Indeed their quest for miniaturization, e.g. to improve memory storage capacity, has reached the limits of current materials technology where "top-down" processes (in which macroscopic materials are made ever smaller) such as laser ablation and molecular beam epitaxy can achieve resolution in the scale of some few nanometers. A currently popular alternative is the "bottom-up" approach in which materials are constructed at a molecular level and may achieve similar properties at the molecular rather than macroscopic scale. In addition, molecular materials possess a number of chemical properties that may provide exciting alternative processing strategies. Not least is their solubility, which allows the preparation of thin films from solution (e.g. Langmuir-Blodgett techniques, electro-coating or even slow evaporation), and their volatility, which may allow thin film formation from the vapor phase (chemical vapor deposition). Because these materials are constructed at the molecular level, it is possible to incorporate a variety of additional features that may give rise to multi-functional properties in which several physical properties are observed which may interact independently, or cooperatively with one another. Electro-optic and photonic magnetic materials are just two examples where the demand for new materials has encountered (in molecular systems) an area with great potential. The scientific motivation is based on the formidable challenge that developing an organic magnetic material represents. To prepare a polyradical whose spin ground state is more than twice that of any metal ion in the periodic table, or to synthesize a magnet based on only carbon, nitrogen, oxygen and hydrogen atoms was beyond common sense not too long ago. Yet both these goals have been achieved in recent years.

While a solid understanding of the many basic principles that govern the magnetism of organic materials has been achieved [1-4], from the strict viewpoint of materials science the area is still far from maturity. To develop an organic compound that becomes a magnet even at only some few degrees Kelvin is a fundamentally important advance but as a material this compound is not particularly useful. This is, in fact, the major challenge for the future: to use all current understanding to develop new organic materials for practical magnetic applications. To achieve this goal, important issues such as chemical stability, magnetic connectivity and control of the solid state structure have to be addressed. In this chapter we discuss some current trends which may provide solutions to these problems, and try to foresee how this area may develop in the future.

II. SUMMARY OF BASIC REQUIREMENTS TO DEVELOP MAGNETIC ORGANIC MATERIALS

Let us summarize, very briefly, the basic conditions that any material must fulfill to be a magnet, emphasizing peculiarities that are characteristic of organic systems.

In the first place, the material must contain molecular units possessing a magnetic moment. In itself this is a nontrivial problem, since the vast majority of organic materials are diamagnetic and many organic radicals associate in the solid state via bond formation. The open shell structure can be stabilized through sterically demanding substituents, but this may have detrimental effects on the exchange interaction (see later sections). The use of electronegative elements (N, O, and S) that provide energetically low-lying antibonding orbitals can assist radical stabilization, as can delocalization of the unpaired electron. The number and variety of available free radicals is large. Moreover, although the syntheses of new open-shell molecules may not be straightforward, they do follow established chemical principles, so there is a steady flow of new, interesting organic free radicals whose magnetic properties deserve to be studied. Alternatively, an open-shell ground state may be achieved through a charge-transfer process. While this approach to organic molecular magnetism has not been so well investigated, the principle of matching the redox properties of donor and acceptor molecules to provide good charge transfer is also well understood.

Second, molecular moments must interact with each other throughout the whole solid. Magnetic exchange pathways are inextricably linked to the solid-state structure, which must be controlled, since it affects the sign, strength, and symmetry of the exchange interactions. The strength of the exchange interaction is very short and normally decays as $1/r^{10}$ or more [5]. Consequently, we require efficient molecular packing if we are to achieve the close intermolecular contacts between radical centers that will lead to the strong exchange interactions required for high magnetic-ordering temperatures. In principle, the closer any orbitals possessing spin density and belonging to neighbor molecules become, the stronger the magnetic interaction will be. Since all the molecules in the solid should eventually interact with one another at the magnetic-ordering transition, it is clear that bulky molecules where the radical center has been sterically protected by diamagnetic groups are not good candidates to develop a magnetic solid. To make the magnetic molecular moments interact with one another in such a way that the interactions extend throughout the whole solid in a cooperative fashion is, indeed, a very demanding goal. It requires an integral design of the material, from the selection of the interacting molecules to control of the crystal lattice. Although molecular packing can be modified by small changes in the molecular geometry, the result is difficult to predict, even for simple organic free radicals. A much better understanding of crystal packing forces, aided by supramolecular synthons (pairs of strongly interacting functional groups on neighboring molecules that give rise to well-defined solid-state geometries) will be essential if we are to tailor their solid-state structures to control their physical properties. Under no circumstances can the structure be affected by the presence of magnetic interactions between neighboring molecules. This is clearly evident by a comparison of the typical energies (10^{-4} eV) involved in magnetic interactions, with the energies of common tools for crystal engineering, namely, the hydrogen bond ($\sim$0.4 eV) and the van der Waals interaction ($\sim$0.1 eV).

Third, in order to form a magnet, the exchange interactions must extend in such a way that spontaneous magnetization results in the solid. This can, at first glance, seem difficult to satisfy if one restricts the appearance of spontaneous magnetization to the formation of a ferromagnet. Ferromagnetic interactions, particularly the intermolecular ones, are still far from being controlled, aside from various elegant attempts to rationalize them [6]. The quest for spontaneous magnetization, however, should not be restricted to just the ferromagnetic state, since there are other spin arrangements related to antiferromagnetic ordering—e.g., weak ferromagnetism and ferrimagnetism—that may also lead to spontaneous magnetization [7].

III. CURRENT ACHIEVEMENTS

There are numerous organic radicals reported in the literature that have been characterized in dilute solution or in inert gas matrices. Many undergo irreversible bond formation in concentrated solution or the solid state, and this may preclude their use as magnetic materials. For example, the reactive carbene functionality, which exhibits a triplet ground state, has been used with some success in constructing high-spin ground state organic molecules, e.g., branched polycarbenes $C_{48}H_{30}$ and $C_{69}H_{42}$ with $S = 6$ and 9, respectively (Scheme 1) [8–10]. Despite their attractive high-spin ground states, these carbene oligomers tend to undergo irreversible bond-forming reactions even at low temperatures.

A number of thermally stable functional groups have been employed as spin-bearing moieties. The most common are the families of nitroxide and nitronylnitroxide radicals. Indeed, these derivatives provided the first examples of organic magnets, when β-p-NPNN and DOTMDAA (Scheme 2) were shown to undergo phase transitions to ferromagnetic states at 0.6 K and 1.8 K, respectively [11,12].

While the spin-polarization method provides a convenient approach to predict intramolecular exchange interactions, the nature of the intermolecular exchange interactions depends on subtle differences in the crystallographic packing [13,14]. The prediction of the solid-state structures of even simple

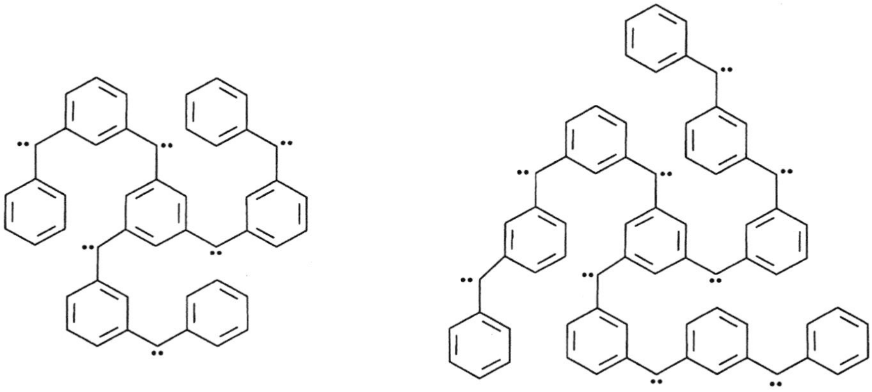

Scheme 1

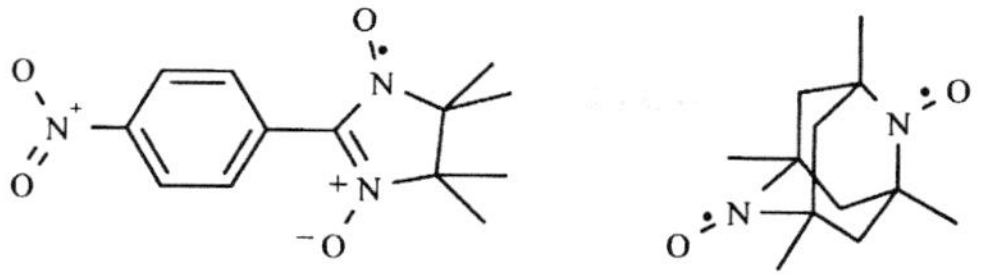

p-NPNN   DOTMDAA

Scheme 2

molecular compounds is still in its infancy, and synthetic chemists rely to a great extent on serendipity. Moreover, in organic chemistry the intermolecular forces between molecules are weak, and polymorphism—i.e., the formation of more than one crystallographic phase—is not uncommon. Indeed, W. C. McCrone has suggested the empirical rule that the number of polymorphs found for a molecule increases with time and money spent in research on that compound [15]! Since magnetic properties are highly dependent on morphology, careful control of crystal growth conditions is required [16]. Indeed, p-NPNN is known in four structural modifications, only one of which undergoes long-range ferromagnetic order. Recently, hydrogen-bonded networks have been utilized to try to control molecular association in the solid state, and the dominant intermolecular exchange interactions in a series of hydroxyphenyl nitronylnitroxide radicals have been modulated between antiferro- and ferromagnetic [6].

While the sign of the exchange interaction is vital to determining the nature of the magnetic order (ferro- vs. antiferromagnetic), it is the magnitude of the exchange interaction that controls the magnetic-ordering temperature. For the majority of organic radicals, the Weiss constants θ are within ± 10 K of (and ordering temperatures T_c are to be expected around or below) liquid helium temperature [13,17]. If we are to achieve organic molecular magnets that have any technological importance, then the T_c must be significantly higher, at least above liquid nitrogen temperature or preferably above room temperature. We believe that this may prove the most difficult task in organic molecular magnetism.

Finally, the nature of the magnetic order must give rise to a spontaneous magnetic moment. Initially, pure ferromagnetism was the sole target of scientists, and the previously cited examples of β-p-NPNN and DOTM-DAA conform to this regime. However, more recently a number of weak ferromagnets have been discovered (such as AOTMP [18] and TOV [19]; Scheme 3), in which canting of antiferromagnetically coupled spins leads to a small net magnetization. Weak ferromagnetism is favored in lower-symmetry structures (in which there is a lack of inversion center between inter-

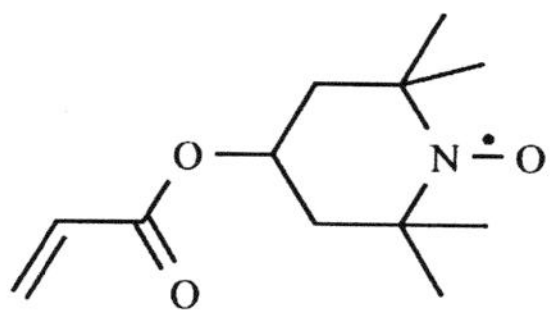

AOTMP 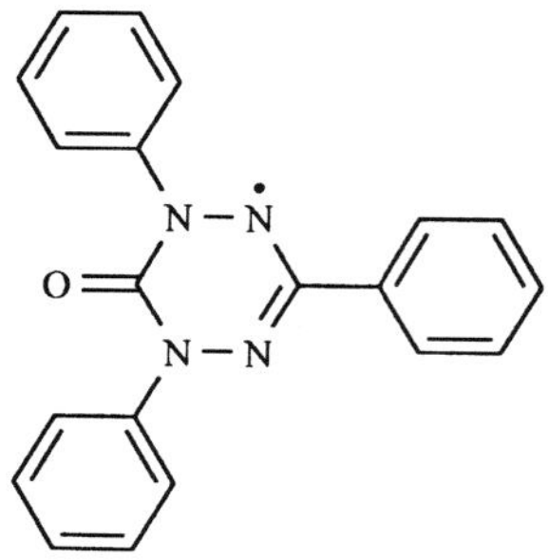   TOV

Scheme 3

acting spins) and is becoming increasingly prevalent in the wider area of molecular magnetism. This is a natural consequence of molecular chemistry in which the inherently lower symmetry of molecular species reduces the crystal symmetry.

IV. FUTURE PROSPECTS

It is always a daunting task to predict the future and, without the aid of a crystal ball, many would say that this is impossible. An analysis of the present literature may, however, give us a guide towards some achievable goals for the early years of the next millenium. As we attempt to predict future strategies, we look again at the requirements for organic magnets, in the light of the current achievements.

One must start with appropriate building blocks for molecular magnets. The most common radical functional group must surely be the nitroxide family, which has generated some elegant models for our fundamental understanding of organic molecular magnetism but whose exchange interactions appear invariably small and unlikely to give rise to high T_c materials. If we wish to increase the magnetic-ordering temperature, we should look for radicals that will promote stronger intermolecular exchange interactions. A recent approach has been to utilize radicals bearing heavier main group elements (P, S, Se) whose more radially diffuse orbitals and less sterically hindered structures may lead to stronger exchange interactions. Indeed, a number of sulfur–nitrogen radicals have been reported with $|\theta|$ up to 102 K, and the β-phase of the sulfur–nitrogen radical p-NCC$_6$F$_4$CNSSN (Scheme 4) becomes weakly ferromagnetic at 36 K [20,21]. Since the

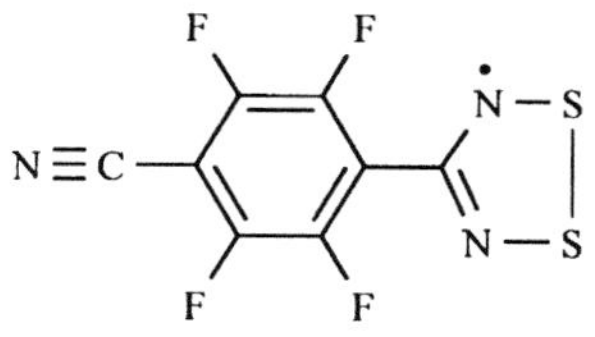

p-NCC$_6$F$_4$CNSSN

Scheme 4

exchange interaction varies as $1/r^{10}$, improvements in T_c may also be made by the application of pressure, which compresses the solid. The application of pressure may decrease intermolecular contacts and thereby increase exchange interactions. Recent studies have shown an increase in T_c of the weak ferromagnet TOV (see Scheme 3), from 4.8 K to 10 K, with pressure [22], although similar studies on β-p-NPNN show a transition from ferromagnetic to antiferromagnetic behavior on the application of pressure [23]. This arises through a change in the dominant exchange interaction caused by subtle changes in intermolecular contacts on compression, as has been observed in model inorganic systems [24,25].

While pure ferromagnets are laudable goals, bulk ferromagnetic order requires stringent conditions, specifically, ferromagnetic exchange throughout the structure. Since nature appears to favor antiferromagnetic exchange, one might wish to consider other approaches to generate a net magnetic moment [7]. We have already mentioned weak ferromagnetism, which relies on the canting of antiferromagnetically coupled spins. Since antiferromagnetic interactions tend to be stronger than ferromagnetic ones, the chances of discovering higher ordering temperatures are better in antiferromagnetically coupled systems than in ferromagnetic ones. For example, the magnetic-ordering temperature rises in verdazyl derivatives from 0.67 K for ferromagnetic p-CDTV (Scheme 5) to 5.4 K for weak ferromagnetic TOV (Scheme 3) [19]. In weak ferromagnetism, the canting angle can be estimated from the g-value, with the greater degree of molecular anisotropy leading to larger canting angles. Weak ferromagnets typically exhibit spontaneous magnetic moments 10^2 or 10^3 times less than pure ferromagnets. Again, incorporation of the heavier main group elements, which exhibit greater spin-orbit coupling, may provide a solution to generate comparatively large spontaneous moments. Alternatively, with the development of high-spin ground state organic molecules, it may be possible to cocrystallize simple $S = 1/2$ and $S > 1/2$ spin systems, with the antiferromagnetic

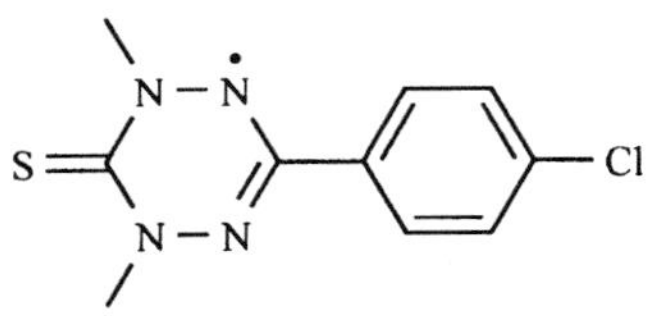

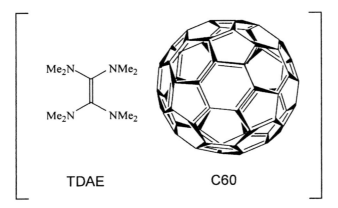

CDTV

Scheme 5

exchange between the two sublattices giving rise to a ferrimagnet. To our knowledge, this has not yet been achieved, but it may provide an exciting target for organic magnets.

Organic charge-transfer salts constitute another group of compounds whose potential possibilities as organic magnets deserve to be explored more systematically. Organometallic materials of this kind have provided a large variety of interesting magnetic materials [26]. Moreover, the fullerene charge-transfer salt $C_{60} \cdot$ TDAE (Scheme 6) was reported to undergo a phase transition to a magnetic state at 16 K [27], which recently has been definitively proved to be ferromagnetic [28]. Very few fullerene-based compounds have been found to show magnetic behavior. For example, $C_{70} \cdot$ TDAE and many different doped $C_{60} \cdot$ TDAE derivatives show no evidence for a magnetically ordered state down to 4 K. The same occurs when TDAE is substituted with other electron donors [29]. Recently, a new fullerene-based derivative showing magnetic ordering below $T_c = 19$ K has been synthesized, namely, (3-aminophenyl)-1H-1,2-methano-[60]fullerene cobaltocene [30].

TDAE   C60

Scheme 6

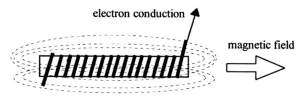

electron conduction

magnetic field

Scheme 7

The interest of these results is that $CoCp_2^+$ has $S = 0$; hence, only the fullerenes are spin-relevant building blocks for the molecular magnetism of these materials. Material architectures based on fullerenes might be quite promising to develop new high-T_c magnetic materials, since the antiferromagnetic state seems to exist in many fullerene materials well above room temperature [31].

A third alternative has been to prepare organic polymers in which well-defined intramolecular exchange interactions are used to generate a magnetic material. A number of ill-defined polymers have been reported to be magnetic, but whose properties appear to be extrinsic rather than intrinsic in nature [32,33].

Finally we propose a rather different strategy, which has not (to our knowledge) been suggested previously. This approach does not rely on any inherent magnetic behavior but returns to an elementary physical phenomenon: the solenoid. When an electric current flows around a coil, it generates a magnetic field within the coil (Scheme 7). The question arises as to how one can design a conducting helix on a molecular scale. And as a cautionary note, just to make the problem a little more difficult, the helices must all be of the same "handedness" in the solid state; otherwise, their magnetic fields will oppose and effectively cancel one another. Under the effects of an electric field gradient, a single crystal of such a molecular conducting helix would act as if it is made up of oriented dipoles and will possess net magnetization. In other words, an electric field can induce the formation of magnetic dipoles, all oriented in the manner as is found for a saturated ferromagnet. Interestingly enough, thermal effects will not tend to disorient these moments in the manner that is found in regular paramagnets.

V. CONCLUSIONS

While the continuing work is fundamental in developing our understanding of organic molecular magnetism, one should not be afraid to look to new molecules, outside current research areas, for inspiration. One should embrace the fundamental research of various groups when designing new stable radicals. If one could ignore the old "common sense," which discounted development of organic materials with very high magnetic moments or the finding of a truly organic magnet containing no metals in it, there would be no a priori reason not to expect an organic magnet at room temperature or above in the near future. One may note that, since the discovery of the first organic magnet with a T_c of 0.6 K in 1991, an increase in T_c by almost two orders of magnitude (36 K) has been achieved. One need only achieve a further increase in T_c by one order of magnitude to achieve the goal of a room-temperature organic ferromagnet. Although this may seem, in principle, contrary to common assumptions that magnetic interactions are weak in organic materials, there are many examples that illustrate how strong antiferromagnetic interactions can be in these systems. There are already examples proving that spontaneous magnetization can arise in certain kinds of antiferromagnets. The future seems fascinating for organic magnetic materials of various sorts, and one may look forward to many unusual phenomena being found as research continues.

REFERENCES

1. Iwamura H. Adv. Phys. Org. Chem. 1990; 26:179, 253.
2. Palacio F. In: Gatteschi D, Kahn O, Miller JS, Palacio F, Eds. Magnetic Molecular Materials. Dordrecht: Kluwer, 1991, vol. 198, p. 1.
3. Kinoshita M. Jpn. J. Appl. Phys. 1994; 33:5718.
4. Lahti P. In: Turnbull M, Sugimoto T, Thompson LK, Eds. Molecule-based Magnetic Materials. Washington, D.C.: American Chemical Society, 1996, vol. 644, p. 218.
5. de Jongh LJ, Block R. Physica 1975; 79B:568.
6. Cirujeda J, Ochando LE, Amigó JM, Rovira C. Rius J, Veciana J. Angew. Chem. Int. Ed. Engl. 1995; 34:55.
7. Palacio F. Mol. Cryst. Liq. Cryst. 1997; 305:385.
8. Nakamura N, Inoue K, Iwamura H. J. Am. Chem. Soc. 1992; 114:1484.
9. Nakamura N, Inoue K, Iwamura H. Angew. Chem. Int. Ed. Engl. 1993; 32:873.
10. Iwamura H. Pure Appl. Chem. 1993; 65:57.
11. Tamura H, Nakazawa Y, Shiomi D, Nowaza K, Hosokoshi Y, Ishikawa M, Takahashi M, Kinoshita M. Chem. Phys. Lett. 1991; 186:401.
12. Chiarelli R, Novak MA, Rassat A, Tholence JL. Nature 1993; 363:147.
13. Deumal M, Cirujeda J, Veciana J, Novoa JJ. Chem. Eur. J. 1998; in press.
14. Deumal M, Cirujeda J, Veciana J, Novoa JJ. Adv. Mat. 1998; 10:1461.
15. Quoted in Dunitz JD, Bernstein J. Accts. Chem. Res. 1995; 28:193.
16. Miller JS. Adv. Mat. 1999; 10:1553.
17. Nakatsuji S, Anzai H. J. Mater. Chem. 1997; 7:2161.

18. Koboyashi TC, Takiguchi M, Hong CU, Amaya K, Kajiwara A, Harada A, Kamachi M. J. Magn. Mat. 1995; 140–144:1447.
19. Takeda K, Mito M, Nakano H, Kawae T, Hitaka M, Takagi S, Deguchi H, Kawasaki S, Mukai K. Mol. Cryst. Liq. Cryst. 1997; 306:431.
20. Banister AJ, Bricklebank N, Lavender I, Rawson JM, Gregory CI, Tanner BK, Clegg W, Elsegood MRJ, Palacio F. Angew. Chem. Int. Ed. Engl. 1996; 35: 2533.
21. Palacio F, Castro M, Antorrena G, Burriel R, Ritter C, Bricklebank N, Rawson J, Smith JNB. Mol. Cryst. Liq. Cryst. 1997; 306:293.
22. Mito M, Hitaka M, Kawae T, Takeda K, Suzuki K, Mukai K. Mol. Cryst. Liq. Cryst. 1999; in press.
23. Mito M, Nakano H, Kawae T, Hitaka M, Takagi S, Deguchi H, Suzuki K, Mukai K, Takeda K. J. Phys. Soc. Jpn. 1997; 66:2147.
24. Morón MC, Palacio F, Clark SM, Paduan-Filho A. Phys. Rev. B 1995; 51: 8660.
25. Morón MC, Palacio F, Clark SM. Phys. Rev. B 1996; 54:7052.
26. Miller JS, Epstein AJ, Reiff WM. Chem. Rev. 1988; 88:201.
27. Allemand PM, Khemani KC, Koch A, Wudl F, Holczer K, Donovan S, Gruner G, Thompson JD. Science 1991; 253:301.
28. Arcon D, Cevc P, Omerzu A, Blinc R. Phys. Rev. Lett. 1998; 80:1529.
29. Klos H, Rystau I, Schutz W, Gotschy B, Skiebe A, Hirsch A. Chem. Phys. Lett. 1994; 224:333.
30. Mihailovic D, Mrzel A, Omerzu A, Umek P, Jaglicic Z, Trontelj Z. Mol. Cryst. Liq. Cryst. 1999; in press.
31. Mihailovic D. Europhys. News 1997; 28:78.
32. Miller JS. Adv. Mater. 1992; 4:298.
33. Miller JS. Adv. Mater. 1992; 4:435.

Index